Chemistry

for the IB DIPLOMA

Christopher Talbot
Richard Harwood
Christopher Coates

HODDER
EDUCATION
AN HACHETTE UK COMPANY

Although every effort has been made to ensure that website addresses are correct at time of going to press, Hodder Education cannot be held responsible for the content of any website mentioned in this book. It is sometimes possible to find a relocated web page by typing in the address of the home page for a website in the URL window of your browser.

Hachette's policy is to use papers that are natural, renewable and recyclable products and made from wood grown in sustainable forests. The logging and manufacturing processes are expected to conform to the environmental regulations of the country of origin.

This material has been developed independently by the publisher and the content is in no way connected with, nor endorsed by, the International Baccalaureate Organization.

Orders: please contact Bookpoint Ltd, 130 Milton Park, Abingdon, Oxon OX14 4SB. Telephone: (44) 01235 827720. Fax: (44) 01235 400454. Lines are open 9.00–5.00, Monday to Saturday, with a 24-hour message answering service. Visit our website at www.hoddereducation.co.uk.

© Christopher Talbot, Richard Harwood, Christopher Coates
First published in 2010 by Hodder Education,
an Hachette UK company,
338 Euston Road
London NW1 3BH

Impression number   5 4 3 2 1
Year   2014 2013 2012 2011 2010

Cover photo © Chris Madeley/Science Photo Library
Illustrations by Ken Vail Graphic Design
Typeset in 10/12pt Goudy and Frutiger families by Ken Vail Graphic Design
Printed in Italy

A catalogue record for this title is available from the British Library

ISBN: 978 0340 98505 2

Contents

Section 4: Glossary, acknowledgements, index and answers to examination questions

Note that the 'B' prefix indicates that these items appear in the printed book.

available on the CD-ROM accompanying this book

Introduction

The International Baccalaureate Diploma programme, a pre-university course for 16- to 19-year-olds, is designed to develop not only a breadth of knowledge, skills and understanding, but well-rounded individuals and engaged world citizens. One of the Diploma's key requirements is concurrent study in six academic areas, at least one of which is an experimental science. Of this chemistry, whether taken at Standard or Higher Level, is the choice of many students. This book is designed to serve them.

Within the IB Diploma programme, the theory content for chemistry is organized into compulsory core topics and options. The organization of this book exactly follows that syllabus sequence:

- **Section 1** is the **common core material** for Standard and Higher Level students: Chapters 1–11
- **Section 2** is the **additional higher level material** for Higher Level students: Chapters 12–20
- **Section 3** consists of the seven **options**: A to G, covered in Chapters 21–27. All the options are available to both Standard and Higher Level students.

The syllabus is presented as topics and options, each of which is the subject of a single chapter in *Chemistry for the IB Diploma*. The topic chapters are provided in the printed book, and the options chapters on the accompanying CD-ROM.

Special features of the chapters of *Chemistry for the IB Diploma* are described below.

- Each chapter begins with 'Starting points' that summarize the essential concepts on which the chapter is based.
- The text is written in straightforward language, uncluttered by phrases or idioms that might confuse students for whom English is a second language. The depth of treatment of topics carefully reflects the objectives and command terms in which the syllabus assessment statements are phrased.
- Photographs and full-colour illustrations are linked to support the relevant text, with annotations included to elaborate the context, function, language or applications of chemistry.
- Throughout the text the IB Diploma chemistry syllabus subtopic assessment statement being addressed is clearly shown, so links between the text and the IB Diploma chemistry syllabus are self-evident.

- The bar at the foot of each page is colour coded to show whether the text is for Standard Level (pink bar), Higher Level (dark red bar), or for both (striped bar).
- Processes of science (science methods) and the history of chemical developments are introduced selectively to aid appreciation of the possibilities and limitations of science.
- At the end of each chapter, typical examination questions, of all types, are given. Full worked answers are provided on the CD-ROM.
- Links to the interdisciplinary Theory of Knowledge (**TOK Link**) element of the IB course are made at appropriate points in most chapters.
- A comprehensive glossary of words and terms is included both in the printed book and on the CD-ROM.
- The CD-ROM also includes Chapter 28, which provides an introduction to IB Diploma chemistry for students and teachers new to the programme, and Chapter 29, which provides detailed chemistry-specific guidance and advice relating to the extended essay (in chemistry).

Using this book

The sequence of chapters in *Chemistry for the IB Diploma* follows the sequence of the syllabus contents. However, the IB Diploma chemistry syllabus is not designed as a teaching syllabus, and the order in which the syllabus content is presented is not necessarily the order in which it should be taught. Different schools and colleges need to design a course delivery model based on individual circumstances. In addition to the study of theory issues on which this book focuses, IB science students are also involved in practical investigations and the Group 4 Project. Investigations are ultimately presented for the internal assessment, based on given internal assessment criteria. How all these components are integrated is also the subject of Chapter 28. This has been written by guest author Gary Seston, an experienced and enthusiastic teacher of the IB Diploma who, importantly, also has examiner experience. Prior to his present post at the United World College of South East Asia in Singapore, Gary also taught at Sotogrande International School, Cadiz. This chapter is an excellent guide of interest to both teachers (especially those new to the IB Diploma) and students.

Author profiles

Christopher Talbot

Chris graduated with honours in Biochemistry from the University of Sussex in the United Kingdom. He has a Masters Degree in Life Sciences (Chemistry) from the National Technological University in the Republic of Singapore. He is currently teaching IB Chemistry and IB Biology at the Overseas Family School, Republic of Singapore. He also taught IB Chemistry and Theory of Knowledge at Anglo-Chinese School (Independent), Republic of Singapore, where he helped prepare students for the International Chemistry Olympiad. He has moderated IB Chemistry coursework.

Richard Harwood

Richard graduated with honours in Chemistry from the University of Manchester Institute of Science and Technology in the UK. He then trained as a teacher and gained a PhD in Medical Biochemistry at the University of Manchester Medical School. After periods of research in biochemistry in both Manchester Medical School and University College, Cardiff, he returned to teaching science as a housemaster at Millfield School in Somerset, England. He subsequently moved to Aiglon College, Switzerland, where he served as Head of Science and Deputy Principal (Academic). He has given many training workshops on science teaching and practical work in various parts of the world.

Christopher Coates

Chris holds a Chemistry degree from the University of Cambridge, and a Master of Education from Monash University, Melbourne. He has previously taught in Suffolk, Yorkshire and Hong Kong at King George V. School, and is currently Head of Science at the Tanglin Trust School, Singapore. He has taught A level and IB Chemistry as well as Theory of Knowledge.

Author's acknowledgements

We are indebted to the following teachers and lecturers who reviewed early drafts of the chapters:
Mr Gordon Woods, formerly Head of Chemistry of Monmouth School (Chapters 1, 3, 4, 13 and 18); the late Bernard Spurgin, formerly Science Master at Wolverhampton Grammar School (Chapter 1); Dr David L. Cooper, University of Liverpool (Chapters 2, 3, 4, 12, 14, 17 and 21); Lorne Schmidt, Deputy Principal (Curriculum), Overseas Family School and College (Chapters 2 and 12); Mr Chooi Khee Wai, Dean of Chemistry, Anglo-Chinese School (Independent), (Chapters 2, 12, 6 and 16); Mrs Usha Penninal Ravi Udayakumar, Anglo-Chinese School (Independent), (Chapters 10 and 20); Dr David Fairley, Overseas Family School (Chapters 5, 7, 15, 17, 21, 23 and the glossary); Dr Sam Logan, University of Ulster at Coleraine, Northern Ireland (Chapters 6 and 16); Professor Norman Billingham, University of Sussex (Chapters 10 and 23); Professor Edward Constable, University of Basel, Switzerland (Chapter 13); Dr Graham L. Patrick, University of Paisley (Chapter 24); Professor Mike Williamson, University of Sheffield (Chapter 21); Dr Jon Iggo, University of Liverpool (Chapter 21); Professor Robin Clark, University College London (Chapter 21); Dr John Moore, University of York (Chapter 21); Professor Richard Compton, University of Oxford (Chapters 7, 17 and 19); Dr Peter Braesicke, University of Cambridge (Chapter 25); Dr Andrew Challinor, University of Leeds (Chapter 25); Dr George Marston, Reading University (Chapter 25); Dr Roger Atkinson, University of California and Professor James Hanson, University of Sussex (Chapters 10 and 27); and Dr Chris Gabbutt, University of Leeds (Chapter 26) and Dr Alistair Chew (Chapter 24). A special word of thanks must go to Nick Lee, experienced chemistry and TOK teacher, workshop leader and examiner, for his most helpful comments on the final drafts.

I am indebted to my colleague, Cesar Reyes Jr, and my father, David Talbot, who are responsible for the majority of the photographs in the book. I also thank Dr Jonathan Hare for providing photographs of analytical instruments for Chapter 21, and Peter O'Byrne for providing the photographs of students for Chapter 28. I am also grateful to Dr Calvin Davidson, University of Sussex (Computational Materials Group) and Dr Chris Ewels, IMN, Université de Nantes, France, who both kindly generated a variety of nano-based structures for Chapter 23. I would also like to thank my colleague Jon Homewood who created the drawings of many of the scientists used in the book.

Finally, I am indebted to all those involved in the production of the book – the International publishing team of Eleanor Miles and Nina Konrad at Hodder Education, freelancers Anne Russell and Anne Trevillion, and Emily Hooton (of Ken Vail Graphic Design).

Chris Talbot
Singapore, January 2010

1 Quantitative chemistry

STARTING POINTS

- Pure substances can be divided into the chemical elements and chemical compounds.
- All substances are composed of atoms.
- Chemical compounds consist of atoms chemically bonded together in a fixed ratio described by a chemical formula.
- The mole is a quantity used to indirectly count atoms, ions, electrons, formula units and molecules.
- A mole contains 6.02×10^{23} particles – this quantity is known as the Avogadro constant (units of mol^{-1}).
- Known amounts of substances (in moles) are prepared by measuring masses of pure solids, liquids or gases. Known amounts of gases can also be prepared by measuring volumes of pure gases (under specified conditions).
- The physical behaviour of gases is described by the three gas laws and summarized by the ideal gas equation. The chemical behaviour of gases is described by Gay-Lussac's and Avogadro's laws.
- Mass is conserved during chemical reactions.
- Titration is a technique, using solutions as reactants, that allows chemists to determine the amounts of substances reacting if the concentration of one of the reacting solutions is known.
- Titrations may involve acid–base, redox or precipitation reactions. A back titration involves two consecutive reactions.
- Amounts are used to deduce chemical formula, amounts of products, percentage purity and percentage yield.

Fundamental concepts of chemistry

States of matter

There are three phases or **states of matter**: **solids**, **liquids** and **gases**. Any substance can exist in each of these three states depending on temperature and pressure.

The simple diagram in Figure 1.1 shows the relationship between these states of matter and the arrangement (idealized, simplified and in two dimensions only) of their particles (ions, atoms or molecules). The arrows represent **physical changes** termed **changes of state**.

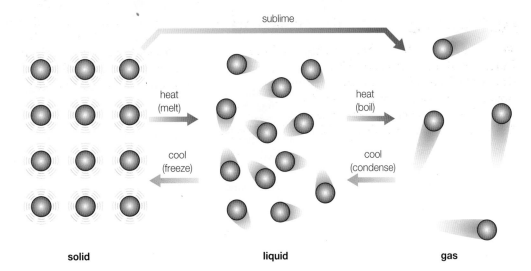

Figure 1.1 The three states of matter and their interconversion

solid   liquid   gas

Elements

The chemical **elements** (Figure 1.2) are the simplest substances and are composed of a single type of atom (Chapter 2). (Many elements exist in two or more slightly different forms of the same element known as isotopes – see Chapter 4). Elements cannot be split up or decomposed into simpler substances by a chemical reaction.

The elements can be classified into three groups based upon the state of matter they exist in at 25 °C. Most of the elements are solids, for example iron, but bromine and mercury are liquids at room temperature and the remainder of the elements are gases, for example oxygen and neon.

The elements can also be classified into two groups: **metals** and **non-metals** (Chapter 4), based on their chemical and physical properties. For example, aluminium is a metal and chlorine is a non-metal.

Many elements exist as atoms, for example metals and the noble gases. However, many non-metals exist as atoms bonded together into **molecules** (Figure 1.3). Examples of non-metal molecules include oxygen, O_2, chlorine, Cl_2, nitrogen, N_2, phosphorus, P_4, and sulfur, S_8. Oxygen, nitrogen and chlorine exist as **diatomic** molecules.

Allotropy is the existence of two or more forms of an element in the same physical state. These different forms are called **allotropes**. Allotropes exist where there is more than one possible arrangement of bonded atoms. For example, solid carbon can exist in three allotropes: diamond, carbon-60 (C_{60}) and graphite (see Chapter 4); oxygen can exist in two allotropes: dioxygen (O_2) and trioxygen (ozone, O_3) (Chapter 25).

Compounds

Many mixtures of elements undergo a chemical reaction when they are mixed together and heated. The formation of a **compound** (Figure 1.4) from its elements is termed **synthesis**. Heat energy is usually released during this reaction (Chapter 5).

Figure 1.2 A sample of the element phosphorus (red allotropic form)

O═O
oxygen molecule O_2

N≡N
nitrogen molecule N_2

H—H
hydrogen molecule H_2

Cl—Cl
chlorine molecule Cl_2

S—S
S   S
| |
S   S
S—S
sulfur molecule S_8

Figure 1.3 Diagram of oxygen, nitrogen, hydrogen, chlorine and sulfur molecules

Figure 1.4 A model showing the structure of the compound calcium carbonate, $CaCO_3$ (black spheres represents carbon, red oxygen and white calcium)

When a mixture of iron and sulfur is heated, large amounts of heat energy are released as the compound iron(II) sulfide, FeS, is formed (Figure 1.5). (Synthesis reactions like this are examples of redox reactions – see Chapter 10.)

Mixtures of elements are easily separated by a physical method, since the atoms of the different elements are not bonded together. For example, iron can be separated from sulfur by the use of a magnet.

However, when a compound is formed the atoms it contains are chemically bonded together, so the compound will have different physical and chemical properties from the constituent elements. For example, iron is magnetic, but the compound iron(II) sulfide is non-magnetic (Figure 1.6). A compound will contain either molecules or ions (Chapter 4).

The splitting of a chemical compound into its constituent elements is termed **decomposition**. This process requires an input of energy, either heat (**thermal decomposition**) or electricity (electrolysis) (Chapter 9).

Figure 1.5 The elements iron and sulfur

Figure 1.6 A sample of iron(II) sulfide and a mixture of iron and sulfur

1.1 The mole concept and Avogadro's constant

1.1.1 **Apply** the mole concept to substances.
1.1.2 **Determine** the number of particles and the amount of substance (in moles).

Introduction

Figure 1.7 The mole concept applied to two solid elements: magnesium and carbon (graphite)

Chemists are interested in the ratios in which chemical elements and compounds react together during chemical reactions. This is important when preparing a pure substance in the laboratory, and even more so in the chemical industry. Using excess reactant, unless necessary, will result in additional costs in order to remove it from the product.

Atoms are very small with very small masses, for example a hydrogen atom ($_1^1H$) weighs only 1.67355×10^{-27} kg. However, the masses of atoms of different elements are different, for example a carbon-12 atom is twelve times more massive than an atom of hydrogen-1.

For this reason, weighing out the same mass of different elements results in different numbers of atoms being present in the samples. It is very difficult for chemists to count large numbers of atoms directly so instead a chemist counts atoms *indirectly* by weighing samples of elements.

For example, 12 grams of carbon-12 atoms and 1 gram of hydrogen-1 atoms both contain the same number of atoms. These samples are described as having the same **amount** of atoms in moles. In this simple example the two samples of elements both contain one **mole** of atoms. The mole concept (Figure 1.7) allows chemists to weigh out samples of substances with equal numbers of particles (atoms, ions or molecules). For elements, one mole of atoms is present when the relative atomic mass (page 6) of the element is weighed out in grams.

The amount of substance (symbol n) is a quantity that is directly proportional to the number of particles in a sample of substance. It is one of the seven base quantities of the SI unit system. The unit of amount is the mole (mol).

A mole of a substance contains 6.02×10^{23} particles of the substance. This is the same number of particles as there are atoms in exactly 12 grams of the isotope carbon-12 ($_6^{12}C$) (see Figure 1.8). The value 6.02×10^{23} mol^{-1} is called the **Avogadro constant** (symbol L).

Figure 1.8 An illustration of the Avogadro constant

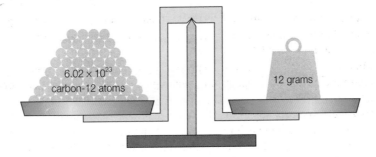

The particles may be atoms (e.g. Ar), molecules (e.g. Br_2), ions (e.g. Na^+), formula units (e.g. NaCl) or electrons (Figure 1.9), but should be specified, for example 1 mol of chlorine atoms or 2 mol of chlorine molecules. (Stoichiometric calculations involving electrons can be found in Chapter 19.)

| 6.02×10^{23} atoms | 6.02×10^{23} molecules | 6.02×10^{23} ions | 6.02×10^{23} formula units | 6.02×10^{23} electrons |

one mole of particles

Figure 1.9 A summary of the mole concept applied to different particles

The mole is simply a convenient counting unit for chemists, large enough to be seen, handled and measured. It is no different from other counting units: a dozen eggs, a gross (144) of nails and a ream (500 sheets) of paper (Figure 1.10). Note that as the objects become *smaller*, the number in a unit amount become *larger*. The value of the Avogadro constant is given on page 2 of the IB *Chemistry data booklet*.

The equation below describes the relationship between the amount of a substance and the number of particles:

$$\text{amount of substance (mol)} = \frac{\text{number of particles}}{6.02 \times 10^{23}\,\text{mol}^{-1}}$$

The formula may be rearranged to make the number of particles the subject.

Figure 1.10 Counting units from left to right: a pair of socks, a ream of paper and a dozen eggs

Worked examples

Calculate the number of molecules of water in 0.01 mol of water.

Number of water molecules = $0.01\,\text{mol} \times 6 \times 10^{23}\,\text{mol}^{-1} = 6 \times 10^{21}$

(Note that units of mol and per mol (mol^{-1}) cancel to leave a pure number. Note also that for simplicity the Avogadro constant is taken to be $6 \times 10^{23}\,\text{mol}^{-1}$.)

Calculate the amount of nitric(V) acid, HNO_3, that contains 9×10^{23} molecules.

$$\text{Amount of nitric(V) acid} = \frac{\text{number of molecules}}{6.02 \times 10^{23}\,\text{mol}^{-1}}$$

$$\text{Amount of nitric(V) acid} = \frac{9 \times 10^{23}}{6 \times 10^{23}\,\text{mol}^{-1}} = 1.5\ \text{mol}$$

Calculate the number of oxygen atoms present in 9×10^{23} molecules of nitric(V) acid, HNO_3.

Each molecule of nitric(V) acid contains three oxygen atoms. Hence, 9×10^{23} molecules of nitric(V) acid contains

$3 \times 9 \times 10^{23} = 2.7 \times 10^{24}$ atoms of oxygen.

TOK Link

Atoms are so small that if a line one metre long were drawn then 6 000 000 000 or 6×10^9 atoms could be lined up end to end. If you were to stand on a sandy beach and look along the beach in both directions, you would not see enough particles to make one mole of grains of sand. The mole concept is analogous to a bank clerk who weighs bags of coins on special scales which effectively count coins (of the same type) by mass. Chapter 23 introduces the science of nanotechnology, whose techniques allow chemists to move and count small numbers of atoms.

■ Extension: Determination of the Avogadro constant

In 1914 William Bragg used X-ray crystallography to determine the Avogadro constant. X-ray crystallography involves passing X-rays through very pure crystals and analysing the scattering patterns to determine the arrangement of particles in the crystal.

General approach

- The spacing of particles in a crystal is first determined.
- Knowing the distance between atoms (or ions) in the crystal, it is then possible to find the volume occupied by one atom.
- The volume of one mole of the substance is then determined.
- Finally, the volume of one mole is divided by the volume of one atom to obtain the Avogadro constant.

Example calculation

Figure 1.11 shows a unit cell of sodium metal, which has a body-centred cubic structure. The unit cell is the simplest arrangement of atoms which, when repeated, will reproduce the same structure.

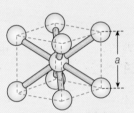

The central atom is located inside the unit cell. The eight atoms at the corners are equally shared between eight unit cells. This means the unit cell effectively contains a total of $(1 + 8 \times \frac{1}{8})$, that is, two atoms.

X-ray diffraction methods show that the width of the sodium unit cell (shown as a in Figure 1.11) is 0.429 nm, or 0.429×10^{-7} cm (1 nm = 10^{-9} m).

Figure 1.11 Unit cell of sodium metal

Thus, the volume of the unit cell (that is, two atoms)

$$= (0.429 \times 10^{-7})^3 \text{ cm}^3$$
$$= 0.0790 \times 10^{-21} \text{ cm}^3$$

Therefore, the volume occupied by one sodium atom = 0.0395×10^{-21} cm^3. The relative atomic mass of sodium = 22.99 and the density of sodium is 0.97 g cm^{-3}.

Using the equation volume $= \dfrac{\text{mass}}{\text{density}}$, the volume of one mole of sodium atoms is given by:

$$\text{volume} = \frac{22.99 \text{ g}}{0.97 \text{ g cm}^{-3}} = 23.70 \text{ cm}^3$$

The Avogadro constant
$$= \frac{\text{volume of one mole of atoms}}{\text{volume of one atom}}$$
$$= \frac{23.70 \text{ cm}^3}{0.0395 \times 10^{-21} \text{ cm}^3}$$
$$= 6 \times 10^{23}$$

1.2 Formulas

1.2.1 **Define** the terms relative atomic mass (A_r) and relative molecular mass (M_r).
1.2.2 **Calculate** the mass of one mole of a species from its formula.

Relative atomic mass, relative formula mass and molar mass

It is very difficult to determine directly the actual masses of individual atoms. However, it is relatively simple to compare the mass of one atom of a chemical element with the mass of atoms of other elements. The relative masses of atoms are determined by the use of a mass spectrometer (Chapter 2). The concept of relative masses of atoms is shown in Figure 1.12.

The **relative atomic mass** of an element is how many times greater the average mass of atoms of that element is than one-twelfth the mass of a carbon-12 atom. The weighted average mass is used since the majority of elements exist as mixtures of isotopes whose masses vary slightly (Chapter 2).

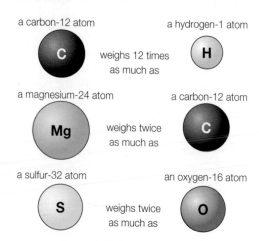

Figure 1.12 Comparing the masses of atoms

$$\text{relative atomic mass} = 12 \times \frac{\text{average mass of one atom of the element}}{\text{mass of one atom of carbon-12}}$$

For example, the average mass of one atom of hydrogen from a large sample of hydrogen atoms is 1.01. (More than 99% of hydrogen atoms have a mass of exactly 1; less than 1% have a mass of exactly 2.)

$$\text{relative atomic mass of hydrogen} = 12 \times \frac{1.01}{12} = 1.01$$

The relative atomic mass expresses masses of atoms as relative values using the carbon-12 atomic mass scale. Relative atomic masses (symbol A_r) are simply pure numbers and do not have units.

Figure 1.13 illustrate the concept of relative atomic mass applied to some isotopes of common elements. Atoms of magnesium-24 are twice as heavy as carbon-12 atoms. Therefore, the relative atomic mass of magnesium-24 atoms is 24. Three helium-4 atoms have the same mass as one carbon-12 atom. Therefore, the relative atomic mass of helium-4 atoms is 4.

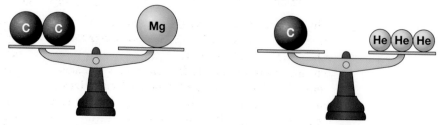

Figure 1.13 Diagrams illustrating the concept of relative atomic mass applied to carbon, magnesium and helium atoms

Relative atomic masses of all the chemical elements are listed on page 6 of the IB *Chemistry data booklet*, but data for the first 20 elements are listed opposite in Table 1.1. The relative atomic masses are reported to two decimal places, but the worked examples employ integer relative atomic masses (except chlorine) in order to simplify the calculations.

Atomic number	Name	Symbol	Relative atomic mass
1	Hydrogen	H	1.01
2	Helium	He	4.00
3	Lithium	Li	6.94
4	Beryllium	Be	9.01
5	Boron	B	10.81
6	Carbon	C	12.01
7	Nitrogen	N	14.01
8	Oxygen	O	16.00
9	Fluorine	F	19.00
10	Neon	Ne	20.18
11	Sodium	Na	22.99
12	Magnesium	Mg	24.31
13	Aluminium	Al	26.98
14	Silicon	Si	28.09
15	Phosphorus	P	30.97
16	Sulfur	S	32.06
17	Chlorine	Cl	35.45
18	Argon	Ar	39.95
19	Potassium	K	39.10
20	Calcium	Ca	40.08

Table 1.1 Atomic data for the first 20 chemical elements

With elements, especially the common gaseous elements, it is important to differentiate between the relative atomic mass and the relative molecular mass. Thus oxygen, for example, has a relative atomic mass of 16.00 but a relative molecular mass of 32.00 because it exists as diatomic molecules (O_2).

The **relative molecular mass** is the sum of the relative atomic masses of all the atoms in one molecule. Relative molecular masses (symbol M_r) are pure numbers and do not have units.

$$\text{relative molecular mass} = 12 \times \frac{\text{average mass of one molecule of the element}}{\text{mass of one atom of carbon-12}}$$

The **relative formula mass** is the sum of the relative atomic masses of all the atoms (in the form of ions) in one formula unit of an ionic compound. Relative formula masses are again pure numbers and do not have units.

The **molar mass** (symbol M) is the mass of one mole (Figure 1.14) of any substance (atoms, molecules, ions or formula units) where the carbon-12 isotope is assigned a value of *exactly* $12 \, g \, mol^{-1}$. It is a particularly useful concept since it can be applied to any chemical entity.

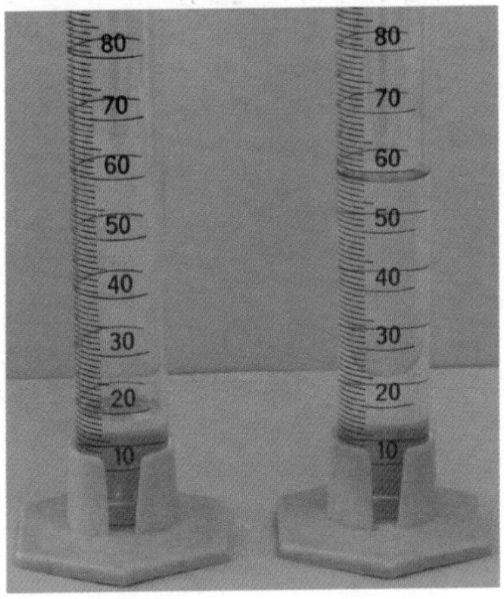

Figure 1.14 (*right*) One mole of ethanol (C_2H_5OH, molar mass = $46 \, g \, mol^{-1}$) and (*left*) one mole of water (H_2O, molar mass = $18 \, g \, mol^{-1}$) in separate measuring cylinders. One mole of different liquids may have very different masses and volumes

Worked examples

Deduce the molar mass of magnesium carbonate, $MgCO_3$.

Molar mass of magnesium carbonate, $MgCO_3 = [24 + 12 + (3 \times 16)] = 84\,g\,mol^{-1}$

The molar mass of magnesium carbonate is $84\,g\,mol^{-1}$.

Deduce the relative molecular mass of carbon dioxide, CO_2.

Molar mass of carbon dioxide, $CO_2 = [12 + (2 \times 16)] = 44\,g\,mol^{-1}$

The relative molecular mass of carbon dioxide is 44.

Deduce the relative formula mass of hydrated iron(II) sulfate crystals, $FeSO_4.7H_2O$.

Molar mass of hydrated iron(II) sulfate crystals, $FeSO_4.7H_2O$

$$= 56 + 32 + (4 \times 16) + 7 \times [(2 \times 1) + 16] = 278\,g\,mol^{-1}$$

The relative formula mass of hydrated iron(II) sulfate is 278.

Note: These are approximate molar masses, as integer values of relative atomic masses have been used to simplify the calculations.

▇ Extension: The relative atomic mass scale

Hydrogen was chosen initially as the standard because chemists realized that the element had the lightest atoms, which could therefore be given a relative atomic mass of 1. Later, when more accurate values for relative atomic masses were obtained, chemists knew that an element could contain atoms of different masses, known as isotopes. It then became necessary to choose a single isotope as the reference standard for relative atomic masses. In 1961, carbon-12 was chosen as the new standard (Figure 1.15). An isotope of carbon was chosen rather than an isotope of hydrogen because carbon is a very common element and, since it is a solid, it is also easier to store and transport than hydrogen, which is a gas.

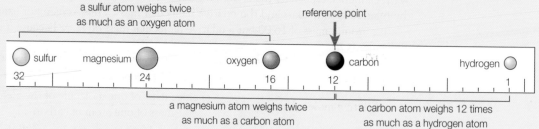

Figure 1.15 A diagram showing the principle of the carbon-12 scale

The equation below describes the relationship between the amount of a substance, its mass (in grams) and its molar mass ($g\,mol^{-1}$):

$$\text{amount of substance (mol)} = \frac{\text{mass (g)}}{\text{molar mass (g}\,mol^{-1})}$$

From this relationship it can be deduced that the mass of one mole of any substance will be equal to its molar mass in grams (Table 1.2).

Formula	Molar mass/$g\,mol^{-1}$	Number of particles	Type of particles
H	1.01	6.02×10^{23}	Atoms
C	12.01	6.02×10^{23}	Atoms
CH_4	16.05	6.02×10^{23}	Molecules
H_2O	18.02	6.02×10^{23}	Molecules
NaCl [Na^+Cl^-]	58.85	$2 \times 6.02 \times 10^{23}$	Ions
$CaCO_3$ [$Ca^{2+}CO_3^{2-}$]	100.09	$2 \times 6.02 \times 10^{23}$	Ions

Table 1.2 Some examples of molar masses

■ **Extension:**   ## Polymers

Samples of polymers (Chapter 23), such as polyethene, often contain a mixture of molecules with slightly different chain lengths centred around an average value. An average molar mass is calculated. However, a form of polymerization, known as living polymerization, can be used to generate polymer chains of identical length and hence identical molar mass.

Calculating quantities

1.2.3   **Solve** problems involving the relationship between the amount of substance in moles, mass and molar mass.

The equation relating the amount of substance in moles, mass and molar mass can be used to calculate any quantity given the values of the other two.

$$\text{amount of substance (mol)} = \frac{\text{mass (g)}}{\text{molar mass (g mol}^{-1})}$$

Determining the amount from mass and molar mass

Worked examples   Calculate the amount of water molecules present in 54 g of water, H_2O.

$$\text{Amount of water molecules} = \frac{54\,g}{18\,g\,mol^{-1}} = 3.0\,mol$$

Calculate the amount of calcium present in 0.500 kg of calcium.

$$\text{Amount of calcium} = \frac{500\,g}{40\,g\,mol^{-1}} = 12.5\,mol$$

Calculate the amount of water present in a drop with a mass of 180 mg.

$$\text{Amount of water} = \frac{0.18\,g}{18\,g\,mol^{-1}} = 0.010\,mol$$

Determining the mass from amount and molar mass

Worked example   Calculate the mass of 0.40 mol of calcium carbonate, $CaCO_3$.

$$\text{Mass of calcium carbonate (g)} = 0.4\,mol \times 100\,g\,mol^{-1} = 40\,g$$

Determining the molar mass from mass and amount

Worked example   0.00200 mol of a substance weighs 1.00 g. Calculate the molar mass of the substance.

$$\text{Molar mass (g mol}^{-1}) = \frac{1.00\,g}{0.00200\,g\,mol^{-1}} = 500\,g\,mol^{-1}$$

Calculating the mass of a single atom or molecule

If the mass of a given number of atoms or molecules is known, then the mass of a single atom or molecule can be calculated.

Worked examples   Calculate the mass of a single atom of carbon.

$$\text{Mass of a single carbon atom} = \frac{12\,g\,mol^{-1}}{6 \times 10^{23}\,mol^{-1}} = 2 \times 10^{-23}\,g$$

Calculate the mass (in grams) of a single molecule of carbon-70, C_{70}.

$$\text{Mass of a single } C_{70} \text{ molecule} = \frac{70 \times 12\,g\,mol^{-1}}{6 \times 10^{23}\,mol^{-1}} = 1.4 \times 10^{-21}\,g$$

Calculating the numbers of atoms of a specific element in a given mass of a molecular compound

The total number of atoms and atoms of a specific chemical element in a given mass of a molecular compound may also be calculated.

Worked example

Calculate the number of carbon and hydrogen atoms and the total number of atoms in 22 g of propane, C_3H_8.

Molar mass of propane, $C_3H_8 = (3 \times 12) + (8 \times 1) = 44 \, \mathrm{g \, mol^{-1}}$

One mole of propane therefore has a mass of 44 g and contains 6×10^{23} molecules of propane.

$$\text{Amount of propane} = \frac{22 \, \mathrm{g}}{44 \, \mathrm{g \, mol^{-1}}} = 0.50 \, \mathrm{mol}$$

Hence, 22 g of propane contains $6 \times 10^{23} \times 0.50 = 3 \times 10^{23}$ molecules of propane and since each molecule of propane contains eleven atoms (three carbon atoms and eight hydrogen atoms), the total number of atoms is $11 \times 3 \times 10^{23} = 3.3 \times 10^{24}$.

The total number of carbon atoms is $3 \times 3 \times 10^{23} = 9 \times 10^{23}$ and the total number of hydrogen atoms is $8 \times 3 \times 10^{23} = 2.4 \times 10^{24}$.

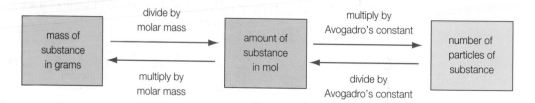

Figure 1.16 Summary of interconversions between amount, mass and number of particles

Language of Chemistry

The name mole (German *Mol*) is attributed to Wilhelm Ostwald (Figure 1.17), who introduced the concept in the year 1902. It is an abbreviation for *Molekulargewicht* (German for molecular weight), which is in turn derived from Latin *moles* 'mass, massive structure'. Ostwald (1853–1932) was a German chemist who was awarded the Nobel Prize in Chemistry in 1909 for his work in physical chemistry. Ironically, he was one of the last prominent chemists to oppose Dalton's atomic theory (Chapter 2). ■

1.2.4 **Distinguish** between the terms **empirical formula** and **molecular formula**.
1.2.5 **Determine** the empirical formula from the percentage composition or from other experimental data.

Molecular and empirical formula

Figure 1.17 Wilhelm Ostwald

The mole concept can be used to calculate the formula of a substance from experimental results. The formula obtained is the simplest possible formula (involving integers) for that compound. It is known as the **empirical formula** of the substance and can be applied to ionic and covalent compounds.

Worked examples

	Sulfur	Oxygen
Combining masses (found from experiment)	32 g	32 g
Amount of atoms	$\dfrac{32 \, \mathrm{g}}{32 \, \mathrm{g \, mol^{-1}}}$	$\dfrac{32 \, \mathrm{g}}{16 \, \mathrm{g \, mol^{-1}}}$
Ratio of moles	1 :	2

The empirical formula is therefore SO_2.

	Copper	Oxygen
Combining masses	4 g	1 g
Amount of atoms	$\dfrac{4\,g}{64\,g\,mol^{-1}}$	$\dfrac{1\,g}{16\,g\,mol^{-1}}$
Ratio of moles	0.0625 :	0.0625
Divide through by the smallest number	$\dfrac{0.0625}{0.0625}$:	$\dfrac{0.0625}{0.0625}$
Ratio of moles	1 :	1

The empirical formula is therefore CuO.

Since ionic compounds exist as giant ionic structures (Chapter 4) the concept of a molecule *cannot* be applied. The formula of an ionic compound is therefore an empirical formula, representing the ions present in their simplest ratio.

The **molecular formula** represents the actual number of atoms in a molecule of a simple covalent substance. The empirical formula and molecular formula may be identical for a molecule or may be different. The empirical formula may be found by dividing the coefficients in the molecular formula by the highest common factor.

For example, the empirical and molecular formula of water are both H_2O; the molecular formula of hydrogen peroxide is H_2O_2, but the empirical formula is HO. The molecular formula of benzene is C_6H_6 and the empirical formula is CH.

Experimental determination of empirical formula

The empirical formula may also be determined from the composition data of the compound. This data is obtained experimentally. Frequently the composition will be expressed as percentages rather than as masses. The method of working is exactly the same because with percentages we are considering the mass of each element in 100 grams of the compound.

Worked example

Determine the empirical formula of a compound containing 85.7% by mass of carbon and 14.3% by mass of hydrogen.

These percentage figures apply to any chosen amount of substance. If you choose 100 grams, then the percentages are simply converted to masses.

	Carbon	Hydrogen
Percentages by mass	85.7%	14.3%
Combining masses in 100 g	85.7 g	14.3 g
Amount of atoms	$\dfrac{85.7\,g}{12\,g\,mol^{-1}}$	$\dfrac{14.3\,g}{1\,g\,mol^{-1}}$
Ratio of moles of atoms	7.14 :	14.3
Ratio of moles of atoms	1 :	2

The empirical formula is therefore CH_2.

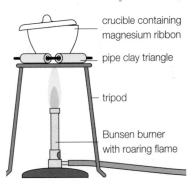

crucible containing magnesium ribbon

pipe clay triangle

tripod

Bunsen burner with roaring flame

Empirical formulas can often be determined by direct determination, for example by chemically converting a weighed sample of one element to the chosen compound and then weighing the compound to find the mass of the second element that chemically combined with the first (Figure 1.18).

An alternative method is to decompose a weighed sample of a compound containing only two elements, so that only one element remains, measure the mass of the remaining element, and then calculate the mass of the element that was originally combined.

Figure 1.18 Apparatus for determining the empirical formula of magnesium oxide (by gravimetric analysis)

Worked example

44.6 grams of an oxide of lead produced 41.4 grams of lead on reduction with hydrogen (to form water). Deduce the empirical formula of the oxide of lead.

	Lead	Oxygen
Combining masses	41.4 g	$(44.6\,g - 41.4\,g) = 3.2\,g$
Amount of atoms	$\dfrac{41.4\,g}{207\,g\,mol^{-1}}$	$\dfrac{3.2\,g}{16\,g\,mol^{-1}}$
Ratio of moles	0.2 :	0.2

The empirical formula is therefore PbO.

Language of Chemistry

Scientific evidence must be empirical, meaning that it is dependent on evidence (raw data) that is observable by the senses. In a related sense 'empirical' in science is synonymous with 'experimental'. Hence, the term 'empirical formula' refers to a formula that is derived from experimental results, often involving weighing of masses. This approach is known as **gravimetric analysis**. ■

A similar approach can be used to determine the empirical formula of a **hydrated** salt (Figure 1.19) whose **water of crystallization** can removed without the **anhydrous** salt undergoing decomposition. In the calculation the water and anhydrous salt are treated as formula units and divided by their molar masses.

Figure 1.19 Blue hydrated copper(II) sulfate crystals and almost colourless anhydrous copper(II) sulfate crystals

Worked example

12.3 grams of hydrated magnesium sulfate, $MgSO_4.xH_2O$, gives 6.0 grams of anhydrous magnesium sulfate, $MgSO_4$, on heating to constant mass. Deduce the value of x.

Mass of water driven off = 12.3 g − 6.0 g = 6.3 g

	$MgSO_4$	H_2O
Combining masses	6.0 g	6.3 g
Amount of atoms	$\dfrac{6\,g}{120\,g\,mol^{-1}}$	$\dfrac{6.3\,g}{18\,g\,mol^{-1}}$
Ratio of moles	0.05 :	0.35
Dividing through by the smallest number	$\dfrac{0.05}{0.05} = 1$:	$\dfrac{0.35}{0.05} = 7$

The empirical formula is therefore $MgSO_4.7H_2O$.

The percentage composition of a hydrocarbon is usually found by combusting a known mass of the pure compound in excess air or oxygen, then finding the masses of both the carbon dioxide (formed from the carbon in the compound) and water (formed from the hydrogen).

Worked example

5.6 grams of a pure hydrocarbon forms 17.6 grams of carbon dioxide and 7.20 grams of water when it undergoes complete combustion. Determine its empirical formula.

Amount of carbon dioxide = $\dfrac{17.6\,g}{44.0\,g\,mol^{-1}}$ = 0.400 mol

Hence, amount of carbon atoms is 0.400 mol, since every carbon dioxide molecule contains one carbon atom.

Amount of water = $\dfrac{7.20\,g}{18.0\,g\,mol^{-1}}$ = 0.400 mol

Hence the amount of hydrogen atoms is 0.800 mol, since every water molecule contains two hydrogen atoms.

The ratio of carbon to hydrogen atoms is 0.400 : 0.800, that is, 1 : 2. Hence the empirical formula is CH_2.

Determining the identity of an element in an empirical formula from percentage by mass data

Worked example Determine the identity of element X in a compound XCO_3 which has 40% by mass of X and 12% by mass of carbon.

In a sample of 100 g of XCO_3 there will be 40 g of element X, 12 g of carbon atoms and 48 g of oxygen (100 g – 40 g – 12 g).

The amount of carbon atoms $= \dfrac{12\,g}{12\,g\,mol^{-1}} = 1\,mol$

The amount of oxygen atoms $= \dfrac{48\,g}{16\,g\,mol^{-1}} = 3\,mol$

Since the empirical formula is XCO_3, the amounts of the three elements must be in a $1:1:3$ ratio. Hence, 40 g of element X contains one mole of that element and the element is therefore calcium, since it has a molar mass of $40\,g\,mol^{-1}$.

Determining the number of atoms of an element in a molecule given its molar mass and the percentage by mass of the element

Worked example An iron-containing protein has a molar mass of $136\,000\,g\,mol^{-1}$. 0.33% by mass is iron. Calculate the number of iron atoms present in one molecule of the protein.

The combined molar mass of the iron $= 0.0033 \times 136\,000 = 448.8\,g\,mol^{-1}$

$$\dfrac{448.8\,g\,mol^{-1}}{56\,g\,mol^{-1}} = 8$$

There are eight iron atoms in each molecule of the protein.

Determining the percentage by mass of an element in a compound of known formula

The experimentally determined percentage composition by mass of a compound is used to calculate the empirical formula of a compound. The reverse process can also be applied and the percentage by mass of a specific element in a compound of known formula can be calculated.

The method may be divided into three steps:
1 Determine the molar mass of the compound from its formula.
2 Write down the fraction by mass of each element (or water of crystallization) and convert this to a percentage.
3 Check to ensure that the percentages sum to 100.

Worked examples Calculate the percentage composition by mass of methane, CH_4 ($M = 16\,g\,mol^{-1}$).

Percentage by mass of carbon $= \dfrac{12}{16} \times 100 = 75\%$

Percentage by mass of hydrogen $= \dfrac{4}{16} \times 100 = 25\%$

Sum of percentages by mass $= (75\% + 25\%) = 100\%$

Calculate the percentage composition by mass of hydrated sodium sulfate, $Na_2SO_4.10H_2O$ ($M = 322\,g\,mol^{-1}$).

Percentage by mass of sodium $= \dfrac{46}{322} \times 100 = 14.3\%$

Percentage by mass of sulfur $= \dfrac{32}{322} \times 100 = 9.9\%$

Percentage by mass of oxygen $= \dfrac{64}{322} \times 100 = 19.9\%$

Percentage by mass of water $= \dfrac{180}{322} \times 100 = 55.9\%$

Sum of percentages by mass $= (14.3\% + 9.9\% + 19.9\% + 55.9\%) = 100\%$

Note that the water (of crystallization) is treated as a separate formula unit and its oxygen is *not* added to that of the sodium sulfate.

Determining the molecular formula

> **1.2.6**　**Determine** the molecular formula when given both the empirical formula and experimental data.

Since the molecular formula is a multiple of the empirical formula, the following relationship holds:

molecular formula = empirical $\times\, n$,　where n represents a small integer

Therefore in order to calculate the molecular formula of a compound it is necessary to know its molar mass. Molar masses can be determined by a variety of physical measurements, including back titrations (for weak acids and bases) (page 41) and weighing gases (page 35). Mass spectrometry is frequently used to determine the molar masses of molecular substances (Chapter 4). Automated instruments for determining the empirical and molecular formulas of organic compounds are available.

Worked example

A compound contains 73.47% carbon, 10.20% hydrogen and 16.33% by mass of oxygen. The compound has a molar mass of $196\,\mathrm{g\,mol^{-1}}$. Calculate the molecular formula.

	Carbon	*Hydrogen*	*Oxygen*
Percentages by mass	73.47	10.20	16.33
Amount of atoms in 100 g	$\dfrac{73.47}{12}$	$\dfrac{10.20}{1}$	$\dfrac{16.33}{16}$
Ratio of moles	6.1225　:	10.20　:	1.020
	$\dfrac{6.1225}{1.020}$	$\dfrac{10.20}{1.020}$	$\dfrac{1.020}{1.020}$
Dividing through by the smallest number	6　:	10　:	1

The empirical formula is therefore $C_6H_{10}O$.

To determine the molecular formula:　$196 = [(6 \times 12) + (10 \times 1) + 16] \times n$

$196 = 98 \times n$

Hence n equals 2 and the molecular formula is $C_{12}H_{20}O_2$.

1.3　Chemical equations

Chemical symbols

Each chemical element is represented by a chemical symbol (Table 1.3). The symbol consists of either one or two letters. The first letter is always a capital or upper case letter and the second letter is always small or lower case. These chemical symbols are international (Figure 1.20).

Name of chemical element	Chemical symbol	Comment
Hydrogen	H	The first letter of the name
Calcium	Ca	The first two letters of the name
Chlorine	Cl	The first letter and one other letter in the name
Sodium	Na	Two letters derived from a non-English name: *natrium* (Latin)

Table 1.3 Selected chemical elements and symbols

Language of Chemistry

A number of chemical elements are named after people, mythical characters or places (Table 1.4). ■

Name and symbol of element	Origin of the name	Additional note
Gallium (Ga)	Named after France (*Gallia*), Latin for France	The discoverer of the metal, Lecoq de Boisbaudran, subtly attached an association with his name. *Lecoq* (rooster) in Latin is *gallus*
Niobium (Nb)	*Niobe*, a mortal woman in Greek mythology	Niobe is a character in the film *Matrix Reloaded*
Vanadium (V)	Scandinavian Goddess *Vanadis* (Freyja)	
Helium (He)	*Helios* is the Greek name for the Sun	Helium was discovered in the Sun before being discovered on Earth
Mendelevium (Md)	Named after Dmitri Mendeleev who formulated the first periodic table in 1869	The element was discovered in 1955 by a team including Glenn Seaborg

Table 1.4 Selected chemical elements describing the origins of their names

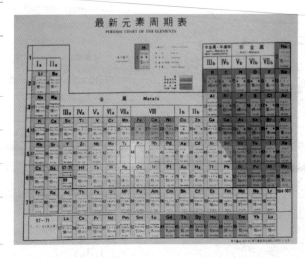

Figure 1.20 A Mandarin periodic table

Chemical formulas

Each chemical compound is represented by a unique chemical formula. The formula of any compound can be determined by performing a suitable experiment. The formulas of many compounds can be deduced using the list of ions shown in Table 1.5. A polyatomic or compound ion is an ion that contains more than two covalently bound atoms with an associated charge; a simple ion is formed by a single element.

Positive ions		Negative ions	
Simple ions	**Formula**	**Simple ions**	**Formula**
Sodium	Na^+	Chloride	Cl^-
Potassium	K^+	Bromide	Br^-
Hydrogen	H^+	Iodide	I^-
		Oxide	O^{2-}
Copper(II)	Cu^{2+}	Sulfide	S^{2-}
Iron(II)	Fe^{2+}		
Magnesium	Mg^{2+}	**Compound or polyatomic ions**	
Calcium	Ca^{2+}	Nitrate	NO_3^-
		Nitrite	NO_2^-
Iron(III)	Fe^{3+}	Sulfate	SO_4^{2-}
Aluminium	Al^{3+}	Sulfite	SO_3^{2-}
		Carbonate	CO_3^{2-}
Compound or polyatomic ions		Phosphate	PO_4^{3-}
Ammonium	NH_4^+	Hydroxide	OH^-

Table 1.5 List of common ions

Figure 1.21 A sample of the compound copper(II) carbonate, $CuCO_3$ [$Cu^{2+}CO_3^{2-}$]

In forming compounds (Figure 1.21) the number of ions used is such that the number of positive charges is equal to the number of negative charges. Ionic compounds are electrically neutral.

Examples of using the charges on ions to deduce the formula of a compound are given below:

- Sodium sulfate is composed of sodium ions, Na^+, and sulfate ions, SO_4^{2-}. Twice as many sodium ions as sulfate ions are necessary in order to have electrical neutrality. Hence, the formula of sodium sulfate is Na_2SO_4 [$2Na^+ SO_4^{2-}$].
- Magnesium hydroxide, is composed of magnesium ions, Mg^{2+}, and hydroxide ions, OH^-. Twice as many hydroxide ions as magnesium ions are necessary in order to have electrical neutrality. Hence, the formula of magnesium hydroxide is $Mg(OH)_2$ [$Mg^{2+} 2OH^-$].

The subscript number after a bracket (as in $(OH)_2$ in the formula for magnesium hydroxide) multiplies all the compound or polyatomic ions inside the bracket.

Chemical equations

> 1.3.1 **Deduce** chemical equations when all reactants and products are given.
> 1.3.2 **Identify** the mole ratio of any two species in a chemical equation.
> 1.3.3 **Apply** the state symbols (s), (l), (g) and (aq).

Chemical reactions are at the centre of chemistry and it is important that the transition from reactants to products is represented with as much precision as possible. Each reaction has an equation. The reaction of iron with chlorine is used as an example to show how to write a correct chemical equation.

- Write down the equations as a word equation, for example:

  iron + chlorine → iron(III) chloride

  The addition sign means 'reacts together' and the arrow means 'yields' and shows the direction of the reaction. (Note that some reactions are reversible, indicated by a double headed arrow ($\rightleftharpoons$), and that both forward and backward reactions will be occurring at the same time (Chapter 7).)
- Insert the correct chemical formulas for the reactant and products.

  $Fe + Cl_2 \rightarrow FeCl_3$

  This equation is unbalanced: the reactants contain (in total) one iron atom and two chlorine atoms, but the products (in total) contain one iron atom and three chlorine atoms.
- Balance the equation by ensuring that the total number of atoms of elements on each side of the equation is equal. This is achieved by inserting integer numbers termed **coefficients** which multiply all the following formula. The chemical formulas should *not* be altered.

  The selection of coefficients is done on a 'trial and error' or inspection basis, although one common approach is to start with any odd numbers in formulas and double them to convert them to even numbers. Elements represented by molecules should be left until last since their coefficients will not unbalance any other molecules. Applying this approach to the example equation gives:

  $Fe + Cl_2 \rightarrow 2FeCl_3$

  followed by:

  $2Fe + Cl_2 \rightarrow 2FeCl_3$

  and finally:

  $2Fe + 3Cl_2 \rightarrow 2FeCl_3$

This equation is now balanced: the total numbers of atoms of each element on both sides of the equation are equal, namely two iron atoms and six chlorine atoms.

The balancing of an equation is a consequence of the **law of conservation of mass**, which states that during a chemical reaction atoms cannot be created or destroyed. The coefficients in a balanced symbol equation indicate the reacting proportions in moles for stoichiometric amounts of the reactants. For example, the equation above indicates that two moles of iron atoms react with three moles of chlorine molecules to produce two moles of iron(III) chloride (formula units).

- Finally, the physical states of reactants and products should be included in small brackets after the chemical formulas.

$$2Fe(s) + 3Cl_2(g) \rightarrow 2FeCl_3(s)$$

Here the state symbol (s) represents a solid, (l) represents a pure liquid, (g) represents a pure gas and (aq) represents an aqueous solution.

- If an element occurs in more than one substance on one side of the equation then leave it to last to balance. Also keep polyatomic ions, for example NO_3^-, SO_4^{2-}, as a unit during balancing.

    Equations may also have additional information that indicate the size of the heat change during the reaction. This will depend on the physical states of the reactants and products, which shows the importance of including state symbols in symbol equations. For example:

$$2Fe(s) + 3Cl_2(g) \rightarrow 2FeCl_3(s) \qquad \Delta H^\ominus = -750\,kJ\,mol^{-1}$$

indicates that when two moles of iron(III) chloride are formed by direct synthesis under standard conditions (25 °C and 1 atmosphere pressure), 1500 kilojoules of heat energy are released (750 kJ for each mole of $FeCl_3$ formed). This is known as a thermochemical equation (Chapter 5).

    An equation can be interpreted at both an atomic or a macroscopic or visible level. The addition of state symbols or an enthalpy change makes the equation macroscopic.

TOK Link

It is good practice to include state symbols in chemical equations and their absence can cause errors. In the absence of state symbols, the reactants and products are assumed to be in their usual physical states at room temperature and pressure.

$$HCl + NaOH \rightarrow NaCl + H_2O$$

The precise interpretation of the equation above is 'one mole of gaseous hydrogen chloride and one mole of solid sodium hydroxide react to give one mole of solid sodium chloride and one mole of liquid water'. However, under anhydrous conditions (in the absence of water), such a chemical reaction would be unlikely to occur.

Presumably, the equation was meant to summarize the neutralization reaction between aqueous solutions of hydrochloric acid and sodium hydroxide:

$$HCl(aq) + NaOH(aq) \rightarrow NaCl(aq) + H_2O(l)$$

State symbols are vital if thermochemical equations (Chapter 5) are written summarizing a chemical equation and its associated energy change. State symbols must be included when writing an equation summarizing a phase change (Chapter 7), for example the sublimation of iodine: $I_2(s) \rightarrow I_2(g)$.

If the focus is purely stoichiometric, that is, on reacting amounts, then state symbols may be redundant, for example:

$$C_6H_5CH_3 + Cl_2 \rightarrow C_6H_5CH_2Cl + HCl$$

This equation shows that one mole of methylbenzene will react with one mole of chlorine to form one mole of chloromethylbenzene and one mole of hydrogen chloride. This reaction will occur regardless of what physical states the reactants are in. Of course, the reaction will be very slow if one or both reactants are solids maintained at low temperatures (Chapter 6).

Additional points about chemical equations

When constructing a balanced equation, ensure that your final set of coefficients are all whole numbers with no common factors other than one. For example, this equation is balanced:

$$4H_2(g) + 2O_2(g) \rightarrow 4H_2O(l)$$

However, all the coefficients have the common factor of two. Divide through by two to eliminate common factors:

$$2H_2(g) + O_2(g) \rightarrow 2H_2O(l)$$

It is allowable, and sometimes necessary, to use fractional coefficients in the balancing process, for example:

$$C_2H_6(g) + \frac{7}{2}O_2(g) \rightarrow 2CO_2(g) + 3H_2O(l)$$

Generally, the fractional coefficient is not retained in the final answer. Multiplying the coefficients through by 2 removes the fraction:

$$2C_2H_6(g) + 7O_2(g) \rightarrow 4CO_2(g) + 6H_2O(l)$$

However, if an equation represents the standard molar enthalpy of combustion, then fractional coefficients may have to be used. The standard molar enthalpy of combustion represents the energy change when one mole of a compound undergoes complete combustion in the presence of excess oxygen (Chapter 5).

Hence the equation:

$$C_2H_6(g) + \frac{7}{2}O_2(g) \rightarrow 2CO_2(g) + 3H_2O(l)$$

correctly represents the standard molar enthalpy of combustion of ethane.

It should also be noted that some reactions do *not* occur, even though balanced equations can be written. Examples include:

$$Cu(s) + H_2SO_4(aq) \rightarrow CuSO_4(aq) + H_2(g)$$

and $\quad Ag(s) + NaCl(aq) \rightarrow AgCl(aq) + Na(aq)$

Hence, the reactivity or electrochemical series (Chapter 9) should be consulted before equations for displacement reactions are written.

Information conveyed by a chemical reaction

Qualitatively, a chemical equation gives the names (via naming rules) of the various reactants and products, and directly gives their physical states. Quantitatively, it expresses the following information:

- the relative number of chemical entities of the reactants and products involved in the chemical reaction
- the relative amounts (in moles) of the reactant and products
- the relative reacting masses of reactants and products
- the relative volumes of gaseous reactants and products.

Consider the following equation:

$$H_2(g) + Cl_2(g) \rightarrow 2HCl(g)$$

Qualitatively, it indicates that hydrogen reacts with chlorine to form hydrogen chloride. The hydrogen, chlorine and hydrogen chloride are all in the gaseous form.

Quantitatively, it conveys the following information:

- one mole of hydrogen molecules reacts with one mole of chlorine molecules to form two moles of molecules of hydrogen chloride
- 2 grams of hydrogen react with 71 grams of chlorine to form 73 grams of hydrogen chloride
- one volume of hydrogen reacts with one volume of chlorine to form two volumes of hydrogen chloride (see Avogadro's law, page 26).

■ **Extension:** Types of chemical reactions

There are several basic types of chemical reactions:

■ **Synthesis**, where two or more elements or compounds may combine to form a more complex compound.

The basic form of this type of reaction is A + X → AX, for example:

$$2Mg(s) + O_2(g) \rightarrow 2MgO(s)$$

and

$$2Na(s) + Cl_2(g) \rightarrow 2NaCl(s)$$

■ **Decomposition**, where a single compound breaks down into its elements or simpler compounds (Figure 1.22).

The basic form of this type of reaction is AX → A + X, for example:

$$CaCO_3(s) \rightarrow CaO(s) + CO_2(g)$$

■ **Displacement**, where a more reactive element takes the place of another element in a compound (Chapters 3 and 9).

The basic form of this type of reaction is A + BX → AX + B or AX + Y → AY + X, for example:

$$Fe(s) + CuSO_4(aq) \rightarrow FeSO_4(aq) + Cu(s)$$

$$2Na(s) + 2H_2O(l) \rightarrow 2NaOH(aq) + H_2(g)$$

and

$$Cl_2(aq) + 2NaBr(aq) \rightarrow 2NaCl(aq) + Br_2(aq)$$

Displacement reactions are all examples of **redox** reactions (Chapter 9).

■ **Precipitation**, when a pair of ions interact to produce an insoluble precipitate. Precipitation reactions are also called double decomposition reactions.

The basic form of this type of reaction is AX + BY → AY + BX, for example:

$$NaCl(aq) + AgNO_3(aq) \rightarrow NaNO_3(aq) + AgCl(s)$$

■ **Acid–base reactions** (Chapter 8), where an acid and a base (metal oxide, metal hydroxide or aqueous ammonia) react to produce a salt and water only, for example:

$$H_2SO_4(aq) + CuO(s) \rightarrow CuSO_4(aq) + H_2O(l)$$

$$HCl(aq) + NaOH(aq) \rightarrow NaCl(aq) + H_2O(l)$$

and

$$NH_4OH(aq) + HCl(aq) \rightarrow NH_4Cl(aq) + H_2O(l)$$

Figure 1.22 Green copper(ii) carbonate undergoing thermal decomposition to form black copper(ii) oxide and carbon dioxide gas

Ionic equations

When a soluble ionic substance is dissolved in water, the ions separate and behave independently. For example, if barium chloride is dissolved in water, hydrated barium and chloride ions are formed:

$$BaCl_2(s) + (aq) \rightarrow BaCl_2(aq) \rightarrow Ba^{2+}(aq) + 2Cl^-(aq)$$

The barium and chloride ions undergo their characteristic reactions regardless of which other ions may be present in the solution. For example, barium ions in solution react with sulfate ions in solution to form a white precipitate of barium sulfate (Figure 1.23).

If a solution of barium chloride, $BaCl_2$, and a solution of sodium sulfate, Na_2SO_4, are mixed, a white precipitate of barium sulfate, $BaSO_4$, is rapidly produced. The following equations describe the precipitate formation:

$$BaCl_2(aq) + Na_2SO_4(aq) \rightarrow BaSO_4(s) + 2NaCl(aq)$$

or

$$Ba^{2+}(aq) + 2Cl^-(aq) + 2Na^+(aq) + SO_4^{2-}(aq) \rightarrow BaSO_4(s) + 2Na^+(aq) + 2Cl^-(aq)$$

Figure 1.23 Precipitate of barium sulfate

The second equation shows that the sodium and chloride ions have *not* undergone any change: they existed as independent ions both before and after the reaction took place. They are termed **spectator ions** and can be removed from the equation to generate a **net ionic equation**:

$$Ba^{2+}(aq) + SO_4^{2-}(aq) \rightarrow BaSO_4(s)$$

This equation may be interpreted to mean that any soluble barium salt will react with any soluble sulfate to produce barium sulfate.

The solubility of common salts is summarized below:

- All sodium, potassium and ammonium salts are soluble.
- All nitrates are soluble.
- All chlorides are soluble *except* silver chloride (Chapter 3) and lead(II) chloride.
- All sulfates are soluble except calcium, barium and lead(II) sulfate.
- Sodium, potassium and ammonium carbonates are soluble *but* all other carbonates are insoluble.

The 'solubility rules' do not need to be memorized.

Net ionic equations must always have the same net charge on each side of the equation. In the net ionic equation above the net charge on both sides of the equation is zero.

Net ionic equations may be written whenever reactions occur in aqueous solution in which some of the ions originally present are removed from solution or when ions not originally present are formed. Ions are removed from solution by the following processes:

- Formation of an insoluble precipitate

$$AgNO_3(aq) + NaCl(aq) \rightarrow AgCl(s) + NaNO_3(aq)$$

$$Ag^+(aq) + Cl^-(aq) \rightarrow AgCl(s)$$

- Formation of molecules containing only covalent bonds

$$HCl(aq) + NaOH(aq) \rightarrow NaCl(aq) + H_2O(l)$$

$$H^+(aq) + OH^-(aq) \rightarrow H_2O(l)$$

- Formation of a new ionic chemical species

$$Zn(s) + CuSO_4(aq) \rightarrow ZnSO_4(aq) + Cu(s)$$

$$Zn(s) + Cu^{2+}(aq) \rightarrow Zn^{2+}(aq) + Cu(s)$$

- Formation of a gas

$$Na_2CO_3(s) + 2HCl(aq) \rightarrow 2NaCl(aq) + CO_2(g) + H_2O(l)$$

$$CO_3^{2-}(aq) + 2H^+(aq) \rightarrow CO_2(g) + H_2O(l)$$

Language of Chemistry

The 'language' of chemistry frequently transcends cultural, linguistic and national boundaries. Although the symbols for the chemical elements are international, the names of the elements are sometimes language dependent, often with the end of the name characterizing the specific language. For example, magnesium changes to *magnésium* in French, *magnesio* in Spanish, *magnesion* in Greek and *magnij* in Russian. In Japanese *katakama* reproduces the sound of the English 'magnesium'. ■

Figure 1.24 The Japanese *kanji* (pictogram) for sulfur (translated as 'yellow substance')

■ Extension: Naming inorganic compounds

Naming ionic compounds

The names of ionic compounds give information about their composition.

- Compounds ending in the suffix *–ide* typically contain two chemical elements chemically bonded together.
  Example: sodium chloride is a compound of sodium and chlorine.
  Exceptions: metal hydroxides, for example sodium hydroxide, which contains the elements sodium, oxygen and hydrogen.
- Compounds ending in the suffixes *–ate* or *–ite* contain oxygen. There is a greater amount of oxygen in the compound ending in *–ate*.
  Examples: sodium sulfate, Na_2SO_4, and sodium sulfite, Na_2SO_3.
- Compounds with the prefix *per-* contain extra oxygen.
  Examples: water (hydrogen oxide), H_2O, and hydrogen peroxide, H_2O_2.
- Compounds with the prefix *thio-* contain extra sulfur in place of an oxygen.
  Examples: sodium sulfate, Na_2SO_4, and sodium thiosulfate, $Na_2S_2O_3$.
- When a metal forms more than one stable positive ion then the name of the ion is the element name and, in parenthesis next to it, a Roman number denoting the charge. The names of ionic compounds with these ions must include these Roman numbers.
  Examples: iron(II) oxide, FeO [Fe^{2+} O^{2-}] and iron(III) oxide, Fe_2O_3 [$2Fe^{3+}$ $3O^{2-}$].

Naming hydrates

Hydrates are substances that include water of crystallization in their formula, for example hydrated copper(II) sulfate, $CuSO_4.5H_2O$. A more precise name would be copper(II) sulfate pentahydrate or copper(II) sulfate-5-water.

This formula indicates that every formula unit [Cu^{2+} SO_4^{2-}] is associated with five water molecules. The dot sign in the formula indicates that although the water is chemically bonded to the formula unit, it can be readily removed by heating (page 11).

$$CuSO_4.5H_2O(s) \rightarrow CuSO_4(s) + 5H_2O(g)$$

■ Extension: The laws of chemical combination

The laws of chemical combination listed below are all consequences of the atomic theory of matter (Chapter 4).

Law of conservation of mass

There is no increase or decrease in mass during a chemical reaction (Figure 1.25). The atoms of a chemical substance cannot be created or destroyed during a chemical reaction. If the reacting substances are weighed before a chemical reaction and the products are accurately weighed after a chemical reaction, the mass is unchanged.

The law applies provided the product(s) do not escape and the mass of all the products is measured. However, if the reaction between calcium carbonate and dilute aqueous acid is performed in an open beaker, then there is a steady decrease in the total mass due to loss of the carbon dioxide, a gaseous product.

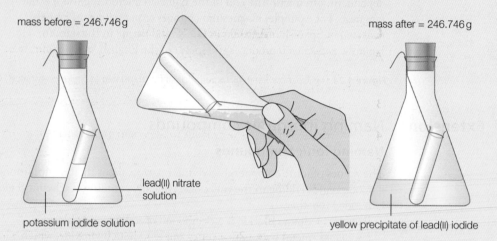

mass before = 246.746 g

mass after = 246.746 g

lead(II) nitrate solution

potassium iodide solution

yellow precipitate of lead(II) iodide

Figure 1.25 An illustration of the law of conservation mass

Law of constant composition

A number of compounds can be prepared by a number of different methods. However, the chemical composition of the compound is identical regardless of the method used. For example, copper(II) oxide can be prepared by heating copper(II) carbonate or copper(II) nitrate.

$$CuCO_3(s) \rightarrow CuO(s) + CO_2(g)$$

$$2Cu(NO_3)_2(s) \rightarrow 2CuO(s) + 4NO_2(g) + O_2(g)$$

The copper(II) oxide produced by these and other reactions can be converted to copper by reaction with hydrogen.

$$CuO(s) + H_2(g) \rightarrow Cu(s) + H_2O(l)$$

Equal masses of copper(II) oxide, produced by different methods, form equal masses of copper when converted to the element.

Law of multiple proportions

The law of multiple proportions applies when two elements form more than one compound, for example, copper(II) oxide, CuO, and copper(I) oxide, Cu_2O. If two elements (A and B) combine together to form more than one compound, then the different masses of A that combine with a fixed mass of B are in a simple ratio. For example, if the masses of two copper oxides are converted to copper by reaction with hydrogen, the masses of copper that combine with 1 gram of oxygen are in the ratio 2:1.

Avogadro's law

Equal volumes of different gases, under the same conditions (page 26) of temperature and pressure, contain the same number of molecules. This law is commonly used in conjunction with Gay-Lussac's law.

Gay-Lussac's law

The volumes of gases reacting and the volumes of gaseous products have a simple numerical relationship to one another, provided all measurements are made at the same temperature and pressure (page 26).

1.4 Mass and gaseous volume relationships in chemical reactions

1.4.1 **Calculate** theoretical yields from chemical equations.

Calculating theoretical yields

Almost all stoichiometric problems can be solved in just four simple steps (Figure 1.26):

1 If necessary, provide a balanced equation.
2 Convert the mass (or volume) units of a given reactant to an amount (in moles).
3 Using the mole ratio from the coefficients in the equation, calculate the amount of the required product.
4 Convert the amount of the product to the appropriate units of mass (or volume).

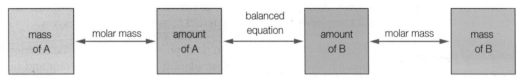

Figure 1.26 Graphical summary of interconversion of mass relationships in a chemical reaction

Determining the mass of a product (from a single reaction)

Worked example

Calculate the mass of calcium oxide that could be obtained by heating 2.5 grams of calcium carbonate, $CaCO_3$. (Assume that the calcium carbonate is pure and that complete decomposition occurs.)

$CaCO_3(s) \rightarrow CaO(s) + CO_2(g)$

Amount of calcium carbonate $= \dfrac{2.5\,\text{g}}{100\,\text{g mol}^{-1}} = 0.025\,\text{mol}$

The coefficients in the equation indicate that one mole of calcium carbonate, $CaCO_3$, decomposes to give one mole of carbon dioxide, CO_2, and one mole of calcium oxide, CaO.

Consequently, 0.025 mol of calcium carbonate, $CaCO_3$, decomposes to give 0.025 mol of carbon dioxide, CO_2, and 0.025 mol of calcium oxide, CaO.

Mass of calcium oxide $= 0.025\,\text{mol} \times 56\,\text{g mol}^{-1} = 1.4\,\text{g}$

Determining the mass of a product (from consecutive reactions), where the product of a reaction is the reactant of a subsequent reaction

Worked example

Calculate the mass of nitric acid that can be produced from 56 grams of nitrogen gas.

$N_2 + 3H_2 \rightarrow 2NH_3$

$4NH_3 + 5O_2 \rightarrow 4NO + 6H_2O$

$2NO + O_2 \rightarrow 2NO_2$

$2H_2O + 4NO_2 + O_2 \rightarrow 4HNO_3$

The coefficients in the equation indicate that: 1 mol of N_2 produces 2 mol of NH_3; 4 mol of NH_3 produces 4 mol of NO; 2 mol of NO produces 2 mol of NO_2; and 4 mol of NO_2 produces 4 mol of HNO_3. So overall, 1 mol of N_2 produces 2 mol HNO_3.

Amount of nitrogen, $N_2 = \dfrac{56\,\text{g}}{28\,\text{g mol}^{-1}} = 2\,\text{mol}$

Amount of nitric acid produced $= 2 \times 2\,\text{mol} = 4\,\text{mol}$

Hence, mass of nitric acid produced $= 4\,\text{mol} \times 63\,\text{g mol}^{-1} = 252\,\text{g}$

1.4.2   **Determine** the limiting reactant and the reactant in excess when quantities of reacting substances are given.

The limiting reactant and the reactant in excess

Frequently during chemical reactions, one of the reactants is present in excess. This means that once the reaction is complete, some of that reactant will be left over. For example, consider the reaction between hydrogen and oxygen to form water:

$$2H_2(g) + O_2(g) \rightarrow 2H_2O(l)$$

Suppose there is a reaction where two moles of hydrogen and two moles of oxygen are available for reaction. The coefficients in the equation indicate that only one mole of oxygen is required to react with two moles of hydrogen. This means that one mole of oxygen will be left over when the reaction is complete.

The amount of water obtained is determined by the amount of reactant that is completely consumed during the reaction. This reactant is termed the **limiting reactant**. The reactant which is not completely consumed is referred to as the reactant in **excess**.

Worked example

Calculate the mass of magnesium that can be obtained from the reaction between 4.8 grams of magnesium and 4.8 grams of sulfur. Identify the limiting reactant and calculate the mass of the unreacted element present in excess.

$$Mg(s) + S(s) \rightarrow MgS(s)$$

Amount of magnesium atoms $= \dfrac{4.8}{24} = 0.20\,\text{mol}$

Amount of sulfur atoms $= \dfrac{4.8}{32} = 0.15\,\text{mol}$

The coefficients in the equation indicate that one mole of magnesium atoms reacts with one mole of sulfur atoms to form one mole of magnesium sulfide. The amounts indicate that sulfur is the limiting reactant and magnesium is present in excess.

Mass of magnesium sulfide formed $= 0.15\,\text{mol} \times 56\,\text{g mol}^{-1} = 8.4\,\text{g}$

Amount of magnesium unreacted $= 0.20\,\text{mol} - 0.15\,\text{mol} = 0.05\,\text{mol}$

Mass of magnesium unreacted $= 0.050\,\text{mol} \times 24\,\text{g mol}^{-1} = 1.2\,\text{g}$

1.4.3   **Solve** problems involving theoretical, experimental and percentage yield.

Percentage and experimental yield

The quantity of product that is calculated to be formed when all the limiting reactant reacts is termed the **theoretical yield**. The mass, volume or amount of a product actually obtained in a chemical reaction is termed the **experimental yield**.

The experimental yield is always less than the theoretical yield for one or more of the following reasons:

- side reactions
- the reaction is reversible and reaches equilibrium (Chapter 7)
- mechanical losses – physical loss of the reactants or products
- impurities present in the reactants.

The **percentage yield** can be calculated from the following expression:

$$\text{percentage yield} = \frac{\text{experimental yield}}{\text{theoretical yield}} \times 100$$

Percentage yields are of particular importance in organic chemistry (Chapter 10) because there are significant side reactions and many organic reactions are reversible.

Worked example

In an experiment to produce a sample of hex-1-ene, 20.4 grams of hexan-1-ol was heated with an excess of phosphoric(V) acid. The phosphoric(V) acid acted as a dehydrating agent, removing water from the alcohol to form hex-1-ene.

$$CH_3CH_2CH_2CH_2CH_2CH_2OH \rightarrow CH_3CH_2CH_2CH_2CH=CH_2 + H_2O$$

 hexan-1-ol   hex-1-ene

After purification of the hex-1-ene, 10.08 grams was produced. Calculate the percentage yield.

From the equation, 1 mol of hexan-1-ol produces 1 mol of hex-1-ene.

$$\text{Amount of hexan-1-ol} = \frac{20.4\,g}{102\,g\,mol^{-1}} = 0.200\,mol$$

Hence, the theoretical amount of hex-1-ene produced is 0.200 mol (since there is excess phosphoric(V) acid).

$$\text{Amount of hex-1-ene} = \frac{\text{mass (g)}}{\text{molar mass (g\,mol}^{-1})} = 0.200\,mol$$

Rearranging, mass of hex-1-ene = $84\,g\,mol^{-1} \times 0.200\,mol = 16.8\,g$

Since only 10.08 g of hex-1-ene was produced, the percentage yield is

$$\frac{10.08\,g}{16.8\,g} \times 100 = 60\%$$

Percentage purity

The percentage purity is the per cent of a specified compound or element in an impure sample. The percentage purity of a substance is often determined by a **titration** or a **back titration** (page 41).

The percentage purity of a sample of a chemical substance can be calculated from the following relationship:

$$\text{percentage purity} = \frac{\text{mass of pure substance in a sample}}{\text{mass of sample}} \times 100$$

Worked example

When 12 grams of impure carbon was burnt in excess oxygen, 33 grams of carbon dioxide was obtained. Calculate the percentage purity of the carbon.

The balanced equation for the reaction is:

$$C(s) + O_2(g) \rightarrow CO_2(g)$$

1 mol   1 mol   1 mol

$$\text{Amount of carbon dioxide formed} = \frac{33\,g}{44\,g\,mol^{-1}} = 0.75\,mol$$

Since one mole of carbon dioxide is obtained from one mole of carbon (in the presence of excess oxygen), the amount of carbon in the original sample must be 0.75 mol.

Therefore, the mass of carbon in the original sample = $0.75\,mol \times 12\,g\,mol^{-1} = 9.0\,g$.

Hence, the percentage purity = $\dfrac{9.0\,g}{12\,g} \times 100 = 75\%$

Reacting volumes of gases

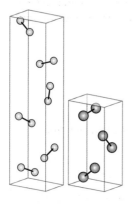

Figure 1.27 Molecular models showing the reaction between hydrogen and chlorine

A French chemist called Gay-Lussac studied chemical reactions between gases. He discovered that in such reactions, the volumes of the reacting gases (measured at the same temperature and pressure) were in a simple whole number ratio to one another and the volumes of the gaseous products. This is known as Gay-Lussac's law. For example, he found that one volume of hydrogen always reacted with exactly the same volume of chlorine to form two volumes of hydrogen chloride (Figures 1.27 and 1.29), and one volume of oxygen always reacted with two volumes of hydrogen to form two volumes of steam.

The Italian chemist Avogadro explained Gay-Lussac's results by suggesting that equal volumes of gases, measured at the same temperature and pressure, contain the same number of molecules. This suggestion is now called **Avogadro's law**. Mathematically, it can be expressed as:

$$V \propto n$$

where V represents the volume of gas and n represents the amount of gas (in moles). For example, if the number of molecules of a mass of gas is doubled, then the volume (at constant temperature and pressure) of the gas is doubled (Figure 1.28).

Using Avogadro's law, Gay-Lussac's observations on the formation of steam by direct synthesis can be interpreted as shown below:

2 volumes of hydrogen + 1 volume of oxygen → 2 volumes of steam

means

2 mol of hydrogen molecules + 1 mol of oxygen molecules → 2 mol of steam (water molecules)

and

2 hydrogen molecules + 1 oxygen molecule → 2 steam (water) molecules

Figure 1.28 Two volumes of hydrogen molecules and one volume of oxygen molecules (an illustration of Avogadro's law)

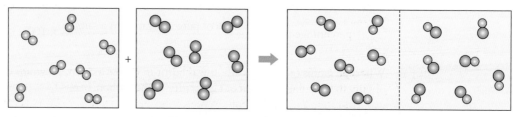

Figure 1.29 A diagrammatic illustration of Avogadro's law for the formation of hydrogen chloride from hydrogen and chlorine

Avogadro's law is a consequence of the large intermolecular distances between molecules in the gaseous state: a gas is mainly *empty space*.

Calculating the volumes of reactants used

Worked examples

Calculate the volume of oxygen needed to burn $200\,cm^3$ of propane.

$$C_3H_8(g) + 5O_2(g) \rightarrow 3CO_2(g) + 4H_2O(l)$$

This equation indicates that one mole of propane reacts with five moles of oxygen to produce three moles of carbon dioxide and four moles of water. Applying Avogadro's law means that one volume of propane reacts with five volumes of oxygen to produce three volumes of carbon dioxide.

(Note that Avogadro's law only applies to gases. Since liquid water is formed you cannot make a statement about its volume.)

The volumes of propane and oxygen must be in a $1:5$ ratio, hence the volume of oxygen needed is $5 \times 200\,cm^3 = 1000\,cm^3$.

Calculate the volume of air needed to burn 200 cm³ of propane. Air is 20% by volume oxygen.

$$C_3H_8(g) + 5O_2(g) \rightarrow 3CO_2(g) + 4H_2O(l)$$

The volume of oxygen required is 1000 cm³. However, air is only $\frac{1}{5}$ oxygen and so you need five times more air than oxygen. Hence, the volume of air required is 5 × 1000 cm³ = 5000 cm³.

Calculating the volumes of products produced

Worked example

Calculate the volume of carbon dioxide produced by the combustion of 0.500 dm³ of butane, C_4H_{10}.

$$2C_4H_{10}(g) + 13O_2(g) \rightarrow 8CO_2(g) + 10H_2O(l)$$

This equation indicates that 2 volumes of butane react with 13 volumes of oxygen to form 8 volumes of carbon dioxide.

The amounts of butane and carbon dioxide are in a 2:8 ratio, hence the volume of carbon dioxide formed is 0.500 dm³ × 4 = 2.00 dm³.

Deducing the molecular formula

Worked example

When 20 cm³ of a gaseous hydrocarbon is reacted with excess oxygen the gaseous products consist of 80 cm³ of carbon dioxide, CO_2, and 80 cm³ of steam, H_2O, measured under the same conditions of pressure and temperature.

Deduce the molecular formula of the hydrocarbon.

20 cm³ hydrocarbon + excess oxygen → 80 cm³ CO_2 + 80 cm³ H_2O

1 molecule of hydrocarbon → 4 molecules of CO_2 + 4 molecules of H_2O

Each molecule of the hydrocarbon must contain four carbon atoms and eight hydrogen atoms. The molecular formula is therefore C_4H_8.

History of Chemistry

Joseph Louis Gay-Lussac (1778–1850) was a French chemist and physicist. He held appointments as Professor of Physics and Professor of Chemistry at three different institutions. Hot air balloons were very popular in France at the end of the 18th century and probably stimulated his research into the physical and chemical properties of gases (Figure 1.30). Gay-Lussac investigated the relationship between temperature and volume of a gas (at constant pressure). The law was first published by Gay-Lussac in 1802, but he referenced unpublished work by French scientist Jacques Charles from around 1787. In English speaking countries this gas law is termed Charles's law but in France it is termed Gay-Lussac's law.

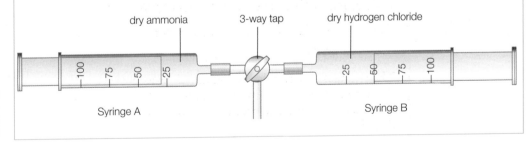

Figure 1.30 Apparatus used for verifying Gay-Lussac's law of combining volumes

Molar volume of a gas

1.4.5 **Apply** the concept of molar volume at standard temperature and pressure in calculations.

It follows from Avogadro's law that the volume occupied by one mole of molecules must be the same for all gases (Figure 1.31). It is known as the gas molar volume and has an approximate value of $22.4\,dm^3$ at $0\,°C$ ($273\,K$) and 1 atmosphere ($1.01 \times 10^5\,Pa$). These conditions are known as standard temperature and pressure (stp).

This relationship, together with Avogadro's constant ($6.02 \times 10^{23}\,mol^{-1}$), allows us to solve the following types of stoichiometry problems.

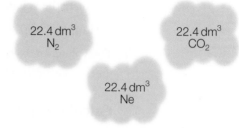

Figure 1.31 An illustration of the molar gas volume (at stp). All samples of the gas contain the same number of particles (atoms or molecule)

Worked examples

Calculate the volume of oxygen in dm^3 at stp that contains 1.35 mol of molecules.

Volume occupied = $1.35\,mol \times 22.4\,dm^3\,mol^{-1} = 30.2\,dm^3$

Calculate the volume of 0.020 g of hydrogen at stp.

Amount of hydrogen, $H_2 = \dfrac{mass\,(g)}{molar\,mass\,(g\,mol^{-1})} = \dfrac{0.020\,g}{2\,g\,mol^{-1}} = 0.010\,mol$

Hence, 0.010 mol of hydrogen, H_2 occupies $0.010\,mol \times 22.4\,dm^3\,mol^{-1} = 0.22\,dm^3$

Calculate the amount of hydrogen gas in $175\,cm^3$ at stp.

Amount of gas = $\dfrac{(175/1000)\,dm^3}{22.4\,dm^3\,mol^{-1}} = 7.81 \times 10^{-3}\,mol$

Calculate the number of molecules present in $2.85\,dm^3$ of carbon dioxide at stp.

Amount of carbon dioxide = $\dfrac{2.85\,dm^3}{22.4\,dm^3\,mol^{-1}} = 0.127\,mol$

Number of molecules = $0.127 \times 6.02 \times 10^{23}\,mol^{-1} = 7.65 \times 10^{22}$

Calculate the density (in grams per cubic decimetre, $g\,dm^{-3}$) of argon gas at stp. The relative atomic mass (from the periodic table) for argon is 39.95. Hence, $22.4\,dm^3$ (1 mol) of argon gas weighs 39.95 g.

Density = $\dfrac{mass}{volume} = \dfrac{39.95\,g}{22.4\,dm^3} = 1.78\,g\,dm^{-3}$

20.8 grams of a gas occupies $7.44\,dm^3$ at stp. Determine the molar mass of the gas.

Amount of gas = $\dfrac{7.44\,dm^3}{22.4\,dm^3\,mol^{-1}} = 0.332\,mol$

Molar mass = $\dfrac{20.8\,g}{0.332\,mol} = 62.7\,g\,mol^{-1}$

When 3.06 grams of potassium chlorate(v), $KClO_3$, is heated it produces $840\,cm^3$ of oxygen (at stp) and leaves a residue of solid potassium chloride, KCl. Deduce the balanced equation. The molar mass of potassium chlorate(v) is $122.5\,g\,mol^{-1}$.

Amount of $KClO_3 = \dfrac{3.06\,g}{122.5\,g\,mol^{-1}} = 0.025\,mol$

Amount of $O_2 = \dfrac{0.84\,dm^3}{22.4\,dm^3\,mol^{-1}} = 0.0375\,mol$

The simplest molar ratio is 2 : 3 and hence the balanced equation must be:

$2KClO_3(s) \rightarrow 2KCl(s) + 3O_2(g)$

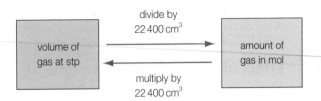

Figure 1.32 Summary of interconversions between the amount of gas and volume (at stp)

History of Chemistry

Amedeo Avogadro (1776–1856) was an Italian physicist who made many early contributions to the concepts of molecularity and relative molecular mass (formerly known as molecular weight) (Figure 1.33). His most critical contribution was making the distinction between atoms and molecules. He trained and practised as a lawyer, but later became Professor of Physics at Turin University. As a tribute to him, the number of particles in one mole of a substance is known as the Avogadro constant (formerly known as the Avogadro number).

Josef Loschmidt (1821–1895) was born in Bohemia, now part of the Czech Republic. He worked as a school teacher, and later as a Professor at the University of Vienna. He calculated the number of molecules in one cubic centimetre of gas at standard temperature and pressure. This quantity is known as the Loschmidt number, but is seldom used today.

Figure 1.33 Caricature of Amedeo Avogadro

Relationship between temperature, pressure and volume of a gas

1.4.6 **Solve** problems involving the relationship between temperature, pressure and volume for a fixed mass of an ideal gas.

In a gas the molecules are completely free to move in all directions and travel in straight lines until they collide with other gas molecules or atoms or bounce off the walls of the container. The *overall* resulting movement is completely random (Figure 1.34).

Forces of attraction between molecules or between single molecules and molecules of the walls of the container are usually so small that they can be neglected. However, when molecules approach close to one another, or to molecules of the walls, their kinetic energy brings them close enough for these forces to become repulsive and no longer negligible – hence they rebound.

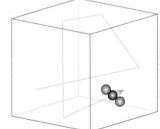

Figure 1.34 The overall random movement of a gas molecule

When a gas is heated the particles move faster and therefore collide more often with each other and the walls of the container. Gases are very compressible because of the large interatomic or intermolecular spaces between the particles. This also causes the density of gases to be very low compared to that of solids and liquids.

Effect of pressure on volume of a gas

The effect of pressure on a fixed volume of gas can be demonstrated (in a qualitative manner) by trapping some air inside a sealed gas syringe connected to a pressure gauge and pushing in the plunger (Figure 1.35): the pressure of the gas increases as the volume decreases. The temperature remains constant during this change (if it is performed slowly, so that the air can remain in temperature equilibrium with the syringe and the surroundings).

Figure 1.35 A syringe of air connected to a pressure gauge

The increase in the pressure of the gas is due to the increase in the frequency of collisions between the gas particles and the container walls. Since the volume is smaller, and the number of particles is constant, they hit the walls more often.

Effect of temperature on volume of a gas

This can be easily demonstrated (in a qualitative manner) by trapping some air inside a sealed gas syringe and placing it in a beaker of hot water: the volume of the gas increases as the temperature increases (Figure 1.36). The pressure remains constant during this change.

Heating the gas makes the particles move (on average) faster so that they hit the walls of the container more frequently and with greater momentum. This pushes the plunger of the syringe outwards, increasing the volume of the gas until this reduces the collision frequency to compensate for the increased speed and the pressure in the trapped gas is equal to the pressure of the atmosphere on the other side of the syringe.

Effect of temperature on pressure of a gas

If a sealed container is heated with gas inside, then once the temperature is sufficiently high the lid will fly off or the container will explode. This shows that gas pressure (at constant volume) increases with temperature.

The effect of changing one of these variables (pressure, temperature and volume) on a fixed mass of gas, while keeping the other one constant, is summarized in Table 1.6.

Figure 1.36 A sealed syringe of air in a beaker of hot water

Table 1.6 Summary of the behaviour of gases

a At constant temperature:

Pressure	Volume
Increase	Decrease
Decrease	Increase

b At constant volume:

Temperature	Pressure
Increase	Increase
Decrease	Decrease

c At constant pressure:

Temperature	Volume
Increase	Increase
Decrease	Decrease

Gas density

The density of a gas is defined as the mass of the gas divided by its volume. The volume of the gas changes with temperature and pressure. Specifically, if the pressure on a gas is increased its volume decreases, resulting in an increase in density. If the temperature of a gas is increased, the volume increases and the density is reduced.

Kinetic theory of gases

The behaviour of gases is explained by the kinetic theory of gases, which makes the following assumptions about the behaviour and properties of particles in a gas:

- the individual molecules or atoms of a gas each have a negligible volume compared to the container
- there are no attractive or repulsive forces operating between the atoms or molecules of the gas, except at collisions between molecules and between individual molecules and the walls of the container (Figure 1.37)
- the collisions between the molecules or atoms with themselves and the walls of the container are perfectly elastic and give rise to gas pressure
- the mean kinetic energy of the molecules or atoms of a gas is directly proportional to its absolute temperature on the thermodynamic scale. The kinetic energy of a gaseous molecule or atom is given by the expression $\frac{1}{2}mv^2$, where m represents the mass of the particle and v represents its speed.

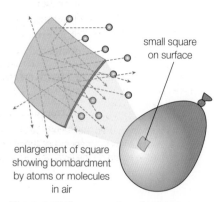

small square on surface

enlargement of square showing bombardment by atoms or molecules in air

Figure 1.37 The generation of gas pressure on the inner surface of a balloon.

This model of a gas is known as the **ideal gas** model and is a good description of the behaviour of most gases, especially at high temperatures and low pressures. However, an ideal gas is a *hypothetical state* since the first two assumptions of the ideal gas model cannot be precisely true. No gas behaves absolutely perfectly as an ideal gas.

The pressure of a gas can be measured using a manometer (Figure 1.38). In its simplest form the manometer is a U-tube about half filled with liquid. With both ends of the tube open, the liquid is at the same height in each leg. When positive pressure is applied to one leg, the liquid is forced down in that leg and up in the other. The difference in height indicates the pressure.

The temperature scale used in kinetic theory is the absolute or thermodynamic (Kelvin) scale. The thermodynamic scale of temperature uses units of kelvin (K), which have the same size as the more familiar degrees Celsius (°C) but whose 'zero' is absolute zero (−273.15 °C) (Figure 1.39). Hence, a temperature change of 1 K is the same as a temperature change of 1 °C. Unlike the Celsius scale, negative numbers are not used in the absolute scale.

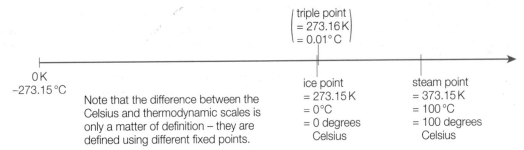

Figure 1.39 also shows:

triple point
= 273.16 K
= 0.01 °C

0 K
−273.15 °C

Note that the difference between the Celsius and thermodynamic scales is only a matter of definition – they are defined using different fixed points.

ice point
= 273.15 K
= 0 °C
= 0 degrees Celsius

steam point
= 373.15 K
= 100 °C
= 100 degrees Celsius

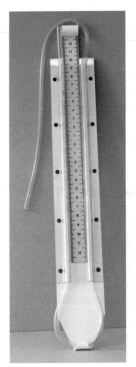

Figure 1.38 Manometer to measure gas pressure

Figure 1.39 The defining temperatures on the absolute or thermodynamic (Kelvin) and Celsius scales

Conversion between the Celsius and thermodynamic scales is governed by the equation:

$$t = T - 273.15$$

where t is the temperature in Celsius and T is the absolute temperature in kelvin. So, for example:

$$60 \,°C = 333.25 \,K - 273.15$$

hence 60 °C is equivalent to 333.15 K.

TOK Link

In 1742 the Swedish astronomer Anders Celsius (1701–1744) created, as a first attempt, an artificial temperature scale where zero represented the boiling point of water (at 1 atm) and 100 represented the freezing point of water (at 1 atm) (Figure 1.40).

It is artificial in the sense that it is based upon the physical properties of water. The melting and boiling points of any other substance could have been chosen and the temperature difference divided up into 100 intervals.

In 1744 the Swedish botanist Carolus Linnaeus reversed Celsius's scale so that zero represented the melting point of ice and 100 represented the water's boiling point. In 1848 William Thomson, later Lord Kelvin, proposed the need for a temperature scale where absolute zero (the complete absence of heat energy) was the scale's null point and which used the degree Celsius for its interval. This is a natural scale since it is independent of the physical properties of any substance and is based on the laws of thermodynamics.

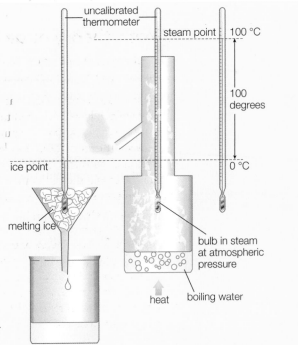

uncalibrated thermometer

steam point — 100 °C

100 degrees

ice point — 0 °C

melting ice

bulb in steam at atmospheric pressure

heat   boiling water

Figure 1.40 Calibration of a mercury thermometer

Language of Chemistry

Chemistry, like all the scientific disciplines, has its own specialized vocabulary. However, the greatest problems tend to arise not from specially invented names, but from widely used words whose meaning does not match their usage in chemistry. For example the word 'ideal' is used to describe a state of perfection. However, an ideal or perfect gas is a simplified or imperfect mathematical model of a gas. 'Ideal laws' simplify the problem of describing phenomena by ignoring the less important features of a system. ■

■ Extension: Entropy

An interesting observation about gases is that they diffuse and mix completely with each other (Figure 1.41). Two gases have never been observed to separate and unmix. This would be a very unlikely or improbable event. This simple observation suggests that one 'driving force' for reactions is an increase in disorder. The degree of disorder in a chemical system can be measured or calculated and is called its entropy (see Chapter 15).

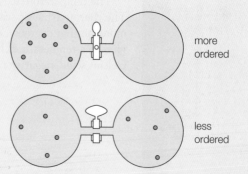

more ordered

less ordered

Figure 1.41 Gases will spread out by diffusion to maximize entropy

The gas laws

Boyle's law

Boyle found that (at constant temperature), the volume of a fixed mass of gas is inversely proportional to its pressure. In other words, if the pressure on a sample of gas is increased, then the volume decreases. This can be expressed mathematically as:

$$P \propto \frac{1}{V} \qquad \text{or as} \qquad P \propto V^{-1}$$

where P represents the pressure and V represents the volume.

Alternatively Boyle's law can be expressed as $P \times V = c$, where c represents a constant that varies with the gas and temperature.

Boyle's law (Figure 1.42) can be used to calculate the new pressure or volume if a fixed mass of gas (at constant temperature) undergoes a change in pressure or volume:

$$P_1 \times V_1 = P_2 \times V_2$$

Figure 1.42 Boyle's law apparatus

where V_1 represents the initial volume, P_1 represents the initial pressure, V_2 represents the final volume and P_2 represents the final pressure.

Worked example

A sample of gas collected in a $350\,cm^3$ container exerts a pressure of $103\,kPa$. Calculate the volume of this gas at a pressure of $150\,kPa$. (Assume that the temperature remains constant.)

Boyle's law: $P_1 \times V_1 = P_2 \times V_2$

$103\,kPa \times 350\,cm^3 = 150\,kPa \times V_2$ $\qquad\qquad 103\,kPa \times \dfrac{350\,cm^3}{150\,kPa} = 240\,cm^3$

■ Extension: Barometers

Barometers are usually calibrated in units of pressure based on the height of the mercury column (in millimetres) that the gas pressure can support. 1 atmosphere (1 atm) = 760 mm Hg. However, the fundamental definition of pressure is defined as the force (in newtons) per unit area (in square metres, m^2):

$$\text{pressure} = \frac{\text{force (N)}}{\text{area (m}^2)}$$

The SI unit of force is the newton per square metre (N/m^2 or $N\,m^{-2}$). It is often called the pascal ($1\,N\,m^{-2} = 1\,Pa$) ($1000\,Pa = 1\,kPa$).

History of Chemistry

Robert Boyle (1627–1691) was an Irish-born British chemist and physicist best known for formulating the law known as Boyle's law. However, his greatest contribution to chemistry was the publication of *The Sceptical Chymist* in 1661. In the book he proposes a simple atomic theory and suggests that chemists should regard elements as substances which cannot be decomposed into two or more simpler substances (though he did not specifically identify any). Previously many philosophers subscribed to the ancient Greek view of four 'elements': earth, air, fire and water. He was regarded as the 'Father of Chemistry' by future generations of chemists.

Charles' law

If the absolute temperature of a gas is doubled then the volume (at constant pressure) doubles. Conversely, if the absolute temperature of a gas is halved (at constant pressure), the volume halves.

This behaviour is known as **Charles' law** (Figure 1.43) and can be expressed mathematically as $V \propto T$, where V represents the volume and T represents the absolute temperature in kelvin. Alternatively Charles' law can be expressed as $V = K \times T$ (where K represents a constant that varies with the gas and pressure).

Note that a doubling of the temperature in degrees Celsius is *not* a doubling of the absolute temperature, for example a doubling of the temperature from 200 to 400 °C is only a rise from (200 + 273) = 473 K to (400 + 273) = 673 K, that is, a ratio of 673/473 or 1.42. Charles' law can be used to calculate the new temperature or volume if a fixed mass of gas (at constant pressure) undergoes a change in temperature or volume. It can be expressed as:

$$\frac{V_1}{T_1} = \frac{V_2}{T_2}$$

where V_1 represents the initial volume, T_1 represents the initial absolute temperature, V_2 represents the final volume and T_2 is the final absolute temperature.

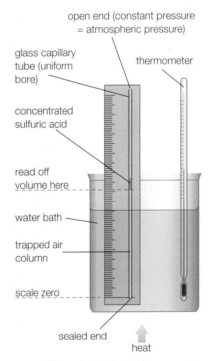

Figure 1.43 Charles' law apparatus

- open end (constant pressure = atmospheric pressure)
- glass capillary tube (uniform bore)
- thermometer
- concentrated sulfuric acid
- read off volume here
- water bath
- trapped air column
- scale zero
- sealed end
- heat

Worked example

A 4.50 dm³ sample of gas is warmed at constant pressure from 300 K to 350 K. Calculate its final volume.

$$V_2 = \frac{V_1 \times T_2}{T_1} = \frac{4.50\,\text{dm}^3 \times 350\,\text{K}}{300\,\text{K}} = 5.25\,\text{dm}^3$$

The pressure law

The **pressure law** states that for a fixed mass of gas (at constant volume) its absolute temperature is directly proportional to pressure. This behaviour can be expressed mathematically as $P \propto T$, where P represents the pressure and T represents the absolute temperature in kelvin. Alternatively the pressure law can be expressed as $P = K \times T$ or $P/T = K$ (where K represents a constant that varies with the gas).

The pressure law can be used to calculate the new pressure or temperature if a fixed mass of gas (at constant volume) undergoes a change in temperature or pressure. This is done using the following equation:

$$\frac{P_1}{T_1} = \frac{P_2}{T_2}$$

where P_1 represents the initial pressure, T_1 represents the initial absolute temperature, P_2 represents the final pressure and T_2 represents the final absolute temperature.

Worked examples

10 dm^3 of a gas is found to have a pressure of 97 000 Pa at 25.0 °C. What would be the temperature required (in degrees Celsius) to change the pressure to 101 325 Pa?

$$\frac{97\,000\,\text{Pa}}{298\,\text{K}} = \frac{101\,325\,\text{Pa}}{T_2}$$

$$T_2 = \frac{101\,325\,\text{Pa} \times 298\,\text{K}}{97\,000\,\text{Pa}}$$

$$T_2 = 311.3\,\text{K} = (311.3 - 273)\,°\text{C} = 38.3\,°\text{C}$$

The temperature of a gas sample is changed from 27 °C to 2727 °C at constant volume. What is the ratio of the final pressure to the initial pressure?

$$\frac{T_2}{T_1} = \frac{P_2}{P_1} = \frac{3000\,\text{K}}{300\,\text{K}} = 10$$

$$\frac{P_2}{P_1} = 10:1$$

Equation of state ('combined gas law')

The equation of state is formed by combining Boyle's law ($P \times V$ = constant) and Charles' law (V/T = constant):

$$\frac{PV}{T} = \text{constant, for a fixed mass of gas}$$

(Note that the constants in the three expressions will all be different.)

This equation is sometimes called the **combined gas law**, but it is properly termed the **equation of state** (for an ideal gas).

This relationship is often written as:

$$\frac{P_1 \times V_1}{T_1} = \frac{P_2 \times V_2}{T_2}$$

Gas volumes are usually compared at stp (see page 29). As with Charles' law, all temperatures must be expressed as absolute temperatures in kelvin.

Worked example

At 60 °C and 1.05×10^5 Pa the volume of a sample of gas collected is 60 cm^3. What would be the volume of the gas at stp?

$$\frac{1.05 \times 10^5\,\text{Pa} \times 60\,\text{cm}^3}{333\,\text{K}} = \frac{1.01 \times 10^5\,\text{Pa} \times V_2}{273\,\text{K}}$$

$$V_2 = \frac{1.05 \times 10^5\,\text{Pa} \times 60\,\text{cm}^3 \times 273\,\text{K}}{1.01 \times 10^5\,\text{Pa} \times 333\,\text{K}} = 51\,\text{cm}^3$$

The ideal gas equation

1.4.7 **Solve** problems using the ideal gas equation, $PV = nRT$.

We have seen that Boyle's law (PV = constant) and Charles' law (V/T = constant) can be combined together to give a combined gas law known as the equation of state:

$$\frac{PV}{T} = \text{constant, for a fixed mass of gas}$$

It follows from Avogadro's law that for one mole of gas (V_m) the constant will be the same for all gases. It is called the gas constant and given the symbol R.

$$\frac{P \times V_m}{T} = R$$

connect to vacuum pump
and then supply of gas X

flask of known
volume

Figure 1.44 Apparatus for determining the relative molecular mass of a gas

which can be rearranged as:

$$PV_m = RT$$

This equation is called the **ideal gas equation** and, for n moles of gas, the equation becomes:

$$PV = nRT,$$

where P represents the pressure in pascals (Pa), V represents the volume in cubic metres (m^3), n represents the amount of gas (mol), R represents the gas constant ($8.314\,J\,K^{-1}\,mol^{-1}$) and T represents the absolute temperature (kelvin). (The ideal gas equation is printed on page 1 of the IB *Chemistry data booklet*.)

It is vital that volumes expressed in dm^3 and cm^3 are converted to cubic metres if the value of the gas constant R given above is used and the pressure is expressed in pascals. $1\,dm^3 = 0.001$ or $10^{-3}\,m^3$ and $1\,cm^3 = 0.000001$ or $10^{-6}\,m^3$.

The ideal gas equation can be used to determine the relative molecular masses of gases (Figure 1.44) or volatile liquids (Figure 1.45).

Figure 1.45 Apparatus used to determine the relative molecular mass of a volatile liquid

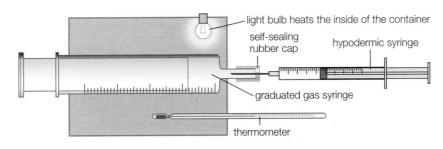

light bulb heats the inside of the container
self-sealing rubber cap
hypodermic syringe
graduated gas syringe
thermometer

Worked example

At 273 K and 101 325 Pa, 12.64 grams of a gas occupy 4.00 dm^3. Calculate the relative molecular mass of the gas.

$$PV = nRT$$

$$n = \frac{PV}{RT}$$

$$n = \frac{101\,325\,Pa \times (4.00 \times 10^{-3})\,m^3}{8.314\,J\,K^{-1}\,mol^{-1} \times 273\,K} = 0.1786\,mol$$

$$\text{Amount of gas (mol)} = \frac{\text{mass (g)}}{\text{relative molecular mass}}$$

$$\text{Relative molecular mass} = \frac{\text{mass (g)}}{\text{amount (mol)}}$$

$$\text{Relative molecular mass (M}_r\text{)} = \frac{12.64}{0.1786} = 70.7$$

(Calculations may also involve measurements involving the density of gas, which should be expressed or converted to SI units of $kg\,m^{-3}$. However, note that $g\,dm^{-3}$ and $kg\,m^{-3}$ are equivalent – no conversion is required.)

Another method of solving this type of problem is to combine the ideal gas equation with the expression for density (d) and the relationship between number of moles (n), mass (m), pressure (P) in kilopascals (kPa) and relative molecular mass (M_r):

$$PV = nRT = \frac{mRT}{M_r} \qquad \text{so} \qquad M_r = \frac{mRT}{PV}$$

$$d = \frac{m}{V} \qquad\qquad \text{so} \qquad V = \frac{m}{d}$$

Substituting for V in the equation for M_r:

$$M_r = \frac{dRT}{P}$$

Worked example

Calculate the relative molecular mass of a gas which has a density of $2.615\,g\,dm^{-3}$ at $298\,K$ and $101\,325$ Pa.

$$M_r = \frac{dRT}{P}$$

$$M_r = \frac{2.615\,kg\,m^{-3} \times 8.314\,J\,K^{-1}\,mol^{-1} \times 298\,K}{101.325\,kPa} = 63.9$$

Graphs relating to the ideal gas equation

1.4.8 Analyse graphs relating to the ideal gas equation.

A gas that obeys the ideal gas equation (and all the gas laws) under all conditions is said to behave as an ideal gas, or to behave ideally. No gas behaves ideally: all gases deviate to some extent from ideal behaviour and are described as **real gases**. The deviation from ideal gas behaviour can be shown by plotting PV/RT against P or PV against P. For a gas behaving ideally these plots would give straight lines (Figure 1.46).

The greatest deviation from ideal behaviour occurs when the gas is subjected to a low temperature and high pressure (Figure 1.47). A real gas deviates from ideal behaviour considerably at high pressures. This is because when gases are put under pressure and compressed the molecules or atoms come sufficiently close together for intermolecular forces (Chapter 4) to operate and the particles to be attracted to each other. In other gases, for example ammonia, stronger hydrogen bonds operate and the deviation from ideal behaviour is even greater.

One of the assumptions of the ideal gas model is that the volume of molecules is negligible compared with the volume occupied by the gas. This is no longer true in a highly compressed gas where the actual volume of the gas molecules becomes significant. At low temperatures deviation from ideal behaviour occurs because the molecules are moving slowly, which significantly strengthens the intermolecular forces operating between neighbouring molecules or atoms.

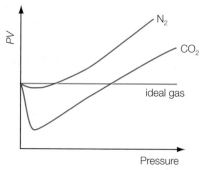

Figure 1.46 Deviation from ideal behaviour at high pressure

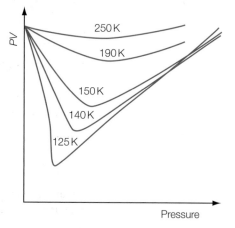

Figure 1.47 Deviation from ideal behaviour at low temperature

1.5 Solutions

1.5.1 **Distinguish** between the terms *solute*, *solvent*, *solution* and *concentration* (g dm⁻³ and mol dm⁻³).

Water dissolves a wide range of different chemical substances. Water is a **solvent** and the substances dissolved in the water are termed **solutes**. The mixture of solvent and solute is termed a **solution**.

When one mole of a solute is dissolved in water and the volume of solution made up to 1000 cm³ (1 dm³), the resulting solution is termed a **molar** solution (1 mol dm⁻³). If two moles of a solute are made up to 1000 cm³ of solution (or one mole to 500 cm³), the solution is described as 2 mol dm⁻³ (Figure 1.48).

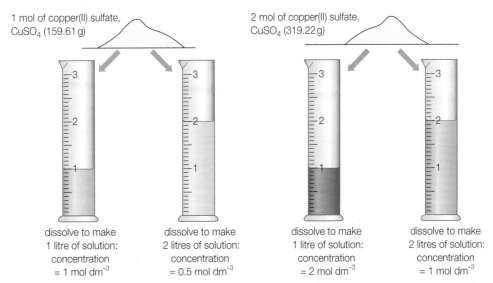

Figure 1.48 A diagram showing the concentration of a solution depends on the amount of solute and the volume of solvent

The **concentration** of a solution is the amount of solute (in moles) contained within one cubic decimetre. The concentration of a solution is given by the following expression:

$$\text{concentration of solution (mol dm}^{-3}) = \frac{\text{amount of solute (mol)}}{\text{volume of solution (dm}^3)}$$

Worked examples

Calculate the concentration of the solution formed when 0.5 mol of glucose is dissolved in 5.0 dm³ of water.

$$\text{Concentration} = \frac{0.5 \text{ mol}}{5.0 \text{ dm}^3} = 0.1 \text{ mol dm}^{-3}$$

Determine the concentration of the solution when 4.00 grams of sodium hydroxide (molar mass 40.0 g mol⁻¹) is dissolved in 200 cm³ of water.

$$\text{Amount of sodium hydroxide} = \frac{4.00 \text{ g}}{40 \text{ g mol}^{-1}} = 0.100 \text{ mol}$$

$$\text{Concentration of sodium hydroxide} = \frac{0.100 \text{ mol}}{0.200 \text{ dm}^3} = 0.50 \text{ mol dm}^{-3}$$

Calculate the mass of hydrated copper(II) sulfate, $CuSO_4.5H_2O$ (molar mass 249.7 g mol⁻¹) present in 25.0 cm³ of a 0.500 mol dm⁻³ solution.

Amount of hydrated copper(II) sulfate = 0.500 mol dm⁻³ × 0.0250 dm³ = 0.0125 mol

Mass of hydrated copper(II) sulfate = 0.0125 mol × 249.7 g mol⁻¹ = 3.12 g

Calculate the concentration (in mol dm⁻³) of a solution of hydrochloric acid containing 14.6 grams of hydrogen chloride in 100 cm³ of solution.

Molar mass of hydrogen chloride = (1.00 g mol⁻¹ + 35.5 g mol⁻¹) = 36.5 g mol⁻¹

$$\text{Amount of hydrogen chloride in 100 cm}^3 = \frac{14.6 \text{ g}}{36.5 \text{ g mol}^{-1}} = 0.400 \text{ mol}$$

Hence, the concentration of hydrogen chloride = 0.400 × 10 = 4 mol dm⁻³

Soluble ionic compounds dissociate into their component ions when dissolved in an excess of water. The concentrations of the individual ions will depend on the amounts of these ions when the substance (salt, base or alkali) dissolves. In a $4.0\,mol\,dm^{-3}$ aqueous solution of aluminium nitrate, for example, the concentration of the aluminium ions is $4.0\,mol\,dm^{-3}$, but the concentration of the nitrate ions is $12.0\,mol\,dm^{-3}$.

$$Al(NO_3)_3(aq) \quad \rightarrow \quad Al^{3+}(aq) \quad + \quad 3NO_3^-(aq)$$
$$1\,mol \qquad\qquad 1\,mol \qquad\qquad 3\,mol$$

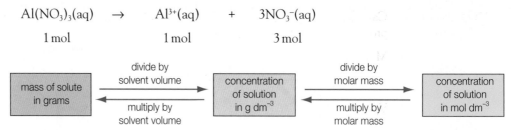

Figure 1.49 Summary of the interconversion between concentration (in moles per cubic decimetre of solution) and concentration in (grams per cubic decimetre of solution)

■ Extension: Additional units for 'concentration'

Mixing ratio

Mixing ratios are frequently used to describe the concentrations of gases in the atmosphere. For example, nitrogen forms approximately 78% by volume of the atmosphere and its concentration, expressed as a mixing ratio, is 780 000 ppmv, where ppmv refers to parts per million by volume. This means that 780 000 out of 1 000 000 particles (atoms or molecules) in the air are nitrogen molecules. Mixing ratios are often used to express the concentrations of pollutant gases (Chapter 25).

Volume strength

Figure 1.50 A stock bottle of hydrogen peroxide

Solutions of hydrogen peroxide (Figure 1.50) are often sold according to their 'volume strength'. The volume strength of a solution of hydrogen peroxide is measured by the number of volumes of oxygen released when it is completely decomposed under standard conditions (0 °C, 1 atmosphere pressure).

For example, 20 volume strength hydrogen peroxide solution means $1\,cm^3$ of the solution will release $20\,cm^3$ of oxygen gas (when completely decomposed). Volume strengths can be converted to other measures of concentration, for example $mol\,dm^{-3}$.

Mass per cent

One method of expressing the concentration of a substance dissolved in water is to express it as a weight or mass per cent. For example, a 5% by mass of aqueous ethanoic acid solution (white vinegar – Figure 1.51) would have 5 g of ethanoic acid for every 100 g of solution.

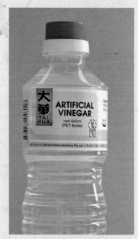

Figure 1.51 Artificial white vinegar (5% by mass ethanoic acid)

Dilution of acids

Acids are supplied as concentrated acids. The solutions required in the laboratory are prepared by diluting the concentrated solutions with water. For safety reasons the dilution is carried out by slowly adding the concentrated acid to the water. Water should *never* be added to concentrated acids. When a concentrated solution is diluted with water, the amount of solute in the solution remains unchanged.

This can be expressed by the following relationship:

$$M_1 \times V_1 = M_2 \times V_2$$

where M_1 represents the initial concentration, M_2 represents the concentration after dilution, V_1 represents the initial volume and V_2 represents the volume after dilution.

Worked example

Calculate the volume to which $25.0\,cm^3$ of $5.0\,mol\,dm^{-3}$ hydrochloric acid must be diluted to produce a concentration of $1.5\,mol\,dm^{-3}$.

$$M_1 \times V_1 = M_2 \times V_2$$

$$5.0\,mol\,dm^{-3} \times 25.0\,cm^3 = 1.5\,mol\,dm^{-3} \times V_2$$

$$V_2 = 25.0 \times \frac{5.0}{1.5} = 83.3\,cm^3$$

Figure 1.52 A selection of apparatus used in a titration: pipette filler, burette and pipette

1.5.2 **Solve** problems involving concentration, amount of solute and volume of solution.

Volumetric chemistry

A solution of known concentration is called a **standard solution**. In volumetric chemistry a series of titrations is carried out, frequently with an acid and a base. In each **titration** (Figure 1.52) a solution is added in small measured quantities, from a burette, to a fixed volume of another solution, measured with a pipette, in the presence of an indicator. The addition of the solution is continued until the indicator just changes colour. At this stage, termed the **end-point**, the two substances are present in **stoichiometric** quantities.

Acid–base titrations

In these titrations acid and base are reacted in the presence of a suitable acid–base indicator (Chapter 18). Uses of acid–base titrations include:

- determining the concentrations of solutions
- determining the percentage purity or molar mass of an acid or base
- deducing the equation for a neutralization reaction
- determining the amount of water of crystallization in a hydrated salt.

Worked examples

Sodium hydroxide reacts with hydrochloric acid according to the following equation:

$$NaOH(aq) + HCl(aq) \rightarrow NaCl(aq) + H_2O(l)$$

Calculate the volume of $0.0500\,mol\,dm^{-3}$ sodium hydroxide solution to react exactly with $25\,cm^3$ of $0.20\,mol\,dm^{-3}$ hydrochloric acid.

Amount of hydrochloric acid $= \dfrac{25.0}{1000}\,dm^3 \times 0.200\,mol\,dm^{-3} = 5.00 \times 10^{-3}\,mol$

The equation's stoichiometry indicates that the alkali and acid react in a $1:1$ molar ratio. Hence the amount of sodium hydroxide is $5.00 \times 10^{-3}\,mol$.

Volume of sodium hydroxide $= \dfrac{1000 \times 5.00 \times 10^{-3}\,mol}{0.0500\,mol\,dm^{-3}} = 100\,cm^3$

0.558 grams of a monobasic aromatic carboxylic acid, HX, was dissolved in distilled water. A few drops of phenolphthalein indicator was added and the mixture was titrated with $0.100\,mol\,dm^{-3}$ sodium hydroxide solution. It took $41.0\,cm^3$ of the alkali to obtain the end-point (with a permanent pink colour). Calculate the molar mass of the organic acid.

Amount of sodium hydroxide $= 0.100\,mol\,dm^{-3} \times \dfrac{41.0}{1000}\,dm^3$

$$= 4.10 \times 10^{-3}\,mol$$

Amount of HX equals 4.1×10^{-3} mol because the acid and base are reacting in a $1:1$ molar ratio:

$$HX(aq) + NaOH(aq) \rightarrow H_2O(l) + X^-Na^+(aq)$$

Hence, molar mass of HX $= \dfrac{\text{mass (g)}}{\text{amount (mol)}} = \dfrac{0.558\,g}{4.10 \times 10^{-3}\,mol} = 136\,g\,mol^{-1}$

$17.5\,cm^3$ of $0.150\,mol\,dm^{-3}$ potassium hydroxide solution react with $20.0\,cm^3$ of phosphoric acid, H_3PO_4 of concentration $0.0656\,mol\,dm^{-3}$. Deduce the equation for the reaction.

Amount of potassium hydroxide $= \dfrac{17.5}{1000}\,dm^3 \times 0.150\,mol\,dm^{-3} = 2.63 \times 10^{-3}\,mol$

Amount of phosphoric acid $= \dfrac{20.0}{1000}\,dm^3 \times 0.0656\,mol\,dm^{-3} = 1.31 \times 10^{-3}\,mol$

The two chemicals react in a $2:1$ molar ratio and hence the equation is:

$$H_3PO_4(s) + 2KOH(aq) \rightarrow K_2HPO_4(aq) + 2H_2O(l)$$

The results of a titration with a solution of known concentration can be used to determine the concentration of the other solution.

Worked example

A $50.0\,cm^3$ sample of concentrated sulfuric acid was diluted to $1.00\,dm^3$. A sample of the diluted sulfuric acid was analysed by titrating with aqueous sodium hydroxide. In the titration, $25.00\,cm^3$ of $1.00\,mol\,dm^{-3}$ aqueous sodium hydroxide required $20.0\,cm^3$ of the diluted sulfuric acid for neutralization. Determine the concentration of the original concentrated sulfuric acid solution.

Steps
i   Construct the equation for the complete neutralization of sulfuric acid by sodium hydroxide.
ii Calculate the amount of sodium hydroxide that was used in the titration.
iii Calculate the concentration of the diluted sulfuric acid.
iv Calculate the concentration of the original concentrated sulfuric acid solution.

Answers
i $2NaOH(aq) + H_2SO_4(aq) \rightarrow Na_2SO_4(aq) + 2H_2O(l)$

ii Amount of sodium hydroxide used in the titration

$$= \frac{25.00}{1000}\,dm^3 \times 1.00\,mol\,dm^{-3} = 0.0250\,mol\,NaOH$$

iii From the equation, amount of H_2SO_4 = amount of NaOH $\div$ 2 = $0.0125\,mol$ in $20.0\,cm^3$, so 'scaling up' to $1000\,cm^3$ to obtain the concentration of diluted sulfuric acid:

$$\frac{1000}{20.0} \times 0.0125\,mol\,dm^{-3} = 0.625\,mol\,dm^{-3}$$

iv 'Scaling up' from 50.0 to $1000\,cm^3$ gives the concentration of the original concentrated sulfuric acid solution: $0.625\,mol\,dm^{-3} \times \dfrac{1000}{50.0} = 12.5\,mol\,dm^{-3}$

Primary standard solutions

Titrations often involve a **primary standard solution**. Its concentration may have been determined by titration with another primary standard solution or by weighing the solute and preparing a solution of known volume (Figure 1.53). The concentrations of primary standard solutions do not change with time. Few chemical substances are suitable for use as primary standards. If a substance is to be weighed accurately enough for use in preparing a primary standard solution, the following criteria must be met:

- The substance must be available in a high state of purity or be easily purified.
- The substance must not be volatile, or some of it would be lost during the weighing process.
- The substance must not react with oxygen, water or carbon dioxide.

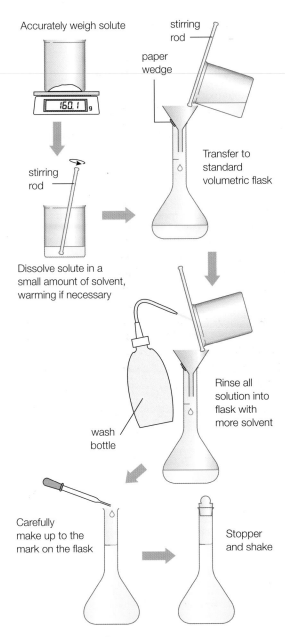

Accurately weigh solute

stirring rod

paper wedge

Transfer to standard volumetric flask

stirring rod

Dissolve solute in a small amount of solvent, warming if necessary

wash bottle

Rinse all solution into flask with more solvent

Carefully make up to the mark on the flask

Stopper and shake

Figure 1.53 Preparing a standard solution

Compounds suitable as primary standards are:

- strong acid – ethanedioic acid (oxalic acid)
- strong base – anhydrous sodium carbonate
- oxidizing agent – potassium dichromate(VI)
- reducing agent – iron(II) sulfate (in the form of hydrated ammonium iron(II) sulfate).

A number of solutions when freshly prepared are used as alternative primary standards, for example sodium and potassium hydroxides, sulfuric and hydrochloric acids and potassium manganate(VII). These cannot be stored since their concentrations change with time due to a chemical reaction (usually with oxygen and/or water in the storage vessel).

Back titration

In the technique known as **back titration**, a known excess of one reagent A is allowed to react with an unknown amount of a reagent B. At the end of the reaction, the amount of A that remains unreacted is found by titration. A simple calculation gives the amount of A that has reacted with B and also the amount of B that has reacted.

In a typical acid–base back titration, a quantity of a base is added to an excess of an acid (or vice versa). All the base and some of the acid react. The acid remaining is then titrated with a standard alkali and its amount determined. From the results, the amount of acid which has reacted with the base can be found and the amount of base can then be calculated. The principle of this type of titration is illustrated in Figure 1.54.

Back titrations are usually used when the determination of the amount of a substance poses some difficulty in the direct titration method, for example insoluble solid substances where the end-point is difficult to detect and volatile substances where inaccuracy arises due to loss of substance during titration.

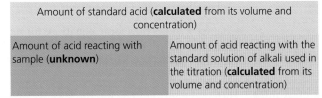

Amount of standard acid (**calculated** from its volume and concentration)	
Amount of acid reacting with sample (**unknown**)	Amount of acid reacting with the standard solution of alkali used in the titration (**calculated** from its volume and concentration)

Figure 1.54 Illustration of the principle of an acid–base back titration

Magnesium oxide is not very soluble in water, and is difficult to titrate directly. Its purity can be determined by use of a 'back titration' method. 4.08 g of impure magnesium oxide was completely dissolved in 100 cm³ of 2.00 mol dm⁻³ aqueous hydrochloric acid. The excess acid required 19.7 cm³ of 0.200 mol dm⁻³ aqueous sodium hydroxide for neutralization. What is the purity of the impure magnesium oxide?

Steps
i Construct equations for the two neutralization reactions.
ii Calculate the amount of hydrochloric acid added to the magnesium oxide.
ii Calculate the amount of excess hydrochloric acid titrated.
iv Calculate the amount of hydrochloric acid reacting with the magnesium oxide.
v Calculate the mass of magnesium oxide that reacted with the initial hydrochloric acid, and hence determine the percentage purity of the magnesium oxide.

Answers

i $MgO(s) + 2HCl(aq) \rightarrow MgCl_2(aq) + H_2O(l)$

$NaOH(aq) + HCl(aq) \rightarrow NaCl(aq) + H_2O(l)$

ii Amount of hydrochloric acid added to the magnesium oxide = $\frac{100}{1000}\,dm^3 \times 2.00\,mol\,dm^{-3}$
= 0.200 mol.

iii Amount of excess hydrochloric acid titrated = $\frac{19.7}{1000}\,dm^3 \times 0.200\,mol\,dm^{-3}$ = 0.003 94 mol HCl, since the mole ratio of NaOH to HCl is 1 : 1.

iv Amount of hydrochloric acid reacting with the magnesium oxide
= 0.200 mol − 0.003 94 mol = 0.196 mol.

v Amount of magnesium oxide that reacted = $\frac{0.196\,mol}{2}$ = 0.098 mol (1 : 2 molar ratio in equation). The molar mass of magnesium oxide is 40.3 g mol^{-1}, hence the mass of magnesium oxide reacting with acid = 0.098 mol × 40.3 g mol^{-1} = 3.95 g and hence percentage purity
= $\frac{3.95\,g}{4.08\,g} \times 100$ = 97%.

Redox titrations

Calculations involving redox titrations (Figure 1.55) are identical to those involving acids and bases. Uses of redox titrations include:

- determining the concentration of a solution
- determining the percentage purity of a salt or other substances, for example an alloy
- determining the charge and relative atomic mass of an ion
- deducing the ionic equation for a reaction
- determining the amount of water of crystallization in a hydrated salt.

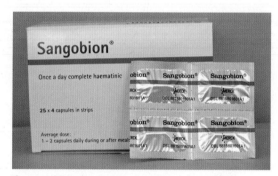

Figure 1.55 Redox titrations can be performed in company laboratories to ensure that materials such as these iron tablets, bleach (sodium chlorate(I)) and vitamin C (ascorbic acid) contain the stated amounts or concentrations of substances in them

Chapter 9 shows how ionic equations for redox titrations can be constructed and examines the principles underlying several types of redox titrations. Two worked examples of common redox titration calculations are presented below.

Worked examples

Hydrated iron(II) sulfate has the formula FeSO$_4$.xH$_2$O. An experiment was performed to determine x, the amount of water of crystallization in hydrated iron(II) sulfate. 50.6 grams of hydrated iron(II) sulfate were dissolved in distilled water to make 250.0 cm^3 of solution. 20.0 cm^3 of this solution reacted completely with 24.0 cm^3 of 0.100 mol dm^{-3} potassium dichromate(VI) solution. Use this data to determine the value of x and hence the formula of hydrated iron(II) sulfate.

$6Fe^{2+}(aq) + 14H^+(aq) + Cr_2O_7^{2-}(aq) \rightarrow 2Cr^{3+}(aq) + 7H_2O(l) + 6Fe^{3+}(aq)$

Amount of $Cr_2O_7^{2-}(aq)$ or $K_2Cr_2O_7(aq)$ = $\frac{24.0}{1000}\,dm^3 \times 0.100\,mol\,dm^{-3}$ = 2.40×10^{-3} mol

Hence amount of $Fe^{2+}(aq)$ or $FeSO_4(aq)$ = 2.40×10^{-3} mol × 6 = 0.0144 mol

The titrated solution consists of 0.0144 mol of $Fe^{2+}(aq)$ or $FeSO_4(aq)$ in 20.0 cm^3.

Hence, concentration of $Fe^{2+}(aq)$ or $FeSO_4(aq)$ = 0.0144 mol × $\frac{1000}{20\,dm^3}$

= 0.720 mol dm^{-3}.

$$FeSO_4.xH_2O(aq) \rightarrow FeSO_4(aq) + xH_2O(l)$$

Amount of $FeSO_4$ present in 50.6 grams of hydrated iron(II) sulfate

$$= 0.0144\,mol \times \frac{250.0}{20.0} = 0.180\,mol$$

Mass of $FeSO_4 = 151.91\,g\,mol^{-1} \times 0.180\,mol = 27.34\,g$

Mass of water $= (50.6\,g - 27.34\,g) = 23.26\,g$

Amount of water $= \dfrac{23.26\,g}{18.02\,g\,mol^{-1}} = 1.29\,mol$

Ratio of H_2O to $FeSO_4 = x = \dfrac{1.29\,mol}{0.180\,mol} = 7.2$

Thus the formula is $FeSO_4.7H_2O$.

Potassium manganate(VII), $KMnO_4$, oxidizes potassium iodide, KI, to iodine, I_2. The iodine liberated is titrated with aqueous sodium thiosulfate, $Na_2S_2O_3$.

$$16H^+(aq) + 2MnO_4^-(aq) + 10I^-(aq) \rightarrow 2Mn^{2+}(aq) + 8H_2O(l) + 5I_2(aq)$$

$$2S_2O_3^{2-}(aq) + I_2(aq) \rightarrow S_4O_6^{2-}(aq) + 2I^-(aq)$$

The iodine produced from $25.0\,cm^3$ of potassium manganate(VII) solution required $26.4\,cm^3$ of $0.500\,mol\,dm^{-3}$ sodium thiosulfate solution for complete reaction. Calculate the concentration of the potassium manganate(VII) solution.

Amount of $S_2O_3^{2-}(aq)$ or $Na_2S_2O_3(aq) = \dfrac{26.4}{1000}\,dm^3 \times 0.500\,mol\,dm^{-3} = 0.0132\,mol$

Using the equations for the reactions:

Amount of $I_2(aq) = \dfrac{0.0132}{2} = 6.6 \times 10^{-3}\,mol$

and hence amount of $MnO_4^-(aq) = \frac{2}{5} \times 6.6 \times 10^{-3}\,mol = 2.6 \times 10^{-3}\,mol$

Concentration of $MnO_4^-(aq) = \dfrac{1000}{25.0}\,dm^3 \times 2.6 \times 10^{-3}\,mol = 0.11\,mol\,dm^{-3}$.

Precipitation titrations

A common type of precipitation titration uses silver nitrate to determine the concentration of chloride ions. Silver nitrate solution is added to a chloride solution in the presence of potassium chromate(VI), which acts as an 'indicator'.

The net ionic equation for a silver nitrate titration is:

$$Ag^+(aq) + Cl^-(aq) \rightarrow AgCl(s)$$

If a chloride solution is acidic, calcium carbonate powder is added to neutralize the acid. One use of silver nitrate titrations is to determine the formula of chlorides (Chapter 13).

Worked example

0.010 mol of an ionic chloride was dissolved in water and found to react completely with $20\,cm^3$ of $1.00\,mol\,dm^{-3}$ silver nitrate solution. Determine the formula, using M to represent the metal.

Amount of silver nitrate $= \dfrac{20}{1000}\,dm^3 \times 1.00\,mol\,dm^{-3} = 0.020\,mol$ of silver ions.

These react with chloride ions in the molar ratio of $1:1$.

$$Ag^+(aq) + Cl^-(aq) \rightarrow AgCl(s)$$

This means that 0.010 mol of the chloride contains 0.020 mol of chlorine, Cl. Hence the formula is MCl_2.

▨ The mole is a chemist's measure of the amount of a chemical substance. The mole is a counting unit used to deal with atoms, ions, electrons and formula units.

▨ One mole (1 mol) of a chemical species contains the same number of particles as there are atoms in exactly 12 grams of the isotope carbon-12 ($^{12}_{6}C$).

▨ The Avogadro constant, L, has the value $6.02 \times 10^{23} \, mol^{-1}$.

▨ Amount of substance (mol) = $\dfrac{\text{number of particles}}{6.02 \times 10^{23} \, mol^{-1}}$

▨ The relative atomic mass of an element, A_r, gives the ratio of the weighted average of the masses of the atoms of an element to the mass of one atom of carbon-12 taken as exactly twelve units. The relative molecular mass, M_r, and relative formula mass are defined similarly. Relative formula mass, relative molecular mass and relative atomic mass are pure numbers and have no units.

▨ The relative atomic mass in grams of any element contains 6.02×10^{23} atoms. The relative molecular mass in grams of any compound contains 6.02×10^{23} molecules.

▨ The molar mass is the mass of one mole of any chemical entity. Molar mass has units of grams per mol ($g \, mol^{-1}$).

▨ Amount of a substance (mol) = $\dfrac{\text{mass of substance (g)}}{\text{molar mass } (g \, mol^{-1})}$

▨ The coefficients in a balanced equation represent the reacting ratios in terms of both numbers of particles (atoms, ions and molecules) and amounts.

▨ The empirical formula is the simplest integer ratio of the atoms or ions in a compound. The molecular formula is the actual formula of a compound. It is identical to the empirical formula or a whole number multiple of the empirical formula.

▨ The masses of a reactant and products in a chemical reaction can be determined experimentally. Converted into amounts of reactant and products, they give the chemical equation for the reaction. Conversely, the chemical equation can be used to calculate the amount and mass of a product from a known mass of reactant.

▨ Gay-Lussac's law of combining volumes states that when gases combine together they do so in volumes that are in a simple whole number ratio to each other and to the product (if it is a gas).

▨ Avogadro's law states that under the same conditions of temperature and pressure, equal volumes of all gases contain the same number of molecules.

▨ One mole (1 mol) of a gas occupies $22.4 \, dm^3$ at standard temperature and pressure, stp (0 °C and 1 atm).

▨ Conversion between gas volumes and amount is given by:

amount = $\dfrac{\text{volume of gas } (dm^3) \text{ at stp}}{\text{molar gas volume } (dm^3 \, mol^{-1})}$

▨ The molecules in a gas are in a constant state of overall random motion.

▨ Boyle's law states that the volume of a fixed mass of gas is inversely proportional to its pressure (at constant temperature). In symbols: $PV = $ constant; $P_1 V_1 = P_2 V_2$.

▨ Charles' law states the volume of a fixed mass of gas is proportional to its absolute temperature (at constant pressure).

$\dfrac{V}{T} = $ constant; $\dfrac{V_1}{T_1} = \dfrac{V_2}{T_2}$

▨ The zero on the absolute scale of temperature is the temperature at which the volume of an ideal gas becomes zero.

▨ The pressure law states that the pressure of a fixed mass of gas is proportional to its absolute temperature (at constant volume).

$\dfrac{P}{T} = $ constant; $\dfrac{P_1}{T_1} = \dfrac{P_2}{T_2}$

▨ An ideal gas can be visualized as a collection of hard, inelastic spheres in constant motion. In an ideal gas the actual volume of the gas particles is negligible and there are no intermolecular forces.

▨ Gas pressure is due to collisions of molecules with the walls of the container. The temperature of a gas is determined by the average kinetic energy of the molecules.

- The ideal gas law is $PV = nRT$, where n represents the amount of gas and R represents the gas constant. The molar mass or relative molecular mass of a gas can be found by weighing a known volume of the gas.
- The concentration of a solution is expressed in grams per cubic decimetre (g dm⁻³) or moles per cubic decimetre ($mol\,dm^{-3}$).
- Concentration ($mol\,dm^{-3}$) = $\dfrac{\text{amount (mol)}}{\text{volume of solution (dm}^3)}$
- Titrations can be used to determine the reacting volumes of solution and, from the volumes and concentrations, the equation for the reaction. Titration of a solution of unknown concentration against a standard solution, with the equation for the reaction, allows the unknown concentration to be calculated.
- A back titration allows the determination of the concentration of a reactant of unknown concentration by reacting it with an excess volume of another reactant of known concentration. The resulting mixture is then titrated, taking into account the excess of reagent which is present. Back titrations are used when the sample under analysis is insoluble in water.
- Many stoichiometry calculations require the following approach:
  - Translate the mass or volume (of gas) of a given reactant into an amount (mol).
  - Use the stoichiometric ratio from the balanced equation to deduce the amount (mol) of the required reactant or product.
  - Reconvert to a mass, volume (for a gas), concentration ($mol\,dm^{-3}$ or g dm⁻³) or percentage purity.

■ *Examination questions – a selection*

Paper 1 IB questions and IB style questions

Q1 Avogadro's number, L, is 6.02×10^{23} and the relative atomic mass of calcium is 40. What is the mass of one mole of calcium atoms?
A $40/L$ grams   **C** $L/40$ grams
B $40L$ grams   **D** 40 grams

Q2 What is the mass in grams of a single molecule of propane, C_3H_8?
A 7.3×10^{-25} g   **C** 6.02×10^{-23} g
B 44 g   **D** 7.3×10^{-23} g

Q3 Which sample has the greatest mass?
A 1.0 mol of N_2H_4 **C** 3.0 mol of NH_3
B 2.0 mol of N_2 **D** 25.0 mol of H_2
    Standard Level Paper 1, May 99, Q1

Q4 Which of the following samples contains the smallest number of atoms?
A 1 g H_2 **C** 1 g S_8
B 1 g O_2 **D** 1 g Br_2

Q5 How many molecules are there in 180 g of H_2O?
A 6.0×10^{22} **C** 6.0×10^{24}
B 6.0×10^{23} **D** 6.0×10^{25}
    Standard Level Paper 1, May 00, Q1

Q6 Hydrogen peroxide, H_2O_2, reacts with manganate(VII) ions, MnO_4^-, in basic solution according to the following equation:

$2MnO_4^-(aq) + 3H_2O_2(aq)$
$\qquad \rightarrow 2MnO_2(s) + 3O_2(g) + 2OH^-(aq) + 2H_2O(l)$

How many moles of hydrogen peroxide would be needed to produce eight moles of water?
A one   **C** three
B two   **D** twelve

Q7 Hydrogen sulfide, H_2S, reacts with oxygen to form sulfur dioxide and water as shown below.

$2H_2S + \_\,O_2 \rightarrow \_\,SO_2 + \_H_2O$

What is the whole number coefficient for oxygen when this equation is balanced?
A 1   **C** 3
B 2   **D** 6
    Standard Level Paper 1, May 99, Q4

Q8 A certain compound has a molecular mass of 56 g mol⁻¹. Which of the following cannot be an empirical formula for this compound?
A BH_3 **C** MgN_2H_4
B C_3H_4O **D** HCl

Q9 Which of the following is an empirical formula?
- **A** N_2F_2
- **B** $C_2H_4O_2$
- **C** C_2H_4O
- **D** C_2N_2

Q10 $2Cl_2$ represents:
- **A** two chlorine molecules
- **B** two chlorine atoms
- **C** two chloride ions
- **D** four free chlorine atoms

Q11 Which one of the following equations correctly represents the combustion of calcium (a member of group 2) in oxygen?
- **A** $Ca(s) + O_2(g) \rightarrow CaO_2(s)$
- **B** $Ca(s) + 2O(g) \rightarrow CaO_2(s)$
- **C** $2Ca(s) + O_2(g) \rightarrow 2CaO(s)$
- **D** $Ca(s) + O(g) \rightarrow CaO(s)$

Q12 Which one of the following equations is **not** correctly balanced?
- **A** $Ca(s) + 2H^+(aq) \rightarrow Ca^{2+}(aq) + H_2(g)$
- **B** $Mg(s) + 2H_2O(l) \rightarrow Mg(OH)_2(aq) + H_2(g)$
- **C** $Fe^{2+}(aq) + Ag^+(aq) \rightarrow Fe^{3+}(aq) + Ag(s)$
- **D** $Fe^{2+}(aq) + Cl_2(g) \rightarrow Fe^{3+}(aq) + 2Cl^-(aq)$

Q13 What is the total number of ions present in the formula, $Fe_2(SO_4)_3$?
- **A** 5
- **B** 3
- **C** 2
- **D** 6

Q14 32.0 grams of sulfur (atomic mass of 32.0) combine with a metal, M (atomic mass of 40.0) to give a product which weighs 52.0 g. What is the empirical formula of the sulfide formed?
- **A** MS
- **B** MS_2
- **C** M_2S
- **D** M_2S_5

Q15 An unknown element M combines with oxygen to form the compound MO_2. If 36.0 g of element M combines exactly with 16.0 g of oxygen, what is the atomic mass of M in grams?
- **A** 12.0
- **B** 16.0
- **C** 24.0
- **D** 72.0

Q16 When 16.00 grams of hydrogen gas reacts with 64.0 grams of oxygen gas in a reaction (atomic masses are H = 1.00, O = 16.00), what will be present in the resulting mixture?
- **A** H_2, H_2O, and O_2
- **B** H_2, H_2O
- **C** O_2, H_2O
- **D** H_2, O_2

Q17 2.4 g of magnesium metal reacted vigorously when heated with excess iron(III) oxide, Fe_2O_3. What mass of metallic iron could be produced in this process?
- **A** 2.8 g
- **B** 3.7 g
- **C** 5.6 g
- **D** 8.4 g

Q18 One molecule of a small protein contains 63 atoms of carbon. The mass percentage of carbon in the protein is 55.74%. What is the molar mass of the protein?
- **A** $1357 \, g \, mol^{-1}$
- **B** $421.7 \, g \, mol^{-1}$
- **C** $821.3 \, g \, mol^{-1}$
- **D** $756.6 \, g \, mol^{-1}$

Q19 Nitrogen(II) oxide, NO, is made from the oxidation of NH_3:

$$4NH_3 + 5O_2 \rightarrow 4NO + 6H_2O$$

An 8.5 g sample of NH_3 gives 15.0 g of NO. What is the percentage yield of NO?
- **A** 40%
- **B** 60%
- **C** 80%
- **D** 100%

Q20 Which one of the following is an incorrect assumption of the kinetic theory of gases?
- **A** Atoms or molecules travel in straight lines between collisions but are in overall random motion.
- **B** Atoms or molecules of a gas are much smaller than the average distances between them.
- **C** Collisions between atoms or molecules of a gas and the containing vessel are perfectly elastic.
- **D** In a given gas, all particles have the same kinetic energy at a given temperature.

Q21 1000 cm³ of hydrogen gas (hydrogen molecules, H_2) contains Z molecules at room temperature and pressure. What will be the number of atoms in 500 cm³ of radon gas (radon atoms) at the same temperature and pressure? (Assume both gases behave ideally.)
- **A** Z
- **B** 2Z
- **C** Z/2
- **D** Z/4

Q22 Which graph shows how the average kinetic energy of the atoms or molecules varies with absolute temperature (in kelvin) for an ideal gas?

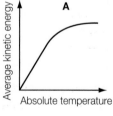

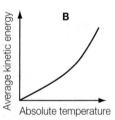

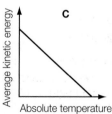

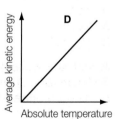

Q23 Under what conditions of temperature and pressure will a real gas behave most like an ideal gas?

	Temperature	Pressure
A	Low	Low
B	High	Low
C	High	High
D	Low	High

Q24 For an ideal gas, which variables are inversely proportional to each other (if all other factors remain constant)?

A P, V C V, T
B P, T D n, P

Q25 When compared at the same pressure and temperature, which one of the following physical properties has the same value for H_2, and for D_2? [D = $^2_1$H].

A average molecular speed
B relative molecular mass
C collision rate between molecules
D average kinetic energy of molecules

Q26 A 350 cm^3 sample of helium gas is collected at 22.0 °C and 99.3 kPa. What volume would this gas occupy at stp?

A 318 cm^3 C 477 cm^3
B 450 cm^3 D 220 cm^3

Q27 A 27.0 g sample of an unknown carbon–hydrogen compound was burned in excess oxygen to form 88.0 g of CO_2 and 27.0 g H_2O. What is the possible molecular formula of the hydrocarbon?

A CH_4 C C_2H_6
B C_4H_6 D C_6H_6

Q28 1000 cm^3 of ammonia gas combines with 1250 cm^3 of oxygen to produce two gaseous compounds with a combined volume of 2500 cm^3, all volumes being measured at 200 °C and 0.500 atm pressure. Which of the following equations fits these facts?

A $4NH_3 + 7O_2 \rightarrow 4NO_2 + 6H_2O$
B $4NH_3 + 5O_2 \rightarrow 4NO + 6H_2O$
C $4NH_3 + 5O_2 \rightarrow 2N_2O_2 + 6H_2O$
D $4NH_3 + 3O_2 \rightarrow 2N_2 + 6H_2O$

Q29 In order to dilute 40.0 cm^3 of 0.600 mol dm^{-3} HCl(aq) to 0.100 mol dm^{-3}, what volume of water must be added?

A 60 cm^3 C 200 cm^3
B 160 cm^3 D 240 cm^3

Q30 What is the concentration of nitrate ions in 0.500 cm^3 of 0.60 mol dm^{-3} $Fe(NO_3)_3$ solution?

A 0.60 mol dm^{-3} C 1.20 mol dm^{-3}
B 0.90 mol dm^{-3} D 1.8 mol dm^{-3}

Q31 If 35.50 cm^3 of a NaOH solution are required for the neutralization of a 25.00 cm^3 sample of 0.200 mol dm^{-3} H_2SO_4, what is the concentration of the NaOH?

A 0.143 mol dm^{-3} C 0.429 mol dm^{-3}
B 0.282 mol dm^{-3} D 0.895 mol dm^{-3}

Q32 How many grams of AgCl would be precipitated if an excess of $AgNO_3$ solution were added to 55.0 cm^3 of 0.200 mol dm^{-3} KCl solution?

[Molar mass of silver chloride = 143.32 g mol^{-1}]

A 1.58 g   C 6.43 g
B 1.11 g   D 7.80 g

Q33 The temperature of an ideal gas sample is changed from 100 °C to 200 °C at constant pressure. What is the ratio of the final volume to the initial volume?

A 1 : 2   C 1.27 : 1
B 4 : 1   D 1 : 1.27

Q34 At stp (i.e. 0 °C and 1 atm pressure (101 kPa)), it was found that 1.15 dm^3 of a gas weighed 3.96 g. What is its molar mass?

A 77 g mol^{-1} C 47 g mol^{-1}
B 39 g mol^{-1} D 4 g mol^{-1}

Q35 A sample of argon gas in a sealed container of fixed volume is heated from 50 to 250 °C. Which quantity will remain constant?

A average speed of the atoms
B pressure of the gas
C average kinetic energy of the atoms
D density of the argon

Paper 2 IB questions and IB style questions

Q1 10 cm^3 of ethene, C_2H_4, is burned in 40 cm^3 of oxygen, producing carbon dioxide and some liquid water. Some oxygen remained.

a Write the equation for the complete combustion of ethene. [2]
b Calculate the volume of carbon dioxide and the volume of oxygen remaining. [2]

Q2 a Write an equation for the formation of zinc iodide from zinc and iodine. [1]
b 100.0 g of zinc is allowed to react with 100.0 g of iodine producing zinc iodide. Calculate the amount (in moles) of zinc and iodine, and hence determine which reactant is in excess. [3]
c Calculate the mass of zinc iodide that will be produced. [1]

Higher Level Paper 2, May 04, Q3

Q3 A balloon, which can hold a maximum of 1000 cm³ of nitrogen before bursting, contains 955 cm³ of nitrogen at 5 °C.

    a Determine whether the balloon will burst if the temperature is increased to 30 °C.   [3]

    b Use the kinetic theory to explain what happens to the molecules of nitrogen inside the balloon as the temperature is increased to 30 °C.   [2]

Q4 Hydrated sodium carbonate has the formula $Na_2CO_3.nH_2O$. An experiment was performed to determine *n*, the amount of water of crystallization. A sample of 50.00 grams of hydrated sodium carbonate was dissolved in 250 cm³ of water. 20.00 cm³ of this solution reacted completely with 13.95 cm³ of 2.00 mol dm⁻³ hydrochloric acid.

$Na_2CO_3(aq) + 2HCl(aq)$
$$\rightarrow 2NaCl(aq) + CO_2(g) + H_2O(l)$$

    a Calculate the amount of hydrochloric acid reacted.   [1]

    b Calculate the amount of sodium carbonate in the 20 cm³ of the solution used in the reaction.   [1]

    c Calculate the concentration of sodium carbonate in the sample.   [1]

    d Calculate the molar mass of the hydrated sodium carbonate.   [1]

    e Calculate the value of *n*.   [2]

Q5 **a** Aqueous XO_4^{3-} ions form a precipitate with aqueous silver ions. Write a balanced equation for the reaction, including state symbols.   [1]

    b When 41.18 cm³ of a solution of aqueous silver ions with a concentration of 0.2040 mol dm⁻³ is added to a solution of ions, 1.172 g of the precipitate is formed.

      i Calculate the amount (in moles) of Ag^+ ions used in the reaction.   [1]

      ii Calculate the amount (in moles) of the precipitate formed.   [1]

      iii Calculate the molar mass of the precipitate.   [2]

      iv Determine the relative atomic mass of X and identify the element.   [2]

Higher Level Paper 2, Nov 03, Q2

2 Atomic structure

STARTING POINTS

- Each chemical element is composed of particles called atoms.
- Atoms are described by their atomic number (Z) (the number of protons) and mass number (A) (the combined number of protons and neutrons).
- Isotopes are atoms of the same chemical element but with different numbers of neutrons in their nuclei.
- Isotopes (of an element) have identical chemical properties but slightly different physical properties.
- Radioactivity (radioactive decay) is the spontaneous breakdown of the nuclei of atoms. Ionizing radiation is released during this process.
- The half-life of a radioisotope (radioactive isotope) is the time it takes for half of the atoms of the isotope to decay.
- Radioisotopes are used in archaeology (radiocarbon dating) and medicine, to treat cancer and act as tracers.
- A mass spectrometer allows the accurate determination of the relative atomic mass of a sample of atoms of a chemical element.
- The relative atomic mass is a weighted average of the relative isotopic masses of the atoms.
- Electrons are arranged in atoms in energy levels. They can move between energy levels via absorption/emission of energy.
- Atomic spectra are obtained by analysing the emission of electromagnetic radiation from excited gaseous atoms (at low pressure).
- The electron arrangement of atoms can be determined from their emission (line) spectra.

2.1 The atom

2.1.1 State the position of protons, neutrons and electrons in the atom.

Atoms are composed of three **sub-atomic particles: protons, neutrons** and **electrons**. Each atom consists of two regions: the **nucleus** and the **electron shells**. The nucleus is a very small dense region located at the centre of the atom. The nucleus contains protons and neutrons. Virtually all of the mass of an atom is due to the protons and neutrons. The electrons occupy the empty space around the nucleus and are arranged in shells. Each shell can hold a specific maximum number of electrons (see page 66). This simple model of the atom is illustrated in Figure 2.1.

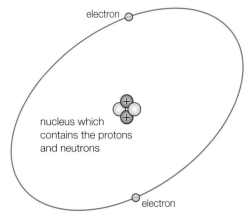

Figure 2.1 A simple model of a helium-4 atom

Language of Chemistry

The word 'atom' is derived from the Greek word *átomos* meaning 'uncuttable' and hence referring to something that cannot be divided. 'Proton' is derived from the Greek word for first. 'Electron' is derived from the Greek word for amber, a substance used in the study of static electricity. 'Neutron' is derived from the Latin root for neutral and the Greek ending *-on* (imitating electron and proton). ■

History of Chemistry

Atomism, the concept that matter is composed of atoms, is thousands of years old. The idea was founded in philosophical reasoning rather than experimentation. The earliest references to atomism date back to India in the sixth century BC. The earliest references in the Western World emerged a century later in Ancient Greece, from Leucippus and his student Democritus.

In 1803, the British chemist **John Dalton** (1766–1844) developed an atomic theory to explain why chemical elements reacted in simple proportions by mass. He proposed that each chemical element consisted of identical atoms and that these atoms could bond together to form chemical compounds. It is not clear whether Dalton was aware of previous ancient ideas about atoms, but his interest in atoms was strongly influenced by the experiments he performed on gases. He did not prove the existence of atoms; he simply demonstrated that his atomic theory was consistent with experimental data.

Unfortunately, some of Dalton's assumptions were later shown to be incorrect. For example, by assuming the formula for water was OH he calculated the relative atomic mass of oxygen to be 7 (where the atomic mass of hydrogen was assigned a value of 1). The critical distinction between atoms and molecules, for example, O and O_2, was not made until 1811, by Avogadro, but it remained ineffective until clarified by Cannizzaro in 1858. Dalton's atomic theory had to be modified in the twentieth century, following the discovery of radioactivity, isotopes and sub-atomic particles, but it is still a useful model for accounting for chemical composition and chemical changes. Dalton's symbols (Figure 2.2) for the elements were later replaced by our modern symbols (Chapter 1) devised by the Swedish chemist Berzelius (1779–1848).

Dalton's chemical theory is often used as an example of a paradigm shift, or revolution in 'scientific thinking'. At the time, the prevalent scientific thought (supported by influential scientists, such as Kelvin and Mach) did not accept the notion of Dalton's atoms. As evidence for Dalton's theory increased, more and more scientists found value in using Dalton's ideas and gradually a shift in thinking about atoms in this manner became more acceptable. This acceptance of a new approach towards viewing a phenomenon is called a 'paradigm shift'.

Figure 2.2 Dalton's symbols for the chemical elements. Note that some of his 'elements' are compounds, for example, magnesia is magnesium oxide

Relative masses and relative charges of protons, neutrons and electrons

2.1.2 **State** the relative masses and relative charges of protons, neutrons and electrons.

Protons have a positive charge and neutrons have no electrical charge and hence are neutral. These particles have almost exactly the same mass. Electrons have a negative charge. The opposite charges of the proton and electron (through electrostatic forces of attraction) hold the atom together. Electrons have a very small mass compared with protons and neutrons. A summary of the characteristics of the sub-atomic particles is given in Table 2.1.

Table 2.1 Characteristics of protons, neutrons and electrons

Sub-atomic particle	Symbol	Relative mass	Relative charge	Nuclide notation
Proton	p	1	+1	$_1^1p$
Neutron	n	1	0	$_0^1n$
Electron	e	5×10^{-4}	−1	$_{-1}^0e$

The charge is measured relative to that of a proton; the mass is measured relative to that of the proton or neutron (as they have nearly the same mass). Atoms are electrically neutral because they contain equal numbers of protons and electrons.

Neutrons help to stabilize the nucleus. They separate the protons, reducing the electrostatic repulsion, and also attract each other and protons (via the strong nuclear force). However, if too many or too few neutrons are present, the nucleus is unstable and will undergo radioactive decay (page 56).

Although electrons are the smallest of the three sub-atomic particles, they control the chemical properties of the chemical elements. Different elements have different chemical properties because they have different numbers of electrons and hence different arrangements in their electron shells. Figure 2.3 gives an idea of the scale of an atom.

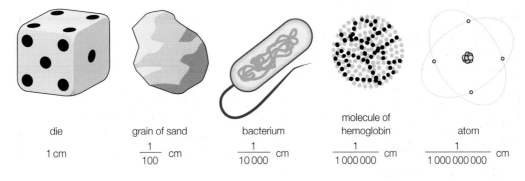

die	grain of sand	bacterium	molecule of hemoglobin	atom
1 cm	$\dfrac{1}{100}$ cm	$\dfrac{1}{10\,000}$ cm	$\dfrac{1}{1\,000\,000}$ cm	$\dfrac{1}{1\,000\,000\,000}$ cm

Figure 2.3 A series of diagrams illustrating the size of atoms

History of Chemistry

The electron was discovered in 1897 by the British physicist **J.J. Thomson** (Figure 2.4) (1856–1940), for which he was awarded the Nobel Prize in Physics in 1906. Thomson made his discovery through a series of experiments involving cathode ray tubes (Crookes tubes) and cathode rays. He found that the beam of cathode rays, produced when a high potential was passed across as gas at low pressure, was composed of negatively charged particles. Thomson measured the ratio of the charge to the mass of the particles in the beam by noting their trajectory under the influence of magnetic and electric fields (Figure 2.5) and found it to be over 1000 times less than that of a hydrogen ion. Thomson identified the particles as electrons.

In further experiments Thomson demonstrated that a hydrogen atom had only one electron. He speculated that atoms consist of a sphere of positive charge with electrons evenly distributed throughout the sphere. This is often referred to as Thomson's 'plum pudding model' (Figure 2.6). Experiments later performed by Ernest Rutherford did not support the 'plum pudding model' and it was replaced with the nuclear model of the atom.

Figure 2.4 Sir John Joseph Thomson

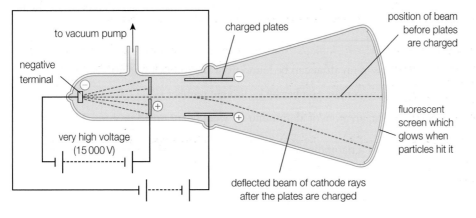

Figure 2.5 Deflection of cathode rays by an electric field

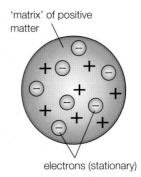

Figure 2.6 Thomson's 'plum pudding' model of the internal structure of the atom

■ Extension: Deflection of sub-atomic particles

Figure 2.7 A cathode ray tube

Moving charged particles such as protons, electrons (cathode rays) and ions are deflected by electric and magnetic fields in an evacuated vacuum tube (Figure 2.7).

This deflection is shown in Figure 2.8. The deflection of positive ions in a magnetic field is a key feature of the mass spectrometer used to study atoms (Section 2.2).

Charged sub-atomic particles and ions that enter a magnetic field are deflected from a straight line to follow an arc of a circle, the radius of which depends on their mass-to-charge ratio (m/z). Positively charged ions with a lower mass-to-charge ratio will be deflected more than those with a higher mass-to-charge ratio. For example, the lighter ion $^{20}Ne^+$ is deflected more than the heavier $^{21}Ne^+$, which in turn is deflected more than the even heavier $^{22}Ne^+$ (Figure 2.9). This is a reflection of *increasing* mass-to-charge ratio. These three ions are derived from three isotopes of neon (page 54).

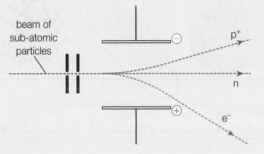
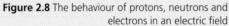

Figure 2.8 The behaviour of protons, neutrons and electrons in an electric field

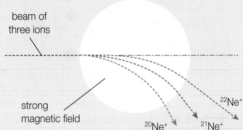

Figure 2.9 Deflection of unipositive ions in a mass spectrometer

Mass number, atomic number and isotopes

2.1.3 **Define** the terms *mass number (A)*, *atomic number (Z)* and *isotopes of an element.*

The **atomic number** (symbol Z) is the number of protons in the nucleus. The atomic number is the same for every atom of a particular element and no two different elements have the same atomic number.

Neutrons and protons have an almost identical mass but electrons have very little mass, so the mass of an atom depends on the number of protons and neutrons in its nucleus. The total number of protons and neutrons is called the **mass number** (symbol A).

Nuclides of a chemical element are described by the following notation $^A_Z X$ where X represents the symbol of the chemical element, Z represents the atomic number and A represents the mass number. For example:

$$\begin{smallmatrix} \text{mass number }(A) \rightarrow 27 \\ \text{atomic number }(Z) \rightarrow 13 \end{smallmatrix} \text{Al}$$

The number of neutrons in an atom can be found from the following relationship:

number of neutrons = mass number (A) − atomic number (Z)

So in the example above, an atom of aluminium-27 would contain 13 protons, 13 electrons and 14 neutrons (27 − 13).

Not all of the atoms in a naturally occurring sample of a chemical element are identical. Atoms of the same element that have different mass numbers are called **isotopes**. Because they are the same element, they will have the same atomic number, but they have different numbers of neutrons. Examples of isotopes are carbon-12, $^{12}_6C$, and carbon-13, $^{13}_6C$.

History of Chemistry

Ernest Rutherford (1871–1937) was a New Zealander who worked with Thomson in the Cavendish Laboratory at Cambridge. Rutherford subsequently held faculty positions at McGill University in Canada, and Manchester and Cambridge Universities in England. He experimented with radium, an alpha-emitting radioisotope isolated by Pierre and Marie Curie. Rutherford's two students Geiger and Marsden fired alpha particles at very thin metal foils to see how they would be deflected or absorbed (Figure 2.10). If Thomson's plum pudding model of the atom was correct, the alpha particles would have mostly been absorbed, and those which did penetrate the foil would have been deflected only slightly. However, over 99% of the alpha particles passed straight through the metal foil, and a very few bounced back. This experiment, known as the 'gold foil' experiment, led Rutherford to discard Thomson's model and propose the idea of a positively charged atomic nucleus. The few large-angle deflections could be explained by repulsion between the positively charged alpha particle and a strong positive charge localized in a small volume, in other words, the atomic nucleus (Figure 2.11). Rutherford was awarded the Nobel Prize in Chemistry in 1908. He famously remarked: 'It was quite the most incredible event that has ever happened to me in my life. It was almost as incredible as if you fired a 15-inch shell at a piece of tissue paper and it came back and hit you.'

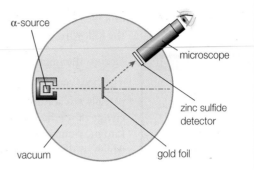

Figure 2.10 Alpha particle scattering experimental apparatus

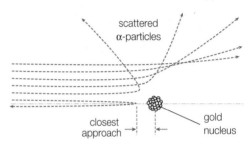

Figure 2.11 Possible trajectories of alpha particles from Rutherford's gold foil experiment

2.1.4 **Deduce** the symbol for an isotope given its mass number and atomic number.

The notation for a specific nuclide can be written if the mass number and atomic number are given. For example, the notation for a sodium atom with a mass number of 23 is written in the following way:

$$^{23}_{11}\text{Na}$$

Chemical elements (names and symbols) and their atomic numbers are listed on page 4 of the IB *Chemistry data booklet*.

Applications of Chemistry

There is considerable debate about the issue of global climate change and predicted future effects (Chapter 25). By drilling ice cores from the polar ice sheets chemists can obtain data on changes in the composition of the atmosphere over the last 100 000 years. The water molecules in the ice are mainly $^{1}\text{H}_2{}^{16}\text{O}$ but it also contains traces of $^{1}\text{H}_2{}^{18}\text{O}$ and $^{1}\text{H}^{2}\text{H}^{16}\text{O}$. The proportions of these heavier molecules depends on the temperature of the atmosphere. Hence, the temperature of the atmosphere in the past can be estimated by measuring the ratio of $^{18}\text{O}/^{16}\text{O}$, or $^{2}\text{H}/^{1}\text{H}$. These measurements are carried out with a mass spectrometer (Section 2.2).

2.1.5 **Calculate** the number of protons, neutrons and electrons in atoms and ions from the mass number, atomic number and charge.

An element forms an **ion** when one or more electrons are added or removed from the atom. A positive ion is formed by the removal of electrons and a negative ion is formed by the addition of electrons. Thus in an ion the number of protons and neutrons remains the same as in the atom –

only the number of electrons changes. The number of protons, neutrons and electrons in an atom or ion can therefore be calculated from the mass number, atomic number and charge, as shown by the following examples.

Worked examples

Deduce the number of electrons, protons and neutrons in $^{31}_{15}P^{3-}$.

The subscript number in the nuclide notation is the atomic number. Hence an atom of phosphorus contains 15 protons and therefore 15 electrons. However, since the ion has a net charge of −3 the ion contains 18 (15 + 3) electrons. The difference between the subscript number (atomic number) and the superscript number (mass number) is equal to the number of neutrons, which in this example is 16 (31 −15).

Deduce the number of electrons in $^{24}_{12}Mg^{2+}$.

Since the atomic number of magnesium is 12, an atom of magnesium contains 12 electrons. However, since the ion has a net charge of +2 the ion contains 10 (12 − 2) electrons. The difference between the subscript number (atomic number) and the superscript number (mass number) is equal to the number of neutrons, which in this example is 12 (24 − 12).

Calculate the total number of electrons in four moles of beryllium atoms.

The atomic number of beryllium is 4, hence each atom of beryllium contains 4 protons and 4 electrons. The total number of electrons in four moles of beryllium atoms is therefore

$$4 \times 4\,\text{mol} \times 6 \times 10^{23}\,\text{mol}^{-1} = 9.6 \times 10^{24}$$

A common mistake is to misread the question and give the answer 16, namely the total number of electrons in four beryllium atoms.

Table 2.2 gives further examples of common nuclides and their symbols and the numbers of their sub-atomic particles.

Name of ion or atom	Symbol of particle	Number of protons	Number of neutrons	Number of electrons
Beryllium atom	$^{9}_{4}Be$	4	5	4
Oxygen atom	$^{17}_{8}O$	8	9	8
Neon atom	$^{21}_{10}Ne$	10	11	10
Fluorine atom	$^{19}_{9}F$	9	10	9
Oxygen atom	$^{18}_{8}O$	8	10	8
Magnesium ion	$^{24}_{12}Mg^{2+}$	12	12	10
Chloride ion	$^{37}_{17}Cl^{-}$	17	20	18
Aluminium ion	$^{27}_{13}Al^{3+}$	13	14	10
Calcium ion	$^{40}_{20}Ca^{2+}$	20	20	18

Table 2.2 Selected nuclides

Properties of isotopes

2.1.6 Compare the properties of the isotopes of an element.

Many elements exist as a mixture of isotopes. Figure 2.12 shows the isotopes of carbon, chlorine and hydrogen. Isotopes of the same element all have the same element symbol and atomic number.

Carbon:
Carbon-12 ($^{12}_{6}C$)
Carbon-13 ($^{13}_{6}C$)

Chlorine:
Chlorine-35 ($^{35}_{17}Cl$)
Chlorine-37 ($^{37}_{17}Cl$)

Hydrogen:
Hydrogen-1 ($^{1}_{1}H$)
Hydrogen-2 ($^{2}_{1}H$)
Hydrogen-3 ($^{3}_{1}H$)

hydrogen-1   hydrogen-2   hydrogen-3

Figure 2.12 Stable isotopes of carbon, chlorine and hydrogen

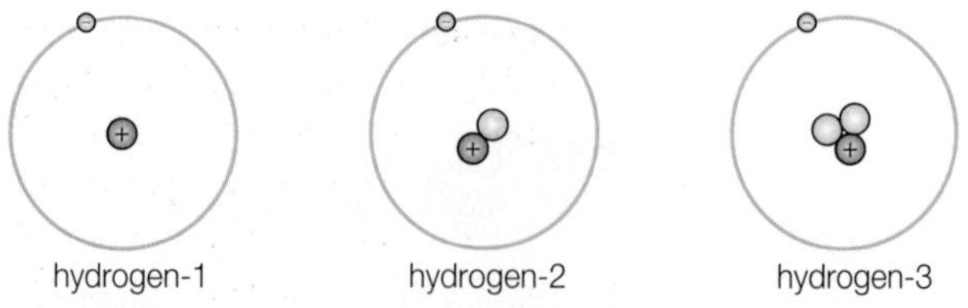

Figure 2.13 The three isotopes of hydrogen: protium, deuterium and tritium

Isotopes of the same chemical element have identical chemical properties but slightly different physical properties. For example, the lighter isotope will always diffuse more rapidly.

Language of Chemistry

The word isotope was coined by F. Soddy to describe different atoms of the same chemical element, because all these isotopes occupy the same place in the periodic table (Chapter 3). Isotope is derived from '*isotopos*', which means 'equal place' in Greek. ∎

∎ Extension: Isotopes in compounds

Compounds will often exist as a mixture of molecules with different relative molecular masses. For example, a sample of carbon consists of the isotopes carbon-12 and carbon-13. A sample of oxygen consists of the isotopes of oxygen-16 and oxygen-17. This means that there will be six types of carbon dioxide molecules:

$$^{12}C^{16}O_2 \qquad ^{12}C^{16}O^{17}O \qquad ^{12}C^{17}O_2 \qquad ^{13}C^{16}O_2 \qquad ^{13}C^{16}O^{17}O \qquad \text{and} \qquad ^{13}C^{17}O_2$$

Applications of Chemistry

Enriched uranium is a type of uranium in which the per cent composition of uranium-235 has been increased through a process of isotope separation. Enriched uranium is a critical step in preparing uranium for nuclear power generation or nuclear weapons. Isotope separation is a very difficult and energy-intensive process. Uranium-235 and uranium-238 have identical chemical properties, and their physical properties are only very slightly different. An atom of uranium-235 is only 1.26% lighter than an atom of uranium-238, which forms 99.3% of natural uranium. The first technique to be developed for uranium enrichment was gaseous diffusion. The uranium was reacted with fluorine to form uranium hexafluoride molecules, UF_6. The uranium hexafluoride was vaporized and the gaseous molecules were then forced through a series of semi-permeable membranes. The rate of diffusion of a gas is proportional to the square root of $1/M$, where M represents the molar mass. Hence, the lighter UF_6 molecules, which contain U-235, diffuse faster than those containing U-238. As a consequence, the front of the diffusing gas becomes enriched in $^{235}UF_6$. The process is repeated many times to achieve a partial separation. The currently preferred method of isotope separation by centrifugation also uses UF_6.

Radioactivity and the uses of radioisotopes

2.1.7   **Discuss** the uses of radioisotopes.

History of Chemistry

In 1896 the French chemist **Henri Becquerel** (1852–1908) discovered that uranium salts released radiation which passed through the wrapping paper around a photographic plate, exposing it (turning it black). The phenomenon was investigated by Pierre (1859–1906) and Marie Curie (1867–1934), who named it radioactivity. Henri Becquerel and the Curies were awarded the Nobel Prize in Physics in 1903. Marie Curie's death was probably caused by prolonged exposure to radiation. At the time its damaging effects were not known and most of her work had been carried out with no safety measures. Her husband died in 1906 in a tragic accident in Paris involving a horse-drawn carriage. Marie Curie was born in Poland but later became a French citizen. The Curies discovered two new elements, radium and polonium, the latter named after Marie's birthplace, Poland.

A number of chemical elements contain unstable nuclides. The nuclei of these chemical elements break up spontaneously with the emission of ionizing radiation (Table 2.3). These unstable nuclides are described as **radioactive** and are called **radioisotopes**. The radiation is of three distinct types and their properties are summarized in Table 2.3. Radiation can be detected and measured using a Geiger–Müller tube (Figure 2.14).

Figure 2.14 Geiger–Müller tube (radiation counter)

Radiation	Relative charge	Relative mass	Nature	Penetration	Deflection by electric field
Alpha particles	+2	4	2 protons and 2 neutrons (He^{2+} ion)	Stopped by a few sheets of paper	Low
Beta particles	−1	$\frac{1}{1837}$	Electron	Stopped by a few mm of plastic or aluminium	High
Gamma rays	0	0	Electromagnetic radiation of very high frequency	Stopped by a few cm of lead	None

Table 2.3 Summary of the properties of alpha and beta particles and gamma rays

When the nucleus of a radioisotope releases an alpha or beta particle, an atom of a new element is formed. For example, carbon-14 and iodine-131 both undergo beta-decay, which can be described by the following nuclear equations:

$$^{14}_{6}C \rightarrow \, ^{14}_{7}N + \, ^{0}_{-1}e \qquad ^{131}_{53}I \rightarrow \, ^{131}_{54}Xe + \, ^{0}_{-1}e$$

Cobalt-60 is another beta emitter, but iodine-125 is a pure gamma emitter. Iodine-131 and cobalt-60 are both gamma emitters (Figure 2.15).

The rate at which nuclei undergo radioactive decay varies between chemical elements. Radioactive decay is an exponential process (Figure 2.16). The rates of radioactive decay are compared using the half-life, which is the time taken for half of the radioactive nuclei to undergo decay. During alpha and beta radioactive decay a more stable isotope is formed.

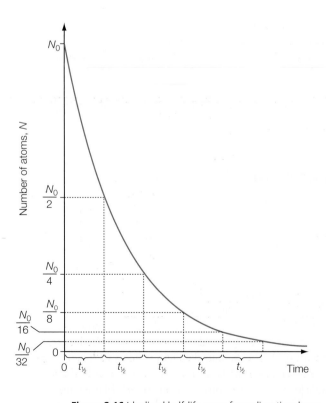

Figure 2.16 Idealized half-life curve for radioactive decay

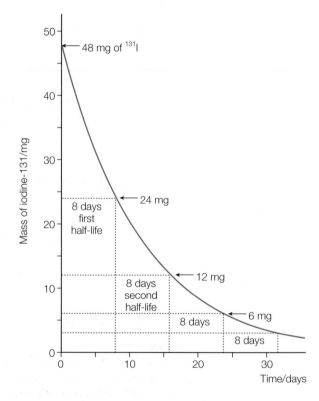

Figure 2.17 A half-life curve for iodine-131

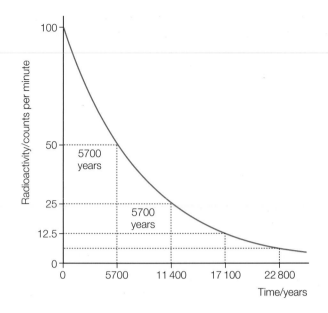

Figure 2.18 The radioactive decay curve of carbon-14

Each radioisotope has its own unique half-life which is unaffected by temperature or pressure. For example, iodine-131 has a half-life of 8 days (Figure 2.17). This means that every 8 days the number of radioactive atoms present halves.

The thyroid gland secretes an iodine-containing hormone known as thyroxine. Iodine-131 is used to treat thyroid cancer and also used to diagnose whether a thyroid gland is functioning normally. Iodine-125 is used to treat prostate cancer and brain tumours. Cobalt-60 is a very powerful gamma emitter and has been used for over 100 years to treat different types of cancer. More recently it has also been used to suppress the body's immune reaction to transplanted human organs. **Radiotherapy** works by damaging the DNA of cancer cells, preventing them from undergoing cell division.

In living animals and plants the percentage of the radioactive isotope carbon-14 remains constant because it is continually being replaced. Animals absorb nutrients from their diet and plants synthesize sugars from carbon dioxide. However, in dead tissue these processes *do not* happen and the carbon-14 undergoes beta-decay.

Hence, by comparing the percentage of carbon-14 in a dead organic sample with that in living tissue, the age of the sample can be estimated. This technique is known as **radiocarbon dating**. For example, if the amount of carbon-14 in a fossil mammal bone was found to be one eighth that in the same bone from a recently killed mammal and the half-life is 5700 years, the estimated age of the fossil bone would be 17 100 years (Figure 2.18). Radiocarbon dating can be used to date organic remains fairly accurately for up to 100 000 years.

2.2 The mass spectrometer

2.2.1 **Describe** and explain the operation of a mass spectrometer.

A **mass spectrometer** (Figure 2.19) allows chemists to determine accurately the relative atomic masses of atoms. It can also be used to determine the relative molecular masses of molecular compounds and establish their structure (Chapter 21).

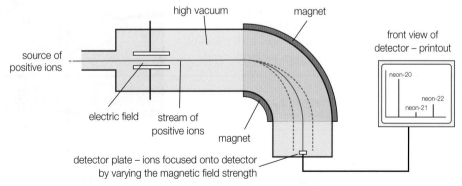

Figure 2.19 Diagram of a single beam mass spectrometer in cross-section

Inside a mass spectrometer there is a vacuum. A sample of the element is vaporized, then introduced into an ionization chamber where it is bombarded with electrons travelling at high speeds. The energetic collisions that take place cause the gaseous atoms to lose one of their electrons and form unipositive ions:

$$M(g) + e^- \xrightarrow{\text{bombarding electron}} M^+(g) + e^- + e^-$$

The beam of positive ions is accelerated by an electric field and then deflected by a powerful magnetic field. The degree of deflection depends on the mass-to-charge ratio of the positive ions. However, since the charge on each ion is the same, the deflection only depends on their masses. The lighter ions which are formed from the lighter isotope atoms are deflected more than the heavier ones.

A detector counts the numbers of each of the different ions that impact upon it, giving a measure of the percentage abundance of each isotope. The counter functions by releasing an electron for every ion it detects; this signal is then amplified. A mass spectrum for chlorine atoms is shown in Figure 2.20. The two peaks are due to detection of $^{35}Cl^+$ and $^{37}Cl^+$ ions. The mass spectrum shows that chlorine is composed of two isotopes: chlorine-35 and chlorine-37 in a 3:1 or 75%:25% ratio by abundance. (The calculation of relative atomic masses is discussed on pages 59–60.)

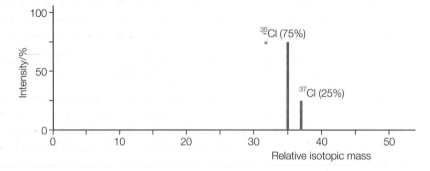

Figure 2.20 Mass spectrum of a sample of naturally occurring chlorine atoms

■ Extension: Interpreting a mass spectrum

The horizontal axis for a mass spectrum is actually mass-to-charge ratio (m/z). However, since all the ions are unipositive, the scale is equivalent to mass since the charge on all the ions is +1. However, if the kinetic energy of the bombarding electrons was increased, then dipositive ions could be formed. For example $^{35}Cl^{2+}$ would appear at 17.5 on the m/z scale.

Applications of Chemistry

In 1976 the unmanned spacecraft Viking 1 landed on the surface of Mars (Figure 2.21). Two mass spectrometers were on board: one was designed to analyse the composition of the upper atmosphere, the other to analyse the Martian soil. The analysis confirmed there was no life on Mars; organic compounds were not present in the soil. The major gas present in the atmosphere was carbon dioxide, with traces of nitrogen, argon, oxygen atoms and oxygen molecules and carbon monoxide.

Figure 2.21 The surface of Mars

History of Chemistry

The mass spectrograph (the forerunner of the mass spectrometer) was invented by British chemist and physicist **Francis William Aston** (1877–1945) in 1919. He identified 212 naturally occurring isotopes. Aston was awarded the Nobel Prize in Chemistry in 1922. A mass spectrograph works on the same principles as a modern day spectrometer, except it records its spectrum of mass values on a photographic plate.

Using mass spectrometry to determine relative atomic mass

Atomic number	10
Element	**Ne**
Atomic mass	20.18

Figure 2.22 The representation of elements in the periodic table of the IB *Chemistry data booklet*. 20.18 is the relative atomic mass of neon. The nuclide notation for neon-20 is $^{20}_{10}$Ne

2.2.2   **Describe** how the mass spectrometer may be used to determine relative atomic mass using the ^{12}C scale.

The chemical elements are listed in a special arrangement called the periodic table (Chapter 3). The periodic table can be found on page 6 of the IB *Chemistry data booklet*. Each chemical element is placed in a box with its chemical symbol (Chapter 1), its atomic number (written above) and its **relative atomic mass** (written below) (Figure 2.22).

The majority of chemical elements in nature exist as a mixture of isotopes in fixed proportions. For example, the mass spectrum of chlorine reveals that a natural sample of chlorine atoms consists of 75% chlorine-35 and 25% chlorine-37.

The relative atomic mass (symbol A_r) is the *weighted average* mass of a sample of naturally occurring atoms on the carbon-12 scale. The relative atomic mass of an element is the weighted average of its isotopes compared to one-twelfth of the mass of one atom of carbon-12:

$$\text{relative atomic mass} = \frac{\text{weighted average mass of the isotopes of the chemical element}}{\frac{1}{12} \times \text{the mass of one atom of carbon-12}}$$

However, since one-twelfth the mass of one atom of carbon-12 is 1, then relative atomic mass of a chemical element is effectively the weighted average isotopic mass divided by 1.

Relative atomic masses can be calculated from a mass spectrum of a chemical element by multiplying the relative isotopic mass of each isotope by its percentage abundance and adding all the values together.

Using chlorine as an example:

$$\text{relative atomic mass of chlorine} = \left(\tfrac{75}{100} \times 35\right) + \left(\tfrac{25}{100} \times 37\right) = 35.5$$

TOK Link

Strong indirect proof for the existence of atoms and molecules came from the kinetic theory of gases (Chapter 1), which assumes that gases consist of many millions of atoms or molecules in rapid random motion. However, the existence of atoms was doubted by a number of notable scientists up until 1900. In 1905, Einstein provided a very convincing mathematical explanation for Brownian motion. This is the random movement of pollen grains or smoke particles visible under a light microscope (Figure 2.23).

Einstein explained this phenomenon by invoking their constant bombardment by gas molecules (Figure 2.24).

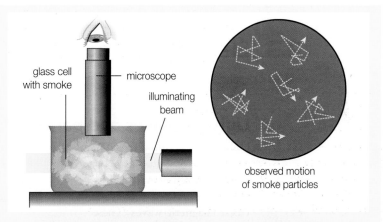

Figure 2.23 Observing Brownian motion in smoke

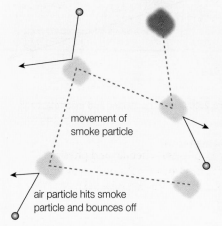

Figure 2.24 A molecular explanation for Brownian motion

Atoms and molecules are too small to see under a light microscope, but now atoms and molecules can be directly visualized as fuzzy dots using a scanning tunnelling microscope (STM). This instrument uses a very fine probe containing a tungsten tip to scan a solid surface. A small potential difference is applied between the probe and the surface. Tiny changes in current are recorded when the surface is uneven.

A computer then generates a contour map of the surface and the outline of individual atoms can be detected. The atoms resemble the hard spheres proposed by Dalton (Figure 2.25), but the STM images are in fact showing the electrons. The fuzziness occurs because the electrons move in a 'cloud' and are not in fixed energy levels or orbits. Previous generations of chemists believed in atoms, but the STM (Chapter 23) provides empirical evidence for the existence of atoms.

Figure 2.25 John Dalton

2.2.3 **Calculate** non-integer relative atomic masses and abundance of isotopes from given data.

Worked example

Rubidium exists as a mixture of two isotopes ^{85}Rb and ^{87}Rb. The percentage abundances are 72.1% and 27.9%, respectively. Calculate the relative atomic mass of rubidium.

Relative atomic mass of rubidium = $(\frac{72.1}{100} \times 85) + (\frac{27.9}{100} \times 87) = 85.6$

The use of simple algebra allows the percentage abundance of one isotope to be calculated given the relative atomic mass of the chemical element and the atomic mass of the other isotope.

Worked example

The relative atomic mass of gallium is 69.7. Gallium is composed of two isotopes: gallium-69 and gallium-71. Calculate the percentage abundance of gallium-69.

Let %Ga-69 = x. Then %Ga-71 = $(100 - x)$, since the two isotopic percentages must sum to one hundred.

$$69.7 = \frac{69x + 71(100 - x)}{100}$$
$$6970 = 69x + 71(100 - x)$$

Expanding the bracket and multiplying all the terms inside by 71:

$6970 = 69x - 71x + 7100$
$6970 = -2x + 7100$ (subtract $71x$ from $69x$)
$-130 = -2x$ (subtract 7100 from 6970)
$$x = \frac{-130}{-2} = 65$$

Hence, the percentage abundance of gallium-69 is 65%.

2.3 Electron arrangement

2.3.1 **Describe** the electromagnetic spectrum.

Light as waves and particles

Light is an important form of energy and is often described using a wave model. According to this theory, light is transmitted in the form of **electromagnetic waves**. These consist of an oscillating electric wave and a magnetic wave which travel together in a sinusoidal pattern (Figure 2.26). The two waves are arranged perpendicularly, that is, at right angles.

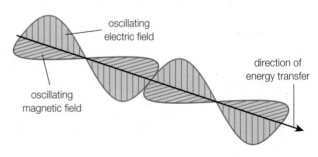

Figure 2.26 Oscillating electric and magnetic fields in an electromagnetic wave

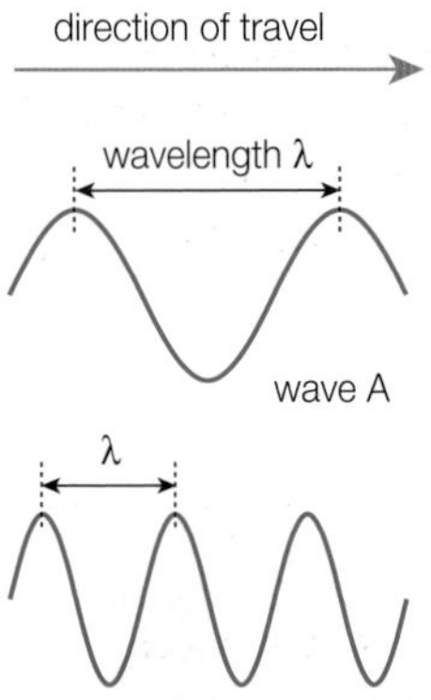

Figure 2.27 Wave A has twice the wavelength of wave B

Light waves and other waves are described by the following terms:

- **Wavelength**
  The wavelength (symbol Greek letter lambda, λ) is defined as the distance between two neighbouring crests or troughs of a wave (Figure 2.27).
- **Frequency**
  The **frequency** (symbol Greek letter f) is defined as the number of waves which pass a point in one second. Its units are **hertz** (Hz). If one wave passes a point every second, then it has a frequency of 1 Hz.

Figure 2.28 A helium–neon laser (632.8 nm wavelength)

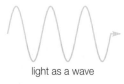

light as a wave

light as a stream of photons (packets of energy)

Figure 2.30 Light and other electromagnetic radiation can be described in two ways: as a wave or as a stream of packets of energy called photons

■ **Speed**

The speed is the distance travelled by a wave in one second. It is denoted by c and is measured in metres per second (m s^{-1}).

The frequency (f) and wavelength (λ) are related to the speed by the **wave equation**:

$$c = f\lambda \qquad \text{or} \qquad f = c / \lambda$$

where c is the speed of light. Light travels in a vacuum at a speed of 3×10^8 metres per second.

Different colours in visible light correspond to electromagnetic waves of different wavelengths and frequencies (Figure 2.28). In addition to visible light there are other types of electromagnetic radiation, such as X-rays, ultraviolet rays, infrared rays, microwaves and radio waves. Figure 2.29 shows the electromagnetic spectrum. It is an arrangement of all the types of electromagnetic radiation in increasing order of wavelength or decreasing order of frequency.

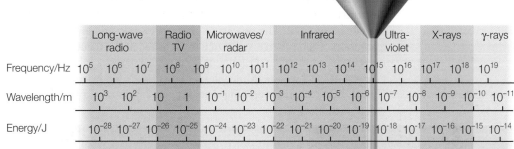

Figure 2.29 The electromagnetic spectrum

Light can also be described by a particle model (Figure 2.30) which treats light as a stream of **photons** or tiny 'packets' of light energy. The two models of light are linked by Planck's equation:

$$E = hf$$

where E represents the energy of a photon (in joules), f represent the frequency of the light (in hertz, Hz, or s^{-1}) and h represents **Planck's constant** (6.63×10^{-34} J s). Planck's equation is given on page 1 of the IB *Chemistry data booklet*; Planck's constant is given on page 2.

Spectra

2.3.2 **Distinguish** between a continuous spectrum and a line spectrum.

If sunlight or light from an electric bulb is formed into a beam by a slit and passed through a prism on to a screen, a rainbow of separated colours is observed. The spectrum of colours formed from white light is composed of visible light of a certain range of wavelengths and is called a **continuous spectrum** (Figure 2.31). A rainbow is an example of a continuous spectrum: there is an infinite number of colours that vary smoothly.

Figure 2.31 Production of a continuous spectrum

History of Chemistry

Isaac Newton (1642–1726) was the first western scientist to show that sunlight is composed of many colours. He used a prism to split light into the spectrum and then used a second prism to re-form white light. He overturned Hooke's theory that colour was a mixture of light and darkness and that prisms coloured light. Newton identified seven colours in the visible spectrum, perhaps inspired by the seven notes in an octave. The Muslim scientist **Ibn al-Haytham** (c. 965–1040) carried out experiments on the dispersion of light before Newton during the Islamic Golden Age.

History of Chemistry

A carbon star (Figure 2.32) is a rare type of star near to the end of its life. Its atmosphere contains more carbon than oxygen. The two elements combine to form carbon monoxide, which leaves the excess carbon to form a variety of unusual chemical species such as C_2, CH, CN and C_3. All these molecules have been detected spectroscopically by analysing the light emitted by these stars, which have a red appearance to the eye. An interest in carbon star chemistry was the impetus for the discovery of carbon-60 (Chapter 4) by Sir Harry Kroto (University of Sussex), who was awarded the Nobel Prize in Chemistry in 1996.

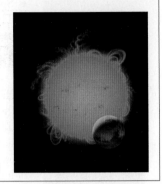

Figure 2.32 Carbon star (red giant)

If gaseous atoms are excited they emit light of certain wavelengths. Excitation occurs when electrons in the atom are raised to a higher energy level, and light is emitted as they return to the unexcited state. The process of electron excitation may be thermal or electrical. Thermal excitation occurs when a substance is vaporized and a flame is formed (Figure 2.33). Electrical excitation occurs when a high voltage is passed across a tube containing a gaseous sample of the element at low pressure. Molecules will be dissociated by the high voltage. Sodium street lamps (Figure 2.34), neon advertising signs (Figure 2.35) and exploding fireworks are all examples of electron excitation.

Figure 2.33 This brick red flame colour is specific for calcium ions

Figure 2.34 Street light (based on sodium vapour lighting)

Figure 2.35 Advertising signs use noble gases

If the light from atoms with excited electrons is passed through a prism, an emission spectrum is formed. Emission spectra consist of a number of separate sets, or series, of narrow coloured lines on a black background. Hence emission spectra are often called **line spectra**. Each chemical element has its own *unique* line spectrum which can be used to identify the chemical element.

The emission spectrum for hydrogen atoms in the visible region is shown in Figure 2.36. This series of lines is called the Balmer series, after Johann Balmer, who first observed these lines. Similar sets of lines are observed in the ultraviolet (Lyman series) and infrared regions of the electromagnetic spectrum (Figure 2.37). An emission or line spectrum (Figure 2.38) differs from a continuous spectrum in two important ways:

1 An emission spectrum is made up of separate lines (coloured if they are in the visible region), that is, it is discontinuous.
2 The lines converge, becoming progressively closer as the frequency or energy of the emission lines increases.

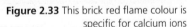

Figure 2.36 The Balmer series of hydrogen

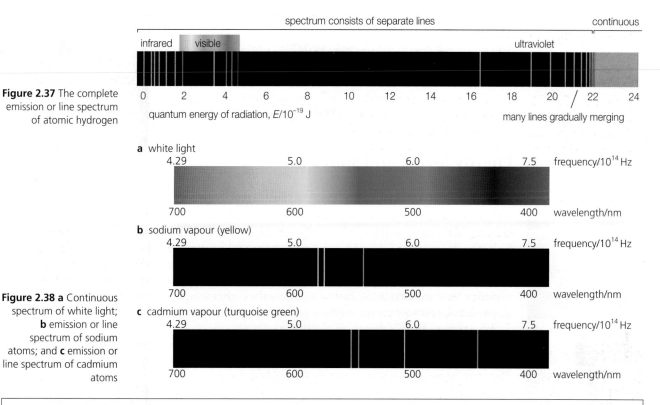

Figure 2.37 The complete emission or line spectrum of atomic hydrogen

quantum energy of radiation, $E/10^{-19}$ J

many lines gradually merging

a white light

b sodium vapour (yellow)

Figure 2.38 **a** Continuous spectrum of white light; **b** emission or line spectrum of sodium atoms; and **c** emission or line spectrum of cadmium atoms

c cadmium vapour (turquoise green)

History of Chemistry

Johann Jakob Balmer (1825–1898) was a Swiss mathematics teacher in a girls' school who studied the visible spectral lines of the emission spectrum of atomic hydrogen. He accurately measured the distances between the lines and devised a mathematical formula to calculate the wavelengths. He accurately predicted the wavelengths of lines close to the convergence limit, where the emission lines merge. The existence of these additional lines in the line spectrum of hydrogen and white stars were confirmed by Ångström. An explanation of why the formula gave the correct wavelengths was only possible when Niels Bohr presented his model of the atom in 1913.

■ Extension: Absorption spectra

If white light is passed through a sample of gaseous atoms and the emerging light is analysed, the light beam will be found to be missing certain wavelengths of light. This is because certain wavelengths have been absorbed by the gaseous atoms. The absorbed energy has caused electron excitation. These absorptions are observed as black lines against the coloured background of the visible spectrum. Figure 2.39 shows the relationship between the **absorption spectrum** (Chapter 21) and the emission spectrum of an excited atom.

Figure 2.39 Relationship between an absorption spectrum and the emission spectrum of the same element: **a** continuous spectrum of white light; **b** absorption spectrum of an element; **c** emission or line spectrum of same element

Language of Chemistry

The strong blue line in the spectrum of indium suggested its name from indigo (one of Newton's seven colours in the visible spectrum). The element thallium was named after the Greek word *Thallos*, a green budding twig. It was discovered in 1861 by the English chemist William Crookes (1832–1919), who was investigating tellurium residues with a **spectroscope**. All the elements release characteristic light when placed in a high voltage emission tube (Figure 2.40). ■

Energy levels and spectra

2.3.3 **Explain** how the lines in the emission spectrum of hydrogen are related to electron energy levels.

Bohr was the first to produce a theory that could account for the complete emission spectrum of hydrogen. He suggested that electrons in atoms moved in energy levels known as **orbits**, where they had certain fixed amounts of potential energy. The further the orbit was away from the nucleus, the greater the amount of potential energy the electron contained. This is analogous to raising an object above the Earth's surface – the higher it is, the greater the amount of gravitational potential energy it contains.

According to **Bohr's theory** an electron moving in one of these orbits does not emit energy. In order to move to an orbit further away from the nucleus, the electron must absorb energy (electrical or thermal energy) to do work against the attraction of the positively charged nucleus. The atom or electron is now said to be in an excited state.

The emission or line spectrum is formed when electrons which have been excited drop back from orbits of high energy to an orbit of lower energy. They emit light of a particular wavelength (Figure 2.41). The energy of the light is equal to the difference between the two energy levels (ΔE) and the frequency of the light is related to the energy difference by Planck's equation:

$$\Delta E = hf$$

Figure 2.40 A high voltage emission tube containing neon gas at low pressure

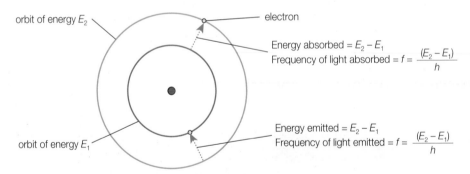

Figure 2.41 The origin of spectral lines

orbit of energy E_2

electron

Energy absorbed = $E_2 - E_1$
Frequency of light absorbed = $f = \dfrac{(E_2 - E_1)}{h}$

orbit of energy E_1

Energy emitted = $E_2 - E_1$
Frequency of light emitted = $f = \dfrac{(E_2 - E_1)}{h}$

Bohr labelled his energy levels or orbits with the letter n and a number. An electron in the lowest energy level (nearest the nucleus) was labelled as $n = 1$; an electron in this orbit is in its **ground state**, the most stable state for a hydrogen atom. The next orbit or energy level is labelled as $n = 2$, and so on. The energy level or orbits correspond to electron shells (page 66).

If an electron receives enough energy to remove it completely from the attraction of the nucleus, the atom is ionized. The energy required to ionize the electron is known as the ionization energy (Chapter 3). It is equivalent to the transition $n = 1$ to $n = \infty$ (infinity).

Language of Chemistry

The n used in Bohr's notation to represent energy levels is known as the **principal quantum number**. It is effectively the 'shell number'. Chemists and physicists use a range of quantum numbers (Chapter 12) to describe electrons in atoms. For example, the spin quantum number (s) indicates whether an electron is spinning clockwise or anticlockwise. ■

Figure 2.42 shows how Bohr's ideas can be used to explain the origin of the Lyman series. The circles represent the energy levels that the electron in a hydrogen atom can occupy. The distances between the circles represent the energy differences between the energy levels. The lines shown in Figure 2.42 are all part of the Lyman series. They are formed as excited electrons from higher energy levels 'fall' from higher energy levels (n = 2, 3, 4, 5 etc.) to the ground state (n = 1). The Balmer series of lines is formed when excited electrons fall from higher energy levels to the second energy level (n = 2).

Figure 2.42 shows that the energy levels become more closely spaced until they converge at high potential energy. This is known as the convergence limit and corresponds to the electron being completely free of the influence of the nucleus of the hydrogen atom. A hydrogen atom that has lost its electrons is said to be ionized. The difference between the convergence limit and the ground state is termed the ionization energy (Figure 2.43).

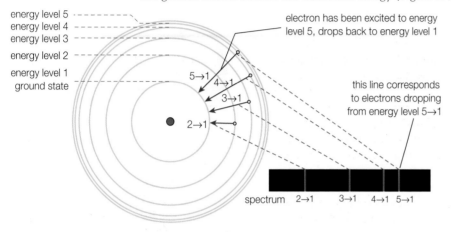

Figure 2.42 How the energy levels in the hydrogen atom give rise to the Lyman series

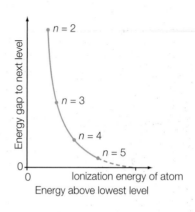

Figure 2.43 A graph showing how the value of ionization energy can be estimated by extrapolation

The diagram in Figure 2.44 is similar to Figure 2.42 except the energy levels have been drawn as straight lines, rather than as circles. In addition, the electron transitions that give rise to the other spectral series have been added.

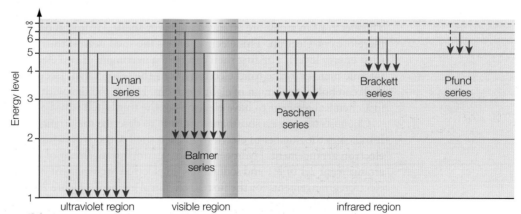

Figure 2.44 The origin of all the major spectral series in the hydrogen emission spectrum

■ Extension: Photon energies

Calculate the energy of photons that give rise to the red line of the Balmer series. The wavelength is 656.3 nm and the speed of light in a vacuum is $3.00 \times 10^8 \, \text{m s}^{-1}$.

$$f = \frac{c}{\lambda} = \frac{3.00 \times 10^8 \, \text{m s}^{-1}}{656.3 \times 10^{-9} \, \text{m}} = 4.57 \times 10^{14} \, \text{s}^{-1}$$

$$\text{energy} = hf = 6.63 \times 10^{-34} \, \text{J s} \times 4.57 \times 10^{14} \, \text{s}^{-1} = 3.03 \times 10^{-19} \, \text{J}$$

History of Chemistry

Niels Bohr (1885–1962) was a Danish physicist who was awarded the Nobel Prize in Physics in 1922 for major contributions to our understanding of atomic structure and in developing quantum mechanics. He was also part of the team that worked on the Manhattan Project, which developed the first atomic bomb. Bohr studied under Rutherford and J.J. Thomson. The Bohr model (1913) of atomic structure is a simple quantum mechanical model: electrons can only have certain values or quantities of energy. The Bohr model was also quantitative and accounted for Balmer's formula. However, Bohr's original theory was flawed and could not account for the spectra of atoms other than the hydrogen atom (or related species, for example, He$^+$) and has been replaced by Schrödinger's quantum mechanical model. 'Anyone who is not shocked by quantum theory has not understood it', is a saying attributed to Bohr.

■ Extension: Sub-atomic structure

The Ancient Greeks believed that atoms were 'uncuttable'; Dalton believed that atoms could not be created or destroyed. However, later in the nineteenth century scientists discovered that atoms were composed of three sub-atomic particles: protons, neutrons and electrons. Physicist continue to study the sub-atomic particles using particle accelerators which are used to study collisions between particles. Protons and neutrons have been found to be made up of smaller particles called quarks (Figure 2.45) which are held together by gluons. Quarks have non-integral charges.

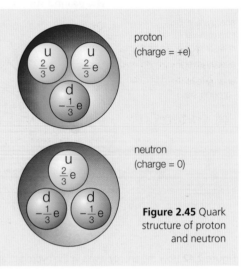

Figure 2.45 Quark structure of proton and neutron

The electron arrangement for atoms and ions

2.3.4 **Deduce** the electron arrangement for atoms and ions up to $Z = 20$.

The electrons in atoms are arranged in energy shells. Hydrogen has an atomic number of 1 and therefore one electron. This electron enters the shell nearest the nucleus. This is the first shell (first energy level). The first shell can hold a maximum of two electrons, so in the lithium atom (atomic number 3) the third electron enters the second shell (second energy level). The second shell can hold a maximum of eight electrons. Hence sodium, with an atomic number of 11, is the first chemical element to have electrons in the third shell (third energy level).

Chemists often use a shorthand notation to describe the arrangement of electrons in shells. It indicates the number of electrons in each shell without drawing the shells. It is known as the **electron arrangement**. Hydrogen has an electron arrangement of 1; lithium has an electron arrangement 2,1 or 2.1 and sodium has an electron arrangement of 2,8,1 or 2.8.1. Table 2.4 lists electron arrangements for the first 20 chemical elements; Figure 2.46 shows the shell structures for selected elements.

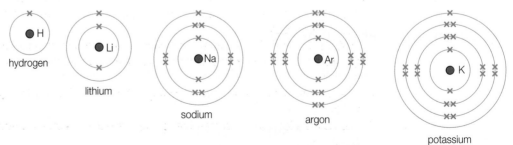

Figure 2.46 Electron arrangements of hydrogen, lithium, sodium, argon and potassium shown as shell structures

Element	Atomic number	Energy shell				Element	Atomic number	Energy shell			
		1st	2nd	3rd	4th			1st	2nd	3rd	4th
Hydrogen	1	1				Sodium	11	2	8	1	
Helium	2	2				Magnesium	12	2	8	2	
Lithium	3	2	1			Aluminium	13	2	8	3	
Beryllium	4	2	2			Silicon	14	2	8	4	
Boron	5	2	3			Phosphorus	15	2	8	5	
Carbon	6	2	4			Sulfur	16	2	8	6	
Nitrogen	7	2	5			Chlorine	17	2	8	7	
Oxygen	8	2	6			Argon	18	2	8	8	
Fluorine	9	2	7			Potassium	19	2	8	8	1
Neon	10	2	8			Calcium	20	2	8	8	2

Table 2.4 Electron arrangements for the first 20 chemical elements

■ Extension: Hybridization

The third shell can hold a maximum of 18 electrons. However, when there are eight electrons in the third shell there is a degree of stability and the next two electrons enter the fourth shell. For the transition metals beyond calcium the additional electrons enter the third shell until it contains a maximum of 18 electrons. In addition, the second and subsequent shells are divided into a number of sub-shells. Atoms (other than hydrogen) also rearrange their electrons before they can form chemical bonds with other atoms. This process is called hybridization (Chapter 14).

■ Extension: Electron shielding

An important concept introduced in Chapter 3 and also in Chapter 12 is electron shielding (Figure 2.47). The electrons in the different shells experience different attractive forces due to the presence of other electrons. The outer electrons experience the most shielding.

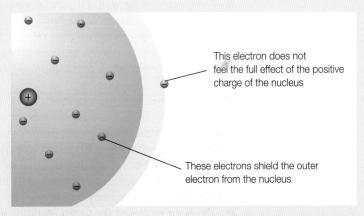

This electron does not feel the full effect of the positive charge of the nucleus

These electrons shield the outer electron from the nucleus

Figure 2.47 Electron shielding

SUMMARY OF KNOWLEDGE

- Atoms consist of a nucleus surrounded by one or more electron shells.
- Atoms contain three sub-atomic particles: electrons, protons and neutrons.
- Protons and neutrons occupy the nucleus of the atom. The electrons move round the nucleus in orbits.
- Protons are positively charged and have a mass almost identical to neutrons. Neutrons are not electrically charged. Electrons carry a negative charge equal in size to that of the proton.
- The sub-atomic particles can be distinguished by their behaviour in electric and magnetic fields.
- Atoms are electrically neutral due to the presence of equal numbers of protons and electrons.

■ The atomic number (*Z*) is the number of protons in an atom of a chemical element. The mass number (*A*) is the number of protons and neutrons in the nucleus in an atom of a chemical element.

■ A nuclide is an atom with a specific number of protons and neutrons. Nuclides are described by the notation $^A_Z X$, where *Z* represents the atomic number, *A* represents the mass number and *X* represents the symbol of the chemical element.

■ Positive ions are formed when atoms lose electrons; negative ions are formed when atoms gain electrons.

■ Isotopes of a chemical element have the same atomic number but different mass numbers. Isotopes of a chemical element have the same chemical properties, but slightly different physical properties.

■ A mass spectrometer is used to show the isotopic composition of a chemical element and determine its relative atomic mass.

■ The relative atomic mass is the mass of a weighted average of a sample of atoms compared to one twelfth of the mass of a carbon-12 atom. The relative isotopic mass is the mass of a nuclide compared to one twelfth of the mass of a carbon-12 atom.

■ A mass spectrometer ionizes gaseous atoms and then deflects a beam of its positive ions. A variable magnetic field is used to deflect the ions and bring them to a detector. Varying the magnetic field strength brings ions of higher mass to the detector, creating a mass spectrum.

■ The isotopes of some chemical elements are radioactive and release ionizing radiation. These isotopes are known as radioisotopes.

■ The three types of ionizing radiation are alpha radiation (helium nuclei), beta radiation (electrons) and gamma radiation (gamma rays).

■ Radioactive decay is an exponential process with a characteristic half-life, which is the time for the rate of radioactive decay to decrease by half. The half-life is independent of the amount of the isotope.

■ Radioisotopes have many uses in medicine (as tracers and radiotherapy) and dating techniques, for example, radiocarbon dating. The uses of radioisotopes depend on their half-life and the radiation they release.

■ Light is a form of energy and can be regarded as having the properties of an electromagnetic wave and a particle. Waves are described by their wavelength, frequency and speed.

■ A continuous spectrum contains all wavelengths from a band of the electromagnetic spectrum.

■ Each line in an emission spectrum of an atom or chemical element corresponds to electrons transitioning from one energy level to another. During electron transitions electromagnetic radiation is emitted or absorbed.

■ The lines in an emission or line spectrum become closer together as wavelength decreases (or frequency increases).

■ The Bohr model of the atom assumes that electrons rotate in circular orbits around the nucleus. The electrons in each orbit have a fixed amount of potential energy.

■ The frequency of the electromagnetic radiation and the gap between two energy levels is given by Planck's equation: $\Delta E = hf$, where *f* represents the frequency and *h* represents Planck's constant.

■ The ground state is the state of an atom in which all electrons have their lowest energies. If electrons are given additional energy and move from a lower to a higher state, then the atom is said to be excited.

■ The Balmer series is caused by excited electrons falling back to the second energy level. The Lyman series is caused by excited electrons falling back to the first energy level.

■ The ionization energy for the hydrogen atom is the energy needed to remove the electron from the ground state of the gaseous atom.

■ *Examination questions – a selection*

Paper 1 IB questions and IB style questions

Q1 Which statement is correct about the isotopes of an element?

A They have the same mass number.

B They have the same numbers of protons and neutrons in the nucleus.

C They have more protons than neutrons.

<u>D</u> They have the same electron arrangement or configuration.

Q2 A chemical element with the symbol X has the electron arrangement 2,8,6. Which chemical species is this chemical element most likely to form?

A the ion X^{3+} <u>C</u> the compound H_2X [$2H^+X^{2-}$]

B the ion X^{6+} D the compound XF_8 [$X^{8+}8F^-$]

Q3 Which of the following particles contains more electrons than neutrons?

   I $_1^1H$ **II** $_{17}^{35}Cl^-$ **III** $_{19}^{39}K^+$

A I only   C I and II only

<u>B</u> II only   D II and III only

              Standard Level Paper 1, May 00, Q6

Q4 What information about the structure of a helium atom can be gained from its emission spectrum?

<u>A</u> Most of the mass of the atom is in its nucleus.

B A helium atom contains two electrons and two protons.

C The electrons in the helium atom are held near the nucleus.

D The electrons may exist in any of several energy levels.

Q5 An element has the electron arrangement 2,8,6. What is the element?

A C   <u>C</u> S   B P   D Ar

Q6 Which is an incorrect statement about the atomic emission spectrum of hydrogen?

A The frequency of each line depends on the difference in energy between the higher and lower energy levels.

B The spectrum consists of several series of lines.

C Electronic transitions to the level $n = 2$ give rise to lines in the visible region.

D It is a continuous spectrum.

Q7 What is the correct number of each particle in a fluoride ion, $^{19}F^-$?

	Protons	Neutrons	Electrons
A	9	10	8
B	9	10	9
<u>C</u>	9	10	10
D	9	19	10

              Standard Level Paper 1, Nov 03, Q5

Q8 What fraction of a radioisotope will remain after three half-lives?

A 1/16   C 1/3

<u>B</u> 1/8   D 3/4

Q9 Why was the Bohr theory of the atom developed?

A To account for changes in gas volumes with temperature.

B To account for the ratios by mass of elements in compounds.

C To account for the emission or line spectrum of hydrogen atoms.

D To account for chemical formulas.

Q10 A particular element consists of two isotopes: 72% of mass number 85 and 28% of mass number 87. What is the expected range of the relative atomic mass?

A less than 85

B between 86 and 87

<u>C</u> between 85 and 86

D more than 88

Q11 How many valence electrons (electrons in the outermost shell) are present in the element of atomic number 14?

<u>A</u> 4   C 2   B 3   D 1

Q12 Which one of the following atoms will have the same number of neutrons as an atom of $_{38}^{88}Sr$?

A $_{39}^{91}Y$ C $_{38}^{89}Sr$ <u>B</u> $_{37}^{87}Rb$ D $_{36}^{84}Kr$

Q13 Which statement is correct for the emission spectrum of the hydrogen atom?

A The lines converge at lower energies.

B The lines are produced when electrons move from lower to higher energy levels.

C The lines in the visible region involve electron transitions into the energy level closest to the nucleus.

D The line corresponding to the greatest emission of energy is in the ultraviolet region.

              Standard Level Paper 1, Nov 03, Q6

Q14 Naturally occurring chlorine consists of the isotopes chlorine-35 and chlorine-37. The relative atomic mass of chlorine is 35.5. Which one of the following statements is true?

A The chlorine-35 and chlorine-37 atoms are present in equal amounts.

B The ratio of chlorine-37 atoms to chlorine-35 atoms is 2 : 1.

C The ratio of chlorine-37 to chlorine-35 atoms is 37/35.

<u>D</u> There are three times as many as chlorine-35 atoms as chlorine-37 atoms.

Q15 Which statement is correct about a line emission spectrum?
  A Electrons neither absorb nor release energy as they move from low to high energy levels.
  B Electrons absorb energy as they move from high to low energy levels.
  C Electrons release energy as they move from low to high energy levels.
  D Electrons release energy as they move from high to low energy levels.
  Standard Level Paper 1, Nov 05, Q6

Q16 Which electronic transition within a hydrogen atom requires the greatest energy?
  A $n = 1 \rightarrow n = 2$ **C** $n = 2 \rightarrow n = 3$
  B $n = 3 \rightarrow n = 5$ **D** $n = 5 \rightarrow n = \infty$

Q17 Which of the following radioisotopes is used in nuclear medicine to image the thyroid gland?
  A iodine-131   **C** fluorine-18
  B carbon-14   **D** uranium-235

Q18 The atomic numbers and mass numbers for four different nuclei are given in the table below. Which two are isotopes?

	Atomic number	Mass number
I	101	258
II	102	258
III	102	260
IV	103	259

  A I and II   **C** III and IV
  B II and III **D** I and IV
  Standard Level Paper 1, Nov 98, Q6

Q19 All isotopes of uranium have the same:
  I   number of protons
  II number of neutrons
  III mass number
  A I only   **C** III only
  B II only   **D** I and III only

Q20 Which is the correct sequence for some of the various stages that typically occur in the analysis of an element during mass spectrometry?
  A vaporization, electron bombardment, acceleration, deflection, detection
  B electron bombardment, vaporization, acceleration, deflection, detection
  C vaporization, electron bombardment, deflection, acceleration, detection
  D deflection, acceleration, electron bombardment, vaporization, detection

Paper 2 IB questions and IB style questions

Q1 The element bromine exists as the isotopes ^{79}Br and ^{81}Br, and has a relative atomic mass of 79.90.
  a Copy and complete the following table to show the numbers of sub-atomic particles in the species shown. [3]

	An atom of ^{79}Br	An ion of ^{81}Br$^-$
Protons		
Neutrons		
Electrons		

  b State and explain which of the two isotopes ^{79}Br and ^{81}Br is more common in the element bromine. [1]
  Standard Level Paper 2, Nov 05, Q3

Q2 The element silver has two isotopes, $^{107}_{47}$Ag and $^{109}_{47}$Ag, and a relative atomic mass of 107.87.
  a Define the term *isotope*. [1]
  b State the number of protons, electrons and neutrons in $^{107}_{47}$Ag$^+$. [2]
  c State the name and the mass number of the isotope relative to which all atomic masses are measured. [1]

Q3 A sample of iridium is analysed in a mass spectrometer. The first and last processes in mass spectrometry are vaporization and detection.
  a i State the names of the second and third processes in the order in which they occur in a mass spectrometer. [2]
   ii Outline what occurs during the second process. [2]
   iii State and explain which one of the following ions will undergo the greatest deflection (under the same conditions in a mass spectrometer):
     ^{191}Ir$^+$ or ^{193}Ir$^+$ [2]
  b The sample of iridium is found to have the following composition of stable isotopes:

Isotope	Ir-191	Ir-193
Relative abundance/%	37.1	62.9

   i Define the term *relative atomic mass*. [2]
   ii Calculate the relative atomic mass of this sample of iridium, giving your answer to two decimal places. [2]
  c Iridium-192 is a short-lived radioisotope used to treat cancer. Define the term radioisotope and name another radioisotope used in nuclear medicine. [2]

Q4 Describe the emission or line spectrum of gaseous hydrogen atoms and explain how this is related to the energy levels in the atom. [3]

3 Periodicity

STARTING POINTS

- ■ The modern (long form) of the periodic table is based upon the work of Mendeleev in 1869.
- ■ Mendeleev grouped the elements according to their chemical properties.
- ■ The periodic table allows predictions to be made about the chemical and physical properties of elements.
- ■ Periodic patterns (trends) of physical, atomic and chemical properties are observed across the periodic table.
- ■ Many of these physical and atomic properties can be accounted for in terms of the balance between the attraction of the nucleus for the electrons and the repulsion (shielding) between electrons.
- ■ Electronegativity can be used as a measure of metallic or non-metallic character.
- ■ Particularly clear trends are observed in elements in group 1 (alkali metals) and group 7 (halogens).
- ■ The elements are arranged in order of increasing atomic (proton) number.
- ■ The arrangement of elements into groups and periods is a reflection of their shell structure.
- ■ The elements can also be classified into four blocks based upon their sub-shell structure.
- ■ The alkali metals are a group of reactive metals that react with water, oxygen and halogens. They act as reducing agents.
- ■ The halogens are a group of reactive, coloured non-metals that exist as diatomic molecules. They act as oxidizing agents and form salts known as halides.
- ■ The reactivity of group 1 elements increases down the group due to an increase in their atomic radii. Group 1 elements lose an electron during ionic bond formation.
- ■ The reactivity of group 7 elements decreases down the group due to an increase in their atomic radii. Group 7 elements gain an electron during ionic bond formation.

3.1 The periodic table

3.1.1 **Describe** the arrangement of elements in the periodic table in order of increasing atomic number.
3.1.2 **Distinguish** between the terms group and period.
3.1.3 **Apply** the relationship between the electron arrangement of elements and their position in the periodic table up to $Z = 20$.
3.1.4 **Apply** the relationship between the number of electrons in the highest occupied energy level for an element and its position in the periodic table.

The arrangement of elements in the periodic table

The chemical elements in the periodic table (Figure 3.1) are arranged in order of increasing atomic number. This arrangement leads to **periodicity** – repeating patterns of chemical and physical properties. These are a reflection of repeating changes in electron configuration. Physical properties such as melting and boiling points, atomic properties such as ionization energy, and chemical properties such as rate of reaction with water, all show periodicity (Sections 3.2 and 3.3)

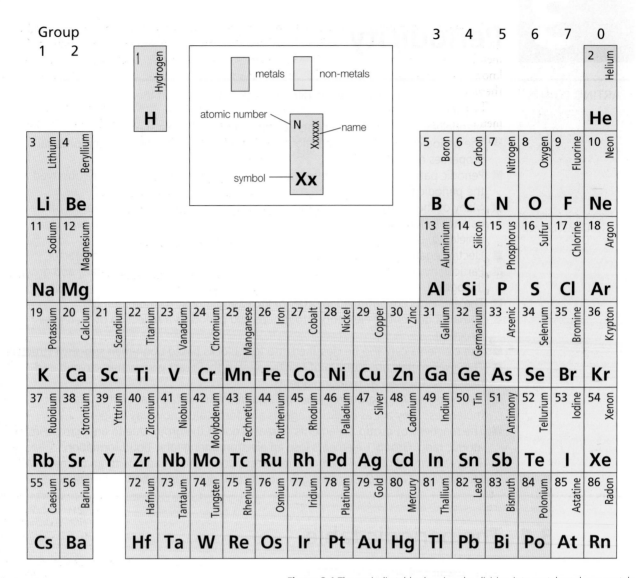

Figure 3.1 The periodic table showing the division into metals and non-metals

A horizontal row of elements in the **periodic table** is called a **period**. There are seven main periods in the periodic table (six are shown in Figure 3.1). Across a period the chemical properties gradually change from those of reactive metals to those of reactive non-metals, ending with an unreactive non-metal (Table 3.1).

Element	Sodium	Magnesium	Aluminium	Silicon	Phosphorus (white)	Sulfur	Chlorine	Argon
Appearance	Silvery metal	Silvery metal	Silvery metal	Silvery solid	White solid	Yellow solid	Greenish gas	Colourless gas
Electron arrangement	2,8,1	2,8,2	2,8,3	2,8,4	2,8,5	2,8,6	2,8,7	2,8,8
Bonding and structure	Giant metallic lattice	Giant metallic lattice	Giant metallic lattice	Giant covalent (three-dimensional)	Simple molecular (P_4)	Simple molecular (S_8)	Simple molecular (Cl_2)	Monatomic Ar
Ion formed	Na^+	Mg^{2+}	Al^{3+}	–	P^{3-}	S^{2-}	Cl^-	–

Table 3.1 Properties of elements in period 3

Figure 3.2 A piece of freshly cut sodium metal

A column of chemical elements in the periodic table is known as a **group**. There are eight groups in the periodic table: 1, 2, 3, 4, 5, 6, 7 and 0 (group 0 is sometimes called group 8). The members of group 1 are known as the **alkali metals** (Figure 3.2) and the elements in group 7 are known as the **halogens** (Section 3.3). The members of group 0 are all unreactive gases known as the **noble gases**.

The **transition metals** are a block of elements that separate groups 2 and 3. The transition metals all have very similar chemical and physical properties. Unlike groups 1 and 2, in the transition metals there are similarities both across and down the block.

The properties of transition metals are summarized below:

- relatively high melting points, high boiling points and high densities
- fairly unreactive towards water, but some react slowly with steam (Chapter 9)
- form more than one stable cation (Chapter 4) and covalently bonded complex ions (Figure 3.3) (Chapter 13)
- often have coloured compounds (Figure 3.4) and coloured solutions (Chapter 21)
- often act as catalysts (Chapters 6 and 7)
- can be combined with other transition metals (and also with other metals) to form a variety of metallic mixtures known as alloys (Figure 3.5) (Chapter 23).

Figure 3.3 From left to right: thiocyanate ions, iron(III) ions and complex ions formed by the reaction between iron(III) and thiocyanate ions

Figure 3.4 A sample of hydrated nickel chloride, $NiCl_2.6H_2O$; nickel is a transition metal

Figure 3.5 British £2 coin showing an outer gold-coloured nickel-brass ring made from 76% copper, 20% zinc and 4% nickel and an inner silver-coloured cupro-nickel disc made from 75% copper and 25% nickel

(Chapter 13 contains further information about transition metals and their compounds.)

Electron arrangement and the periodic table

Group	1	2	3	4	5	6	7	0 (or 8)
Period								
1	1 **H** 1							2 **He** 2
2	3 **Li** 2,1	4 **Be** 2,2	5 **B** 2,3	6 **C** 2,4	7 **N** 2,5	8 **O** 2,6	9 **F** 2,7	10 **Ne** 2,8
3	11 **Na** 2,8,1	12 **Mg** 2,8,2	13 **Al** 2,8,3	14 **Si** 2,8,4	15 **P** 2,8,5	16 **S** 2,8,6	17 **Cl** 2,8,7	18 **Ar** 2,8,8
4	19 **K** 2,8,8,1	20 **Ca** 2,8,8,2						

Figure 3.6 The short form of the periodic table, showing the first 20 chemical elements and their electron arrangements

It is the electrons in the outer or valence shell that determine the chemical and physical properties of the chemical element. The position of a chemical element in the periodic table is related to its electron arrangement. The period number indicates the number of shells in the atom of the element. All chemical elements in the same period have the same number of shells. The group number (except 0) indicates the number of valence electrons in the atom of the element.

Figure 3.6 shows how the electron arrangement of a chemical element is related to its group and period number. This so-called 'short form' of the periodic table omits the transition elements.

Based on the electron arrangements of the elements (Chapter 12), the periodic table can be divided into four blocks of elements:

- s-block elements
- p-block elements
- d-block elements
- f-block elements.

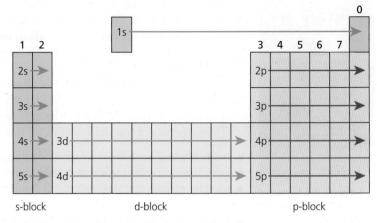

Figure 3.7 A diagram showing electron sub-shell filling in periods 1 to 5

The s-block consists of hydrogen, helium and groups 1 (alkali metals) and 2. All the s-block elements have a half-filled s orbital (s^1) or a completely filled s orbital (s^2) in the outermost shell.

The p-block consists of groups 3 to 0. The s- and p-blocks are collectively called the main group elements. Each p-block element has an outer electron configuration which varies from s^2p^1 (group 3), s^2p^2 (group 4) through to s^2p^6 (the noble gases in group 0).

The d-block (Chapter 13) consists of three series of metals. Each series of d-block metals contains ten metals with outer electron configurations ranging from d^1s^2 to $d^{10}s^2$.

There are two series of metals at the bottom of the periodic table known as the f-block metals because they contain f orbitals which are being filled. The two f-block series, known as the lanthanides and actinides, each contain 14 elements.

The block classification of elements is emphasized in a simple three-dimensional periodic table (Figure 3.8).

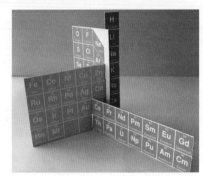

Figure 3.8 Simple three-dimensional periodic table

Language of Chemistry

The periodic table is said to show **periodicity** of physical and chemical properties. Periodicity is a general term that refers to an event that happens at regular intervals. For example, a freely swinging pendulum exhibits periodicity. The periodic table shows chemical periodicity in both the groups (columns) and periods (rows). The periodic table was generated from the periodic law which states that many of the physical and chemical properties, for example melting points of the elements (Figure 3.9), tend to vary in a regular manner with increasing atomic number. Moving from the lowest atomic number to the highest atomic number atoms, the properties of the elements are similar at regular intervals of 2, 8, 18 and 32. These numbers correspond to the filling of the first four shells of electrons. The term 'periodic trend' describes the way in which a property increases or decreases along a series of elements in the periodic table. This can refer to the changes in properties down a group or across a period. ■

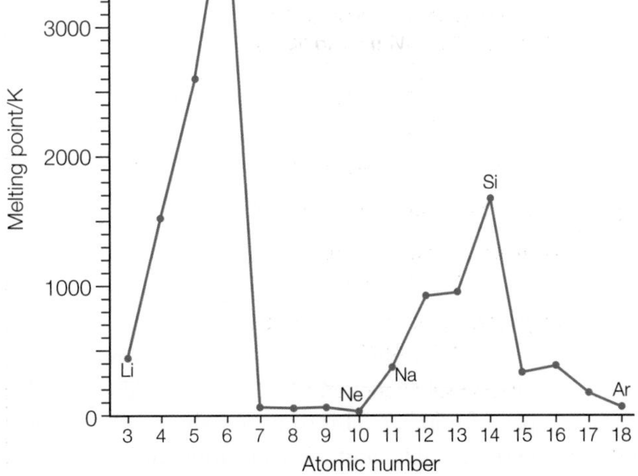

Figure 3.9 A graph showing periodicity in the melting points of elements

History of Chemistry

Döbereiner's triads

In 1829 the German chemist **Johann Döbereiner** (1780–1849) noticed that, where groups of three similar chemical elements occurred, the relative atomic mass of the middle element came about half-way between those of the other two. Two of Döbereiner's triads are shown in Figure 3.10.

Newlands' octaves

In 1864 the British chemist **John Newlands** (1837–1898) found that if the elements were arranged in order of increasing relative atomic mass then a pattern appeared (Figure 3.11). Starting at any given element, the eighth one from it was, as he phrased it, 'a kind of repetition of the first'. Because of the similarity to a musical scale he called it the Law of Octaves. Newlands' octaves place some very different elements in the same column, for example, phosphorus and manganese and iron and sulfur. The pattern breaks down if the list of elements is extended. It was widely ridiculed at the time but laid the foundations for later work by Mendeleev.

Li lithium 6.9	Cl chlorine 35.5
Na sodium 23.0	Br bromine 79.9
K potassium 39.1	I iodine 126.9

Figure 3.10 Two of Döbereiner's triads

H	Li	Be	B	C	N	O
F	Na	Mg	Al	Si	P	S
Cl	K	Ca	Cr	Ti	Mn	Fe

Figure 3.11 Newlands' octaves

Figure 3.12 Miniature set including a stamp (in the middle part) commemorating the publication of Mendeleev's first periodic table in 1869

Mendeleev's periodic table

Dimitri Mendeleev (1834–1907) (Figure 13.12) was a Russian chemist who arranged the known chemical elements into a table in order of increasing relative atomic mass in a similar manner to Newlands **but**, in order to obtain better chemical periodicity, he left gaps for undiscovered elements. He made predictions for the chemical and physical properties of five 'missing' chemical elements based upon the properties of neighbouring elements. These predictions were later proved to be accurate, following the discovery of germanium (Table 3.2), gallium, scandium, francium and technetium.

	Predicted properties of *eka*-silicon, Es (predicted by Mendeleev in 1871)	Properties of germanium, Ge (discovered by Winkler in 1886)
Relative atomic mass	72	72.6
Density	5.5 g cm^{-3}	5.47 g cm^{-3}
Appearance	Dirty grey metal	Lustrous (shiny) grey metal
With air	Will form a white powder, EsO_2, on heating	Gives a white powder, GeO_2, on heating
With acids	Slight reaction only	No reaction with dilute sulfuric or hydrochloric acid
Properties of oxide, MO_2	Very high melting point, 4.7 times denser than water	Very high melting point, 4.7 times denser than water
Properties of chloride, MCl_4	Liquid, boiling point less than 100 °C	Liquid, boiling point 86 °C

Table 3.2 A selection of predictions about germanium made by Mendeleev

TOK Link

Mendeleev is a good example of a 'risk-taker'; he invited chemists to test and potentially falsify his predictions. The requirement that a scientific hypothesis be falsifiable was proposed by the philosopher Karl Popper to be the 'criterion of demarcation' of the empirical sciences because it sets apart scientific knowledge from other forms of knowledge. A hypothesis that is not subject to the possibility of empirical falsification does not belong in the realm of science.

Mendeleev's periodic table (Figure 3.13) is known as the 'short form' and is still used in Russia and former communist countries (Figure 3.14). His table is divided into eight groups and, with the exception of the eighth, each group is divided into two sub-groups A and B. This form of the periodic table suffers from some drawbacks, for example manganese is classified with the halogens, with which it has little in common. It should be noted that the structure of atoms was not known at this time and many elements had not been discovered, most notably the noble gases. Mendeleev also had to correct some relative atomic masses which had been determined incorrectly.

Group	1		2		3		4		5		6		7		8
Sub-group	A	B	A	B	A	B	A	B	A	B	A	B	A	B	
1st period	H														
2nd period	Li		Be			B	C		N		O		F		
3rd period	Na		Mg			Al	Si		P		S		Cl		
4th period	K	Cu	Ca	Zn	–	–	Ti	*	V	As	Cr	Se	Mn	Br	Fe Co Ni
5th period	Rb	Ag	Sr	Cd	Y	In	Zr	Sn	Nb	Sb	Mo	Te	–	I	Ru Rh Pd
6th period	Cs	Au	Ba	Hg	La	Tl	–	Pb	Ta	Bi	W	–	–	–	Os Ir Pt
7th period	–		–		–		Th		–		U				

Figure 3.13 Mendeleev's periodic table (modernized form)

Figure 3.14 A commemorative version of Mendeleev's periodic table

The periodic table shown in Figure 3.14 is on the end wall of the four storey building in which Mendeleev worked from 1893 onwards. It was created in 1934 to celebrate the centenary of Mendeleev's birth. The title is written in Russian Cyrillic script and reads '*Periodic System of Elements, D.I. Mendeleev*'. The elements whose symbols are blue were discovered between Mendeleev's death in 1907 and 1934. Blanks were left for francium and astatine, still to be discovered. J represents iodine and A represented argon until 1958. At the bottom of the periodic table are the group formulas for the hydrides and oxides, emphasizing that the table is based on *chemical* properties.

Mendeleev's periodic table was based on relative atomic masses and the chemical properties of the elements. His work was done prior to knowledge about the electronic structure of atoms. In 1869 two elements had to be listed in the *wrong* order according to their relative atomic masses so they could be fitted into the correct group based on their chemical properties (Figure 3.15).

Iodine has a lower relative atomic mass than tellurium and hence should be placed in group 6. However, Mendeleev placed it in group 7 since it clearly has similar properties to the other halogens. The discovery of the noble gases introduced a similar reversal of relative atomic mass between argon and potassium.

18	19
Ar	**K**
39.95	39.10

52	53
Te	**I**
127.60	126.90

Figure 3.15 Atomic data for argon and potassium and tellurium and iodine

History of Chemistry

A new basis for the order of elements in the periodic table was established by the English chemist **Henry Moseley** (1887–1915)(Figure 3.16), who studied the X-rays released when different atoms of metallic elements were bombarded by electrons. He discovered a simple relationship between the frequency of the X-rays and the atomic (proton) number (Figure 3.17).

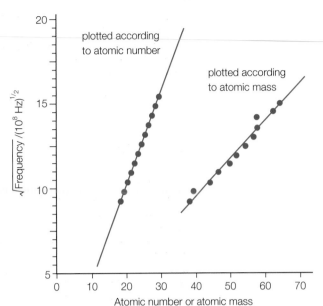

Figure 3.16 Henry Moseley

Figure 3.17 Some of Moseley's results for the X-ray spectra of various metals

From these results Moseley suggested that one proton (and therefore one electron) was added to the atom on going from one element to the next. Atomic number was therefore a more fundamental property of atoms than relative atomic mass. When the elements are arranged in order of atomic number the problems of elements being in the 'wrong order', for example iodine and tellurium, are removed. Moseley published his findings in 1914, but was to die in the First World War.

■ **Extension:** Position of hydrogen

In the IB *Chemistry data booklet* hydrogen is placed in group 1 with the alkali metals. It has one electron in its atom and can form a unipositive charged ion (H^+), like the alkali metals in group 1. However, like the halogens in group 7, it only needs to gain one electron to attain the electronic structure of the nearest noble gas, helium. Hydrogen can form the hydride ion, H^-, when bonded to a reactive metal such as sodium. It also shows some unique properties that make it difficult to classify in any group.

3.2 Physical properties

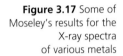

3.2.1 **Define** the terms *first ionization energy* and *electronegativity*.

First ionization energy

The **first ionization energy** is the minimum energy required to remove one mole of electrons from one mole of gaseous atoms (under standard thermodynamic conditions of 25 °C and 1 atm).

In general: $X(g) \rightarrow X^+(g) + e^-$

For example, the first ionization energy of hydrogen is given by the following equation:

$$H(g) \rightarrow H^+(g) + e^- \qquad \Delta H = +1310\,kJ\,mol^{-1}$$

The amount energy required to carry out this process for a mole of hydrogen atoms is 1310 kilojoules.

Atoms of each element have different values of first ionization energy. The factors that control the values of first ionization energy are discussed in Chapter 12.

Electronegativity

The **electronegativity** of an atom is the ability or power of an atom in a covalent bond to attract shared pairs of electrons to itself. The greater the electronegativity of an atom, the greater its ability to attract shared pairs of electrons to itself.

Electronegativity value are based on the Pauling scale. A value of 4.0 is given to fluorine, the most electronegative atom. The least electronegative elements, caesium and francium, both have an electronegativity value of 0.7. The values for all the other elements lie between these two extremes. Note that electronegativity values are pure numbers with no units.

History of Chemistry

Linus Pauling (1901–1994) was an American chemist who was awarded two Nobel Prizes: Chemistry (1954) and Peace (1962). His early work was centred on chemical bonding and intermolecular forces. He developed the concepts of hybridization and resonance (Chapter 14). He also studied biological molecules and correctly proposed the α-helix and β-sheets as common secondary structures in proteins. However, he incorrectly predicted a triple helix structure for DNA. Later in his life he began controversial research into the use of vitamin C as an anti-cancer compound, both as a preventive measure and to treat it.

Trends in the physical properties of the elements in group 1 and group 7

3.2.2 **Describe** and **explain** the trends in atomic radii, ionic radii, first ionization energies, electronegativities and melting points for the alkali metals (Li → Cs) and the halogens (F → I).

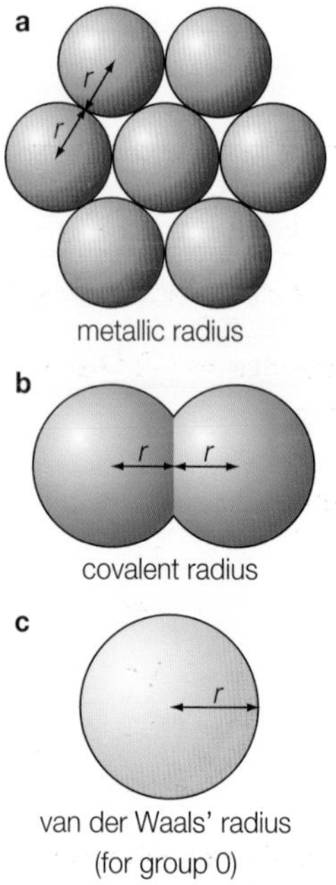

Trends in atomic and ionic radii

At the right of the periodic table the atomic radius is defined as half the distance between the nuclei of two covalently bonded atoms (Figure 3.18). For example, the bond length for chlorine atoms in a chlorine molecule (distance between two chlorine nuclei) is 0.199 nm. Therefore the atomic radius of chorine is $\frac{1}{2} \times 198 = 99$ pm (1 picometre (pm) = 10^{-12} m; 1 nanometre (nm) = 10^{-9} m). At the left of the periodic table, the atomic radius is that of the atom in the metal lattice (the metallic radius). For the noble gases the atomic radius is that of an isolated atom (the van der Waals' radius).

The atomic radius of an atom is determined by the balance between two opposing factors:

- the shielding effect by the electrons of the inner shell(s) – this makes the atomic radius larger. The shielding effect is the result of repulsion between the electrons in the inner shell and those in the outer or valence shell
- the nuclear charge (due to the protons) – this is an attractive force that pulls all the electrons closer to the nucleus. With an increase in nuclear charge, the atomic radius becomes smaller.

However, when moving down a group in the periodic table, there is an *increase* in the atomic radius as the nuclear charge increases (Tables 3.3 and 3.4 and Figures 3.19 and 3.20). This is the result of two factors:

- the increase in the number of complete electron shells between the outer (valence) electrons and the nucleus
- the increase in the shielding effect of the outer electrons by the inner electrons.

Moving down a group, both the nuclear charge and the shielding effect increase. However, the outer electrons enter new shells. So, although the nucleus gains protons, the electrons are not only further away, but also more effectively screened by an additional shell of electrons.

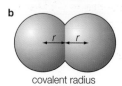
metallic radius

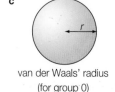
covalent radius

c
van der Waals' radius
(for group 0)

Figure 3.18 Atomic radius

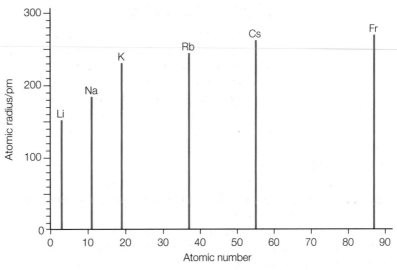

Atom	Atomic number	Atomic radius/pm
Li	3	152
Na	11	186
K	19	231
Rb	37	244
Cs	55	262
Fr	87	270

Table 3.3 The variation of atomic radii in group 1

Figure 3.19 Bar chart showing the variation of atomic radii in group 1

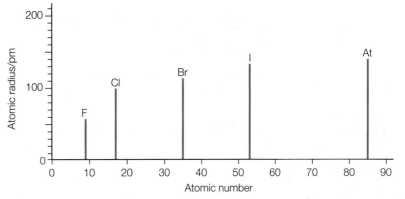

Atom	Atomic number	Atomic radius/pm
F	9	58
Cl	17	99
Br	35	114
I	53	133
At	85	140

Table 3.4 The variation of atomic radii in group 7

Figure 3.20 Bar chart showing the variation of atomic radii in group 7

Ionic radii for ions of the same charge also increase down a group for the same reason (Tables 3.5 and 3.6). Ionic radii are the radii for ions in a crystalline ionic compound (Figure 3.22).

Ion	Atomic number	Ionic radius/pm
Li^+	3	68
Na^+	11	98
K^+	19	133
Rb^+	37	148
Cs^+	55	167
Fr^+	87	No data

Table 3.5 The variation of ionic radii in group 1

Ion	Atomic number	Ionic radius/pm
F^-	9	133
Cl^-	17	181
Br^-	35	196
I^-	53	219
At^-	85	No data

Table 3.6 The variation of ionic radii in group 7

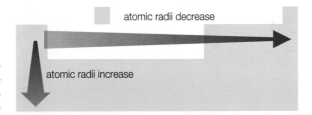

Figure 3.21 Summary of trends in periodicity in atomic radii in the periodic table

atomic radii decrease

atomic radii increase

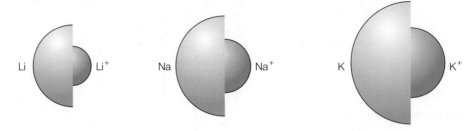

Figure 3.22 The relative sizes of the atoms and ions of group 1 metals

Trends in first ionization energy

On moving down a group, the atomic radius increases as additional electron shells are added. This causes the shielding effect to increase. The further the outer or valence shell is from the nucleus, the smaller the attractive force exerted by the protons in the nucleus. Hence, the more easily an outer electron can be removed and the lower the ionization energy. So, within each group, the first ionization energies decrease down the group. This is shown in Table 3.7 and Figure 3.23.

Atom	Atomic number	First ionization energy/kJ mol⁻¹
Li	3	519
Na	11	494
K	19	418
Rb	37	402
Cs	55	376

Table 3.7 The variation of first ionization energy in group 1

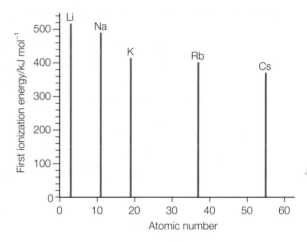

Figure 3.23 Bar graph showing the variation of first ionization energy in group 1

■ Extension: Effective nuclear charge

An alternative way to account for changes in ionization energies is to use the concept of **effective nuclear charge** (Figure 3.24). This is the nuclear charge experienced by the electrons after taking into account the shielding effect of electrons. For example, in the atoms of group 2 the effective nuclear charge is +2, which is calculated by adding the charges of the protons and shielding electrons. However, moving down group 2 the outer or valence electrons are held less strongly, being further away from the same effective nuclear charge.

Figure 3.24 Shielding in beryllium, magnesium and calcium atoms

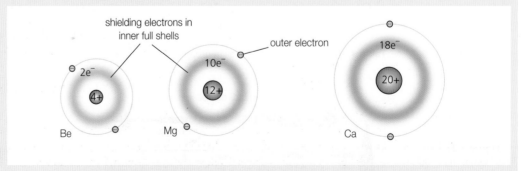

Trends in electronegativity

Electronegativity values generally decrease down a group. Clear decreasing trends in electronegativity can be found in group 1 (the alkali metals) (Table 3.8) and group 7 (the halogens) (Table 3.9). Electronegativity can be interpreted as a measure of non-metallic or metallic character. Decreasing electronegativity down a group indicates a decrease in non-metallic character and an increase in metallic character.

The decrease in electronegativity down groups 1 and 7 can be explained by the increase in atomic radius. There is therefore an increasing distance between the nucleus and shared pairs of electrons. Hence the attractive force is decreased. Although the nuclear charge increases down a group, this is counteracted by the increased shielding due to additional electron shells.

The trends in electronegativity can be used to explain the redox properties of groups 1 and 7. Reducing power decreases down group 1; oxidizing power increases up group 7 (Chapter 9).

Atom	Atomic number	Electronegativity
Li	3	1.0
Na	11	0.9
K	19	0.8
Rb	37	0.8
Cs	55	0.7
Fr	87	0.7

Table 3.8 The variation of electronegativity in group 1

Atom	Atomic number	Electronegativity
F	9	4.0
Cl	17	3.0
Br	35	2.8
I	53	2.5
At	85	2.2

Table 3.9 The variation of electronegativity in group 7

Trends in melting point

■ Group 1

The melting points of the alkali metals decrease down the group (Table 3.10 and Figure 3.25). Metals are held together in the solid and liquid states by metallic bonding (Chapter 4). Metals are composed of a lattice of positive ions surrounded by delocalized electrons which move between the ions. The delocalized electrons are valence electrons shed by the metal atoms as they enter the lattice.

The melting points decrease down the group because the strength of the metallic bonding decreases. This occurs because the attractive forces between the delocalized electrons and the nucleus decrease owing to the increase in distance. The increase in nuclear charge is counteracted by the increase in shielding.

Atom	Atomic number	Melting point/K
Li	3	454
Na	11	371
K	19	337
Rb	37	312
Cs	55	302
Fr	87	300

Table 3.10 The variation of melting point in group 1

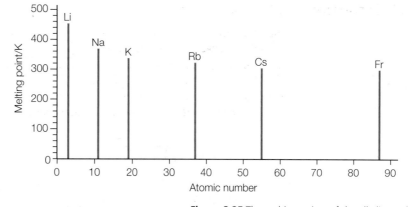

Figure 3.25 The melting points of the alkali metals

■ Group 7

In contrast to the alkali metals, the melting and boiling points of the halogens *increase* down the group (Table 3.11 and Figure 3.26). This is because as the molecules become large, the attractive forces between them increase. These shorter-range attractive forces are known as van der Waals' forces and increase with the number of electrons in atoms or molecules (Chapter 4).

Atom	Atomic number	Melting point/K
F	9	54
Cl	17	172
Br	35	266
I	53	387
At	85	575

Table 3.11 The variation of melting point in group 7

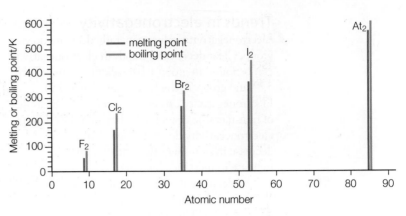

Figure 3.26 Melting and boiling points of the halogens

Trends in physical properties of elements across period 3

3.2.3 **Describe** and **explain** the trends in atomic radii, ionic radii, first ionization energies and electronegativities for elements across period 3.

Trends in atomic radii

There is a gradual decrease in atomic radius across period 3 from left to right (Table 3.12 and Figure 3.27). When moving from group to group across a period, the number of protons and the number of electrons increases by one. Since the electrons are added to the same shell, there is only a slight increase in the shielding effect across the period. At the same time additional protons are added to the nucleus, increasing the nuclear charge. The effect of the increase in nuclear charge more than outweighs the small increase in shielding and consequently all the electrons are pulled closer to the nucleus. Hence, atomic radii decrease across period 3. The same effect is observed in other periods.

Atom	Atomic radius/pm
Na	186
Mg	160
Al	143
Si	117
P	110
S	104
Cl	99
Ar	No data

Table 3.12 The atomic radii in period 3

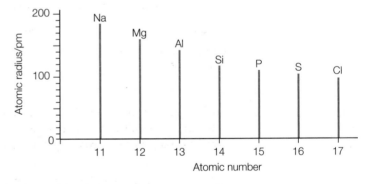

Figure 3.27 Bar graph of the atomic radii in period 3

Trends in ionic radii

The data in Table 3.13 shows the following trends in ionic radii across period 3.

- The radii of positive ions decrease from the sodium ion, Na^+ to the aluminium ion, Al^{3+}.
- The radii of negative ions decrease from the phosphide ion, P^{3-} to the chloride ion, Cl^-.
- The ionic radii increase from the aluminium ion, Al^{3+} to the phosphide ion, P^{3-}.

Element	Sodium	Magnesium	Aluminium	Silicon	Phosphorus	Sulfur	Chlorine
Ion	Na^+	Mg^{2+}	Al^{3+}	(Si^{4+} and Si^{4-})	P^{3-}	S^{2-}	Cl^-
Ionic radius/pm	98	65	45	(42 and 271)	212	190	181

Table 3.13 The ionic radii in period 3

The data for the silicon ions are calculated values. Silicon does not form simple ions and its bonding is covalent.

Isoelectronic species

Isoelectronic species are atoms and ions that have the same number of electrons. For a specific number of electrons, the higher the nuclear charge, the greater the forces of attraction between the nucleus and the electrons. Hence, the smaller the atomic or ionic radius.

Ions of sodium, magnesium and aluminium are isoelectronic species (Table 3.14). The nuclear charge increases from the sodium ion to the aluminium ion. The higher nuclear charge pulls all the electron shells closer to the nucleus. Hence, the ionic radii decrease.

Similarly, the nuclear charge increases from the phosphide ion to the chloride ion. The higher nuclear charge causes the electron shells to be pulled closer to the nucleus. Again, the ionic radii decrease (Table 3.15).

Species	Na^+	Mg^{2+}	Al^{3+}
Nuclear charge	+11	+12	+13
Number of electrons	10	10	10
Ionic radius/pm	98	65	45

Table 3.14 Atomic data for sodium, magnesium and aluminium ions

Species	P^{3-}	S^{2-}	Cl^-
Nuclear charge	+15	+16	+17
Number of electrons	18	18	18
Ionic radius/pm	212	190	181

Table 3.15 Atomic data for phosphide, sulfide and chloride ions

The large increase in size from the aluminium ion to the phosphide ion is due to the presence of an additional electron shell. This causes a large increase in the shielding effect and as a result the ionic radius increases.

Trends in first ionization energy

The first ionization energies of the elements in period 3 are listed in Table 3.16. The general trend is an increase in first ionization energy across the periodic table. When moving across a period from left to right the nuclear charge increases but the shielding effect only increases slightly (since electrons enter the same shell). Consequently, the electron shells are pulled progressively closer to the nucleus and as a result first ionization energies increase.

Element	Sodium	Magnesium	Aluminium	Silicon	Phosphorus	Sulfur	Chlorine
First ionization energy/kJ mol^{-1}	494	736	577	786	1060	1000	1260

Table 3.16 First ionization energies for the elements in period 3

However, the increase in first ionization energy is not uniform and there are two *decreases* – between magnesium and aluminium and between phosphorus and sulfur. These decreases can only be explained by reference to sub-shells and orbitals (see Chapter 12).

Comparing electronegativity values

3.2.4 **Compare** the relative electronegativity values of two or more elements based on their positions in the periodic table.

The electronegativities of the elements in period 3 are listed in Table 3.17. The general trend is an increase in first ionization energy across the periodic table. When moving across a period from left to right the nuclear charge increases but the shielding effect only increases slightly (since electrons enter the same shell). Consequently, the electron shells are pulled progressively closer to the nucleus and as a result electronegativity values increase.

Element	Sodium	Magnesium	Aluminium	Silicon	Phosphorus	Sulfur	Chlorine
Electronegativity	0.9	1.2	1.5	1.8	2.1	2.5	3.0

Table 3.17 Electronegativity values for the elements in period 3

Generally, the electronegativity values of chemical elements increase across a period and decrease down a group (Figure 3.28). This observation can be used to compare the relative electronegativity values of two elements in the periodic table. To do this, find the positions of the elements in the periodic table. Then simply see which one is further up and to the right; that is the more electronegative element (Figure 3.29). The further apart the two elements are in the periodic table, the larger the difference will be in their electronegativities. This is important in determining the type of bonding between the two elements (Chapter 4).

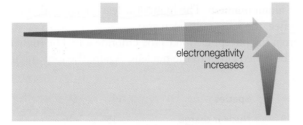

Figure 3.28 Trends in electronegativity for s- and p-block elements

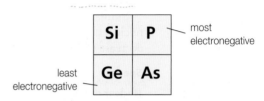

Figure 3.29 Relative values of electronegativity of elements in the periodic table

■ **Extension:** ## Diagonal relationships

Electronegativity increases across a period and decreases down a group. This results in what are known as diagonal relationships, where a pair of elements have similar chemical properties. The most important pairs are lithium and magnesium, beryllium and aluminium, and boron and silicon.

History of Chemistry
Dimitri Mendeleev was born in 1834 in Tobolsk, Siberia, the youngest of 17 children. When Dimitri was 13 years old, his father died and his mother's glass-making factory burnt down. In 1849 the family relocated to St Petersburg (formerly Leningrad) and he later became Professor of Chemistry at the University of St Petersburg. In 1862 he married Feozva Nikitichna Leshcheva. This marriage ended in divorce and in 1882 Mendeleev married one of his students, Anna Popova. He was dismissed from the University in 1890 for supporting the causes of students against the authorities. In 1893 he was appointed the Director of the Bureau of Weights and Measures and helped to formulate new standards for measures such as mass and length. Mendeleev was nominated for the 1906 Nobel Prize in Chemistry, but narrowly lost to Frenchman Henri Moissan, who had isolated fluorine. He probably would have been awarded the 1907 Nobel Prize in Chemistry, but died early in 1907 from influenza. Nobel Prizes cannot be awarded posthumously (after death).

3.3 Chemical properties

3.3.1 **Discuss** the similarities and differences in the chemical properties of elements in the same group.

The alkali metals

The alkali metals are a group of very reactive metals. The first three members of the group are lithium, sodium and potassium Their atomic and physical properties are summarized in Table 3.18. The electrode potentials are a measure of reducing strength (Chapter 19). The more negative the value, the greater the tendency for the atom to lose an electron (in aqueous solution).

Element	Lithium	Sodium	Potassium
Electron arrangement	2,1	2,8,1	2,8,8,1
Electron configuration	$1s^22s^1$	$1s^22s^22p^63s^1$	$1s^22s^22p^63s^23p^64s^1$
Chemical symbol	Li	Na	K
First ionization energy/kJ mol^{-1}	519	494	418
Atomic radius/nm	0.152	0.186	0.231
Melting point/K	454	371	337
Boiling point/K	1600	1156	1047
Density/g cm^{-3}	0.53	0.97	0.86
Standard electrode potential, $E^\ominus$ M$^+$(aq) $\mid$ M(s)/V	−3.03	−2.71	−2.92

Table 3.18 The atomic and physical properties of three alkali metals

Figure 3.30 Reaction between sodium and water

Sodium

Sodium is a soft silvery-white metal and an excellent conductor of heat and electricity. It rapidly corrodes in moist air, initially to form sodium oxide, Na_2O. When placed in water sodium floats but immediately reacts with the water (Figure 3.30) to form a solution of sodium hydroxide and hydrogen gas:

$$2Na(s) + 2H_2O(l) \rightarrow 2NaOH(aq) + H_2(g)$$

The heat energy produced by this exothermic reaction (Chapter 5) is sufficient to melt the sodium, but not usually to ignite the hydrogen (unless the sodium is not allowed to move). The sodium burns with a brilliant golden-yellow flame.

Sodium hydroxide is a strong alkali (Chapter 8). It is completely ionized in water and forms a strongly alkaline solution of sodium hydroxide with a high pH:

$$2Na(s) + 2H_2O(l) \rightarrow 2Na^+(aq) + 2OH^-(aq) + H_2(g)$$

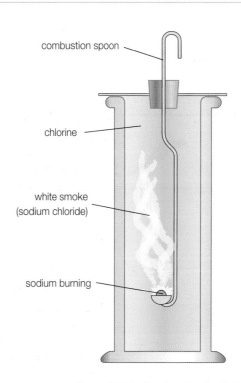

Figure 3.31 Sodium burning in chlorine

This reaction is an example of a redox reaction (Chapter 9), in which the sodium acts as a reducing agent.

When a piece of hot sodium is lowered into a gas jar of chlorine, the metal continues to burn, forming a white smoke of sodium chloride (Figure 3.31). Similar reactions occur with bromine and iodine to form sodium bromide and sodium iodide, but the reactions are slower and less heat is released.

$$2Na(s) + Cl_2(g) \rightarrow 2NaCl(s)$$

Applications of Chemistry

Sodium is used as a coolant in some types of nuclear reactors because of its high thermal conductivity. It carries heat away from the core to a steam generator, where water is converted to steam to drive the turbines of an electrical generator. Another advantage of using sodium, rather than water, is the lack of corrosion to the steel used in the construction.

Potassium and lithium

Potassium is a soft silvery metal that, like sodium, is a good conductor of heat and electricity. The reactions of potassium are less vigorous than corresponding reactions of sodium (partly due to its lower first ionization energy), but the reactions are otherwise identical. Its reaction with water is sufficient to raise the temperature of the hydrogen to its ignition point; the metal burns with a lilac (pale purple) flame. Lithium is a hard silver metal that has identical reactions to sodium, but slower (partly due to its higher first ionization energy). Lithium and potassium also react with chlorine: the reaction with potassium is faster and more exothermic (compared to sodium); the reaction with lithium is slower and less exothermic (compared to sodium).

History of Chemistry

Francium, atomic number 87, is the most unstable of the first 101 elements. The longest-lived isotope has a half-life of 22 minutes. It is an alkali metal element situated at the bottom of group 1 just below caesium. In 1871 the Russian chemist Mendeleev had predicted the existence of an alkali metal that he called *eka*-caesium. The element was discovered in 1939 by the French chemist **Marguerite Perey** (1909–1975), a former assistant of Marie Curie, working at the Curie Institute in Paris. She isolated francium from the radioactive element actinium. Small amounts of francium were prepared by co-precipitation with caesium salts and by paper chromatography. Francium was the last naturally occurring element to be discovered. Weighable amounts have not been prepared and it currently has no uses. However, studies of francium atoms have confirmed some predictions of quantum theory.

Chemistry and Literature

Mario Petrucci (1958–) is an Italian living in London. A graduate in physics from Cambridge University, Petrucci is a freelance poet and essayist. His poem 'Last Wish' was inspired by the Chernobyl disaster and is informed by historical research and by the author's knowledge as a scientist.

In April 1986, one of the reactors at the Chernobyl nuclear power plant in Ukraine exploded. Further explosions and fires spread radioactive debris over a wide area. ■

Last Wish (Chernobyl 1986)

*You bury me in concrete. Bury me
in lead. Rather I was buried
with a bullet in the head.*

*You seal me in powder. Cut the hair
last. Then take the trimmings
and seal them in glass.*

*You wrap me in plastic. Wash me
in foam. Weld the box airless
and ram the box home.*

*For each tomb that's hidden a green
soldier turns. None decomposes.
Nothing for worms.*

*A buckle. A pencil. Break one thing
I left. Give some small part of me
ordinary death.*

The halogens

The halogens are a group of very reactive non-metals. The first three members of the group are chlorine, bromine and iodine. Their atomic and physical properties are summarized in Table 3.19.

Element	Chlorine	Bromine	Iodine
Chemical formula	Cl_2	Br_2	I_2
Structure	Cl–Cl	Br–Br	I–I
Electron arrangement	2,8,7	2,8,18,7	2,8,18,18,7
Detailed outer shell arrangement	$3s^2 3p^5$	$4s^2 4p^5$	$5s^2 5p^5$
State at room temperature and pressure	Gas	Liquid	Solid
Colour	Pale green	Red-brown	Black
Melting point/K	172	266	387; 458 (sublimes)
Boiling point/K	239	332	
Standard electrode potential, $E^\ominus$ $X_2(aq) \mid X^-(aq)/V$	1.36	1.09	0.54

Table 3.19 The atomic and physical properties of the halogens

Figure 3.32 Saturated bromine water and gaseous iodine

All the halogens have an outer or valence shell with seven electrons. A full shell or noble gas configuration is obtained by the addition of one extra electron (from a metal) to form a **halide** ion, or by the sharing of electrons to form covalent bonds and hence molecules.

All the halogens exist as diatomic molecules where two halogen atoms are held together by a single covalent bond (a shared pair of electrons). Diatomic molecules are present in all three physical states.

All the halogens (Figure 3.32) are coloured, with the colour becoming progressively darker as you move down the group. The volatility of the halogens decreases down the group as boiling and melting points increase. This decrease correlates with an increase in the strength or extent of van der Waals' forces operating between molecules (Chapter 4). These are weak attractive forces that operate between neighbouring molecules in the liquid and solid states.

■ Extension: Properties of the halogens

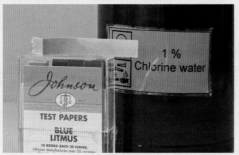

Figure 3.33 The reaction between blue litmus paper and chlorine gas

Figure 3.34 Iodine added to ethanol (on the left) and water (on the right)

Solubility

Halogens are absorbed into organic solvents, such as tetrachloromethane ('carbon tetrachloride') or hexane. In these non-polar solvents chlorine is colourless, bromine is red and iodine is violet. In polar organic solvents such as ethanol ('alcohol') and propanone ('acetone'), bromine and iodine give brownish solutions.

Chlorine is moderately soluble in water, forming a solution known as chlorine water. It contains a mixture of hydrochloric and chloric(I) acids in equilibrium with chlorine molecules. The position of the equilibrium is pH dependent and a low pH (acidic conditions) favours chlorine molecules (Chapter 7).

$$Cl_2(aq) + H_2O(l) \rightleftharpoons HCl(aq) + HOCl(aq)$$

$$HCl(aq) \rightarrow H^+(aq) + Cl^-(aq); \quad HOCl(aq) \rightleftharpoons H^+(aq) + OCl^-(aq)$$

Chlorine gas turns moist blue litmus paper red and then decolorizes it (Figure 3.33). The bleaching properties of chlorine water are due to the presence of chlorate(I) ions.

Bromine undergoes a similar reaction to form bromine water. Iodine is slightly soluble in water, but readily soluble in ethanol (Figure 3.34). This is an illustration of the 'like dissolves like' principle (Chapter 4): iodine is non-polar so is more soluble in ethanol than water, due to the lower polarity of ethanol.

Applications of Chemistry

Household 'chlorine bleach' is a dilute solution of sodium chlorate(I) (sodium hypochlorite). It is prepared by absorbing chlorine gas into cold sodium hydroxide solution. More concentrated solutions are used to disinfect drinking water and swimming pools. Bleach should never be mixed with other household cleaners. With bleach, acid-based cleaners produce chlorine and ammonia-based products produce toxic chloramines, for example NH_2Cl.

■ **Extension:** ## Standard electrode potential

The standard electrode potential (Chapter 19) is a measure of how much tendency a chemical species in solution has to lose or gain electrons. Positive numbers indicate a chemical species (molecule, ion or atom) which is an oxidizing agent – a species which has a high tendency to accept electrons. Negative numbers indicate a chemical species (molecule, ion or atom) which is a reducing agent – a species which has a high tendency to donate electrons.

The decrease in standard electrode potentials indicates that the halogens become progressively less powerful as oxidizing agents as you move down the group, that is, they have a decreasing tendency to accept electrons:

$$X_2(aq) + 2e^- \rightarrow 2X^-(aq)$$

Reactions of the halogens

Displacement reactions

When chlorine water is added to an aqueous solution of potassium bromide, KBr, the solution becomes yellow-orange owing to the formation of bromine:

$$Cl_2(aq) + 2Br^-(aq) \rightarrow Br_2(aq) + 2Cl^-(aq)$$

Chlorine also reacts with potassium iodide solution to form a brown solution of iodine:

$$Cl_2(aq) + 2I^-(aq) \rightarrow I_2(aq) + 2Cl^-(aq)$$

The two reactions shown above for chlorine are known as **displacement reactions** and involve a more reactive halogen, chlorine, displacing or 'pushing out' a less reactive halogen from its salt.

These are redox reactions – the halogen acts as an oxidizing agent and the halide ion acts as a reducing agent (Chapter 9). There is a transfer of electrons from the iodide ions and bromide ions to the chlorine molecules. Going down group 7 the halogens become more weakly oxidizing and the halide ions become more strongly reducing.

Bromine water will give a displacement reaction with a solution of an iodide:

$$Br_2(aq) + 2I^-(aq) \rightarrow I_2(aq) + 2Br^-(aq)$$

However, as bromine is less reactive than chlorine, it is unable to displace chloride ions and no reaction occurs. Iodine, being the most unreactive halogen, is unable to displace bromide or chloride ions and no reaction occurs.

■ **Extension:** ## Explaining trends in the behaviour of the halogens

The trends in oxidizing and reducing power for the halogens and the halide ions can be easily explained in terms of the relative sizes of the halogen atoms and halide ions (Figure 3.35). A halide ion is oxidized by the removal of one of its outer eight electrons. In a large halide ion, the outer electrons are more easily removed as they are further from the nucleus and more effectively shielded from its attraction by the inner electrons. Small halide ions have their outer electrons located closer to the nucleus and less effective shielding occurs, hence their affinity for electrons is higher. A similar argument explains why a small halogen atom can attract an extra electron with a greater affinity than a larger halogen atom.

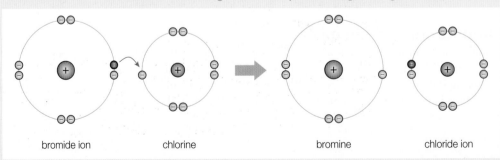

Figure 3.35 The reaction between a halide ion and a halogen atom

bromide ion   chlorine   bromine   chloride ion

Reactions of the halide ions

The term halide ions collectively refers to fluoride, F^-, chloride, Cl^-, bromide, Br^- and iodide, I^-, ions which are present in metal salts, for example, sodium chloride, NaCl [Na^+ Cl^-].

Halide ions are colourless, but the four halide ions may be distinguished from each other in solution by the use of silver nitrate solution (acidified with nitric acid).

With a solution of a chloride salt, silver nitrate gives a white precipitate of silver chloride. For example:

$$NaCl(aq) + AgNO_3(aq) \rightarrow NaNO_3(aq) + AgCl(s)$$

or ionically (Figure 3.36):

$$Cl^-(aq) + Ag^+(aq) \rightarrow AgCl(s)$$

The silver chloride rapidly turns purple in sunlight due to photodecomposition:

$$2AgCl(s) \rightarrow 2Ag(s) + Cl_2(g)$$

Bromides and iodides give cream and yellow precipitates of silver bromide and silver iodide (Figure 3.37), respectively:

$$Br^-(aq) + Ag^+(aq) \rightarrow AgBr(s)$$

$$I^-(aq) + Ag^+(aq) \rightarrow AgI(s)$$

(Fluorides do not give any precipitate with acidified silver nitrate solution since silver fluoride is soluble.)

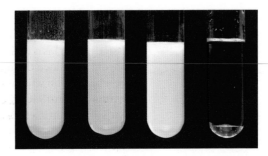

Figure 3.37 The colours of the silver halides – from left to right, silver iodide, silver bromide, silver chloride and silver fluoride

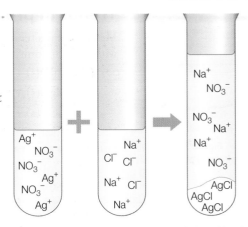

Figure 3.36 The precipitation of silver chloride

■ Extension: The hydrogen halides

The hydrogen halides can be prepared by direct synthesis from the elements (in the presence of ultraviolet light) or by the action of concentrated sulfuric acid on a metal halide. For example:

$$Cl_2(g) + H_2(g) \rightarrow 2HCl(g)$$

$$NaCl(s) + H_2SO_4(l) \rightarrow NaHSO_4(s) + HCl(g)$$

The hydrogen halides are all colourless gases that fume in moist air. Boiling points rise from hydrogen chloride to hydrogen iodide owing to the increase in van der Waals' forces of attraction (Chapter 4). The boiling point of hydrogen fluoride is unexpectedly high owing to the presence of hydrogen bonding.

The hydrogen halides are all soluble in water and form acidic solutions. For example, hydrogen chloride reacts with water to form hydrochloric acid. It is a strong acid and ionization is complete (Chapter 8).

$$HCl(g) + (aq) \rightarrow H^+(aq) + Cl^-(aq)$$

Hydrogen fluoride reacts with water to form hydrofluoric acid. However, this is a reversible reaction since it is a weak acid:

$$HF(g) + (aq) \rightleftharpoons H^+(aq) + F^-(aq)$$

One reason for the weak acid strength of hydrofluoric acid is the relatively high bond strength of the H–F bond in the hydrogen fluoride molecule.

The presence of a hydrogen halide can be confirmed by reacting the gas with ammonia gas. White fumes of the ammonium salt are formed, for example:

$$NH_3(g) + HCl(g) \rightarrow NH_4Cl(s) \qquad [NH_4^+ \ Cl^-]$$

Astatine

Astatine (At) is a radioactive element and the least reactive of the halogens. Studies of astatine are difficult since the most stable isotope has a half-life of only 8.3 hours. Astatine is expected to be a black solid which can be displaced from its salts by all the other halogens. Astatine was first synthesized in 1940 by Dale Corson, K. MacKenzie and Emilio Segré, who bombarded bismuth with alpha particles. Astatine-211 is being investigated for use in radiotherapy to treat cancer.

3.3.2 **Discuss** the changes in nature, from ionic to covalent and from basic to acidic, of the oxides across period 3.

Trends in properties of the oxides in period 3

Figure 3.38 Partially hydrolysed phosphorus(v) oxide, P_4O_{10}

Metallic oxides tend to be ionic and hence **basic**. The more reactive metals form oxides that react with water to form alkaline solutions:

$$Na_2O(s) + H_2O(l) \rightarrow 2NaOH(aq)$$
$$MgO(s) + H_2O(l) \rightarrow Mg(OH)_2(aq)$$
$$\text{or} \quad O^{2-}(s) + H_2O(l) \rightarrow 2OH^-(aq)$$

Non-metallic oxides tend to be covalent and **acidic**. The more reactive non-metals (Figure 3.38) form oxides that react with water to form acidic solutions.

$$P_4O_{10}(s) + 6H_2O(l) \rightarrow 4H_3PO_4(aq)$$
$$SO_3(g) + H_2O(l) \rightarrow H_2SO_4(aq)$$
$$\text{or} \quad SO_3(g) + H_2O(l) \rightarrow H_2SO_4(aq) \rightleftharpoons H^+(aq) + HSO_4^-(aq)$$

(A more detailed and complete discussion about period 3 oxides (and chlorides) can be found in Chapter 13.)

Acid rain

Figure 3.39 Gravestones eroded by carbonic acid and acid rain

Pure rain water is slightly acidic and has a pH of about 5.6. This acidity is caused by carbon dioxide in the atmosphere reacting with rain droplets to form carbonic acid. Rain water with a pH of less than 5.6 is termed **acid rain**. The main acids present in acid rain are sulfuric acid (H_2SO_4) and nitric acid (HNO_3).

The sulfuric acid in acid rain is formed from sulfur dioxide in the atmosphere. Sulfur dioxide is released from volcanoes, but the majority comes from the burning of sulfur-containing fuels, primarily coal in power stations. Car exhaust emissions and the smelting of metals, such as zinc, also contribute to sulfur dioxide pollution. The sulfur dioxide undergoes oxidation to form sulfur trioxide which reacts with water to form sulfuric acid. Sulfur dioxide also reacts with water to form sulfurous acid, H_2SO_3.

The nitric acid present in acid rain is formed from oxides of nitrogen, nitrogen monoxide, NO, and nitrogen dioxide, NO_2. These two oxides are produced during combustion processes, especially those in car engines and in power stations. Nitrogen monoxide is rapidly oxidized by air to nitrogen dioxide, which reacts with water in the presence of oxygen to form nitric acid.

Acid rain causes direct and indirect damage to the environment. In lakes it can directly kill a variety of organisms, such as young fish and insect larvae. Acidic water releases aluminium ions from rocks and soil which are washed into lakes. Aluminium ions are toxic and interfere with the gills of fish, preventing them from extracting dissolved oxygen from the water.

Trees, especially, those at high altitudes, are prone to damage by both acid rain and gaseous sulfur dioxide. The trees drop their leaves and can no longer photosynthesize. Ozone also plays a role in damaging trees and in catalysing the formation of sulfur trioxide from sulfur dioxide.

Acid rain can also cause damage to building materials and historical monuments. This is because the sulfuric acid in the rain chemically reacts with the calcium carbonate (Figure 3.39)

in limestone or marble to create calcium sulfate, which then flakes off. Acid rain also reacts with iron and promotes its oxidation.

$$CaCO_3(s) + H_2SO_4(aq) \rightarrow CaSO_4(aq) + CO_2(g) + H_2O(l)$$

Acid rain is a form of acid deposition (Chapter 25).

History of Chemistry

Robert Angus Smith (1817–1884) was a Scottish chemist who carried out research into a number of environmental issues. He is especially famous for his work on air pollution, during the course of which he discovered what he termed 'acid rain' (Figure 3.40). In 1852 Smith found acid rain in Manchester (centre of the Industrial Revolution in England) and deduced the relationship between acid rain and sulfur-based atmospheric pollution. He was trained to be a minister in the Church of Scotland, but left before graduating and spent two years studying chemistry under Justus von Liebig (Chapter 8).

Figure 3.40 The effects of acid rain and acid gases

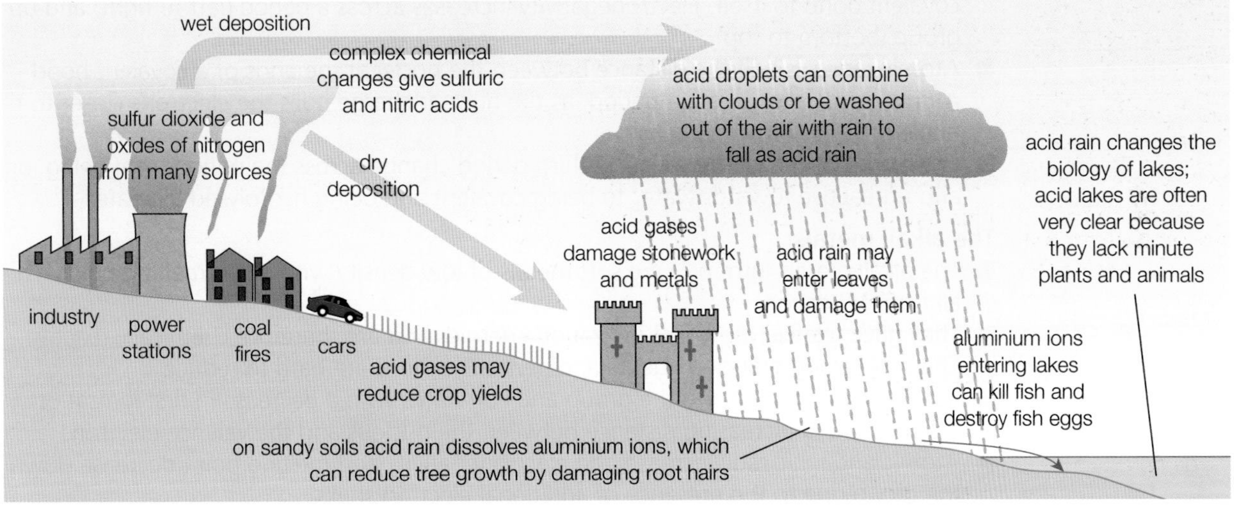

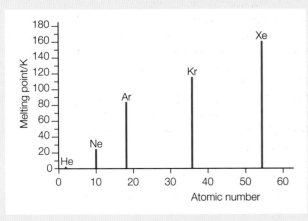

■ Extension: Use of the noble gases

The noble gases form the last group of the periodic table. They are all colourless, odourless and unreactive gases. The first noble gas compounds, XeF_2, XeF_4 and XeF_6, were formed in 1962. The melting and boiling points of the elements increase down the group (Figure 3.41) owing to an increase in van der Waals' forces of attraction (Chapter 4). Neon is used in advertising signs (Chapter 2) and argon is used to fill light bulbs (Figure 3.42). Krypton and xenon are used in lighthouse and projector bulbs. Helium is used to fill advertising airships (a trivial use of a rare and non-renewable resource) and is used in research at very low temperatures.

Figure 3.41 Melting points of the noble gases

Figure 3.42 Light bulbs are filled with argon, which will not react with the hot filament

SUMMARY OF
KNOWLEDGE

- In the periodic table, chemical elements are arranged in order of increasing atomic (proton) number.
- Chemical elements with similar chemical properties are placed under each other in groups. Members of a group all have the same number of electrons in their outer shells.
- Members of group 1 are the alkali metals, members of group 7 are the halogens and members of group 0 are the noble gases. Groups 2 and 3 are separated by the transition metals block.
- The elements in the third period gradually change across the period from metallic to non-metallic. Members of a period all have the same number of electron shells.
- The elements in the periodic table can be classified into four blocks based upon the arrangement of the electrons in the outer sub-shell. There are four blocks: s, p, d and f.
- The first ionization energy of an atom is the energy required to remove completely a mole of electrons from a mole of gaseous atoms: $M(g) \rightarrow M^+(g) + e^-$.
- The electronegativity of an atom is a measure of its ability to attract the electrons in a covalent bond to itself. Electronegativity increases across a period (left to right) and up a group (bottom to top).
- Atomic radius is half the distance between the nuclei at the ends of a covalent bond. Atoms get smaller across a period, as the nuclear charge pulls the electrons closer to the nucleus.
- The oxides of the elements in the third period change across the period from being ionic and unreactive towards water to being covalent and being hydrolysed by water.

The alkali metals
- The alkali metals (group 1) are soft metals of low density with a low melting point. They all form M^+ cations.
- They have relatively low first ionization energies and are therefore chemically reactive. They are strong reducing agents and hence their ions are hard to reduce.
- Reactivity increases down the group and correlates with a decrease in first ionization energy, due to increasing distance between the nucleus and the valence electron.
- Atomic and ionic radii increase and electronegativity and melting point decrease down the group due to the presence of additional electron shells.
- Key reactions of group 1 metals:
  - oxygen with heated metal: $2M(s) + \frac{1}{2}O_2(g) \rightarrow M_2O(s)$
  - halogen with heated group 1 metal: $M(s) + \frac{1}{2}X_2(g) \rightarrow MX(s)$
  - water with metal: $M(s) + H_2O(l) \rightarrow MOH(aq) + \frac{1}{2}H_2(g)$

The halogens
- The halogens, X_2, are a group of reactive non-metals in group 7. They all form X^- ions.
- Reactivity increases up the group. This correlates with an increase in first electron affinity due to decreasing distance between the nucleus and the incoming electron.
- Key reactions of group 7 elements:
  - displacement reactions: $X_2(aq) + 2Y^-(aq) \rightarrow 2X^-(aq) + Y_2(aq)$, where X represents a more reactive halogen (more powerful oxidizing agent) than Y
  - reaction with water: $X_2(aq) + H_2O(l) \rightleftharpoons HOX(aq) + H^+(aq) + X^-(aq)$
  - reaction with group 1 metals: $\frac{1}{2}X_2(g) + M(s) \rightarrow MX(s)$
  - precipitation reactions: $X^-(aq) + Ag^+(aq) \rightarrow AgX(s)$

■ *Examination questions – a selection*

Paper 1 IB questions and IB style questions

Q1 Which element shows chemical behaviour similar to calcium?
- **A** strontium
- **B** chlorine
- **C** sodium
- **D** boron

Q2 The following are three statements concerning the periodic table.
- **I** The horizontal rows are called periods and the vertical columns are called groups.
- **II** Electronegativity decreases down any group and across a period from left to right.
- **III** Reactivity increases down all groups.

Which of the above is/are true?
- **A** I, II and III
- **B** I and II only
- **C** II and III only
- **D** I only

Q3 Which is the correct trend (left to right) across period 3 for the oxides?
- **A** basic to acidic
- **B** acidic to basic
- **C** increasingly basic
- **D** neutral to acidic

Q4 What happens when chlorine water is added to an aqueous solution of potassium iodide?
- **A** No reaction occurs because chlorine is less reactive than iodine.
- **B** Chlorine molecules are oxidized to chloride ions.
- **C** Iodide ions are oxidized to iodine molecules.
- **D** A purple precipitate of iodine is formed.

Q5 Which of the following determines the order in which the elements are arranged in the modern form of the periodic table?
- **A** relative atomic mass
- **B** mass number
- **C** atomic number
- **D** chemical reactivity

Q6 Which is a correct statement about the element with an atomic number of 20?
- **A** It is in group 4.
- **B** It is in group 2.
- **C** It is a transition metal.
- **D** It is in group 7 and is a halogen.

Q7 In general, atomic radii decrease:
- **A** within a group from lower to higher atomic number
- **B** within a period from lower to higher atomic number
- **C** with an increase in the number of isotopes of an element
- **D** with an increase in the shielding of the nuclear charge

Q8 When the elements are listed in order of increasing reactivity with air, the correct order is:
- **A** Na, K, Cs
- **B** Cs, K, Na
- **C** Cs, Na, K
- **D** K, Cs, Na

Q9 For which type of isoelectronic ions do ionic radii decrease with increasing nuclear charge?
- **A** positive ions only
- **B** negative ions only
- **C** neither positive or negative ions
- **D** both positive and negative ions

Q10 Which properties are typical of most non-metals in period 3 (Na to Ar)?
- **I** They form ions by gaining one or more electrons.
- **II** They are poor conductors of heat and electricity.
- **III** They have high melting points.

- **A** I and II only
- **B** I and III only
- **C** II and III only
- **D** I, II and III

Standard Level Paper 1, Nov 05, Q7

Q11 On the periodic table, groups of elements show similarities in their chemical properties. This can be best explained by the:
- **A** differences in the number of protons in the nucleus of the atoms
- **B** similarities in the results of emission spectrum analysis of gaseous samples of a group
- **C** similarities in the electronic structures of the atoms
- **D** differences in the number of neutrons in the nucleus of the atoms

Q12 Which atom has the smallest atomic radius?
- **A** $_{31}Ga$
- **B** $_{20}Ca$
- **C** $_{35}Br$
- **D** $_{37}Rb$

Q13 Which one of the following series represents the correct size order for the various iodine species?
- **A** $I < I^- < I^+$
- **B** $I < I^+ < I^-$
- **C** $I^+ < I < I^-$
- **D** $I^- < I < I^+$

Q14 Which one of the following will be observed as the atomic number of the elements in a single group of elements on the periodic table increases?
- **A** an increase in atomic radius
- **B** an increase in ionization energy and hence decrease in reactivity
- **C** a decrease in ionic radius
- **D** an increase in electronegativities

Q15 Which of the following properties of the halogens increase from F to I?
  I   atomic radius
  II melting point
  III electronegativity
  A I only   **C** I and III only
  B I and II only   **D** I, II and III
Standard Level Paper 1, Nov 03, Q7

Q16 In general, how do ionization energies vary as the periodic table is crossed from left to right?
  A They remain constant.
  B They increase.
  C They increase to a maximum and then decrease.
  D They decrease.

Q17 0.01 mol samples of the following oxides were added to separate 1 dm³ portions of water. Which will produce the most acidic solution?
  A $Al_2O_3(s)$ **C** $Na_2O(s)$
  B $SiO_2(s)$ **D** $SO_3(g)$

Q18 Which property increases with increasing atomic number for both the alkali metals and the halogens?
  A melting points
  B first ionization energies
  C electronegativities
  D atomic radii

Q19 Which one of the following elements has the lowest first ionization energy?
  A Li   **C** B   **B** Na   **D** Mg

Q20 Barium is an element in group 2 of the periodic table (below strontium with atomic number 56). Which of the following statements about barium is not correct?
  A Its first ionization energy is lower than that of strontium.
  B It has two electrons in its outermost energy level.
  C Its atomic radius is smaller than that of strontium.
  D It forms a chloride with the formula $BaCl_2$.

Paper 2 IB questions and IB style questions
(IB *Chemistry data booklet* required)

Q1 a i Define the term *ionization energy*. [2]
  ii Write an equation, including state symbols, for the process occurring when measuring the first ionization energy of aluminium. [1]
  b Explain why the first ionization energy of magnesium is greater than that of sodium. [3]
  c Lithium reacts with water. Write an equation for the reaction and state two observations that could be made during the reaction. [3]
Standard Level Paper 2, Nov 05, Q4

Q2 a i Explain why the ionic radius of bromine is less than that of selenium. [2]
  ii Explain what is meant by the term *electronegativity* and explain why the electronegativity of fluorine is greater than that of chlorine. [3]
  b For each of the following reactions in aqueous solution, state one observation that would be made, and deduce the equation.
  i The reaction between chlorine and potassium iodide. [2]
  ii The reaction between silver ions and bromide ions. [2]
  c Deduce whether or not each of the reactions in **b** is a redox reaction, giving a reason in each case. [4]

Q3 a What factors determine the size of an atom or ion? [3]
  b i Explain why the ionic radius of sodium is much smaller than its atomic radius. [2]
  ii Explain why the cations of group 1 increase in size with increasing atomic number. [2]
  c Explain why the ionic radius of Mg^{2+} is less than that of Na^+. [2]
  d Arrange the following species in order of increasing size:
  i N, N^{3-} [1]
  ii Fe, Fe^{2+} and Fe^{3+} [1]

Q4 Describe and explain the variation in ionic radius of the elements across period 3 from sodium to chlorine. [6]

4 Bonding

STARTING POINTS

- There are three types of chemical bonding: metallic, ionic and covalent bonding.
- There are many examples of substances with bonding intermediate in nature.
- Some substances contain both covalent and ionic bonding.
- Each type of bonding results in the formation of substances with characteristic physical properties.
- The type of bond formed largely depends on the difference in electronegativity between the atoms.
- When atoms react to form chemical bonds, only the electrons in the outer or valence shell are involved.
- Noble gas configurations are important in the Lewis theory of chemical bonding.
- The formation of chemical bonds is driven by the decrease in potential energy.
- The electrons in the outer shells of atoms, ions and molecules are represented by Lewis diagrams, which usually represent electrons as dots and crosses.
- Covalent bonding usually occurs between non-metal atoms and involves the sharing of pairs of electrons.
- The directional nature of covalent bonds gives rise to small molecules having definite, fixed shapes.
- Ionic bonding occurs between metals and non-metals and involves the complete transfer of electrons from the metal to the non-metal.
- Metallic bonding in metals involves delocalized valence electrons spread out within a regular three-dimensional array of positive ions.
- Various types of intermolecular forces of attraction occur between molecules.
- Giant covalent and simple covalent substances have very different properties.

4.1 Ionic bonding

> **4.1.1** **Describe** the ionic bond as the electrostatic attraction between oppositely charged ions.

Ionic bonding occurs when one or more electrons are transferred from the outer shell of one atom to the outer shell of another atom. The atom losing an electron or electrons forms a positively charged ion (cation) and the atom gaining an electron or electrons forms a negatively charged ion (anion). An ionic bond is the electrostatic attraction between oppositely charged ions (Figure 4.1).

Ionic bonding is described as non-directional since each ion is attracted to every other ion of opposite charge. In contrast, covalent bonding involves the sharing of electrons between atoms and is directional.

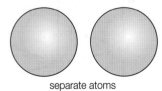
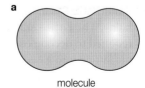
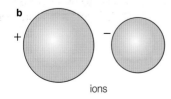

Figure 4.1 Electron rearrangement during **a** covalent bonding and **b** ionic bonding

separate atoms   molecule   ions

Language of Chemistry

Ionic bonding is also referred to as electrovalent bonding. However, this is a less satisfactory term since it is a vague reference to electrical attraction. The prefix *co-* means sharing, so covalent bonding refers to the sharing of pairs of electrons. ■

4.1.2   **Describe** how ions can be formed as a result of electron transfer.

Formation of ions by electron transfer

The formation of an ionic compound typically involves the reaction between a metal and a non-metal. An example of ionic bond formation involves the reaction between sodium and chlorine to form sodium chloride.

The electron arrangements of the sodium and chlorine atoms are:

    sodium atom, Na   2,8,1
    chlorine atom, Cl   2,8,7

The ionic bonding in sodium chloride occurs when the valence electron from the third shell of the sodium atom is transferred to the chlorine atom.

The electron arrangements of the sodium and chloride ions are:

    sodium ion, Na⁺   2,8
    chloride ion, Cl⁻   2,8,8

These ions have stable noble gas electron arrangements: the sodium ion has the electron arrangement of neon and the chloride ion has the electron arrangement of argon.

Lewis diagrams can be used to represent the transfer of electrons that occurs during the formation of ionic bonds. For example, the reaction between sodium and chlorine atoms is described in Figure 4.2 using Lewis diagrams.

Figure 4.2 Ionic bonding in sodium chloride, NaCl showing **a** all electrons and **b** only the outer or valence electrons. The curved arrow indicates the transfer of an electron from the sodium atom to the chlorine atom

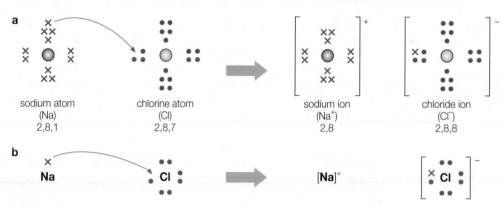

a

| sodium atom (Na) 2,8,1 | chlorine atom (Cl) 2,8,7 | sodium ion (Na⁺) 2,8 | chloride ion (Cl⁻) 2,8,8 |

b

Na   Cl   [Na]⁺   [Cl]⁻

The ions will be arranged into a regular arrangement (Figure 4.3) known as a **lattice** (page 102). Within the lattice oppositely charged ions attract and ions of the same charge repel each other. However, there is an overall, or net, attractive force. The strength of an ionic lattice is measured by its lattice enthalpy. The lattice enthalpy is the energy required to decompose one mole of an ionic lattice into gaseous ions (Chapter 15).

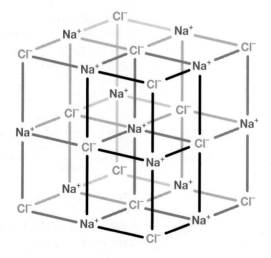

Figure 4.3 Ionic lattice for sodium chloride

Figure 4.4 shows the electron transfer that takes place during the formation of calcium fluoride. A calcium atom (2,8,8,2) obtains a full outer shell by losing two electrons. These are transferred, one to each of the fluorine atoms (2,7). A compound is formed containing two fluoride ions, F^-, for each calcium ion, Ca^{2+}. The formula of the compound is CaF_2. Ionic compounds are always electrically neutral.

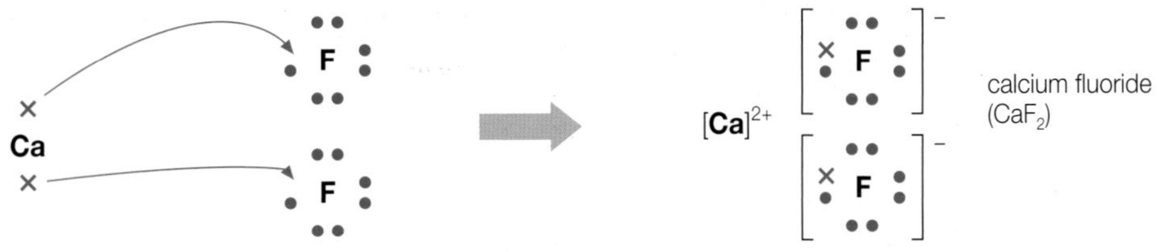

Figure 4.4 Simplified diagram of the ionic bonding in calcium fluoride, CaF_2

calcium fluoride (CaF_2)

Language of Chemistry

Positive ions are often referred to as cations because they move towards the cathode (negative electrode) when placed in an electric field. Negative ions move towards the anode (positive electrode), so they are termed anions. ■

History of Chemistry

All noble gas atoms (except helium) have eight valence electrons. In 1920 the American chemist **Gilbert N. Lewis** (1875–1946) observed that atoms of the elements in groups 1 to 7 tended to lose, gain or share the appropriate number of electrons in order to obtain a full outer shell of eight electrons. This tendency for atoms in compounds to achieve a stable noble gas configuration with eight valence electrons is termed the **octet rule**. An **octet** is four pairs of valence electrons in an atom. The octet rule applies to ionic and covalent compounds formed by the elements of periods 2 and 3.

Lewis made many contributions to physical chemistry. In 1916 he proposed the idea that a covalent bond consisted of a shared pair of electrons. In 1923 he formulated the electron pair theory of acid–base reactions, now known as Lewis theory (Chapter 8). He also made contributions to thermodynamics and was the first scientist to prepare 'heavy water', $^2_1H_2O$.

■ Extension: Why ionic compounds form

It should be emphasized that ionic bonding is *not* driven by the transfer of electrons in order for ions to achieve stable noble gas configurations. The removal of electrons from atoms and ions requires energy, so ionization is always an endothermic process (Chapter 3). The driving force for the formation of ionic compounds is that when ions are brought close together in an ionic crystal the favourable electrostatic forces of attraction more than outweigh the energy changes required for ion formation (in the gas phase) (see Chapter 15 for a discussion of lattice enthalpy and the Born–Haber cycle).

Formation of ions by elements in groups 1, 2 and 3

4.1.3 **Deduce** which ions will be formed when elements in groups 1, 2 and 3 lose electrons.

The elements in groups 1, 2 and 3 have only 1, 2 or 3 electrons in their outer shell. These elements at the beginning of a period *lose* electrons to form positive ions (cations). The resulting simple ions obey the octet rule and have an electron arrangement like the noble gas at the end of the previous period.

Examples: $Na \rightarrow Na^+ + e^-$; $Mg \rightarrow Mg^{2+} + 2e^-$; $Al \rightarrow Al^{3+} + 3e^-$; Na^+, Mg^{2+} and Al^{3+} all have the same electronic structure as Ne.

Formation of ions by elements in groups 5, 6 and 7

> **4.1.4   Deduce** which ions will be formed when elements in groups 5, 6 and 7 gain electrons.

The elements in groups 5, 6 and 7 have 5, 6 or 7 electrons in their outer shell. These elements near the end of the period gain electrons to form negative ions (anions). The resulting simple ions obey the octet rule and have an electron arrangement like the noble gas at the end of the period.

$$Examples: \quad P + 3e^- \rightarrow P^{3-} \qquad\qquad P^{3-}, S^{2-} \text{ and } Cl^- \text{ all have}$$
$$S + 2e^- \rightarrow S^{2-} \qquad\qquad \text{the same electronic}$$
$$Cl + e^- \rightarrow Cl^- \qquad\qquad\quad \text{structure as Ar}$$

The elements in group 0 have full outer shells of electrons. This is a stable electron arrangement and these elements only form compounds with the most reactive elements, notably fluorine. The first two elements in group 4, carbon and silicon, have outer shells which are half full. These two elements generally do not form simple ions but instead form covalent bonds (Section 4.2). (However, carbon reacts with metals to form a number of metal carbides.) Table 4.1 shows the electron arrangements of the atoms and simple ions of the elements in period 3 of the periodic table.

Table 4.1 Electron arrangements of the atoms and simple ions of the elements in period 3

Group	1	2	3	4	5	6	7	0
Element	Sodium	Magnesium	Aluminium	Silicon	Phosphorus	Sulfur	Chlorine	Argon
Electron arrangement	2,8,1	2,8,2	2,8,3	2,8,4	2,8,5	2,8,6	2,8,7	2,8,8
Number of electrons in outer shell	1	2	3	4	5	6	7	8
Common simple ion	Na^+	Mg^{2+}	Al^{3+}	–	P^{3-} (phosphide)	S^{2-} (sulfide)	Cl^- (chloride)	–
Electron arrangement of ion	2,8	2,8	2,8	–	2,8,8	2,8,8	2,8,8	–

Ions of the transition elements

> **4.1.5   State** that transition elements can form more than one ion.

The transition elements form more than one stable positive ion. For example, copper forms copper(I), Cu^+, and copper(II), Cu^{2+}, and iron forms iron(II), Fe^{2+}, and iron(III), Fe^{3+}. The Roman number indicates the oxidation number of the transition metal (Chapter 9). The charges on the simple positive ions must be learnt and those most commonly encountered are summarized in Table 4.2. A charge of positive two is the most common charge on a transition metal simple ion.

Name of transition metal	Simple positive ions
Silver	Ag^+
Iron	Fe^{2+}, Fe^{3+}
Copper	Cu^+, Cu^{2+}
Manganese	Mn^{2+}, Mn^{3+} and Mn^{4+}
Chromium	Cr^{3+} and Cr^{2+} (not stable in air)

Table 4.2 Charges of selected transition element ions

Language of Chemistry

The formulas of ionic compounds are *empirical* formulas (Chapter 1). For example, sodium chloride consists of a lattice containing a large number of sodium and chloride ions in a 1 : 1 ratio. Each sodium ion is attracted to every chloride ion; each chloride ion is attracted to every sodium ion. However, no molecules are present and thus only an empirical formula, NaCl, can be written. If the ionic nature is to be emphasized, then the formula may be written as $[Na^+Cl^-]$. ∎

Applications of Chemistry

Ionic compounds are usually solids at room temperature and pressure. They form crystals and melt at relatively high temperatures. One of the first liquid ionic salts was synthesized when an organic salt, alkylpyridinium chloride, was heated with aluminium chloride. A clear, colourless, ionic liquid was formed. Ionic liquids have a number of potentially useful 'green' properties. In particular, they are non-volatile and non-toxic. Chemists are now investigating the possibility of using ionic liquids instead of the toxic and volatile organic solvents currently used in many industrial processes.

■ Extension: Electron configurations of selected atoms and ions

Gallium is a group 3 metal with the following electron configuration:

$$1s^2 2s^2 2p^6 3s^2 3p^6 3d^{10} 4s^2 4p^1$$

The gallium(III) ion is formed by the loss of three valence electrons. The electron configuration of the gallium(III) ion, Ga^{3+}, is:

$$1s^2 2s^2 2p^6 3s^2 3p^6 3d^{10}$$

A noble gas core with an outer d^{10} configuration is known as a pseudo-noble gas configuration. A d^{10} cation is stable because the third shell is completely filled with 18 electrons, with 10 electrons in the d sub-shell.

The iron(II) ion (Fe^{2+}) is unstable in solution, whereas the iron(III) ion (Fe^{3+}) is stable. Iron(II) compounds are readily oxidized to iron(III) compounds. In contrast, the manganese(II) ion (Mn^{2+}) is stable, whereas the manganese(III) ion (Mn^{3+}) is unstable.

Fe^{2+} $1s^2 2s^2 2p^6 3s^2 3p^6 3d^6$

Fe^{3+} $1s^2 2s^2 2p^6 3s^2 3p^6 3d^5$

Mn^{2+} $1s^2 2s^2 2p^6 3s^2 3p^6 3d^5$

Mn^{3+} $1s^2 2s^2 2p^6 3s^2 3p^6 3d^4$

The iron(III) and manganese(II) ions are stable because of the special stability associated with the half-filled 3d sub-shell (d^5 configuration) (Chapter 13).

Language of Chemistry

A sodium atom and a sodium ion have *very different* properties. For example, sodium ions dissolve in water without a chemical reaction. In contrast, sodium atoms react with water to form sodium ions. These differences occur because the sodium ion is charged and has a stable electron arrangement. A sodium ion has the same electron arrangement as an argon atom, but they have different properties because the sodium ion is charged and has a different number of protons in its nucleus. ■

Predicting the type of bonding from electronegativity values

4.1.6 **Predict** whether a compound of two elements would be ionic from the position of the elements in the periodic table or from their electronegativity values.

4.2.5 **Predict** whether a compound of two elements would be covalent from the position of the elements in the periodic table, or from their electronegativity values.

Ionic bonding between two elements typically occurs when a metal is chemically bonded with a non-metal. Hence, the bonding in the compound barium fluoride, BaF_2, is predicted to be ionic since barium is a metal and fluorine a non-metal.

Ionic bonding is favoured if the metal and non-metal elements are reactive. The reactivity of metals and non-metals can be assessed using **electronegativity** values (Chapter 3). Ionic bonding is most likely when there is a large difference in electronegativity values between the two elements.

The electronegativity (Table 4.3) of an atom is the ability or power of an atom in a covalent bond to attract shared pairs of electrons to itself. The greater the electronegativity of an atom, the greater its ability to attract shared pairs of electrons to itself. The most electronegative elements are highly reactive non-metals and the least electronegative elements are the reactive metals.

H 2.1																	He
Li 1.0	Be 1.5											B 2.0	C 2.5	N 3.0	O 3.5	F 4.0	Ne
Na 0.9	Mg 1.2											Al 1.5	Si 1.8	P 2.1	S 2.5	Cl 3.0	Ar
K 0.8	Ca 1.0	Sc 1.3	Ti 1.5	V 1.6	Cr 1.6	Mn 1.5	Fe 1.8	Co 1.8	Ni 1.8	Cu 1.9	Zn 1.6	Ga 1.6	Ge 1.8	As 2.0	Se 2.4	Br 2.8	Kr
Rb 0.8	Sr 1.0	Y 1.2	Zr 1.4	Nb 1.6	Mo 1.8	Tc 1.9	Ru 2.2	Rh 2.2	Pd 2.2	Ag 1.9	Cd 1.7	In 1.7	Sn 1.8	Sb 1.9	Te 2.1	I 2.5	Xe
Cs 0.7	Ba 0.9	La 1.1	Hf 1.3	Ta 1.5	W 1.7	Re 1.9	Os 2.2	Ir 2.2	Pt 2.2	Au 2.4	Hg 1.9	Tl 1.8	Pb 1.8	Bi 1.9	Po 2.0	At 2.2	Rn
Fr 0.7	Ra 0.9	Ac 1.1															

Table 4.3
Electronegativity values
(Pauling scale)

Electronegativity generally increases on passing across a period, owing to the increasing nuclear charge and decreasing atomic radius. Electronegativity decreases on moving down a group since the combined effects of increasing atomic size and the shielding effect outweigh the increase in nuclear charge (Chapter 3).

There are some general rules for predicting the type of chemical bond based upon the electronegativity differences.

- If the difference in electronegativity values is greater than 1.8, then the bond is ionic.
- If the difference in electronegativity values is 0, then the bond is non-polar covalent.
- If the difference in electronegativity values is greater than 0 but less than 1.8, then the bond is polar covalent.

Polar covalent bonds are covalent bonds with ionic character (partial electron transfer). Ionic and covalent bonding are extremes forms of bonding: polar bonds are intermediate in nature. The larger the difference in electronegativity between the atoms, the greater the polarity of the bond and the greater the ionic character (Figure 4.5).

Figure 4.5 The spectrum of bonding from ionic to covalent via polar covalent. The delta symbols shown for polar covalent bonding represent fractional charges on the two atoms

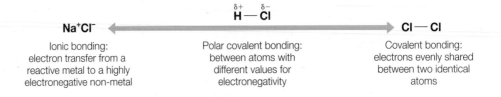

Na$^+$Cl$^-$	$\overset{\delta+}{H}$—$\overset{\delta-}{Cl}$	Cl—Cl
Ionic bonding: electron transfer from a reactive metal to a highly electronegative non-metal	Polar covalent bonding: between atoms with different values for electronegativity	Covalent bonding: electrons evenly shared between two identical atoms

Worked example

Using electronegativity values from Table 4.3, predict the type of bonding in fluorine molecules (F_2), hydrogen iodide (HI) and lithium fluoride (LiF).

Using the values in the table:

Fluorine, F_2 — Difference in electronegativity = 4.0 − 4.0 = 0
Non-polar covalent bond, F—F

Hydrogen iodide, HI — Difference in electronegativity = 2.5 − 2.1 = 0.4
Polar covalent bond, $^{\delta+}$H—I$^{\delta-}$

Lithium fluoride, LiF — Difference in electronegativity = 4.0 − 1.0 = 3.0
Ionic bond, Li$^+$F$^-$

■ Extension: Relationship between electronegativity difference and ionic character

Table 4.4 and Figure 4.6 show how the proportion of ionic character introduced into a covalent bond depends on the difference in electronegativity between the two atoms. Where covalent bond polarization occurs, then intermolecular forces of attraction will be generated. These forces affect the compound's physical properties, such as solubility, melting and boiling points (Section 4.3).

Electronegativity difference	Percentage of ionic character/%
0	0
0.5	6
1.0	22
1.5	43
2.0	63
2.5	79
3.0	96

Table 4.4 Percentage of ionic character in covalent bonds in relation to the electronegativity difference between the two elements

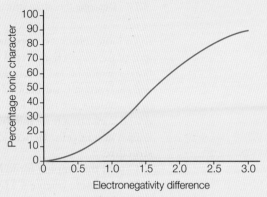

Figure 4.6 Graph of the relationship between percentage of ionic character in a covalent bond and the electronegativity difference between the two elements

Worked example

Deduce the percentage ionic character in the O—H bond.

The difference in electronegativity is 1.4. Hence from Table 4.4 the percentage ionic character is approximately 39%.

Polyatomic ions

4.1.7 **State** the formula of common polyatomic ions formed by non-metals in periods 2 and 3.

Many ions contain more than one atom. These types of ions are known as polyatomic ions. Table 4.5 summarizes the names, formulas and structures of commonly encountered polyatomic ions. A number of these ions are stabilized by resonance (π delocalization; see Chapter 14).

Name of ion	Formula	Structure of polyatomic ion	Example of compound
Ammonium	NH_4^+	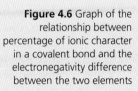	NH_4Cl, ammonium chloride
Oxonium or hydroxonium	H_3O^+		$H_3O^+Cl^-$, hydrochloric acid (Chapter 8)
Sulfate	SO_4^{2-}		$MgSO_4$, magnesium sulfate
Hydrogencarbonate	HCO_3^-		$KHCO_3$, potassium hydrogencarbonate

Name of ion	Formula	Structure of polyatomic ion	Example of compound
Nitrate	NO_3^-		$AgNO_3$, silver nitrate
Phosphate	PO_4^{3-}		K_3PO_4, potassium phosphate
Hydroxide	OH^-		$NaOH$, sodium hydroxide
Carbonate	CO_3^{2-}		Na_2CO_3, sodium carbonate

Table 4.5 Common polyatomic ions

In compounds such as magnesium sulfate, $MgSO_4$ or $[Mg^{2+}SO_4^{2-}]$, the bonding within the sulfate ions is covalent, but it is ionic between the sulfate and magnesium ions.

Structures of giant ionic compounds

4.1.8 Describe the lattice structure of ionic compounds.

In an ionic substance, oppositely charged ions attract each other and ions of the same charge repel each other. Hence, each cation in a lattice is surrounded by a number of anions as its closest neighbours, and each anion is surrounded by a number of cations. Two cations, or two anions, are never located next to each other. A single ionic crystal contains a huge number of ions in regular repeating units, known as unit cells. Hence, ionic solids are said to possess a **giant structure**.

One of the simplest ionic lattices is the lattice adopted by sodium chloride (Figure 4.7). It is a simple cubic lattice, also referred to as the rock salt structure. Each sodium ion is surrounded by six chloride ions; each chloride ion is surrounded by six sodium ions. The structure of ionic lattices was established from X-ray diffraction studies.

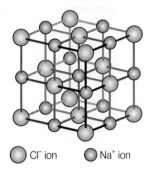

○ Cl⁻ ion   ● Na⁺ ion

Figure 4.7 Ionic lattice for sodium chloride

■ **Extension:** Coordination number

The number of ions that surround another of the opposite charge in an ionic lattice is called the **coordination number**. The sodium chloride lattice is known as a 6:6 lattice since each ion is surrounded by six oppositely charged ions (Figure 4.8). The coordination number of an ionic lattice depends on the relative sizes of the ions and their relative charges (Chapter 15).

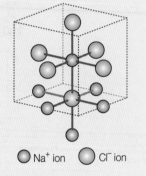

Figure 4.8 The sodium chloride lattice showing the 6:6 coordination

○ Na⁺ ion   ○ Cl⁻ ion

■ **Extension:** Analysis of crystals by X-ray diffraction

The vast majority of solids are crystalline solids and consist of a regular three-dimensional arrangement of atoms, molecules or ions, known as a lattice. The wavelengths of X-rays and the distances between the particles in a crystal are of a similar size. Hence X-rays are diffracted (scattered) if they strike the crystal lattice at a shallow angle (Figure 4.9).

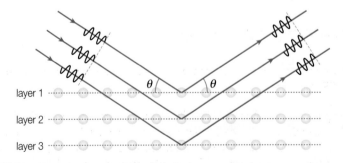

Figure 4.9 Strong X-ray reflections from a set of crystal planes will occur if the waves arrive and leave in phase

layer 1

layer 2

layer 3

The X-ray diffraction pattern is photographed and computer software is used to calculate the positions of the particles within the lattice. This can be used to generate an electron density map (Figure 4.10) of the molecule in a molecular lattice. Each contour line connects points of the same electron density. Bond lengths and bond angles of the molecule may be obtained from the electron density map.

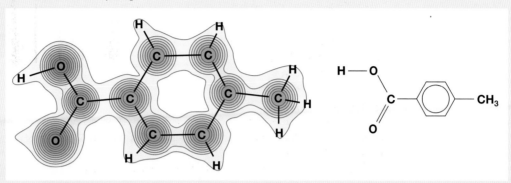

Figure 4.10 The structure of the 4-methyl benzoic acid molecule superimposed on its electron density map

■ Extension: Non-crystalline solids

Some solids are non-crystalline, or amorphous. In non-crystalline solids, the particles are not arranged in a lattice (Figure 4.11). Many non-crystalline solids, for example glass (Figure 4.12), are often called 'supercooled liquids'. Glass is a compound of silicon, oxygen and sodium. There is some order, but a lattice is not present. However, if the glass is shattered then the glass will crystallize. A number of polymers, for example, nylon may have crystalline and amorphous regions (Chapter 23).

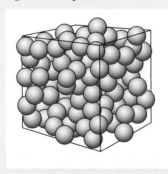

Figure 4.11 The structure of a non-crystalline (amorphous) solid

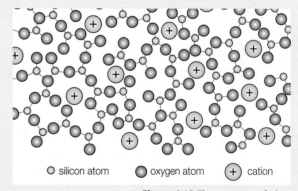

○ silicon atom   ● oxygen atom   ⊕ cation

Figure 4.12 The structure of glass

4.2 Covalent bonding

4.2.1 **Describe** the covalent bond as the electrostatic attraction between a pair of electrons and positively charged nuclei.

The simplest covalently bonded molecule is the hydrogen molecule, H_2. The two hydrogen atoms are held together because their nuclei are both attracted to the electron pair which is shared between them. Both the atoms are identical so the electrons are shared equally – a single non-polar covalent bond is formed. This simple electrostatic model is summarized in Figure 4.13.

However, very often the two atoms bonded will have different sizes. The smaller atom will attract the shared pair(s) of electrons more strongly since its nucleus will be closer to the electrons and will experience less shielding (Figure 4.14). The smaller atom is more electronegative. The resulting covalent bond is a single polar covalent bond.

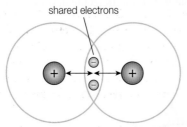

Both nuclei are attracted to the same pair of shared electrons. This holds the nuclei together.

Figure 4.13 A simple electrostatic model of the covalent bond in the hydrogen molecule, H_2

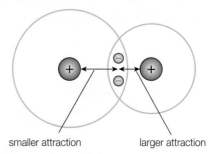

smaller attraction   larger attraction

Figure 4.14 The unequal sharing of electrons in a polar covalent bond

■ Extension: A simple model of covalent bonding

Consider the energy changes involved when a single covalent bond is formed from two hydrogen atoms (Figure 4.15). As the two hydrogen atoms approach each other, each nucleus starts to electrostatically attract the other atom's electron. The covalent bond starts to form and energy is released. However, if the two hydrogen atoms came too close together, there would be considerable repulsion between the nuclei and the potential energy of the system would rise. The covalent bond in the hydrogen molecule represents a position of equilibrium or balance in which the forces of attraction between the nuclei and the bonding electrons exactly match the repulsive forces between the two nuclei. This simple 'spring model' of covalent bonds is used in infrared spectroscopy (Chapter 21). The distance between the two bonded hydrogen nuclei is known as the bond length and the energy required to separate the atoms in the bond is known as the bond dissociation enthalpy (Chapter 5).

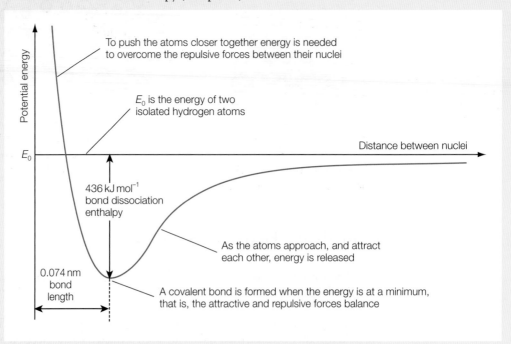

Figure 4.15 Variation in the potential energy of the two hydrogen atoms as the distance between them is varied

Formation of covalent bonds

4.2.2 **Describe** how the covalent bond is formed as a result of electron sharing.

Using Lewis structures to describe the formation of covalent bonds

In a single bond, each atom contributes one electron to the shared pair of electrons. Through the sharing of a pair of electrons, each atom now achieves the configuration of a noble gas. When two or more atoms are joined by covalent bonds, the particle or chemical species that results is known as a **molecule**. Covalent compounds are composed of molecules and each molecule is a group of bonded atoms held together by covalent bonds. Covalent bonds are usually formed between non-metallic elements.

A diatomic molecule is a molecule that consists of two identical atoms joined together by covalent bonds. For example, hydrogen gas exists as diatomic molecules, H_2. The structure of a hydrogen molecule, H_2, can be shown by using a Lewis (electron dot diagram) structure (Figure 4.16). A pair of electrons can be represented by dots, crosses, a combination of dots and crosses or by a line. Many chemists prefer to use a combination of dots and crosses so that it is clear which atom contributed the electrons.

Figure 4.16 Lewis structures (electron dot diagrams) for the hydrogen molecule, H_2

H⦂H   H⦂H   H⦂H   H — H

Lewis structures (electron dot diagrams) are shown in Figure 4.17 for the chlorine molecule, Cl_2. Note that the lone (unshared) pairs of electrons must be represented and that Cl–Cl is *not* a Lewis structure since it does not display the lone pairs of the two chlorine atoms. Lewis structures only include the outer or valence electrons since these are the electrons involved in bonding.

Figure 4.17 Lewis structures (electron dot diagrams) for the chlorine molecule, Cl_2

⦂Cl⦂Cl⦂   ⦂Cl⦂Cl⦂   ⦂Cl⦂Cl⦂ |Cl̄—Cl̄|

In the case of the elements oxygen and nitrogen, two and three pairs of electrons respectively must be shared between the two atoms of their molecules to achieve a stable noble gas electron arrangement. The oxygen molecule has a double bond and the nitrogen molecule has a triple bond (Figure 4.18).

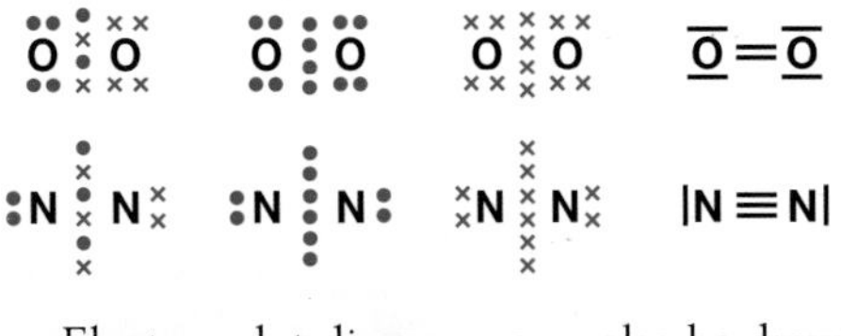

Figure 4.18 Lewis structures (electron dot diagrams) for oxygen and nitrogen molecules

O⦂O   O⦂O   O⦂O   Ō=Ō

⦂N⦂N⦂   ⦂N⦂N⦂   ⦂N⦂N⦂ |N≡N|

Electron dot diagrams may also be drawn for molecules of compounds. Figure 4.19 gives the electron dot diagrams for hydrogen chloride, methane, water, ammonia, ethane, ethene and carbon dioxide molecules. All the diagrams represent molecules in which the bonded atoms have achieved the electron arrangement of a noble gas.

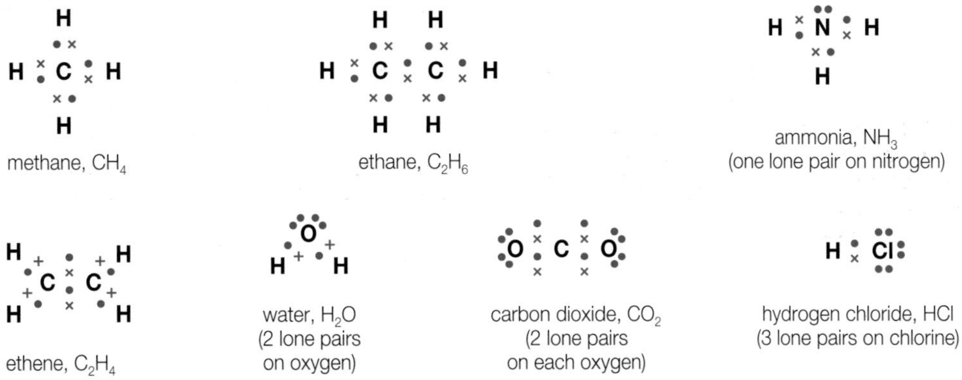

Figure 4.19 Lewis structures (electron dot diagrams) for a selection of simple covalent compounds

methane, CH_4

ethane, C_2H_6

ammonia, NH_3
(one lone pair on nitrogen)

ethene, C_2H_4

water, H_2O
(2 lone pairs on oxygen)

carbon dioxide, CO_2
(2 lone pairs on each oxygen)

hydrogen chloride, HCl
(3 lone pairs on chlorine)

Lewis structures (electron dot diagrams) may also be drawn to show the formation of ions in ionic compounds (Figure 4.20).

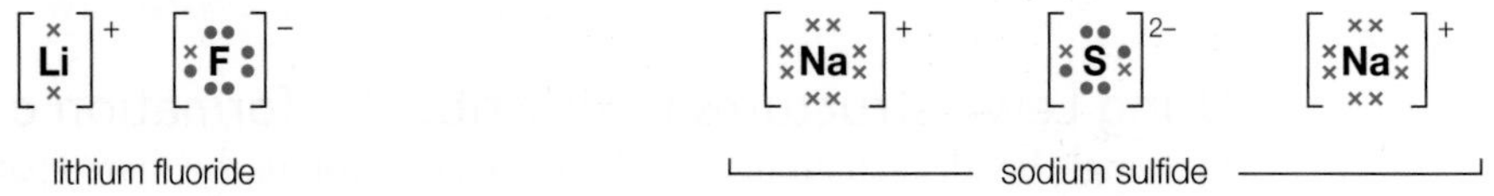

Figure 4.20 Lewis structures (electron dot diagrams) for lithium fluoride, LiF, and sodium sulfide, Na₂S

lithium fluoride   sodium sulfide

The only difference between Lewis structures (electron dot diagrams) of polyatomic ions (Figure 4.21) and molecules is that with ions you must consider the charge when counting valence electrons. Since electrons are negative, ions with a negative charge have extra electrons (sometimes denoted by a square); ions with a positive charge are short of electrons. A number of polyatomic ions have dative bonds, where both electrons in the covalent bond are donated by *one* of the two atoms that form the bond.

Figure 4.21 Lewis structures (electron dot diagrams) for the hydroxide, cyanide and carbonate ions

hydroxide ion   cyanide ion

carbonate ion

Lewis structures may also be drawn with dots and crosses including Venn diagrams (Figure 4.22). This approach is used for complicated molecules, as it allows an easy check to be made of exactly which bonds the electrons are in.

Figure 4.22 Lewis structures including Venn diagrams for the ethane, phosphorus trifluoride and hydrogen peroxide molecules

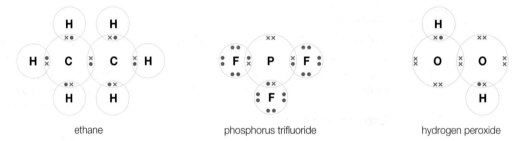

ethane   phosphorus trifluoride   hydrogen peroxide

History of Chemistry

Valency is a measure of the number of chemical bonds formed by the atoms of a given element. The concept of valency, or 'combining power', originated with the English chemist **Edward Frankland** (1825–1899). The concept of valency is usually applied to the formation of molecules (Figure 4.23) by non-metallic elements. Common valencies are: carbon, 4; nitrogen, 3; hydrogen, 1 and chlorine, 1. These valencies (Table 4.6) can be related to the number of electrons in the outer shells of the elements. For example, nitrogen has five electrons in its outer shell and hence needs to share three more electrons to gain an octet. Valency has largely been replaced by the concept of oxidation number (Chapter 9).

Figure 4.23 Models of molecules of nitrous acid (HONO), chloric(I) acid, HOCl and hydrogen cyanide, HCN

Group

Period	1	2											3	4	5	6	7
1							**H** 1										
2	**Li** 1	**Be** 2											**B** 3	**C** 4	**N** 3	**O** 2	**F** 1
3	**Na** 1	**Mg** 2											**Al** 3	**Si** 4	**P** 3,5	**S** 2,4,6	**Cl** 1
4	**K** 1	**Ca** 2	**Sc** 3	**Ti** 4	**V** 3,4,5	**Cr** 3,6	**Mn** 2,4,7	**Fe** 2,3	**Co** 2	**Ni** 2	**Cu** 1,2	**Zn** 2					**Br** 1
5	**Rb** 1	**Sr** 2									**Ag** 1			**Sn** 2,4			**I** 1
6	**Cs** 1	**Ba** 2												**Pb** 2,4			

Table 4.6 Some common valencies of selected elements

■ Extension: Molecules that do not obey the octet rule

There are a number of molecules that do not obey the octet rule. There are three types of exceptions:

- molecules in which a central atom has an incomplete octet
- molecules in which a central atom has an expanded octet
- molecules with an odd number of electrons.

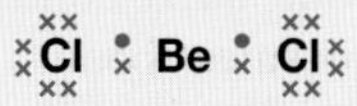

Incomplete octet

In the beryllium chloride molecule ($BeCl_2(g)$), the beryllium atom has only four electrons in its valence shell (Figure 4.24). The molecule is described as electron deficient.

Another example of an electron-deficient molecule is aluminium trichloride, $AlCl_3$. The aluminium atom has only six electrons in its valence shell (Figure 4.25).

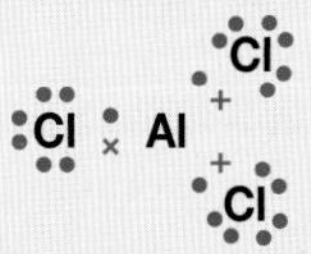

beryllium chloride

Figure 4.24 Lewis diagram of the beryllium chloride molecule

aluminium trichloride

Figure 4.25 Lewis diagram of the aluminium trichloride molecule

Expanded octet

Molecules with an expanded octet have a central atom which has more than eight electrons in its valence shell (Figure 4.26). Examples are phosphorus pentafluoride, PF_5 (Chapter 13) and sulfur hexafluoride, SF_6 (Chapter 14). In phosphorus pentafluoride, there are five covalent bonds (that is, 10 electrons around the phosphorus atom) and in sulfur hexafluoride, SF_6, there are six covalent bonds (that is, 12 electrons around the sulfur atom).

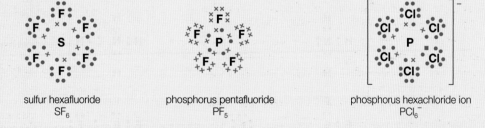

sulfur hexafluoride SF_6

phosphorus pentafluoride PF_5

phosphorus hexachloride ion PCl_6^-

Figure 4.26 Lewis structures (electron dot diagrams) of sulfur hexafluoride, SF_6, phosphorus pentafluoride, PF_5 and PCl_6^-.

The outer shells of the elements phosphorus and sulfur in period 3 of the periodic table can hold up to 18 electrons. The elements in period 2 do not form compounds with more than eight electrons in the outer shell because the second shell can only hold up to eight electrons. Hence, nitrogen trifluoride, NF_3, is a stable species but nitrogen pentafluoride, NF_5, is unknown. However, phosphorus trifluoride, PF_3, and phosphorus pentafluoride, PF_5, are both stable molecules.

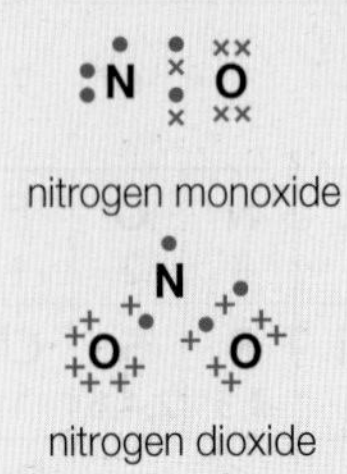

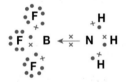

nitrogen monoxide

nitrogen dioxide

Figure 4.27 Lewis structures (electron dot diagrams) of nitrogen monoxide and nitrogen dioxide molecules

Odd electron molecules

In most stable molecules, the number of electrons is even and complete pairing of electrons occurs. However, a small number of molecules and ions contain an odd number of valence electrons. Most odd electron molecules have a central atom from an odd-numbered group, such as nitrogen and chlorine. Nitrogen monoxide, •NO, and nitrogen dioxide, •NO$_2$, are examples of odd electron molecules (Figure 4.27). The oxy-chlorine radical •OCl is an intermediate formed during the destruction of ozone by chlorofluorocarbons (Chapter 25). Molecules and ions with an unpaired electron are known as **free radicals**; their existence can be explained by the molecular orbital (MO) theory of bonding (Chapter 14). Both molecules exist as resonance hybrids of two Lewis structures.

carbon monoxide

Figure 4.28 Lewis structure for the carbon monoxide molecule, CO

Coordinate (dative) bonding

In some molecules and polyatomic ions, both electrons to be shared come from the same atom. The covalent bond formed is known as a **coordinate** or **dative covalent bond**. In Lewis structures (electron dot diagrams), a coordinate or dative bond is often denoted by an arrow pointing from the atom which donates the lone pair to the atom which receives it.

For example, the carbon monoxide molecule, CO (Figure 4.28), contains one dative bond. Once formed, the dative bond is indistinguishable from the other two single covalent bonds.

Dative bonding may also be found in molecular addition compounds (or adducts), such as boron trifluoride ammonia, BF$_3$.NH$_3$ (Figure 4.29).

The boron atom in boron trifluoride has only six electrons in its outer shell and so can accept an additional two electrons to fill the shell and obey the octet rule. The nitrogen atom on the ammonia molecule donates its lone pair of electrons to form the dative bond between the nitrogen and the boron atom.

The formation of dative bonds between a pair of reacting chemical species is the basis of a theory of acidity known as Lewis theory (Chapter 8). Dative bonds are also involved in the formation of transition metal complex ions (Chapter 13). Dative bond formation is often part of organic reaction mechanisms (Chapter 20). Aqueous solutions of acids contain the oxonium ion, H$_3$O$^+$, a datively bonded species (Chapter 8).

Dative bonding is also present in some common polyatomic ions. Three examples are shown in Figure 4.30. Ammonia and hydrogen chloride react together rapidly to form ammonium chloride. This is a white ionic solid with the formula NH$_4$Cl [NH$_4$$^+$ Cl$^-$]. When a fluoride ion shares a lone pair with the boron atom in boron trifluoride, a tetrafluoroborate ion, BF$_4$$^-$, is formed. In the nitrate ion, NO$_3$$^-$, the nitrogen atom achieves an octet by forming a dative bond with one of the oxygen atoms.

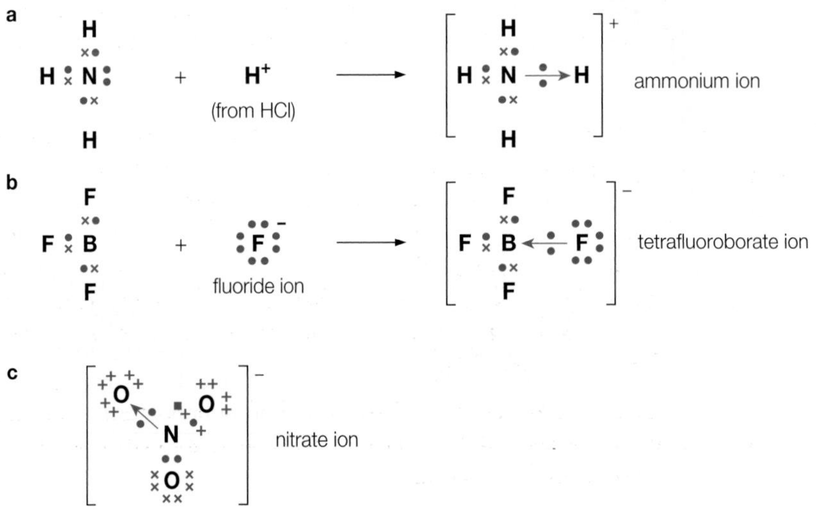

boron trifluoride ammonia

Figure 4.29 Lewis structure for the boron trifluoride ammonia, BF$_3$.NH$_3$ molecule

a

ammonium ion

(from HCl)

b

tetrafluoroborate ion

fluoride ion

c

nitrate ion

Figure 4.30 Formation of the **a** ammonium and **b** tetrafluoroborate ions; **c** structure of the nitrate ion

Extension: Bonding and electron orbitals

The Lewis model (electron dot model) is a very useful but simplistic approach to chemical bonding. However, a deeper understanding of chemical bonding and chemical reactions is obtained if atomic orbitals are considered (Chapter 12). Covalent bonds are formed when orbitals overlap to form molecular orbitals (Chapter 14). In hydrogen the two 1s atomic orbitals overlap and merge to form a σ (sigma) bond (Figure 4.31). All single bonds are σ bonds. The second bond of a double bond, known as a π (pi) bond (Figure 4.32), is formed by the sideways overlap of p orbitals. The π bond is weaker than the single bond. This explains why ethene, which has a carbon–carbon double bond, is relatively reactive (Chapter 20).

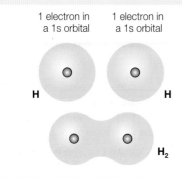

Figure 4.31 The formation of a covalent bond by the overlap of 1s atomic orbitals

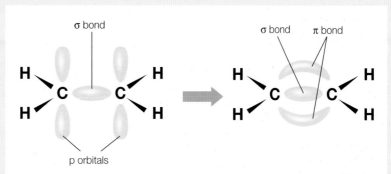

Figure 4.32 The formation of the first bond (σ bond) and the second bond (π bond) in the ethene molecule

Drawing Lewis structures for molecules and ions

4.2.3 **Deduce** the Lewis (electron dot) structures of molecules and ions for up to four electron pairs on each atom.

In general the following steps are followed when writing Lewis structures for molecules and ions:

- Calculate the total number of valence electrons for all the atoms in the molecule or ion. The number of valence electrons is deduced from its group in the periodic table.
- Arrange all the atoms surrounding the central atom by using a pair of electrons per bond. The central atom is most often the atom which is least electronegative. Hydrogen is never a central atom.
- Assign the remaining electrons to the terminal atoms so that each terminal atom has eight electrons. However, if hydrogen is the terminal atom, then it can only hold two electrons.
- Place any electrons left over on the central atom. In the case of period 3 elements, such as sulfur and phosphorus, the central atom may have more than eight electrons.
- Form multiple bonds if there are not enough electrons to give the central atom an octet of electrons.

Worked example

Write the Lewis structure (electron dot diagram) for hydrogen cyanide, HCN.

- The total number of valence electrons is ten: one from the hydrogen atom; four electrons from the carbon atom and five electrons from the nitrogen atom.
- The bonds around the central atom use four electrons: **H ⁝ C ⁝ N**
- There are six electrons left to place. The hydrogen cannot take any more in its outer shell, as it already has two electrons, so these can be placed around the nitrogen atom, which can accomodate eight electrons in its outer shell. There are no more electrons for the central carbon atom. **H ⁝ C ⁝ N̈⁝**
- The central carbon does not have enough electrons to form an octet, so move two lone pairs of electrons from the nitrogen atom to the carbon atom to form a triple bond between the two atoms. **H ⁝ C ⁝ N̈⁝**

Extension: Alternative Lewis structures (formal charge)

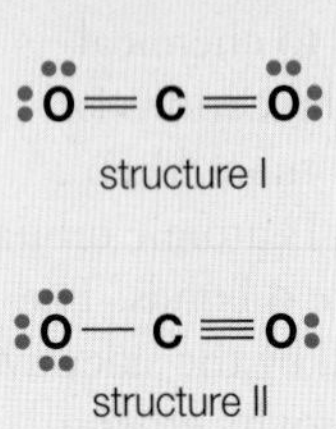

$:\overset{..}{O}=C=\overset{..}{O}:$

structure I

$:\overset{..}{\underset{..}{O}}-C\equiv O:$

structure II

Figure 4.33 Two possible structures for carbon dioxide

It is sometimes possible to write two different Lewis structures for a molecule that have different arrangements of the electrons. For example, there are two ways of writing the Lewis structure of carbon dioxide (Figure 4.33).

In both Lewis structures the octet rule is obeyed. One method of determining which is the most stable structure is based on the concept of **formal charge**.

The formal charge on an atom in a Lewis structure is the charge it would have if the bonding electrons were shared equally. The formal charge of an atom is the number of valence electrons in the free atom minus the number of electrons assigned to that atom.

$$\text{formal charge} = \left[\begin{array}{c}\text{number of valence}\\\text{electrons in free atom}\end{array}\right] - \left[\begin{array}{c}\text{number of electrons}\\\text{assigned to atom}\end{array}\right]$$

The electrons are assigned to the atoms in a Lewis structure according to the following rules:
- All lone pairs are assigned to the atom on which they are found.
- Half of the bonding electrons are assigned to each atom in the bond.

Hence, the definition of formal charge can be rewritten:

$$\text{formal charge} = \left[\begin{array}{c}\text{number of valence}\\\text{electrons in free atom}\end{array}\right] - \frac{1}{2}\left[\begin{array}{c}\text{number of}\\\text{bonding electrons}\end{array}\right] - \left[\begin{array}{c}\text{number of}\\\text{lone pairs}\end{array}\right]$$

As a general rule, when several Lewis structures are possible, the most stable Lewis structure is the one with no formal charges, or the Lewis structure in which the atoms bear the smallest formal charges and the negative charges appear on the more electronegative atoms. The sum of the formal charges of the atoms in a Lewis structure must equal zero for a neutral molecule, or equal the charge for a polyatomic ion.

The method for finding the most stable structure is illustrated below for carbon dioxide.

Calculating formal charges for Lewis structure I for carbon dioxide

Left-hand oxygen atom
Valence electrons of oxygen atom = 6
Electrons assigned to the oxygen atom = $(\frac{1}{2} \times 4) + 4 = 6$
Hence, formal charge on the oxygen atom = $(6 - 6) = 0$

(The right-hand oxygen atom is identical to the left-hand oxygen in Lewis structure I).

Carbon atom
Valence electrons of carbon atom = 4
Number of electrons assigned to carbon atom = $(\frac{1}{2} \times 8) + 0 = 4$
Hence, formal charge on the carbon atom = $(4 - 4) = 0$

Summary　　　　$:\overset{..}{O}=C=\overset{..}{O}:$

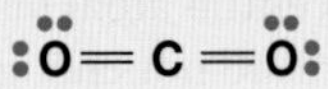

formal charges　　0　　0　　0

Calculating formal charges for Lewis structure II for carbon dioxide

Left-hand oxygen atom
Valence electrons of oxygen atom = 6
Electrons assigned to the oxygen atom = $(\frac{1}{2} \times 2) + 6 = 7$
Hence, formal charge on the oxygen atom = -1

Right-hand oxygen atom
Valence electrons of oxygen atom = 6
Electrons assigned to the oxygen atom = $(\frac{1}{2} \times 6) + 2 = 5$
Hence, formal charge on the oxygen atom = $+1$

Carbon atom

Valence electrons of carbon atom = 4

Number of electrons assigned to carbon atom = $(\frac{1}{2} \times 2) + (\frac{1}{2} \times 6) = 4$

Hence, formal charge on the carbon atom = 0

Summary

$$:\overset{\cdot\cdot}{\underset{\cdot\cdot}{O}} - C \equiv O:$$

formal charges   −1   0   +1

Hence, structure I is the stable Lewis structure because the atoms have no formal charges. (The concept of formal charges can also be used select the major resonance form (Chapter 14).)

Bond strength

4.2.4   **State** and **explain** the relationship between the number of bonds, bond length and bond strength.

Double bonds are stronger than single bonds (for the same pair of atoms) because there are more pairs of shared electrons between the two nuclei. Triple bonds are stronger than double bonds (for the same pair of atoms). This observation can be accounted for by the increase in attraction by the nuclei for the shared pairs of electrons (based on the simple classical model of covalent bonding – see pages 103–104).

Consequently, bond strengths (as measured by bond enthalpies) *increase* and bond lengths *decrease* from single to double to triple bonds (for the same pair of atoms). This is illustrated in Table 4.7 for the element carbon.

This relationship is typically observed for atoms in other covalent bonds. For example, ethanoic acid, CH_3—COOH, contains two carbon–oxygen bonds, one single and one double. The double bond, C=O (length 0.122 nm), is significantly shorter than the single bond, C—O (0.143 nm).

Bond type	Bond enthalpy/ kJ mol^{-1}	Bond length/ nm
Single (C—C)	348	0.154
Double (C=C)	612	0.134
Triple (C≡C)	837	0.120

Table 4.7 The bond enthalpies and bond lengths of carbon–carbon bonds

■ **Extension:**   ## Bond lengths in benzene

The benzene molecule has carbon–carbon bond lengths and strengths intermediate between carbon–carbon single and carbon–carbon double bonds. This indicates the molecule is a resonance hybrid (Chapter 14).

Bond polarity

4.2.6   **Predict** the relative polarity of bonds from electronegativity values.

We have seen how electronegativity values can be used to predict whether a bond is ionic or covalent (page 99). The relative polarities of covalent bonds can also be predicted from electronegativity values. The larger the difference between the electronegativites of the atoms forming the covalent bond, the more unequal the sharing will be and the more polar the bond.

The electronegativities of selected elements are shown in Table 4.8.

Element	F	O	N	Cl	C	H
Electronegativity	4.0	3.5	3.0	3.0	2.5	2.1

Table 4.8 Electronegativity values of six elements

The values quoted show that C—Cl and C—O bonds are both polar. However, the C—O bond (electronegativity difference of 1.0) is more polar than the C—Cl bond (electronegativity difference of 0.5). The N—Cl bond, however, is non-polar because the electronegativity difference between nitrogen and chlorine is zero. The C—H bond (electronegativity difference of 0.4) has very low polarity.

Applications of Chemistry

Computer software can be used to generate an electrostatic potential surface (Figures 4.34 and 4.35) which shows the variation in electrostatic potential around a molecule. The software uses different colours to represent the different values of the electrostatic potential on this surface. Red is used to colour the regions of most negative electrostatic potential and blue is used to colour the regions of most positive electrostatic potential. Intermediate colours represent intermediate values so that the potential increases in the order red < orange < yellow < green < blue. Production of these surfaces allows chemists to predict the chemical and physical properties of the molecules. Chemical reactions are usually associated with highly charged sites; the most highly charged site is usually the most reactive.

Figure 4.34 Water molecule with superimposed potential map (mesh)

Figure 4.35 Hydrogen fluoride molecule with superimposed potential map (solid)

Valence shell electron pair repulsion theory

> **4.2.7** **Predict** the shape and bond angles for species with four, three and two negative charge centres on the central atom using the valence shell electron pair repulsion theory (VSEPR).

The shapes of molecules and ions can be predicted by the **valence shell electron pair repulsion theory (VSEPR)**. If the Lewis structure is drawn for a molecule or a polyatomic ion, the shape of this molecule or ion can be predicted using this theory.

The VSEPR theory states that:

- the electron pairs around the central atom repel each other
- bonding pairs and lone pairs of electrons arrange themselves to be as far apart as possible. Bonding (shared) and lone (unshared) pairs are termed negative charge centres.

The molecule or polyatomic ion adopts the shape that minimizes the repulsion between the bonding and lone pairs of electrons. The shapes of the molecules and polyatomic ions are therefore determined by the electron pairs rather than by the atoms.

Basic molecular shapes

Three of the five basic molecular shapes are linear, trigonal planar and tetrahedral. Table 4.9 shows the arrangement of the electron pairs (charge centres) that results in minimum repulsion and the basic shapes of the molecules. The two other basic shapes adopted by molecules, trigonal bipyramidal and octahedral, are discussed in Chapter 14.

Molecule shape	Number of electron pairs	Description
	2	Linear
	3	Triangular planar (trigonal planar)

Table 4.9 Basic molecular shapes

Molecule shape	Number of electron pairs	Description
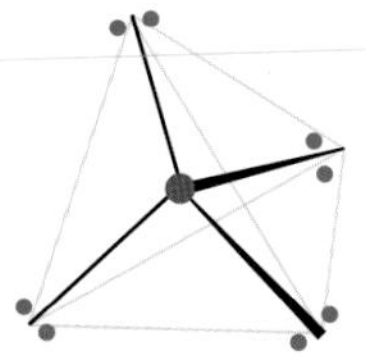	4	Tetrahedral
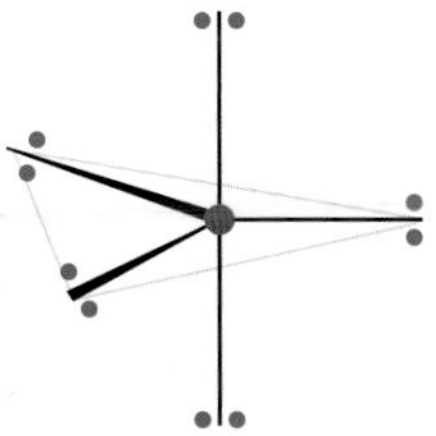	5	Triangular bipyramidal (trigonal bipyramidal)
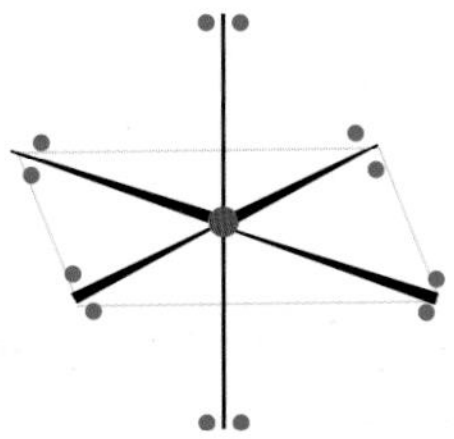	6	Octahedral

Table 4.9 (cont.)

Shapes of molecules and bond angles

The shapes and bond angles of molecules and ions are primarily determined by the number of electron pairs. However, the number of electron pairs alone does not account completely for the shapes and bond angles. In the VSEPR theory, a lone pair of electrons repels other electron pairs more strongly than a bonding pair. This is because the region in space occupied by a lone pair of electrons is closer to the nucleus of an atom than a bonding pair. Bonding pairs of electrons are spread out between the nuclei of the atoms which they bind together. Thus, a lone pair can exert a greater repelling effect than a bonding pair. The order of the repulsion strength of lone pairs and bond pairs of electron is:

lone pair–lone pair repulsion > lone pair–bond pair repulsion > bond pair–bond pair repulsion

strongest   weakest

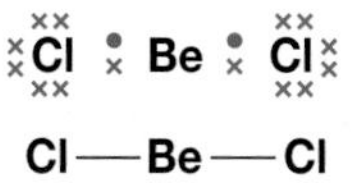

Two electron pairs

Consider the gaseous beryllium chloride molecule, $BeCl_2(g)$. The Lewis structure of the molecule shows there are only two electron pairs (two negative charge centres) in the valence shell of the beryllium atom (Figure 4.36).

These two pairs of electrons try to separate as far as possible from each other so as to minimize electron repulsion. Thus, the beryllium chloride molecule adopts a linear shape with a bond angle of 180°, because the electron pairs are farthest apart when they are on opposite sides of the beryllium atom.

Three electron pairs

Boron is an element in group 3 of the periodic table. Therefore, it has three valence electrons. The Lewis structure (Figure 4.37) of the boron trifluoride molecule, BF_3, shows there are only three electron pairs (three negative charge centres) in the valence shell of the boron atom.

These three bonding pairs repel each other equally, with the result that the boron trifluoride molecule is a trigonal planar (flat) molecule. The three boron–fluorine bonds point towards the three corners of an equilateral triangle. The bond angles are all equal at 120°.

Cl—Be—Cl

beryllium chloride

Figure 4.36 Lewis structure and molecular shape of beryllium chloride

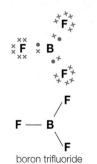

boron trifluoride

Figure 4.37 Lewis structure and molecular shape of boron trifluoride

Four electron pairs

The methane molecule, CH_4, has four bonding pairs of electrons (four negative charge centres) located in the valence shell of the central carbon atom. The repulsion between the bonding pairs of electrons is minimized when the angle between the electron pairs is 109.5° (the tetrahedral angle). Wedges and tapers are used to show the directions of the bonds (Figure 4.38).

Figure 4.38 Lewis structure and molecular shape of the methane molecule

—— bond in the plane of the paper
·····ıı bond behind the plane of the paper
—— bond in front of the plane of the paper

It is important to distinguish between the orientation or arrangement of electron pairs in the outer shell and the shapes of molecules. The shape of a molecule or ion refers to the positions of the atoms or groups of atoms around the central atom and not the orientation of the electron pairs. The shape of a molecule or ion depends not only on the number of electron pairs, but also on whether these electron pairs are bonding pairs of electrons or lone pairs of electrons.

Therefore, there are three molecular shapes corresponding to four electron pairs that are arranged tetrahedrally around the central atom. These molecular shapes are described as tetrahedral, pyramidal and V-shaped/non-linear or bent. For example the methane molecule, CH_4, is tetrahedral in shape; the ammonia molecule, NH_3, is trigonal pyramidal; and the water molecule, H_2O, is V-shaped/non-linear or bent (Figure 4.39) – even though each of these molecules has four pairs of electrons arranged tetrahedrally around the central atoms, namely carbon, nitrogen and oxygen, respectively.

Figure 4.39 Lewis structures and molecular shapes of the ammonia and water molecules

The bond angles of the methane, ammonia and water molecules are shown in Table 4.10. The bond angles decrease as the number of lone pairs of electrons increases (Figure 4.40).

Molecule	Number of lone pairs	Bond angles
Methane, CH_4	0	109.5°
Ammonia, NH_3	1	107.0°
Water, H_2O	2	104.5°

Table 4.10

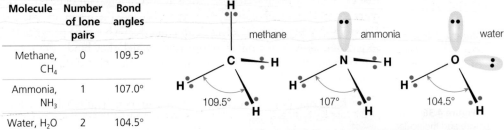

Figure 4.40 The shapes and bond angles of the methane, ammonia and water molecules

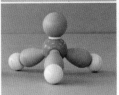

Figure 4.41 Models of hydrogen fluoride (three lone pairs), water (two lone pairs) and ammonia molecules (one lone pair)

The bond angles of H—C—H bonds in the methane molecule are 109.5°. The predicted bond angles of H—N—H bonds in ammonia would be 109.5°, *if* all the electron pairs repelled each other equally as they do in methane. However, the experimentally determined bond angles (via X-ray diffraction) in the ammonia molecule are 107°, 2.5° smaller than the predicted tetrahedral angle. An even smaller bond angle is shown by the water molecule.

The progressive decrease in the bond angle is caused by the lone pair electron repulsion being greater than the bond pair electron repulsion. This is because the electrons of a lone pair are closer to the nucleus than a bonding pair. This greater repulsive effect tends to push the bonding pairs in the ammonia molecule closer together, so that the ammonia molecule is a slightly distorted tetrahedron with a smaller than expected bond angle. The effect is even greater in the water molecule, where the additional repulsion between the two lone pairs causes a greater deviation in bond angle from the tetrahedral bond angle.

Multiple bonds

VSEPR theory can also be used to explain the shapes of molecules or ions that contain a double or triple bond. A double or triple bond has the same effect as a single bond because all the bonding pairs of electrons are located between the two atoms forming a covalent bond. A double or triple bond is therefore counted as one bonding pair (one negative charge centre) when predicting the shapes of molecules and ions.

Thus the carbon dioxide molecule, CO_2, has a linear structure like the beryllium chloride molecule (Figure 4.42), and the ethene molecule, C_2H_4, is trigonal planar around each of the two carbon atoms (Figure 4.43). It is a planar molecule.

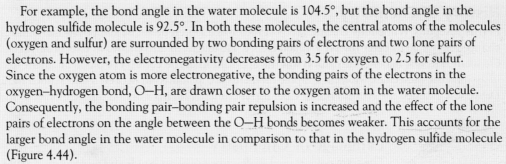

Figure 4.42 Lewis structure and molecular shape of the carbon dioxide molecule

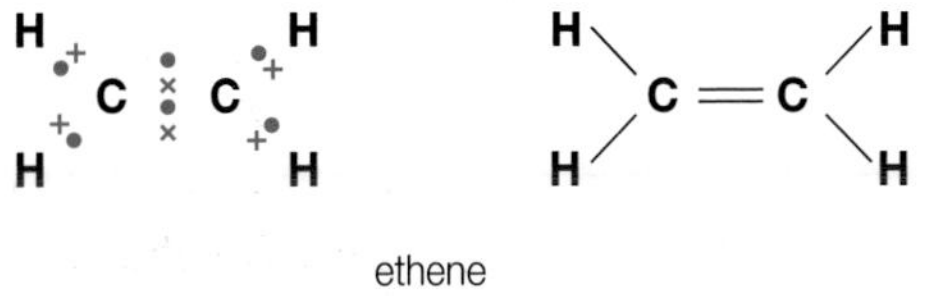

Figure 4.43 Lewis structure and molecular shape of the ethene molecule

■ Extension: Effect of electronegativity on molecular shape

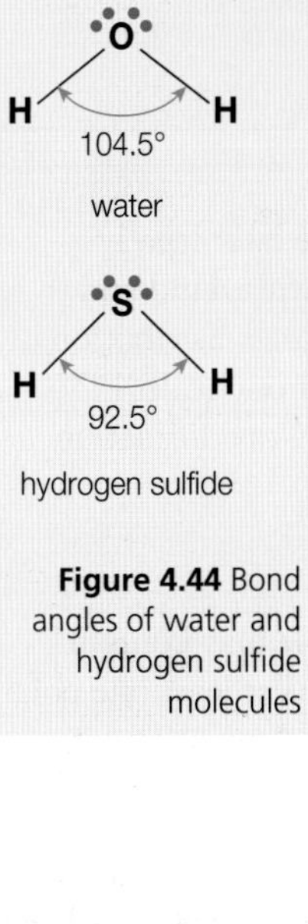

The shapes of molecules and ions are determined mainly by the number of electron pairs located around the central atom of the molecule or ion, and the effects of any lone pairs of electrons on the bond angles. Another factor that affects the bond angle is the electronegativity of the central atom.

For example, the bond angle in the water molecule is 104.5°, but the bond angle in the hydrogen sulfide molecule is 92.5°. In both these molecules, the central atoms of the molecules (oxygen and sulfur) are surrounded by two bonding pairs of electrons and two lone pairs of electrons. However, the electronegativity decreases from 3.5 for oxygen to 2.5 for sulfur. Since the oxygen atom is more electronegative, the bonding pairs of the electrons in the oxygen–hydrogen bond, O—H, are drawn closer to the oxygen atom in the water molecule. Consequently, the bonding pair–bonding pair repulsion is increased and the effect of the lone pairs of electrons on the angle between the O—H bonds becomes weaker. This accounts for the larger bond angle in the water molecule in comparison to that in the hydrogen sulfide molecule (Figure 4.44).

Figure 4.44 Bond angles of water and hydrogen sulfide molecules

4.2.8 **Predict** whether or not a molecule is polar from its molecular shape and bond polarities.

Bond polarity and dipole moment

Non-polar and covalent bonds

Diatomic molecules of elements, such as hydrogen, oxygen, nitrogen and halogens, consist of two identical atoms covalently bonded together. The bonding electrons are symmetrically arranged around the two nuclei and are attracted equally to both nuclei. This is because the two atoms in the bond are identical and have the same electronegativity values. This type of bond is called a non-polar bond.

When two atoms with *different* electronegativity values form a covalent bond, the shared pairs of bonding electrons will be attracted more strongly by the more electronegative element. This results in an asymmetrical distribution of the bonding electrons.

For example, in the hydrogen chloride molecule, the more electronegative chlorine atom attracts the bonding pair more strongly than the hydrogen atom does. Consequently, the chlorine atom has a partial or fractional negative charge and the hydrogen atom has a partial or fractional positive charge. The hydrogen chloride molecule is described as being **polar** and the bond in the hydrogen chloride molecule is described as **polar covalent**.

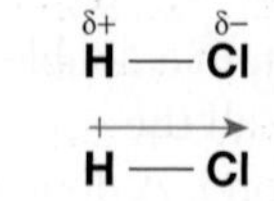

hydrogen chloride

Figure 4.45 Bond polarity in the hydrogen chloride molecule

In polar molecules, the centres of the positive and negative charges do not coincide. One end of a polar molecule has a partial positive charge, and the other a partial negative charge. The degree of bond polarity depends on the difference in the electronegativity values of the two elements.

The polarity of the hydrogen chloride molecule can be indicated in two ways (Figure 4.45). By convention the arrow always points from the positive charge to the negative charge. The symbols '$\delta+$' and '$\delta-$' represent the partial positive and negative charges on the hydrogen and chlorine atoms. The arrow denotes the shift in electron density towards the more electronegative chlorine atom. The crossed end of the arrow represents a plus sign that designates the positive end (the less electronegative atom).

The separation of charge in a polar bond is termed **polarization**. When two electrical charges of opposite sign are separated by a small distance, a **dipole** is established. The size of a dipole is measured by its dipole moment.

■ Extension: Dipole moment

The size of a dipole is measured by its **dipole moment**. If two charges of equal magnitude but opposite in sign, Q^+ and Q^-, are separated by a distance r, the system is said to have a dipole moment of magnitude given by the following formula:

$$\text{dipole moment } (\mu) = \text{charge} \times \text{distance} = Q \times r$$

Dipole moments are usually measured in debyes (D). $1\,D = 3.34 \times 10^{-30}\,C\,m$, where C stands for coulomb and m for metre.

History of Chemistry

Petrus Josephus Wilhelmus Debijie (1884–1966) was a Dutch physicist and physical chemist. He later legally changed his name to Peter Joseph William Debye. He taught in universities in Czechoslovakia, Switzerland and Germany. His first major contribution was the application of the concept of dipole moment to charge distribution in asymmetric molecules. Molecular dipole moments are measured in debyes, a unit named in his honour. In 1936 Debye was awarded the Nobel Prize in Chemistry, primarily for his work on dipole moments and X-ray diffraction.

Language of Chemistry

The term polar is used to describe covalent bonds and molecules. A molecule may contain polar bonds, but be non-polar. A molecule cannot be polar unless it contains at least one polar bond. Hence, the term polar should be referenced to either bond or molecular polarity. ■

Polarity of molecules

Diatomic molecules that contain two atoms of different electronegativities are described as polar molecules. However, the polarity of a molecule containing more than two atoms depends on both the polarities of the bonds and the shape of the molecule. Molecules that are very polar have large dipole moments. Non-polar molecules have a zero dipole moment (Table 4.11).

A molecule containing more than two atoms of different electronegativities may be non-polar even though there are polar bonds in the molecule. Molecules with polar bonds are non-polar because these molecules are symmetrical, that is, the central atom is symmetrically surrounded by identical atoms.

Bond dipoles and dipole moments are vector quantities, which means they have both a magnitude

Name of molecule	Formula	Polarity of molecule
Hydrogen chloride	HCl	Polar
Water	H_2O	Polar
Ammonia	NH_3	Polar
Benzene	C_6H_6	Non-polar
Boron trichloride	BCl_3	Non-polar
Methane	CH_4	Non-polar
Bromobenzene	C_6H_5Br	Polar
Carbon dioxide	CO_2	Non-polar
Sulfur dioxide	SO_2	Polar
Tetrachloromethane	CCl_4	Non-polar

Table 4.11 Polarity of some molecules

(size) and direction. The overall dipole moment of a polyatomic molecule is the sum of its bond dipoles. In the carbon dioxide molecule the two bond dipoles, although equal in magnitude, are exactly opposite in direction. Hence, the dipoles vectorially cancel each other and the overall dipole moment is zero (Figure 4.46).

The water molecule has a bent shape. The two polar bonds are identical, so the bond dipoles are equal in magnitude. However, the bond dipoles are not directly opposite one another and therefore do not cancel each other. Consequently, the water molecule is a polar molecule (Figure 4.46) because it has an overall dipole moment.

The tetrachloromethane molecule is non-polar. Based on the electronegativity difference between carbon and chlorine, the carbon–chlorine bonds are polar. The resultant dipole moment is zero, which means that the dipoles must be oriented in such a way that they cancel each other. The tetrahedral arrangement of the four chlorine atoms around the central carbon atom provides the symmetrical distribution of bond dipoles that leads to this vectorial cancellation (Figure 4.46). Consequently, the tetrachloromethane molecule is non-polar.

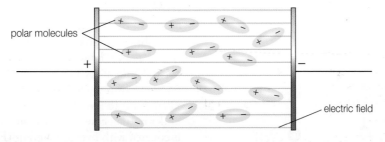

carbon dioxide   water   tetrachloromethane

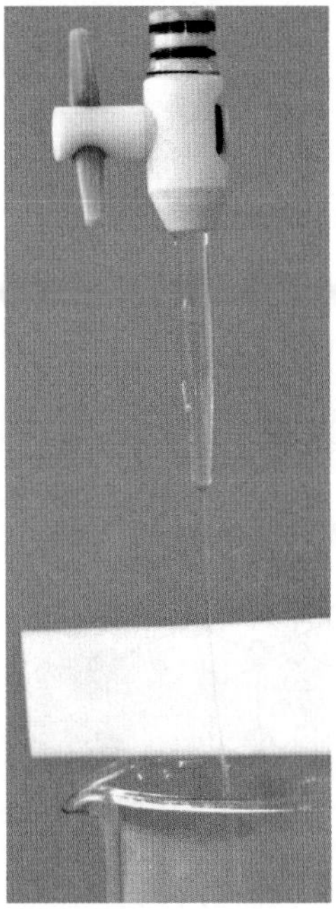

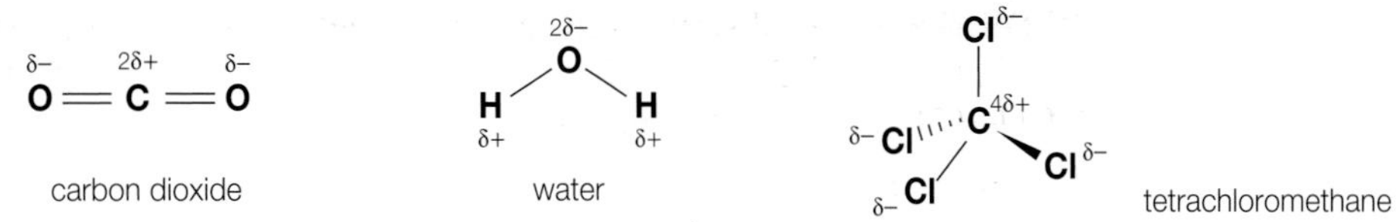

Figure 4.46 The polarity of carbon dioxide, water and tetrachloromethane molecules

One simple way to test whether a liquid is polar or non-polar is to use a charged rod, as shown in Figure 4.47. When a charged rod is brought close to the stream of a liquid running from the jet of a burette, a polar liquid will be deflected from its vertical path towards the charged rod but a non-polar liquid will not be affected. The greater the polarity of the liquid, the greater the deflection (under the same experimental conditions).

When polar molecules are placed in an electric field (Figure 4.48) the electrostatic forces will line up the molecules with the electric field. However, the order is disrupted by random movements due to the kinetic energy of the molecules.

Figure 4.47 The effect of a charged polyethene rod on water

polar molecules

electric field

Figure 4.48 Polar molecules in an electric field

Giant covalent lattices

4.2.9 **Describe** and compare the structure and bonding in the three allotropes of carbon (diamond, graphite and C_{60} fullerene).

Giant covalent lattices usually consist of a three-dimensional lattice of covalently bonded atoms. These atoms can be either all of the same type as in silicon and carbon (diamond and graphite), or of two different elements, such as silicon dioxide.

Diamond and graphite

Pure carbon exists in three allotropic forms: diamond (Figure 4.49), graphite and a family of related molecules known as the **fullerenes**. **Allotropes** are two (or more) crystalline forms of the same element, in which the atoms (or molecules) are bonded differently.

Figure 4.49 Model of a diamond lattice

The properties of diamond, graphite and the fullerene carbon-60 (C_{60}) are summarized in Table 4.12. The differences in physical properties are due to the large differences in the bonding between the carbon atoms in the three allotropes.

Allotrope	Diamond	Graphite	Carbon-60 (C_{60})
Colour	Colourless and transparent	Black and opaque	Black (in large quantities)
Hardness	Very hard	Very soft and slippery	Soft
Electrical conductivity	Very poor – a good insulator	Good – along the plane of the layers	Very poor – a good insulator
Density	$3.51\,g\,cm^{-3}$	$2.23\,g\,cm^{-3}$	$1.72\,g\,cm^{-3}$
Melting point/K	3823	Sublimation point 3925–3970	Sublimation point 800
Boiling point/K	5100	Sublimation point 3925–3970	Sublimation point 800

Table 4.12 The physical properties of diamond, graphite and carbon-60

In diamond (Figure 4.50), each carbon atom is tetrahedrally bonded to four other carbon atoms by single, localized covalent bonds. A very rigid three-dimensional network is formed. In diamond the bond angles are 109° and each carbon atom has a coordination number of four because there are four neighbouring carbon atoms near to it.

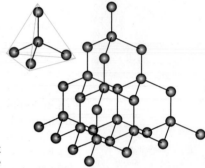

Figure 4.50 Structure of diamond: unit cell and lattice

■ Extension: Diamond

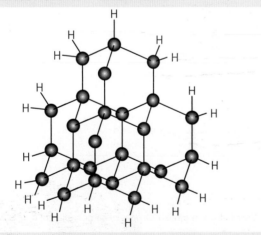

Figure 4.51 The hydrogen atoms on the edge of a diamond crystal

The carbon atoms inside a diamond crystal are all bonded to four other carbon atoms. However, those atoms on the flat surface of the side of a crystal only have three carbon atoms. They were predicted to have a spare valency. However, recently the surfaces of very clean diamond crystals have been studied using photoelectron spectroscopy and the scanning tunnelling microscope (STM) (Chapter 23). It was discovered that under normal conditions the surface of a diamond crystal is covered with hydrogen atoms (Figure 4.51). Diamond is therefore a hydrocarbon with a very high carbon to hydrogen ratio.

Applications of Chemistry

Naturally occurring diamonds are very poor conductors of electricity, but excellent conductors of heat owing to their ability to transmit atomic vibrations. The best conducting diamonds are 'isotopically pure' crystals composed almost totally of carbon-12. Carbon-13 is an impurity in naturally occurring diamonds and slows down the heat transfer. Pure crystals can conduct heat up to 50% more efficiently than natural diamond. The heat conducting properties of diamond crystals is used in lasers, communications and electronics. These three technologies require the removal of large amounts of unwanted heat.

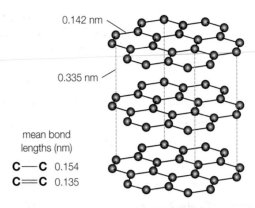

Figure 4.52 Sample of graphite crystals and a model of graphite   **Figure 4.53** Structure of graphite

In graphite, each carbon atom is covalently bonded to only three other carbon atoms. A two-dimensional network is formed consisting of hexagonal rings of carbon atoms. A graphite crystal (Figure 4.52) is composed of many layers of hexagonally arranged carbon atoms, stacked on top of one another (Figure 4.53). Each sheet can be regarded as a single molecule of carbon. There is no covalent bonding between the layers of carbon atoms, but extensive van der Waals' forces operate due to the relatively large surface area.

Each carbon atom has a spare electron which becomes delocalized along the plane, resulting in two-dimensional metallic bonding (Figure 4.54). The presence of delocalized electrons accounts for the ability of graphite to conduct electricity along the plane of the crystal when a voltage is applied.

Within the graphite layers, the carbon–carbon bond length is in between that of a single and a double carbon–carbon bond, suggesting there is a partial double bond character between carbon atoms in the layers.

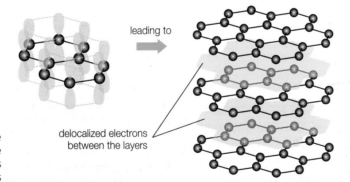

Figure 4.54 The structure of graphite showing the delocalization of electrons between the layers

Fullerenes

In 1985 a new allotropic form of pure carbon, known as carbon-60, was discovered. It is a simple molecular form of carbon and was first prepared by very rapidly condensing vapour consisting of carbon atoms, produced from graphite using a high power laser. The synthesis of carbon-60 was performed in an inert atmosphere of helium maintained at low pressure.

The atoms in a molecule of carbon-60 are arranged into the shape of a truncated icosahedron (Figure 4.55). This is the mathematical name given to the football or soccer ball. This structure has 60 vertices (corners) and 32 faces: 12 pentagons and 20 hexagons. The pentagons are 'isolated' – no pentagons are adjacent.

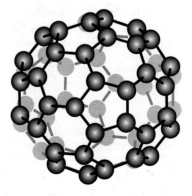

Figure 4.55 Structure of carbon-60 (C_{60})

Figure 4.57 The structure of carbon-60 (showing the alternating single and double carbon–carbon bonds)

a

b

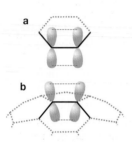

Figure 4.58 Simplified diagrams showing p orbital overlap in **a** benzene (planar) and **b** carbon-60 (non-planar)

Language of Chemistry

Carbon-60 was named buckminsterfullerene after the American engineer and architect Richard Buckminster Fuller who designed geodesic domes (Figure 4.56). A colloquial name for carbon-60 molecules is 'bucky balls'. ∎

Figure 4.56 Spaceship Earth (full sphere geodesic dome) (EPCOT Centre), Disneyland, Florida

The bonding in a molecule of carbon-60 is shown in Figure 4.57. This is a series of alternating carbon–carbon double and single bonds. This arrangement of bonds is known as a conjugated system (Chapter 21) and would be expected to give carbon-60 similar chemical properties to benzene (Chapter 27). However, the p orbital overlap inside and outside the curved surface is poor (Figure 4.58). Inside the orbital lobes are too close and repulsion occurs; outside the orbital lobes are too far away from each other for effective overlap.

The molecule's carbon–carbon double bonds therefore behave like those of an alkene and it undergoes a variety of addition reactions (Chapter 10), for example, bromination. Carbon-60 is stable in air, but is degraded by ultraviolet light and reacts with ozone. Freshly prepared samples dissolve in methylbenzene and other non-polar solvents to yield a purple solution.

Carbon-60 is the most abundant member of a family of related closed carbon cages (C_{32} upwards), known as **fullerenes**. Each of the fullerenes has an even number of carbon atoms and contains 12 five-membered rings and a variable number of six-membered rings. Related to the fullerenes are nanotubes which can be regarded as a sheet of rolled up graphite capped at each end by 'half' a fullerene (Chapter 23).

History of Chemistry

In September 1985 British chemist **Harold Kroto** (Figure 4.59) of the University of Sussex collaborated with Americans Richard E. Smalley, Robert F. Curl, James R. Heath and Sean C. O'Brien at Rice University in Houston, Texas in some experiments on graphite. Kroto had an interest in molecules found in interstellar space and had wanted to show that molecules containing long chains of carbon atoms could be formed under the conditions believed to be typical of the outer atmospheres of stars known as red giants. Smalley had developed a cluster beam apparatus which could vaporize small samples of solid graphite.

Using mass spectrometry (Chapter 2) they detected long chains of carbon atoms, but in addition all the mass spectra showed the presence of a stable C_{60} species. After building some models prompted by Kroto's recollection of the geodesic dome at Expo 67 in Montreal, they proposed the truncated icosahedron as its structure. In 1990 American physicist Donald Huffman and his German colleague Wolfgang Krätschmer found that soot enriched in carbon-60 could be readily formed by passing a large electric current through graphite rods in helium. The carbon-60 could be extracted from the soot by dissolving it in methylbenzene and then filtering to remove the insoluble soot.

Figure 4.59 Professor Sir Harry Kroto standing in front of a carbon-60 arc generator

Kroto, Smalley and Curl were awarded the Nobel Prize in Chemistry in 1996.

4.2.10 **Describe** the structure of and bonding in silicon and silicon dioxide.

Silicon dioxide

Silicon is another element which exists as a giant covalent structure. The most common form of silicon dioxide (silica) is quartz, which has a structure similar to diamond in which tetrahedral SiO_4 groups are bonded together by Si—O—Si bonds (Figure 4.60). Silicon dioxide has physical properties that are very similar to diamond. It is hard, transparent and has a high melting and boiling point. A common impure form of silicon dioxide is sand, which is coloured yellow by the presence of iron(III) oxide.

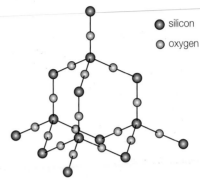

● silicon
○ oxygen

Figure 4.60 Structure of quartz (silicon dioxide)

Silicon

The element silicon has a giant covalent structure similar to that of diamond in which each silicon atom is bonded to four others by single covalent bonds. Silicon is less hard than diamond owing to the larger size of the silicon atoms, which results in longer and hence weaker bonds. Like diamond, silicon is an insulator, but can be made to conduct small electric currents after adding small amounts of other atoms in a process known as doping (Chapter 23).

Applications of Chemistry

The second hardest known substance, after diamond, is silicon carbide (SiC). It occurs very rarely in nature but can be synthesized by heating silicon dioxide and graphite in a furnace at a temperature of 2500 °C. Pure silicon carbide is colourless. Artificial silicon carbide is known commercially as carborundum and is used as an abrasive for cutting and polishing. It has a structure related to diamond.

4.3 Intermolecular forces

4.3.1 **Describe** the types of intermolecular forces (attractions between molecules that have temporary dipoles, permanent dipoles or hydrogen bonding) and explain how they arise from the structural features of molecules.
4.3.2 **Describe** and **explain** how intermolecular forces affect the boiling points of substances.

Van der Waals' forces

Substances composed of non-polar molecules, such as oxygen, carbon dioxide, nitrogen, the halogens and the noble gases, can all be liquefied and then solidified by cooling. This observation suggests that there are attractive forces operating between molecules and atoms in the liquid and solid states. These short-range attractive forces are known as **van der Waals' forces** and are due to the formation of temporary dipoles (Figure 4.61).

Temporary dipoles are caused by the temporary and random fluctuations in the electron density in a molecule or an atom. Over an averaged period of time, the electron density is spread evenly around the nucleus or nuclei. However, at a given instant, the electron density distribution may be asymmetrical, giving the atom or molecule a temporary dipole (Figure 4.62). A **dipole** is a separation of charge.

$\delta+$ $\delta-$

CI   CI

More of the electron cloud is at one end of the molecule; the Cl_2 molecule has an instantaneous dipole.

Figure 4.62 The formation of temporary dipole in a chlorine molecule

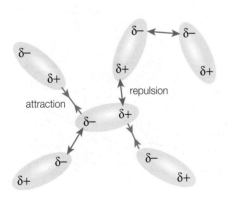

Figure 4.61 Attraction between molecules with dipoles

The formation of a temporary dipole in one atom or molecule causes electrons in a neighbouring atom or molecule to be displaced, resulting in the formation of another temporary dipole. This process is termed induction and the newly formed dipole an induced dipole. The formation of induced dipoles is rapidly transmitted through the liquid or solid. The forces of attraction between temporary or induced dipoles (Figure 4.63) are known as van der Waals' forces.

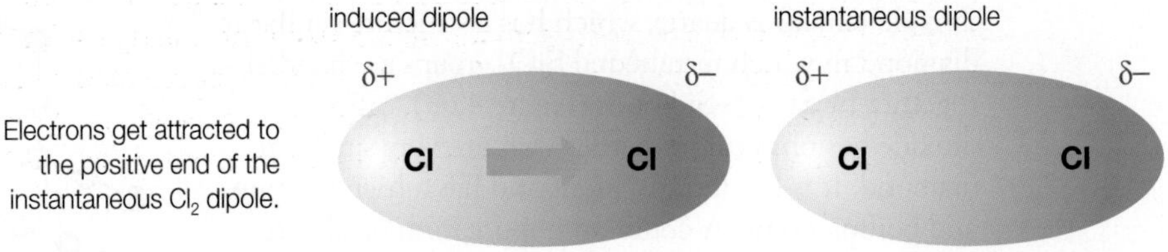

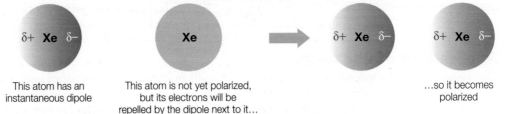

Figure 4.63 An instantaneous dipole-induced dipole attraction

Van der Waals' forces are generally small, but the exact size or extent depends on the polarizability of the atom. This is a measure of the ease with which the electron density of an atom or molecule can be distorted by an electric field. The larger the molecule or atom, the greater the volume occupied by the electrons and the greater the polarization and the larger the size of the temporary or induced dipole (Figure 4.64). Hence, van der Waals' forces increase with relative atomic or relative molecular mass.

Figure 4.64 The polarization of xenon atoms leads to production of induced dipole forces between atoms

This atom has an instantaneous dipole

This atom is not yet polarized, but its electrons will be repelled by the dipole next to it...

...so it becomes polarized

Factors which influence van der Waals' forces

The strength of van der Waals' forces is influenced by two factors:

- molecular size
- molecular shape.

Table 4.13 shows the boiling points of the halogens. The molecules increase in size and contain a greater number of electrons as group 7 is descended. In addition, the electrons become located further away from the nucleus and hence are less strongly attracted. Consequently, the electron cloud can be distorted increasingly easily. In other words, the polarizability of the bond increases, together with the size of the induced dipole. This results in stronger and more extensive van der Waals' forces. Therefore, the boiling points increase from fluorine to bromine. A similar trend is observed in the noble gases (Figure 4.65).

Note that it is the number of electrons that govern van der Waals' forces, not the force of gravity, which is negligible.

Molecule	Boiling point/K	Molar mass/g mol^{-1}
F_2	85	38
Cl_2	239	71
Br_2	332	160

Table 4.13 The boiling points of halogens

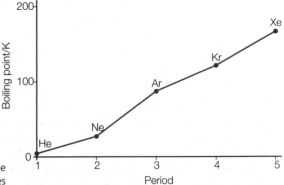

Figure 4.65 The boiling points of the noble gases

Molecular shape is also important in determining the size or extent of van der Waals' forces of attraction. Molecules with a large surface area allow a closer contact between molecules. This gives rise to greater or more extensive van der Waals' forces of attraction than in molecules of similar molecular mass but with more compact shapes due to branching.

For example, pentane has a higher boiling point than its isomer 2,2-dimethylpropane. Both molecules have the same molecular formulas, but different structures (Chapter 10). The van der Waals' forces between pentane molecules in the liquid or solid states are stronger because the linear molecules have a larger surface area for interaction. In contrast, its isomer, 2,2-dimethylpropane, is more compact (adopting a roughly spherical shape) owing to its extensive branching, and hence has a smaller surface area for interaction (Figure 4.66).

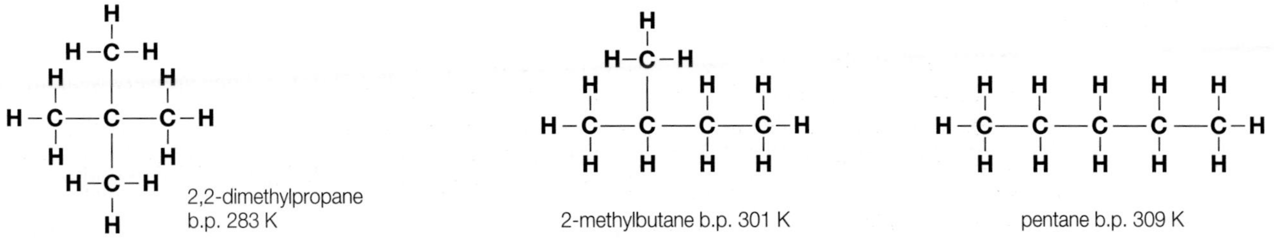

2,2-dimethylpropane
b.p. 283 K

2-methylbutane b.p. 301 K

pentane b.p. 309 K

Figure 4.66 Branched chain alkanes have lower boiling points than straight chain isomers

In comparing the relative strengths of intermolecular forces, the following generalizations are useful:

- When molecules have very different molecular masses, van der Waals' forces are more significant than dipole–dipole forces. The molecule with the largest relative molecular mass has the strongest intermolecular attractions.
- When molecules have similar molecular masses, dipole–dipole forces are more significant. The most polar molecule has the strongest intermolecular attractions.

Table 4.14 shows the increase in melting and boiling points of the hydrogen halides from hydrogen chloride to hydrogen iodide. All the molecules have a linear shape.

Table 4.14 The melting and boiling points of the hydrogen halides

Molecule	Molecular mass	Melting point/K	Boiling point/K
Hydrogen chloride, HCl	36.5	159	188
Hydrogen bromide, HBr	81.0	186	207
Hydrogen iodide, HI	128.0	222	238

The hydrogen chloride molecule is the most polar molecule since chlorine is the most electronegative of the three halogen atoms considered. However, hydrogen chloride has the lowest boiling and melting points of these three hydrogen halides. The data shows that the intermolecular forces of attraction are strongest in hydrogen iodide molecules. Thus the influence of van der Waals' forces is more significant than dipole–dipole forces when comparing molecules of very different molecular masses.

Van der Waals' forces are responsible for the soft and slippery properties of graphite (page 118). Van der Waals' forces of attraction also account for the deviations of the noble gases and the halogens from ideal gas behaviour (Chapter 1). They are also partly responsible for the solubility of covalent compounds, especially organic compounds, in organic solvents (Section 4.5).

History of Chemistry

Johannes Diderik van der Waals (1837–1923) was a Dutch scientist who is best known for his research into establishing the relationship between pressure, volume and temperature of fluids (gases and liquids). He modified the ideal gas equation to generate the van der Waals equation, which took into account that molecules are not point masses and do attract each other. He initially worked as a school teacher but later was appointed Professor of Physics at the University of Amsterdam. He was the first scientist to propose the concept of intermolecular forces between atoms and non-polar molecules. He was awarded the Nobel Prize in Physics in 1910.

Dipole–dipole forces

A molecule that contains polar bonds may be polar or non-polar, depending on the shape of the molecule. In a polar molecule, the molecules have permanent dipole moments. A dipole–dipole force exists between polar molecules because the positive end of the dipole of one molecule will electrostatically attract the negative end of the dipole of another molecule (Figure 4.67).

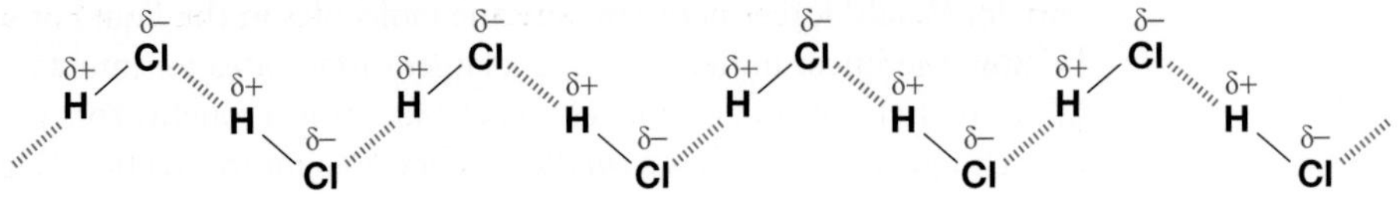

Figure 4.67 Dipole–dipole forces in solid hydrogen chloride

Dipole–dipole forces are often called permanent dipole–dipole forces because they only occur between molecules with permanent dipole moments. Dipole–dipole forces are only effective when the polar molecules are very close together, in the solid and liquid states. They are very weak in comparison to ionic or covalent bonds.

The strength of a dipole–dipole force depends on the size of the dipole moment of the molecule involved. The larger the dipole moment, the more polar the molecules of the substance and the greater the strength of the dipole–dipole force. For polar substances with similar relative molecular masses, the higher the dipole moment, the stronger the dipole–dipole attractions and the higher the boiling point, as shown in Table 4.15.

Name of substance	Formula	Relative molecular mass	Dipole moment/D	Boiling point/K
Propane	$CH_3CH_2CH_3$	44	0.1	231
Methoxymethane	CH_3OCH_3	46	1.3	249
Ethanenitrile	CH_3CN	41	3.9	355

Table 4.15 Dipole moments and boiling points for molecules having similar relative molecular masses

Hydrogen bonding

If two molecules of hydrogen fluoride are close to one another, the hydrogen atom of one molecule will be attracted to the fluorine atom of the other molecule (Figure 4.68). This occurs because of the electrostatic attraction between the partial positive charge on the hydrogen atom and the partial negative charge on the fluorine atom. This charge separation or dipole exists because fluorine is more electronegative than hydrogen. The electrostatic attraction that holds the hydrogen atom of one molecule to the fluorine atom of another molecule is an example of a **hydrogen bond.** Hydrogen bonds are often represented by long dotted (or dashed) lines, as shown in Figure 4.68. In the solid and liquid states, hydrogen fluoride consists of zigzag chains of hydrogen fluoride molecules. The neighbouring hydrogen fluoride molecules are held together by hydrogen bonds.

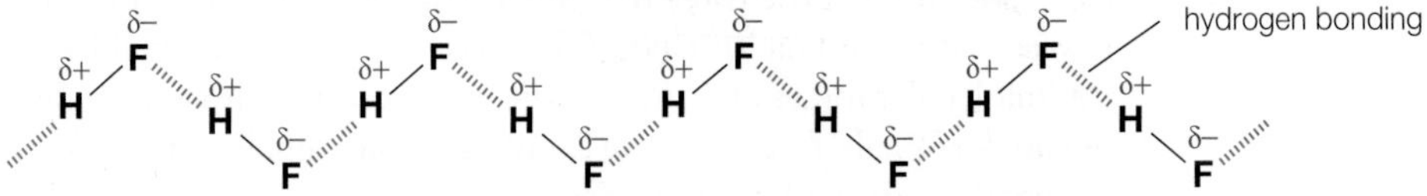

Figure 4.68 Hydrogen bonding in liquid hydrogen fluoride

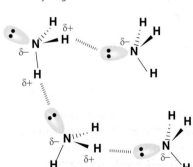

Figure 4.69 Hydrogen bonding in ammonia

Hydrogen bonding is a strong permanent dipole–dipole attraction between molecules which contain a hydrogen atom covalently bonded to a fluorine, oxygen or nitrogen atom. These three atoms are small and highly electronegative. The essential requirements for the formation of a hydrogen bond are a hydrogen atom directly attached to oxygen, nitrogen or fluorine and a lone pair of electrons on the electronegative atom.

In the ammonia molecule, the nitrogen atom has one lone pair of electrons. This means that each ammonia molecule can form one hydrogen bond (Figure 4.69). Nitrogen is larger and less electronegative than fluorine and hence the resulting hydrogen bonding in ammonia is weaker than the hydrogen bond formed by hydrogen fluoride.

Each water molecule has two lone pairs which can form hydrogen bonds with two other water molecules. This helps to explain the three-dimensional lattice structure in ice.

If we consider the overall effect of the hydrogen bonds in water and hydrogen fluoride, the collective strength of the hydrogen bonds in water is greater than the strength of the hydrogen bonds in hydrogen fluoride. This is because each oxygen atom (with two lone pairs) in the water molecule can form two hydrogen bonds with two other water molecules, whereas each fluorine atom in the hydrogen fluoride molecule can form only one hydrogen bond with another hydrogen fluoride molecule.

History of Chemistry

Fritz London (1900–1954) was a German-born American theoretical physicist. He made fundamental contributions to the theories of chemical bonding and intermolecular forces. Van der Waals' forces are often termed London dispersion forces, or simply London forces. His early work, in conjunction with Heitler, showed how quantum mechanics could be used to explain the formation of the hydrogen molecule. He was also interested in the properties of liquid helium and superconductors: conductors which have virtually no resistance.

Effects of hydrogen bonding on physical properties

Hydrogen bonding affects:

- the boiling points of water, ammonia, hydrogen fluoride and other molecules
- the solubility of simple covalent molecules such as ammonia, methanol and ethanoic acid in water
- the density of water and ice
- the viscosity of liquids, for example, the alcohols.

Effect of hydrogen bonding on boiling point

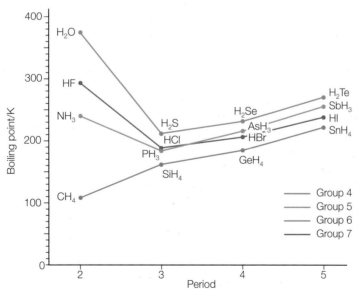

Figure 4.70 The boiling points of the hydrides of elements in groups 4 to 7

Figure 4.70 compares the boiling points of the hydrides of elements in groups 4 to 7 and allows the following conclusions to be drawn:

- The hydrides of group 4 elements (methane, CH_4, silane, SiH_4, germane, GeH_4 and stannane, SnH_4) display 'normal' behaviour, that is, the boiling points increase regularly when the relative molecular mass increases. This is because the van der Waals' forces of attraction increase as the molecular size increases.
- With the exception of ammonia, water and hydrogen fluoride, hydrogen bonds are not present in the hydrides of elements in groups 5, 6 and 7. The increase in the boiling points for the

hydrides of each periodic group is therefore due to the increase in van der Waals' forces as the molecular size increases.

- The boiling points of ammonia, water and hydrogen fluoride are anomalously high compared to those of the hydrides of other elements in groups 5, 6 and 7 of the periodic table. This is evidence for the existence of hydrogen bonds which are appreciably stronger than the van der Waals' forces that exist between molecules.

Effect of hydrogen bonding on the solubility of simple covalent compounds

Water is a good solvent for liquids and gases consisting of small polar molecules that can form hydrogen bonds with water molecules. For example, ammonia is a simple covalent compound. In general, simple covalent compounds are insoluble in water. Ammonia and amines are soluble in water because ammonia and amine molecules can form hydrogen bonds with water molecules. Similarly, alcohols and carboxylic acids are soluble in water because the alcohol groups in the organic compounds can form hydrogen bonds with water molecules.

However, not all organic compounds that contain primary amine groups, $-NH_2$, or alcohol groups, $-OH$, are totally soluble in water. As the relative molecular mass increases, the polar amine or alcohol functional group represents a progressively smaller portion of the molecule, but the hydrocarbon-based portion of the molecule becomes increasingly larger. Since hydrocarbons are insoluble in water, the solubility decreases as the relative molecular mass of the amine, phenol or alcohol increases (Figure 4.71).

Phenol, C_6H_5OH, is slightly soluble in water and slightly soluble in non-polar solvents at room temperature. Phenol has a polar hydroxyl group, $-OH$, which allows it to hydrogen bond to water molecules when dissolved in water. However, the solubility is limited by the bulky non-polar benzene ring which interacts with molecules of non-polar solvents, for example, benzene.

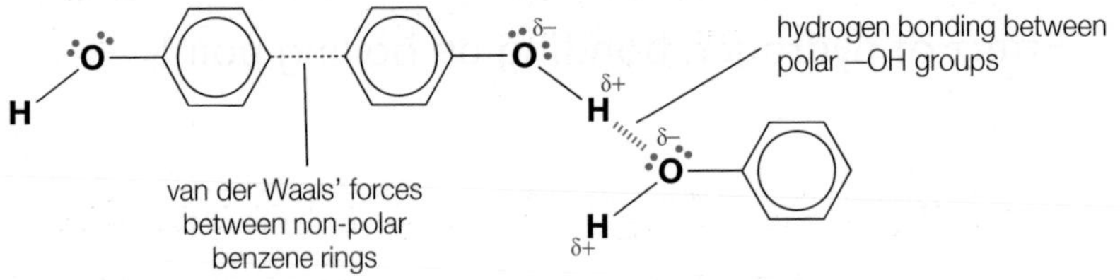

Figure 4.71 Accounting for the solubility of phenol in polar and non-polar solvents

Hydrogen bonding is responsible for many of the characteristic properties of water and ice. For example, hydrogen bonding is responsible for the high boiling point of water and the low density of ice. Hydrogen bonding is also responsible for the high surface tension of water (Figure 4.72). Surface tension – the 'skin' on the surface of water – arises because the molecules on the surface are pulled in to the bulk of the liquid strongly but there is no force pulling them out so the surface tends to adopt a minimum area.

Figure 4.72 Through hydrogen bonding, molecules on the surface of water form *very temporary* hexagonal arrays which are responsible for the high surface tension

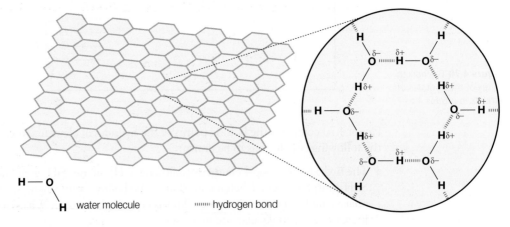

Hydrogen bonding is responsible for the unusual relative densities of water and ice. Water is one of the few substances that is less dense as a solid than it is as a liquid. In ice, each water molecule is surrounded tetrahedrally by four other water molecules joined by intermolecular hydrogen bonding. The water molecules arrange themselves into a lattice (similar to that of diamond) to *maximize* the number of hydrogen bonds and hence *minimize* the energy. The lattice (Figure 4.73) has a relatively large amount of space between the molecules. The 'open' structure of ice accounts for the fact that ice is less dense than water at 0 °C.

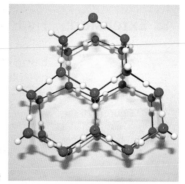

Figure 4.73 The structure of ice

■ Extension: The anomalous expansion of water

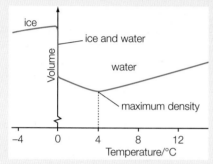

Figure 4.74 The volume of a fixed mass of water versus temperature

When ice melts, some of the hydrogen bonds are broken which allows some of the water molecules to be packed closer together, resulting in a decrease in volume. Hence, water has a higher density than ice. With further heating, additional hydrogen bonds are broken as more free water molecules are produced, so that just above the melting point, the density of water increases with temperature. At the same time, water expands as it is heated (due to increased molecular movement), which causes its density to decrease. These two processes – the reduction in volume due to melting and the thermal expansion – act in opposite directions. From 0 to 4 °C, the process of reduction in volume predominates, and water becomes progressively denser. Beyond 4 °C, the thermal expansion predominates, and the density of water decreases with increasing temperature. Figure 4.74 shows the variation in density with temperature for water.

Hydrogen bonding in biological molecules

Proteins are polymers with long chains consisting of repeating units with the structure —RCCONH—. Hydrogen bonding can occur between the carbonyl groups (>C=O) and the amine groups (H—N<). Hydrogen bonding can occur between the chains or within the chain to form secondary structures known as the β-sheet and the α-helix (Chapter 22).

In chromosomal DNA the four bases form pairs (via hydrogen bonding – Figure 4.75) with specific partners, allowing DNA replication to occur. The hydrogen bonds are relatively weak and can be broken by appropriate enzymes. Thymine always hydrogen bonds to adenine; cytosine always hydrogen bonds to guanine. These are known as base pairs (Chapter 22).

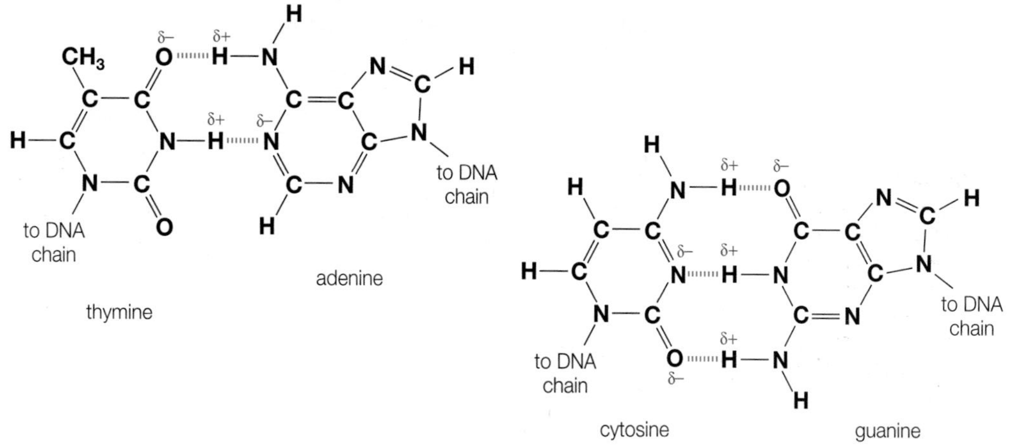

Figure 4.75 Base pairing in DNA

Hydrogen bonding in ethanoic acid

When dissolved in non-polar solvents or heated to just above its boiling point, ethanoic acid has an apparent molar mass of $120\,g\,mol^{-1}$ instead of the $60\,g\,mol^{-1}$ that would be expected from its molecular formula of CH_3COOH. These observations can be accounted for by the formation of a hydrogen-bonded dimer (Figure 4.76).

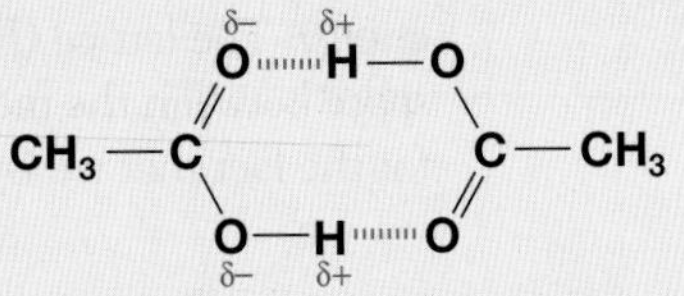

Figure 4.76 The formation of a hydrogen-bonded dimer of ethanoic acid

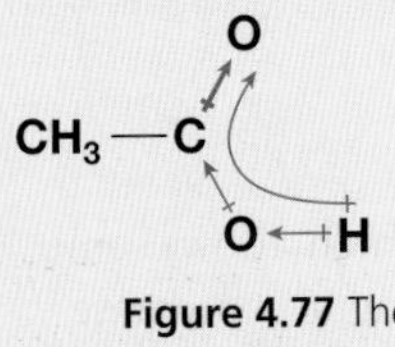

$$2\ \textbf{CH}_3\textbf{COOH} \longrightarrow (\textbf{CH}_3\textbf{COOH})_2$$

M_r 60   120

Figure 4.77 The effect of introducing a carbonyl group into an alcohol molecule

When dissolved in water, such an association cannot occur because the ethanoic acid molecules will be hydrogen bonded to the water molecules instead. At high temperature (well above its boiling point), the dimer will dissociate into individual ethanoic molecules.

The hydrogen bonding between the ethanoic molecules is relatively strong owing to the electron-withdrawing effect (Figure 4.77) of the carbonyl group, which increases the size of the positive charge on the hydrogen atom of the —OH group. This effect is not present in alcohols and hence carboxylic acids form stronger hydrogen bonds compared to alcohols (of similar molar mass).

Intramolecular hydrogen bonding

Hydrogen bonding that occurs between atoms of the same molecule is termed *intramolecular* hydrogen bonding. This is exemplified by 2-nitrophenol. In 2-nitrophenol, the hydrogen atom of the alcohol group, —OH, can form an intramolecular hydrogen bond with the oxygen atom of the nitro group, —NO_2 (Figure 4.78).

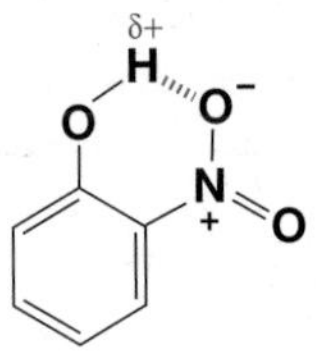

In 4-nitrophenol, the hydrogen atom of the alcohol group, —OH, cannot form an intramolecular hydrogen bond with the oxygen atom in the nitro group, —NO_2, because they are too far apart. In contrast, 4-nitrophenol forms *intermolecular* hydrogen bonds (Figure 4.79). Consequently, the melting point of 4-nitrophenol is higher than that of 2-nitrophenol, which is mainly associated via van der Waals' forces.

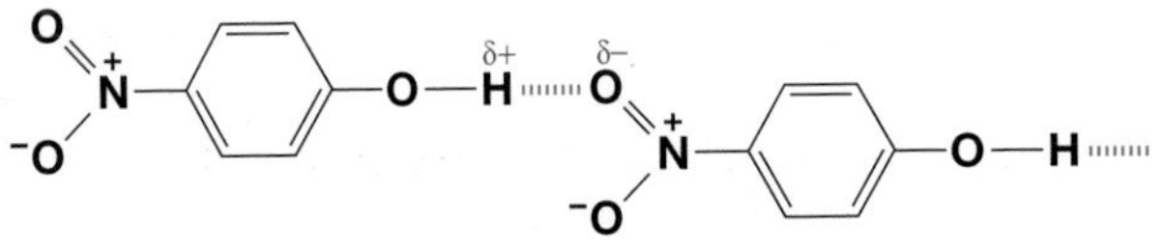

Figure 4.78 Hydrogen bonding in 2-nitrophenol

Figure 4.79 Hydrogen bonding in 4-nitrophenol

4.4 Metallic bonding

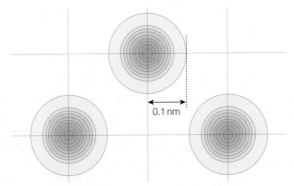

0.1 nm

In metals, the valence electrons are no longer associated with a particular metal atom, but are free to move throughout the metal. The mobile valence electrons are described as **delocalized**. Hence, the metal atoms are effectively ionized. This description is confirmed by X-ray analysis (Figure 4.80) which shows that metal crystals are composed of positive ions (cations) surrounded by the delocalized valence electrons.

Figure 4.80 An electron density map of aluminium

Metallic bonding (Figure 4.81) is the electrostatic attraction between the metal ions and the delocalized electrons. Metallic bonding is *non-directional*: all of the valence electrons are attracted to the nuclei of all the metal ions. The presence of delocalized electrons accounts for the physical properties of metals. This model of metallic bonding is often known as the electron-sea model, where the delocalized electrons form the 'sea'.

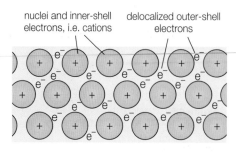

Figure 4.81 The structure of a metallic lattice

4.4.2 **Explain** the electrical conductivity and malleability of metals.

Explaining the physical properties of metals

Metals are **ductile** and **malleable** and are excellent conductors of heat and electricity. In a metal, the valence electrons do not belong to any particular atom. Hence, if sufficient force is applied to the metal, one layer of metal atoms can slide over another without disrupting the metallic bonding (Figure 4.82). The metallic bonding in a metal is strong and flexible and so metals can be hammered into thin sheets (malleability) or drawn into long wires (ductility) without breaking.

However, if atoms of other elements are added by alloying, the layers of ions will not slide over each other so readily. The alloy is thus less malleable and ductile and consequently harder and stronger (Chapter 23).

Figure 4.82 The application of a shear force to a metal lattice: adjacent layers can slide over each other

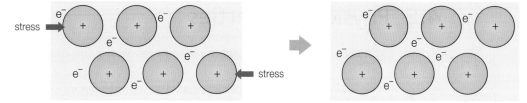

When a voltage (potential difference) is applied across a metal the delocalized electrons are repelled from the negative electrode and move towards the positive electrode. This orderly flow of electrons constitutes an electric current.

The delocalized electrons can also conduct heat by carrying kinetic energy (in the form of vibrations) from a hot part of the metal lattice to a colder part of the lattice. The presence of delocalized electrons in metals accounts for the thermal and electrical conductivity of metals.

■ **Extension:** ## The melting points of metals

The melting point is an approximate measure of the strength of the metallic bonding in a metal lattice. The higher the melting point, the stronger the bonding.

The delocalized electrons in the valence shell of the metal atoms are responsible for the metallic bond, and therefore for its strength. The other factor controlling the strength of metallic bonding is the size of the metal ion. The smaller the ionic radius, the stronger the metallic bonding. Thus:

$$\text{strength of metallic bond} \propto \frac{\text{number of valence electrons per atom}}{\text{metallic radius}}$$

The strength of metallic bonding therefore increases in period 3 from sodium, through magnesium to aluminium (Figure 4.83) as the number of valence electrons per atom increases from one, to two and then three. The ionic radius decreases from sodium to aluminium, as the large increase in nuclear charge outweighs the small increase in shielding (Chapter 3). Consequently, the melting points increase from sodium to aluminium (Table 4.16). The same trend is exhibited by the metallic elements in period 2.

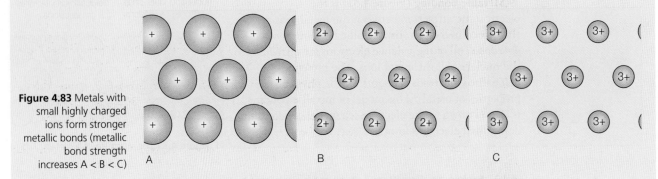

Figure 4.83 Metals with small highly charged ions form stronger metallic bonds (metallic bond strength increases A < B < C)

Element	Sodium	Magnesium	Aluminium
Melting point/K	371	922	936
Boiling point/K	1156	1363	2740
Metallic radius (10^{-12} m)	98	65	45

Table 4.16 Melting and boiling points of period 3 metals

The boiling points of metals are considerably higher than their melting points. This implies that most of the metallic bonding still exists in the liquid state. However, when the liquid changes into a gas (vapour), the atoms must be separated to large distances which involves breaking the metallic bonds.

4.5 Physical properties

4.5.1 **Compare** and explain the properties of substances resulting from different types of bonding.

Table 4.17 compares the properties of solid crystalline substances with different types of bonding.

Type of bonding:	Metallic	Giant covalent	Simple molecular	Ionic
Examples	Sodium, aluminium, iron, mercury and brass (copper and zinc)	Diamond, polyethene, nylon, silicon dioxide and graphite	Iodine, methane, hydrogen chloride, water, benzene carboxylic acid, ethanol, ammonia and fullerenes	Sodium chloride, magnesium oxide, calcium fluoride and sodium carbonate
Composition	Metal atoms	Non-metallic atoms	Molecules	Ions
Nature of bonding	Cations attracted to delocalized valence electrons	Atoms bonded by strong covalent bonds	Covalently bonded molecules held together by weak intermolecular forces	Strong electrostatic attraction between oppositely charged ions
Physical state at room temperature and pressure	Solids (except mercury)	Solids	Gases, liquids and solids	Solids
Hardness	Usually hard (but group 1 metals are soft)	Extremely hard	Soft (if solids)	Hard and brittle; undergo cleavage
Melting point	Usually high (except group 1 and mercury)	Very high	Very low or low	High
Electrical conductivity (in molten state)	Conductor	Non-conductors (except graphite)	Usually non-conductors	Conductor
Solubility	Nil, but dissolve in other metals to form alloys	Totally insoluble in all solvents	Usually soluble in non-polar solvents; usually less soluble in polar solvents	Usually soluble in polar solvents; insoluble in non-polar solvents

Table 4.17 Comparing and contrasting the properties of solid crystalline substances

Solubility

Solubility of solids

When a solution is formed, the particles from the simple molecular solid mix freely with those from the liquid. The process of dissolution may be thought of as occurring in three stages.

1 The solid's lattice must be broken. This process will be endothermic, that is heat is absorbed.
2 The intermolecular forces in the liquid, whether van der Waals' forces, hydrogen bonds or dipole–dipole attractions, must be disrupted to some extent. Again, this is an endothermic process, as attractive forces are being broken and this requires energy.
3 New bonds are formed between the molecules in the solid and the liquid. This is an exothermic process.

Generally, a solid is more likely to dissolve in a liquid if the overall enthalpy change is exothermic. High solubility is therefore *more likely* if:

strength of the attraction between the molecular solid and liquid molecules in the solution	>	combined strengths of the attractions between molecules in the pure solid and between molecules in the pure liquid

Although the thermodynamics of solubility are rather more complex than this (involving a consideration of *entropy changes* as described in Chapter 15), this simple 'rule of thumb' often helps us to account for patterns in solubility. It can be summarized in the phrase 'like dissolves like'.

Worked example

Account for the observation that iodine is soluble in hexane but not in water (Figure 4.84).

Iodine and hexane are non-polar substances. When mixed together a solution is formed because:

strength of the iodine/ hexane attraction in solution (van der Waals' forces)	>	combined strengths of the attractions in iodine solid (van der Waals' forces) and hexane liquid (van der Waals' forces)

Figure 4.84 Iodine introduced to hexane and water

Also, water is a polar solvent, with its molecules forming hydrogen bonds. When mixed with non-polar iodine molecules, nearly all of the water molecules continue to hydrogen bond with each other. Thus, the resulting iodine/water attractions are extremely weak in comparison to the combined strength of the hydrogen bonds in water and the van der Waals' forces in iodine. Consequently, iodine is virtually insoluble in water.

Solubility of liquids

The dissolving of one liquid in another may be explained in a similar way to the dissolving of a solid in a liquid. For example, water will mix with polar liquids such as ethanol (C_2H_5OH) and propanone ((CH_3)$_2$CO). The oppositely charged ends of the different molecules attract one another and hydrogen bonds are formed (Figure 4.85). The hydrogen bonds formed between the water and ethanol molecules are stronger than the hydrogen bonds formed between the molecules in the pure liquids. Entropy also plays an important role in determining the solubility of liquids in liquids.

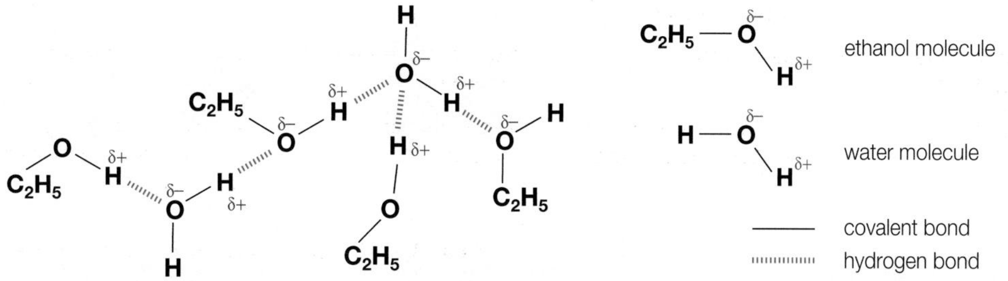

Figure 4.85 A mixture of water and ethanol

However, when water is added to an unreactive and non-polar liquid, such as tetrachloromethane, CCl_4, two layers separate out. The water molecules attract each other strongly, via hydrogen bond formation, but have no tendency to mix with the molecules of tetrachloromethane (Figure 4.86). It is not energetically favourable to replace the strong hydrogen bonds formed between water molecules with the weaker van der Waals' forces formed between water and tetrachloromethane molecules. The 'like dissolves like' principle holds.

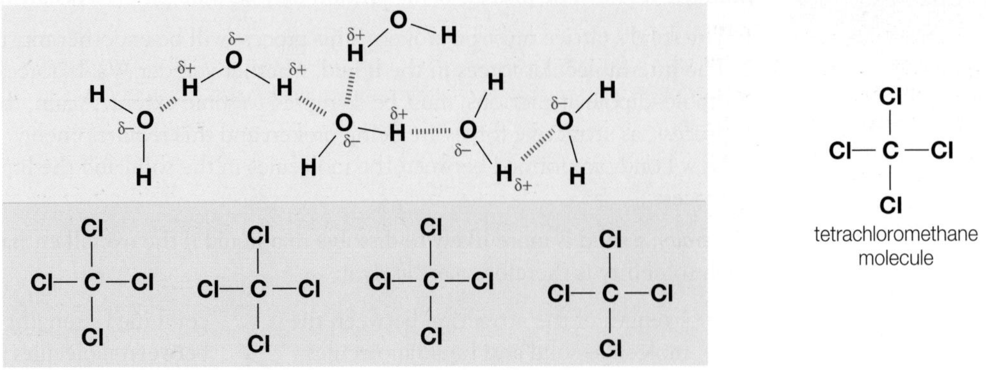

Figure 4.86 The interface between water and tetrachloromethane

Solubility of gases

Gases are generally only slightly soluble in water. Examples of gases in this category include oxygen (Chapter 25), hydrogen, nitrogen and the noble gases. A small number of gases are highly soluble in water because they react with water to release ions.

For example, sulfur dioxide, reacts with water to form a solution of hydrogen and hydrogen sulfate ions:

$$SO_2(g) + H_2O(g) \rightleftharpoons H^+(aq) + HSO_3^-(aq)$$

This solution is known as sulfurous acid and is a major component of acid rain (Chapters 3 and 25).

Hydrogen chloride reacts with water to form hydrochloric acid. The force of attraction between the negatively charged oxygen atoms of the water molecules and the positively charged hydrogen atoms of the hydrogen chloride molecules is sufficient that the hydrogen–chlorine bond is polarized and broken. The chlorine atom retains both electrons of this bond, while the oxygen atom of the water molecule uses one of its lone pairs to form a dative bond with the hydrogen ion (Figure 4.87). Virtually all of the hydrogen chloride molecules ionize in this way and therefore the hydrochloric acid solution is a strong acid (Chapter 8).

Figure 4.87 The reaction between water and hydrogen chloride molecules

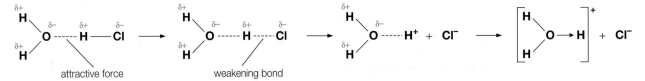

Physical properties of ionic compounds

The brittle nature of an ionic crystal, and its ability to be cleaved along planes, results from the mutual repulsion between layers when ions are displaced as seen in Figure 4.88.

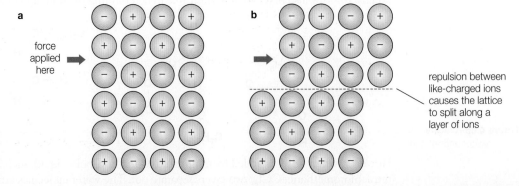

Figure 4.88 Cleavage in ionic solids: a layer of ions **a** before and **b** after cleavage

Figure 4.89 An impure sample of corundum: a very hard, crystalline form of the ionic compound aluminium oxide

The attraction between oppositely charged ions is much greater than the repulsive forces operating between ions of the same charge. Hence, the lattice is relatively strong and large amounts of heat energy are needed to break it up and separate the ions. Ionic solids are therefore hard (Figure 4.89) and have high melting and boiling points. The strong electrostatic attractive forces operating between ions of opposite charge remain when the solid melts to form a liquid, so boiling points are relatively high.

Since the ions are held rigidly in the lattice by electrostatic forces of attraction, the solids cannot conduct electricity when a voltage is applied. However, when molten or dissolved in water to form an aqueous solution (if the compound is soluble), the lattice is broken up and the hydrated ions are free to move. When an ionic solid dissolves in water, the mobile water molecules interact with the ions on the surface of the lattice and bond to the ions (Figure 4.90). This process is known as **hydration** and the ions are said to be **hydrated**.

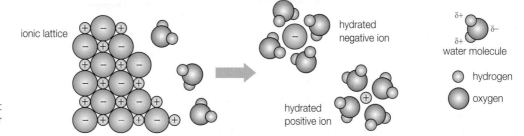

Figure 4.90 An ionic solid dissolving in water

■ Extension: Ion–dipole interactions

The bonds formed between the ions and water molecules are known as ion–dipole interactions. Energy is released during the hydration process and hence hydration is an exothermic process. The hydrated ions are no longer attracted to the oppositely charged ions so they enter the water. Eventually, provided water is present in significant excess, the lattice is completely broken down and all the ions are hydrated. The hydrated cations and anions are not attracted to each other owing to the presence of the water molecules.

There are two types of ion–dipole interactions:

■ **Ion–dipole bonds**
These are the electrostatic forces of attractions which exist between an ion and the oppositely charged region of a water molecule. Ion–dipole forces are formed between simple metal ions from groups 1 and 2 and anions.

■ **Dative covalent bonds**
If a metal ion has empty low energy 3d and 4s orbitals it can form dative covalent bonds with water molecules and a complex ion is formed. d-block metals and metals from period 3 onwards, for example lead ions, aluminium ions (Figure 4.91) and tin ions, can all form complex ions with water (Chapter 13). A lone pair of electrons on the oxygen atom is used to form a dative bond with the central metal ion.

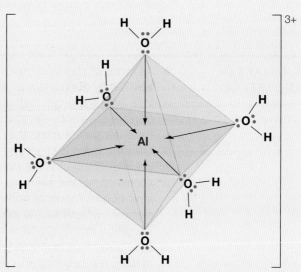

Figure 4.91 Structure of hexaaquaaluminium ion

Physical properties of simple molecular compounds

Simple molecular compounds are formed by the covalent bonding of a relatively small number of atoms. The bonds holding the atoms together in molecules are relatively *strong* covalent bonds. However, the molecules are associated together in the solid and liquid states by relatively *weak* intermolecular forces (Figure 4.92). Therefore, under standard conditions, simple molecular compounds are either gases or liquids or soft solids with low melting points. The melting points are low because of the weak intermolecular attractions that exist between molecules in the liquid and solid states.

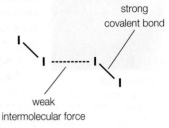

Figure 4.92 Strong covalent and weak intermolecular forces (van der Waals' forces)

Most simple covalent compounds whose intermolecular forces are van der Waals' forces, for example iodine (Figure 4.93) and the halogenoalkanes (Chapter 20), are poorly soluble in water, but are soluble in less polar or non-polar solvents. Simple covalent compounds whose intermolecular forces are hydrogen bonds are often soluble in water, for example amines, carboxylic acids, amides and sugars, provided they have relatively low molar mass or can form multiple hydrogen bonds.

Generally, simple molecular compounds do not conduct electricity when molten. This is because they do not contain ions but molecules. Molecules are electrically neutral and are not attracted to charged electrodes.

However, a number of simple molecular compounds are soluble in water and undergo a chemical reaction with water (hydrolysis) to release ions. The molecular substance is completely or partially converted into ions. Examples of such substances include chlorine, Cl_2, ammonia, NH_3, and hydrogen chloride, HCl. The first two reactions are equilibrium reactions, but the last reaction goes to completion.

$$Cl_2(g) + H_2O(l) \rightleftharpoons HCl(aq) + HOCl(aq)$$

$$NH_3(g) + H_2O(l) \rightleftharpoons NH_4^+(aq) + OH^-(aq)$$

$$HCl(g) + H_2O(l) \rightarrow H_3O^+(aq) + Cl^-(aq)$$

In the last reaction H_3O^+ is the oxonium ion, present in aqueous solutions of acid. It is formed when a water molecule forms a dative bond to a proton.

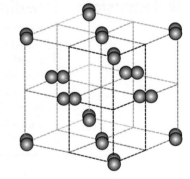

Figure 4.93 Structure of iodine: a simple molecular substance

---

SUMMARY OF KNOWLEDGE

- All forms of chemical bonding result from the electrostatic attraction of one atom's electron by the nucleus of another atom.
- Ionic bonding occurs between metal atoms and non-metal atoms and involves the transfer of valence electrons from the metal to the non-metal. Both atoms usually gain full outer shells (unless they are transition metals).
- The ionic bond is an electrostatic attraction between oppositely charged ions.
- In ionic compounds the cation and anion charges sum to zero.
- Atoms of metallic elements in groups 1, 2 and 3 lose 1, 2 and 3 electrons when forming ionic bonds. Atoms of non-metallic elements in groups 5, 6 and 7 will gain 3, 2 and 1 electrons when forming ionic bonds.
- Transition metals can form two or more stable simple ions.
- Ionic bonding is favoured when the metal is reactive (low electronegativity) and the non-metal is reactive (high electronegativity).
- Polyatomic ions contain two or more atoms covalently bonded together. Many polyatomic ions contain oxygen.
- An ionic lattice is a regular repeating arrangement of closely packed ions.

- A covalent bond is formed by the sharing of electron pair(s) between two atoms. In a dative covalent bond, the shared pair of electrons comes from only one of the atoms.
- A covalent bond can be described as the electrostatic attraction between a pair of shared electrons and the nuclei of the two atoms. Covalent bonds are directional.
- Covalent bonds have optimum or equilibrium lengths. Too far apart and the electrons cannot interact with each other's nuclei; too close together and the bond is destabilized by inter-nuclear repulsion.
- A single bond is composed of one shared pair of electrons; a double bond is composed of two shared pairs and a triple bond is composed of three shared pairs.
- The normal valency of an atom (in periods 2 and 3) is equal to the number of outer electrons if that number is four or less. Otherwise, the valency is equal to eight minus the number of outer electrons.
- Covalent bonds can be represented by lines in structural formulas of molecules.
- Pairs of electrons (bonding and non-bonding) can be represented in Lewis structures by dots and/or crosses or bars.
- Many atoms obey the octet rule and form stable molecules or ions where they have a full outer shell of eight electrons.
- In a dative covalent bond both electrons of a bonding pair are donated by the same atom.
- Covalent bond strength increases and bond length decreases with the number of shared electron pairs.
- Covalent bonds are formed when both atoms have similar values in electronegativity. Covalent bonding is favoured when both atoms are non-metals.
- In covalent compounds, the larger the energy of a bond, the more difficult it is to break that bond and the lower the reactivity of the compound.
- In covalent compounds, a difference in electronegativity between the bonded atoms makes the bond polar, giving a degree of ionic character.
- If two bonded atoms differ in electronegativity, the shared electron pair will be closer to the more electronegative atom. This results in the bond having a dipole: a separation of charge.
- Molecules can have dipole moments if their bond dipoles do not cancel out.
- Polar molecules in the liquid state are attracted towards a charged rod.
- Diamond has a three-dimensional giant covalent structure based upon a tetrahedral arrangement of carbon bonds. Graphite has a two-dimensional giant covalent layered structure based on a trigonal planar arrangement of carbon bonds. Carbon-60 has a simple molecular structure. Silicon and silicon dioxide have giant structures related to diamond.
- The shapes of covalent molecules and relative sizes of bond angles are explained by the valence shell electron pair repulsion theory (VSEPR): the shape of a molecule is determined by the number of regions of high electron density around the central atom of a molecule.
- The shape adopted by the molecule is that which minimizes the repulsion between the regions of high electron density.
- Regions of high electron density include: single bonds, multiple bonds and lone pairs. The order of repulsion is: lone pair-lone pair > lone pair-bonding pair > bonding pair-bonding pair.
- There are three main types of intermolecular forces: van der Waals' forces (due to the presence of temporary dipoles); dipole–dipole forces (due to the presence of permanent dipoles) and hydrogen bonds (due to the presence of large permanent dipoles). Their strengths increase in this order (for molecules of similar molar mass and molecular shape). Intermolecular forces are generally much weaker than covalent bonds.
- Hydrogen bonds are a relatively strong, directional intermolecular force that occurs between molecules that have hydrogen covalently bonded to the small and electronegative atoms of the elements nitrogen, oxygen or fluorine. The hydrogen bonds formed can be regarded as a weak ionic bond between the hydrogen of one molecule and the nitrogen, oxygen or fluorine atom on another molecule.

■ A dipole–dipole force is the attraction between oppositely charged regions of polar molecules. Dipole–dipole forces involve permanent dipoles.

■ Van der Waals' forces of attraction arise due to the random movement of electrons which create temporary dipoles which induce dipoles in adjacent molecules. The result is a very weak force of attraction between now oppositely charged regions of the two molecules.

■ Van der Waals' forces depend on a molecule's size and shape. Van der Waals' forces are maximized in large molecules with linear shapes.

■ In metals, individual valence electrons separate from the individual atoms and become delocalized. Metallic bonding is the electrostatic attraction between the nuclei of metal ions and the delocalized valence electrons.

■ The strength of metallic bonding increases with the number of valence electrons and decreasing ionic radius.

■ The nature of metallic bonding accounts for the electrical conductivity of metals: the delocalized electrons flow when a voltage is applied.

■ The nature of metallic bonding accounts for the malleability of metals: the layers of ions slide over each other when a force is applied.

■ The physical properties of a compound or element are related to the type of bonds and/ or intermolecular forces present. The type of bonding in a compound or element can be deduced from its physical properties.

■ Substances with a simple molecular structure have low melting points, high volatility and are electrical insulators. Some simple molecular substances hydrolyse with water.

■ Substances with a giant molecular structure have very high melting points, low volatility, and are electrical insulators under all conditions.

■ Ionic substances have high melting points, are brittle (when stressed), and only conduct electricity when molten or in aqueous solution (if soluble).

■ Graphite is an electrical conductor due to the presence of delocalized electrons in its layers.

■ Hydration is the process by which water molecules penetrate a solid lattice and attach themselves to the particles (atoms, ions or molecules).

■ Solvent molecules can be attached to particles by (i) van der Waals' forces, (ii) dipole– dipole interactions, (iii) hydrogen bonds, (iv) ion–dipole forces or (v) dative covalent bonds.

■ *Examination questions – a selection*

Paper 1 IB questions and IB style questions

Q1 Which compound contains ionic bonds?
A magnesium chloride, $MgCl_2$
B dichloroethane, CH_2Cl_2
C ethanoic acid, CH_3COOH
D silicon tetrabromide, $SiBr_4$

Q2 When CH_4, NH_3, H_2O are arranged in order of **decreasing bond angle**, what is the correct order?
A CH_4, NH_3, H_2O
B NH_3, H_2O, CH_4
C NH_3, CH_4, H_2O
D H_2O, NH_3, CH_4

Q3 In which of the following pairs does the second substance have the lower boiling point?
A Cl_2, Br_2
B H_2O, H_2S
C C_3H_8, C_4H_{10}
D CH_3OCH_3, CH_3CH_2OH

Q4 A group 1 element, X, bonds with a group 7 element, Y. What is the most likely formula and type of bonding in this compound?
A XY_2 covalent
B XY ionic
C XY covalent
D X_2Y ionic

Q5 In which of the following is there at least one double bond?
- **I** O_2
- **II** CO_2
- **III** C_2F_4

A I only   **C** II and III only
B III only   **D** I, II and III

Q6 According to VSEPR theory, which molecule would be expected to have the smallest bond angle?
A H_2O **C** SiH_4
B H_2CO **D** NH_3

Q7 In which of the following substances would hydrogen bonding would be expected to occur?
- **I** C_2H_6
- **II** CH_3CH_2COOH
- **III** CH_3OCH_3

A II only   **C** II and III only
B I and III only   **D** I, II and III

Q8 Why is the boiling point of ethane greater than that of neon?
A The ethane molecule is polar.
B Hydrogen bonds form between ethane molecules but are not present in liquid neon.
C More electrons are present in ethane than in neon.
D A molecule of ethane has a greater mass than a neon atom.

Q9 Which molecule has the greatest polarity?
A fluorine
B hydrogen fluoride
C hydrogen iodide
D tetrafluoromethane

Q10 Which is the best description of metallic bonding?
A The attraction between positive and negative ions.
B The attraction between protons and electrons.
C The attraction between the nuclei of positive ions and delocalized valence electrons.
D The attraction between nuclei and shared electron pairs.

Q11 Which compound is the most soluble in water?
A ethane   **C** propan-1-ol
B propane   **D** hexan-1-ol

Q12 Which statement is **not** true about metallic bonding?
A It is present in mixtures of metals (alloys).
B It results from the transfer of electrons from metal atoms to non-metal atoms.
C It involves the delocalization of electrons.
D It is electrostatic in nature.

Q13 Element X is in group 3 and element Y is in group 6 of the periodic table. Which is the most likely formula of the compound formed when X and Y react together?
A XY   **B** X_2Y_3 **C** X_3Y_2 **D** XY_2

Q14 Which molecule contains a multiple bond?
A H_2 **B** H_2O **C** C_2F_4 **D** C_2F_6

Q15 Which is **not** present in $C_2H_5OC_2H_5$ in the liquid state?
A covalent bonding
B van der Waals' forces
C dipole–dipole attractions
D hydrogen bonding

Q16 Chlorine has a lower boiling point than bromine. Which property of the two elements is responsible for this observation?
A ionization energies
B bond enthalpies
C bond polarities
D number of electrons

Q17 Which of the following molecules is planar?
A NCl_3 **B** C_2H_4 **C** C_3H_6 **D** SF_6

Q18 A solid has a high melting point, does not conduct electricity as a solid, but does when it is dissolved in water. What type of substance is the solid?
A ionic   **C** giant molecular
B simple molecular   **D** metallic

Q19 When the Lewis structure for $HCOOCH_3$ is drawn, how many bonds and how many lone pairs of electrons are present?
A 8 and 4   **C** 5 and 5
B 7 and 5   **D** 7 and 4

Q20 The angle between the two carbon–carbon bonds in CH_3CHCF_2 is closest to:
A 180°   **B** 120°   **C** 109°   **D** 90°

Q21 The compounds X, Y, Z, have approximately the same molar mass.
- **X** C_5H_{12}
- **Y** $CH_3CH_2CH_2CH_2OH$
- **Z** $CH_3OCH_2CH_2CH_3$

When these compounds are arranged in order of increasing boiling point (lowest boiling point first), the correct order is:
A X, Z, Y   **C** Y, Z, X
B X, Y, Z   **D** Z, Y, X

Q22 What is the formula for the compound formed by strontium and nitrogen?
A SrN   **B** Sr_2N **C** Sr_2N_3 **D** Sr_3N_2

Q23 The molar masses of C_2H_6, CH_3OH and CH_3F are similar. How do their boiling points compare?
A $C_2H_6 < CH_3OH < CH_3F$
B $CH_3F < CH_3OH < C_2H_6$
C $CH_3OH < CH_3F < C_2H_6$
D $C_2H_6 < CH_3F < CH_3OH$

Q24 Which intermolecular forces exist in dry ice, $CO_2(s)$?
A dipole–dipole interactions
B covalent bonds
C van der Waals' forces
D hydrogen bonds

Q25 Which one of the following molecules would be expected to be linear?
A H_2O_2 **B** NO_2 **C** SO_3 **D** CO_2

Q26 Which of the compounds H_2O, H_2S, H_2Se and HCl has the highest boiling point?
A H_2O **B** H_2S **C** H_2Se **D** HCl

Q27 Which is an incorrect statement about carbon-60 (C_{60})?
A It is a giant molecular substance.
B It is a soft powder.
C The surface of its molecules is composed of rings of five and six carbon atoms.
D Van der Waals' forces of attraction hold the molecules in a lattice.

Q28 In general, the strengths of the following intermolecular forces and bonds increase in the order:
A covalent bonds, hydrogen bonds, van der Waals' forces
B covalent bonds, van der Waals' forces, hydrogen bonds
C hydrogen bonds, covalent bonds, van der Waals' forces
D van der Waals' forces, hydrogen bonds, covalent bonds

Q29 Given the following electronegativity values:
H: 2.1   N: 3.0   O: 3.5   F: 4.0
which bond has the greatest polarity?
A O–H in H_2O **C** N–O in NO_2
B N–F in NF_3 **D** N–H in NH_3

Q30 Which one of the following species has a triangular pyramidal geometry?
A BCl_3 **B** NCl_3 **C** H_2Se **D** C_2H_2

Paper 2 IB questions and IB style questions

Q1 Describe the variation in melting points and electrical conductivities of the elements sodium to argon, and explain these variations in terms of their structures and bonding.   [6]

Q2 a Draw electron dot structures for N_2 and F_2 and explain why F_2 is much more reactive than N_2.   [3]
b Compare the polarity of the bonds N–F and C–F. Are the molecules NF_3 and CF_4 polar or non-polar? In all your answers give your reasons.   [5]
Standard Level Paper 2, May 99, Q5

Q3 Explain at the molecular level why ethanol (C_2H_5OH) is soluble in water, but cholesterol ($C_{27}H_{45}OH$) and ethane (C_2H_6) are not.   [4]
Standard Level Paper 2, May 01, Q6

Q4 The elements potassium and fluorine and the compound potassium fluoride can be used to show the connection between bonding, structure and physical properties.
a Describe the type of bonding in potassium metal and explain why potassium is a good conductor of electricity.   [4]
b Draw a Lewis structure for fluorine. Name and describe the bonding within and between the molecules in liquid fluorine.   [4]
c Write the electronic structures of both potassium and fluorine and describe how the atoms combine to form potassium fluoride.   [4]
d Explain why potassium fluoride does not conduct electricity until it is heated above its melting point.   [1]

Q5 a The letters A, B, C and D represent four consecutive elements in the periodic table. The number of electrons in the highest occupied energy levels are:
A: 3   B: 4   C: 5   D: 6
Write the formula for:
i an ionic compound formed from A and C, showing the charges   [2]
ii a covalent compound containing B and D.   [1]
b State the number of protons, neutrons and electrons in the ion $^{12}_{6}C_{60}{}^{6-}$.   [2]
c State the type of bonding in the compound SiF_4. Draw the Lewis structure for this compound.   [3]
d Outline the principles of the valence shell electron pair repulsion (VSEPR) theory.   [3]
e i Use the VSEPR theory to predict and explain the shape and bond angle of each of the molecules SBr_2 and C_2Br_2.   [6]
ii Deduce whether or not each molecule is polar, giving a reason for your answer.   [3]

5 Energetics

STARTING POINTS

- Chemical reactions involve the transfer of energy in the form of heat and work.
- Chemical reactions are either exothermic (heat released) or endothermic (heat absorbed).
- Enthalpy (energy) changes, ΔH, are expressed in units of kilojoules per mole (kJ mol^{-1}).
- The standard enthalpy change, $\Delta H^{\ominus}$, is the heat energy transferred under standard conditions (1 atmosphere pressure and 298 K).
- Only enthalpy changes can be determined. The enthalpy, H, of a substance cannot be determined absolutely.
- Enthalpy (energy) changes are measured using calorimeters.
- Molecules contain potential energy (in their bonds) and kinetic energy (due to their movement).
- Average bond enthalpies (bond energies) can be used to calculate enthalpy changes.
- Average bond enthalpies can also be used to understand the structure and bonding in molecules and understand the mechanisms of chemical reactions.
- Chemical reactions obey the law of conservation of energy: energy cannot be created or destroyed.
- In a chemical reaction involving a number of steps the enthalpy changes of the individual steps can be summed to give the overall enthalpy change.

5.1 Exothermic and endothermic reactions

5.1.1 **Define** the terms *exothermic reaction*, *endothermic reaction* and *standard enthalpy change of reaction* ($\Delta H^{\ominus}$).
5.1.2 **State** that combustion and neutralization are exothermic processes.
5.1.3 **Apply** the relationship between temperature change, enthalpy change and the classification of a reaction as endothermic or exothermic.
5.1.4 **Deduce,** from an enthalpy level diagram, the relative stabilities of reactants and products and the sign of the enthalpy change for the reaction.

Enthalpy changes

Figure 5.1 A burning magnesium sparkler (an exothermic reaction)

Chemical reactions involve a transfer of energy. Chemical substances contain chemical energy, a form of potential energy. Many chemical reactions involve a transfer of chemical energy into heat.

For example, when methane (the major component of natural gas) burns in excess oxygen, chemical energy is transferred to the surroundings as heat. The products of this **combustion** reaction are water and carbon dioxide.

$$CH_4(g) + 2O_2(g) \rightarrow CO_2(g) + 2H_2O(l)$$

The majority of chemical reactions release heat energy to their surroundings (Figures 5.1 and 5.2). This type of reaction is known as an **exothermic reaction**. A few chemical reactions absorb heat energy from their surroundings: these reactions are known as **endothermic reactions**. An example of an endothermic reaction is the thermal decomposition of calcium carbonate to form calcium oxide and carbon dioxide.

Figure 5.2 Burning Camping Gaz (compressed butane)

$$CaCO_3(s) \rightarrow CaO(s) + CO_2(g)$$

When an exothermic reaction transfers heat energy to the surroundings the chemical reactants lose potential energy. The products have less potential energy than the reactants. This potential energy stored in the chemical bonds is known as **enthalpy** and is given the symbol H. The transfer of heat energy that occurs (*at constant pressure*) during a chemical reaction from the reaction mixture (known as the **system**) to the surroundings is known as the enthalpy change, ΔH, where the Greek letter delta means 'change in'. Enthalpy changes can be shown using **enthalpy level diagrams** (Figures 5.3 and 5.4).

Thermochemical equations are equations which show the associated enthalpy changes.

$$CH_4(g) + 2O_2(g) \rightarrow CO_2(g) + 2H_2O(l) \qquad \Delta H^\ominus = -890\,kJ\,mol^{-1}$$

$$CaCO_3(s) \rightarrow CaO(s) + CO_2(g) \qquad \Delta H^\ominus = +180\,kJ\,mol^{-1}$$

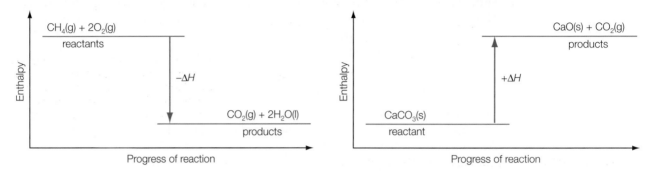

Figure 5.3 Enthalpy level diagram for an exothermic reaction – the combustion of methane

Figure 5.4 Enthalpy level diagram for an endothermic reaction – the thermal decomposition of calcium carbonate

For an exothermic reaction the enthalpy change, ΔH, is described as negative because potential energy has been lost from the reaction mixture to the surroundings, in the form of heat. The products are at a lower energy or enthalpy level than the reactants.

Endothermic reactions absorb heat energy from the surroundings. The products of an endothermic reaction contain more potential energy or enthalpy than the reactants. For this type of reaction the enthalpy change is positive, because the chemical reactants gain heat from their surroundings.

■ Extension: Activation energy

All chemical reactions, both endothermic and exothermic, have an activation energy barrier (Chapter 6). The activation energy of a reaction is usually denoted by E_a, and given in units of kilojoules per mole. The activation energy barrier controls the rate of the reaction: the smaller the value of the activation energy, the greater the rate of the reaction.

Figure 5.5 Matches and match box. The match head consists of potassium chlorate(v), sulfur and phosphorus trisulfide; the frictional heat when the match is struck against the side of the box is sufficient to give the reactants enough energy to overcome the activation barrier

Language of Chemistry

The negative and positive signs in enthalpy changes do *not* represent 'positive' and 'negative' energy. They simply indicate the direction of the flow of heat energy (Figure 5.6). The term exothermic is derived from the Greek word *exo*, meaning outside and *thermo* meaning heat. Endothermic is derived from the Greek word *endon*, meaning within. The term enthalpy is derived from the Greek word *thalpo*, to heat. ■

Enthalpy changes are usually measured in units of kilojoules per mole ($kJ\,mol^{-1}$). The values of the enthalpy change for a particular reaction will vary with the conditions, especially concentration of chemicals.

Hence, standard enthalpy changes, $\Delta H^{\ominus}$, are measured under standard conditions:

- a pressure of 1 atmosphere or 1.013×10^5 Pa
- a temperature of $25\,°C$ ($298\,K$)
- concentrations of $1\,mol\,dm^{-3}$
- the most thermodynamically stable allotrope (which in the case of carbon is graphite).

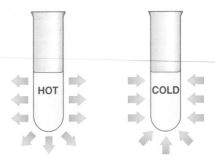

Exothermic reactions give out heat. This warms the mixture and then heat is lost to the surroundings

Endothermic reactions take in heat. This cools the mixture at first and then heat is gained from the surroundings

Figure 5.6 The directions of heat flow during exothermic and endothermic reactions

■ Extension: Alternative energy sources

Currently our main source of energy is the combustion of fossil fuels: coal, oil and natural gas (methane). They release large amounts of heat energy when burnt but they are a non-renewable source and the carbon dioxide they release contributes to the enhanced greenhouse effect and hence global warming (Chapter 25). Coal contains sulfur and sulfur compounds. When these are burnt, they release sulfur dioxide which causes acid rain (Chapters 3 and 25).

The issues of pollution, climate change and the increasing cost of fossil fuels has encouraged research into the use of alternative, renewable sources of energy. These include hydroelectric power, tidal and wave power, wind power, geothermal power, solar power (Figure 5.7) and nuclear power. Most of the alternative energy sources rely on the Sun. Some fuels, known as biofuels, can be made from renewable resources, for example ethanol can be made by fermentation. Hydrogen is another renewable energy resource and is pollution free since water is formed when it undergoes combustion. Hydrogen can also be used in fuel cells (Chapter 23). One drawback to the use of hydrogen is transport and safe storage, but it is hoped that the use of metal hydrides may overcome these problems.

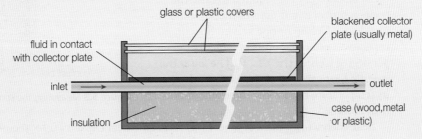

glass or plastic covers

fluid in contact with collector plate

inlet

insulation

blackened collector plate (usually metal)

outlet

case (wood, metal or plastic)

Figure 5.7 Structure of a solar energy panel

Further points about enthalpy changes

In a reversible reaction (Chapter 7), if the forward reaction is exothermic, then the reverse reaction is endothermic, for example:

$$N_2(g) + 3H_2(g) \rightleftharpoons 2NH_3(g) \qquad \Delta H^{\ominus} = -92\,kJ\,mol^{-1}$$
$$2NH_3(g) \rightleftharpoons N_2(g) + 3H_2(g) \qquad \Delta H^{\ominus} = +92\,kJ\,mol^{-1}$$

Similarly, if the forward reaction is endothermic, the reverse reaction is exothermic.

The enthalpy change depends on the amounts of reactants used. If the coefficients of the thermochemical equation are multiplied or divided by a common factor, the value of the enthalpy change is changed by the same factor. For example:

$$\text{if} \quad CO(g) + \tfrac{1}{2}O_2(g) \rightarrow CO_2(g) \qquad \Delta H^{\ominus} = -283\,kJ\,mol^{-1}$$
$$\text{then} \quad 2CO(g) + O_2(g) \rightarrow 2CO_2(g) \qquad \Delta H^{\ominus} = (2 \times -283) = -566\,kJ\,mol^{-1}$$

Thermochemical equations are often manipulated according to these rules when solving problems using Hess's law (Section 5.3).

■ **Extension:** **Measuring enthalpy changes**

It is not possible to measure the total enthalpy of a chemical– it is only possible to measure enthalpy *changes*. There are no *absolute* enthalpies. To use an analogy, when you climb to the top of a building that is 200 metres high, you do not know how far away the centre of the Earth is, but you know it is 200 metres further away than it was before you climbed the building.

5.2 Calculation of enthalpy changes

5.2.1 **Calculate** the heat energy change when the temperature of a pure substance is changed.
5.2.2 **Design** suitable experimental procedures for measuring the heat energy changes of reactions.
5.2.3 **Calculate** the enthalpy change for a reaction using experimental data on temperature changes, quantities of reactants and mass of water.
5.2.4 **Evaluate** the results of experiments to determine enthalpy changes.

Specific heat capacity

When a substance is heated, the temperature of the substance increases. The size of the increase depends on the heat capacity of the substance. The heat capacity of the substance is the amount of heat energy required to raise the temperature of a substance by one degree Celsius or one kelvin. Heat capacity has units of joules per degree Celsius ($J\,°C^{-1}$) or joules per kelvin ($J\,K^{-1}$).

The specific heat capacity (Figure 5.8) is the amount of heat required to raise the temperature of a unit mass of the substance by one degree Celsius or one kelvin. Specific heat capacity, c, often has units of joules per gram per degree Celsius ($J\,g^{-1}\,°C^{-1}$). The *lower* the specific heat capacity of a substance, the *greater* its temperature rise for the same amount of heat absorbed.

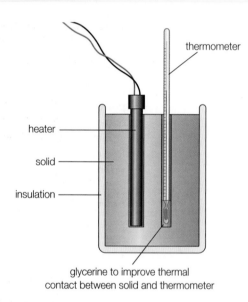

Figure 5.8 Measuring the specific heat capacity of a solid

heat energy (q) = mass of object (m) × specific heat capacity (c) × temperature rise (ΔT)

Worked examples

Calculate the heat capacity of 80.0 grams of water. The specific heat capacity of water is $4.18\,J\,g^{-1}\,°C^{-1}$.

Heat capacity = $80.0\,g \times 4.18\,J\,g^{-1}\,°C^{-1} = 334.4\,J\,°C^{-1}$

How much heat energy is required to increase the temperature of 20 grams of nickel (specific heat capacity $440\,J\,kg^{-1}\,°C^{-1}$) from 50 °C to 70 °C?

$q = mc\Delta T$

$q = 0.02\,kg \times 440\,J\,kg^{-1}\,°C^{-1} \times 20\,°C = 176\,J$

Measuring enthalpy changes

5.2.2 **Design** suitable
experimental
procedures for
measuring the
heat energy
changes of
reactions.

Enthalpy changes are usually measured by their effect on a known volume of water in a container known as a calorimeter. A chemical reaction involving known amounts of chemical dissolved in the water may be performed and the temperature increase or decrease measured. Alternatively, a combustion reaction may be performed and the temperature increase in the water bath recorded.

The heat produced or absorbed may be calculated from the following expression:

$$\text{heat change} = \begin{array}{c}\text{total mass of water} \\ \text{or solution}\end{array} \times \begin{array}{c}\text{specific heat capacity} \\ \text{of water}\end{array} \times \begin{array}{c}\text{temperature} \\ \text{change}\end{array}$$

In symbols:

$$q\ (\text{J}) = m\ (\text{g}) \times c\ (\text{J}\,\text{g}^{-1}\,{}^{\circ}\text{C}^{-1}) \times \Delta T\ ({}^{\circ}\text{C})$$

The calculations of heat transferred to the water are based on the following assumptions.

- The reaction is assumed to occur sufficiently rapidly for the maximum temperature to be achieved before the reaction mixture begins to cool to room temperature. This occurs if the next condition is completely fulfilled.
- There is no heat transfer between the solution, thermometer, the surrounding air and the calorimeter.

(In practice neither of these conditions is completely fulfilled, but the rate of heat transfer in or out of the calorimeter is tracked and extrapolated back to the moment the reaction began.)

- The solution is sufficiently dilute that its density and specific heat capacity are taken to be equal to that of water, namely, $1\,\text{g}\,\text{cm}^{-3}$ and $4.18\,\text{J}\,\text{g}^{-1}\,{}^{\circ}\text{C}^{-1}$.

The heat change is for a specific amount of chemicals used in the reaction. This is usually less than one mole, so by simple proportion or 'scaling up' the heat change is then calculated for the amount of chemicals shown in the chemical equation.

Worked example

50.00 cm³ of 0.100 mol dm⁻³ silver nitrate solution was put in a calorimeter and 0.200 g of zinc powder added. The temperature of the solution rose by 4.3 °C. Deduce which reagent was in excess and then calculate the enthalpy change for the reaction (per mole of zinc that reacts). Assume that the density of the solution is $1.00\,\text{g}\,\text{cm}^{-3}$ and the specific heat capacity of the solution is $4.18\,\text{J}\,\text{g}^{-1}\,{}^{\circ}\text{C}^{-1}$. Ignore the heat capacity of the metals and dissolved ions.

$$q = 50\,\text{g} \times 4.18\,\text{J}\,\text{g}^{-1}\,{}^{\circ}\text{C}^{-1} \times 4.3\,{}^{\circ}\text{C} = 898.7\,\text{J}$$

$$\text{Amount of silver nitrate} = \frac{50.0}{1000}\,\text{dm}^3 \times 0.100\,\text{mol}\,\text{dm}^{-3} = 0.005\,\text{mol}$$

$$\text{Amount of zinc} = \frac{0.200\,\text{g}}{65.37\,\text{g}\,\text{mol}^{-1}} = 0.0031\,\text{mol}$$

$$2\text{AgNO}_3(\text{aq}) + \text{Zn}(\text{s}) \rightarrow \text{Zn}(\text{NO}_3)_2(\text{aq}) + 2\text{Ag}(\text{s})$$

Zinc is the *excess* reactant (Chapter 1) and hence the temperature change and the enthalpy change are determined by the *limiting* reactant, the silver nitrate. Therefore:

$$\Delta H = -0.8987\,\text{kJ}/0.005\,\text{mol} = -180\,\text{kJ}\,\text{mol}^{-1}$$

The enthalpy changes of reactions in solution can be easily measured with the simple apparatus shown in Figure 5.9. A lid may be fitted to minimize heat transfer. More accurate measurements could be performed in a calorimeter based around a Thermos or vacuum flask.

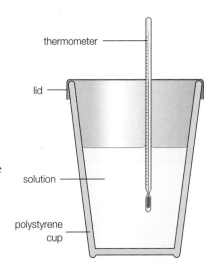

Figure 5.9 A simple calorimeter: polystyrene
cup, lid and thermometer

Problems with calorimetry

There are three problems associated with the use of calorimeters:

- Not having the desired reaction occur. This is relevant to enthalpies of combustion where incomplete combustion occurs.

- Loss of heat to the surroundings (Figure 5.10) (exothermic reactions); absorption of heat from the surroundings (endothermic reaction). This unwanted flow of heat can be reduced by 'lagging' the calorimeter to ensure it is well insulated.

- Using an incorrect specific heat capacity in the calculation of the heat change. If a copper can is used as a calorimeter during an enthalpy change of combustion investigation, then its specific heat capacity needs to be taken into account during the calculation.
  For example:

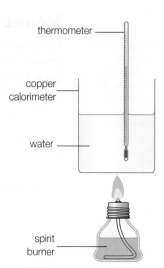

Figure 5.10 Simple apparatus used to measure enthalpy changes of combustion of liquids

heat transferred = [mass of water × specific heat capacity of water × temperature change]
+ [mass of copper × specific heat capacity of copper × temperature change]

Temperature corrections

Accurate results can be obtained by using simple calorimeters (e.g. a polystyrene cup fitted with a lid) for fast reactions, such as neutralizations or precipitations. However, for slower reactions such as metal ion displacement, the results will be less accurate with the same apparatus. This is because there is heat loss to the surroundings; this will increase if the reaction is slow because the heat will be lost over a longer period of time. Consequently, the temperature rise observed in the calorimeter is not as great as it should be. However, an allowance can be made for this by plotting a temperature–time graph (or cooling curve). The method is described below.

One reagent is placed in the polystyrene cup and its temperature recorded at, say, 1 minute intervals for, say, 4 minutes, stirring continuously. At a known time, say 4.5 minutes from the start, the second reagent is added, stirring continuously, and the temperature recorded until the maximum temperature is reached. As the reaction mixture starts to cool, temperature recording and stirring are continued for at least 5 minutes. A graph of temperature against time is then plotted.

The lines are extrapolated to the time of mixing to determine the temperature change that would have occurred had mixing of the reagents been instantaneous with no heat loss to the surroundings. Graphs are given for an exothermic (Figure 5.11) and an endothermic reaction (Figure 5.12).

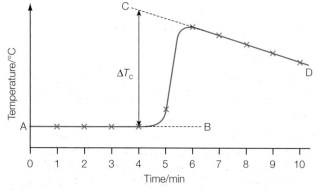

Figure 5.11 A temperature correction curve for an exothermic reaction

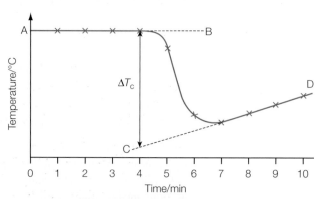

Figure 5.12 A temperature correction curve for an endothermic reaction

Enthalpy change of combustion

The standard enthalpy change of combustion for a substance is the heat energy released when one mole of a pure substance is completely burnt in excess oxygen under standard conditions.

An example of the enthalpy change of combustion is the combustion of methane. The reaction can be described by the following thermochemical equation:

$$CH_4(g) + 2O_2(g) \rightarrow CO_2(g) + 2H_2O(l) \qquad \Delta H_c^{\ominus} = -698\,kJ\,mol^{-1}$$

Enthalpy changes of combustion are always negative as heat is released during combustion processes.

Measuring enthalpy changes of combustion of fuels

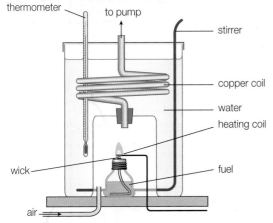

Figure 5.13 Measuring the enthalpy change of combustion of a liquid fuel using a flame combustion calorimeter

Enthalpy changes for the combustion of liquids can be measured in a flame combustion calorimeter (Figure 5.13). The investigation is carried out as follows:

1　Place a known volume and hence known mass of water into the calorimeter.
2　Stir and record the temperature of the water.
3　Record the mass of the spirit burner.
4　Turn on the pump so that there is steady flow of air and hence oxygen through the copper coil.
5　Use the electrically operated heating coil to light the wick.
6　Slowly stir the water throughout the experiment.
7　Allow the spirit burner to heat up the water.
8　Record the maximum temperature of the water.
9　Reweigh the spirit burner to determine the mass of liquid fuel combusted.

Worked example

Some example results are given below for the combustion of methanol.

Volume of water = 100 cm³

Temperature rise = 34.5 °C

Mass of methanol burned = 0.75 g

Specific heat capacity of water = 4.18 J g⁻¹ °C⁻¹

These results can be used to calculate the molar enthalpy change of combustion.

Heat energy transferred = $100\,g \times 4.18\,J\,°C^{-1}g^{-1} \times 34.5\,°C = 14421\,J$

Amount of methanol burnt = $\dfrac{0.75\,g}{32\,g\,mol^{-1}} = 0.023\,mol$

Amount of energy released per mole of methanol = $\dfrac{14421\,J}{0.023\,mol} = 627\,000\,J\,mol^{-1}$

Hence the enthalpy change of combustion of methanol is −627 kJ mol⁻¹.

The experimental literature value for the standard enthalpy change of combustion of methanol is −726 kJ mol⁻¹. The absolute error is (726 − 627), that is, 99 kJ mol⁻¹. The percentage error is (726–630)/726 × 100, that is 14% (Chapter 11).

The error is due to the large heat losses that occur during the use of the flame combustion calorimeter. Heat losses to the surrounding air are relatively large, despite the use of heat shields. Also, heat energy from the flame heats up the material of the calorimeter itself, as well as the water. A correction can be made for the heat losses to the calorimeter if the specific heat capacity of copper is known.

Applications of Chemistry

Spacecraft, like the Space Shuttle, must carry a mixture of a fuel and oxidizing agent (Chapter 9) to provide oxygen. These are known as propellants. The Space Shuttle has two booster rockets containing a solid propellant consisting of aluminium powder and ammonium chlorate(VII) which react together according to the following equation:

$$3Al(s) + 3NH_4ClO_4(s) \rightarrow Al_2O_3(s) + AlCl_3(s) + 3NO(g) + 6H_2O(g)$$

The exhaust gases leave the Shuttle at a speed of almost 1600 metres per second and at a temperature of almost 2000 °C. The aluminium-containing products form the dense white clouds (Figure 5.14) observed during take-off of the Shuttle.

Figure 5.14 The Space Shuttle Discovery at take-off

■ **Extension:** ### The bomb calorimeter

The experimental literature values for standard enthalpy changes of combustion are obtained using a more accurate bomb calorimeter (Figure 5.15). These can be obtained for fuels as well as dried foodstuffs (Chapter 26).

■ A known mass of the solid or liquid fuel is placed inside the stainless steel container, the 'bomb'. This container is filled with oxygen under pressure.
■ The fuel is ignited electrically and the heat produced is transferred directly to the surrounding water; the temperature rise is measured.
■ The determination of the standard enthalpy change of combustion is identical to the calculation outlined for the flame combustion calorimeter. The heat losses in this investigation are minimized

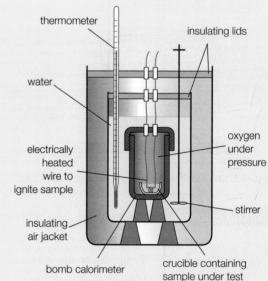

Figure 5.15 Measuring the enthalpy change of combustion of a fuel using a bomb calorimeter

because the water completely surrounds the bomb and the apparatus is well insulated. In addition, complete combustion is promoted by the use of pure oxygen.

Enthalpy change of neutralization

The standard enthalpy change of neutralization is the enthalpy change that takes place when one mole of hydrogen ions is *completely* neutralized by an alkali under standard conditions.

An example of the enthalpy change of neutralization is the reaction between sodium hydroxide solution and hydrochloric acid. The reaction can be described by the following thermochemical equation:

$$NaOH(aq) + HCl(aq) \rightarrow NaCl(aq) + H_2O(l) \qquad \Delta H^{\ominus} = -57 \, kJ \, mol^{-1}$$

The enthalpy change of neutralization of a strong acid with a strong alkali is almost the same for all strong acids and strong alkalis. This is because strong acids and strong alkalis undergo complete ionization or dissociation in water:

$$NaOH(s) + (aq) \rightarrow Na^+(aq) + OH^-(aq)$$
$$HCl(g) + (aq) \rightarrow H^+(aq) + Cl^-(aq)$$

The reaction between a strong base and a strong acid is the combination of hydrogen and hydroxide ions to form water molecules. The other ions are spectator ions (they take no part in the reaction). The reaction can be described by the following ionic equation:

$$H^+(aq) + OH^-(aq) \rightarrow H_2O(l) \qquad\qquad \Delta H^\ominus = -57\,kJ\,mol^{-1}$$

For sulfuric acid, a dibasic acid, the enthalpy of neutralization equation is

$$\tfrac{1}{2}H_2SO_4(aq) + KOH(aq) \rightarrow \tfrac{1}{2}K_2SO_4(aq) + H_2O(l) \quad \Delta H^\ominus = -57\,kJ\,mol^{-1}$$

and *not*

$$H_2SO_4(aq) + 2KOH(aq) \rightarrow K_2SO_4(aq) + 2H_2O(l)$$

This is because during the neutralization process each sulfuric acid molecule releases two hydrogen ions:

$$H_2SO_4(aq) \rightarrow 2H^+(aq) + SO_4^{2-}(aq)$$

However, where neutralizations involve a *weak* acid, a *weak* base, or both, then the enthalpy of neutralization will be *smaller* in magnitude than $-57\,kJ\,mol^{-1}$, that is, slightly less exothermic.

$$CH_3COOH(aq) + NaOH(aq) \rightarrow CH_3COONa(aq) + H_2O(l)$$

For example, the enthalpy of neutralization for ethanoic acid and sodium hydroxide is $-55.2\,kJ\,mol^{-1}$ because some of the energy released on neutralization is used to ionize or dissociate the acid.

$$CH_3COOH(aq) \rightarrow CH_3COO^-(aq) + H^+(aq)$$

This can be shown in the form of an enthalpy level cycle (Figure 5.16).

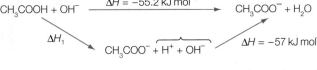

Figure 5.16 Enthalpy level cycle for the neutralization of ethanoic acid

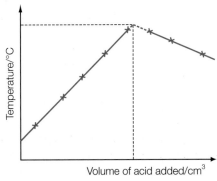

Figure 5.17 A graph of temperature versus volume of acid for an investigation to determine the enthalpy change of neutralization for an acid

A simple method for determining the enthalpy change of neutralization involves mixing equal volumes of dilute solutions of a strong acid and a strong base and measuring the temperature rise. A thick plastic cup fitted with a lid makes a cheap and effective calorimeter.

The maximum temperature can be deduced from the graph (see Figure 5.17) by extrapolating both lines back to find out where they intersect.

Worked example

$50.00\,cm^3$ of $1.0\,mol\,dm^{-3}$ hydrochloric acid was added to $50.00\,cm^3$ of $1.0\,mol\,dm^{-3}$ sodium hydroxide solution. The temperature rose by $6.8\,°C$. Calculate the enthalpy change of neutralization for this reaction. Assume that the density of the solution is $1.00\,g\,cm^{-3}$ and the specific heat capacity of the solution is $4.18\,J\,g^{-1}\,°C^{-1}$.

$$q = mc\Delta T$$

$$q = 100\,g \times 4.18\,J\,g^{-1}\,°C^{-1} \times 6.8\,°C = 2842\,J$$

$$NaOH(aq) + HCl(aq) \rightarrow NaCl(aq) + H_2O(l)$$

$$\text{Amount of hydrochloric acid} = \frac{50.00}{1000}\,dm^3 \times 1.0\,mol\,dm^{-3} = 0.050\,mol$$

$$\text{Amount of sodium hydroxide} = \frac{50.00}{1000}\,dm^3 \times 1.0\,mol\,dm^{-3} = 0.050\,mol$$

$$\text{Enthalpy change of neutralization} = \frac{-2.842\,kJ}{0.050\,mol} = 56.8\,kJ\,mol^{-1} = 57\,kJ\,mol^{-1}$$

(The answer is expressed to two significant figures because $6.8\,°C$ has only two significant figures.)

■ Extension: Enthalpy change of solution

The standard enthalpy change of solution is the enthalpy change that occurs when one mole of a solute dissolves in a large excess of water, so that no further heat change occurs when more water is added to the solution.

For example, the enthalpy change of solution when dissolving sodium chloride in water is the enthalpy change of the following reaction:

$$NaCl(s) + (aq) \rightarrow Na^+(aq) + Cl^-(aq) \qquad \Delta H^\ominus = +3.9 \, kJ \, mol^{-1}$$

This enthalpy change of solution is positive, that is, the reaction is endothermic. This reaction proceeds because a large positive entropy change occurs during the dissolving process (Chapter 15). Many ionic compounds have enthalpies of solution that are negative, that is, the reactions are exothermic.

Worked example

0.848 grams of anhydrous lithium chloride, LiCl, are added to 36.0 grams of water at 25 °C in a polystyrene cup acting as a calorimeter. The final temperature of the solution was 29.8 °C. Calculate the enthalpy change of solution for one mole of lithium chloride.

$$\text{Amount of lithium chloride} = \frac{0.848 \, g}{42.4 \, g \, mol^{-1}} = 0.0200 \, mol$$

$$\text{Amount of water} = \frac{36.0 \, g}{18.02 \, g \, mol^{-1}} = 2.00 \, mol$$

$$\text{Therefore,} \frac{\text{amount of LiCl}}{\text{amount of H}_2\text{O}} = \frac{1}{100}$$

$$\Delta H = 36.0 \, g \times 4.18 \, J \, g^{-1} \, °C^{-1} \times 4.8 \, °C = -0.72 \, kJ$$

'Scaling up' to molar quantities:

$$\Delta H = -0.72 \, kJ \times \frac{1 \, mol}{0.020 \, mol} = -36 \, kJ$$

Therefore:

$$LiCl(s) + (aq) \rightarrow LiCl(aq) \qquad \Delta H^\ominus_{sol} = -36 \, kJ \, mol^{-1} \text{ (to two significant figures)}$$

5.3 Hess's law

Hess's law and enthalpy change

5.3.1 Determine the enthalpy change of a reaction that is the sum of two or three reactions with known enthalpy changes.

Hess's law states that if a reaction consists of a number of steps, the overall enthalpy change is equal to the sum of the enthalpy changes for all the individual steps. Hess's law (Figure 5.18) states that the overall enthalpy change in a reaction is constant and not dependent on the pathway taken.

When reactant A is converted directly into product D by route 1, or indirectly by route 2 (via intermediates B and C), then according to Hess's law the enthalpy change in route 1 will equal the enthalpy changes of the reactions in route 2.

In symbols:

$$\Delta H_1 = \Delta H_2 + \Delta H_3 + \Delta H_4$$

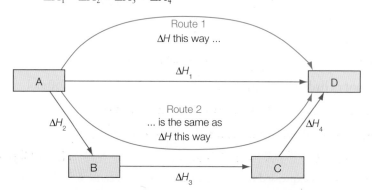

Figure 5.18 An illustration of the principle of Hess's law

History of Chemistry

Germain Henri Hess (1802–1850) was a Swiss-born Russian chemist and doctor who formulated what is now known as Hess's law in 1840. It was originally known as the law of constant heat summation. Hess's interest in chemistry was increased following a meeting with Berzelius, a famous Swedish chemist and developer of our modern chemical symbols (Chapter 1). In 1830 he began full time researching and teaching chemistry and later became a professor at the Saint Petersburg Technological Institute.

Using Hess's law to calculate the enthalpy change of a reaction

Hess's law can be used to calculate the enthalpy change of a reaction. Consider the following reaction:

$$C(s) + \tfrac{1}{2}O_2(g) \rightarrow CO(g)$$

The enthalpy change of this reaction *cannot* be found directly by experiment because carbon dioxide is always formed when carbon reacts with even a limited amount of oxygen. This is an unavoidable reaction. However, the enthalpy changes of combustion of carbon and carbon monoxide *can* be found experimentally.

The reactions and their enthalpy changes of reaction can be linked using Hess' law, as shown in Figure 5.19. There are two pathways from carbon to carbon dioxide: a *direct* pathway (route 1) and an *indirect* pathway (route 2), where the carbon is burnt to form carbon monoxide and then burnt to produce carbon dioxide.

Route 1: $C(s) + O_2(g) \rightarrow CO_2(g)$ $\Delta H_1^\ominus = -394\,\text{kJ}\,\text{mol}^{-1}$

Route 2: $C(s) + \tfrac{1}{2}O_2(g) \rightarrow CO(g)$ $\Delta H_2^\ominus = ?$

$CO(g) + \tfrac{1}{2}O_2(g) \rightarrow CO_2(g)$ $\Delta H_3^\ominus = -283\,\text{kJ}\,\text{mol}^{-1}$

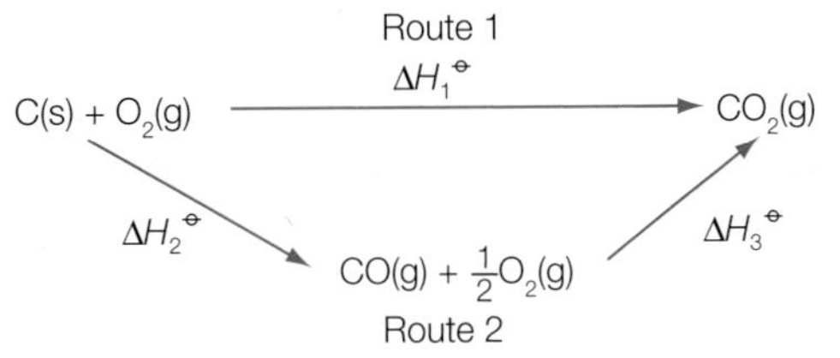

Route 1
$\Delta H_1{}^\ominus$
$C(s) + O_2(g) \xrightarrow{\hspace{3cm}} CO_2(g)$

$\Delta H_2{}^\ominus$ $\Delta H_3{}^\ominus$

$CO(g) + \tfrac{1}{2}O_2(g)$
Route 2

Figure 5.19 An example of Hess's law

Applying Hess's law, the enthalpy change in route 1 equals the enthalpy change in route 2. In symbols:

$$\Delta H_1 = \Delta H_2 + \Delta H_3$$
$$\Delta H_2 = \Delta H_1 - \Delta H_3 = -394 - (-283) = -111\,\text{kJ}\,\text{mol}^{-1}$$

So the enthalpy change for the combustion of carbon to form carbon monoxide is $-111\,\text{kJ}\,\text{mol}^{-1}$.

Reactions in aqueous solutions

The conversion of solid sodium hydroxide into sodium chloride solution illustrates the use of a Hess's law energy cycle for reactions in aqueous solution. Figure 5.20 shows two pathways for this reaction.

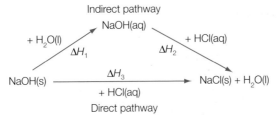

Indirect pathway
NaOH(aq)
$+ H_2O(l)$ $+ HCl(aq)$
ΔH_1 ΔH_2

NaOH(s) $\xrightarrow{\hspace{2cm}}$ NaCl(s) + $H_2O(l)$
ΔH_3
$+ HCl(aq)$
Direct pathway

Figure 5.20 A further example of Hess's law

The first pathway (indirect pathway) involves two steps:

Step 1: solid sodium hydroxide dissolved in water:

$$NaOH(s) + (aq) \rightarrow NaOH(aq) \qquad\qquad \Delta H_1 = -43\,\text{kJ}\,\text{mol}^{-1}$$

Step 2: sodium hydroxide solution neutralized by hydrochloric acid

$$NaOH(aq) + HCl(aq) \rightarrow NaCl(aq) + H_2O(l) \qquad \Delta H_2 = -57 \text{kJ mol}^{-1}$$

The second pathway (direct pathway) involves one step.

Step 3: solid sodium hydroxide is added directly to hydrochloric acid.

$$NaOH(s) + HCl(aq) \rightarrow NaCl(aq) + H_2O(l)$$
$$\Delta H_3 = \Delta H_1 + \Delta H_2 = (-43) + (-57) = -100 \text{kJ mol}^{-1}$$

Decomposition of calcium carbonate

Hess's law can be used to determine the value of an endothermic reaction, for example the thermal decomposition of calcium carbonate:

$$CaCO_3(s) \rightarrow CaO(s) + CO_2(g)$$

The reaction is slow and a high temperature is required to bring it to completion. Direct measurement of the temperature is therefore not practical. Instead, two reactions that take place readily at room temperature are carried out and their enthalpy changes used to find the enthalpy of decomposition of calcium carbonate. These reactions are the reactions of calcium carbonate and calcium oxide with dilute hydrochloric acid:

$$CaCO_3(s) + 2HCl(aq) \rightarrow CaCl_2(aq) + H_2O(l) \qquad \Delta H^\ominus = -17 \text{kJ mol}^{-1}$$
$$CaO(s) + 2HCl(aq) \rightarrow CaCl_2(aq) + H_2O(l) \qquad \Delta H^\ominus = -195 \text{kJ mol}^{-1}$$

A Hess's law cycle (Figure 5.21) can be drawn to indicate the direct and indirect routes or pathways. $\Delta H^\ominus + (-195) = -17$ and $\Delta H^\ominus = -17 - (-195) = +178 \text{kJ mol}^{-1}$.

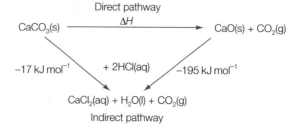

Figure 5.21 Hess's law cycle for the decomposition of calcium carbonate

Enthalpy of hydration of an anhydrous salt

Hess's law can also be used to determine the enthalpy of hydration of an anhydrous salt. For example, anhydrous copper(II) sulfate:

$$CuSO_4(s) + 5H_2O(l) \rightarrow CuSO_4.5H_2O (s)$$

The enthalpy of hydration of anhydrous copper(II) cannot be found directly. This is because if five moles of water are added to anhydrous copper(II) sulfate, hydrated copper(II) sulfate is not produced in a controlled way. It can only be produced by crystallization from a solution. The enthalpy change can be found *indirectly* by determining the enthalpy of solution of both anhydrous and hydrated copper(II) sulfates (Figure 5.22).

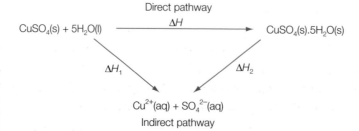

Figure 5.22 Hess's law cycle to find the enthalpy change when anhydrous copper(II) sulfate crystals are hydrated

An algebraic method

Hess's law problems do *not* have to be solved by drawing enthalpy cycles (although they are preferred for complicated energetic calculations). They can also be solved by an 'algebraic method' that involves manipulating the equations so that when added together they give the required enthalpy change.

Worked example

From the following data at 25 °C and 1 atmosphere pressure:

Equation 1: $2CO_2(g) \rightarrow 2CO(g) + O_2(g)$ $\Delta H^\ominus = 566\,kJ\,mol^{-1}$
Equation 2: $3CO(g) + O_3(g) \rightarrow 3CO_2(g)$ $\Delta H^\ominus = -992\,kJ\,mol^{-1}$

Calculate the enthalpy change calculated for the conversion of oxygen to one mole of ozone (O_3), i.e. for the reaction:

$$\tfrac{3}{2}O_2(g) \rightarrow O_3(g)$$

$\tfrac{3}{2}O_2$ is required in the reactant side and O_3 is required on the product side. Reversing equation 1 and multiplying it by $\tfrac{3}{2}$ gives:

$$\tfrac{3}{2}O_2(g) + 3CO(g) \rightarrow 3CO_2\,(g) \qquad\qquad \Delta H^\ominus = -849\,kJ\,mol^{-1}$$

(Equation 3)

Reversing equation 2 gives:

$$3CO_2(g) \rightarrow 3CO(g) + O_3\,(g) \qquad\qquad \Delta H^\ominus = +992\,kJ\,mol^{-1}$$

(Equation 4)

Adding equations 3 and 4 gives the desired reaction (as the CO and CO_2 molecules cancel out), and adding the $\Delta H^\ominus$ values which gives the desired final $\Delta H^\ominus$.

$$\tfrac{3}{2}O_2(g) \rightarrow O_3\,(g) \qquad\qquad \Delta H^\ominus = -849 + 992 = +143\,kJ\,mol^{-1}$$

One of the most important uses of Hess's law is to calculate enthalpy changes that are difficult to measure experimentally.

Worked example

Calculate the enthalpy change for the conversion of graphite to diamond under standard thermodynamic conditions.

$$C(s,\,graphite) + O_2(g) \rightarrow CO_2(g) \qquad \Delta H_1^\ominus = -393\,kJ\,mol^{-1}$$
$$C(s,\,diamond) + O_2(g) \rightarrow CO_2(g) \qquad \Delta H_2^\ominus = -395\,kJ\,mol^{-1}$$

The problem may be solved via use of an energy cycle (Figure 5.23) or via algebraic manipulation.

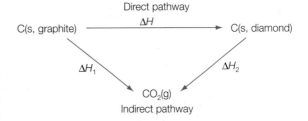

Figure 5.23 Energy cycle showing direct conversion of graphite to diamond and indirect conversion via carbon dioxide

Using an energy cycle

$$\Delta H = \Delta H_1 - \Delta H_2 \qquad \Delta H = (-393) - (-395) \qquad \Delta H^\ominus = +2\,kJ\,mol^{-1}$$

Using an algebraic method

$$C(s,\,graphite) + O_2(g) \rightarrow CO_2(g) \qquad \Delta H^\ominus = -393\,kJ\,mol^{-1}$$

Then reversing the original second equation, also reversing the sign of the enthalpy change:

$$CO_2(g) \rightarrow C(s,\,diamond) + O_2(g) \qquad \Delta H^\ominus = +395\,kJ\,mol^{-1}$$

The two equations are added together and the oxygen and carbon dioxide molecules cancelled.

$$C(s,\,graphite) \rightarrow C(s,\,diamond) \qquad \Delta H^\ominus = +2\,kJ\,mol^{-1}$$

It should also be noted that the value and sign of the enthalpy change (ΔH) do *not* indicate whether a reaction occurs or not. There are spontaneous exothermic and endothermic reactions. The Gibbs free energy change (ΔG) determines whether a reaction occurs or not (under standard conditions) (Chapter 15). In addition, the size and sign of the enthalpy change do *not* give any indication of the rate of reaction, since a fast rate of reaction can be prevented by a large activation energy barrier.

■ Extension: Enthalpy changes during changes of state

It is essential to specify the physical states of the substances involved when writing thermochemical equations to represent an enthalpy change. This is because any change in physical state (Chapter 1) has its own enthalpy change (Chapter 17).

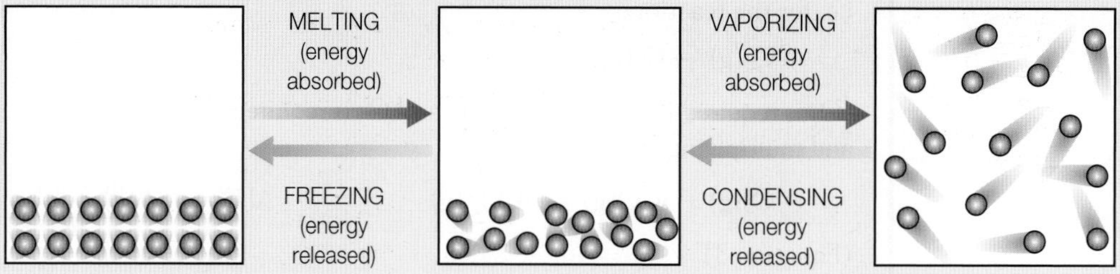

Figure 5.24 Changes of state and energy changes

MELTING (energy absorbed) →
← FREEZING (energy released)

VAPORIZING (energy absorbed) →
← CONDENSING (energy released)

The **enthalpy of fusion** of ammonia would be the enthalpy change for the reaction:

$$NH_3(s) \rightarrow NH_3(l)$$

and the **enthalpy of vaporization** of ammonia would be the enthalpy change for the reaction:

$$NH_3(l) \rightarrow NH_3(g)$$

Enthalpies of vaporization and sublimation are always endothermic since intermolecular forces of attraction need to be overcome. The enthalpy of sublimation for iodine would be the enthalpy change for the reaction:

$$I_2(s) \rightarrow I_2(g)$$

The values of these enthalpy changes will vary with the strength of intermolecular forces (Chapter 4).

■ Extension: Dissolving of ionic compounds

When an ionic compound dissolves in water, the process may be exothermic or endothermic, depending on the substance concerned. The value for the enthalpy change, ΔH, for such reactions is the sum of two factors:

$$\Delta H = + \begin{bmatrix} \text{energy to overcome the} \\ \text{electrostatic forces of attraction} \\ \text{between ions in the lattice} \end{bmatrix} - \begin{bmatrix} \text{energy released when the} \\ \text{ions attract water molecules} \\ \text{around themselves} \end{bmatrix}$$

The sign of the enthalpy change, ΔH, for the reaction depends on which of these two quantities is larger. The dissolving of ionic solids is discussed in Chapters 4 and 15.

History of Chemistry

Gunpowder was discovered by Chinese alchemists in the 9th century. The Chinese (Figure 5.25) used it in fireworks, but also in some weapons including guns in the 13th century. The Chinese used fireworks to frighten away evil spirits with their loud sound (*bian pao*) and also to pray for happiness and prosperity. Between the 11th and 13th centuries gunpowder spread from China to the Islamic World and then medieval Europe. The arrival of gunpowder in Europe and the subsequent wars changed the political and social structure of the continent. Gunpowder, also known as black powder, is prepared from finely powdered sulfur, charcoal (carbon) and potassium nitrate to provide oxygen. It burns rapidly, producing large volumes of gas and hot solids.

The equation below is close to the stoichiometric reacting proportions, but is simplified and not a complete description of the burning process.

$$10KNO_3(s) + 8C(s) + 3S(s) \rightarrow 2K_2CO_3(s) + 3K_2SO_4(s) + 6CO_2(g) + 5N_2(g)$$

Figure 5.25 This drawing shows arrows fired by gunpowder and is from a 17th century Chinese treatise on the art of war

5.4 Bond enthalpies

5.4.1 **Define** the term *average bond enthalpy*.

5.4.2 **Explain,** in terms of average bond enthalpies, why some reactions are exothermic and others are endothermic.

The bond enthalpy (bond energy) is the amount of energy required to break one mole of a specific covalent bond between two atoms in one mole of gaseous molecules. Measurement of bond enthalpies can be performed using a mass spectrometer (Chapter 2). The concept of bond enthalpy is illustrated in Figure 5.26 using the hydrogen molecule.

For the hydrogen molecule, the thermochemical equation describing the bond dissociation enthalpy is:

$$H_2(g) \rightarrow 2H(g) \qquad \Delta H^\ominus = +436\,kJ\,mol^{-1}$$

The bond enthalpy for hydrogen is $436\,kJ\,mol^{-1}$. Because energy is required to overcome or break the attractive forces between the shared pair of electrons and the nuclei, the bond breaking process is endothermic (that is, heat energy is absorbed from the surroundings). It should be noted that if the H—H bond had been formed, then 436 kilojoules of heat energy would have been released to the surroundings. This is a simple application of Hess's law (page 148). Bond breaking is *always* an endothermic process; bond formation is *always* an exothermic process.

The strength of a covalent bond is indicated by the size of the bond dissociation enthalpy. The larger the bond enthalpy, the stronger the covalent bond. Also, bond enthalpy is inversely proportional to bond length.

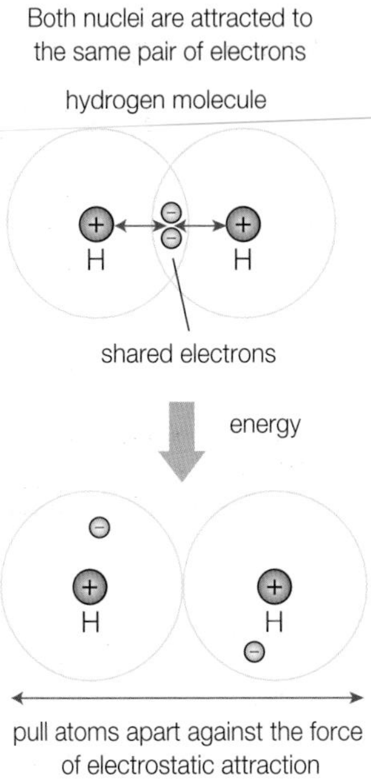

Many bond enthalpies are *average* bond enthalpies. For example, the C—H bond enthalpy is based upon the average bond energies in methane, alkanes and other hydrocarbons.

Because bond enthalpies are often average values it means that enthalpy changes calculated using bond enthalpies will *not* be exactly equal to an accurate experimentally determined value.

Both nuclei are attracted to the same pair of electrons

hydrogen molecule

shared electrons

energy

pull atoms apart against the force of electrostatic attraction

Figure 5.26 To break a covalent bond the attractive forces between the shared pair or pairs of electrons and the nuclei of the two atoms needs to be overcome

---

■ **Extension:**

Average bond enthalpies

A selection of bond enthalpies and bond lengths is given in Table 5.1. A more extensive table of bond enthalpies is given in Table 10 on page 11 of the IB *Chemistry data booklet*.

Factors affecting average bond enthalpies

■ **Effect of bond length**

The larger the atoms joined by a particular bond, the longer the bond length. Large atoms have more electrons than smaller atoms and this results in an increase in repulsion between the electron shells of each atom. In addition the nucleus of each atom is more effectively shielded (Chapter 12). Both of these effects lead to a weakening of the bond. For example, in the halogens the bond strength weakens in the order: chlorine, bromine and iodine. Fluorine, however, has a surprisingly low bond enthalpy, which is accounted for by lone pair–lone pair repulsion.

■ **Effect of number of bonding electrons**

The more electrons present in a bond, the greater the strength of the bond. This is because an increasing number of electrons leads to an increase in electrostatic forces of attraction. Hence triple bonds are expected to be stronger than double bonds, which should be stronger than single (for the same element). This can clearly be observed with carbon (Table 5.1).

Bond	$\Delta H^\ominus$ /$kJ\,mol^{-1}$	Bond length /nm
H–H	436	0.07
C–C	348	0.15
C=C	612	0.13
C≡C	837	0.12
N–N	163	0.15
N=N	409	0.12
O–O	146	0.15
O=O	496	0.12
F–F	158	0.14
Cl–Cl	242	0.20
Br–Br	193	0.23
I–I	151	0.27
C–H	412	0.11
O–H	463	0.10
C≡N	890	0.12
H–F	562	0.09
N–H	388	0.10
O–H	463	0.10
C=O	743	0.12

Table 5.1 A selection of average bond enthalpies (at 298 K)

■ **Effect of bond polarity**
Bonds become more polar as the difference in electronegativity (Chapter 3) between the two bonded atoms increases (Chapter 4). This increases the ionic character of the bond and reinforces the covalent bond. This can be observed with N—H, O—H and F—H where bond strength increases with increasing polarity.

Using bond dissociation enthalpies to calculate enthalpy changes of reaction

Bond enthalpies can be used to determine the enthalpy change for a particular reaction involving molecules in the gaseous state, for example the combustion of methane.

$$CH_4(g) + 2O_2(g) \rightarrow CO_2(g) + 2H_2O(g)$$

The reaction can be regarded as occurring in two steps: first, all of the bonds in the reactants have to be broken to form atoms. This is an endothermic process and heat energy has to be absorbed from the surroundings. In the second step, bond formation occurs. This is an exothermic process and releases heat energy to the surroundings. The overall reaction is exothermic since the energy released during bond formation is greater than the energy absorbed during bond breaking (Figure 5.27).

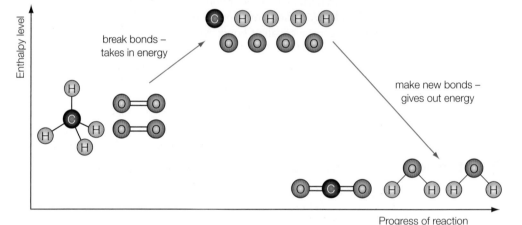

Figure 5.27 Breaking and forming of bonds during the combustion of methane

Bond enthalpy data can be used to calculate the enthalpy change for this reaction.

Bond breaking
Breaking 4 C—H bonds in a methane molecule = 4 × 412 = 1648 kJ
Breaking 2 O=O bonds in two oxygen molecules = 2 × 496 = 992 kJ
Total amount of energy required to break all these bonds = (1648 + 992) = 2640 kJ

Bond making
Making 2 C=O bonds in a carbon dioxide molecule = 2 × 743 = 1486 kJ
Making 4 O—H bonds in two water molecules = 4 × 463 = 1852 kJ
Total amount of energy released to surroundings when these bonds are formed
= (1486 + 1852) = 3338 kJ

The enthalpy change of this reaction is:

$$\Delta H = \sum \binom{\text{energy required}}{\text{to break bonds}} - \sum \binom{\text{energy released when}}{\text{bonds are formed}}$$

$$= (2640 - 3338)$$

$$= -698 \, \text{kJ mol}^{-1}$$

Since more energy is released when the new bonds in the products are formed than is needed to break the bonds in the reactants to begin with, there is an overall release of energy in the form of heat. The reaction is exothermic. In a reaction which is endothermic the energy absorbed by bond breaking is greater than the energy released by bond formation.

The bond breaking and making processes can be represented by a Hess's law cycle (see Figure 5.28). This calculated value is slightly different from the real value because the bond enthalpies are averages. In addition, the water that is produced (in the calculation) is *not* in its standard state, as a liquid. The gaseous state is always used when performing bond enthalpy calculations.

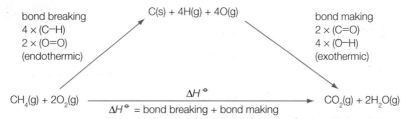

Figure 5.28 Using a Hess's law cycle to represent the bond breaking and bond making processes for the complete combustion of methane

Average bond enthalpies can be used to calculate the enthalpy change for any reaction involving molecules in the gaseous state. This is done by assuming that an alternative route for all reactions can be achieved theoretically via the gaseous atoms involved in the compounds (Figure 5.29).

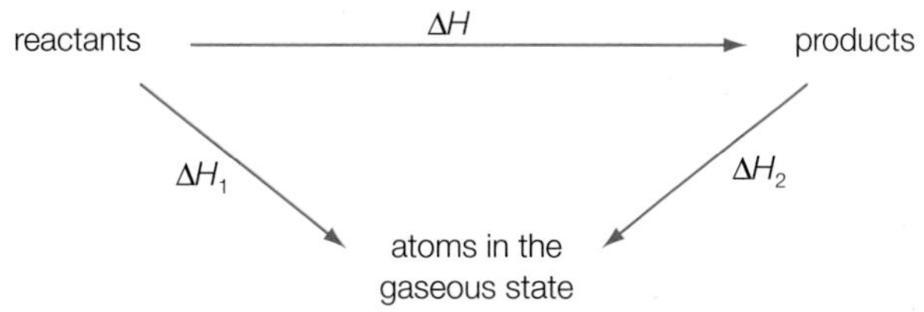

Figure 5.29 Generalized energy cycle to determine an enthalpy change from bond energies

ΔH_1 = sum of the average bond enthalpies of the reactants

ΔH_2 = sum of the average bond enthalpies of the products

Applying Hess's law gives:

$$\Delta H = \Delta H_1 - \Delta H_2$$

This leads to the expression:

$$\Delta H = \sum \left(\begin{array}{c} \text{average bond enthalpies} \\ \text{of the reactants} \end{array} \right) - \sum \left(\begin{array}{c} \text{average bond enthalpies} \\ \text{of the products} \end{array} \right)$$

i.e. $\Delta H = \sum (\text{bonds broken}) - \sum (\text{bonds made})$

Bond enthalpy data can also be used to determine an unknown bond enthalpy provided that the enthalpy change and all the other bond enthalpies are known.

Worked example

The bond enthalpies for $H_2(g)$ and $HCl(g)$ are $435\,kJ\,mol^{-1}$ and $431\,kJ\,mol^{-1}$ respectively for the reaction

$$H_2(g) + Cl_2(g) \rightarrow 2HCl(g)$$

and the enthalpy change of reaction is $-184\,kJ\,mol^{-1}$. Calculate the bond enthalpy of chlorine.

Enthalpy change $= \sum (\text{bonds broken}) - \sum (\text{bonds made})$

$$-184 = (435 + Cl-Cl) - (2 \times 431)$$
$$-184 = (435 + Cl-Cl) - 862$$
$$Cl-Cl = 243\,kJ\,mol^{-1}$$

Chemical Demonstration

A British twenty pound note is immersed in a 1 : 1 mixture of ethanol and water (by volume). The ethanol will burn, but the money does not (Figure 5.30). The heat of combustion for this reaction is $-1367\,kJ\,mol^{-1}$, which is sufficient to cause the combustion of the paper. The function of the water is to absorb some of the energy of combustion as the water heats to its boiling point and is vaporized. The energy available from the combustion of 25 g of ethanol (0.543 mol) is approximately 743 kJ. Heating 25 g of H_2O from 20 °C to 100 °C requires 8.4 kJ and the added step of vaporizing the water requires 56.5 kJ, making a total of 65 kJ absorbed in these procedures.

Figure 5.30 Twenty pound note burning in a 1 : 1 mixture of ethanol and water (note the demonstrator should be wearing a lab coat!)

■ Extension: Some exceptions

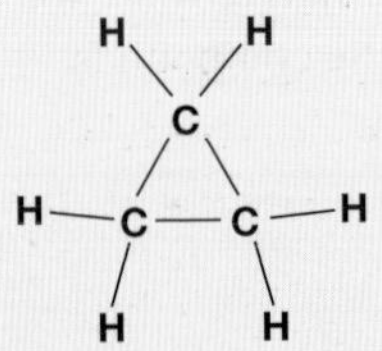

Calculations involving bond enthalpies generally give values of enthalpy changes that are close to experimentally determined values. However, with a small number of molecules the calculated enthalpy change is *significantly* different from the real value.

For example, cyclopropane (see Figure 5.31) is much *less* energetically stable than a bond energy calculation predicts. This is due to the 'strain energy' placed into the molecule during its formation. The angles between the carbon–carbon bonds forming the ring are 60°, whereas the *preferred* bond angle is 109° (Chapter 4).

In contrast the benzene molecule is *more* energetically stable than a bond energy calculation would suggest. This is because the molecule is a hybrid (Chapter 14) of the two resonance structures shown in Figure 5.32. Its carbon–carbon bonds are *intermediate* between single and double bonds, but because the π electrons (Chapter 14) move around the ring they stabilize the molecule.

The benzene molecule is a cyclic or ring system, like cyclopropane, but does *not* suffer from strain energy.

Figure 5.31 Structure of cyclopropane

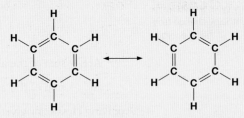

Figure 5.32 Resonance structures of the benzene molecule

■ Extension: Enthalpies of combustion of alkanes

Bond energies are by and large additive, which means that the specific bond energies are approximately constant for a range of related molecules, for example alkanes and alcohols.

Consider the alkanes, a group of hydrocarbons that are derived from methane, CH_4, by progressively adding methylene, $-CH_2-$ units (Figure 5.33).

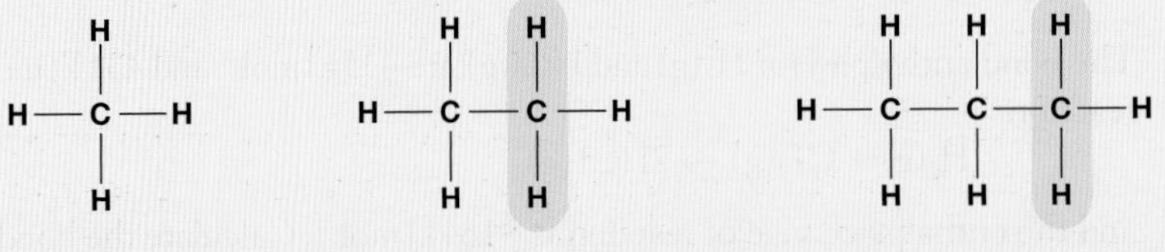

Figure 5.33 The formation of alkanes by the progressive addition of methylene, $-CH_2-$ units

If a series of hydrocarbons is combusted the addition of each extra methylene group will be responsible for an additional enthalpy change:

$$-CH_2- + 1\tfrac{1}{2}O_2(g) \rightarrow CO_2(g) + H_2O(l)$$

The enthalpy change, ΔH, for this process can be calculated using bond enthalpies :

Breaking bonds: $1 \times C–C, 2 \times C–H, 1\tfrac{1}{2}\,O=O$
Making bonds: $2 \times C=O, 2 \times O–H$

Using the values from Table 10 on page 11 of the IB *Chemistry data booklet*:

	Breaking bonds:	$1 \times C–C$	348
		$2 \times C–H$	2×412
		$1\frac{1}{2} O=O$	$1\frac{1}{2} \times 496$
Total energy		1916 kJ	
	Making bonds:	$2 \times C=O$	2×743
		$2 \times O–H$	2×463
Total energy			2412 kJ

Enthalpy change = Σ(bonds broken) – Σ (bonds made) = $(1916) + (-2412) = -496\,kJ\,mol^{-1}$

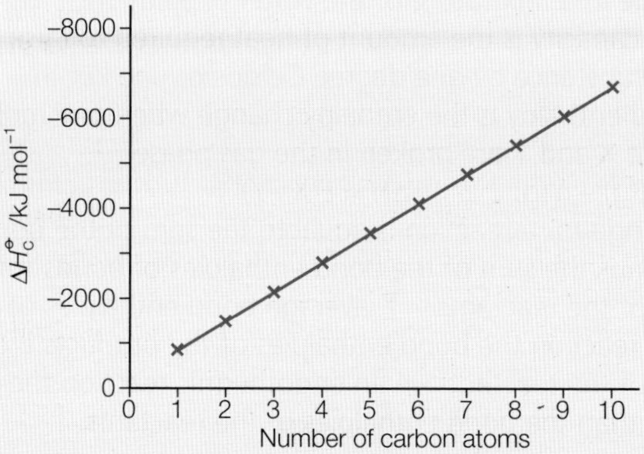

Figure 5.34 Graph of the standard enthalpies of combustion, $\Delta H_c^{\ominus}$, of the straight-chain alkanes plotted against the number of carbon atoms in the molecule

Hence the additional enthalpy of combustion for each additional methylene unit, $–CH_2–$, is $-496\,kJ\,mol^{-1}$. This type of simple calculation predicts that there should be an approximate linear relationship between the enthalpy change of combustion of an alkane and the number of carbon atoms (Figure 5.34). Experimental values confirm this prediction.

■ Extension: Feasibility of reactions

There are many examples of reactions which are spontaneous. The vast majority of these reactions are exothermic. Hence it appears that the enthalpy change, ΔH, is a reliable guide to which direction a reaction will go. However, there are examples of endothermic reactions that occur without the need for heat to initiate the reaction, for example, the reaction between citric acid and a solution of sodium hydrogencarbonate. Some salts dissolve endothermically in water. Chapter 15 introduces a factor, known as entropy, that, in conjunction with enthalpy and temperature, determines whether or not reactions occur at a specified temperature.

SUMMARY OF KNOWLEDGE

■ Atoms have a lower potential energy when bonded than when uncombined. When atoms form bonds potential energy is converted into heat.

■ Enthalpy (H) is quantity which is a related to the potential energy of the system. All reactions are accompanied by enthalpy changes, ΔH.

■ An enthalpy change (ΔH) is the heat exchanged with the surroundings when the reaction occurs at constant pressure.

■ Standard enthalpy changes, $\Delta H^{\ominus}$, refer to standard conditions, 298 K and 1 atmosphere pressure.

■ Exothermic reactions have negative enthalpy changes: heat flows from the chemicals into the surroundings. Endothermic reactions have positive enthalpy changes: heat flows from the surroundings into the chemicals.

■ Many chemical reactions are exothermic. Neutralization and combustion reactions are always exothermic. Thermal decomposition reactions are endothermic.

■ Enthalpy changes can be summarized by enthalpy diagrams: potential energy is on the vertical axis and extent of reaction along the horizontal axis. Exothermic reactions involve a change from reactants with high enthalpy to products with lower enthalpy. Endothermic reactions involve a change from reactants with low enthalpy to products with higher enthalpy.

■ Hess's law states that the enthalpy change in a reaction depends only on the enthalpies of the reactant and products, and is therefore independent of the route or path taken.

■ Chemical equations can be manipulated like algebraic equations: equations can be added or subtracted to eliminate chemical species. If a chemical reaction is reversed the sign of the enthalpy change is reversed.

■ Enthalpy changes can be directly determined by measuring the heat released or absorbed during a reaction, using a calorimeter. Practical methods of measuring enthalpy changes for liquids and solutions involve letting the reaction mixture heat itself. For combustion reactions the reaction is allowed to heat a known quantity of water.

■ Enthalpy changes can be derived using the following formula: $q = mc\Delta T$, where m represents the mass of water or solution, c represents the specific heat capacity of water and ΔT the change in temperature of the water or solution.

■ The specific heat capacity is the amount of heat required to raise the temperature of a unit mass of the substance by one degree Celsius or one kelvin.

■ The average bond enthalpy is the enthalpy change when one mole of covalent bonds between atoms of X and Y are broken in the gas phase:
$$X{-}Y(g) \rightarrow X(g) + Y(g); \qquad \Delta H = E(X{-}Y)$$

■ The enthalpy change for a reaction is equal to the sum of the bond enthalpies of the bonds broken minus the sum of the bond enthalpies of bonds formed: $\Delta H = \Sigma$ average bond enthalpies of the reactants $- \Sigma$ average bond enthalpies of the products.

■ In an exothermic reaction the bond enthalpies of the products are greater than the bond enthalpies of the reactants. In an endothermic reaction the bond enthalpies of the products are less than the bond enthalpies of the reactants.

■ Examination questions – a selection

Paper 1 IB questions and IB style questions

Q1 Which of the following must have a negative value for an exothermic reaction?
A voltage of a voltaic cell **C** enthalpy change
B change in state   **D** equilibrium constant

Q2 Which of the following is observed when the change in enthalpy is positive for the dissolving of a salt in water in an insulated copper beaker?
A Heat is evolved to the surroundings and the beaker feels cold.
B Heat is evolved to the surroundings and the beaker feels warm.
C Heat is absorbed from the surroundings and the beaker feels warm.
D Heat is absorbed from the surroundings and the beaker feels cold.

Q3 What is the specific heat capacity of an alcohol in $Jg^{-1}K^{-1}$ if 560.0 J of heat are required to raise the temperature of a 64.0 g sample of ethanol from 295.0 K to 310.0 K?
A 0.583 **B** 0.194 **C** 8.75 **D** 0.292

Q4 The following equation shows the formation of calcium oxide from calcium metal.
$$2Ca(s) + O_2(g) \rightarrow 2CaO(s) \qquad \Delta H^\circ = -1270\,kJ$$
Which of these statements is true?

A 1270 kJ of energy are released for every mol of calcium reacted.
B 635 kJ of energy are absorbed for every mol of calcium oxide formed.
C 635 kJ of energy are released for every mol of oxygen gas reacted.
D 1270 kJ of energy are released for every 2 mol of calcium oxide formed.

Q5 Which of the following processes is/are endothermic?
   I $H_2O(s) \rightarrow H_2O(g)$
   II $CO_2(g) \rightarrow CO_2(s)$
   III $N_2(g) \rightarrow 2N(g)$
A I only   **C** I and II only
B III only   **D** I and III only

Q6 All the following processes are exothermic **except**:
A $2C_2H_5(g) \rightarrow C_4H_{10}(g)$
B $F_2(g) \rightarrow 2F(g)$
C $Cl(g) + e^- \rightarrow Cl^-(g)$
D $4Fe(s) + 3O_2(g) \rightarrow 2Fe_2O_3(s)$

Q7 Which of the following reactions would you expect to provide the largest amount of heat?
A $C_2H_6(l) + 7O_2(l) \rightarrow 4CO_2(g) + 6H_2O(g)$
B $C_2H_6(l) + 7O_2(g) \rightarrow 4CO_2(g) + 6H_2O(g)$
C $C_2H_6(g) + 7O_2(g) \rightarrow 4CO_2(g) + 6H_2O(g)$
D $C_2H_6(g) + 7O_2(g) \rightarrow 4CO_2(g) + 6H_2O(l)$

Q8 Why does the temperature of boiling water remain constant even though heat is supplied at a constant rate?
A Heat is lost to the surroundings.
B The heat is used to break the covalent bonds in the water molecules.
C Heat is also taken in by the container.
D The heat is used to overcome the intermolecular forces of attraction between water molecules.

Standard Level Paper 1, Nov 05, Q14

Q9 When 0.050 mol of nitric acid is reacted with 0.050 mol of potassium hydroxide in water, the temperature of the system increases by 13.7 °C. Calculate the enthalpy of reaction in kJ mol⁻¹.

$$HNO_3(aq) + KOH(aq) \rightarrow KNO_3(aq) + H_2O(l)$$

Assume that the heat capacity of the system was 209.2 J °C⁻¹.
A +57.3 kJ mol⁻¹   **C** −2.87 kJ mol⁻¹
B +2.87 kJ mol⁻¹   **D** −57.3 kJ mol⁻¹

Q10 What can be deduced about the relative stability of the reactants and products and the sign of ΔH, from the enthalpy level diagram below?

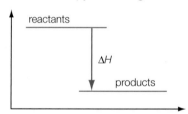

	Relative stability	Sign of ΔH
A	products more stable	−
B	products more stable	+
C	reactants more stable	−
D	reactants more stable	+

Standard Level Paper 1, May 99, Q16

Q11 The specific heat capacities of some metals are given below.

Metal	Specific heat capacity (J g⁻¹ K⁻¹)
copper	0.385
magnesium	1.020
mercury	0.138
platinum	0.130

If 100 kJ of heat is added to 10.0 g samples of each of the metals above, which are all at 25 °C, which metal will have the lowest temperature?
A copper   **C** mercury
B magnesium   **D** platinum

Q12 Consider the following bond energies:
C=C   615 kJ mol⁻¹
C—F   484 kJ mol⁻¹
C—C   348 kJ mol⁻¹
F—F   158 kJ mol⁻¹
Which one of the following gives the enthalpy change (in kJ mol⁻¹) for the addition reaction between fluorine and ethene (in the gaseous state)?
A −615 − 158 + 348 + 2(484)
B −615 − 158 − 348 − 2(484)
C 615 + 158 − 348 − 2(484)
D 615 + 158 + 348 + 2(484)

Q13 Which one of the following statements is correct?
A Breaking covalent bonds absorbs energy and making ionic bonds absorbs energy.
B Bond breaking is endothermic and bond making is exothermic.
C Bond breaking is exothermic and bond making is endothermic.
D Breaking bonds releases energy and making bonds absorbs energy.

Q14 The bond energy for the H—F bond is equal to the enthalpy change for which process?
A $H^+(g) + F^-(g) \rightarrow HF(g)$
B $HF(g) \rightarrow H(g) + F(g)$
C $\frac{1}{2}F_2(g) + \frac{1}{2}H_2(g) \rightarrow HF(g)$
D $HF(g) \rightarrow \frac{1}{2}F_2(g) + \frac{1}{2}H_2(g)$

Q15 When a sample of a pure hydrocarbon (melting point 85 °C) cools, the temperature is observed to remain constant as it solidifies. Which statement accounts for this observation?
A The heat released in the change of state equals the heat loss to the surroundings.
B The temperature of the system has fallen to room temperature.
C The solid which forms insulates the system, preventing heat loss.
D Heat is gained from the surroundings as the solid forms, maintaining a constant temperature.

Q16 Consider the following equation:

$$6CO_2(g) + 6H_2O(l) \rightarrow C_6H_{12}O_6(s) + 6O_2(g)$$
$$\Delta H = 2824\,kJ\,mol^{-1}$$

What is the enthalpy change associated with the production of 100.0 g of $C_6H_{12}O_6$?
A 157 kJ   **C** 508 kJ
B 282 kJ   **D** 1570 kJ

Q17 $N_2(g) + O_2(g) \rightarrow 2NO(g)$ $\Delta H = 180.4 \, kJ \, mol^{-1}$

$N_2(g) + 2O_2(g) \rightarrow 2NO_2(g)$ $\Delta H = 66.4 \, kJ \, mol^{-1}$

Use the enthalpy values to calculate ΔH for the reaction:

$NO(g) + \frac{1}{2}O_2(g) \rightarrow NO_2(g)$

A −57 kJ   C 57 kJ
B −114 kJ   D 114 kJ

Standard Level Paper 1, May 00, Q18

Q18

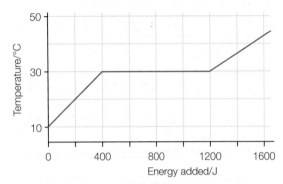

The heating curve for 10 g of a substance is given above. How much energy would be required to melt completely 40 g of the substance that is initially at 10 °C ?

A 4800 J   C 1600 J   B 2400 J   D 800 J

Q19 The bond energies for H_2, I_2 and HI are 432, 149 and 295 kJ mol⁻¹ respectively. From these data, what is the enthalpy change (in kJ) for the reaction below?

$H_2(g) + I_2(g) \rightarrow 2HI \, (g)$

A +9   C −286
B +286   D −9

Q20 Consider the following reaction:

$N_2(g) + 3H_2(g) \rightarrow 2NH_3(g)$

Bond enthalpies (in kJ mol⁻¹) involved in the reaction are:

N≡N	a
H−H	b
N−H	c

Which expression could be used to calculate the enthalpy of reaction?

A $a + 3b − 6c$ C $a − 3b + 6c$
B $6c − a + 3b$ D $a + 3b − 2c$

Paper 2 IB questions and IB style questions

Q1 a i Explain what is meant by the term *standard enthalpy of reaction*.   [3]

ii Describe an experiment to determine the enthalpy change of the reaction between dilute hydrochloric acid and aqueous sodium hydroxide. Show how the value

of ΔH would be calculated from the data obtained.   [9]

iii Draw an enthalpy diagram for the neutralization reaction above. Indicate on your diagram the enthalpy change of the reaction and hence compare the relative stabilities of reactant and products.   [4]

b Explain, giving one example, the usefulness of Hess's law in determining ΔH values.   [4]

Standard Level Paper 2, May 00, Q5

Q2 a i Define the term *average bond enthalpy*. [3]

ii Explain why the fluorine molecule, F_2, is not suitable as an example to illustrate the term average bond enthalpy.   [1]

b i Using values from Table 10 of the IB *Chemistry data booklet*, calculate the enthalpy change for the following reaction:

$CH_4(g) + F_2(g) \rightarrow CH_3F(g) + HF(g)$ [3]

ii Sketch an enthalpy diagram for the reaction.   [2]

iii Without carrying out a calculation, suggest, with a reason, how the enthalpy change for the following reaction compares with that of the previous reaction.

$CH_3F(g) + F_2(g) \rightarrow CH_2F_2(g) + HF(g)$ [2]

Q3 In aqueous solution, lithium hydroxide and hydrochloric acid react as follows.

$LiOH(aq) + HCl(aq) \rightarrow LiCl(aq) + H_2O(l)$

The data below is from an experiment to determine the standard enthalpy change of this reaction.

50.0 cm³ of a 0.500 mol dm⁻³ solution of LiOH was mixed rapidly in a glass beaker with 50.0 cm³ of a 0.500 mol dm⁻³ solution of HCl.

Initial temperature of each solution = 20.6 °C
Final temperature of the mixture = 24.1 °C

a State, with a reason, whether the reaction is exothermic or endothermic.   [1]

b Explain why the solutions were mixed rapidly.   [1]

c Calculate the enthalpy change of this reaction in kJ mol⁻¹. Assume that the specific heat capacity of the solution is the same as that of water.   [4]

d Identify the major source of error in the experimental procedure described above. Explain how it could be minimized.   [2]

e The experiment was repeated but with an HCl concentration of 0.520 mol dm⁻³ instead of 0.500 mol dm⁻³. State and explain what the temperature change would be.   [2]

6 Kinetics

STARTING POINTS

- Chemical reactions proceed at different rates.
- The relative rate of formation of products or consumption of reactants is related to the stoichiometry.
- The rate of a chemical reaction is the change in concentration with time.
- The rate of a chemical reaction can be measured by following a chemical or physical property that changes as the reaction proceeds.
- Collision theory (based on kinetic theory) and the Maxwell–Boltzmann distribution explain the molecular processes involved in chemical reactions.
- The activation energy acts as a barrier to a chemical reaction.
- Catalysts increase the rates of chemical reactions.
- There are two very different types of catalysis: homogeneous catalysis and heterogeneous catalysis.
- A chemical reaction often proceeds by a number of simple chemical reactions and intermediates: these steps are known as the mechanism.

6.1 Rates of reaction

Introduction

6.1.1 Define the term *rate of reaction*.

6.1.2 Describe suitable experimental procedures for measuring rates of reactions.

6.1.3 Analyse data from rate experiments.

The branch of chemistry concerned with reaction rates and the sequence of elementary steps by which a chemical reaction occurs is called **reaction kinetics** or **chemical kinetics**. The study of kinetics allows chemists to:

- determine how quickly a reaction will take place
- determine the conditions required for a specific reaction rate
- propose a reaction mechanism.

Reaction rates

Some reactions are very fast, for example neutralization and precipitation reactions (Chapter 1). Some reactions are slow, for example the enzymatic browning of fruits (Chapter 26), and some very slow, for example rusting (Chapter 9).

The **rate** of a chemical reaction is a measure of the 'speed' of the reaction: those reactions which are complete in a relatively short space of time are said to have high rates. The rate refers to the change in the amount (if it is a liquid or solid) or concentration (if it is a gas or solution) of a reactant or product. The rate is defined as the change in concentration or amount of a reactant or product with time, t:

$$\text{rate} = \frac{\text{change in concentration}}{\text{change in time}} \quad \text{or} \quad \text{rate} = \frac{(\text{concentration at time } t_2 - \text{concentration at time } t_1)}{(\text{time } t_2 - \text{time } t_1)}$$

Figure 6.1 shows a graphical method of visualizing how reactant and product concentrations are related to time. [A] represents reactant concentrations and [C] represents product concentrations. A 1:1 molar ratio exists between A consumed and C formed. $[C]_\infty$ represents the concentration of the product after infinite time has passed and the reaction has stopped.

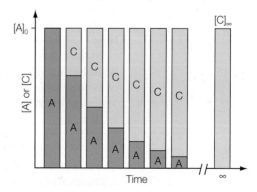

Figure 6.1 The relationship between reactant and product concentrations

In symbols: rate = $\dfrac{\Delta[A]}{\Delta t}$ or, using calculus notation, rate = $\dfrac{d[A]}{dt}$

where [A] represents the concentration or amount of a reactant or product, t_2 is greater than t_1 and Δ, the Greek letter delta, and d, the first differential, both indicate a small change in a quantity. The usual units for reaction rate are moles per cubic decimetre per second ($mol\,dm^{-3}\,s^{-1}$).

Worked examples

0.04 mol of a substance is produced in a 2.5 dm^3 vessel in 20 seconds. What is the rate of reaction?

Determine the amount produced in 1.0 dm^3: $\quad$ concentration = $\dfrac{0.04\,mol}{2.5\,dm^3}$ = 0.016 $mol\,dm^{-3}$

Determine the amount produced per second: $\quad$ rate = $\dfrac{0.016\,mol\,dm^{-3}}{20\,s^{-1}}$ = $8 \times 10^{-4}\,mol\,dm^{-3}\,s^{-1}$

22 grams of carbon dioxide is produced in 15 seconds in a vessel of capacity 4 dm^3. What is the rate of reaction?

Amount of CO_2 = $\dfrac{22\,g}{44\,g\,mol^{-1}}$ = 0.5 mol

Again, adjusting the volume to 1 dm^3: $\dfrac{0.5\,mol}{4\,dm^3}$ = 0.125 mol dm^{-3}

Again, adjusting for the time: rate = $8.33 \times 10^{-3}\,mol\,dm^{-3}\,s^{-1}$

Acidified hydrogen peroxide and aqueous potassium iodide react according to the following equation:

$$2H^+(aq) + H_2O_2(aq) + 2I^-(aq) \rightarrow I_2(aq) + 2H_2O(l)$$

It was found that the concentration of iodine was 0.06 $mol\,dm^{-3}$ after allowing the reactants to react for 30 seconds. Calculate the average rate of formation of iodine during this time.

average rate = $\dfrac{(\text{concentration at time } t_2 - \text{concentration at time } t_1)}{(\text{time } t_2 - \text{time } t_1)}$

average rate = $\dfrac{(0.06\,mol\,dm^{-3} - 0\,mol\,dm^{-3})}{(30\,s - 0\,s)}$ = $2 \times 10^{-3}\,mol\,dm^{-3}\,s^{-1}$

This means that *on average* 2×10^{-3} moles of iodine are being formed per cubic decimetre (litre) every second.

If the balanced equation for the overall reaction is known, then the rates of change in concentrations in all reactants and products are related to each other via the coefficients in the balanced equation (Chapter 1).

Any reaction can be represented by the following general equation:

aA $\quad +\quad$ bB $\quad \rightarrow \quad$ cC $\quad + \quad$ dD
reactants $\qquad\qquad\qquad\qquad$ products

The relative rates of reaction are given by the following expression:

$$\text{rate} = -\dfrac{1}{a}\dfrac{d[A]}{dt} = -\dfrac{1}{b}\dfrac{d[B]}{dt} = +\dfrac{1}{c}\dfrac{d[C]}{dt} = +\dfrac{1}{d}\dfrac{d[D]}{dt}$$

The *negative* sign indicates that the concentrations of the reactants A and B *decrease* with time, whereas the *positive* sign indicates that the concentrations of the products C and D *increase* with time.

In the example above for the reaction between acidified hydrogen peroxide and aqueous potassium iodide, the average rate of appearance of water is twice the average rate of appearance of iodine, that is, $4 \times 10^{-3}\,mol\,dm^{-3}\,s^{-1}$. This is because two water molecules are formed for every iodine molecule formed.

In symbols:

$$\text{rate} = \dfrac{1}{2}\dfrac{\Delta[I_2(aq)]}{\Delta t} = \dfrac{\Delta[H_2O(l)]}{\Delta t}$$

The rate of disappearance or consumption of hydrogen peroxide is the same as the rate of appearance of iodine, that is, $2 \times 10^{-3}\,mol\,dm^{-3}\,s^{-1}$.

In symbols:

$$\text{rate} = \frac{-\Delta[\text{H}_2\text{O}_2(\text{aq})]}{\Delta t} = \frac{\Delta[\text{I}_2(\text{aq})]}{\Delta t}$$

This is because one molecule of iodine is formed for every hydrogen peroxide molecule consumed. The negative sign in this expression indicates a decrease in the peroxide concentration with time.

Figure 6.2 shows a graph of the amount or concentration of a reactant against time. (This form of graph is obtained in most reactions, with the exception of autocatalysis or zero order reactions (Chapter 16).)

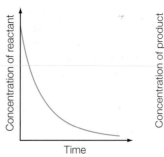

Figure 6.2 Graph of concentration or amount of reactant against time

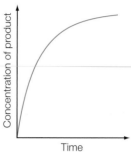

Figure 6.3 Graph of concentration or amount of product against time

You can see that the gradient of the graph continually decreases with time and, hence, the rate of reaction decreases with time. The reaction rate is zero when the reactants are all consumed and the reaction stops.

Figure 6.3 shows a graph of the concentration or amount of product against time. The gradient of the graph decreases progressively with time because the rate of reaction decreases with time as the reactants are consumed.

The graphs show that rate of a reaction is not constant but varies with time due to changes in the concentrations of reactants. Since the rate of a reaction varies with time, it is often appropriate to express the rate at a particular time instead of over a time period.

History of Chemistry

The basis of chemical kinetics is the measurement of the rates of chemical reactions. The first measurements of a rate of a reaction were performed in 1850 by the German chemist **Ludwig Wilhelmy** (1812–1864), who demonstrated that the hydrolysis of sucrose into glucose and fructose depends on the first power of the sucrose concentration. This is known as a first-order reaction (Chapter 16). He also examined the influence of temperature on the reaction. He followed the conversion using a polarimeter. However, since he did not have a thermostatted water bath, he had to note the variation in temperature during the day and make small corrections to the observed rate.

The **instantaneous rate of reaction** can be determined graphically (see Figures 6.4 and 6.5) from a graph of product or reactant concentration or amount against time. The instantaneous rate is the rate of a reaction at a particular time, unlike the **average rate**, which is the average rate over a particular time interval.

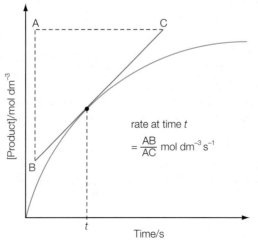

Figure 6.4 Concentration–time graph for the formation of a product. The rate of formation of product at time *t* is the gradient (or slope) of the curve at this point

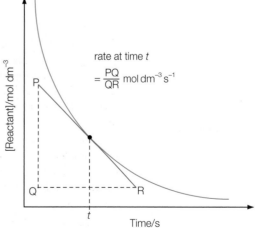

Figure 6.5 Concentration–time graph for the disappearance of a reactant. The rate of loss of reactant at time *t* is the gradient (or slope) of the curve at this point

The instantaneous rate of reaction at any time is equal to the gradient or slope of the graph at that time. The rate will either be positive or negative, depending on whether the y-axis shows the concentration of a product or a reactant.

The steeper the gradient of the graph, the faster the reaction, and the higher its rate. When the graph is horizontal, that is, the gradient is zero, the rate of reaction is zero, indicating the reaction has finished.

In practical situations, raw data is typically collected of some property that changes with time. This raw data may be directly proportional to the reactant concentration, for example absorbance, or may have a more complicated relationship, for example pH, which is a logarithmic function. This raw data can be converted into a rate expressed in moles per cubic decimetre.

It is important in kinetics to discover how the rate of reaction varies with concentration of the reactants. It allows chemists to deduce the order and the rate expression for the reaction (Chapter 16). One simple approach is to draw a number of tangents on a concentration–time graph (Figure 6.6) and then plot a graph of the rates (the numerical value of the gradients) against concentration. Many reactants show a directly proportional relationship between concentration and rate (which is known as first-order (Chapter 16)).

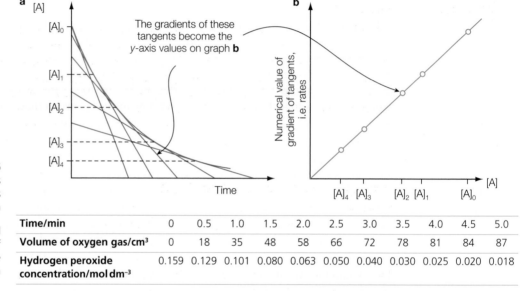

Figure 6.6
a Concentration versus time graph; **b** rate versus concentration graph

Table 6.1 The catalysed decomposition of hydrogen peroxide solution

Time/min	0	0.5	1.0	1.5	2.0	2.5	3.0	3.5	4.0	4.5	5.0
Volume of oxygen gas/cm³	0	18	35	48	58	66	72	78	81	84	87
Hydrogen peroxide concentration/mol dm⁻³		0.159	0.129	0.101	0.080	0.063	0.050	0.040	0.030	0.025	0.020	0.018

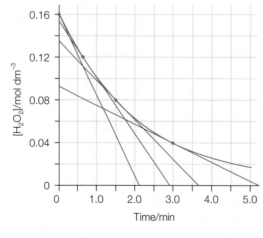

Figure 6.7 A graph of concentration of hydrogen peroxide against time

Table 6.1 shows data for the catalysed decomposition of hydrogen peroxide solution into oxygen gas and water.

A graph of the concentration of hydrogen peroxide solution against time (Figure 6.7) allows the gradients at various points on the curve to be calculated. For example, at the start of the reaction the gradient is given by:

$$\text{gradient} = \frac{0.16\,\text{mol dm}^{-3}}{2.1\,\text{min}}$$

$$= -0.076\,\text{mol dm}^{-3}\,\text{min}^{-1}$$

Thus the rate of decomposition at the start of the reaction is $0.076\,\text{mol dm}^{-3}\,\text{min}^{-1}$. Similar graphical calculations at other points produce additional rate values (Table 6.2).

Table 6.2 Hydrogen peroxide concentrations and corresponding values of instantaneous rate of decomposition

Hydrogen peroxide concentration/mol dm⁻³	0.16	0.12	0.08	0.04
Rate/mol dm⁻³ min⁻¹	0.076	0.049	0.033	0.019

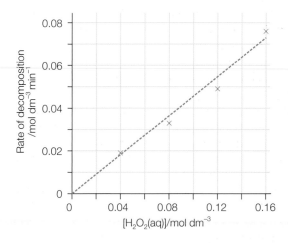

The relationship between the concentration and the rate of decomposition of hydrogen peroxide is illustrated in the graph of rate versus concentration in Figure 6.8. The dotted line of best fit shows that, within the limits of experimental error, the rate of decomposition of hydrogen peroxide is directly proportional to its concentration.

Figure 6.8 A graph of rate of hydrogen peroxide decomposition against concentration

Extension: Initial rates

Measuring rates from a concentration–time graph involves drawing a graph and measuring the gradients to the curve at a number of points, at least five. This can be an inaccurate process towards the end of the reaction when it is slowing down and the change in rate is relatively small. This problem can be avoided by measuring the initial rates in a series of investigations where the initial concentration of the reactant under investigation is varied (Chapter 16). In this approach the amount of a product (or loss of a reactant) is measured over a small period of time. Since the reactant concentration changes very slightly, the rate will be approximately constant and the initial rate is calculated by dividing the change in concentration by the time taken.

Measuring rates of reaction

To measure the rate of reaction the reactants need to be mixed together so that the reaction begins. The concentration of one of the reactants (or products) is then measured against time. The temperature (and for gaseous reactions, the pressure) must be controlled and kept constant.

There are many different ways in which the rate of reaction can be measured for a particular reaction. All of them measure either directly or indirectly a change in the concentration of either a reactant or product. Suitable changes include:

- colour
- formation of a precipitate
- change in mass, for example a gas produced, causing a loss of mass
- volume of gas produced
- time taken for a given mass of a product to appear
- pH
- temperature.

Reactions that produce gases

Reactions that produce gases are most easily investigated by collecting and measuring the volume of gas produced in a gas syringe. The volume of gas collected will increase as the concentration of the reactants decreases. (The rate of increase of volume of gas (tangent to the volume–time curve) can be used as a measure of reaction rate.)

Figure 6.9 shows apparatus suitable for investigating the reaction between calcium carbonate and dilute hydrochloric acid. This arrangement ensures the two reactants are kept separate while the apparatus is set up so that the start time can be accurately recorded.

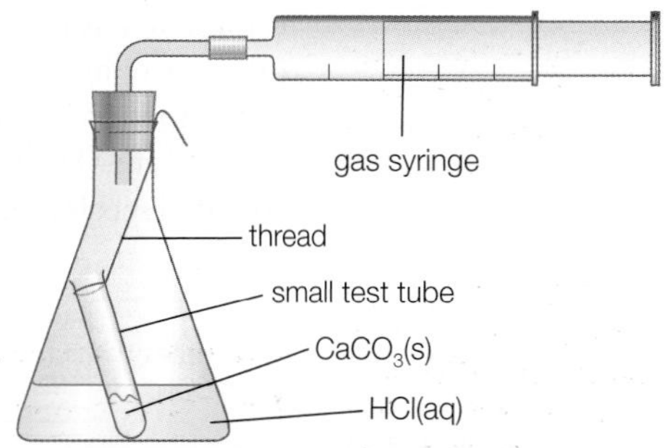

gas syringe

thread

small test tube

$CaCO_3(s)$

$HCl(aq)$

Figure 6.9 Apparatus used to study the rate of a reaction that releases a gas

Figure 6.10 shows a typical graph (line II) of the results where total volume of carbon dioxide gas is plotted against time. The reaction is fastest when the graph is steepest at the beginning of the reaction. The reaction finishes when the graph becomes horizontal, that is, when there is no further production of hydrogen.

The other line (line I) indicates data from a second similar experiment using the same amounts of calcium carbonate and hydrochloric acid but with conditions changed so the reaction is faster. This might be done by using acid at a higher temperature, using acid of a higher concentration or using powdered calcium carbonate rather than lumps.

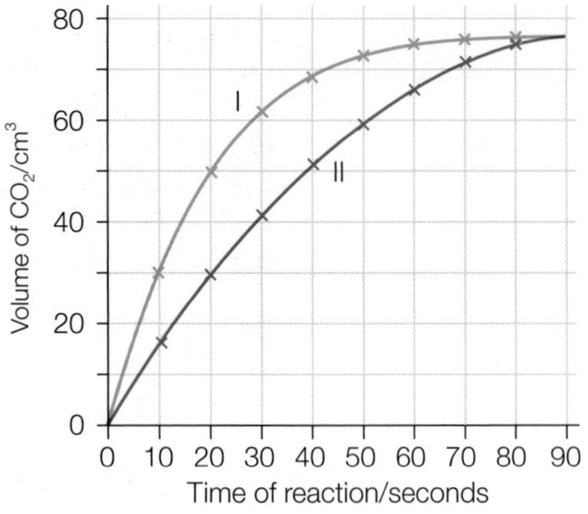

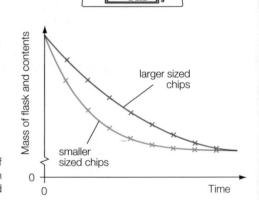

Figure 6.10 A graph of total volume of carbon dioxide collected against time

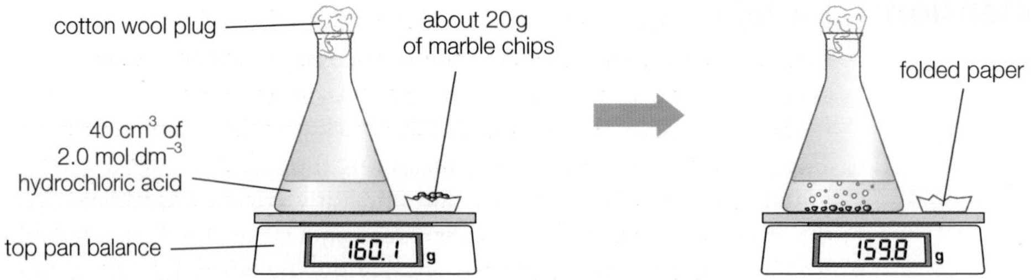

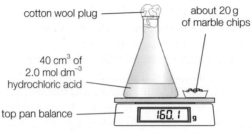

Figure 6.11 Following the rate of reaction between marble chips (calcium carbonate) and hydrochloric acid

cotton wool plug
about 20 g of marble chips
40 cm^3 of 2.0 mol dm^{-3} hydrochloric acid
top pan balance
160.1 g
folded paper
159.8 g

Alternatively, a reaction producing a gas can be performed in an open flask placed on an electronic balance. The reaction can then be investigated by recording the loss of mass as the gas is produced. This is shown in Figure 6.11 for the reaction between calcium carbonate and hydrochloric acid. Figure 6.12 shows the effect of chip size on a graph of mass of flask and contents versus time.

Figure 6.12 The effect of chip size on a graph of mass of flask and contents versus time for the reaction between calcium carbonate and hydrochloric acid

larger sized chips
smaller sized chips
Mass of flask and contents
Time

Reactions that produce a colour change

If one of the reactants or products of a reaction is highly coloured, the intensity of the colour can be used to measure the rate of reaction. If a reactant is coloured, the colour of the reaction mixture will fade during the reaction. If a product is coloured, then reaction mixture will gradually become more intensely coloured as the reaction proceeds. An instrument known as a colorimeter can measure the colour intensity. Later, experiments can be done to find the relationship between the colour intensity and the concentration of either the reactant or product.

In a colorimeter (see Figure 6.13), a narrow beam of light passes through the reaction mixture towards a sensitive photocell. The current generated within the photocell depends on the intensity of light that was transmitted through the reaction mixture, which in turn depends on the concentration of the coloured product or reactant. The colorimeter is used to measure absorbance against time (Figure 6.14) because absorbance is directly proportional to the concentration of the coloured species (provided the concentration is relatively low) (Chapter 21).

Figure 6.13 Block diagram of a colorimeter or spectrometer

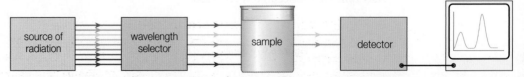

source of radiation
wavelength selector
sample
detector

An example of a reaction that can be investigated using a colorimeter is the iodination of propanone in the presence of dilute acid. In this reaction the iodine is the only coloured species and its colour varies from a pale yellow to a darker orange or brown colour, depending upon its concentration. The lower the light absorbance, the further the reaction has progressed. Data logging may be used to investigate this reaction.

$$CH_3COCH_3(aq) + I_2(aq) \rightarrow CH_3COCH_2I(aq) + HI(aq)$$

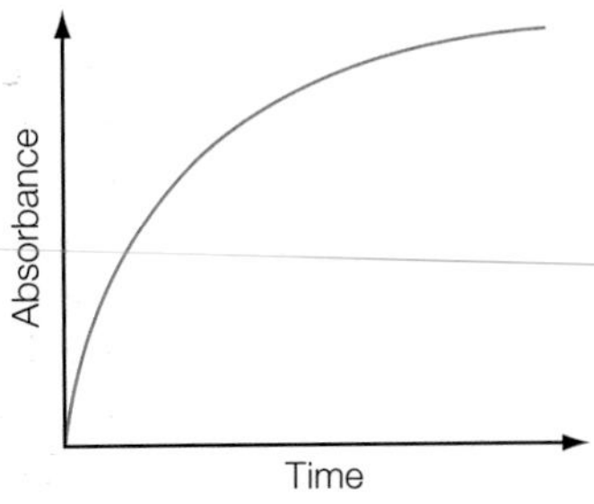

Figure 6.14 A typical graph of absorbance against time. This is equivalent to the volume-against-time graph in Figure 6.10

Reactions that involve a change in ion concentration

If one of the reactants or products in a reaction is either hydroxide ions, $OH^-(aq)$ or hydrogen ions, $H^+(aq)$ (oxonium ions $H_3O^+(aq)$), then there will be a change in pH (Chapter 8). This can be followed with a pH probe and meter.

If there is an overall change in the number of ions during a reaction, then there will a change in conductivity that can be measured using a conductivity probe and meter (Figure 6.15). Conductivity increases in the reaction mixture if there is an overall increase in the number of ions during a reaction; it decreases if there is an overall decrease in the number of ions. The conductivity also varies with the size of the ions: generally, smaller ions move faster and have higher conductivities than larger, slower moving ions.

For example, the alkaline hydrolysis of bromoethane can be followed using conductivity measurements. As this reaction proceeds the small and fast moving hydroxide ions are consumed and replaced by slower moving bromide ions. Therefore, the electrical conductivity decreases as the reaction proceeds.

$$C_2H_5Br(l) + OH^-(aq) \rightarrow C_2H_5OH(aq) + Br^-(aq)$$

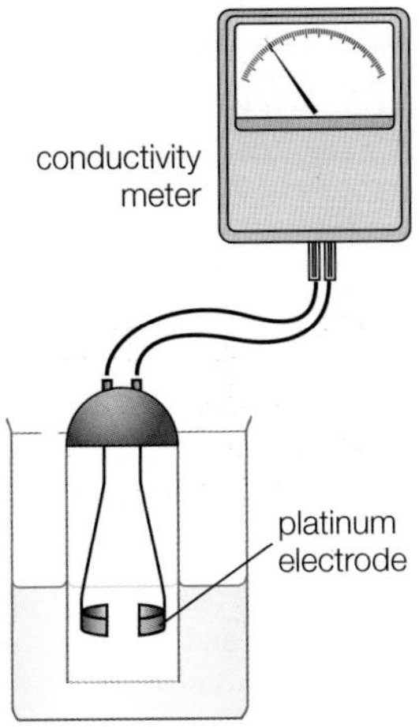

Figure 6.15 A conductivity cell and meter

Pressure and volume changes

Some reactions involving liquids show a small change in volume that can be measured and recorded. Reactions between gases are often investigated by measuring and recording changes in volume (at constant pressure) (Figure 6.16) or pressure (at constant volume). This technique can only be used if the number of moles of reactants is different from the number of products, for example:

$$2NO(g) + O_2(g) \rightarrow 2NO_2(g)$$

However, if the number of moles of reactants is equal to the number of moles of products, there will be no change in volume and hence pressure, for example:

$$H_2(g) + I_2(g) \rightarrow 2HI(g)$$

It is important to ensure that temperature is kept constant during these reactions so that pressure and volume changes are not, in part, due to temperature changes.

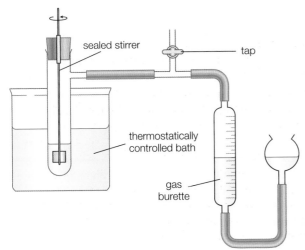

Figure 6.16 Apparatus for measuring changes in gas volume in a reaction maintained at constant pressure

Withdrawal of samples and titration

For some reactions small samples of the reaction mixture can be removed and then analysed by performing an acid–base or redox titration (Chapter 1) with a standard solution (Figure 6.17). This will allow the amount of a particular reactant remaining to be determined. The small sample

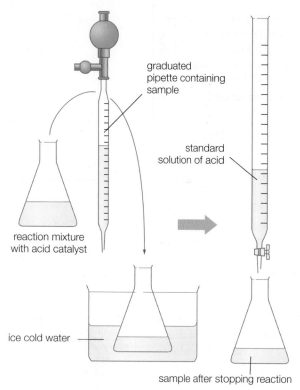

Figure 6.17 Following the course of an acid-catalysed reaction via sampling and titration with alkali

graduated pipette containing sample

standard solution of acid

reaction mixture with acid catalyst

ice cold water

sample after stopping reaction

of the reaction mixture is usually added to a large volume of cold water so the reaction is stopped, or at least slowed down. It suffers from being a destructive technique. This method can be used to measure the rate of saponification (alkaline hydrolysis) of ethyl ethanoate (Chapter 1):

$$C_2H_5COOC_2H_5(l) + NaOH(aq)$$
$$\rightarrow C_2H_5COO^-Na^+(aq) + C_2H_5OH(l)$$

At regular intervals during the reaction, a sample of the reaction mixture is taken and titrated against hydrochloric acid using a suitable indicator (Figure 6.17). This allows the concentration of the sodium hydroxide remaining in the reaction mixture to be calculated (Chapter 1). The smaller the volume of hydrochloric acid solution required for neutralization, the further the reaction has progressed.

$$NaOH(aq) + HCl(aq) \rightarrow NaCl(aq) + H_2O(l)$$

The reaction between iodine and propanone can also be followed by chemical analysis. At regular time intervals, samples of the reaction mixture are quenched with sodium hydrogencarbonate solution to remove the acid catalyst and stop the reaction. The iodine in the quenched mixture is then titrated with sodium thiosulfate using starch as the indicator.

$$I_2(aq) + 2S_2O_3^{2-}(aq) \rightarrow 2I^-(aq) + S_4O_6^{2-}(aq)$$

Rotation of the plane of polarized light

Some organic molecules, particularly sugars and amino acids, rotate the plane of polarized light (Chapter 10). The direction of rotation and the angle can be measured with a polarimeter. Changes in the concentrations of these optically active molecules causes a change in the amount of rotation.

For example, sucrose is hydrolysed in acidic solution:

$$C_{12}H_{22}O_{11}(aq) + H_2O(l) \rightarrow C_6H_{12}O_6(aq) + C_6H_{12}O_6(aq)$$
sucrose   water   glucose   fructose

During this reaction a change in the optical rotation of the solution occurs. Although both the reactant and the two products are optically active, the sizes and directions in which they rotate plane-polarized light differ. The overall change in optical activity can therefore be measured by means of a polarimeter (Chapter 20).

Using time as a measure of rate

In some reactions it is easy to measure the time it takes for a particular stage of a reaction to be reached. For example, in the reaction between sodium thiosulfate solution and dilute acid, a precipitate of sulfur is produced.

$$S_2O_3^{2-}(aq) + 2H^+(aq) \rightarrow SO_2(g) + H_2O(l) + S(s)$$

The effect of changing the temperature or the concentrations of two reactants can be investigated by carrying out the reaction in a conical flask placed on a cross drawn on a piece of paper (Figure 6.18). The time from the start of the reaction until there is sufficient sulfur to hide the cross when it is looked at from above the solution is recorded. The same total volume of solutions and the same cross have to be used in each experiment.

The graphs in Figures 6.19 and 6.20 summarize the results of kinetic investigations into the rate of reaction between sodium thiosulfate and hydrochloric acid. The volume of sodium thiosulfate used is proportional to its concentration and so the reciprocal of the time ($\frac{1}{t}$ or t^{-1}) or 'rate' taken for the obscuring of the cross is directly proportional to the **initial rate** in $mol\,dm^{-3}\,s^{-1}$.

Figure 6.18 The production of a colloidal suspension of sulfur from the reaction between thiosulfate and hydrogen ions

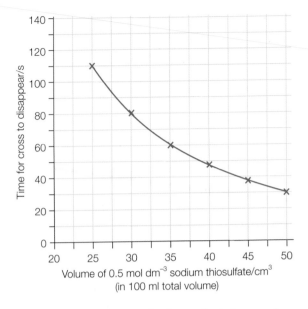

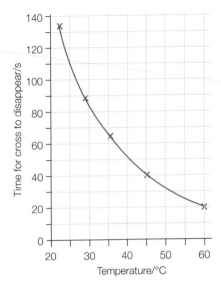

Figure 6.19 Sample raw data for the reaction between sodium thiosulfate and acid at different concentrations of sodium thiosulfate

Figure 6.20 Sample raw data for the reaction between sodium thiosulfate and acid at different temperatures

Extension: Clock reactions

Clock reactions are reactions in solutions that can be followed by a particular technique. Suppose a reactant A reacts to give products B and C. Two substances P and Q are added. P reacts with and removes B; only a small amount of P is added. Q interacts with B to produce a colour change, and so once all the P is consumed there is a sharp colour change. Clearly, it is vital that neither P nor Q interferes with the reaction of A.

Probably the most famous clock reaction is an iodine clock reaction, a redox reaction involving hydrogen peroxide and acidified iodide ions. The iodine is not seen when the reactants are first mixed because it is being converted to colourless iodide ions in a reaction with another reactant. The reaction of the acidified iodide ions with hydrogen peroxide is:

$$H_2O_2(l) + 2I^-(aq) + 2H^+(aq) \rightarrow I_2(aq) + 2H_2O(l)$$

In kinetic investigations of this reaction, sodium thiosulfate of known concentration together with a little starch solution are added to the mixture of hydrogen peroxide and acidified iodide ions. The iodine produced by the main reaction immediately reacts with thiosulfate ions:

$$I_2(aq) + 2S_2O_3^{2-}(aq) \rightarrow 2I^-(aq) + S_4O_6^{2-}(aq)$$

When all the thiosulfate has been used up, the iodine will be produced very rapidly and the reaction mixture suddenly turns blue-black (Figure 6.21) if starch is present (or brown if the starch is absent). In this clock reaction, if B were iodine then P would be thiosulfate and Q would be starch.

The time t (known as the induction period) from mixing the reactants to the appearance of the blue-black colour of

Figure 6.21 An iodine clock reaction

the starch–iodine complex is the time for a fixed amount of iodine to be formed. The appearance of the iodine indicates when a particular amount of iodine has been formed, regardless of the time required for this to occur. We can therefore simply use $1/t$ as measure of the initial rate of reaction.

Another common version of the iodine clock is that between peroxodisulfate(VI) and iodide ions:

$$S_2O_8{}^{2-}(aq) + 2I^-(aq) \rightarrow 2SO_4{}^{2-}(aq) + I_2(aq)$$

Investigations into kinetics are again performed by adding solutions of thiosulfate ions of different concentrations and some starch solution to reaction mixtures containing peroxodisulfate and iodide ions. The time from mixing to the appearance of the blue-black starch–iodine complex is again measured and $1/t$ is used as a measure of the initial reaction rate.

Another clock reaction is one where A is a halogenoalkane, RX, reacting by an S_N1 mechanism (Chapter 20) to give an alcohol, ROH, and halide ions, X⁻. P would be sodium hydroxide solution and Q an acid–base indicator, showing when the alkali was all used up (Chapter 20).

Extension: Fast and slow reactions

Some reactions that involve oppositely charged ions in solution have very high rates of reaction. Examples include precipitation (Chapter 1); for example, when barium ions are added to sulfate ions an almost instantaneous white precipitate of barium sulfate is formed:

$$Ba^{2+}(aq) + SO_4{}^{2-}(aq) \rightarrow BaSO_4(s)$$

Neutralization reactions between acids and alkalis (Chapter 8) are also very rapid, the essential reaction being:

$$H^+(aq) + OH^-(aq) \rightarrow H_2O(l) \qquad \text{or} \qquad H_3O^+(aq) + OH^-(aq) \rightarrow 2H_2O(l)$$

Redox reactions (Chapter 9) involving electron transfer are also usually fast; for example, the reaction between iodine and sodium thiosulfate:

$$I_2(aq) + 2S_2O_3{}^{2-}(aq) \rightarrow 2I^-(aq) + S_4O_6{}^{2-}(aq)$$

Reactions that involve covalent molecules tend to occur at much slower rates, for example esterification (Chapter 20):

$$CH_3COOH(l) + C_2H_5OH(l) \rightleftharpoons CH_3COOC_2H_5(aq) + H_2O(l)$$

This reaction is catalysed by hydrogen ions, $H^+(aq)$, released from concentrated sulfuric acid.

Slow reactions also occur between ions of the *same* charge, for example the reaction between iodide ions and peroxodisulfate ions:

$$S_2O_8{}^{2-}(aq) + 2I^-(aq) \rightarrow 2SO_4{}^{2-}(aq) + I_2(aq)$$

This reaction can be catalysed by iron(II) ions (Chapter 16).

6.2 Collision theory

6.2.1 **Describe** the kinetic theory in terms of the movement of particles whose average energy is proportional to temperature in kelvin.

6.2.2 **Define** the term *activation energy*, E_a.

6.2.3 **Describe** the collision theory.

6.2.4 **Predict** and **explain**, using the collision theory, the qualitative effects of particle size, temperature, concentration and pressure on the rate of a reaction.

6.2.5 **Sketch** and **explain** qualitatively the Maxwell–Boltzmann energy distribution curve for a fixed amount of gas at different temperatures and its consequences for changes in reaction rate.

6.2.6 **Describe** the effect of a catalyst on a chemical reaction.

6.2.7 **Sketch** and **explain** Maxwell–Boltzmann curves for reactions with and without catalysts.

Simple **collision theory** states that before a chemical reaction can occur, the following requirements must be met:

■ The reactants (ions, atoms or molecules) of the reactants must physically collide and come into direct contact with each other.

- For many reacting molecules **steric factors** are involved: the molecules must collide in the correct relative positions so their reactive atoms or functional groups are aligned. This is known as collision geometry (see Figure 6.22).

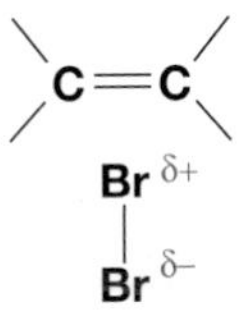

Figure 6.23 A bromine molecule undergoing polarization as it approaches an ethene molecule

Figure 6.22 Two nitrogen dioxide molecules approaching with sufficient kinetic energy to overcome the activation energy barrier must collide in the correct orientation in order to form dinitrogen tetroxide

no reaction   reaction

An example of the influence of steric factors is provided by the bromination of ethene (Figure 6.23) where the bromine molecule has to approach the pi bond of the double bond 'sideways on' (Chapter 20).

Another example from organic chemistry (Chapter 20) is provided by the reaction between 1-bromobutane molecules and hydroxide ions. This mechanism is known as an S_N2 mechanism and a successful collision involves the hydroxide ion approaching the carbon atom from behind. This is known as backside attack (Figure 6.24).

- Each of the colliding particles must be travelling at sufficient velocity so that when they collide there is enough kinetic energy to enable the reaction to occur. This fixed amount of kinetic energy that the particles need is required to overcome an endothermic 'energy barrier' (see Figure 6.25) whose value is the **activation energy**.

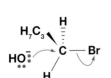

Figure 6.24 Backside attack on the carbon atom of a 1-bromobutane molecule by a hydroxide ion

Values of barrier heights vary widely between chemical reactions and control how rapidly reactions take place and how their rates respond to changes in temperature. (Most reactions involve a number of steps and the activation energy may not correspond to the height of any individual activation energy barrier. It is a concept best applied to the individual elementary steps of a multi-step reaction.)

Fast reactions are associated with low values of energy barriers (and hence activation energies) and slow reactions are associated with high values of energy barriers (and activation energies).

If the colliding species do not possess sufficient kinetic energy to surmount the energy barrier and/or the correct collision geometry then an ineffective collision will occur and the reacting species will not undergo a chemical reaction.

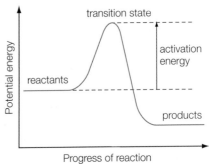

Figure 6.25 Concept of barrier height and activation energy (for an exothermic reaction)

Concentration

The term concentration refers to the numbers of particles (usually ions) present in a particular volume of solution. It is usually expressed in moles per decimetre cubed ($mol\,dm^{-3}$) (Chapter 1).

It is *generally* found that the greater the concentration of the reactants (A and B), the greater the rate of reaction. This is because increasing the concentration of the reactants (A and B) increases the number of collisions between particles of A and B and, therefore, increases the rate of reaction (Figure 6.26). In particular, a doubling in the concentration of one of the reactants *usually* doubles the rate of reaction. This is because the collision rate involving that reactant has been doubled.

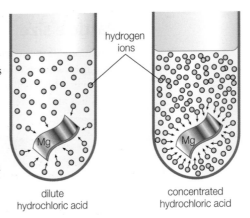

Figure 6.26 The effect of concentration on the rate of a reaction between magnesium and hydrochloric acid (hydrogen ions)

This also explains why the greatest rate of reaction occurs as soon as the reacting solutions are mixed, that is, when they are at their highest concentrations. As the reaction proceeds the concentrations of the chemicals decrease and the rate of reaction decreases because there are fewer collisions.

Pressure

When one or more of the reactants are gases an increase in pressure can lead to an increase in the rate of reaction (provided the order – see Chapter 16 – is positive). The increase in pressure forces the particles closer together which causes an increase in the collision rate and hence an increase in the rate of reaction.

An increase in pressure can be regarded as an increase in 'concentration' since more gas molecules will be present in a particular volume of space (Figure 6.27). Since liquids and solids undergo little change in volume when the pressure is increased (Chapter 1), their reactions rates are little affected by changes in pressure.

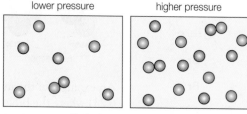
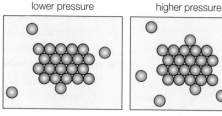

Figure 6.27 The effect of pressure on gaseous reactions and between a gas and a solid

Temperature

When particles in gases, liquids or solutions are heated, they move with higher velocities.

This has two consequences: firstly, they travel a greater distance in a given time and so will be involved in more collisions and hence an increase in rate (Figure 6.28). However, more importantly, at a higher temperature a larger proportion of the colliding species will have kinetic energies equal to or exceeding the energy barrier.

Frequently, a rise of 10 °C approximately doubles the initial rate of reaction (Chapter 16). (However, note that this relationship is a simplification and does not hold for all reactions over all temperatures.)

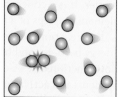
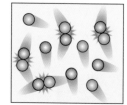

cold – slow movement, few collisions, little kinetic energy

hot – fast movement, more collisions, more kinetic energy

Figure 6.28 The effect of temperature on gaseous molecules

Particle size

Figure 6.29 The effect of particle size on the surface area of a solid reactant

When one of the reactants is a solid, the reaction takes place on the surface of the solid. If the solid is broken up into smaller pieces or particles, the surface area is increased, giving a greater area over which collisions can occur (see Figure 6.29). A number of important industrial catalysts are solids and the reactions they catalyse occur on the surface of the catalyst.

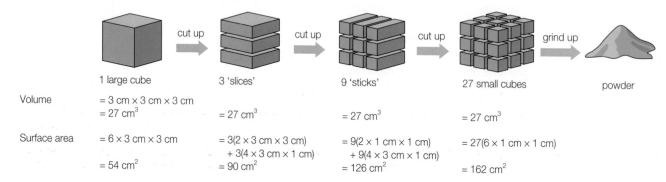

	1 large cube	3 'slices'	9 'sticks'	27 small cubes	powder
Volume	$= 3\,cm \times 3\,cm \times 3\,cm$ $= 27\,cm^3$	$= 27\,cm^3$	$= 27\,cm^3$	$= 27\,cm^3$	
Surface area	$= 6 \times 3\,cm \times 3\,cm$ $= 54\,cm^2$	$= 3(2 \times 3\,cm \times 3\,cm)$ $+ 3(4 \times 3\,cm \times 1\,cm)$ $= 90\,cm^2$	$= 9(2 \times 1\,cm \times 1\,cm)$ $+ 9(4 \times 3\,cm \times 1\,cm)$ $= 126\,cm^2$	$= 27(6 \times 1\,cm \times 1\,cm)$ $= 162\,cm^2$	

Light

The rates of some reactions are greatly increased by exposure to light. For example, the silver halides (Chapter 3), silver nitrate, hydrogen peroxide and nitric acid are all **photosensitive** and undergo partial decomposition (to form radicals, often in the form of reactive atoms) in the presence of sunlight.

$$2AgX(s) \rightarrow 2Ag(s) + X_2(g), \text{ where X represents a halogen}$$
$$2AgNO_3(s) \rightarrow 2Ag(s) + 2NO_2(g) + O_2(g)$$
$$2H_2O_2(aq) \rightarrow 2H_2O(l) + O_2(g)$$
$$4HNO_3(aq) \rightarrow 2H_2O(l) + 4NO_2(g) + O_2(g)$$

Mixtures of hydrogen and bromine, or methane and chlorine (Chapter 10) do not react in the dark, but in the presence of light a very rapid reaction takes place.

$$H_2(g) + Br_2(g) \rightarrow 2HBr(g)$$
$$CH_4(g) + Cl_2(g) \rightarrow CH_3Cl(g) + HCl(g)$$

Table 6.3 Summary of the factors affecting rates of reaction

Factor	Reactions affected	Change made in conditions	Usual effect on the initial rate of reaction
Temperature	All	Increase	Increase
		Increase by 10 K	Approximately doubles
Concentration	All	Increase	Usually increases (unless zero order)
		Doubling of concentration of one of the reactants	Usually exactly doubles (if first order)
Light	Generally those involving reactions of mixtures of gases including the halogens	Reaction in sunlight or ultraviolet light	Very large increase
Particle size	Reactions involving solids and liquids, solids and gases or mixtures of solids	Powdering the solid, resulting in a large increase in surface area	Very large increase

The concept of reaction order is discussed in Chapter 16.

The Maxwell–Boltzmann distribution

Theoretical calculations and experimental measurements both suggest that the translational (and vibrational) kinetic energies of gas molecules in an ideal gas are distributed over a range known as a **Maxwell–Boltzmann distribution** (see Figure 6.30). Similar distributions of kinetic energies are present in the particles in solutions and liquids. The activation energy, E_a, is defined as the minimum amount of kinetic energy which colliding or vibrating molecules require in order to react.

The total area under the curve is directly proportional to the total number of molecules and the area under any portion of the curve is directly proportional to the number of molecules with kinetic energies in that range. When the temperature of a gas, liquid or solution is increased a number of changes occur in the shape of the Maxwell–Boltzmann distribution (see Figure 6.31).

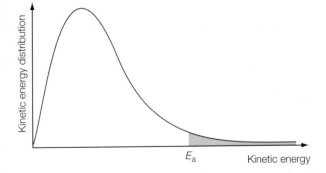

Figure 6.30 Maxwell–Boltzmann distribution of kinetic energies in an ideal gas

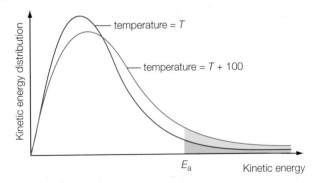

Figure 6.31 Maxwell–Boltzmann distribution of kinetic energies in a solution or gas at two different temperatures

■ The peak of the curve moves to the right so the most likely value of kinetic energy for the molecules increases.

■ The curve flattens so the total area under it and, therefore, the total number of molecules remains constant.

■ The area under the curve to the right of the activation energy, E_a, increases. This means that at higher temperatures, a greater percentage of molecules have energies equal to or in excess of the activation energy, E_a.

When the temperature is increased the collision rate increases because the average speeds of particles in the gas, liquid or solution are increased. *However*, this has only a very minor effect on the rate of reaction and cannot account for the rapid increase in reaction as the temperature increases. For many reactions when the temperature of the reactants is increased by 10 °C, the collision rate increases by about 2%, but the rate of reaction increases by about 100% because of the increased number of molecules that possess the activation energy.

■ **Extension:** ## Rate of reaction in solution

Simple collision theory can be modified and extended to reactions in solution. In solutions, which contain solvated molecules and ions rather than simple molecules or atoms, interactions are known as encounters rather than collisions. It would be expected that encounter rates should be smaller than collision frequencies because the solvent molecules reduce the collision rate between reactants. However, encounters may be more likely than collisions where molecules are trapped in a temporary 'cage' of solvent molecules (Figure 6.32).

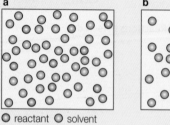
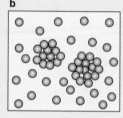

○ reactant ○ solvent

Figure 6.32 The effect of a solvent 'cage' on the rate of reaction in a solvent

---

Catalysts

A **catalyst** is a substance that can increase the rate of a reaction but remains chemically unchanged at the end of the reaction. Catalysts are important in many industrial processes, where they are frequently transition metals or their compounds (Chapters 3, 13 and 23). Catalysts increase the rate of reactions by providing a new alternative pathway or **mechanism** for the reaction that has a lower barrier height than the uncatalysed pathway (see Figure 6.33). (This statement is simplistic since most reactions consist of a number of steps, each of which has its own associated activation energy.) There are two types of catalysts: homogeneous and heterogeneous (Chapter 16).

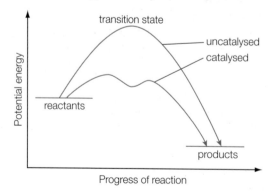

Figure 6.33 General enthalpy level diagram for uncatalysed and catalysed reactions of an exothermic reaction

Catalysts increase the rates of reversible and irreversible reactions. Catalysts do *not* alter the position of equilibrium (Chapter 7), they only increase the rate at which equilibrium is achieved. In other words, the presence of a catalyst does not increase the yield of products but increases the rate of their production. This is because a catalyst lowers both the forward and backward

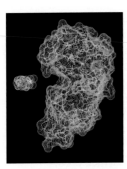

Figure 6.34 A computer model showing a small molecule entering the active site of a protein molecule

activation energy barrier heights, E_a, increasing the rates of both forward and backward reactions to the same degree. (This effect of a catalyst on forward and backward energy barriers is known as the principle of microscopic reversibility.) Hence, to find a good catalyst for a particular reaction it is sufficient to look for a good catalyst for the reverse reaction.

However, catalysts have no effect on reactions that are not thermodynamically spontaneous; in other words, they are not able to catalyse reactions that are not energetically favourable. This is because catalysts do *not* alter the enthalpy change, ΔH, or Gibbs free energy change, ΔG (Chapter 15), that occurs during the reaction.

Biological catalysts are known as enzymes (Chapter 23) and consist of proteins often associated with metal ions. A substance that decreases the rate of a reaction is called an **inhibitor**. An example of an inhibitor is the 'anti-knock' compound, tetraethyl lead(IV), used to prevent pre-ignition of 'leaded' petrol vapour (Chapter 10). There are many specific and general inhibitors known for enzymes: many nerve gases and poisons, for example cyanides, operate as enzyme inhibitors, often by interacting with the active site of the enzyme (Figure 6.34).

Catalysts are widely used in the chemical industry: examples include finely divided iron in the Haber process for making ammonia and platinum in the Contact process (Chapter 7). A complex organo-metallic catalyst, known as a Ziegler–Natta catalyst, is used in the production of polymers synthesized from alkenes (Chapter 23).

Aqueous hydrogen peroxide decomposes to water and oxygen. Solid manganese(IV) oxide ('manganese dioxide') acts as a catalyst (Figure 6.35):

$$2H_2O_2(aq) \rightarrow 2H_2O(l) + O_2(g)$$

The insoluble manganese dioxide can be filtered off, washed and dried before being reused as a catalyst. The decomposition of hydrogen peroxide can also be demonstrated using finely chopped pieces of fresh liver or blood which release the enzyme catalase (Chapter 22).

Figure 6.35 The production of an oxygen-filled foam from the manganese(IV) oxide catalysed decomposition of hydrogen peroxide (note that the demonstrator should be wearing a lab coat)

Another example of catalysis involves the oxidation of potassium sodium tartrate (potassium sodium 2,3-dihydroxybutanedioic acid) by hydrogen peroxide solution to give a mixture of oxygen and carbon dioxide gases. The reaction is catalysed by cobalt(II) chloride. As the experiment proceeds the pink colour of the aqueous cobalt(II) ions changes to green, revealing the presence of a cobalt(III) complex intermediate, before reverting to the original pink colour, indicating 'regeneration' of the catalyst (Figure 6.36). These two reactions demonstrate the two types of catalysis: homogeneous and heterogeneous catalysis (Chapter 16).

Figure 6.36 The reaction between tartrate ions and hydrogen peroxide in the presence of cobalt(II) ions, acting as a catalyst. The pink solution on the left contains cobalt(II) ions, the green solution in the middle contains a temporary green intermediate containing cobalt(III) ions, and the pink solution on the right contains regenerated cobalt(II) ions

SUMMARY OF KNOWLEDGE

▓ Rate = $\dfrac{\text{change in concentration or amount}}{\text{change in time}}$

The SI units of rate are $\text{mol}\,\text{dm}^{-3}\,\text{s}^{-1}$.

▓ For the reaction $aA + bB \rightarrow cC + dD$

$$\text{rate} = -\frac{1}{a}\frac{d[A]}{dt} = -\frac{1}{b}\frac{d[B]}{dt} = \frac{1}{c}\frac{d[C]}{dt} = \frac{1}{d}\frac{d[D]}{dt}$$

■ Rates are determined by measuring the change in concentration of a reactant or product over a period of time. Methods include titrations, pressure measurement (for gases), colour changes and mass changes.

■ The rate of a reaction can be affected by: concentration of reactants in solution, surface area of solid reactants, temperature, light, pressure (of gaseous reactants) and presence of catalysts.

■ The activation energy is the minimum kinetic energy that reactants need to form products. The activation energy is the energy barrier to a reaction (or elementary step) and controls the rate of the reaction.

■ Collision theory accounts for the variation in rate with temperature, surface area, concentration of reactants and use of catalysts.
  – Increasing the temperature. This increases the collision rate but a more significant factor is the increase in the proportion of reactants having a combined kinetic energy greater to or equal to the activation energy.
  – Increasing concentrations. This increases the chances of the reactant molecules colliding and reacting.
  – Increasing the surface area (for reactions that involve solids). This increases the surface area over which the reaction can occur.
  – Using a catalyst. This increases the number of successful collisions.

■ Requirements for a chemical reaction to occur: the reactants must collide together with sufficient kinetic energy; for many reactions the reactants must have the correct collision geometry (steric factor).

■ The Maxwell–Boltzmann distribution is a graph of the proportion of a sample of molecules that has a specific value of kinetic energy. Graphs at different temperatures show that the proportion of molecules with high kinetic energy increases with temperature.

■ A catalyst is a chemical substance that increases the rate of a reaction without itself undergoing any permanent chemical change. Catalysts provide a new reaction mechanism with a lower activation energy than the uncatalysed reaction.

■ *Examination questions – a selection*

Paper 1 IB questions and IB style questions

Q1 The reaction between *excess* magnesium carbonate and hydrochloric acid can be followed by measuring the volume of carbon dioxide produced with time. The results of one such reaction are shown below. How does the rate of this reaction change with time and what is the main reason for this change?

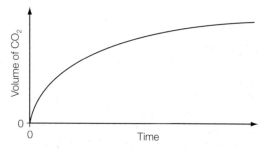

A The rate decreases with time because the acid becomes more dilute.

B The rate increases with time because the acid becomes more dilute.

C The rate decreases with time because the magnesium carbonate particles get smaller.

D The rate increases with time because the magnesium carbonate particles get smaller.

Q2 Zinc reacts with sulfuric acid as shown below.

$$Zn(s) + H_2SO_4(aq) \rightarrow ZnSO_4(aq) + H_2(g)$$

Two identical samples of zinc powder were reacted with separate samples of *excess* acid as follows:

Reaction 1: zinc added to $0.5\,mol\,dm^{-3}$ sulfuric acid
Reaction 2: zinc added to $2.0\,mol\,dm^{-3}$ sulfuric acid

What is the same for reactions 1 and 2?

A total mass of hydrogen formed
B total reaction time
C average rate of production of hydrogen gas
D initial reaction rate

Q3 The reaction between nitrogen and chlorine in the atmosphere under normal conditions is extremely slow.

Which statement best explains this?

A The concentration of nitrogen is too high, which slows the reaction.

B The molar mass of nitrogen molecules is less than that of chlorine molecules.

C Nitrogen and chlorine molecules are both non-polar molecules.

D Very few nitrogen and chlorine molecules have sufficient kinetic energy to react.

Q4 What is the action of a catalyst?

A provides a new mechanism or pathway for the reaction

B increases the enthalpy change, ΔH, for the reaction

C changes the sign of the enthalpy change, ΔH, for the reaction

D decreases the activation energy, E_a, for the forward reaction only

Q5 Which statement is correct for a collision between reactant particles leading to a reaction?

A Colliding particles must have different energies.

B All reactant particles must have the same energy.

C Colliding particles must have a kinetic energy higher than the activation energy.

D Colliding particles must have the same velocity.

Standard Level Paper 1, Nov 05, Q19

Q6 Which change of condition will decrease the rate of the reaction between excess calcium granules and dilute hydrochloric acid?

A increasing the amount of calcium

B increasing the concentration of the hydrochloric acid

C decreasing the temperature of the acid

D converting the calcium granules into powder

Q7 *Excess* magnesium was added to a beaker of aqueous hydrochloric acid on a balance. A graph of the mass of the beaker and contents was plotted against time (line 1).

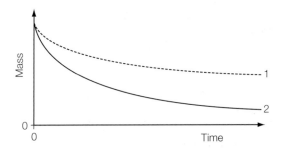

What change in the experiment could give line 2?

I the same mass of magnesium but in smaller pieces

II the same volume of a more concentrated solution of hydrochloric acid

III a lower temperature

A I only

B II only

C III only

D none of the above

Standard Level Paper 1, Nov 03, Q19

Q8 The rate of a reaction between two gases increases when the temperature is increased and a catalyst is added. Which statements are correct for the effect of these changes on the reaction?

	Increasing the temperature	Adding a catalyst
A	collision frequency increases	activation energy increases
B	activation energy increases	activation energy does not change
C	activation energy does not change	activation energy decreases
D	activation energy increases	collision frequency increases

Standard Level Paper 1, Nov 03, Q20

Q9 Hydrogen peroxide undergoes decomposition in the presence of a manganese(IV) oxide catalyst to produce oxygen and water. The graph below shows how the total volume of oxygen generated varies with time.

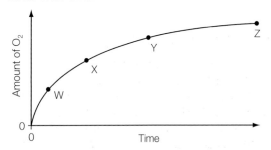

Which point on the graph represents the fastest rate of reaction?

A W   **B** X   **C** Y   **D** Z

Q10 For the reaction:

$$2HCl(aq) + CaCO_3(s) \rightarrow CaCl_2(aq) + H_2O(l) + CO_2(g)$$

which gives the fastest rate?

A $1.0\,mol\,dm^{-3}$ HCl and $CaCO_3$ chips

B $2.0\,mol\,dm^{-3}$ HCl and $CaCO_3$ chips

C $2.0\,mol\,dm^{-3}$ HCl and $CaCO_3$ powder

D $4.0\,mol\,dm^{-3}$ HCl and $CaCO_3$ powder

Q11 In the reaction shown below, which species may be acting as a catalyst?

$O_3 + Br \rightarrow BrO + O_2$
$BrO + O \rightarrow Br + O_2$

A O_2　　**B** Br　　**C** O　　　**D** BrO

Q12 An increase in the rate of a reaction is generally brought about by increases in which of the following?

I reactant concentration
II particle size
III temperature

A I and II only　　**C** I, II and III
B I and III only　　**D** II and III only

Q13 All of the following statements are correct **except**:

A The smaller the size of the reacting particles, the faster the reaction.
B The rate of a chemical reaction can be decreased by decreasing the temperature.
C Increasing the pressure often decreases the rates of reaction involving gases.
D The rates of most chemical reactions decrease with time.

Q14 When $100\,cm^3$ of $1.0\,mol\,dm^{-3}$ methanoic acid, HCOOH, are added to 1 gram of magnesium turnings at $20\,°C$, hydrogen gas is slowly produced. All of the following will increase the initial rate of hydrogen production **except**:

A substituting $1.0\,mol\,dm^{-3}$ hydrochloric acid for $1.0\,mol\,dm^{-3}$ methanoic acid
B using $300\,cm^3$ of $1.0\,mol\,dm^{-3}$ methanoic acid instead of $100\,cm^3$
C substituting powdered magnesium in the place of magnesium turnings
D increasing the temperature of the $1.0\,mol\,dm^{-3}$ methanoic acid to $60\,°C$

Q15 The rates of many reactions increase rapidly with small increases in temperature. Which of the following best accounts for this behaviour?

A The bonds become weaker.
B The activation energy decreases.
C The number of molecular collisions increases.
D The number of molecules with the necessary kinetic energy increases.

Q16 A catalyst:

A creates another reaction pathway
B is consumed in a chemical reaction
C is always a transition element or compound
D is always a solid

Q17 All of the following substances are observed to react with aqueous ethanoic acid at room temperature. Which one probably reacts fastest?

A a strip of zinc metal
B magnesium oxide powder
C marble chips (calcium carbonate)
D an aqueous solution of sodium hydroxide

Q18 What are the usual units for rate?

A $mol^{-1}\,(dm^3)^2\,s^{-1}$　　**C** $mol\,dm^{-3}\,s^{-1}$
B $mol\,dm^{-3}$　　　　　**D** $mol\,s^{-1}$

Q19 For which one of the following reactions would pressure measurements be the least sensitive for measuring the initial rate of reaction?

A $2H_2O_2(l) \rightarrow 2H_2O(l) + O_2(g)$
B $N_2O_4(g) \rightarrow 2NO_2(g)$
C $H_2(g) + I_2(g) \rightarrow 2HI(g)$
D $2NCl_3(l) \rightarrow N_2(g) + 3Cl_2(g)$

Q20 Which statement explains why the speed of some chemical reactions is increased when the surface area of the reactant is increased?

A This change increases the density of the reactant particles.
B This change increases the concentration of the reactant.
C This change exposes more reactant particles to a possible collision.
D This change alters the electrical conductivity of the reactant particles.

Paper 2 IB questions and IB style questions

Q1 **a** Define the term *rate of reaction*. [1]
　　b The reaction between gases **C** and **D** is slow at room temperature.
　　　i Suggest **two** reasons why the reaction is slow at room temperature. [2]
　　　ii A relatively small increase in temperature causes a relatively large increase in the rate of this reaction. State **two** reasons for this. [2]
　　　iii Suggest **two** ways of increasing the rate of reaction between **C** and **D** other than increasing temperature. [2]
　　　　　Standard Level Paper 2, Nov 05, Q2

Q2 The following results were obtained when 2.40 grams of magnesium ribbon were added to a large excess of dilute hydrochloric acid while the temperature was kept constant (by means of a water bath). The hydrogen gas was collected in a large gas syringe.

Time/seconds	Volume of hydrogen gas evolved/cm³
0	0
20	900
40	1400
60	1720
80	1950
100	2100
120	2240
140	2240

a What was the volume of gas produced in
  i the first 20 second interval (from 0 seconds to 20 seconds) [1]
  ii the second 20 second interval (from 20 seconds to 40 seconds) [1]
  iii the third 20 second interval (from 40 seconds to 60 seconds)? [1]
b Explain why the volume of hydrogen gas changes in each of these 20 second intervals. [2]
c Why is the volume the same after 140 seconds and 120 seconds? [1]
d How would the initial rate of production of gas change if
  i the temperature were increased [1]
  ii the volume of acid were diluted with an equal volume of water [1]
  iii the same mass of magnesium powder were used instead of magnesium ribbon? [1]
e How would the final volume of gas change if
  i a greater mass of magnesium were used with the same volume of acid [1]
  ii a larger volume of acid were used with the same mass of magnesium? [1]

f An alternative method of monitoring the change of rate of this reaction is to perform the reaction in an open flask on an electronic balance. Suggest why this method is less preferable. [2]
g State and explain what the relationship is between the rate of consumption of hydrochloric acid and the rate of formation of hydrogen gas. [2]
h Deduce the ionic equation and explain why the reaction is very unlikely to proceed directly via this reaction in one step. [2]

Q3 The graph below was obtained when calcium carbonate reacted with dilute hydrochloric acid, under two different conditions, X and Y.

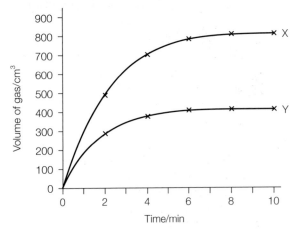

a **i** Name the gas produced in the reaction. [1]
  ii Write a balanced equation for the reaction occurring. [2]
b Identify the volume of gas produced and the time taken for the reaction under condition X to be complete.
c **i** Explain the shape of curve Y in terms of collision theory. [2]
  ii Suggest two possible reasons for the differences between curves X and Y. [2]

7 Equilibrium

STARTING POINTS

■ Certain physical systems can reach a state of equilibrium in which two opposing processes become balanced and there is no further overall change in the composition of the system. To achieve equilibrium the system must be closed. A closed system is one in which neither matter nor energy can be lost or gained.

■ Chemical reactions do not always go to 100% completion; in some cases the 'products' react with each other and the reverse reaction occurs at the same time as the forward reaction. Such reactions are said to be 'reversible reactions'.

■ Where reversible reaction systems take place in a closed system the reaction reaches a point where the rate of the reverse reaction is equal to that of the forward reaction. This situation is referred to as a 'dynamic equilibrium'.

■ Once this situation is achieved the concentrations of the reactants and products in the equilibrium mixture remain constant unless the conditions are changed. However, individual molecules in the mixture continue to react and the two opposing reactions continue to take place.

■ The effects of changing the physical conditions of the system or in some way altering the 'closed' nature of the system can be predicted by applying Le Châtelier's principle.

■ Some key industrial processes are based on reversible chemical reactions. These include the key reactions of the Haber process for ammonia synthesis and the Contact process for making sulfuric acid.

■ Le Châtelier's principle is helpful in establishing the most productive conditions for these industrial processes, but considerations of reaction kinetics and safety need also to be taken into account in setting up an economically viable process.

7.1 Dynamic equilibrium

There are some chemical reactions that very obviously go virtually to completion. For example, spectacular firework displays involve a range of colourful and eye-catching reactions that help us to celebrate significant occasions (Figure 7.1). However, there is a significant number of reactions that do not go even close to completion. In these cases a reverse reaction is set up that runs in 'competition' with the forward reaction.

Such reversible reactions are key to certain industrially important processes such as the synthesis of ammonia and the production of sulfuric acid. They are also crucial in our body chemistry. The reversible nature of so many metabolic reactions in our biochemistry means that these reactions can be controlled by subtle changes in conditions. For instance, reversible effects control how oxygen binds to, and is released from, the hemoglobin in red blood cells as they flow through our lungs and other tissues of our body. The complexity of our body chemistry is dependent on the fine control that is possible where the metabolic pathways consist of sequences of reversible reactions.

Figure 7.1 Fireworks at the New Year celebrations over Sydney Harbour

7.1.1 Outline the characteristics of chemical and physical systems in a state of equilibrium.

Some of the essential features of a dynamic equilibrium are best illustrated by considering physical systems. Such systems can give clear-cut examples of basic dynamic changes taking place while the overall properties of different parts, or components, of the system remain constant.

Physical equilibria

We are all familiar with the phenomenon of evaporation. Puddles of water disappear after a rain shower. Propanone disappears from the cupped palm of your hand even as you watch, and your hand feels cold as the liquid evaporates. These are open systems: once evaporated, the molecules in the vapour escape and mix with the air. Evaporation continues as the molecules gain enough kinetic energy from the surroundings to escape the surface of the liquid and enter the atmosphere. Eventually all the liquid disappears into the air. This process has many uses, including the evaporation of water from salt pans (Figure 7.2).

Figure 7.2 The Inca salt pans at Salinas Ollantaytambo, Peru

If you place some water in a sealed container a different situation arises. To start with, water will still vaporize. However, the air in the container will become saturated with water vapour until it can hold no more water. Equilibrium will be established between the liquid water and the water-saturated air above it. Some water molecules will still have sufficient kinetic energy to escape the surface and enter the vapour phase. Simultaneously some molecules in the vapour will condense back into the liquid. At equilibrium, the rate of vaporization is equal to the rate of condensation.

$$H_2O(l) \rightleftharpoons H_2O(g)$$

Experiments can be carried out on this type of system. Figure 7.3 represents two containers (A and B) with the same amount of water in them, but the surface area in A is twice that in B. The rates of evaporation and condensation are both twice as fast in A as in B, but the position of equilibrium is unchanged. The vapour pressure of water in the two systems is the same.

This type of physical liquid–vapour equilibrium can be visually demonstrated using bromine. Bromine is the one non-metallic element that is a liquid at room temperature. It is a volatile liquid (boiling point 332 K). When placed in a sealed container, the orange-brown vapour collects over the deep red-brown liquid. As the liquid slowly evaporates over a period of time, the colour of the vapour becomes more intense. Eventually, the intensity of colour of the vapour as it sits over the liquid remains constant (Figure 7.4).

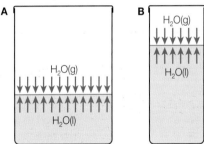

Figure 7.3 The effect of surface area on the evaporation of water in a closed container

The unchanging colour of the vapour in the flask suggests that a position of balance has been reached. Some of the bromine has formed a vapour and some of the bromine remains as a liquid. A position of equilibrium has been reached between bromine liquid and bromine gas. This equilibrium can be summarized as:

$$Br_2(l) \rightleftharpoons Br_2(g)$$

At this point the rate of evaporation and the rate of condensation are the same. There is no net change in the amounts of bromine liquid and vapour present.

Figure 7.4 A sealed flask containing bromine demonstrates a physical equilibrium between the liquid and its vapour

The equilibrium sign ($\rightleftharpoons$) is used to show that both bromine liquid and bromine gas are present in the flask. Do all the liquid bromine molecules remain in the liquid while all the gaseous molecules stay as vapour (a **static equilibrium**)? Or is there an exchange of molecules, with some liquid molecules entering the vapour state while an equal number of vapour molecules condense to liquid (a **dynamic equilibrium**)? Experiments show that liquid and gas molecules move around rapidly and randomly, giving rise to our ideas of the kinetic theory of matter (Chapter 1). Given these ideas it seems likely that a dynamic rather than a static equilibrium is set up in the flask. In which case, the rate at which molecules leave the liquid surface and enter the vapour is equal to the rate at which other molecules in the vapour return to the liquid. Random molecular activity occurs even after all the obvious visual signs of change have disappeared.

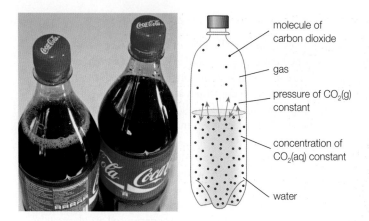

Figure 7.5 Dynamic equilibrium in a sealed bottle of fizzy drink. The bottle on the left has had its cap slightly opened, and then closed again

The solubility of gases in water extends these ideas to another physical situation. There are many fizzy drinks that involve carbon dioxide dissolved under pressure, forming CO_2(aq) (Figure 7.5). The pressure needs to be maintained if the carbon dioxide is to stay dissolved as CO_2(aq). This is done by keeping the lid screwed on tightly. Once we release the pressure by slightly unscrewing the lid the carbon dioxide begins to come out of solution. A stream of bubbles is produced which can be slowed down by re-tightening the cap on the bottle (Figure 7.5). This process is said to be a **reversible reaction**, which means it can go in either direction.

$$\text{increase pressure}$$
$$CO_2(g) + (aq) \rightleftharpoons CO_2(aq)$$
$$\text{decrease pressure}$$

Again the double-headed arrow symbol, $\rightleftharpoons$, represents a dynamic equilibrium or a 'balanced state'. This can be explained in the following way. In the bottle of fizzy drink, the concentration of the carbon dioxide molecules dissolved in water and the concentration of carbon dioxide molecules in the gas phase are constant. If you examined the situation further then you would see that carbon dioxide molecules are constantly moving back and forth between the liquid and the gas phase (Figure 7.5). The rate of movement of carbon dioxide molecules from the liquid to the gas phase is the same as the rate of movement of carbon dioxide from the gas phase to the water. So even though molecules of carbon dioxide are moving between the two environments, no apparent change is taking place.

Language of Chemistry

The terms 'gas' and 'vapour' are often used interchangeably and in most circumstances this is quite acceptable. The term 'vapour' has the more limited and exact meaning and refers to gases formed by evaporation of substances that are usually liquids or solids at room temperature. Thus, in the case we looked at just now, the orange-brown bromine 'gas' is more usually referred to as bromine vapour; it is directly in contact with bromine liquid. Physicists will use the term 'vapour' to refer to any substance in the gas phase under conditions where it can be liquefied simply by increasing the pressure. In these cases they are gases below their critical temperature (Chapter 17). ∎

Chemical equilibria

A chemical equilibrium can only occur when the chemical system is closed. One reaction that gives a visual demonstration of aspects of a chemical equilibrium is based on a chemical test for iron(III) ions (Fe^{3+} ions) in solution. Aqueous iron(III) ions react with thiocyanate ions (SCN^- ions) to produce a blood red colour (Figure 7.6). The red colour is due to the soluble complex ion, $[Fe(SCN)]^{2+}$.

$$Fe^{3+}(aq) + SCN^-(aq) \rightleftharpoons [Fe(SCN)]^{2+}(aq)$$
pale yellow   colourless   deep red

Figure 7.6 The equilibrium between Fe^{3+}(aq) ions and SCN^- (aq) ions can be studied by colorimetry. A soluble blood red coloured complex is formed between the two ions

The system forms an equilibrium mixture containing unreacted Fe^{3+} ions, unreacted SCN^- ions and the complex ion product $[Fe(SCN)]^{2+}$.

It is possible to study the nature of the equilibrium set up by this reaction at room temperature by looking at the effect of adding various ions on the intensity of the red colour of the solution. Provided sufficiently dilute solutions are used, this can be done using a colorimeter. In this way we avoid any bias involved in simply using our own eyesight. If a few drops of a solution containing a soluble iron(III) salt are added to an equilibrium solution the colour of the solution becomes darker. A new state of equilibrium has been quickly achieved in which the concentration of $[Fe(SCN)]^{2+}$ is greater than before. Increasing the concentration of $Fe^{3+}(aq)$ has increased the concentration of the complex ion.

In a similar way, the concentration of $[Fe(SCN)]^{2+}(aq)$ can also be increased when a few drops of potassium thiocyanate solution are added to the equilibrium solution. The red colour again intensifies. In this way we can see that an equilibrium mixture has been set up by a chemical reaction. This dynamic balance between the ions in the mixture can be disturbed by some simple additions to the mixture.

Reversible chemical reactions reach this balanced state of equilibrium when the rates of both the forward and reverse reactions are equal. If this is true then, under given conditions, the same equilibrium mixture should be reached whether we start with the chemicals on one side of the equation or the other. For instance, the gas phase reaction

$$H_2(g) + I_2(g) \rightleftharpoons 2HI(g)$$

has been studied under various conditions. The reaction can be stopped quickly (quenched) by cooling and the amount of iodine present in the equilibrium mixture found by titration with sodium thiosulfate. Figure 7.7 illustrates the type of results obtained in studies of this kind. Note that the same equilibrium concentrations of $H_2(g)$, $I_2(g)$ and $HI(g)$ are reached whichever direction the equilibrium is approached from.

Figure 7.7 a Graph of the concentration of reactants and products with time when reacting equal amounts of hydrogen gas and iodine vapour **b** Graph showing the achievement of the same equilibrium state by decomposition of hydrogen iodide vapour at the same temperature

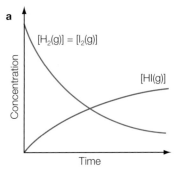
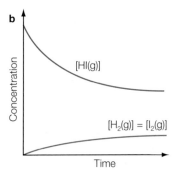

TOK Link

By its very nature the essential features of a physical or chemical equilibrium state are at a sub-microscopic level beyond our sight. In order to make sense of what we understand is happening at this level it is often necessary to use analogies or thought experiments to create mental 'pictures' of what is occurring in these situations.

Various analogies have been suggested to describe dynamic equilibria at the molecular level. One of the most frequently used is that of a person walking or running up a 'down' escalator at the same rate as the moving staircase is descending. The person is moving, the escalator is moving too, but the overall effect is that the person remains at the same point up the incline (Figure 7.8). A similar analogy from the fitness club would be a person running or walking on a treadmill set at an appropriate speed.

Other analogies that have been used include private house parties where the same number of guests move between a limited number of rooms; or indeed two neighbours 'fighting' over the windfall apples from a tree that is on the boundary between their gardens – they toss the fallen fruit back and forth between them, reaching a point of 'equilibrium' in terms of the number of apples on each garden lawn, dependent on their relative fitness.

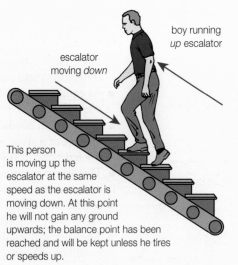

This person is moving up the escalator at the same speed as the escalator is moving down. At this point he will not gain any ground upwards; the balance point has been reached and will be kept unless he tires or speeds up.

Figure 7.8 One useful analogy of dynamic equilibrium is that of a person trying to go up the wrong escalator

Such analogies always have their weaknesses but they do try to convey the idea of continuous change producing a stable situation where certain overall properties do not alter. One 'thought experiment' that you can try is to imagine that you are one of the atoms in a chemical mixture at equilibrium, a nitrogen atom in an ammonia molecule, for instance. Imagine that you detach yourself from the ammonia molecule and become part of a nitrogen molecule, N_2 – but only if another nitrogen atom comes the other way, from the nitrogen molecule to the ammonia molecule. Then you can imagine being part of a dynamic equilibrium.

The equivalent of such thought experiments can be carried out using radioactive or 'heavy' isotopes (Chapter 2) of an element involved in the reaction. In this way we can 'tag' certain atoms and show that movement continues to occur even though a system is in equilibrium. An experiment can be set up to show that there is dynamic exchange of atoms between the molecules in an equilibrium mixture. A heavy isotope is one having an extra neutron or neutrons in the nucleus of the atom. For example, deuterium (^2_1H), sometimes given the symbol D, is an isotope of hydrogen in which the nucleus of each atom contains a neutron as well as a proton. In studies on the ammonia equilibrium, some of the hydrogen is replaced by an equal amount of 'heavy hydrogen', D_2. The D_2 molecules behave chemically in exactly the same way as H_2 molecules and will take part in the reaction (Figure 7.9). When the new equilibrium mixture is subsequently analysed using a mass spectrometer (Chapter 2) some NH_2D, NHD_2, ND_3 and HD will be detected. This finding can only occur if there is an exchange of atoms between the molecules of ammonia, hydrogen and deuterium in the equilibrium mixture.

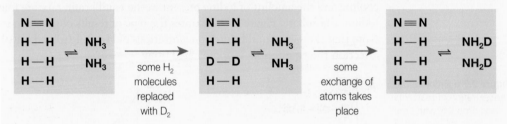

Figure 7.9 Incorporation of deuterium into ammonia within the ammonia, hydrogen and deuterium equilibrium mixture

7.2 The position of equilibrium

From the discussion so far, the basic features of dynamic equilibria can be summarized as follows:

- A dynamic equilibrium can only be established in a closed system. There can be no loss or gain of material to or from the surroundings.
- At equilibrium, macroscopic properties such as the amounts of the various substances involved are unchanging under the given conditions. The amounts of reactants and products remain constant.
- The equilibrium is dynamic, not static. At the sub-microscopic level the particles present continue to take part in the forward and reverse processes.
- At equilibrium the rates of the forward and reverse reactions are equal so that no net change takes place.
- Under given conditions the equilibrium position can be achieved from either direction. A mixture of a given equilibrium composition can be made starting with either the substances on the left-hand or right-hand side of the equation for the reversible reaction.
- The dynamic nature of these equilibria means that they are stable under fixed conditions but sensitive to alterations in these conditions. This immediate sensitivity to changes in conditions, such as alterations in temperature, pH or the concentration of a reactant, can be taken as an indication that the system was indeed at equilibrium.
- The chemical equilibria we will study in these chapters are all examples of homogeneous equilibria. The reactants and products are all in the same physical phase – either in the gaseous or liquid state, or in aqueous solution.

The equilibrium constant

7.2.1 **Deduce** the equilibrium constant expression (K_c) from the equation for a homogeneous reaction.

The quantitative study of many chemical equilibria has shown that the following equilibrium law applies to such systems. The position of the equilibrium for a particular reversible reaction can be defined by a constant which has a numerical value found by relating the equilibrium concentrations of the products to those of the reactants. This constant is known as the **equilibrium constant**, K_c, for the reaction.

We can write an equilibrium expression for K_c for a general reaction:

$$aA + bB \rightleftharpoons cC + dD$$

where a, b, c and d represent the amounts of each substance (in moles) in the equation.

$$K_c = \frac{[C]^c_{eqm} [D]^d_{eqm}}{[A]^a_{eqm} [B]^b_{eqm}}$$

It is important to note that the concentration values fed into the expression *must* be those that occur at equilibrium, not the starting values. For this reason the terms $[C]_{eqm}$ etc. are used, and strictly speaking this is how the expression should be written. However, this can make the expression look rather cluttered and generally the expression is written as summarized below:

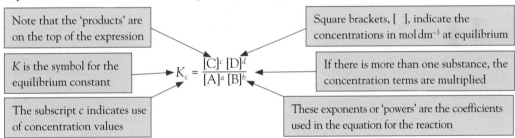

Note that the 'products' are on the top of the expression	Square brackets, [ ], indicate the concentrations in $mol\,dm^{-3}$ at equilibrium
K is the symbol for the equilibrium constant	If there is more than one substance, the concentration terms are multiplied
The subscript c indicates use of concentration values	These exponents or 'powers' are the coefficients used in the equation for the reaction

$$K_c = \frac{[C]^c [D]^d}{[A]^a [B]^b}$$

It should be noted that the equilibrium constant, K_c, is constant for a given temperature. If the temperature changes, then the value of K_c will change. This is a general expression of the **equilibrium law**. The general expression is then adapted to the particular reaction being studied.

Worked example

Use the figures given in Table 7.1 to demonstrate that using this type of relationship and the equilibrium concentrations for a particular reaction produces a constant value.

$$N_2(g) + 3H_2(g) \rightleftharpoons 2NH_3(g) \qquad K_c = \frac{[NH_3]^2}{[N_2] [H_2]^3}$$

	$[N_2]/mol\,dm^{-3}$	$[H_2]/mol\,dm^{-3}$	$[NH_3]/mol\,dm^{-3}$
Experiment 1	0.922	0.763	0.157
Experiment 2	0.399	1.197	0.203
Experiment 3	2.59	2.77	1.82

Table 7.1 Results from three experiments on the ammonia synthesis reaction at 500 °C

Substituting values from Table 7.1:

Experiment 1 $K_c = \dfrac{(0.157)^2}{0.922 \times (0.763)^3} = 0.0602$

Experiment 2 $K_c = \dfrac{(0.203)^2}{0.399 \times (1.197)^3} = 0.0602$

Experiment 3 $K_c = \dfrac{(1.82)^2}{2.59 \times (2.77)^3} = 0.0602$

This is a constant value for all three experiments.

In applying the equilibrium law to a particular reaction it is important to first write down the equation for the reaction studied. Using an example we looked at earlier:

$$H_2(g) + I_2(g) \rightleftharpoons 2HI(g)$$

The concentrations of the substances on the right-hand (product) side of the equation are written in the numerator (the upper part of the fraction), while the concentrations of those substances on the left-hand (reactant) side are placed as the denominator (the lower part of the fraction). So the equilibrium expression for this reaction is:

$$K_c = \frac{[HI]^2}{[H_2][I_2]}$$

It is important to always quote K_c with the relevant chemical equation for the reaction being considered. The value of K_c is a number that indicates the extent to which the equilibrium lies to the right-hand side of the equation we have written. It is crucial to be clear about the equation to which a K_c applies. Suppose the value for K_c above was X at a particular temperature, what would be the value of K_c for the reaction written as the decomposition of hydrogen iodide?

$$2HI(g) \rightleftharpoons H_2(g) + I_2(g)$$

For this reverse reaction, the equilibrium constant, K_c', at the same temperature is:

$$K_c' = \frac{[H_2][I_2]}{[HI]^2} = \frac{1}{X}$$

In general, the values, at the same temperature, for the forward and reverse equations for an equilibrium system are related as follows:

$$K_c' = \frac{1}{K_c} \qquad \text{or} \qquad K_c^{-1}$$

i.e. the value of the equilibrium constant for the reverse reaction is the reciprocal of that for the forward reaction.

The equilibrium law is very much an experimentally determined one. The basic structure of the expression for a particular reaction is confirmed by the value of K_c remaining constant for a series of experiments at a given temperature. Table 7.2a shows the values for K_c obtained in such a series of experiments when hydrogen and iodine are reacted in a sealed container at 700K. These results show that a constant value of 54 is obtained when the equilibrium concentrations are fed into the expression:

$$K_c = \frac{[HI]^2}{[H_2][I_2]}$$

The results for further experiments where the equilibrium at 700K is approached from the other direction are given in Table 7.2b. These results demonstrate the reciprocal relationship of the equilibrium constants found from the two sets of experiments.

	$[H_2]_{eqm}$ /10^{-3} mol dm^{-3}	$[I_2]_{eqm}$ /10^{-3} mol dm^{-3}	$[HI]_{eqm}$ /10^{-3} mol dm^{-3}	K_c
Experiment 1	4.56	0.74	13.49	54
Experiment 2	3.56	1.25	15.50	54
Experiment 3	2.25	2.34	16.86	54

Table 7.2a The results for a series of experiments on the reaction $H_2(g) + I_2(g) \rightleftharpoons 2HI(g)$

	$[H_2]_{eqm}$ /10^{-3} mol dm^{-3}	$[I_2]_{eqm}$ /10^{-3} mol dm^{-3}	$[HI]_{eqm}$ /10^{-3} mol dm^{-3}	K_c'	K_c (1/K_c')
Experiment 4	0.48	0.48	3.52	0.0186	54
Experiment 5	0.50	0.50	3.67	0.0186	54

Table 7.2b The experimental results for the reaction $2HI(g) \rightleftharpoons H_2(g) + I_2(g)$

Worked example

Write the equilibrium expressions for the following reversible reactions. They are all examples of homogeneous equilibria.

a $2SO_2(g) + O_2(g) \rightleftharpoons 2SO_3(g)$

b $Fe^{3+}(aq) + SCN^-(aq) \rightleftharpoons [Fe(SCN)]^{2+}(aq)$

c $4NH_3(g) + 5O_2(g) \rightleftharpoons 4NO(g) + 6H_2O(g)$

d i The following reaction is an esterification reaction producing ethyl ethanoate:

$CH_3CO_2H(l) + C_2H_5OH(l) \rightleftharpoons CH_3CO_2C_2H_5(l) + H_2O(l)$

ii The value of K_c for the above reaction at 25 °C is 4.0. This equilibrium can be approached experimentally from the opposite direction. What is the value for K_c for this reaction, the hydrolysis of ethyl ethanoate, at 25 °C?

$CH_3CO_2C_2H_5(l) + H_2O(l) \rightleftharpoons CH_3CO_2H(l) + C_2H_5OH(l)$

a Remember to keep in mind the written equation. The concentration of sulfur trioxide will be on top in the equilibrium expression. Remember also to include the balancing coefficients from the equation as powers in the expression for K_c.

$$K_c = \frac{[SO_3(g)]^2}{[SO_2(g)]^2 [O_2(g)]}$$

You will see the state symbols for each substance included in the equilibrium expression, emphasizing the fact that this is a homogeneous equilibrium. Unless you are specifically asked to include them they can be omitted. It means that the expression looks less cumbersome. Thus, an answer to this question would read:

$2SO_2(g) + O_2(g) \rightleftharpoons 2SO_3(g)$ $\qquad K_c = \dfrac{[SO_3]^2}{[SO_2]^2 [O_2]}$

b $Fe^{3+}(aq) + SCN^-(aq) \rightleftharpoons [Fe(SCN)]^{2+}(aq)$ $\qquad K_c = \dfrac{[Fe(SCN)]^{2+}}{[Fe^{3+}] [SCN^-]}$

c This is the first step in the industrial conversion of ammonia into nitric acid. Do not be put off by the complexity of the powers involved.

$4NH_3(g) + 5O_2(g) \rightleftharpoons 4NO(g) + 6H_2O(g)$ $\qquad K_c = \dfrac{[NO]^4[H_2O]^6}{[NH_3]^4 [O_2]^5}$

d i $CH_3CO_2H(l) + C_2H_5OH(l) \rightleftharpoons CH_3CO_2C_2H_5(l) + H_2O(l)$ $\quad K_c = \dfrac{[CH_3CO_2C_2H_5] [H_2O]}{[CH_3CO_2H] [C_2H_5OH]}$

ii K_c for the esterification reaction is 4.0. Therefore, the value for the hydrolysis reaction is $\dfrac{1}{4.0} = 0.25$

How far will a reaction go?

7.2.2 Deduce the extent of a reaction from the magnitude of the equilibrium constant.

The magnitude of K_c gives us a useful indication of how far a reaction has gone towards completion under the given conditions. The higher the value of K_c, the further to the right the equilibrium position will lie at that temperature. This relationship arises from the structure of the equilibrium expression.

If the value of K_c is high, then this shows that at equilibrium there is a high proportion of products compared to reactants. In this case we say that the equilibrium lies well over to the right, as the equation is written. If the K_c value is low, then this indicates that only a small fraction of the reactants have been converted into products, and the equilibrium lies over to the left.

Two seemingly very similar reactions are those of chlorine and iodine with hydrogen to form the respective hydrogen halides.

$H_2(g) + I_2(g) \rightleftharpoons 2HI(g)$ $\qquad K_c = 2$ at 277 °C

$H_2(g) + Cl_2(g) \rightleftharpoons 2HCl(g)$ $\qquad K_c = 10^{18}$ at 277 °C

The large difference in magnitude for the values of K_c illustrates how stable hydrogen chloride is at this temperature compared to hydrogen iodide. The reaction between hydrogen and chlorine has gone virtually to completion to produce hydrogen chloride (HCl) molecules that do not readily decompose with heat. The bonding in hydrogen iodide is weaker and therefore the reverse reaction is more evident, with an equilibrium being established.

As a general rule, if $K_c \gg 1$ then the reaction is said to have gone virtually to completion. There has been an almost complete conversion of reactants to products. If $K_c \ll 1$ the reaction has hardly taken place at all, very little of the reactants have been converted to products. Thus the equilibrium constant indicates the extent of a reaction at a particular temperature (see Table 7.3). Do note though that it gives *no information* at all about how fast the equilibrium state is achieved.

Table 7.3 The relationship between the value of K_c and the extent of a reaction

Reaction hardly goes	'Reactants' predominate at equilibrium	Equal amounts of reactants and products	'Products' predominate at equilibrium	Reaction goes virtually to completion
$K_c < 10^{-10}$	$K_c = 0.01$	$K_c = 1$	$K_c = 100$	$K_c > 10^{10}$

It is also important to realize that the value of K_c is not altered by the addition of more reactants, the removal of some product from the system, or any other adjustment of the components of the mixture. Consider the general equation we met before:

$$aA + bB \rightleftharpoons cC + dD$$

for which:

$$K_c = \frac{[C]^c \, [D]^d}{[A]^a \, [B]^b}$$

If we added more A, while keeping the temperature constant, the value of the bottom part of the equation would increase and the overall value of the expression would no longer equal K_c. However, the reaction re-adjusts itself. Some of the extra A reacts with B to form more C and D. The concentrations adjust to establish a new equilibrium mixture so that the overall value for the expression again equals K_c.

The K_c for a given reversible reaction at equilibrium only changes if the temperature changes.

Language of Chemistry

When discussing reversible reactions and predicting the adjustments that take place in an equilibrium mixture when conditions change, it is important to give the equation for the reaction clearly. Even though the reaction is reversible, the species on the right of the equation are often referred to as 'products', while those on the left are still termed the 'reactants'. The structure of the equilibrium expression depends on the direction in which the equation is written, as does the value of K_c that we derive from it. ∎

Referring to the reaction below, it is the species on the 'right' of the equation that goes on the top of the equilibrium expression.

$$2CrO_4^{2-}(aq) + H^+(aq) \rightleftharpoons Cr_2O_7^{2-}(aq) + OH^-(aq)$$

yellow   orange

$$K_c = \frac{[Cr_2O_7^{2-}][OH^-]}{[CrO_4^{2-}]^2[H^+]}$$

The shifts in equilibrium of this reaction can be easily seen because of the different colours of the ions involved (Figure 7.10). The reaction can be used to illustrate the language of direction and 'shifts' that we frequently use in this context.

Figure 7.10 The equilibrium between chromate(VI) and dichromate(VI) ions in solution can be manipulated by the addition of acid or alkali

When potassium chromate(VI) is dissolved a yellow solution is produced. Adding a few drops of acid (H^+ ions) sets up the equilibrium, though well to the left. Further additions of acid disturb the equilibrium. The equilibrium shifts to the right to remove the additional hydrogen ions, H^+, and the colour of the solution turns orange. The presence of orange dichromate(VI) ions is favoured by low pH.

If hydroxide ions (OH^-) ions are added to this orange solution the equilibrium shifts to the left to remove these ions. The solution turns yellow again; the formation of chromate(VI) ions is favoured.

■ Extension: The units of the equilibrium constant (K_c)

In order to calculate a standard value of the equilibrium constant, K_c, for a reaction the values of the concentrations entered in the equilibrium expression *must* be in $mol\,dm^{-3}$.

However, strictly speaking the values entered should be 'activity' values rather than concentrations. For gases and relatively dilute solutions the concentration values are sufficiently close to the 'activity' values that their use does not introduce any significant errors into the calculation. Since the 'activity' values for substances are simply a number – they do not have any units – it follows that *any K_c value will also simply be a number without units* no matter which reaction you are studying. Certainly in the IB examination you will not be asked for any units relating to K_c values.

■ Extension: The relationship between the equilibrium constant (K_c) and the reaction quotient (Q_c)

In the discussion above we have emphasized that the concentrations entered in the equilibrium expression when calculating K_c must be those occurring once equilibrium has been established. However, the same general expression, this time with non-equilibrium concentration values fed into it, can be of use in predicting how a given mixture will react. In these cases the value calculated is known as the **reaction quotient, Q_c**.

The following reaction illustrates the meaning of this.

$$N_2O_4(g) \rightleftharpoons 2NO_2(g) \qquad K_c \text{ at } 500\,K = 41$$

Suppose a mixture of dinitrogen tetroxide ($2\,mol\,dm^{-3}$) and nitrogen(IV) oxide ($6\,mol\,dm^{-3}$) is contained in a sealed tube at $500\,K$, will the reaction move the composition of the mixture to the left or the right? Or, indeed, will the composition of the mixture change at all? Entering these concentrations into the expression for the reaction quotient, Q_c, we get the following value:

$$Q_c = \frac{[NO_2]^2}{[N_2O_4]} = 18$$

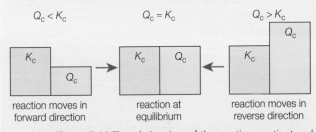

$Q_c < K_c$ $Q_c = K_c$ $Q_c > K_c$

reaction moves in forward direction   reaction at equilibrium   reaction moves in reverse direction

Figure 7.11 The relative sizes of the reaction quotient and equilibrium constant indicate the direction in which a reaction mixture tends to change

So, in this case, Q_c is less than K_c and therefore the reaction will move towards the right to produce more nitrogen(IV) oxide and eventually achieve equilibrium.

This example illustrates the fact that by comparing the value of Q_c with the equilibrium constant, K_c, the direction of reaction can be predicted (see Figure 7.11).

- If $Q_c < K_c$ the reaction will move to the right, generating more product to reach equilibrium.
- If $Q_c = K_c$ the reaction is at equilibrium.
- If $Q_c > K_c$ the reaction will move to the left, generating more reactants to reach equilibrium.

The value of Q_c in relation to K_c thus indicates the direction in which any net reaction must proceed as the system moves towards its equilibrium state. It is worth noting, however, that this prediction makes no comment on the *rate* at which the equilibrium may be achieved. Certainly some reactions only reach equilibrium very slowly. For instance, the esterification reaction mentioned in the worked example on page 187 can take several weeks to achieve equilibrium in the absence of a catalyst. The stalactites that hang from the roofs of caves in limestone regions take geological periods of time to form. So it is worth remembering that in some cases the prediction based on Q_c may not be realized as the reaction may be so slow in both directions that equilibrium is never reached.

History of Chemistry

Henri Louis Le Châtelier (1850–1936) was an influential French/Italian chemist of the late 19th and early 20th centuries. He is most famous for devising Le Châtelier's principle, used by chemists to predict the effect of a change in conditions on a chemical equilibrium. Although his initial training was as an engineer, Le Châtelier chose to teach chemistry rather than pursue a career in industry. He taught first at the *École des Mines* in Paris and, after a series of other appointments, he taught at the Sorbonne, where he succeeded Henri Moissan. He is most famous for the law on chemical equilibrium which bears his name, Le Châtelier's principle. This is summarized as follows:

When a system at equilibrium is disturbed, the equilibrium position will shift in the direction which tends to minimize, or counteract, the effect of the disturbance.

This principle can be applied to a wide range of chemical and physical situations. Le Châtelier was named '*chevalier*' (knight) of the *Légion d'honneur* in 1887, and was eventually awarded the title of '*grand officier*' (Knight Grand Officer) in 1927.

Le Châtelier's principle

7.2.3 Apply Le Châtelier's principle to predict the qualitative effects of changes of temperature, pressure and concentration on the position of equilibrium and on the value of the equilibrium constant.

When the conditions under which a chemical equilibrium has been established are changed there is an effect on 'the position of the equilibrium'. Is it possible to predict the effect of such changes? Indeed we can, on the basis of a principle first proposed by a French chemist Henri Le Châtelier. (Do note that for IB you do not need to learn a statement of the principle, as some published versions use quite complex language. However, it can be very useful to have a familiar version in mind when tackling questions.) This principle is a descriptive statement of what happens when a dynamic equilibrium is disturbed by a change in conditions; it is not an explanation as to why the change happens. Put simply, the system responds to negate the change by responding in the opposite way. For instance, if we add more of a reactant, the system will react to remove it; if we remove a product, the system will react to replace it.

The possible changes in conditions that we need to consider are:

- changes in the concentration of either reactants or products
- changes in pressure for gas phase reactions
- changes in temperature
- the presence of a catalyst.

This general principle is of importance industrially as it allows chemists to alter the reaction conditions to produce an increased amount of the product and, therefore, increase the profitability of a chemical process.

Note that in all cases, once the equilibrium has been re-established after the change, the value of K_c will be unaltered except when there is a change in temperature (Table 7.4).

Change made	Effect on 'position of equilibrium'	Value of K_c
Concentration of one of the components of the mixture	Changes	Remains unchanged
Pressure	Changes if the reaction involves a change in the total number of gas molecules	Remains unchanged
Temperature	Changes	Changes
Use of a catalyst	No change	Remains unchanged

Table 7.4 A broad summary of the effects of changing conditions on the position of an equilibrium

Changes in concentration

Consider the esterification reaction we looked at earlier:

$$CH_3CO_2H(l) + C_2H_5OH(l) \rightleftharpoons CH_3CO_2C_2H_5(l) + H_2O(l)$$
ethanoic acid   ethanol   ethyl ethanoate   water

If we remove some of the water from the equilibrium mixture then we would predict that more carboxylic acid and alcohol would react to replace it, producing more ester. This can in fact be done by adding a few drops of concentrated sulfuric acid to the mixture. The sulfuric acid provides hydrogen ions, H⁺, which act as a catalyst for the reaction, but also acts as a dehydrating agent, removing water from the mixture. The addition of the acid favours the production of the ester since, by removing water, it shifts the position of equilibrium to the right.

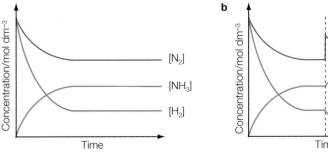

Step 2: equilibrium shifts to right to replace H_2O

$$CH_3CO_2H(l) + C_2H_5OH(l) \rightleftharpoons CH_3CO_2C_2H_5(l) + H_2O(l)$$
ethanoic acid   ethanol   ethyl ethanoate   water

Step 1: product removed

One massively important industrial process that depends on a reversible reaction is the production of ammonia (NH_3). The reaction involved is:

$$N_2(g) + 3H_2(g) \rightleftharpoons 2NH_3(g)$$

Because of its economic significance this reaction has been extensively studied. What, for instance, would be the effect on the equilibrium of increasing the concentration of the nitrogen gas in the reaction mixture?

Before more nitrogen gas is added into the system, the equilibrium concentrations of the reactants would be constant (Figure 7.12a). Then for a short time immediately after the addition of the extra nitrogen the system is no longer at equilibrium. To return the system to equilibrium some of the added nitrogen gas has to be used up and converted to more ammonia. A new equilibrium is established with different concentrations of each of the reactants and products. Most importantly, the equilibrium concentration of the ammonia has increased (Figure 7.12b).

Figure 7.12 a An equilibrium is established between N_2, H_2 and NH_3. **b** After the equilibrium has been disturbed, a new equilibrium position is established containing more ammonia

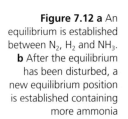
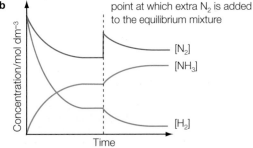

The addition of more nitrogen gas to the system has had the effect predicted by Le Châtelier's principle. The system has counteracted the change by using up some of the added nitrogen to produce more ammonia gas.

Generally:

■ increasing the concentration of a reactant will move the position of equilibrium to the right, favouring the forward reaction and increasing the equilibrium concentrations of the products.

Conversely, the opposite of this is also true:

■ the addition of more product to an equilibrium mixture would shift the position of the equilibrium to the left; the reverse reaction would be favoured.

Note that when the new equilibrium concentrations are substituted into the equilibrium expression the value of K_c remains unchanged.

■ Extension: The link between Le Châtelier's principle and the reaction quotient, Q_c

As we have stressed earlier, Le Châtelier's principle provides a very useful descriptive tool that helps us predict the outcome of a change in conditions on a chemical equilibrium. However, it does not provide an explanation for these effects. Use of the reaction quotient, Q_c, can give an insight into why changing the concentration of a component of an equilibrium mixture gives rise to the effect it does. Take the following reaction as an example:

$$N_2(g) + 3H_2(g) \rightleftharpoons 2NH_3(g)$$

At equilibrium $\qquad Q_c = K_c = \dfrac{[NH_3]^2}{[N_2][H_2]^3}$

If more hydrogen is added to the equilibrium mixture this will increase the value of the denominator in Q_c. Its value will no longer be equal to K_c – it will have a lower value – and therefore the reaction will adjust to increase Q_c by producing more ammonia (see page 191).

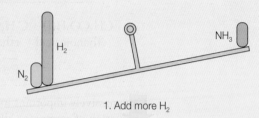

1. Add more H_2

The argument outlined here can be illustrated using the analogy of the 'see-saw' as shown in Figure 7.13. Here the angle of the beam represents the composition of the equilibrium mixture (the pictorial equivalent of K_c). At a particular temperature the value of K_c is constant. So the system must respond to any change in the composition of the mixture in a way that restores the angle of the beam.

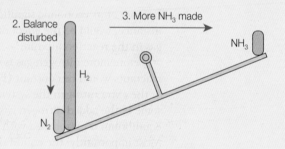

2. Balance disturbed

3. More NH_3 made

System responds to restore position of see-saw beam – the amounts of N_2, H_2 and NH_3 are different, but the position is the same

Figure 7.13 Diagrammatic representation of the effect of adding more hydrogen to the equilibrium mixture involved in ammonia synthesis

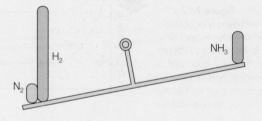

Changing the pressure for a gas phase reaction

Changing the pressure under which a chemical reaction is carried out only affects reactions that involve gases. The solid and liquid phases are essentially non-compressible in this context, and so reactions involving these phases are unaffected by changes in pressure. Indeed, gas phase reactions will only be affected by a change of pressure if the reaction involves a change in the number of molecules on the two sides of the equation.

There is a direct relationship between the number of molecules of a gas in a container and the pressure the gas exerts. So if a reaction at equilibrium is subjected to an increase in pressure, then the system will respond by favouring the side of the equation with the smaller number of molecules. In doing this the pressure of the mixture will be reduced. A useful reaction to study in this context is the decomposition of dinitrogen tetroxide (N_2O_4) because the change can be followed by the colour change observed. Dinitrogen tetroxide is a colourless gas that decomposes to brown nitrogen dioxide gas.

$$N_2O_4(g) \rightleftharpoons 2NO_2(g)$$
colourless   brown

An equilibrium mixture of these two gases can be set up in a gas syringe at a particular temperature (Figure 7.14). You can change the pressure in the reaction mixture by pushing the syringe piston in, or by sharply pulling it out, and compare the resulting colour of the gas mixture with the original. The interpretation of the changes observed is complicated by the simple effects on colour intensity caused by any volume change.

When the pressure is increased by pushing in the piston there is an initial darkening of the colour of the gas mixture. This is due to the increase in concentration of the mixture. However, this is quickly followed by a lightening of the colour as the equilibrium mixture adjusts to its new composition in which there is a higher concentration of colourless N_2O_4. This produces a mixture involving fewer molecules and therefore reduces the pressure in the syringe, thus counteracting the increase in externally applied pressure.

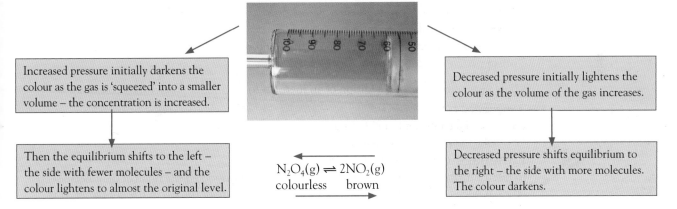

Increased pressure initially darkens the colour as the gas is 'squeezed' into a smaller volume – the concentration is increased.

Decreased pressure initially lightens the colour as the volume of the gas increases.

Then the equilibrium shifts to the left – the side with fewer molecules – and the colour lightens to almost the original level.

$$N_2O_4(g) \rightleftharpoons 2NO_2(g)$$
colourless   brown

Decreased pressure shifts equilibrium to the right – the side with more molecules. The colour darkens.

Figure 7.14 The effect of changing pressure on the gaseous equilibrium involving nitrogen dioxide can be followed by observing the changes in colour intensity

Alternatively, if the pressure in the syringe is lowered by pulling out the piston, the effect of reduced pressure can be seen. After an initial lightening of the colour, as the new equilibrium is established the gas mixture finishes up darker than it was originally. Decreased pressure favours the side of the equation that involves more molecules. This new mixture involves a total of more gas molecules, counteracting the decrease in applied pressure.

For gas phase reactions where there are different numbers of molecules on the two sides of the equation:

- increased pressure shifts the equilibrium position to the side of the equation with fewer molecules
- decreased pressure shifts the equilibrium position to the side with more molecules.

None of these changes results in a change in the value of K_c.

If there is no change in the number of molecules during the course of a reaction then changes in pressure will have no effect on the equilibrium position of the reaction at a given temperature. This is illustrated by the reaction of hydrogen and iodine to form hydrogen iodide:

$$H_2(g) + I_2(g) \rightleftharpoons 2HI(g)$$

At a given temperature the position of this equilibrium, and others like it, cannot be manipulated by changing the external pressure applied to the mixture. If the pressure is altered by changing the volume then the concentrations of all the species change by the same factor, leaving K_c unchanged.

■ Extension: Adding an inert gas to the system

One possible way of altering the pressure of a system is to introduce an inert gas into the mixture. The gas added can be a noble gas such as argon or any gas that does not react with those involved in the reaction. The effect produced depends on the different conditions involved; namely, whether the volume or pressure of the system is kept constant.

Addition of an inert gas at constant volume

When an inert gas is added to an equilibrium mixture under conditions where the volume of the system is kept constant, the total pressure of the system will increase. However, the concentrations of the different reacting components of the mixture will be unchanged. Hence, under these conditions, there will be no effect on the position of the equilibrium.

Addition of an inert gas at constant pressure

If an inert gas is added to a system at equilibrium such that the pressure is kept constant, then there will be a resulting increase in volume. Because of this the concentrations of each of the reactants and products in the mixture will be reduced. In accordance with Le Châtelier's principle the system will respond by shifting the equilibrium position to the side of the equation that has the greater number of molecules. For example, if argon were added to the equilibrium:

$$2SO_2(g) + O_2(g) \rightleftharpoons 2SO_3(g)$$

at constant pressure, the equilibrium position would shift to the left. An equilibrium such as:

$$H_2(g) + I_2(g) \rightleftharpoons 2HI(g)$$

would be unaffected by the addition of an inert gas to the mixture under these conditions.

Language of Chemistry

In discussing the effects of changing pressure, we have considered the number of molecules on each side of the equation and related this to the pressure of the gas in the container. It is possible also to link the number of molecules to the volume that the gas would occupy. This can be done because one mole of any gas has the same volume if the conditions of temperature and pressure are the same (Chapter 1).

In applying Le Châtelier's principle, therefore, it is possible to use the language of 'volume', and expansion or contraction, in discussing the changes predicted. Thus for the reaction involving the decomposition of dinitrogen tetroxide:

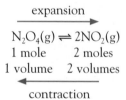

An increase in external pressure will favour the side of the reaction that occupies less volume. In this case the equilibrium will shift to the left, the direction of the contraction in volume.

A decrease in pressure will favour the decomposition to nitrogen dioxide, NO_2, as this side of the reaction occupies a greater volume. The equilibrium will shift to the right, the direction of the expansion in volume. ■

Changing the temperature

Le Châtelier's principle can be used to predict the effect of a temperature change on the position of an equilibrium. The key factor to be considered here is whether the forward reaction is exothermic (a negative ΔH value) or endothermic (a positive ΔH value)(see Chapter 5). Remember that, in a reversible reaction, the reverse reaction has an enthalpy change that is equal and opposite to that of the forward reaction.

When the temperature is increased, the equilibrium position will shift in the direction that will tend to lower the temperature, that is, the endothermic direction. If the temperature is lowered the equilibrium will shift in the exothermic direction so as to generate heat and raise the temperature. These effects are summarized in Table 7.5.

Nature of forward reaction (sign of ΔH)	Change in temperature	Shift in the position of equilibrium	Effect on value of K_c
Endothermic (positive ΔH)	Increase	To the right	K_c increases
Endothermic (positive ΔH)	Decrease	To the left	K_c decreases
Exothermic (negative ΔH)	Increase	To the left	K_c decreases
Exothermic (negative ΔH)	Decrease	To the right	K_c increases

Table 7.5 The effects of temperature changes on chemical equilibria

Consider the reaction:

$$N_2O_4(g) \rightleftharpoons 2NO_2(g) \qquad \Delta H^\ominus = +57\,kJ\,mol^{-1}$$
colourless   brown

When an enthalpy value is quoted alongside an equilibrium equation like this it refers to the forward reaction. So in this case the decomposition of N_2O_4 is endothermic. If an equilibrium mixture is set up in a sealed container at room temperature it will have a certain intensity of colour.

If the mixture is then placed in an ice bath (Figure 7.14a) its colour will lighten as a new equilibrium mixture containing more N_2O_4 is established. A decrease in temperature causes the equilibrium position to shift to the left. The value of K_c decreases as a result of these changes.

Alternatively, if the original mixture were placed in a hot water bath (Figure 7.14b), then the colour will darken as the new equilibrium mixture will contain more NO_2. An increase in temperature causes the equilibrium position to shift to the right. The value of K_c increases as a result of these changes (Table 7.6a).

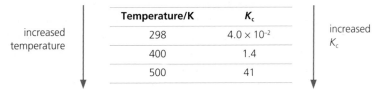

increased temperature →

Temperature/K	K_c
298	4.0×10^{-2}
400	1.4
500	41

→ increased K_c

Table 7.6a Data on the change in the equilibrium constant K_c for the **endothermic** reaction $N_2O_4(g) \rightleftharpoons 2NO_2(g)$ at different temperatures

increased temperature →

Temperature/K	K_c
298	4.2×10^8
400	4.5×10^4
500	62

→ decreased K_c

Table 7.6b Data on the change in the equilibrium constant K_c for the **exothermic** reaction $N_2(g) + 3H_2(g) \rightleftharpoons 2NH_3(g)$ at different temperatures

Figure 7.14 An equilibrium mixture of N_2O_4 and NO_2 in a sealed gas syringe is placed firstly in **a** an ice bath and then in **b** a hot water bath

It is important to note that, unlike changing the concentration or pressure, a change in temperature will also change the value of K_c. For endothermic reactions, an increase in temperature results in an increase in the concentration of products in the equilibrium mixture and therefore an increased K_c. The opposite will be true for exothermic reactions (see Table 7.6b).

In summary, for a chemical equilibrium:

- an increase in temperature always favours the endothermic process
- a decrease in temperature always favours the exothermic process.

Worked example

Draw up a table showing how the position of equilibrium in reactions A, B and C would be affected by the following changes:

i increased temperature

ii increased pressure.

Reaction A: the interconversion of oxygen and ozone.

$$3O_2(g) \rightleftharpoons 2O_3(g) \qquad\qquad \Delta H^\ominus = +285\,kJ\,mol^{-1}$$

Reaction B: the reaction between sulfur dioxide and oxygen in the presence of a platinum/rhodium catalyst.

$$2SO_2(g) + O_2(g) \rightleftharpoons 2SO_3(g) \qquad\qquad \Delta H^\ominus = -197\,kJ\,mol^{-1}$$

Reaction C: the reaction between hydrogen and carbon dioxide.

$$H_2(g) + CO_2(g) \rightleftharpoons H_2O(g) + CO(g) \qquad \Delta H^\ominus = +41\,kJ\,mol^{-1}$$

Use the applications of Le Châtelier's principle to create a table:

Reaction	Effect of increased temperature on equilibrium position	Effect of increased pressure on equilibrium position
Reaction A	Shift to the right – more ozone K_c increased	Fewer molecules on the right; therefore, shift to right – more O_3
Reaction B	Shift to the left – less SO_3 K_c decreased	Fewer molecules on the right; therefore, shift to right – more SO_3
Reaction C	Shift to the right – more CO K_c increased	No change as there are the same number of molecules on both sides

■ Extension: Reaction sequences and K_c values

Reactions between the oxides of nitrogen are also of interest in demonstrating how the overall K_c value for a sequence of reactions relates to the individual values for the different steps.

Worked example

Show that the K_c value for the reaction

$$N_2(g) + 2O_2(g) \rightleftharpoons 2NO_2(g) \qquad\qquad \text{Reaction 3 } (K_{c3})$$

is given by $K_{c3} = K_{c1} \times K_{c2}$ where reactions 1 and 2 are as follows:

$$N_2(g) + O_2(g) \rightleftharpoons 2NO(g) \qquad\qquad \text{Reaction 1 } (K_{c1})$$

$$2NO(g) + O_2(g) \rightleftharpoons 2NO_2(g) \qquad\qquad \text{Reaction 2 } (K_{c2})$$

Reactions 1 and 2 are sequential reactions that added together give the overall equation for reaction 3.

$$N_2(g) + O_2(g) \rightleftharpoons 2\cancel{NO(g)}$$
$$\underline{2\cancel{NO(g)} + O_2(g) \rightleftharpoons 2NO_2(g)}$$
$$N_2(g) + 2O_2(g) \rightleftharpoons 2NO_2(g)$$

Now we need to establish the expressions for K_{c1}, K_{c2} and K_{c3} respectively.

$$K_{c1} = \frac{[NO]^2}{[N_2][O_2]} \qquad K_{c2} = \frac{[NO_2]^2}{[NO]^2[O_2]} \qquad K_{c3} = \frac{[NO_2]^2}{[N_2][O_2]^2}$$

Now we need to work out the expression for $K_{c1} \times K_{c2}$.

$$K_{c1} \times K_{c2} = \frac{\cancel{[NO]^2}}{[N_2][O_2]} \times \frac{[NO_2]^2}{\cancel{[NO]^2}[O_2]} = \frac{[NO_2]^2}{[N_2][O_2]^2}$$

This is the expression for K_{c3}.

Therefore $K_{c3} = K_{c1} \times K_{c2}$

In general, for reactions that take place in sequence, the overall K_c value is equal to the product of the individual K_c values for each of the individual reaction steps. 'Coupled' reactions of this type are particularly important in biochemistry where it is possible for ten or so reactions to be linked in this way (see the comment at the beginning of the chapter).

The role of catalysts

7.2.4 **State** and **explain** the effect of a catalyst on an equilibrium reaction.

A **catalyst** is a substance that increases the rate of a chemical reaction by providing an alternative reaction pathway of lower activation energy (E_a)(see Chapter 6). This means that more particles in a reaction mixture have sufficient kinetic energy on collision to react with each other. Industrially catalysts are of significance, as they allow reactions to occur at reasonable rates under milder, and therefore more economic, conditions.

In fact the presence of a catalyst has no effect on the position of a chemical equilibrium. Figure 7.15 shows how a catalyst lowers the activation energy of a reaction. However, the effect is applicable to the E_a values of both the forward and reverse reactions: both values are reduced by the same amount. Consequently the presence of a catalyst increases the rate of the forward and reverse reactions equally. There is no change in the position of the equilibrium or the value of K_c.

However, the advantage of using a catalyst is that its presence reduces the time required for the equilibrium to be established. This effect is demonstrated by the esterification reaction we discussed earlier.

$$CH_3CO_2H(l) + C_2H_5OH(l) \rightleftharpoons CH_3CO_2C_2H_5(l) + H_2O(l)$$
$$\text{ethanoic acid} \quad \text{ethanol} \quad \text{ethyl ethanoate} \quad \text{water}$$

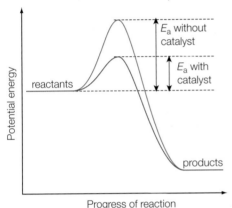

Figure 7.15 Effect of a catalyst in lowering the activation energy for both forward and reverse reaction

The uncatalysed reaction takes many weeks to reach equilibrium. However, the addition of hydrogen ions, H^+, as a catalyst reduces that time to a few hours.

Equilibrium in industrial processes

7.2.5 **Apply** the concepts of kinetics and equilibrium to industrial processes.

The phenomena of reversible reactions and dynamic equilibria are widespread and relevant to many areas of chemistry. Consider, for instance, the whole field of acid–base chemistry (Chapters 8 and 18) where these ideas are crucial to our understanding of what an acid is, as well as to the use of indicators and buffers. In a similar way our conceptual grasp of electrochemistry is very much dependent on the interplay of reversible reactions (Chapter 19). We have commented earlier on how certain important industrial processes are dependent on some key reversible reactions. The ability to predict the effects of changes in physical conditions provided by Le Châtelier's principle is very useful indeed in establishing the best conditions to use for these processes. Such considerations help us to adapt conditions so as to maximize the yield of product. However, these are not the only considerations to be kept in mind. The rate at which a given yield is produced is also important economically, so the time taken to achieve a particular equilibrium is also of significance. Quite often these different considerations work in opposite directions and a compromise set of conditions is employed which gives an acceptable yield in an economically viable time.

The Haber process for ammonia manufacture

Important though nitrogen is for plant growth, most plants cannot 'fix' nitrogen directly from the air. Only certain plants, such as peas for example, can convert nitrogen directly into a usable chemical form because of *Rhizobium* bacteria present in their root nodules. To promote the growth of other crop plants, a nitrogen-containing compound has to be spread as a fertilizer. The important fertilizers include urea, $CO(NH_2)_2$, ammonium sulfate, $(NH_4)_2SO_4$, and ammonium dihydrogen phosphate(v), $NH_4H_2PO_4$. All of these involve the use of ammonia in their manufacture. Ammonia is also used to make nitric acid, some polymers (polyamides such as nylon) and explosives.

The chemical process that produces the ammonia gas can be represented by the following equation:

$$N_2(g) + 3H_2(g) \rightleftharpoons 2NH_3(g) \qquad\qquad \Delta H^{\ominus} = -92\,kJ\,mol^{-1}$$

Nitrogen gas, from the air, is mixed with hydrogen gas, obtained from the reaction of methane with steam (steam reforming). The nitrogen and hydrogen are fed into the main reaction vessel in the ratio of $1:3$ by volume.

The production of cheap nitrogen and hydrogen gases in the correct ratio is an essential part of the whole process. Most ammonia plants use methane (as natural gas), air and water as starting materials (Figure 7.16). Any sulfur in the methane must first be removed otherwise it would poison the surface of the catalyst, reducing the efficiency of adsorption of the reacting gases. The methane is mixed with steam in the presence of a nickel catalyst at 750 °C and the following equilibrium reaction takes place in the 'primary reformer':

$$CH_4(g) + H_2O(g) \rightleftharpoons CO(g) + 3H_2(g) \qquad \Delta H^{\ominus} = +206\,kJ\,mol^{-1}$$

Air is introduced into the mixture to provide the nitrogen, but the oxygen present needs to be removed from the mixture. This is done in the 'secondary reformer' where the oxygen reacts with some of the hydrogen that has just been produced. The unwanted carbon monoxide produced is removed from the mixture in the 'shift reactor' by reaction with more steam. After removal of carbon dioxide, the final mixture consists of nitrogen and hydrogen in a ratio of $1:3$. This mixture is then compressed and fed into the reaction vessel (Figure 7.16).

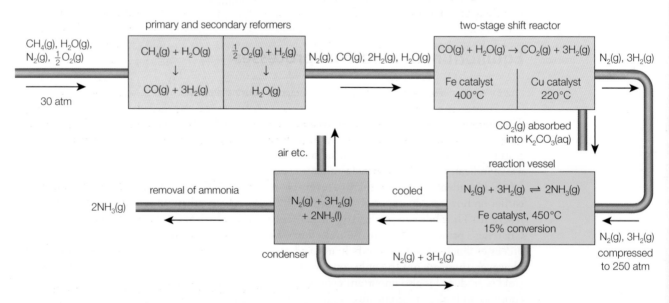

Figure 7.16 Flow diagram showing the stages of the Haber process

The conditions used industrially in the main reaction vessel are arrived at by consideration of both Le Châtelier's principle and kinetic factors (Chapter 6). The aim is to achieve a satisfactory yield of ammonia at a reasonable and economic rate.

The application of Le Châtelier's principle to the reaction

$$N_2(g) + 3H_2(g) \rightleftharpoons 2NH_3(g) \qquad \Delta H^\ominus = -92\,kJ\,mol^{-1}$$

concludes that the highest yield of ammonia is obtained by using low temperatures and high pressures. This is shown in Figure 7.17.

In reality, for economic reasons, it is the rate at which equilibrium is achieved that proves the determining factor, rather than simply the percentage of ammonia present at equilibrium. Equilibrium is reached most quickly at relatively high temperatures, high pressures and in the presence of a catalyst. The reasons for these conditions are explained in turn below.

The choice of pressure

The Haber process is markedly affected by changes in pressure. If the pressure is increased then, in accordance with Le Châtelier's principle, the position of the equilibrium will shift to the right as there are fewer molecules on that side of the equation.

$$N_2(g) + 3H_2(g) \rightleftharpoons 2NH_3(g)$$

On the left-hand side of the equation there are four moles of gas, on the right-hand

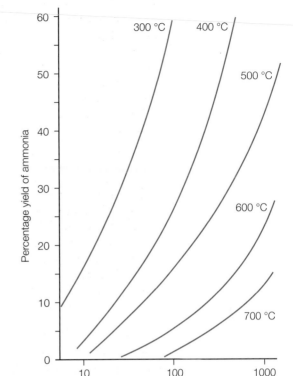

Figure 7.17 Equilibrium percentage yield of ammonia at different temperatures and pressures

side there are two moles of gas. This means that higher pressures will move the position of the equilibrium to the right, producing more ammonia gas. This can be seen in Figure 7.17, which shows the percentage of ammonia in the equilibrium mixture at different pressures.

The use of higher pressures is also favoured for kinetic reasons and most industrial plants operate at 200 or 250 atmospheres. Some plants do operate at pressures up to 1000 atmospheres but these very high pressures demand a large expenditure of energy for compression. More importantly, the very thick walls needed for the reaction vessels (special chromium steel is used) so that such pressures can be safely contained are very costly. The decision is one of balancing the high initial set-up costs against the eventual higher profits resulting from increased yield.

The choice of temperature

In the Haber process the forward reaction is exothermic ($\Delta H^\ominus = -92\,kJ\,mol^{-1}$). This means that the production of ammonia will be favoured by lower temperatures. Increased temperature will result in less ammonia in the equilibrium mixture (see Figure 7.17). By this consideration it would follow that the Haber process should be carried out at low temperatures.

Industrially, however, a temperature of 450 °C is actually used. Three reasons justify the use of this relatively high temperature:

■ Firstly, at low temperatures the reaction is very slow and would take a long time to reach equilibrium. Even the most efficient catalyst, working at high pressures, does not work fast enough to obtain a reasonable conversion at room temperature. A compromise temperature is used – one that gives a reasonable percentage conversion while achieving equilibrium at a fast enough rate.

■ The second reason is that the catalyst used for the reaction has an optimum operating temperature range (Figure 7.18a) and the process is generally carried out at the upper end of this range. The temperature sensitivity of the catalytic action relates to the mechanism by which the heterogeneous catalyst works. The catalysed reaction mechanism depends on the nitrogen and hydrogen molecules adsorbing to the irregular metal surface. The attachment to the surface is by a process known as chemisorption in which electron density is donated from nitrogen and hydrogen atoms into vacant d orbitals in the iron atoms (Figure 7.18b). This adsorption results in a weakening of the bonds in the nitrogen and hydrogen molecules; particularly significant in the case of the triple-bonded nitrogen molecules (N≡N). The weakening of the bonds in the reacting molecules means they can react more readily and ammonia molecules are then formed on the metal surface. The final step of the catalytic sequence is desorption of the ammonia from the metal. The catalytic activity is critically dependent on the strength of the interactions involved in the adsorption of the gas molecules to the surface. These interactions must be strong enough for the molecules to attach to the metal, but weak enough to allow the subsequent release of the product molecules.

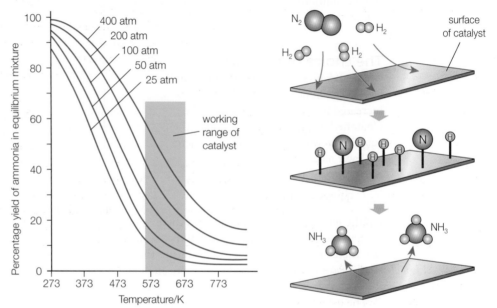

Figure 7.18   **a**   The effective working conditions for the iron catalyst in the Haber Process   **b**   The mechanism of chemisorption of nitrogen and hydrogen molecules, their dissociation into atoms and subsequent reaction on the iron surface of the catalyst in the Haber process

■ Thirdly, because ammonia is easily condensed out of the equilibrium mixture, the hydrogen and nitrogen are readily recycled. This means that a low yield is less of a concern as the unused reactants are not wasted.

Catalysts

The Haber process uses a catalyst of freshly produced, finely divided iron. This is obtained by reducing iron oxide (magnetite, iron(II,III) oxide, Fe_3O_4) with hydrogen. A catalyst is used in this process as it reduces the time needed to reach equilibrium.

The action of the iron catalyst is modified by the presence, in trace quantities, of a promoter. A promoter is a substance that may improve the performance of the catalyst. In some Haber process plants traces of molybdenum are used as the promoter, whilst in others potassium or aluminium oxides or potassium hydroxide are used.

A typical industrial plant for the Haber process operates at 250 atmospheres pressure and 450 °C and produces an actual conversion to ammonia of about 15%. The ammonia can be removed by liquefaction, and then the unchanged gases are recycled through the converter. This recycling ensures the almost complete use of the nitrogen and hydrogen fed into the plant.

History of Chemistry

Fritz Haber (1868–1934), one of Germany's most famous chemists, is also one of the most complex figures to evaluate in the history of science. His life and career were inextricably linked with the political upheaval in Europe that led to two world wars. The moral and ethical issues that coloured that period of turmoil impacted on his life and the choices he personally made leave us with a highly equivocal view of his achievements. His was a life of personal striving, immense scientific achievement and personal tragedy.

He studied at several universities, receiving a PhD in Chemistry in 1891. After a few apparently aimless years working for his father and trying to progress in academic life, he gained a position as a lab assistant at the University of Karlsruhe in 1894. He then rose through the ranks to become a professor. He was passionately patriotic and this, together with career interests, led him to renounce Judaism and convert to Christianity.

Figure 7.19 Fritz Haber (1868–1934)

Haber's scientific legacy is unquestionable. He devised a method for the direct synthesis of ammonia from nitrogen and hydrogen. His process used high pressure and temperature, together with an osmium catalyst. This method was then adapted for use with an iron catalyst by Carl Bosch (1874–1949), and the process was scaled up for industrial production. The achievement of industrial nitrogen fixation was crucial for the development of inexpensive fertilizers and revolutionized food production worldwide. Haber received the Nobel Prize in Chemistry in 1918 for his work on ammonia synthesis.

Even this seemingly beneficial contribution to scientific progress had a double edge. Continuing his research, Haber developed a process for converting ammonia into nitric acid. Nitric acid was then used as the basis for synthesizing a variety of insecticides and producing nitrate high explosives. The novel production of explosives significantly helped the German effort in the First World War (1914–18), enabling Germany to negate the effect of the Allied blockades of nitrates from Chile. Haber became increasingly involved in that national war effort; specifically in the military use of gases such as chlorine to subdue (via suffocation) enemy troops in the trenches (Figure 7.20). His idea seems to have been to use chlorine gas to temporarily incapacitate enemy soldiers and take them out of the war, not to maim or kill them. However, the effects of chlorine have been vividly portrayed in Wilfred Owens's poem 'Dulce et Decorum Est' written just prior to Owen's death on the battlefield in 1914.

GAS! GAS! Quick boys! – An ecstasy of fumbling,
Fitting the clumsy helmets just in time,
But someone still was yelling out and stumbling
And floundering like a man in fire or lime –
Dim through the misty panes and thick green light,
As under a green sea, I saw him drowning.
In all my dreams before my helpless sight
He plunges at me, guttering, choking, drowning.

Figure 7.20 Trench warfare was fought in horrific conditions of endless shelling and physical deprivation

Chlorine was soon supplanted by phosgene ($COCl_2$) and then by a far worse agent, mustard gas (bis[2-chloroethyl] sulfide). But Haber's involvement in chemical warfare was to have tragic personal consequences. His first wife, Clara, a research chemist in her own right, committed suicide at the height of Haber's connection with the war effort. She appears to have been pushed over the edge by Haber's decision to continue in the gas warfare programme.

Haber lived for science, both for its own sake and also for its influence in shaping human life, culture and civilization. His talents were wide-ranging, and he possessed an astonishing knowledge of politics, history, economics, science and industry, meaning he might have succeeded equally well in other fields. He continued to work on a number of areas of chemistry after World War I, and developed friendships with other key scientific figures of the era; Einstein and Max Planck, for instance.

However, despite Haber's clear loyalty to his country, his Jewish ancestry was at odds with the rising tide of anti-semitism in Nazi Germany, making his presence in the country undesirable to the authorities. In 1933, he was forced to leave Germany, and he died of heart problems in Switzerland in 1934. Ironically and as a final tragic twist in this complex story, Zyklon B, a development from the hydrogen cyanide (HCN) insecticide Haber had originally introduced, was used to kill prisoners in the Nazi concentration camps. Reportedly, among the victims were some of Haber's relatives.

Little has been written about Haber's life until recently, because his papers had been kept locked away by those wishing to protect his reputation. When the papers were made publicly available in the early 1990s, they served to show the triumphs, failings and tragedy of a man whose life bore out the contradictions of the time in which he lived. Haber was one of the greatest scientists of his generation and yet he has also been described as one of science's greatest scoundrels.

History of Chemistry

Carl Bosch (1874–1940) was the son of a plumber in Cologne, Germany. After studying chemistry at university, Bosch joined the major chemical company, BASF, in 1899 and quickly gained a reputation as a brilliant chemical engineer. Bosch was responsible for developing an industrial plant to manufacture ammonia from Haber's laboratory process. The first factory for the industrial production of ammonia opened in 1913. The use of ammonia-based fertilizers completely transformed world food production. In 1931, Bosch was awarded the Nobel Prize in Chemistry, jointly with Friedrich Bergius, for their contributions to the invention and development of chemical high-pressure methods.

The Contact process for the manufacture of sulfuric acid

Sulfuric acid (H_2SO_4) is the single most produced chemical world-wide and is now almost entirely produced by the Contact process. Some 150 million tonnes are manufactured globally each year, with the main uses being in the manufacture of the following:

- fertilizers
- paints and pigments
- detergents and soaps
- dyestuffs.

The Contact process consists of three stages (see Figure 7.21).

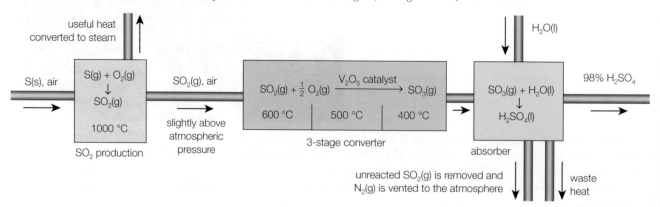

Figure 7.21 Flow diagram showing the stages of the Contact process

Stage 1: In the first stage, sulfur is burnt in air at 1000 °C.

$$S(s) + O_2(g) \rightarrow SO_2(g)$$

Stage 2: The sulfur dioxide produced is then mixed with more air and passed over a vanadium(V) oxide, V_2O_5, catalyst at about 450 °C.

$$2SO_2(g) + O_2(g) \rightleftharpoons 2SO_3(g) \qquad \Delta H^\ominus = -197 \, kJ \, mol^{-1}$$

It is this key second stage (known as the Contact step) that is reversible and requires the consideration of kinetic and equilibrium ideas. To improve the yield of sulfur trioxide the equilibrium is displaced to the right by mixing the sulfur dioxide with an excess of air – about three times more than is necessary from the equation.

Le Châtelier's principle would suggest that the highest yield of sulfur trioxide would be obtained using high pressures and low temperatures. However, in practice a very low temperature cannot be used as the rate of achieving equilibrium would be so slow as to be uneconomic. Another factor here is that the vanadium(V) oxide catalyst only becomes effective at temperatures above 400 °C. To improve the percentage conversion, the reacting gases are passed through a sequence of four separate catalyst beds. The starting temperature of each bed is 450 °C, but the exothermic reaction raises the temperature by several hundred degrees each time. Therefore the gases are cooled back to 450 °C before passage to the next bed. This repetitive technique produces a 99.5% conversion of sulfur dioxide to sulfur trioxide.

Given this high rate of conversion it is unnecessary, in practice, to use a high pressure to increase the yield of sulfur trioxide, SO_3. The cost of using high pressures is uneconomic and a pressure of 1–2 atmospheres is used to ensure the gases circulate freely through the catalyst beds.

Stage 3: In the third stage the sulfur trioxide produced is dissolved in 98% sulfuric acid and then water is added. Carrying out the production of the acid solution in this way avoids the potentially violent and highly exothermic reaction that would occur if the gas were passed directly into water.

$$SO_3(g) + H_2SO_4(l) \rightarrow H_2S_2O_7(l)$$
$$\text{oleum}$$

$$H_2S_2O_7(l) + H_2O(l) \rightarrow 2H_2SO_4(aq)$$

SUMMARY OF KNOWLEDGE

- Reversible physical processes, such the evaporation–condensation of liquids, can establish themselves in equilibrium provided they are in a closed vessel.
- Such systems reach a position of dynamic equilibrium which remains unchanged so long as the physical conditions remain constant.
- While many chemical reactions do go to virtual completion there is a substantial number that do not. Such reactions are reversible reactions, where the new 'products' react with each other and the reverse reaction occurs at the same time as the forward reaction.
- Where such chemical reactions occur in a closed system, a dynamic equilibrium is set up where the rates of the forward and reverse reactions are equal.
- At equilibrium the two reactions continue to take place but there is no overall change in the concentration of the components of the equilibrium mixture under the given physical conditions.
- The point of equilibrium for a particular reaction can be achieved starting from either direction under a given set of conditions.
- The equilibrium law describes how the equilibrium constant, K_c, can be determined for a particular reaction. The equilibrium constant relates directly to a particular chemical equation and indicates the extent of reaction (or position of the equilibrium) at a given temperature.
- The equilibrium constant, K_c, for a general reaction: $aA + bB \rightleftharpoons cC + dD$
  is given by

  $$K_c = \frac{[C]^c \, [D]^d}{[A]^a \, [B]^b}$$

  (where a, b, c and d are the balancing numbers (coefficients) in the above chemical equation; and [C] etc. represent the equilibrium concentrations of the reactants and products).

- The value of K_c for a given reaction is constant provided the temperature remains unchanged.
- The effect on the equilibrium position of changing the physical conditions of concentration, pressure, or temperature for a particular reaction can be predicted by Le Châtelier's principle.
- A change in concentration of any component of the equilibrium mixture will result in a shift in the equilibrium position. Thus, if one of the products is removed from the mixture the equilibrium will shift to the right to replace the substance removed.
- The equilibrium positions of certain gas phase reactions are affected by changes in external pressure. These are reactions where there is a change in the amount (number of moles) of substance during the course of the reaction. For instance, an increase in pressure will shift the position of an equilibrium towards the side of the reaction that involves fewer molecules.
- An increase in temperature will always favour the endothermic process and shift the position of an equilibrium in that direction. Lowering the temperature will always favour the exothermic process. A change of temperature will alter the value of K_c.
- A catalyst has no effect on the position of an equilibrium at a particular temperature as it increases the rate of the forward and reverse reactions equally. It will, however, decrease the time it takes the system to reach equilibrium.
- The only change of physical conditions that alters the value of K_c is a change of temperature. In all other cases the equilibrium concentrations adjust so that the value of K_c remains constant.
- Reversible reactions are central to certain key industrial processes such as the Haber process for ammonia synthesis and the Contact process for the manufacture of sulfuric acid.
- Considerations based on Le Châtelier's principle are important in establishing the industrial conditions used to achieve the best yield of product in these processes. Such considerations must be linked to ideas relating to the rate of reaction so that the yield of product is achieved in an economically viable way.
- For ammonia production the conditions most often used are high pressure (200 atmospheres) and an optimum temperature of 450 °C in the presence of a finely divided iron catalyst. Although the conversion rate is about 15%, the ease of condensing the ammonia produced and recycling the unreacted nitrogen and hydrogen means that the overall process is very efficient.
- In the major stage of the Contact process sulfur dioxide and air are passed over a series of catalyst beds (containing vanadium(v) oxide) at a temperature of around 450 °C. Although Le Châtelier's principle predicts that high pressures would increase the yield of sulfur trioxide, this is found to be unnecessary industrially as the conversion rate is already about 98% using a pressure of 1–2 atmospheres.

Examination questions – a selection

Paper 1 IB questions and IB style questions

Q1 Which statements are correct for a reaction at equilibrium?

   I The forward and reverse reactions both continue.
   II The rates of the forward and reverse reactions are equal.
   III The concentrations of reactants and products are equal.

A I and II only   **C** II and III only
B I and III only   **D** I, II and III

           Standard Level Paper 1, May 05, Q21

Q2 Which statement is always true for a chemical reaction that has reached equilibrium?
A The yield of product(s) is less than 50%.
B The rate of the reverse reaction is greater than the rate of the forward reaction.
C The amounts of reactants and products do not change.
D Both forward and reverse reactions have stopped.

Q3 Which statement(s) is/are true for a mixture of ice and water at equilibrium?

   I The rates of melting and freezing are equal.
   II The amounts of ice and water are equal.
   III The same position of equilibrium can be reached by cooling water or heating ice.

A I only   **C** I and III only
B II only   **D** III only

Q4 Which of the following is a correct statement for the effect of a catalyst on a reversible chemical reaction?
A It increases the amount of product.
B It increases the activation energy of the reaction.
C It allows the chemical reaction to reach equilibrium more quickly.
D It only increases the rate of the forward reaction.

Q5 Which statement is true about chemical reactions at equilibrium?
A The forward and reverse reactions proceed at equal rates.
B The forward and reverse reactions have stopped.
C The concentrations of the reactants and products are equal.
D The forward reaction is exothermic.

Q6 What changes occur when the temperature is increased in the following reaction at equilibrium?

$$Br_2(g) + Cl_2(g) \rightleftharpoons 2BrCl(g) \qquad \Delta H^\ominus = +14\,kJ\,mol^{-1}$$

	Position of equilibrium	Value of equilibrium constant
A	Shifts towards the products	Decreases
B	Shifts towards the reactants	Decreases
C	Shifts towards the reactants	Increases
D	Shifts towards the products	Increases

Q7 Which changes will shift the position of equilibrium to the right in the following reaction?

$$2CO_2(g) \rightleftharpoons 2CO(g) + O_2(g)$$

   I adding a catalyst
   II decreasing the oxygen concentration
   III increasing the volume of the container

A I, II and III   **C** II and III only
B I and II only   **D** I and III only

Q8 The key reaction in the manufacture of sulfuric acid can be represented by the equation below.

$$2SO_2(g) + O_2(g) \rightleftharpoons 2SO_3(g) \qquad \Delta H^\ominus = -197\,kJ\,mol^{-1}$$

What happens when a catalyst is added to an equilibrium mixture from this reaction?
A The rate of the forward reaction increases and that of the reverse reaction decreases.
B The rates of both forward and reverse reactions increase.
C The value of $\Delta H^\ominus$ increases.
D The yield of sulfur trioxide increases.

Q9 $I_2(g) + 3Cl_2(g) \rightleftharpoons 2ICl_3(g)$

What is the equilibrium constant expression for the reaction above?

A $K_c = \dfrac{[I_2][Cl_2]^3}{[ICl_3]^2}$ **B** $K_c = \dfrac{2[ICl_3]}{[I_2] + 3[Cl_2]}$

C $K_c = \dfrac{2[ICl_3]}{3[I_2][Cl_2]}$ **D** $K_c = \dfrac{[ICl_3]^2}{[I_2][Cl_2]^3}$

Q10 $2SO_2(g) + O_2(g) \rightleftharpoons 2SO_3(g) \qquad \Delta H^\ominus = -197\,kJ\,mol^{-1}$

According to the above information, what temperature and pressure conditions produce the greatest amount of SO_3?

	Temperature	Pressure
A	Low	Low
B	Low	High
C	High	High
D	High	Low

Q11 The volume of the reaction vessel containing the following equilibrium mixture

$$SO_2Cl_2(g) \rightleftharpoons SO_2(g) + Cl_2(g)$$

is increased. When equilibrium is re-established, which of the following will have occurred?
- **A** The amount of $SO_2Cl_2(g)$ will have increased.
- **B** The amount of $SO_2Cl_2(g)$ will have decreased.
- **C** The amount of $Cl_2(g)$ will have remained unchanged.
- **D** The amount of $Cl_2(g)$ will have decreased.

Q12 In which of the following reactions does the position of equilibrium remain unaffected by change in pressure?
- **A** $2O_3(g) \rightleftharpoons 3O_2(g)$
- **B** $2NO_2(g) \rightleftharpoons N_2O_4(g)$
- **C** $2NO(g) + Cl_2(g) \rightleftharpoons 2NOCl(g)$
- **D** $N_2(g) + O_2(g) \rightleftharpoons 2NO(g)$

Q13 For a gaseous reaction, the equilibrium constant expression is:

$$K_c = \frac{[O_2]^5[NH_3]^4}{[NO]^4 [H_2O]^6}$$

Which equation corresponds to this equilibrium expression?
- **A** $4NH_3 + 5O_2 \rightleftharpoons 4NO + 6H_2O$
- **B** $4NO + 6H_2O \rightleftharpoons 4NH_3 + 5O_2$
- **C** $8NH_3 + 10O_2 \rightleftharpoons 8NO + 12H_2O$
- **D** $2NO + 3H_2O \rightleftharpoons 2NH_3 + \frac{5}{2}O_2$

Standard Level Paper 1, May 02, Q21

Q14 The following equilibrium can be set up in a closed container.

$$\begin{array}{ccc} N_2O_4(g) & \rightleftharpoons & 2NO_2(g) \\ \text{colourless} & & \text{brown} \end{array} \qquad \Delta H^{\ominus} = +57\,kJ\,mol^{-1}$$

Which of the following changes would produce a darkening of the colour of the gaseous mixture in the container?
- **A** an increase in temperature
- **B** adding a catalyst
- **C** an increase in pressure
- **D** a decrease in temperature

Q15 Which of the following changes will shift the position of equilibrium of this reaction in the forward direction?

$$N_2(g) + O_2(g) \rightleftharpoons 2NO(g) \qquad \Delta H^{\ominus} = +181\,kJ\,mol^{-1}$$

- **I** increasing the pressure
- **II** adding a catalyst
- **III** increasing the temperature
- **A** I only
- **C** III only
- **B** II only
- **D** I and III

Standard Level Paper 1, Nov 01, Q21

Q16 $2H_2(g) + CO(g) \rightleftharpoons CH_3OH(g)$

Methanol is made in industry by means of the reaction above. The equilibrium expression for this reaction is:

A $K_c = \dfrac{[CH_3OH]}{2[H_2][CO]}$ **B** $K_c = \dfrac{[CH_3OH]}{[H_2]^2 [CO]}$

C $K_c = \dfrac{2[H_2][CO]}{[CH_3OH]}$ **D** $K_c = \dfrac{[H_2]^2[CO]}{[CH_3OH]}$

Q17 $N_2(g) + 3H_2(g) \rightleftharpoons 2NH_3(g) \qquad \Delta H^{\ominus} = -91.8\,kJ\,mol^{-1}$

The industrial synthesis of ammonia is based on the reaction above. Which factor(s) will increase the equilibrium concentration of ammonia?
- **I** increase in pressure
- **II** increase in temperature
- **A** I only
- **C** neither I nor II
- **B** both I and II
- **D** II only

Q18 N_2O_4 and NO_2 produce an equilibrium mixture according to the equation below:

$$N_2O_4(g) \rightleftharpoons 2NO_2(g) \qquad \Delta H^{\ominus} = +57\,kJ\,mol^{-1}$$

An increase in the equilibrium concentration of NO_2 can be produced by increasing which of the factors below?
- **I** pressure
- **II** temperature
- **A** neither I nor II
- **C** I only
- **B** both I and II
- **D** II only

Q19 The hydration of ethene to ethanol occurs according to the following equation:

$$C_2H_4(g) + H_2O(g) \rightleftharpoons C_2H_5OH(g)$$

If this reaction is exothermic, which of the following sets of conditions would give the best equilibrium yield of ethanol?

	Temperature/°C	Pressure/atm
A	1000	3
B	750	2
C	250	10
D	500	4

Q20 The smaller an equilibrium constant, K_c:
- **A** the slower the reaction rate
- **B** the lower the concentration of products at equilibrium
- **C** the more endothermic the reaction
- **D** the faster the reactants are converted to products

Paper 2 IB questions and IB style questions

Q1 a The diagrams below represent equilibrium mixtures for the reaction:

$$Y + X_2 \rightleftharpoons XY + X$$

at 350 K and 550 K, respectively. Deduce and explain whether the reaction is exothermic or endothermic. [2]

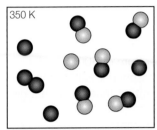

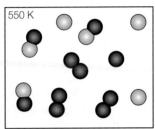

●X   ○Y

b The equation for the main reaction in the Haber process is:

$$N_2(g) + 3H_2(g) \rightleftharpoons 2NH_3(g) \qquad \Delta H^\circ = -92\,kJ\,mol^{-1}$$

  i State **two** characteristics of a reversible reaction at equilibrium. [2]
  ii This reaction is described as *homogeneous*. State what is meant by the term *homogeneous*. [1]
  iii Write the equilibrium constant expression for the reaction. [2]

c When nitrogen and hydrogen are mixed together at room temperature and atmospheric pressure the reaction is very slow. In industry, typical values of pressure and temperature used can be 250 atmospheres and 450 °C.

  i State the effects on both the rate of reaction and the value of the equilibrium constant of increasing the temperature. [2]
  ii State the effects on both the rate of reaction and the value of the equilibrium constant of increasing the pressure. [2]
  iii Suggest why a pressure of 1000 atmospheres is not used. [1]
  iv State and explain what will happen to the position of the equilibrium if some of the ammonia is removed. [2]

d Name the catalyst used in the Haber process. State and explain its effect on the value of the equilibrium constant. [3]

e Use the collision theory to explain the effect of increasing the temperature on the rate of reaction between nitrogen and hydrogen. [3]

Q2 For the reversible reaction:

$$H_2(g) + I_2(g) \rightleftharpoons 2HI(g) \qquad\qquad \Delta H > 0$$

the equilibrium constant $K_c = 60$ at a particular temperature.

a Give the equilibrium expression. [1]
b For this reaction, what information does the value of K_c provide about the **relative** concentrations of the product and reactants at equilibrium? [1]
c What effect, if any, will an increase in pressure have on the **equilibrium position**? [1]
d Explain why an increase in temperature increases the value of the **equilibrium constant** for the above reaction. [1]

Q3 The equation for one reversible reaction involving oxides of nitrogen is shown below:

$$N_2O_4(g) \rightleftharpoons 2NO_2(g) \qquad \Delta H^\circ = +57\,kJ\,mol^{-1}$$

Experimental data for this reaction can be represented on the following graph:

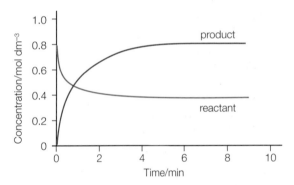

a Write an expression for the equilibrium constant, K_c, for the reaction. Explain the significance of the horizontal parts of the lines on the graph. State what can be deduced about the magnitude of K_c for the reaction, giving a reason. [4]
b Use Le Châtelier's principle to predict and explain the effect of increasing the temperature on the position of equilibrium. [2]
c Use Le Châtelier's principle to predict and explain the effect of increasing the pressure on the position of equilibrium. [2]
d State and explain the effects of a catalyst on the forward and reverse reactions, on the position of equilibrium and on the value of K_c. [6]

Standard Level Paper 2, Nov 05, Q6(a)

Q4 a The following equilibrium is established at 1700°C.

$$CO_2(g) + H_2(g) \rightleftharpoons H_2O(g) + CO(g)$$

If only carbon dioxide gas and hydrogen gas are present initially, sketch on a graph a line representing rate against time for **i** the forward reaction *and* **ii** the reverse reaction until shortly after equilibrium is established. Explain the shape of each line. [7]

b K_c for the equilibrium reaction is determined at two different temperatures. At 850°C, $K_c = 1.1$ whereas at 1700°C, $K_c = 4.9$. On the basis of these K_c values explain whether the reaction is exothermic or endothermic. [3]

Q5 The table below gives information about the percentage yield of ammonia obtained in the Haber process under different conditions.

Pressure/atm	Temperature/°C			
	200	300	400	500
10	50.7	14.7	3.9	1.2
100	81.7	52.5	25.2	10.6
200	89.1	66.7	38.8	18.3
300	89.9	71.1	47.1	24.4
400	94.6	79.7	55.4	31.9
600	95.4	84.2	65.2	42.3

a From the table, identify which combination of temperature and pressure gives the highest yield of ammonia. [1]

b The equation for the main reaction in the Haber process is:

$$N_2(g) + 3H_2(g) \rightleftharpoons 2NH_3(g) \qquad \Delta H \text{ is negative}$$

Use this information to state and explain the effect on the yield of ammonia of increasing
i pressure [2]
ii temperature. [2]

c In practice, typical conditions used in the Haber process are a temperature of 500°C and a pressure of 200 atmospheres. Explain why these conditions are used rather than those that give the highest yield. [2]

d Write the equilibrium constant expression, K_c, for the production of ammonia. [1]

e i Suggest why this reaction is important for humanity. [1]
ii A chemist claims to have developed a new catalyst for the Haber process, which increases the yield of ammonia. State the catalyst normally used for the Haber process, and comment on the claim made by this chemist. [2]

8 Acids and bases

- Acids and bases are chemically complementary.
- Acidity is due to the presence of excess hydrogen ions.
- Alkalinity is due to the presence of excess hydroxide ions.
- Acidity and alkalinity are both measured on the pH scale.
- pH and hydrogen ion concentration have an inverse logarithmic relationship.
- Neutralization is the reaction between an acid and a base.
- Acids and bases are defined according to the Brønsted–Lowry theory which views acid–base reactions as proton transfers.
- Acids and bases can also be defined under the more general Lewis theory which views acid–base reactions in terms of dative bond formation.

8.1 Theories of acids and bases

Brønsted–Lowry theory

8.1.1 **Define** acids and bases according to the Brønsted–Lowry and Lewis theories.

8.1.2 **Deduce** whether or not a species could act as a Brønsted–Lowry and/or a Lewis acid or base.

8.1.3 **Deduce** the formula of the conjugate acid (or base) of any Brønsted–Lowry base (or acid).

The **Brønsted–Lowry theory** of acids and bases involves the transfer of protons or hydrogen ions within an aqueous solution. An **acid** is defined as a molecule or ion that acts as a proton donor and a **base** is defined as a molecule or ion that acts as a proton acceptor. For example, when hydrogen chloride gas is dissolved in water it reacts to form hydrochloric acid. The following equilibrium is established:

$$HCl(g) + H_2O(l) \rightleftharpoons H_3O^+(aq) + Cl^-(aq)$$

In the forward reaction (left to right) the hydrogen chloride molecule is acting as an acid because it donates a proton or hydrogen ion, H^+, to the water molecule, which is acting as a base since it forms an oxonium or hydronium ion, $H_3O^+(aq)$.

In the reverse or backward reaction (right to left) the hydronium or oxonium ion acts as an acid by donating a hydrogen ion to the chloride ion to form hydrogen chloride. The chloride ion is acting as a base. The equation above can be split into two 'half-equations' which more clearly show the proton transfer:

$$\underset{\text{acid}}{HCl(aq)} \rightleftharpoons \underset{\text{conjugate base}}{Cl^-(aq)} + H^+(aq)$$

This reaction shows that when a species loses a proton, the product has to be a base since the process is reversible (to a varying degree depending on the acid – Section 8.3). The chloride ion is described as the **conjugate base** of the hydrogen chloride molecule.

$$\underset{\text{base}}{H_2O(l)} + H^+(aq) \rightleftharpoons \underset{\text{conjugate acid}}{H_3O^+(aq)}$$

This reaction shows that when a species gains a proton, the product has to be an acid since the process is reversible. The hydronium or oxonium ion is described as the **conjugate acid** of the water molecule. An acid–base reaction always involves (at least) two **conjugate pairs** that differ by H^+.

Brønsted–Lowry theory can also be applied to the behaviour of bases in aqueous solution. For example, when ammonia gas is dissolved in water the following chemical equilibrium is established:

$$\underset{\text{base}}{NH_3(g)} + \underset{\text{acid}}{H_2O(l)} \rightleftharpoons \underset{\text{conjugate acid}}{NH_4^+(aq)} + \underset{\text{conjugate base}}{OH^-(aq)}$$

The ammonia is acting as a base by accepting a proton from the water. Water is acting as an acid here, in contrast to its behaviour with acids when it acts as a base. Species that are able to act as both acids and bases (proton donors and proton acceptors), depending on the species they are reacting with, are termed **amphiprotic**.

Language of Chemistry

The word acid is derived from the Latin word *acidus* meaning sour. The concept of a base in chemistry was first introduced by the French chemist Guillaume François Rouelle in 1754. He noted that acids, which in those days were mostly volatile liquids (for example, ethanoic acid), converted into solid salts only when combined with specific substances known as bases. ■

History of Chemistry

Antoine Lavoisier (Chapter 9) suggested that acids were substances containing oxygen. Berthollet (Chapter 5) proved in 1787 that 'prussic acid' (hydrocyanic acid, HCN) contained only hydrogen, carbon and nitrogen. **Sir Humphry Davy** (1778–1829) was an English chemist who discovered several group 1 and 2 metals. He suggested that all acids contained hydrogen as the essential element. In 1838 the German chemist **Justus von Liebig** (1803–1873) (Figure 8.1) proposed that acids were substances that can react with metals to produce hydrogen.

Figure 8.1 Justus von Liebig

It is important to note that in the equations below for ethanoic acid and hydrochloric acid the *competition* is between the base and its conjugate for a proton, H^+.

$$HCl(g) + H_2O(l) \rightleftharpoons H_3O^+(aq) + Cl^-(aq)$$
$$\text{acid} \qquad \text{base} \qquad \text{conjugate acid} \qquad \text{conjugate base}$$

$$CH_3COOH(l) + H_2O(l) \rightleftharpoons H_3O^+(aq) + CH_3COO^-(aq)$$
$$\text{acid} \qquad \text{base} \qquad \text{conjugate acid} \qquad \text{conjugate base}$$

In the case of hydrochloric acid the water molecule is a much stronger base than the chloride ion. In other words, the water molecule has a much greater tendency to accept a proton, $H^+(aq)$, than does the chloride ion. Consequently, the position of the equilibrium will lie on the right and virtually all of the hydrogen chloride molecules will be ionized or dissociated. In general, *strong acids* produce relatively *weak conjugate bases* in aqueous solutions (Chapter 18).

In the case of ethanoic acid the ethanoate ion is a much stronger base than the water molecule. In other words, the ethanoate ion has a much greater tendency to accept a proton, $H^+(aq)$, than does the water molecule. In general, *weak acids* produce relatively *strong conjugate bases* in aqueous solutions (Figure 8.2).

Similarly, *strong bases* produce *weak conjugate acids* in aqueous solutions and *weak bases* produce *strong conjugate acids* in aqueous solutions (Table 8.1). The strengths of acid, bases and their respective conjugates can be measured and expressed in terms of K_a or pK_a (Chapter 18).

if equilibrium lies to the *right* then strong acid but weak conjugate base

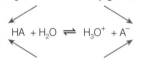

$$HA + H_2O \rightleftharpoons H_3O^+ + A^-$$

weak acid, stronger conjugate base if the equilibrium lies to the *left*

Figure 8.2 The relationship between an acid and its conjugate base

Acid	Strength	Base	Strength
H_2SO_4	very strong	HSO_4^-	very weak
HCl		Cl^-	
HNO_3		NO_3^-	
H_3O^+	fairly strong	H_2O	weak
HSO_4^-		SO_4^{2-}	
CH_3COOH	weak	CH_3COO^-	less weak
H_2CO_3		HCO_3^-	
NH_4^+		NH_3	
HCO_3^-	very weak	CO_3^{2-}	fairly strong
H_2O		OH^-	

Table 8.1 Some common acids and conjugate bases in order of their strengths

Acids (Figure 8.3) that have a single proton to donate are said to be **monoprotic**. Common examples include hydrochloric, HCl(aq), nitric, HNO₃(aq), nitrous, HNO₂(aq) and ethanoic acids, CH₃COOH(aq). Acids that have two protons to donate are said to be **diprotic**. Common examples include 'carbonic acid', H₂CO₃(aq), sulfuric acid, H₂SO₄(aq), and sulfurous acid, H₂SO₃(aq). The only common triprotic acid is phosphoric acid, H₃PO₄(aq).

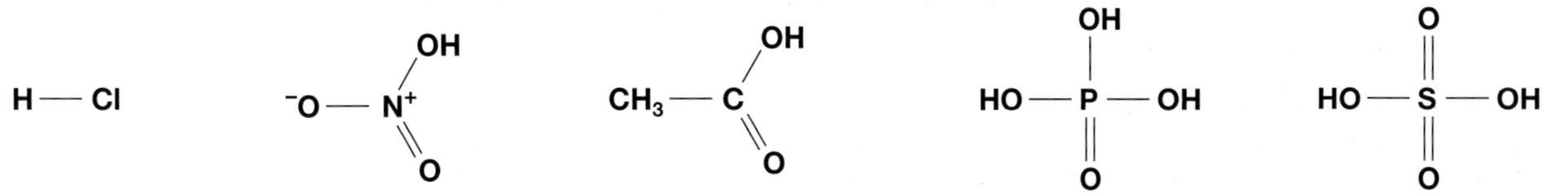

$$H\!-\!Cl$$

hydrogen chloride

nitric(v) acid

ethanoic acid

phosphoric(v) acid

sulfuric acid

Figure 8.3 Structural formulas for hydrochloric, sulfuric, nitric(v), phosphoric(v) and ethanoic acids

For a substance to be an acid the hydrogen usually has to be attached to oxygen or a halogen. This accounts for the monoproticity of ethanoic acid: only the hydrogen atom attached to the oxygen atom is acidic and replaceable by a metal ion. The other three hydrogen atoms of ethanoic acid are attached to a carbon atom and are therefore *not* acidic.

Care must be taken when using the term 'conjugate' when referring to diprotic or triprotic acids. For example, consider the ionization or dissociation of the weak acid, sulfurous acid (Chapter 25), H₂SO₃(aq):

$$\underset{\text{acid}}{H_2SO_3(aq)} + \underset{\text{base}}{H_2O(l)} \rightleftharpoons \underset{\text{conjugate acid}}{H_3O^+(aq)} + \underset{\text{conjugate base}}{HSO_3^-(aq)}$$

$$\underset{\text{acid}}{HSO_3^-(aq)} + \underset{\text{base}}{H_2O(l)} \rightleftharpoons \underset{\text{conjugate acid}}{H_3O^+(aq)} + \underset{\text{conjugate base}}{SO_3^{2-}(aq)}$$

According to the first equation the hydrogensulfite ion, HSO₃⁻(aq) is the *conjugate base* of sulfurous acid, *but* according to the second equation it is the *conjugate acid* of the sulfite ion, SO₃²⁻(aq). The two equations illustrate the fact that 'conjugate' is a relative term and only links a specific pair of acids and bases. The hydrogensulfite ion, like the water molecule, is another example of an amphiprotic species.

The terms 'acid' and 'base' are also *relative* terms. If two concentrated acids are reacted together, then the weaker acid of the two will be 'forced' to act as a base. For example, when concentrated nitric and sulfuric acids are reacted together in a 1 : 2 molar ratio, a so-called nitrating mixture is formed which contains a cation known as the nitronium ion, NO₂⁺. This cation is involved in the nitration of organic compounds (Chapter 27).

The first equilibrium to be established in the nitrating mixture is shown below:

$$\underset{\text{base}}{HNO_3(aq)} + \underset{\text{acid}}{H_2SO_4(aq)} \rightleftharpoons \underset{\text{conjugate acid}}{H_2NO_3^+(aq)} + \underset{\text{conjugate base}}{HSO_4^-(aq)}$$

The Brønsted–Lowry model can be extended to reactions that occur in the gas phase. An example of a gas-phase acid–base reaction is encountered when hydrogen chloride and ammonia gases react to form white fumes of ammonium chloride (Figure 8.4):

$$HCl(g) + NH_3(g) \rightarrow NH_4Cl(s)$$

This reaction involves the transfer of a hydrogen ion (H⁺) from hydrogen chloride to ammonia and is therefore a Brønsted–Lowry acid–base reaction, even though it occurs in the gas phase.

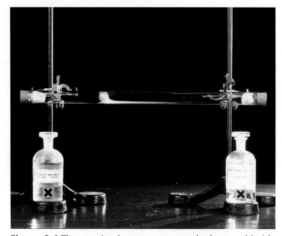

Figure 8.4 The reaction between gaseous hydrogen chloride and gaseous ammonia

History of Chemistry

Johannes Brønsted (Figure 8.5) (1879–1947) was a Danish physical chemist. Following a degree in chemical engineering he was appointed as professor of inorganic and physical chemistry at Copenhagen University. In 1923 he introduced his proton-based theory of acid–base reactions (Figure 8.6). The English chemist **Thomas Lowry** (Figure 8.5) (1874–1936) published an identical theory in the same year, both realizing that water had an active role in acidity.

Figure 8.5 Johannes Brønsted (left) and Thomas Lowry (right)

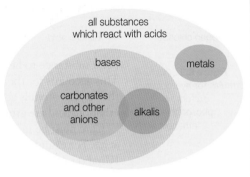

Figure 8.6 Venn diagram showing the Brønsted–Lowry classification of chemicals which react with acids

Lewis theory

The Brønsted–Lowry theory defines an acid as a proton donor and a base as a proton acceptor. However, when a base accepts a proton it donates an electron pair to the proton:

In general:

$$\text{B:} \quad\curvearrowright\quad H^+ \rightarrow {}^+BH$$

where B: represents a base (with one or more lone pairs of electrons), H^+ represents the proton, BH^+ represents the conjugate acid and the curly arrow represents the movement of an electron pair to form a dative or coordinate covalent bond (Chapter 4). Simple examples include the reaction between an ammonia molecule and a proton to form the ammonium ion.

$$H_3N: \quad\curvearrowright\quad H^+ \rightarrow {}^+NH_4$$

and between a water molecule and a proton to form the hydronium or oxonium ion:

$$H_2\ddot{O}: \quad\curvearrowright\quad H^+ \rightarrow H_3O^+$$

The movement of an electron pair during an acid–base reaction is the basis of the **Lewis theory** of acidity developed by Gilbert Lewis (Chapter 4). A **Lewis acid** is defined as a substance that can accept a pair of electrons from another atom to form a coordinate or dative covalent bond. A **Lewis base** is defined as a substance that can donate a pair of electrons to another atom to form a dative covalent (coordinate) bond. In the simple examples above, the proton (H^+) is the Lewis acid and the ammonia molecule and water molecules are the Lewis bases.

However, the Lewis theory is *more general* than the Brønsted–Lowry theory: some reactions are classified as acid–base reactions under the Lewis definitions that are not regarded as acid–base reactions under the Brønsted–Lowry theory. Therefore the terms Lewis acid and Lewis base are often reserved for species which are Lewis acids and bases, but which are *not* Brønsted–Lowry acids and bases. In these reactions no protons are involved, water is absent and the reactions frequently occur in the gas phase. For example, the gases ammonia and boron trifluoride react together to form an adduct called ammonia boron trifluoride:

$$H_3N: \quad\curvearrowright\quad BF_3 \rightarrow NH_3BF_3$$

The ammonia is the Lewis base (electron pair donor) and boron trifluoride is the Lewis acid (electron pair acceptor). The reaction is driven, in part, by the need for the boron in boron trichloride to overcome its electron deficiency: it has only six electrons in its outer shell. In the **adduct** the boron has acquired a full outer shell of eight electrons (an octet).

An interesting example of Lewis acid–base behaviour is illustrated by the aluminium chloride dimer (Chapter 13) formed when aluminium chloride undergoes sublimation. Each of the two aluminium chloride molecules forms a dative or coordinate bond to the aluminium of the other molecule, using one of its lone pairs on its chlorine atoms, while accepting a lone pair of electrons from the chlorine of the other molecule. Each aluminium chloride molecule is thus acting simultaneously as both a Lewis acid and a Lewis base.

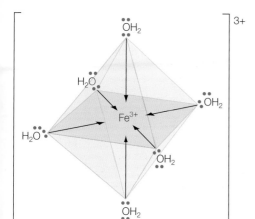

Common examples of Lewis bonding are found in the complex ions formed by the transition metals (Chapter 13). Charged metal ions become surrounded by water molecules in aqueous solution; for example, many transition metal ions become surrounded by six water molecules. $Fe^{3+}(aq)$ exists as the hexaaquairon(III) ion $[Fe(H_2O)_6]^{3+}$ (Figure 8.7).

The six water molecules in this complex each donate a lone pair of electrons from the oxygen atoms of their water molecules to the empty 3d orbitals of the central iron(III) ion. The water molecules, known as ligands, are acting as Lewis bases (electron pair donors) and the iron(III) ion is acting as a Lewis acid (electron pair acceptor).

Figure 8.7 The structure of the hexaaquairon(III) ion $[Fe(H_2O)_6]^{3+}$

8.2 Properties of acids and bases

8.2.1 **Outline** the characteristic properties of acids and bases in aqueous solution.

Acids

Common acids found in the laboratory include ethanoic acid, $CH_3COOH(aq)$, sulfuric acid, $H_2SO_4(aq)$, hydrochloric acid, $HCl(aq)$, and nitric acid, $HNO_3(aq)$. Acids are a group of compounds that exhibit the following properties when dissolved in water to form a dilute solution.

pH
Acids have a pH value less than 7 and turn the indicator blue litmus paper red (Section 8.4). The pH value is a measure of the acidity of the solution and indicators (Figure 8.8) are dyes that change colour according to the pH of the solution.

Figure 8.8 Blue and red litmus paper

Conductivity
Acids are electrolytes (Chapter 9), meaning they undergo chemical decomposition when an electric current is passed through their aqueous solutions.

Reaction with metals

Most dilute acids react to give hydrogen gas (Figure 8.9) and a solution of a salt when a reactive metal (Chapter 9) such as magnesium, iron or zinc is added. For example:

$$Mg(s) + 2HCl(aq) \rightarrow MgCl_2(aq) + H_2(g)$$
$$Mg(s) + 2HNO_3(aq) \rightarrow Mg(NO_3)_2(aq) + H_2(g)$$
$$Mg(s) + H_2SO_4(aq) \rightarrow MgSO_4(aq) + H_2(g)$$
$$Mg(s) + 2CH_3COOH(aq) \rightarrow (CH_3COO)_2Mg(aq) + H_2(g)$$

or, ionically,

$$Mg(s) + 2H^+(aq) \rightarrow Mg^{2+}(aq) + H_2(g)$$

In general,

reactive metal + dilute acid → salt + hydrogen

The more unreactive metals, for example copper and lead, do *not* react with dilute acids (Chapter 9).

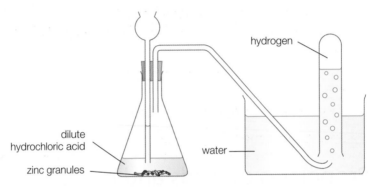

Figure 8.9 Apparatus for collecting the hydrogen produced by the reaction of zinc and hydrochloric acid

Reaction with metal carbonates

Dilute acids react to give carbon dioxide gas when a metal carbonate or metal hydrogencarbonate is added (Figure 8.10). For example,

$$CaCO_3(s) + 2HCl(aq) \rightarrow CaCl_2(aq) + H_2O(l) + CO_2(g)$$
$$NaHCO_3(s) + HCl(aq) \rightarrow NaCl(aq) + H_2O(l) + CO_2(g)$$

or ionically,

$$CO_3^{2-}(aq) + 2H^+(aq) \rightarrow H_2O(l) + CO_2(g)$$
$$HCO_3^-(aq) + H^+(aq) \rightarrow H_2O(l) + CO_2(g)$$

In general,

Figure 8.10 The reaction between calcium carbonate and dilute hydrochloric acid

metal carbonate or metal hydrogencarbonate + dilute acid → salt + water + carbon dioxide

The reaction between calcium carbonate and dilute sulfuric acid is *slow* because an almost insoluble layer of calcium sulfate, $CaSO_4$, protects the calcium carbonate from further attack by the acid.

The presence of carbon dioxide can be confirmed by bubbling the gas through limewater (a solution of calcium hydroxide). The solution initially turns cloudy, but then clears if excess carbon dioxide is passed through the limewater:

$$CO_2(g) + Ca(OH)_2(aq) \rightarrow CaCO_3(s) + H_2O(l)$$
$$CaCO_3(s) + H_2O(l) + CO_2(g) \rightarrow Ca(HCO_3)_2(aq)$$

Reaction with bases

Bases include metal oxides, metal hydroxides and aqueous ammonia. A base is a substance that reacts with an acid to form a salt and water *only*. This reaction is known as a **neutralization**.

Alkalis are bases which are soluble in water (Figure 8.11). They include the group 1 hydroxides, barium hydroxide and aqueous ammonia, $NH_3(aq)$, sometimes called 'ammonium hydroxide', $NH_4OH(aq)$. Alkalis have a soapy feel (they react with oils and fats in the skin) and have a bitter taste.

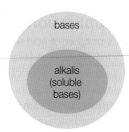

Figure 8.11 The relationship between alkalis and bases

Figure 8.12 The reaction between excess copper(II) oxide and hydrochloric acid to form green copper(II) chloride solution

Reaction with metal oxides

Dilute acids react to give a salt and water when a metal oxide is added (Figure 8.12). For example:

$$CuO(s) + H_2SO_4(aq) \rightarrow CuSO_4(aq) + H_2O(l)$$
$$CuO(s) + 2HNO_3(aq) \rightarrow Cu(NO_3)_2(aq) + H_2O(l)$$
$$CuO(s) + 2HCl(aq) \rightarrow CuCl_2(aq) + H_2O(l)$$
$$CuO(s) + 2CH_3COOH(aq) \rightarrow Cu(CH_3COO)_2(aq) + H_2O(l)$$

or ionically,

$$O^{2-}(s) + 2H^+(aq) \rightarrow H_2O(l)$$

In general,

metal oxide + dilute acid → salt + water

Reaction with metal hydroxides

Dilute acids react to give a salt and water when a metal hydroxide or aqueous ammonia is added. For example:

$$NaOH(aq) + HNO_3(aq) \rightarrow NaNO_3(aq) + H_2O(l)$$
$$NH_3(aq) + HNO_3(aq) \rightarrow NH_4NO_3(aq)$$

or ionically,

$$OH^-(aq) + H^+(aq) \rightarrow H_2O(l)$$
$$NH_3(aq) + H^+(aq) \rightarrow NH_4^+(aq)$$

In general,

metal hydroxide + dilute acid → salt + water

A summary of the reactions of acids in dilute aqueous solution is given in Figure 8.13.

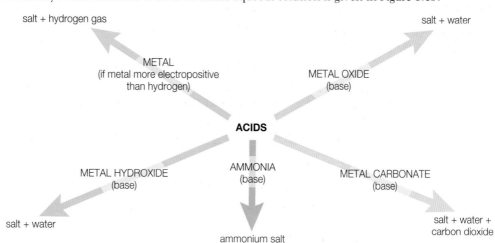

Figure 8.13 Summary of the reactions of acids

Applications of Chemistry

A number of acid and bases are used in the home as cleaning agents (Figure 8.14). Rust removers often contain phosphoric(V) acid which forms a protective layer of iron(III) phosphate to help prevent further rusting. Rust stains on clothing can be removed using ethanedioic (oxalic) acid. The reaction involves complexing of iron ions (Chapter 13). Hard water contains a high concentration of calcium ions and forms deposits in kettles and hot water pipes. The carbonate deposits can be removed with acid, for example vinegar (ethanoic acid).

Oven cleaners usually contain sodium hydroxide (Figure 8.15), which convert oils and fats into water-soluble products (propane-1,2,3-triol and carboxylate ions). Ammonia and sodium carbonate are also present in many liquid cleaners. They are both weaker bases and hence less corrosive to the skin and eyes. Sodium carbonate is present in dishwasher crystals. Carbonate ions undergo hydrolysis with water molecules to release excess hydroxide ions (Chapter 18).

Figure 8.14 Denture cleaning tablets. Active ingredients include the salts sodium hydrogencarbonate, sodium perborate and citric acid

Figure 8.15 Oven cleaner pads

■ Extension: Salts

A **salt** is an ionic compound formed when the replaceable hydrogen of an acid is completely or partly replaced by a metal (ion). The number of replaceable hydrogen atoms in an acid is termed the basicity or **proticity** of the acid. Table 8.2 gives the basicity or proticity of some common acids.

For example, $HCl(aq)$ $\rightarrow$ $NaCl(aq)$
                 hydrochloric acid (an acid)   sodium chloride (a salt)

Name of acid	Formula	Basicity or proticity
Hydrochloric acid	HCl	1
Nitric acid	HNO$_3$	1
Ethanoic acid	CH$_3$COOH	1
Sulfuric acid	H_2SO$_4$	2
Carbonic acid	H_2CO$_3$	2
Phosphoric(V) acid	H_3PO$_4$	3

Table 8.2 Basicity or proticity of some common acids

In the case of a diprotic or dibasic acid, containing more than one replaceable hydrogen atom, salts can be formed when all or some of the hydrogen is replaced. Salts formed by replacing all of the hydrogen are termed **normal salts**; those formed by replacing only part of the hydrogen are termed **acid salts**. Table 8.3 gives examples of the sodium salts formed by common acids.

Acid	Salt	Example
Hydrochloric acid, HCl	Chlorides	Sodium chloride, NaCl
Nitric acid, HNO$_3$	Nitrates	Sodium nitrate, NaNO$_3$
Ethanoic acid, CH$_3$COOH	Ethanoates	Sodium ethanoate, CH$_3$COONa
Sulfuric acid, H$_2$SO$_4$	Sulfates (normal salts) and hydrogensulfates (acid salts)	Sodium sulfate, Na$_2$SO$_4$, and sodium hydrogensulfate, NaHSO$_4$
Carbonic acid, H$_2$CO$_3$	Carbonates (normal salts) and hydrogencarbonates (acid salts)	Sodium carbonate, Na$_2$CO$_3$, and sodium hydrogencarbonate, NaHCO$_3$

Table 8.3 Examples of sodium salts formed by common acids

Aqueous solutions of salts may be neutral, acidic or alkaline (Chapter 18). Salts that form acidic or alkaline solutions have undergone hydrolysis with water. Salts may be prepared in a variety of ways, depending on their solubility (Figure 8.16).

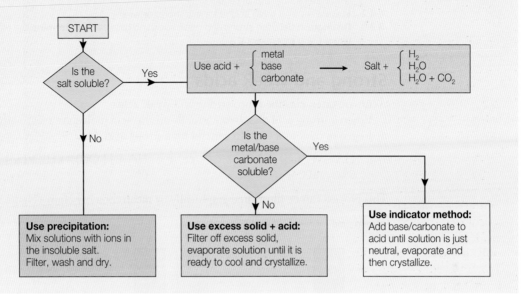

Figure 8.16 Summary of the preparation of salts

The importance of water

Pure or anhydrous acids do *not* behave as acids and do not exhibit the characteristic properties of acids described previously. These properties are only shown after the acids have been reacted and dissolved in water to form dilute aqueous solutions.

The importance of water in acid solutions can be demonstrated by dissolving hydrogen chloride gas, HCl, in both water and an organic liquid such as methylbenzene, $C_6H_5CH_3$. Solutions are formed in both cases, *but* only the aqueous solution exhibits typical acidic properties (Table 8.4).

Test	Solution of hydrogen chloride in water	Solution of dry hydrogen chloride in dry methylbenzene
Dry universal indicator paper	Turns red – acidic solution formed	Remains green – neutral solution formed
Addition of calcium carbonate	Carbon dioxide gas produced	No reaction
Electrical conductivity	Good conductor	Non-conductor
Temperature change on formation of solution	Rise in temperature	Little change in temperature

Table 8.4 Reactions of hydrogen chloride with water and methylbenzene

The results are accounted for by suggesting that when hydrogen chloride dissolves in water, a chemical reaction occurs and ions are formed, resulting in the formation of a hydrochloric acid:

$$HCl(g) + (aq) \rightarrow H^+(aq) + Cl^-(aq)$$

It is the hydrogen ions that are responsible for all the acidic properties of acids previously described, and these are only formed in the presence of water. When hydrogen chloride is dissolved in methylbenzene, no hydrogen ions are formed and undissociated or un-ionized hydrogen chloride molecules, HCl(solv), are present.

8.3 Strong and weak acids and bases

8.3.1 **Distinguish** between strong and weak acids and bases in terms of the extent of dissociation, reaction with water and electrical conductivity.
8.3.2 **State** whether a given acid or base is strong or weak.
8.3.3 **Distinguish** between *strong* and *weak* acids and bases, and **determine** the relative strengths of acids and bases, using experimental data.

Strong and weak acids

Acids are often classified into **strong** and **weak acids**.

When a strong acid dissolves, virtually all the acid molecules react with the water to produce hydrogen or oxonium ions. In general for a strong acid, HA:

$$HA \rightarrow H^+(aq) + A^-(aq) \qquad \text{or} \qquad HA + H_2O(l) \rightarrow H_3O^+(aq) + A^-(aq)$$
$$0\% \quad 100\% \qquad\qquad\qquad\qquad 0\% \qquad\qquad 100\%$$

This process can be illustrated graphically by means of a bar chart (Figure 8.17).

strong acid

Figure 8.17 Graphical representation of the behaviour of a strong acid in aqueous solution

The four common strong acids are hydrochloric, nitric, sulfuric and chloric(VII) (perchloric).

$$HCl(g) + H_2O(l) \rightarrow H_3O^+(aq) + Cl^-(aq)$$

$$HNO_3(l) + H_2O(l) \rightarrow H_3O^+(aq) + NO_3^-(aq)$$

$$H_2SO_4(l) + H_2O(l) \rightarrow H_3O^+(aq) + HSO_4^-(aq)$$

$$HClO_4(l) + H_2O(l) \rightarrow H_3O^+(aq) + ClO_4^-(aq)$$

Monoprotic organic acids are usually weak. When a weak acid dissolves in water, only a small percentage of its molecules (typically 1%) react with water molecules to release hydrogen or oxonium ions. An equilibrium is established, with the majority of the acid molecules not undergoing ionization or dissociation. In other words, the equilibrium lies on the left-hand side of the equation.

In general for a weak acid, HA:

$$HA \rightleftharpoons H^+(aq) + A^-(aq) \qquad \text{or} \qquad HA + H_2O(l) \rightleftharpoons H_3O^+(aq) + A^-(aq)$$
$$99\% \quad 1\% \qquad\qquad\qquad\qquad 99\% \qquad\qquad 1\%$$

This process can be illustrated graphically by means of a bar chart (Figure 8.18).

weak acid

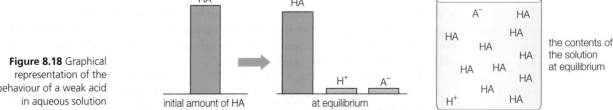

Figure 8.18 Graphical representation of the behaviour of a weak acid in aqueous solution

Examples of common weak acids are ethanoic acid and aqueous carbon dioxide or 'carbonic acid':

$$CH_3COOH(l) + H_2O(l) \rightleftharpoons CH_3COO^-(aq) + H_3O^+(aq)$$

and

$$CO_2(g) + 2H_2O(l) \rightleftharpoons HCO_3^-(aq) + H_3O^+(aq)$$

The term carbonic acid is in quotation marks, since the compound carbonic acid, H_2CO_3, does not actually exist and cannot be isolated. Attempts to isolate it result in the formation of carbon dioxide and water.

History of Chemistry

Boric acid (Figure 8.19) was first prepared by **Wilhelm Homberg** (1652–1715) from borax, by the action of mineral acids, and was given the name *sal sedativum Hombergi* (sedative salt of Homberg). Boric acid was originally extracted from steam vents (known as *soffioni*) in Tuscany in Italy, where it occurs in the steam at 0.06 % by mass. Boric acid is a poisonous substance, but since it is a powerful germicide it was used as food preservative.

Figure 8.19 Boric acid, H_3BO_3 (a weak inorganic acid)

It is also used as an antiseptic and insecticide. Borax ($Na_2B_4O_7$), the sodium salt of boric acid, is another important boron compound. In medieval Europe the merchants of Venice had a monopoly on the supply of borax, which was used at that time to treat bacterial and fungal infections. The early European supplies came from high-altitude lakes on the plateaus of Tibet, a country which at that time was closed to foreigners on pain of death.

Strong and weak acids (see Table 8.5) of the *same concentration*, such as hydrochloric and ethanoic acids, can be easily distinguished:

- A weak acid has a lower concentration of hydrogen ions and hence a higher pH than a strong acid of the same concentration. This can be established using narrow range universal indicator paper or, preferably, a pH probe and meter.
- A weak acid, because of its lower concentration of hydrogen ions, will be a much poorer electrical conductor than a strong acid of the same concentration.
- Weak acids react more slowly with reactive metals, metal oxides, metal carbonates and metal hydrogencarbonates than strong acids of the same concentration. This is again due to a lower concentration of hydrogen ions in the weak acid since it is the hydrogen ions that are responsible for the typical chemical properties of acids.
- Strong and weak acids can also be distinguished by measuring and comparing their enthalpies of neutralization (Chapter 5).

	0.1 mol dm⁻³ HCl(aq)	0.1 mol dm⁻³ CH₃COOH(aq)
[H⁺(aq)]	$0.1 \, mol \, dm^{-3}$	$\approx 0.0013 \, mol \, dm^{-3}$
pH	1.00	2.87
Electrical conductivity	high	low
Relative rate of reaction with magnesium	fast	slow
Relative rate of reaction with calcium carbonate	fast	slow

Table 8.5 Comparison of a weak and strong acid of the same concentration

It is important not to confuse the terms strong and weak with dilute and concentrated (Chapter 1). The term concentrated, as applied to acids, means that a relatively large amount of the pure acid (weak or strong) has been dissolved in a relatively small volume of water. Therefore, a 0.1 mol dm⁻³ solution of hydrochloric acid can be described as a dilute solution of a strong acid and a 0.1 mol dm⁻³ solution of ethanoic acid can be described as a dilute solution of a weak acid. A concentrated solution of a weak acid such as ethanoic acid might contain a greater concentration of hydrogen ions than a very dilute solution of a strong acid, such as hydrochloric acid. Acid strength does not change as the acid is diluted (at constant temperature).

Language of Chemistry

The terms strong and weak acids are not entirely satisfactory since they are qualitative descriptions. Strong and weak are absolute terms and some acids, such as phosphoric(v) acid, $H_3PO_4(aq)$, are described as moderately strong. In Chapter 18, a quantitative measure of acid strength that does not vary with dilution, known as the acid dissociation constant, will be introduced. ∎

Strong and weak bases

In addition to strong and weak acids there are strong and weak bases. A **strong base** undergoes almost 100% ionization or dissociation when in dilute aqueous solution. Strong bases include the metal hydroxides of group 1 and barium hydroxide. Strong bases have high pH values and high conductivities (Table 8.6).

In general for a strong ionic base, BOH:

$$BOH + (aq) \rightarrow OH^-(aq) + B^+(aq)$$
$$\quad 0\% \qquad\qquad 100\%$$

The three common strong bases are sodium hydroxide, potassium hydroxide and barium hydroxide:

$$NaOH(s) + (aq) \rightarrow Na^+(aq) + OH^-(aq)$$

$$KOH(s) + (aq) \rightarrow K^+(aq) + OH^-(aq)$$

$$Ba(OH)_2(s) + (aq) \rightarrow Ba^{2+}(aq) + 2OH^-(aq)$$

	0.1 mol dm⁻³ NaOH(aq)	**0.1 mol dm⁻³ NH₃(aq)**
[OH⁻(aq)]	0.1 mol dm⁻³	≈ 0.0013 mol dm⁻³
pH	13	11–12
Electrical conductivity	high	low

Table 8.6 Comparison of a weak and strong base of the same concentration

All bases are weak except the hydroxides of groups 1 and 2. **Weak bases** are composed of molecules that react with water molecules to release hydroxide ions. In general for a weak molecular base, BOH:

$$BOH + (aq) \rightleftharpoons OH^-(aq) + B^+(aq)$$

An equilibrium is established, with the majority of the base molecules not undergoing ionization or dissociation. In other words, the equilibrium lies on the left-hand side of the equation. Weak bases have low pH values and low conductivities (Table 8.6).

Examples of weak bases include caffeine (Chapter 25), the bases of nucleic acids (Chapter 22), aqueous ammonia and ethylamine (Chapter 20):

$$NH_3(g) + (aq) \rightleftharpoons NH_4^+(aq) + OH^-(aq)$$

$$C_2H_5NH_2(g) + (aq) \rightleftharpoons C_2H_5NH_3^+(aq) + OH^-(aq)$$

Despite the equilibria lying on the left there is a sufficiently high concentration of hydroxide ions in dilute aqueous ammonia (and ethylamine) to precipitate many transition metal hydroxides from aqueous solutions of the metal salt (Chapter 13).

Calcium hydroxide is a strong base, but it is very dilute because it is only slightly soluble in water.

$$Ca(OH)_2(s) + (aq) \rightarrow Ca^{2+}(aq) + 2OH^-(aq)$$

Applications of Chemistry

Rain water is slightly acidic with a pH of 5.6, due to the presence of carbonic acid. Rain water (Chapter 3 and Chapter 25) in polluted areas may have a pH between 5 and 2. The main acid present in acid rain is sulfurous acid and the major sources are cars and coal-fired power stations.

Acid rain has direct and indirect effects on living organisms. It can directly kill young fish and invertebrates. Acid rain also damages the leaves of trees, and releases aluminium ions from rocks and soil, which wash into lakes and rivers. Aluminium ions are toxic and interfere with gas exchange in the gills of fish. Acid rain also slowly dissolves away buildings and objects, for example statues, made from marble or limestone (calcium carbonate).

One approach to tackling acidified lakes and rivers is to add large amounts of limestone to the lakes. This is a cheap base which neutralizes some of the excess acid. Acid rain is a specific example of acid deposition: the deposition of wet (rain, snow, sleet, fog, cloudwater, dew) and dry (acidifying particles and gases) acidic components.

8.4 The pH scale

8.4.1 **Distinguish** between aqueous solutions that are acidic, neutral or alkaline using the pH scale.
8.4.2 **Identify** which of two or more aqueous solutions is more acidic or alkaline, using pH values.
8.4.3 **State** that each change of one pH unit represents a 10-fold change in the hydrogen ion concentration [$H^+(aq)$].
8.4.4 **Deduce** changes in [$H^+(aq)$] when the pH of a solution changes by more than one pH unit.

The pH scale

The pH scale is a number scale (Figure 8.20) that is used to describe the acidity, alkalinity or neutrality of an aqueous solution. The scale runs from 0 to 14:

- an aqueous solution with a pH below 7 is acidic
- an aqueous solution with a pH of above 7 is alkaline
- an aqueous solution with a pH of exactly 7 is neutral.

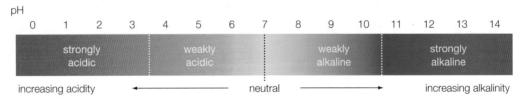

Figure 8.20 The pH scale

As the pH numbers below 7 become progressively *smaller*, the solution becomes *increasingly acidic*. Conversely, as the pH numbers above 7 become progressively larger, the solution becomes increasingly alkaline.

The pH of aqueous solutions can be measured by using universal indicator, either in the form of a solution or as paper. This is a mixture of indicators that has different colours in solutions of different pH. The exact colours usually correspond to a 'rainbow' sequence as the pH increases (Figure 8.21).

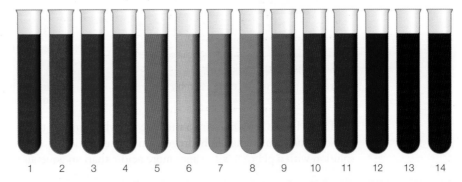

Figure 8.21 pH scale and the colours of universal indicator

Figure 8.22 shows the pH of some common laboratory chemicals and household substances.

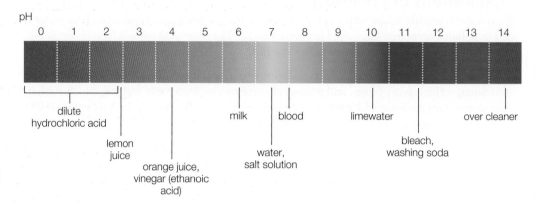

Figure 8.22 The pH of common substances

A more accurate method of measuring pH involves using a pH probe and meter (Figure 8.23).

Figure 8.23 pH probe

The pH probe contains a very thin glass bulb filled with acid of known concentration and therefore pH. When the probe is placed in an aqueous solution, a voltage difference is established between the acid inside the bulb and the solution outside the bulb (Figure 8.24). This voltage is then converted to a pH reading by the pH meter.

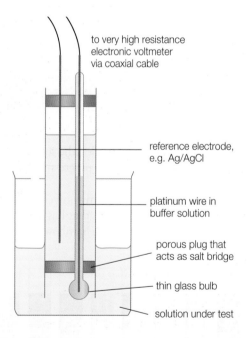

Figure 8.24 pH meter electrode assembly

pH	Concentration of hydrogen ions, $H^+(aq)/mol\,dm^{-3}$
0	$1\times10^0 = 1.0$
1	$1\times10^{-1} = 0.1$
2	$1\times10^{-2} = 0.01$
3	$1\times10^{-3} = 0.001$
4	$1\times10^{-4} = 0.0001$
5	$1\times10^{-5} = 0.00001$
6	$1\times10^{-6} = 0.000001$
7	$1\times10^{-7} = 0.0000001$
8	$1\times10^{-8} = 0.00000001$
9	$1\times10^{-9} = 0.000000001$
10	$1\times10^{-10} = 0.0000000001$
11	$1\times10^{-11} = 0.00000000001$
12	$1\times10^{-12} = 0.000000000001$
13	$1\times10^{-13} = 0.0000000000001$
14	$1\times10^{-14} = 0.00000000000001$

Table 8.7 The relationship between pH and $H^+(aq)$ concentration

The pH scale is *logarithmic* to the base 10, which means that every change in one unit on the pH scale means a change in hydrogen ion concentration of 10 units. For example, an aqueous solution with a pH of 4 is 10 times more acidic than an aqueous solution with a pH of 5 and

100 times (10 × 10) more acidic than an aqueous solution with a pH of 6. Additionally, an aqueous solution with a pH of 8 is 100 times (10 × 10) less alkaline than an aqueous solution with a pH of 10. pH is directly related to the concentration of hydrogen ions present in the solution (Table 8.7).

This table clearly shows that if the pH of a solution is given the value x, then the corresponding hydrogen ion concentration is $10^{-x}\,mol\,dm^{-3}$. For now, we shall use only integral values of x, but expressions involving non-integers can be evaluated using a calculator (Chapter 18). However, note again that the pH scale is logarithmic, which means that although a pH of 5.5, as a number, is half-way between pH 5 and 6, $10^{-5.5}$ is *not* half-way between 10^{-5} and 10^{-6}.

Table 8.7 indicates that distilled water (pH 7) and alkalis, for example $1\,mol\,dm^{-3}$ aqueous sodium hydroxide (pH 14), despite being neutral and alkaline, respectively, contain hydrogen ions, albeit at low concentrations. Hydrogen ions are present in neutral and alkaline aqueous solutions because water itself is very slightly dissociated into hydrogen and hydroxide ions (Chapter 18).

$$H_2O(l) \rightleftharpoons H^+(aq) + OH^-(aq)$$

However, in distilled water (pH 7) the concentrations of hydroxide and hydrogen ions are equal, that is, $[H^+(aq)] = [OH^-(aq)]$. In an acidic solution, the concentration of hydrogen ions will be larger than the concentration of hydroxide ions, that is, $[H^+(aq)] > [OH^-(aq)]$ and in alkaline solution, the concentration of hydroxide ions will be larger than the concentration of hydrogen ions, that is, $[OH^-(aq)] > [H^+(aq)]$ (Table 8.8).

Solution	$[H^+(aq)]/mol\,dm^{-3}$	$[OH^-(aq)]/mol\,dm^{-3}$
Acidic	$>10^{-7}$	$<10^{-7}$
Neutral	10^{-7}	10^{-7}
Basic	$<10^{-7}$	$>10^{-7}$

Table 8.8 Ion concentrations (at 25 °C)

Applications of Chemistry

Hydrofluoric acid, HF, is a weak acid which is formed when the gas hydrogen fluoride reacts with water. It is used as a source of fluorine and to make the polymer Teflon (PTFE) used in non-stick frying pans (Figure 8.25) and woks. Hydrogen fluoride is also used to etch glass. Although a weak acid, it is highly corrosive and a contact poison. It penetrates the skin very rapidly and forms calcium fluoride when it comes into contact with calcium ions in the bones. A hydrofluoric acid burn covering less than 2% of the body can be fatal and there is a delay between exposure and symptoms.

Figure 8.25 Non-stick frying pan

Language of Chemistry

The concept of pH was first introduced by Danish chemist S. P. L. Sørensen (Figure 8.26) at the Carlsberg Laboratory in 1909. The pH scale is logarithmic to base 10. The p in pH comes from the German word *potenz* and Danish word *potenus*, which both mean 'power'. The H stands for hydrogen. ∎

Figure 8.26 S. P. L. Sørensen

SUMMARY OF
KNOWLEDGE

- Under the Brønsted–Lowry theory for defining acidity and basicity, an acid is defined as a proton donor, while a base is a proton acceptor. The reaction between an acid and a base involves the transfer of a proton (H^+).
- Under the Lewis theory for defining acidity and basicity, an acid is defined as an electron pair acceptor, while a base is an electron pair donor.
- Acids have polarized covalent bonds to hydrogen atoms which carry a partial positive charge. These bonds break in aqueous solution to release protons.

$$HA \rightarrow H^+ + A^-$$

- Bases have a lone pair or lone pairs of electrons. These electrons receive the donated proton and a dative bond is formed.

$$B + H^+ \rightarrow BH^+$$

- Equations showing generalized acid and base behaviour:

$$HA(aq) + H_2O(l) \rightleftharpoons \quad H_3O^+(aq) \quad + \quad A^-(aq)$$
acid   base   conjugate acid   conjugate base

$$B(aq) + H_2O(l) \rightleftharpoons \quad BH^+(aq) \quad + \quad OH^-(aq)$$
base   acid   conjugate   conjugate
                 acid of B   base of water

- A pair of species differing by a single proton is called a conjugate pair.
- A strong acid is almost completely ionized in aqueous solution. Strong acids include sulfuric, nitric and hydrochoric acids. Weak acids are only slightly ionized in solution. Organic acids, such as ethanoic acid, are all weak acids.
- A strong base is almost completely ionized in aqueous solution. Strong bases include potassium, sodium and barium hydroxides. Weak bases are only slightly ionized in solution. Aqueous ammonia solution and organic bases, such as the amines, are all weak bases.
- Strong acids have a higher electrical conductivity, lower pH and faster reactions with bases than weak acids (of the same concentration and temperature). Strong bases have a higher electrical conductivity and a higher pH than weak bases (of the same concentration and temperature).
- A strong acid is a good proton donor, hence its conjugate base is a poor proton acceptor. So strong acids will have weak conjugate bases, and conversely, weak acids will have strong conjugate bases.
- Acids react with reactive metals (those above hydrogen in the reactivity series) to release hydrogen and form a salt. Acids react with metal carbonates to form water, a salt and carbon dioxide. An acid reacts with a metal oxide to form a salt and water only. This reaction is termed neutralization.

■ *Examination questions – a selection*

Paper 1 IB questions and IB style questions

Q1 Which one of the following descriptions defines a strong acid?
  A It is concentrated.
  B It does dissociate in water.
  C It absorbs water from the air.
  D It almost completely dissociated in water.

Q2 Which of the following represents the reaction between zinc powder and a dilute aqueous solution of sulfuric acid?
  A $Zn + 2H_2SO_4 \rightarrow 2ZnS + 2H_2O + 3O_2$
  B $4Zn + H_2SO_4 \rightarrow 4ZnO + H_2S$
  C $Zn + H_2SO_4 \rightarrow ZnSO_4 + H_2$
  D $Zn + H_2SO_4 \rightarrow ZnH_2 + SO_2 + O_2$

Q3 When the following $1.0\,mol\,dm^{-3}$ solutions are listed in increasing order of pH (lowest first), what is the correct order?

- **A** $HNO_3 < H_2CO_3 < NH_3 < Ba(OH)_2$
- **B** $NH_3 < Ba(OH)_2 < H_2CO_3 < HNO_3$
- **C** $Ba(OH)_2 < H_2CO_3 < NH_3 < HNO_3$
- **D** $HNO_3 < H_2CO_3 < Ba(OH)_2 < NH_3$

Standard Level Paper 1, Nov 05, Q24

Q4 What is the pH of pure distilled water?

- **A** 0
- **B** 4
- **C** 7
- **D** 6

Q5 Which one of the following represents the reaction between calcium hydroxide and dilute hydrochloric acid?

- **A** $Ca(OH)_2 + HCl \rightarrow CaOCl + H_2O$
- **B** $CaOH + HCl \rightarrow CaCl_2 + H_2O$
- **C** $CaOH + 2HCl \rightarrow Cl_2 + CaOH_2$
- **D** $Ca(OH)_2 + 2HCl \rightarrow CaCl_2 + 2H_2O$

Q6 The amino acid alanine has the structure:

$H_2N{-}CH(CH_3){-}COOH$

Which of the following species represents its conjugate acid?

- **A** $^+NH_3CH(CH_3)COOH$
- **B** $^+NH_3CH(CH_3)COOH_2{}^+$
- **C** $^+NH_3CH(CH_3)COO^-$
- **D** $NH_2CH(CH_3)COO^-$

Q7 Four flasks labelled A, B, C and D contain equal volumes of hydrochloric acid at different concentrations. When equal volumes of $1\,mol\,dm^{-3}$ sodium hydroxide are added to each flask the pH values below are produced.

Which flask has the most concentrated acid?

Flask	A	B	C	D
pH	2	6	8	12

- **A** flask A
- **B** flask B
- **C** flask C
- **D** flask D

Q8 In the equilibrium below, which species represents a conjugate acid–base pair?

$CH_2ClCOOH(aq) + H_2O(l)$
$\rightleftharpoons CH_2ClCOO^-(aq) + H_3O^+(aq)$

- **A** $CH_2ClCOOH/H_2O$
- **B** CH_2ClCOO^-/H_3O^+
- **C** H_2O/CH_2ClCOO^-
- **D** H_2O/H_3O^+

Q9 Hydrogen chloride dissolved in water reacts with magnesium. Hydrogen chloride dissolved in ethanol does not react with magnesium. Which statement accounts for this observation?

- **A** Ethanol accepts hydrogen ions (protons) and water does not.
- **B** Water is a hydrogen ion/proton acceptor.
- **C** Magnesium is very soluble in ethanol but insoluble in water.
- **D** Hydrogen chloride does not form ions in water.

Q10 Which of the following $1.00\,mol\,dm^{-3}$ aqueous solutions would have the highest pH value?

- **A** ammonia
- **B** ethanoic acid
- **C** sulfuric acid
- **D** sodium hydroxide

Q11 A dilute aqueous solution of benzenecarboxylic acid (an organic acid) is a poor conductor of electricity. Which of the following statements accounts for this observation?

- **A** Benzenecarboxylic acid solution has a high concentration of ions.
- **B** Benzenecarboxylic acid is only slightly dissociated in water.
- **C** Benzenecarboxylic acid is completely dissociated in water.
- **D** It is a strong acid.

Q12 Which is the correct description for an aqueous solution with a pH of 9.5?

- **A** alkaline
- **B** acidic
- **C** neutral
- **D** amphoteric

Q13 Methanoic acid, $HCOOH$, is a stronger acid than propanoic acid, CH_3CH_2COOH. Which one of the statements about these acids is correct?

- **A** Propanoic acid is more dissociated in water than methanoic acid.
- **B** Magnesium will react with methanoic acid but not with propanoic acid.
- **C** A $1.0\,mol\,dm^{-3}$ solution of methanoic acid will turn blue litmus red, but a $1.0\,mol\,dm^{-3}$ solution of propanoic acid will turn red litmus blue.
- **D** The pH of a solution of $1\,mol\,dm^{-3}$ propanoic acid is higher than that of $1\,mol\,dm^{-3}$ methanoic acid.

Q14 A decrease in the pH of an aqueous solution corresponds to:

- **A** a decrease in the H^+ concentration and an increase in the OH^- concentration
- **B** an increase in the H^+ concentration and a decrease in the OH^- concentration
- **C** a decrease in the H^+ concentration with no change in the OH^- concentration
- **D** an increase in the OH^- concentration and an increase in the H^+ concentration

Q15 Calcium oxide is added to a lake to neutralize the effects of acid rain. The pH value of the lake water rises from 4 to 6. What is the change in concentration of $[H^+(aq)]$ in the lake water?

- **A** an increase by a factor of 2
- **B** an increase by a factor of 100
- **C** a decrease by a factor of 2
- **D** a decrease by a factor of 100

Q16 Which chemical can behave as a Brønsted–Lowry base and as a Brønsted–Lowry acid?
A CO_3^{2-} **B** HSO_4^- **C** NO_3^-
D Such a species does not exist.

Q17 The pH of solution X is 1 and that of Y is 2. Which statement is correct about the hydrogen ion concentrations in the two solutions?
A $[H^+]$ in X is half that in Y.
B $[H^+]$ in X is twice that in Y.
C $[H^+]$ in X is one tenth of that in Y.
D $[H^+]$ in X is ten times that in Y.
Standard Level Paper 1, May 05, Q23

Q18 In which one of the following reactions does the nitric acid molecule act as a base?
A $HNO_3 + H_2O \rightarrow H_3O^+ + NO_3^-$
B $HNO_3 \rightarrow H^+ + NO_3^-$
C $HNO_3 + CH_3COOH \rightarrow CH_3COOH_2^+ + NO_3^-$
D $HNO_3 + 2H_2SO_4 \rightarrow NO_2^+ + 2HSO_4^- + H_3O^+$

Q19 Which one of the following species is not amphoteric (amphiprotic), that is, not capable of showing acidic and basic properties in aqueous solution?
A H_2O **C** NH_4^+
B $H_2PO_4^-$ **D** NH_2CHCH_3COOH

Q20 Which of the following would exactly neutralize $100\,cm^3$ of $1\,mol\,dm^{-3}$ sulfuric acid?
A $0.1\,mol$ of $Ba(OH)_2$ **C** $0.2\,mol$ of Na_2CO_3
B $0.1\,mol$ of KOH **D** $0.1\,mol$ of NH_3

Q21 Which of the following compounds containing hydrogen acts as an acid in aqueous solution?
A hydrogen chloride   **C** methane
B ammonia   **D** ethene

Q22 Which is not a strong acid?
A nitric acid   **C** carbonic acid
B sulfuric acid   **D** hydrochloric acid
Standard Level Paper 1, Nov 06, Q23

Q23 Which equation correctly describes phosphoric(v) acid behaving as a monoprotic acid in aqueous solution?
A $H_3PO_4(aq) \rightarrow H^+(aq) + H_2PO_4^-(aq)$
B $H_2PO_4^-(aq) \rightarrow H^+(aq) + HPO_4^{2-}(aq)$
C $H_3PO_4(aq) \rightarrow 3H^+(aq) + PO_4^{3-}(aq)$
D $H_3PO_4(aq) \rightarrow 2H^+(aq) + HPO_4^{2-}(aq)$

Q24 According to the Lewis theory, a base:
A is a proton acceptor
B is a proton donor
C makes available a share in a pair of electrons
D accepts a share in a pair of electrons

Q25 Ammonia molecules in aqueous solution can be considered as:
A a Lewis acid (only)
B a Lewis base (only)
C a Brønsted–Lowry base (only)
D both a Brønsted–Lowry base and a Lewis base

Paper 2 IB questions and IB style questions

Q1 a The pH values of solutions of three organic acids of the same concentration were measured.
acid X   pH = 5
acid Y   pH = 2
acid Z   pH = 3

   i Identify which solution is the least acidic.   [1]
   ii Deduce how the $[H^+]$ values compare in solutions of acids Y and Z.   [1]
   iii Arrange the solutions of the three acids in decreasing order of electrical conductivity, starting with the greatest conductivity, giving a reason for your choice.   [2]

Q2 Carbonic acid (H_2CO_3) is described as a weak acid and hydrochloric acid (HCl) is described as a strong acid.
   a Explain, with the help of equations, what is meant by strong and weak acid using the above acids as examples.   [4]
   b Outline **two** ways, other than using pH, in which you could distinguish between carbonic acid and hydrochloric acid of the same concentration.   [2]
   c A solution of hydrochloric acid, HCl(aq), has a pH of 1 and a solution of carbonic acid, H_2CO_3(aq), has a pH of 5. Determine the ratio of the hydrogen ion concentrations in these solutions.   [2]
   d The relative strengths of the two acids can be illustrated by the following equation:

   $HCO_3^-(aq) + HCl(aq) \rightleftharpoons H_2CO_3(aq) + Cl^-(aq)$

   i Identify the acid and its conjugate base and the base and its conjugate acid in the above equation.   [2]
   ii Name the theory that is illustrated in **d i**.   [1]
   Standard Level Paper 2, Nov 02, Q5

9 Oxidation and reduction

STARTING POINTS
- Many chemical reactions involve oxidation and reduction.
- Redox reactions involve simultaneous reduction and oxidation processes.
- An oxidation process involves either the addition of oxygen, the removal of hydrogen, the loss of electrons or an increase in oxidation number.
- A reduction process involves either the loss of oxygen, the addition of hydrogen, the gain of electrons or a decrease in oxidation number.
- Oxidation numbers are a system used by chemists to keep track of electrons during redox reactions. They make the assumption that all compounds are composed of ions.
- Oxidizing agents cause a substance to be oxidized; reducing agents cause a substance to be reduced. Redox reactions involve a reaction between a reducing and an oxidizing agent.
- Displacement reactions occur when a more reactive element replaces a less reactive element; displacement reactions are redox reactions.
- The reactivity series lists metals and non-metals in order of their ability to displace metal ions in aqueous solution.
- Metal ion displacement reactions are the basis for voltaic cells (simple batteries). The voltage depends on how far apart the two metals are in the reactivity series.
- Electrolysis is the use of electricity to decompose an ionic compound.

9.1 Introduction to oxidation and reduction

9.1.1 **Define** *oxidation* and *reduction* in terms of electron loss and gain.

Oxidation

Many chemical reactions involve **oxidation**. This was originally defined as:

- the addition of oxygen to a substance

  or
- the loss or removal of hydrogen from a substance.

An example of the first type of oxidation involves the burning or combustion of magnesium (Figure 9.1) in air or oxygen:

The magnesium has gained oxygen and we say that the magnesium has been oxidized.

$$2Mg(s) + O_2(g) \rightarrow 2MgO(s)$$

An example of the second type of oxidation involves the reaction between manganese(IV) oxide ('manganese dioxide') and concentrated aqueous hydrochloric acid:

$$MnO_2(s) + 4HCl(aq) \rightarrow MnCl_2(aq) + 2H_2O(l) + Cl_2(g)$$

Figure 9.1 The combustion of magnesium to form magnesium oxide

The hydrochloric acid loses hydrogen and is therefore oxidized. Later, we will see why two *apparently very different* reactions are both regarded as oxidation reactions.

Strictly speaking, both these reactions are correctly described as **redox reactions** (Section 9.2) since they involve both oxidation and reduction. The formation of rust (hydrated iron(III) oxide) (Figure 9.2) is a very familiar redox reaction. It involves the reaction between iron, oxygen and liquid water (Chapter 19).

Figure 9.2 A rusting railway trolley (showing pitting)

History of Chemistry

Joseph Priestley (1733–1804) was an English chemist and non-conformist clergyman who independently discovered oxygen in 1774. He used a magnifying glass to focus the rays of the sun on a sample of a substance he called the 'red calx of mercury', now known as mercury(II) oxide, HgO. The result was silvery globules of mercury and a colourless gas (oxygen) which he termed dephlogisticated air. He found that a candle burned more brightly and for a longer time in this gas than it would have in air.

Priestley knew that air was required for rusting and burning and accepted the phlogiston theory, which held that all substances that burn and all metals that rust contain a substance called phlogiston. The phlogiston theory was proposed in the late 17th century by two German chemists, Johann Becher and Georg Stahl.

According to the theory, burning and rusting both represent the escape of phlogiston, and air is necessary for both processes because phlogiston is absorbed into it. When the air becomes saturated with phlogiston, the phlogiston has no place to go and the flame goes out or the rusting stops.

Although the theory made qualitative sense and helped explain burning and rusting, it suffered from a quantitative defect: it could not adequately account for the observed changes in mass that accompany burning and rusting. It was known as early as 1630 that when a piece of iron rusts, the rust formed weighs more than the original iron. A few phlogistonists tried to explain this by asserting that phlogiston had negative mass However, when a lump of charcoal (carbon) burns, again presumably with the loss of phlogiston, its mass decreases.

The problem could not be resolved until the French chemist Antoine Lavoisier discovered the true role of oxygen in both burning and rusting. He repeated Priestley's experiments with mercury(II) oxide and showed the mass of the mercury formed equals the original mass of the oxide minus the mass of the oxygen generated.

Reduction

Many chemical reactions involve **reduction**. Reduction is the reverse of oxidation and was originally defined as:

- the loss or removal of oxygen from a substance
  or
- the addition of hydrogen to a substance.

An example of the first type of reduction involves the reaction between hydrogen gas and heated copper(II) oxide:

$$CuO(s) + H_2(g) \rightarrow Cu(s) + H_2O(l)$$

The copper(II) oxide loses oxygen and we therefore say that the copper(II) oxide has been reduced. An example of the second type of reduction involves the reaction between ethene and hydrogen to form ethane in the presence of a hot metal catalyst (Chapter 10):

$$C_2H_4(g) + H_2(g) \rightarrow C_2H_6(g)$$

The ethene gains hydrogen and we therefore say that the ethene has been reduced. This reaction is also an example of hydrogenation (Chapter 10), where hydrogen is added to an organic or carbon-containing compound.

Strictly speaking, both of these reactions are correctly described as redox reactions (Section 9.2). The term reduction should be restricted to what happens to the copper(II) oxide and ethane, respectively.

History of Chemistry

The French chemist **Antoine Lavoisier** (1743–1794) recognized and named the element oxygen. He also introduced the term oxidation for any reaction of a substance with oxygen. Reduction referred to the removal of oxygen. In 1837 the German organic chemist **Justus von Liebig** (1803–1873) proposed another definition for oxidation and reduction: reduction was the removal of oxygen or the addition of hydrogen.

O xidation

I s

L oss

R eduction

I s

G ain

Figure 9.3 Mnemonic for redox reactions and electron transfer

Later, when electrolysis (Section 9.5) was discovered and modern theories of atomic structure (Chapter 2) and chemical bonding (Chapter 4) were developed, the terms oxidation and reduction were redefined in terms of electrons. Specifically, oxidation was defined as the loss of electrons from a substance; reduction was defined as the gaining of electrons by a substance (Figure 9.3).

Note that these modern definitions include many of the oxidation and reduction reactions previously defined in terms of loss and gain of oxygen and hydrogen. For example:

$$2Mg(s) + O_2(g) \rightarrow 2MgO(s)$$

can be rewritten to emphasize the loss and gain of electrons that occur during this reaction.

Magnesium oxide, MgO, is an ionic compound [$Mg^{2+} O^{2-}$], so the magnesium atom has lost two electrons to form a magnesium ion, Mg^{2+}. Oxygen is a molecular substance, so each oxygen atom has gained two electrons to become an oxide ion, O^{2-}.

These two processes can be described by the following equations:

Oxidation: $Mg \rightarrow Mg^{2+} + 2e^-$

Reduction: $O_2 + 4e^- \rightarrow 2O^{2-}$

Since during reduction one substance gains electrons, there must be a second process involving oxidation where a substance is losing electrons. Reduction and oxidation processes must therefore occur together simultaneously. Such processes are called **redox** (<u>red</u>uction–<u>ox</u>idation) reactions. They are the basis for voltaic cells (Section 9.4).

The two equations are known as **half-equations** since they only describe one of the two reactions that must occur together. The ionic equation to describe the redox reaction is obtained by adding the two half-equations together and cancelling the electrons that appear on both sides of the equation:

Half-equations: $Mg \rightarrow Mg^{2+} + 2e^-$; $O_2 + 4e^- \rightarrow 2O^{2-}$

The first half-equation has to be multiplied through by two so the number of electrons is the same as that in the second half-equation:

$$2Mg \rightarrow 2Mg^{2+} + 4e^-$$

Sum of the two half-equations:

$$2Mg + O_2 + 4e^- \rightarrow 2Mg^{2+} + 4e^- + 2O^{2-}$$

Cancelling of electrons:

$$2Mg + O_2 \rightarrow 2[Mg^{2+} O^{2-}]$$

The new definitions of oxidation and reduction, as electron loss and electron gain respectively, also include many examples of redox reactions which do *not* involve oxygen or hydrogen. For example, the burning of sodium metal in chlorine gas (Chapter 3) to form sodium chloride:

$$2Na(s) + Cl_2(g) \rightarrow 2NaCl(s)$$

The sodium atoms lose electrons to form sodium ions: $2Na \rightarrow 2Na^+ + 2e^-$
and are therefore oxidized.

The chlorine molecules gain electrons to form chloride ions: $Cl_2 + 2e^- \rightarrow 2Cl^-$
and are therefore reduced.

Two half-equations can only be added together if the numbers of electrons in both are the same. If they are not, then one or both of the equations needs to be multiplied through by an appropriate coefficient.

For example, half-equations: $Na \rightarrow Na^+ + e^-$; $O_2 + 4e^- \rightarrow 2O^{2-}$

Multiplying the first half-equation through by four: $4Na \rightarrow 4Na^+ + 4e^-$

Summing the two half-equations: $4Na + O_2 + 4e^- \rightarrow 4Na^+ + 2O^{2-} + 4e^-$

Cancelling of electrons and conversion of the ionic formula to a 'molecular formula':

$$4Na + O_2 \rightarrow 2Na_2O$$

Language of Chemistry

Some of the conceptual difficulties related to redox involve language. Although oxidation originally meant reaction with oxygen, the modern understanding of oxidation as a loss of electrons means that oxygen need not be involved. Similarly, the term reduction originally described the process of reducing an ore to its metal content, in other words, obtaining a metal from its ore (usually an oxide). Today, in chemistry reduction means a gain of electrons, which seems inconsistent with the historical term. ■

Oxidation numbers

9.1.2 **Deduce** the oxidation number of an element in a compound.

The electron transfer approach to oxidation (electron loss) and reduction (electron gain) is useful, *but* it does have some drawbacks and limitations. For example, consider the combustion of sulfur in air or oxygen to form sulfur dioxide:

$$S(s) + O_2(g) \rightarrow SO_2(g)$$

According to the historical definition of oxidation, as the addition of oxygen, the sulfur has been oxidized. However, neither the two reactants, nor the product, are ionic, and there is no obvious transfer of electrons. There is a large number of reactions that fit the historical definitions of oxidation and reduction, such as the hydrogenation of ethene, but *not* the modern definitions in terms of electron loss and gain.

One way to overcome this problem is to develop new definitions of oxidation and reduction that cover both the historical definitions in terms of oxygen and hydrogen and the modern definitions in terms of electrons. The concept of an **oxidation number**, which consists of a number and a sign, allows chemists to avoid the problems associated with using two separate and sometimes conflicting definitions for oxidation and reduction.

The concept applies equally well to reactions involving ionic compounds, where there is obvious electron transfer, and reactions involving covalent compounds, where there is no obvious transfer of electrons. Oxidation is now defined as an increase in oxidation number and reduction is defined as a decrease in oxidation number.

There are some simple rules (based on the electronegativities of the elements and their bonding) that we can apply to find oxidation numbers:

■ The oxidation number of any uncombined element is zero.
  Example: $O_2(g)$ the oxidation number of oxygen is zero.
■ For a simple ion, the oxidation number of the ion is equal to the charge on the ion.
  Examples: Cl^- and Fe^{2+}, the oxidation numbers are -1 and $+2$, respectively. Note the different order for the sign and number of the oxidation number compared to the ion.
■ For a compound, the sum of the oxidation numbers of the elements is zero.
  Example: $NaCl$ [Na^+ Cl^-] the sum of the oxidation numbers ($+1$ and -1) is 0.
■ For an oxoanion, the sum of the oxidation numbers of the elements is equal to the charge on the ion.
  Example: in the sulfate ion, SO_4^{2-}, the sum of the oxidation numbers is -2, that is, [$+6 + (-2 \times 4)$].
■ The oxidation number of hydrogen is $+1$ (except where is it is combined with a reactive metal, for example, sodium hydride, NaH [Na^+ H^-], where it is -1).
■ The oxidation number of oxygen is -2 (except in hydrogen peroxide, H_2O_2, where it is -1 and oxygen difluoride, OF_2, where it is $+2$).

Worked examples

Deduce the oxidation number of sulfur in sulfur dioxide, SO_2.

(O.N. of S) + (2 × O.N. of O) = 0

(O.N. of S) + (2 × –2) = 0

O.N. of S = +4

Deduce the oxidation number of nitrogen in the nitrate ion, NO_3^-.

(O.N. of N) + (3 × O.N. of O) = –1

(O.N. of N) + (3 × –2) = –1

O.N. of N = +5

In a covalent bond in a molecule, the more **electronegative** element is given a negative oxidation number and the less electronegative element, a positive oxidation number. For example, in chlorine fluoride, ClF, the chlorine atom is less electronegative than the fluorine atom (the most electronegative element). Therefore chlorine is assigned an oxidation number of +1 and fluorine that of –1. By convention, the less electronegative element appears first in the formula of a binary compound, so chlorine fluoride has the formula ClF and not FCl.

(In some compounds whose formulae were established before the concept of oxidation number was developed, the convention is not observed. For example, the formula of ammonia is NH_3, though nitrogen is more electronegative than hydrogen.)

In sulfur trioxide, SO_3, the oxygen is the more electronegative element (it is above sulfur in the periodic table) and is therefore assigned the negative oxidation number. The three oxygen atoms, each of oxidation number –2, are balanced by the one sulfur atom with an oxidation number of +6. The sum of the oxidation numbers [(–2 × 3) + 6] equals zero.

The idea of an oxidation number is an artificial concept since it considers all compounds, even covalent ones, to be ionic. An alternative, but equivalent, expression is that it is assumed that the atom with the greater electronegativity 'owns' or 'controls' all the bonding or shared electrons of a particular covalent bond.

For example, the sulfur trioxide molecule, SO_3, is assumed to be $[S^{6+} 3O^{2-}]$ and the water molecule, H_2O, is assumed to be $[2H^+ O^{2-}]$, as shown in Figure 9.4.

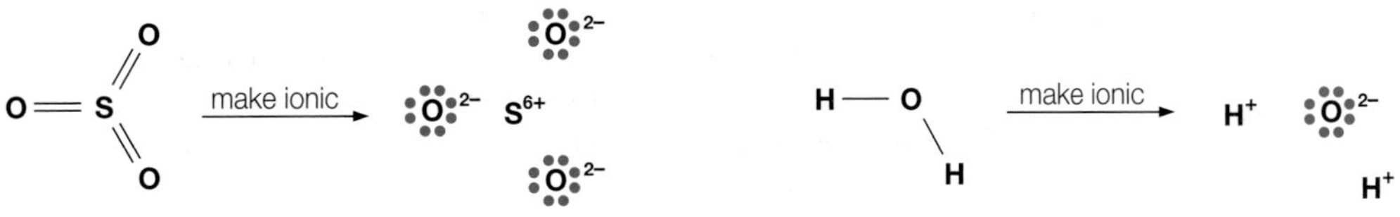

Figure 9.4 The ionic formulations of the sulfur trioxide and water molecules

The oxidation numbers for the central sulfur and oxygen in these two species are +6 and –2, respectively.

A negative sign for an oxidation number means the atom has gained 'control' of the electrons (compared to the element) and a positive sign for an oxidation number means that the atom has lost 'control' of electrons (compared to the element). The numerical value of an oxidation number indicates the number of electrons over which electron 'control' has changed compared to the situation in the element.

So in the two examples above, SO_3 and H_2O, sulfur has lost 'control' of six electrons and oxygen has gained 'control' of two electrons, compared to their elements.

(The oxidation number is written as a Roman numeral if included in the name of a compound (page 235). This is common practice for transition metal compounds: for example, the correct name for the permanganate ion, MnO_4^-, is manganate(VII), pronounced 'manganate seven'.)

TOK Link

Manganese in the manganate(VII) ion, MnO_4^-, has an oxidation number of +7. This implies the presence of Mn^{7+}. It must be remembered that the oxidation state of an atom does not represent the 'real' charge on that atom. This is particularly true of high oxidation states, where the ionization energy required to produce a highly charged positive ion is far greater than the energies available in chemical reactions. The manganese–oxygen bonds in the manganate(VII) ion are polar covalent bonds. The assignment of electrons between atoms in calculating an oxidation state is purely a set of useful but artificial beliefs for the understanding of many chemical reactions. Oxidation numbers are not 'real' – they are simply mathematical constructs.

History of Chemistry

In the early days of chemistry, compounds were given trivial or non-systematic names. Many come from historic usage, frequently alchemy. Trivial names often derive from a property of the compound. For example, silver nitrate was known as lunar caustic. The solid form was used as a cauterizing agent to ensure that bacteria did not cause infection following surgery. Alchemists associated silver with the moon or *Luna* (its Latin name).

The Latin name for silver, *argentum*, derives from 'white, shining'. Sodium sulfate was known as Glauber's salt and is named after the German **Johann Rudolf Glauber** who discovered it in the 17th century in spring water. The replacement of trivial names with systematic Stock names has resulted in a loss of historical knowledge about the discovery of compounds and a shift away from the substance's properties to its formula.

■ Extension: Oxidation numbers and the periodic table

Oxidation numbers of elements in compounds generally increase regularly as you move across periods 2 and 3 in the periodic table (Table 9.1). For example, the maximum oxidation number of the elements in oxides increases from +1 in sodium to +7 for chlorine (Chapter 13).

Na	Mg	Al	Si	P	S	Cl
Na_2O	MgO	Al_2O_3	SiO_2	P_4O_{10}	SO_3	Cl_2O_7
+1	+2	+3	+4	+5	+6	+7

Table 9.1 Formulas of the highest oxides and their oxidation numbers in the elements of period 3

The maximum oxidation number corresponds to the number of electrons in the outer shell, all of which are involved in bonding in the highest oxide. The maximum oxidation number also corresponds to the group number in the periodic table (Table 9.2)

Na	Mg	Al	Si	P	S	Cl
2,8,1	2,8,2	2,8,3	2,8,4	2,8,5	2,8,6	2,8,7
1	2	3	4	5	6	7

Table 9.2 Electron arrangements and group numbers of the elements in period 3

The term 'highest oxide' is used since phosphorus, sulfur and chlorine can all exhibit lower oxidation states: sulfur dioxide, SO_2 (oxidation number of sulfur +4); tetraphosphorus hexaoxide, P_4O_6 (oxidation number of phosphorus +3) and chlorine monoxide, Cl_2O (oxidation number of chlorine +1).

Similar trends in oxidation states (with the exception of sulfur) are observed for the chlorides of period 3 (Chapter 13) (Table 9.3).

Na	Mg	Al	Si	P	S
NaCl	$MgCl_2$	$AlCl_3$/ Al_2Cl_6	$SiCl_4$	PCl_5	SCl_4
+1	+2	+3	+4	+5	+4

Table 9.3 Formulas of the highest chlorides and their oxidation numbers in the elements of period 3

With the exception of sulfur, the maximum oxidation number again corresponds to the number of electrons in the outer shell, all of which are involved in bonding in the highest chloride. The maximum oxidation number again corresponds to the group number of the periodic table.

The term 'highest chloride' is used since phosphorus and sulfur can exhibit lower oxidation states: phosphorus trichloride, PCl_3 (oxidation number of phosphorus +3); sulfur dichloride, SCl_2 (oxidation number of sulfur +2) and disulfur dichloride, S_2Cl_2 (Cl—S—S—Cl) (oxidation number of sulfur +1).

■ Extension: Trends in the redox properties of the elements

Period 3

The trends in the redox properties of the elements in period 3 are summarized in Table 9.4. Chlorine is a strong oxidizing agent and will oxidize all the other elements in period 3. A trend is observed from strong reducing agent to strong oxidizing agent. This correlates with an increase in electronegativity (Chapter 4), which is a measure of the ability of an atom to attract a pair of electrons in a covalent bond. Note that oxidation numbers for non-metals generally differ by two, as a consequence of covalent bonds containing pairs of electrons.

Property/Element	Sodium, Na	Magnesium, Mg	Aluminium, Al	Silicon, Si	Phosphorus, P	Sulfur, S	Chlorine, Cl
Oxidation numbers	+1 only	+2 only	+3 only	+4 (−4 rarely)	+5, +3, −3	+6, +4, +2, −2	+7, +5, +3, +1, −1
Examples of compounds in these oxidation states	NaBr	$MgSO_4$	Al_2O_3	SiO_2, SiH_4	PCl_5 PCl_3 PH_3	SO_3 SO_2 SCl_2 H_2S	$HClO_4$ $NaClO_3$ $NaClO_2$ $NaClO$ HCl
Redox properties (all reactions of elements are redox)	Strong reducing agent; chemistry is summarized by $Na \rightarrow Na^+ + e^-$	Strong reducing agent; chemistry is summarized by $Mg \rightarrow Mg^{2+} + 2e^-$	Strong reducing agent; chemistry is summarized by $Al \rightarrow Al^{3+} + 3e^-$	Usually a reducing agent	Usually a reducing agent	A reducing agent, but can be an oxidizing agent with hydrogen and reactive metals	An oxidizing agent, especially in solution. Can be a reducing agent with fluorine and water
Standard electrode potential, $E^\ominus$/V	−2.71	−2.36	−1.66	*	*	*	1.36
Electronegativity of element	0.9	1.2	1.5	1.8	2.1	2.5	3.0
Type of element	Metal	Metal	Metal	Metalloid	Non-metal	Non-metal	Non-metal

Table 9.4
Summary of redox properties of the elements in period 3

The **standard electrode potential** is a measurement of the reducing power of an element in aqueous solution under standard conditions (Chapter 19). The more negative the value, the greater the reducing power of the element concerned.

The equations below describe typical redox reactions exhibited by some of the elements in period 3. Oxidation numbers have been calculated to indicate the nature of the redox reaction.

Sodium

$$2\underline{Na}(s) + 2H_2O(l) \rightarrow 2\underline{Na}OH(aq) + H_2(g)$$

    0   +1

Magnesium

$$2\underline{Mg}(s) + CO_2(g) \rightarrow 2\underline{Mg}O(s) + C(s)$$

    0   +2

All the reactions involving metallic elements involve loss of electrons from the metal atoms to form positive ions. The metals are acting as reducing agents.

Phosphorus

$$4\underline{P}(s) + 8H_2SO_4(aq) \rightarrow 4H_3\underline{P}O_4(aq) + S(s) + 7SO_2(g) + 2H_2O(l)$$

    0   +5

$$\underline{P}(s) + 3Na(s) \rightarrow Na_3\underline{P}(s)$$

    0   −3

In the first reaction phosphorus is acting as a reducing agent, but in the second it is acting as an oxidizing agent.

Chlorine

$$8NH_3(aq) + 3\underline{Cl}_2(g) \rightarrow N_2(g) + 6NH_4\underline{Cl}(aq)$$
$$0 -1$$

$$\underline{Cl}_2(g) + F_2(g) \rightarrow 2\underline{Cl}F(g)$$
$$0 +1$$

In the first reaction chlorine is acting as an oxidizing agent, but in the second it is acting as a reducing agent.

Transition metals

The fact that the transition metals form a variety of relatively stable oxidation states is due, in part, to the availability of 3d and 4s electrons for ion and covalent bond formation (Chapter 13). The common oxidation states of the first row transition metals are shown in Table 9.5.

d-block metal	Ti	V	Cr	Mn	Fe	Co	Ni	Cu
Common oxidation states	+3	+2	+2	+2	+2	+2	+2	+1
	+4	+3	+3	+4	+3	+3		+2
		+4	+6	+6				
		+5		+7				
Examples of ions in these oxidation states	Ti^{3+}	V^{2+}	Cr^{2+}	Mn^{2+}	Fe^{2+}	Co^{2+}	Ni^{2+}	Cu^+
	Ti^{4+}	V^{3+}	Cr^{3+}	Mn^{4+}	Fe^{3+}	Co^{3+}		Cu^{2+}
		VO^{2+}	CrO_4^{2-}	MnO_4^{2-}				
		VO_2^+	$Cr_2O_7^{2-}$	MnO_4^-				

Table 9.5 Examples of compounds showing common oxidation states of the first row transition metals

Scandium at the beginning of the first row of the d-block and zinc at the end are not classified as transition metals since they only exhibit one stable oxidation state in their compounds, +3 (Sc^{3+}) and +2 (Zn^{2+}), respectively. Notice that the lower oxidation states correspond to simple or atomic ions, for example, Mn^{2+} (+2), while the higher oxidation states correspond to covalently bonded oxoanions, for example, MnO_4^- (+7).

■ Extension: Difficulties with oxidation numbers

Bonds between elements

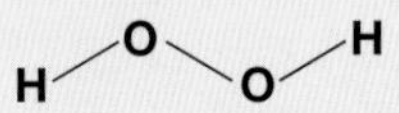

Figure 9.5 Structural formula of the hydrogen peroxide molecule

Another rule for determining oxidation numbers is that bonds between atoms of the same element do not count towards the oxidation number. So, for example, in hydrogen peroxide (Figure 9.5), the oxidation numbers of hydrogen and oxygen are +1 and –1, respectively. Since hydrogen peroxide is a molecule, the sum of the oxidation numbers $[(+1 \times 2) + (-1 \times 2)]$ will be zero.

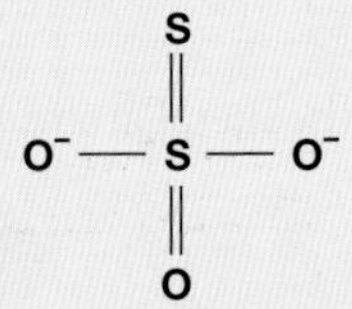

Sodium thiosulfate (Figure 9.7) contains the thiosulfate ion (Figure 9.6), $S_2O_3^{2-}$, which has the systematic name thiosulfate(VI) to indicate its relationship to the sulfate(VI) ion, SO_4^{2-}. However, this is incorrect if the rule for calculating oxidation numbers in the case of element–element bonds is applied (cf. H_2O_2). Applying this rule we find that that the sulfur atom bonded only to the other sulfur has an oxidation number of zero and the central sulfur atom has an oxidation number of +4.

Figure 9.6 Structural formula of the thiosulfate ion

Figure 9.7 Photograph of sodium thiosulfate crystals

Organic chemistry

Because of the large number of element–element bonds present in organic (carbon-containing) compounds, the concept of oxidation numbers is not very helpful and adds very little to the understanding of organic chemistry. However, the 'old-fashioned' definitions of oxidation and reduction in terms of hydrogen and oxygen are frequently applied to oxygen-containing compounds such as aldehydes, ketones, carboxylic acids and alcohols (Chapter 10).

For example:

$$CH_3CH_2OH \xrightarrow{[2H]} CH_3CHO \xrightarrow{[O]} CH_3COOH$$

ethanol   ethanal   ethanoic acid
(an alcohol)   (an aldehyde)   (a carboxylic acid)

The first conversion involves the loss of hydrogen and is hence classified as oxidation. The second conversion involves the addition of oxygen and is again classified as an oxidation. The [O] and [H] are a shorthand for oxidation and reduction; they do *not* imply that oxygen and hydrogen atoms are intermediates in the oxidation process.

The oxidation of ethanol to ethanoic acid can be brought about by heating ethanol with an acidified aqueous solution of potassium dichromate(VI), $K_2Cr_2O_7$. Half-equations can also be written for the two oxidations shown above:

$$CH_3CH_2OH \rightarrow CH_3CHO + 2H^+ + 2e^-$$

$$CH_3CHO + H_2O \rightarrow CH_3COOH + 2H^+ + 2e^-$$

The electrons will be accepted by the dichromate(VI) ion.

Another common oxidizing agent used in organic chemistry is acidified or alkaline potassium manganate(VII). A reducing reagent used in organic chemistry is hydrogen gas and a heated metal catalyst (Chapter 6).

Naming inorganic compounds

9.1.3 **State** the names of compounds using oxidation numbers.

The concept of oxidation number is used in the modern chemical naming of ionic inorganic substances. This system of nomenclature, or naming, is called the **Stock notation**. In this system the oxidation number is inserted immediately after the name of an ion. Roman numerals are inserted after the name or symbol of the element.

For example:

$FeCl_2$	$[Fe^{2+} \ 2Cl^-]$	iron(II) chloride
$FeCl_3$	$[Fe^{3+} \ 3Cl^-]$	iron(III) chloride

This notation is only used for the transition metals and tin and lead from group 4 (IV) of the periodic table where variable or multiple oxidation states are exhibited. For the metals from groups 1 (I), 2 (II) and 3 (III) it is not usually necessary to indicate the oxidation state of the metal, for example calcium chloride rather than calcium(II) chloride. Some compounds contain two cations, for example the 'mixed oxide' of lead:

Pb_3O_4 $[2Pb^{2+} \ Pb^{4+} \ 4O^{2-}]$ dilead(II) lead(IV) oxide

The Stock system is also used to name complex ions (Chapter 13). For example $[Fe(CN)_6]^{3-}$, which consists of an iron(III) ion surrounded by six cyanide ions, that is, $[Fe^{3+} \ 6CN^-]$, is named as the hexacyanoferrate(III) ion. Stock names are used for the following oxoanions:

Chromate(VI)	CrO_4^{2-}		Dichromate(VI)	$Cr_2O_7^{2-}$
Manganate(VII)	MnO_4^-		Manganate(VI)	MnO_4^{2-}
Chlorate(I)	ClO^-		Chlorate(III)	ClO_2^-
Chlorate(V)	ClO_3^-		Chlorate(VII)	ClO_4^-

Figure 9.8 Photograph of crystals of sodium chlorate(V), a powerful oxidizing agent

This is because, for example, the names 'chlorate' (Figure 9.8) and 'manganate' are not precise enough and potentially refer to more than one species. For compounds between non-metals the Stock notation is generally not used and the actual numbers of the atoms in the molecular formula are shown in the name. For example, dinitrogen oxide, N_2O, rather than nitrogen(I) oxide and sulfur hexafluoride, SF_6, rather than sulfur(VI) fluoride.

Identifying redox reactions

9.1.4 **Deduce** whether an element undergoes oxidation or reduction in reactions using oxidation numbers.

Redox reactions or redox equations are easily recognized by:

■ deducing all of the oxidation numbers of the atoms in the chemical species present in the molecular, ionic or half-equation. Note that the equation does not have to be balanced in order to do this.
■ examining the numbers to see if the oxidation number of any atom has changed. If it has, the reaction is a redox reaction. An increase in oxidation number is oxidation and a decrease in oxidation number is reduction.

If there are *no* changes in oxidation numbers during the chemical reaction, then the reaction is *not* a redox reaction. Examples of non-redox reactions include acid–base reactions (Chapter 8), precipitation reactions (Chapter 1) and complex ion formation (Chapter 13).

Consider the following reactions.

$$2FeCl_2(s) + Cl_2(g) \rightarrow 2FeCl_3(s)$$

The oxidation numbers of iron are +2 and +3, respectively and the oxidation numbers for chlorine are 0 and –1, respectively. The iron has undergone oxidation and the chlorine has undergone reduction.

$$Mn(NO_3)_2(s) \rightarrow MnO_2(s) + 2NO_2(g)$$

The oxidation numbers of manganese are +2 and +4, respectively, and the oxidation numbers for nitrogen are +5 and +4, respectively. The manganese has undergone oxidation and the nitrogen has undergone reduction.

$$(NH_4)_2Cr_2O_7(s) \rightarrow Cr_2O_3(s) + 4H_2O(g) + N_2(g)$$

The oxidation numbers of chromium are respectively +6 and +3, and the oxidation numbers for nitrogen are –3 and 0, respectively. The chromium has undergone reduction and the nitrogen has undergone oxidation.

$$MgO(s) + 2HCl(aq) \rightarrow MgCl_2(aq) + H_2O(l)$$

The oxidation numbers of magnesium, chlorine, hydrogen and oxygen remain unchanged at +2, –1, +1 and –2. Acid–base reactions are therefore *not* redox reactions.

$$[Cu(H_2O)_6]^{2+}(aq) + 4NH_3(aq) \rightarrow [Cu(NH_3)_4(H_2O)_2]^{2+}(aq) + 4H_2O(l)$$

The oxidation number of copper is +2 in both of the complex ions. There is no change in oxidation number and this reaction, an example of ligand displacement (Chapter 13), is therefore *not* a redox reaction.

Disproportionation

Disproportionation occurs when a single species is both oxidized and reduced simultaneously (Figure 9.9). An example of disproportionation is the catalytic decomposition of hydrogen peroxide (Chapter 6).

One of the oxygen atoms in the hydrogen peroxide molecule becomes part of an oxygen molecule and during this change the oxidation number increases from –1 to zero. Hence, this is oxidation. The other oxygen atom in the hydrogen peroxide molecule become part of a water molecule and during this change the oxidation number decreases from –1 to –2. Hence, this is reduction.

Another example of disproportionation is the reaction between chlorine and water (Chapter 3) to form a mixture of hydrochloric and chloric(I) ('hypochlorous') acids:

$$\underline{Cl}_2(g) + H_2O(l) \rightarrow HO\underline{Cl}(aq) + H\underline{Cl}(aq)$$
$$0+1-1$$

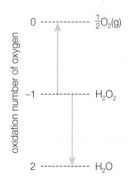

$H_2O_2(aq) \rightarrow H_2O(l) + \frac{1}{2}O_2(g)$

oxidation number of oxygen

0 ------------ $\frac{1}{2}O_2(g)$

–1 ------------ H_2O_2

2 ------------ H_2O

Figure 9.9 The disproportionation of hydrogen peroxide

One of the chlorine atoms in the chlorine molecule becomes a chloride ion, and during this change the oxidation number decreases from 0 to −1. Hence, this is reduction. The other chlorine atom in the chlorine molecule becomes part of the chlorate(I) ion, and during this change the oxidation number increases from 0 to +1. Hence, this is oxidation.

Other examples of disproportionation include:

- the reaction between chlorine and cold dilute aqueous sodium hydroxide (Chapter 23)
- the reaction between soluble copper(I) compounds, such as copper(I) sulfate, and water (Chapter 13)
- the overall cell reaction that occurs in a lead–acid car battery during discharge (Chapter 23).

9.2 Redox equations

Constructing half-equations

9.2.1 **Deduce** simple oxidation and reduction half-equations given the species involved in a redox reaction.

Many of the oxidizing and reducing agents previously described only bring about oxidation and reduction in an acidified aqueous solution. Their half-equations frequently involve water molecules and hydrogen ions. The following procedure describes how such half-equations can be constructed.

1 Write down the formulae of the reactant and products, for example:

$$Cr_2O_7^{2-} \rightarrow Cr^{3+}$$

2 Balance with respect to the chromium:

$$Cr_2O_7^{2-} \rightarrow 2Cr^{3+}$$

3 Balance the oxygen atoms of the dichromate(VI) ion with water molecules:

$$Cr_2O_7^{2-} \rightarrow 2Cr^{3+} + 7H_2O$$

4 Balance the hydrogen atoms present in the water with hydrogen ions:

$$14H^+ + Cr_2O_7^{2-} \rightarrow 2Cr^{3+} + 7H_2O$$

5 Determine the total charges on both sides of the almost completed half-equation:

LHS: $+14 + -2 = +12$;   RHS: $(2 \times +3) = +6$

6 Balance the two charges by adding electrons to the side of the equation with the most positive value:

$$6e^- + 14H^+(aq) + Cr_2O_7^{2-}(aq) \rightarrow 2Cr^{3+}(aq) + 7H_2O(l)$$
LHS: $+14 + -2 + -6 = +6$;   RHS: $(2 \times +3) = +6$

An identical process is used to construct half-equations for reducing agents that operate in an aqueous acidic solvent. The one difference is that the electrons will appear on the right-hand side of the half-equation.

1 Write down the formulae of the reactant and products, for example:

$$HNO_2 \rightarrow NO_3^-$$ (the equation is already balanced with respect to the nitrogen)

2 Balance the oxygen of the nitrous acid (nitric(III) acid) with a water molecule:

$$H_2O + HNO_2 \rightarrow NO_3^-$$

3 Balance the hydrogen present in the water and nitrous acid with hydrogen ions:

$$H_2O + HNO_2 \rightarrow NO_3^- + 3H^+$$

4 Determine the total charges on both sides of the almost completed half-equation:

LHS: $= 0$;   RHS: $-1 + (3 \times +1) = +2$

5 Balance the two charges by adding electrons to the side of the equation with the most positive value:

$$H_2O(l) + HNO_2(aq) \rightarrow NO_3^-(aq) + 3H^+(aq) + 2e^-$$
LHS: 0;   RHS: $-1 + (3 \times +1) + -2 = 0$

Forming redox equations

Redox equations are written by combining two half-equations: one describing the action of an oxidizing agent and the other describing the action of a reducing agent. Often one or both of the two half-equations must be multiplied by suitable coefficients so that the number of electrons gained by the oxidizing agent equals the number of electrons lost by the reducing agent. The electrons can then be cancelled from both sides of the equations and, if necessary, the number of water molecules and hydrogen ions (if present) simplified.

Worked example

Write a redox equation for the reduction of acidified manganate(VII) ions and the oxidation of methanol using the balanced half-equations below:

$$2H_2O(l) + CH_3OH(l) \rightarrow CO_2(g) + H_2O(l) + 6H^+(aq) + 6e^-$$
$$MnO_4^-(aq) + 8H^+(aq) + 5e^- \rightarrow Mn^{2+}(aq) + 4H_2O(l)$$

Multiplying through the top half-equation by five and the bottom half-equation by six:

$$10H_2O(l) + 5CH_3OH(l) \rightarrow 5CO_2(g) + 5H_2O(l) + 30H^+(aq) + 30e^-$$
$$6MnO_4^-(aq) + 48H^+(aq) + 30e^- \rightarrow 6Mn^{2+}(aq) + 24H_2O(l)$$

Adding the two half-equations together:

$$10H_2O(l) + 5CH_3OH(l) + 6MnO_4^-(aq) + 48H^+(aq) + 30e^-$$
$$\rightarrow 5CO_2(g) + 5H_2O(l) + 30H^+(aq) + 30e^- + 6Mn^{2+}(aq) + 24H_2O(l)$$

Cancelling electrons:

$$10H_2O(l) + 5CH_3OH(l) + 6MnO_4^-(aq) + 48H^+(aq)$$
$$\rightarrow 5CO_2(g) + 5H_2O(l) + 30H^+(aq) + 6Mn^{2+}(aq) + 24H_2O(l)$$

Simplifying the number of water molecules on the right-hand side of the equation:

$$10H_2O(l) + 5CH_3OH(l) + 6MnO_4^-(aq) + 48H^+(aq)$$
$$\rightarrow 5CO_2(g) + 30H^+(aq) + 6Mn^{2+}(aq) + 29H_2O(l)$$

The consumption of 10 water molecules and the production of 29 water molecules is equivalent to the production of 19 molecules:

$$5CH_3OH(l) + 6MnO_4^-(aq) + 48H^+(aq) \rightarrow 5CO_2(g) + 30H^+(aq) + 6Mn^{2+}(aq) + 19H_2O(l)$$

The consumption of 48 hydrogen ions and the production of 30 hydrogen ions is equivalent to the consumption of 18 hydrogen ions:

$$5CH_3OH(l) + 6MnO_4^-(aq) + 18H^+(aq) \rightarrow 5CO_2(g) + 6Mn^{2+}(aq) + 19H_2O(l)$$

Redox titrations

Redox titrations are similar to acid–base titrations (Chapter 1). Acid–base titrations involve the transfer of one or more hydrogen ions from the acid to the base. A redox titration involves the transfer of one or more electrons from a reducing agent to an oxidizing agent.

As demonstrated above, a redox reaction can be described by two half-equations: in one half-equation the reducing agent loses electrons and in the other half-equation the oxidizing agent gains electrons. The overall or stoichiometric equation for a redox titration can be obtained by combining the two half-equations, so that the number of electrons lost by the reducing agent equals the number of electrons gained by the oxidizing agent. For example:

$$MnO_4^-(aq) + 8H^+(aq) + 5e^- \rightarrow Mn^{2+}(aq) + 4H_2O(l)$$
and $$H_2O_2(aq) \rightarrow O_2(g) + 2H^+(aq) + 2e^-$$

The bottom half-equation is multiplied through by five and the top half-equation by two so that they both contain the same number of electrons. The two equations are then added together and simplified:

$$2MnO_4^-(aq) + 16H^+(aq) + 10e^- \rightarrow 2Mn^{2+}(aq) + 8H_2O(l)$$
and $$5H_2O_2(aq) \rightarrow 5O_2(g) + 10H^+(aq) + 10e^-$$

Adding together:

$$2MnO_4^-(aq) + 16H^+(aq) + 5H_2O_2(aq) \rightarrow 2Mn^{2+}(aq) + 8H_2O(l) + 5O_2(g) + 10H^+(aq)$$

Simplifying the numbers of hydrogen ions:

$$2MnO_4^-(aq) + 6H^+(aq) + 5H_2O_2(aq) \rightarrow 2Mn^{2+}(aq) + 8H_2O(l) + 5O_2(g)$$

Common oxidizing and reducing agents used in redox titrations, together with their appropriate half-equations are shown below.

Oxidizing agents for redox titrations

Acidifed manganate(VII) ions

$$MnO_4^-(aq) + 8H^+(aq) + 5e^- \rightarrow Mn^{2+}(aq) + 4H_2O(l)$$

Manganate(VII) ions are purple in colour but the reduced form, manganese(II) ions, is almost colourless. Solutions of potassium manganate(VII) are *not* primary standards (Chapter 1) because potassium manganate(VII) is difficult to prepare pure and it reacts slowly with water to form manganese(IV) oxide, especially in the presence of light.

Acidified dichromate(VI) ions

$$14H^+(aq) + Cr_2O_7^{2-}(aq) + 6e^- \rightarrow 2Cr^{3+}(aq) + 7H_2O(l)$$

Dichromate(VI) ions are orange in colour but the reduced form, chromium(III), is green. Solutions of potassium dichromate(VI) can be used as primary standards.

Iron(III) ions or salts

$$e^- + Fe^{3+}(aq) \rightarrow Fe^{2+}(aq)$$

Iodine

$$I_2(aq) + 2e^- \rightarrow 2I^-(aq)$$

Iodine (in potassium iodide solution) is red brown in colour, but colourless when in the reduced form as the iodide ion.

Indicators are not needed for titrations that involve manganate(VII) ions, dichromate(VI) ions or iodine since there is a colour change. At the end-points of these titrations, adding a slight excess of the reducing agent will produce a permanent colour change in the solution.

However, the sensitivity of the iodine colour change is often improved by adding starch solution as an indicator. This gives a deep blue-black coloured complex in the presence of iodine. The complex disappears at the end-point when all the iodine is converted to iodide.

Acidified hydrogen peroxide

$$H_2O_2(aq) + 2H^+(aq) + 2e^- \rightarrow 2H_2O(l)$$

Hydrogen peroxide is a moderately powerful oxidizing agent; however, when in the presence of a more powerful oxidizing agent it is 'forced' to act as a reducing agent.

Reducing agents for redox titrations

Figure 9.10 Photograph of ethanedioic acid crystals

Iron(II) salts or iron(II) ions

$$Fe^{2+}(aq) \rightarrow Fe^{3+}(aq) + e^-$$

Ethanedioic (oxalic) acid and ethanedioate (oxalate) ions

$$(COOH)_2(aq) \rightarrow 2CO_2(g) + 2H^+(aq) + 2e^-$$
$$(COO^-)_2(aq) \rightarrow 2CO_2(g) + 2e^-$$

This autocatalytic reaction (Chapter 16) is carried out at 80 °C since the reaction is relatively slow at room temperature.

Ethanedioic acid (Figure 9.10) and its salts are primary standards and are frequently used to standardize solutions of potassium manganate(VII), that is, determine their concentration to a high degree of accuracy.

Hydrogen peroxide

$$H_2O_2(aq) \rightarrow 2H^+(aq) + O_2(g) + 2e^-$$

This reaction occurs when hydrogen peroxide is in the presence of a more powerful oxidizing agent, such as dichromate(VI) or manganate(VII) ions.

Iodide ions

$$2I^-(aq) \rightarrow I_2(aq) + 2e^-$$

Sodium thiosulfate(VI) or thiosulfate(VI) ions

$$2S_2O_3^{2-}(aq) \rightarrow S_4O_6^{2-}(aq) + 2e^-$$

A common redox reaction is the reaction of an oxidizing agent with excess potassium iodide solution to form iodine. The iodine is then titrated with sodium thiosulfate solution, using starch as an indicator (Figure 9.11).

The overall equation for the reaction is:

$$2S_2O_3^{2-}(aq) + I_2(aq) \rightarrow S_4O_6^{2-}(aq) + 2I^-(aq)$$

Figure 9.11 The different colours observed during a thiosulfate titration: **a** aqueous iodine, **b** addition of the indicator, **c** deep blue-black colour of indicator with aqueous iodine and **d** the colourless end-point

9.2.3 **Define** the terms *oxidizing agent* and *reducing agent*.
9.2.4 **Identify** the oxidizing and reducing agents in redox equations.

Oxidizing agents

An **oxidizing agent** is defined as a substance that brings about the oxidation of substances by accepting electrons from the substance they oxidize. Oxidizing agents undergo a process of reduction.

Some common oxidizing agents, some examples of their reactions and their appropriate half-equations are described below. The strengths of oxidizing agents are described by their standard electrode potentials (Section 19.1). (Oxidizing agents for redox titrations have been described on page 239.)

Oxygen

During the reaction in the gas phase oxygen molecules gain electrons to form oxide ions:

$$\tfrac{1}{2}O_2(g) + 2e^- \rightarrow O^{2-}(g)$$

Ozone (trioxygen)

Ozone (Chapter 25) is an extremely powerful oxidizing agent in acidic solution:

$$O_3(aq) + 2H^+(aq) + 2e^- \rightarrow O_2(g) + H_2O(l)$$

It produces iodine from neutral or alkaline potassium iodide solution:

$$2KI(aq) + O_3(aq) + H_2O(l) \rightarrow 2KOH(aq) + O_2(g) + I_2(aq)$$

Chlorine

During the reaction chlorine molecules gain electrons to form chloride ions.

$$Cl_2(aq) + 2e^- \rightarrow 2Cl^-(aq)$$

This reaction occurs in both the gas phase and in acidic solution.

Similar reactions occur with the other halogens, but with a decreasing tendency as you go down the group.

Acidified potassium manganate(VII)

During the reaction purple manganate(VII) ions are converted, under strongly acidic conditions, to pale pink manganese(II) ions:

$$MnO_4^-(aq) + 8H^+(aq) + 5e^- \rightarrow Mn^{2+}(aq) + 4H_2O(l)$$

Acidified aqueous potassium dichromate(VI)

During the reaction the orange solution containing dichromate(VI) ions (Figure 9.12) is converted to a solution containing green chromium(III) ions:

$$Cr_2O_7^{2-}(aq) + 14H^+(aq) + 6e^- \rightarrow 2Cr^{3+}(aq) + 7H_2O(l)$$

(This reaction and its associated colour change was the basis for the 'breathalyser' formerly used by many police forces around the world to detect and measure alcohol levels in the breath of drivers (Chapters 21 and 24).)

Acidified hydrogen peroxide

$$H_2O_2(aq) + 2H^+(aq) + 2e^- \rightarrow 2H_2O(l)$$

Figure 9.12 Potassium dichromate(VI) crystals

Metal ions

Metal ions or cations of unreactive metals can behave as weak oxidizing agents, for example:

$$Ag^+(aq) + e^- \rightarrow Ag(s)$$

Hydrogen ions ('protons')

Hydrogen ions from dilute aqueous solutions of acids, for example the reaction between hydrochloric acid and magnesium:

$$Mg(s) + 2HCl(aq) \rightarrow MgCl_2(aq) + H_2(g)$$

The half-equation for the reduction of the hydrogen ion is:

$$2H^+(aq) + 2e^- \rightarrow H_2(g)$$

This reaction can be regarded as a displacement reaction where the hydrogen ions are displaced from the acid by the magnesium atoms.

Manganese(IV) oxide

Manganese dioxide or manganese(IV) oxide in acidic solution:

$$MnO_2(s) + 4H^+(aq) + 2e^- \rightarrow Mn^{2+}(aq) + 2H_2O(l)$$

Reducing agents

A reducing agent is defined as a substance that brings about the reduction of a substance by donating electrons to the substance it reduces. Reducing agents undergo a process of oxidation (Figure 9.13).

Some common reducing agents and examples of their reactions are described below. The strengths of reducing agents, like oxidizing agents, are described by their standard electrode potentials (Section 19.1). (Reducing agents for redox titration have been described on pages 239–240.)

Hydrogen

For example, with lead(II) oxide

$$PbO(s) + H_2(g) \rightarrow H_2O(l) + Pb(s)$$

The half-equation is:

$$O^{2-}(s) + H_2(g) \rightarrow H_2O(l) + 2e^-$$

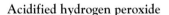

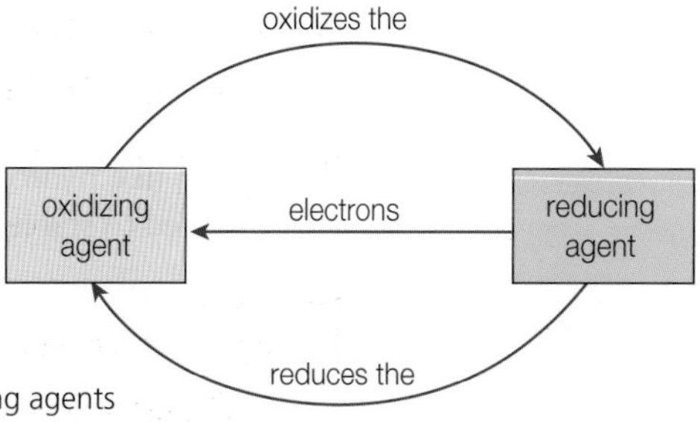

Figure 9.13 The relationship between oxidizing and reducing agents

Carbon

For example, with lead(II) oxide

$$PbO(s) + C(s) \rightarrow Pb(s) + CO(g)$$

The half-equation is:

$$O^{2-}(s) + C(s) \rightarrow CO(g) + 2e^-$$

Carbon monoxide

When carbon monoxide acts as a reducing agent, carbon dioxide is produced, for example with iron(III) oxide (at high temperature):

$$Fe_2O_3(s) + 3CO(g) \rightarrow 2Fe(l) + 3CO_2(g)$$

The half-equation is:

$$O^{2-}(g) + CO(g) \rightarrow CO_2(g) + 2e^-$$

This is the basis of the extraction of iron on an industrial scale in a blast furnace (Chapter 23). Carbon monoxide is a more powerful reducing agent at low temperatures than carbon, but at high temperatures carbon is the more powerful reducing agent.

Metals

The more reactive metals are strong reducing agents. For example, zinc acts as a reducing agent with aqueous copper(II) sulfate solution:

$$CuSO_4(aq) + Zn(s) \rightarrow ZnSO_4(aq) + Cu(s)$$

Ionically, this can be written as:

$$Cu^{2+}(aq) + Zn(s) \rightarrow Zn^{2+}(aq) + Cu(s)$$

after removing the 'spectator' sulfate ions.

This and similar reactions involving metals and metal ions are known as displacement reactions and occur when a more reactive metal reacts with the ions of a less reactive metal.

■ Extension: Substances that can act as both oxidizing and reducing agents

The terms oxidizing agent and reducing agent, like the terms acid and base, are *relative terms*. A weak reducing agent will be 'forced' to act as an oxidizing agent in the presence of a more powerful reducing agent. Conversely, a weak oxidizing agent will be 'forced' to act as a reducing agent in the presence of a more powerful oxidizing agent.

For example, with acidified aqueous potassium iodide, hydrogen peroxide acts as an oxidizing agent and converts iodide ions to iodine:

$$H_2O_2(aq) + 2H^+(aq) + 2I^-(aq) \rightarrow 2H_2O(l) + I_2(aq)$$

The hydrogen peroxide is reduced to water during the reaction:

$$H_2O_2(aq) + 2H^+(aq) + 2e^- \rightarrow 2H_2O(l)$$

However, in the presence of acidified potassium manganate(VII), a stronger oxidizing agent than hydrogen peroxide, hydrogen peroxide is 'forced' to act as a reducing agent.

$$5H_2O_2(aq) + 2MnO_4^-(aq) + 6H^+(aq) \rightarrow 5O_2(g) + 2Mn^{2+}(aq) + 8H_2O(l)$$

The hydrogen peroxide is oxidized to water and oxygen:

$$H_2O_2(aq) \rightarrow O_2(g) + 2H^+(aq) + 2e^-$$

Substances like hydrogen peroxide that are able to act as both oxidizing and reducing agents can be converted to stable compounds that have higher and lower oxidation states.

9.3 Reactivity

9.3.1 **Deduce** a reactivity series based on the chemical behaviour of a group of oxidizing and reducing agents.

Reactions of metals with metal ions in solution

Figure 9.14 The reaction between zinc and copper(II) ions to form zinc ions and copper atoms

It has previously been stated that metals often act as reducing agents and that the greater the chemical reactivity of the metal, the greater its ability to bring about reduction. A group of metals can be readily sorted into order of reactivity, and hence reducing power, by performing a number of simple experiments involving the metals and aqueous solutions of their ions (Figure 9.14).

This approach to establishing a reactivity series is tabulated below for copper, lead, iron, magnesium, zinc and tin. Test tubes are filled with a small volume of the following aqueous solutions (each of which have the same concentration): copper(II) nitrate; lead(II) nitrate, iron(II) sulfate, magnesium nitrate, zinc nitrate and tin(II) chloride. (The nitrate, sulfate and chloride ions are 'spectator' ions and do not participate in any reactions that occur. These solutions can therefore be regarded as aqueous solutions of the metal ions.)

Into each of these solutions is placed a small piece of freshly cleaned magnesium ribbon. The surface of the magnesium is observed for several minutes for any colour changes indicative of a chemical reaction.

The process is then repeated, in turn, with fresh solutions and pieces of the other metals in the place of magnesium. The results are tabulated, where a tick indicates a reaction has occurred and a cross indicates no observable reaction has taken place (Table 9.6).

Ion in solution/ Metal	Cu^{2+}(aq)	Pb^{2+}(aq)	Fe^{2+}(aq)	Mg^{2+}(aq)	Zn^{2+}(aq)	Sn^{2+}(aq)
Copper	✗	✗	✗	✗	✗	✗
Lead	✓	✗	✗	✗	✗	✗
Iron	✓	✓	✗	✗	✗	✓
Magnesium	✓	✓	✓	✗	✓	✓
Zinc	✓	✓	✓	✗	✗	✓
Tin	✓	✓	✗	✗	✗	✗

Table 9.6 Summary of results for a series of reactions between selected metals and their ions

Metal	Number of displacement reactions
Mg	5
Zn	4
Fe	3
Sn	2
Pb	1
Cu	0

Table 9.7 A reactivity series for selected metals based on displacement reactions

Each tick represents a chemical reaction and by summing the number of reactions that each metal has produced, as shown in Table 9.7, a reactivity or activity series can be constructed.

Up the **reactivity** or **activity series** the metals become increasingly chemically reactive and their reducing power, or ability to donate electrons, increases.

The reactions that occur, as indicated by the ticks in Table 9.6, are known as **displacement reactions** since they involve a more reactive metal displacing, or 'pushing out', a less reactive metal from its salt. Molecular, ionic reactions and half-equations can be written for all the displacement reactions, for example:

Formula equation: $Mg(s) + CuSO_4(aq) \rightarrow MgSO_4(aq) + Cu(s)$

Rewriting in terms of ions: $Mg(s) + Cu^{2+}(aq) + SO_4^{2-}(aq) \rightarrow Mg^{2+}(aq) + SO_4^{2-}(aq) + Cu(s)$

Cancelling spectator ions: $Mg(s) + Cu^{2+}(aq) \rightarrow Mg^{2+}(aq) + Cu(s)$

Rewriting in terms of half-equations:
$Mg(s) \rightarrow Mg^{2+}(aq) + 2e^-$
$Cu^{2+}(aq) + 2e^- \rightarrow Cu(s)$

Displacement reactions can also be carried out in the solid state using powdered samples of metals and metal compounds. For example, if iron(III) oxide and aluminium are heated together, a very exothermic reaction known as the thermite reaction occurs, resulting in the formation of aluminium oxide and molten iron (Figure 9.15).

$$Fe_2O_3(s) + 2Al(s) \rightarrow 2Fe(l) + Al_2O_3(s)$$

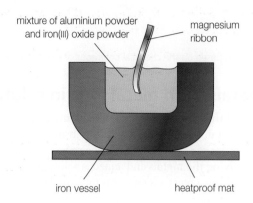

Figure 9.15 Thermite reaction mixture apparatus

The thermite reaction (Figure 9.16) occurs because aluminium is a more powerful reducing agent than iron and has a stronger tendency to lose its electrons.

Some metals react with water and these can be regarded as displacement reactions with hydrogen being displaced from water. For example, the reaction between sodium and water:

Figure 9.16 The thermite experiment

$$2Na(s) + 2H_2O(l) \rightarrow 2NaOH(aq) + H_2(g)$$

The relevant half-equations are:

$$2Na(s) \rightarrow 2Na^+(aq) + 2e^- \qquad 2H_2O(l) + 2e^- \rightarrow 2OH^-(aq) + H_2(g)$$

Using the reactivity series

9.3.2 **Deduce** the feasibility of a redox reaction from a given reactivity series.

When metals are placed in a reactivity series (shown opposite), their order is very similar to an arrangement based on standard electrode potentials (Chapter 19).

Potassium
Sodium
Calcium
Magnesium
Aluminium
(Carbon)
Zinc
Iron
Tin
Lead
(Hydrogen)
Copper
Silver
Gold

■ **Extension:** Aluminium

The position of aluminium may be somewhat surprising since everyday experience suggests that aluminium is a relatively unreactive metal that does not undergo corrosion. The apparent low reactivity of aluminium is accounted for by the presence of an extremely thin protective layer of aluminium oxide present on the metal surface that prevents the metal underneath from oxidizing further.

The inclusion of the non-metals carbon and hydrogen extends the usefulness of the activity or reactivity series. In these reactions hydrogen is behaving like a metal, since aqueous solutions of acids contain positively charged hydrogen ions (cf. positive ions of metals in salts). Metals above hydrogen, for example zinc, will displace hydrogen from dilute acids, but metals below it, for example copper, will not displace hydrogen from dilute acids.

Molecular equation: $Zn(s) + 2HCl(aq) \rightarrow ZnCl_2(aq) + H_2(g)$

Ionic equation: $Zn(s) + 2H^+(aq) \rightarrow Zn^{2+}(aq) + H_2(g)$

Half-equations: $Zn(s) \rightarrow Zn^{2+}(aq) + 2e^-$

$2H^+(aq) + 2e^- \rightarrow H_2(g)$

Metals above carbon in the reactivity series, such as sodium and aluminium, cannot be produced by reduction of metal oxides with carbon; instead electrolysis has to be used. Metals below carbon, such as iron and zinc, can be produced by reduction of metal oxides with carbon.

$ZnO(s) + C(s) \rightarrow Zn(g) + CO(g)$

The reactions of metals with their ions, water (Figure 9.17), dilute acid, carbon and hydrogen are summarized in Table 9.8.

Reactivity series	Reaction with dilute acid	Reaction with air/oxygen	Reaction with water	Ease of extraction
Potassium (K)	Producing H$_2$ with decreasing vigour	Burn very brightly and vigorously	Produce H$_2$ with decreasing vigour with cold water	Difficult to extract
Sodium (Na)				
Calcium (Ca)		Burn to form an oxide with decreasing vigour	React with steam with decreasing vigour	Easier to extract
Magnesium (Mg)				
Aluminium (Al)				
Zinc (Zn)				
Iron (Fe)				
Lead (Pb)		React slowly to form the oxide		
Copper (Cu)	Do not react with dilute acids		Do not react with cold water or steam	
Silver (Ag)				
Gold (Au)		Do not react		Found as the element (native)
Platinum (Pt)				

Table 9.8 Reactivity series of metals

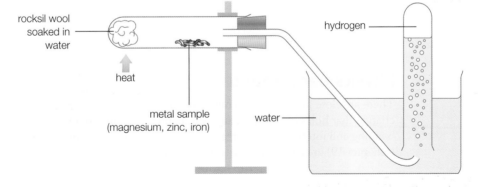

Figure 9.17 The apparatus used to test the action of steam on a metal

rocksil wool soaked in water

heat

metal sample (magnesium, zinc, iron)

water

hydrogen

Displacement reactions also occur with non-metals, in particular the halogens where a more reactive halogen will displace a less reactive halogen from one of its compounds (Chapter 3). A reactivity or activity series can be written for the halogens, which corresponds to the positions of the elements of the periodic table:

Fluorine

Chlorine

Bromine

Iodine

As you move up the reactivity or activity series the halogens become increasingly chemically reactive and their oxidizing power, or ability to receive electrons and form halide ions, increases.

For example, when chlorine gas or chlorine water is added to an aqueous solution of potassium bromide, the chlorine (being higher up the reactivity series) displaces the less reactive bromine. The colourless solution of potassium bromide turns orange as the bromine is produced. The bromine is more easily identified if a small volume of non-polar organic solvent, such as tetrachloromethane (carbon tetrachloride), is added to the reaction mixture. The bromine, being non-polar, will enter the organic layer and due to its higher concentration will be more visible. Molecular equations, ionic equations and half-equations can be written for this displacement reaction, for example:

Molecular equation: $2KBr(aq) + Cl_2(g) \rightarrow 2KCl(aq) + Br_2(aq)$

Rewriting in terms of ions (and ignoring any reactions between the halogens and water):

$$2K^+(aq) + 2Br^-(aq) + Cl_2(g) \rightarrow 2K^+(aq) + 2Cl^-(aq) + Br_2(aq)$$

Cancelling spectator ions: $2Br^-(aq) + Cl_2(g) \rightarrow 2Cl^-(aq) + Br_2(aq)$

Rewriting in terms of half-equations: $2Br^-(aq) \rightarrow 2e^- + Br_2(aq)$

$$2e^- + Cl_2(g) \rightarrow 2Cl^-(aq)$$

You can see that the bromide ions have undergone oxidation and the chlorine has undergone reduction. Chlorine has behaved as an oxidizing agent and bromide ions have behaved as a reducing agent. No reaction of course occurs if iodine solution is added to potassium bromide solution because iodine is lower down the reactivity or activity series than bromine and is therefore a less powerful oxidizing agent.

The reactivity and hence oxidizing power of the halogens is correlated with the size of their atoms (Chapter 3). As the halogen atoms get larger, the nucleus has decreasingly less electrostatic attraction for the electrons in the outer shell and becomes progressively less able to attract an extra electron to complete its outer shell: hence the oxidizing power of the halogens decreases from fluorine to iodine.

Conversely, the larger halide ions, such as iodide, have weaker electrostatic attraction for the electrons in their outer shell, compared to smaller halide ions such as fluoride. Consequently, iodide ions give up their extra electrons very readily; they are easily oxidized. The smaller fluoride ions have stronger electrostatic attraction for their outer electrons and are much less readily oxidized.

9.4 Voltaic cells

9.4.1   **Explain** how a redox reaction is used to produce electricity in a voltaic cell.
9.4.2   **State** that oxidation occurs at the negative electrode (anode) and reduction occurs at the positive electrode (cathode).

Physics background

There are close links between chemistry and electricity. It is the attraction of opposite electric charges that holds electrons in atoms (Chapter 2). This attraction is also the basis for chemical bonding and intermolecular forces (Chapter 4). Electrolysis and voltaic cells (simple 'batteries') (Chapter 19) are part of a branch of chemistry termed electrochemistry.

Electric charge

There are two types of electric charge: positive and negative. Like charges repel, so a negative ion is repelled by a positively charged surface. Opposite charges attract, so a negative ion is attracted to a positively charged surface. Electric charge is measured in **coulombs**, symbol C. An electron has a charge of 1.6×10^{-19} coulombs. In an electric circuit, electric charge flows through wires. The charge is carried by electrons, which therefore flow from the negative terminal to the positive terminal.

Electric current

The rate at which electric charge flows through a circuit is called the **current** and is measured in **amperes** (usually abbreviated to amps), symbol A. A large current could be produced by a large amount of charge moving slowly or a small amount of charge moving quickly. A current of one amp is a flow of charge of one coulomb per second.

Potential difference

Electric current flows through a circuit *if* there is a difference in electric potential between two points in a circuit. This is analogous to a ball rolling down a ramp: at the top of the ramp it has high potential energy, at the bottom it has less potential energy. This potential difference gives electrical energy to the charge. Potential difference is measured in **volts**, symbol V, and is often loosely termed voltage.

One volt gives a charge of one coulomb one joule of energy. The charge then transfers the energy into other forms, for example heat, light or chemical energy (in a chargeable battery). The charge is not used, it simply travels round the circuit carrying energy. An analogy is a postman (charge) who picks up letters (energy) at the post office (power supply or battery). He delivers the letters (transfers their energy) as he follows his route and then returns to the post office to pick up more letters (Figure 9.18).

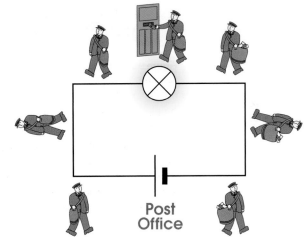

Figure 9.18 Model of an electrical circuit showing the flow of electrons

A simple **voltaic cell**, known as the **Daniell cell** (Figure 9.19), can be constructed by placing a zinc electrode in a solution of zinc sulfate and a copper electrode in a solution of copper(II) sulfate. The two electrodes are connected via wires and a high-resistance voltmeter. This is known as the external circuit and allows electrons to flow. The circuit is completed by a salt bridge which allows ions to flow in order to maintain electrical neutrality (Chapter 19). A simple salt bridge consists of a filter paper soaked in saturated potassium nitrate. Potassium and nitrate ions are chosen because they will not react with the other ions in solution or with the electrodes.

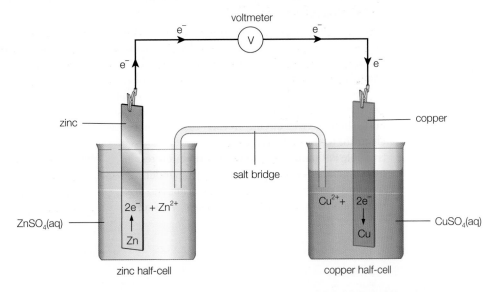

Figure 9.19 A Daniell cell

Since zinc is higher in the reactivity series (see page 243) it will undergo oxidation and release electrons onto the surface of the zinc electrode (making it negative). The zinc ions produced dissolve into the water (Figure 9.20).

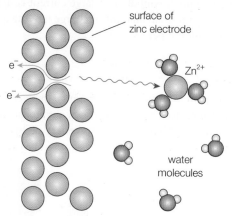 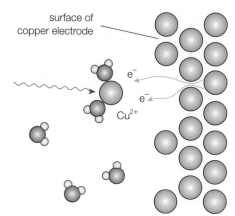

Figure 9.20 Zinc atoms forming hydrated zinc ions on the surface of the zinc electrode of a Daniell cell

Figure 9.21 Hydrated copper(II) ions forming copper atoms on the surface of the copper electrode of a Daniell cell

The electrons flow from the surface of the zinc electrode through the external circuit to the surface of the copper electrode. Copper(II) ions on the surface of the copper electrode accept the electrons and undergo reduction (Figure 9.21).

The process continues until either all the zinc electrode or all the copper(II) ions are consumed. The zinc is acting as a reducing agent and copper(II) ions are acting as an oxidizing agent. By definition, the **anode** in a voltaic cell is the electrode at which oxidation occurs and the **cathode** is the electrode at which reduction occurs.

The relevant half-equations are:

Anode: $\qquad\qquad\qquad\qquad$ $Zn(s) \rightarrow Zn^{2+}(aq) + 2e^-$

Cathode: $\qquad\qquad\qquad\qquad$ $Cu^{2+}(aq) + 2e^- \rightarrow Cu(s)$

Overall ionic equation: $Zn(s) + Cu^{2+}(aq) \rightarrow Cu(s) + Zn^{2+}(aq)$

The overall chemical change is the same as that which occurs when zinc is placed in copper(II) sulfate solution. Heat energy is released in that situation, but the arrangement in the voltaic cell, where the two reactions are physically separated, enables the release of electrical energy.

An electric current flows from the anode to the cathode because there is a difference in electric potential energy between the electrodes. Experimentally the difference in electric potential between the anode and the cathode is measured by a voltmeter and the reading (in volts) is called the **cell potential**.

The voltage of a cell depends not only on the nature of the electrodes and the ions, but also on the concentrations of the ions and the temperature at which the cell is operated. Voltaic cell voltages are normally measured under standard conditions (Chapter 19). The cell potential of the Daniell cell under standard conditions is 1.1 volts.

The Daniell cell is one example of a simple voltaic cell. Similar voltaic cells can be made from two different metals in contact with an aqueous solution of their ions and connected by a salt bridge and external circuit. In each case the more reactive metal forms the anode which supplies electrons to the cathode.

Table 9.9 summarizes some experimental results from a number of voltaic cells operating under standard conditions. These cell potentials can also be calculated from standard electrode potentials (Chapter 19).

Metal electrodes	Cell potential/V
Copper and magnesium	2.70
Copper and iron	0.78
Lead and zinc	0.64
Lead and iron	0.32

Table 9.9 Selected voltaic cells and cell potentials

Of these metals, copper and magnesium are *furthest apart* in the reactivity series. This combination of electrodes gives the *highest* cell potential. Lead and iron are the *closest* in the reactivity series and give the *lowest* voltage. Hence, the further apart the two metals are in the reactivity series, the higher the cell potential.

Applications of Chemistry

The most obvious method of preventing corrosion is to keep oxygen and water away from the surface of the metal by covering it with a layer of paint, grease or enamel. Another method is to electroplate the object with a thin layer of metal. The most effective approach is an electrochemical approach known as cathodic protection.

If a metal higher in the reactivity series is connected to iron, it acts as the anode and undergoes oxidation while the iron remains intact. The other metal is 'sacrificed' to protect the iron. Examples of metals used in this way are zinc and magnesium. The use of zinc blocks on the hull of a ship is an example of a sacrificial method (Figure 9.22).

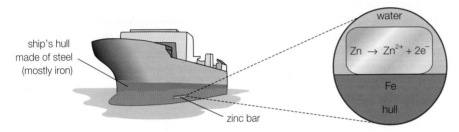

Figure 9.22 Sacrificial protection of an iron ship

Zinc, being more reactive than iron, is oxidized to form zinc ions:

$$Zn(s) \rightarrow Zn^{2+}(aq) + 2e^-$$

The electrons released reduce dissolved oxygen molecules to form hydroxide ions, $OH^-(aq)$:

$$O_2(g) + 2H_2O(l) + 4e^- \rightarrow 4OH^-(aq)$$

9.5 Electrolytic cells

9.5.1 **Describe,** using a diagram, the essential components of an electrolytic cell.
9.5.2 **State** that oxidation occurs at the positive electrode (anode) and reduction occurs at the negative electrode (cathode).
9.5.3 **Describe** how current is conducted in an electrolytic cell.

Conductors and insulators

A substance that allows electricity to pass through itself is called a conductor. A substance that does not allow electricity to pass through itself is called an insulator. Common insulators include non-metallic elements (with the exception of graphite), dry samples of covalent compounds and solid samples of ionic substances.

Some substances, for example silicon and germanium, conduct electricity very slightly and are known as semi-conductors (Chapter 23). Conductors include metals and graphite (Chapter 4), aqueous solutions (Figure 9.23) of acids and alkalis (Chapter 8), and ionic compounds, when they are dissolved in water or molten.

In order for a substance to conduct electricity it must contain electrically charged particles that are free to move when the substance is subjected to a

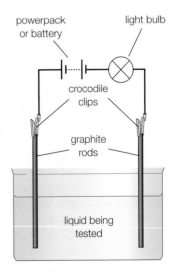

Figure 9.23 A circuit for testing the conductivity of liquids and solutions

potential difference or voltage. In metals, in both the solid and liquid states, the charged particles are the valence electrons (Chapter 4). It is a flow of these valence electrons through the metal that constitutes an electric current (Figure 9.24).

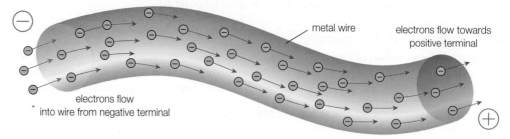

metal wire

electrons flow towards positive terminal

electrons flow into wire from negative terminal

Figure 9.24 Electrons flowing along a metal wire form an electric current

All the other substances mentioned above in the list of conductors contain negatively and positively charged ions that are free to move through the substance. These ions, in effect, constitute an electric current since electrons are transported from one electrode to another.

Ionic solids do *not* conduct electricity because the ions are firmly held in the lattice by powerful electrostatic forces and cannot move. Only when the ionic substance is molten or dissolved in water are the ions released from the lattice and free to move.

The electrolysis of a molten salt

An important and fundamental difference between conduction in a metal (or graphite) and in an aqueous solution of an ionic compound or a molten sample is that when an electric current passes through a metal, the metal itself is chemically unaffected. However, when an electric current passes through an ionic substance, either molten or in solution, the compound undergoes chemical decomposition. A substance that conducts electricity and is decomposed by the passage of an electric current is known as an **electrolyte** (Figure 9.25). The process of decomposing an electrolyte with an electric current is called **electrolysis**. It is an important technique used on the industrial scale to prepare aluminium, chlorine, sodium hydroxide and hydrogen (Chapter 23).

When electricity is passed through an electrolyte, the electricity enters and leaves via electrical conductors known as electrodes, which are usually made of graphite or metal. The electrode connected to the positive terminal of the battery or direct current (dc) power supply is known as the **anode** and the negative electrode is known as the **cathode**. Negative ions, or **anions**, are attracted towards the anode and positive ions, or **cations**, are attracted towards the cathode. When the ions reach the surface of the electrodes they undergo redox reactions.

Figure 9.25 Lead(II) bromide, which acts as an electrolyte when molten

The simplest form of electrolysis is the electrolysis of a molten binary salt, such as lead(II) bromide, $PbBr_2$ [Pb^{2+} $2Br^-$]. Inert or chemically unreactive graphite or metal electrodes are used (Figure 9.26) and the decomposition products are molten lead and bromine vapour. The overall reaction is:

$$PbBr_2(l) \rightarrow Pb(l) + Br_2(g)$$

The lead is formed at the cathode and the bromine vapour is formed at the anode.

At the anode: Negatively charged bromide ions are electrostatically attracted towards the positively charged anode. At the anode they lose electrons and form bromine molecules. The relevant half-equations describing the oxidation of bromide ions are shown below.

$$2Br^-(l) \rightarrow Br_2(g) + 2e^-$$
$$\text{or} \quad Br^-(l) \rightarrow Br(g) + e^-; \quad 2Br(g) \rightarrow Br_2(g)$$

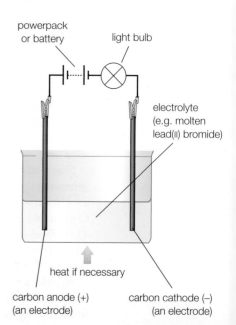

powerpack or battery

light bulb

electrolyte (e.g. molten lead(II) bromide)

heat if necessary

carbon anode (+) (an electrode)

carbon cathode (−) (an electrode)

Figure 9.26 Experimental apparatus for electrolysis

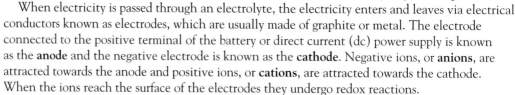

At the cathode: Positively charged lead(II) ions are attracted towards the negatively charged cathode. At the cathode they gain electrons and form lead atoms. The relevant half-equation describing the reduction of lead(II) ions is shown below.

$$Pb^{2+}(l) + 2e^- \rightarrow Pb(l)$$

During electrolysis, each lead(II) ion accepts two electrons from the cathode and at the same time two bromide ions each release an electron to the anode. The overall effect of these two processes is equivalent to two electrons flowing through the liquid lead(II) bromide from the cathode to the anode.

9.5.4 **Deduce** the products of the electrolysis of a molten salt.

Electrolysis of other compounds

All ionic compounds undergo electrolysis in the molten state and obey two simple rules:

- Metals always form positively charged ions, or cations, which migrate to the cathode and are discharged as atoms.
- Non-metals always form negatively charged ions, or anions, which migrate to the anode and are discharged as molecules.

Examples of the products of the electrolysis of molten or fused electrolytes are shown in Table 9.10.

Table 9.10 Examples of electrolysis of molten electrolytes

Electrolyte	Overall decomposition	Cathode half-equation	Anode half-equation
Sodium chloride, NaCl (Figure 9.27)	$2NaCl \rightarrow 2Na + Cl_2$	$Na^+ + e^- \rightarrow Na$	$2Cl^- \rightarrow Cl_2 + 2e^-$
Potassium iodide, KI	$2KI \rightarrow 2K + I_2$	$K^+ + e^- \rightarrow K$	$2I^- \rightarrow I_2 + 2e^-$
Copper(II) chloride, $CuCl_2$	$CuCl_2 \rightarrow Cu + Cl_2$	$Cu^{2+} + 2e^- \rightarrow Cu$	$2Cl^- \rightarrow Cl_2 + 2e^-$

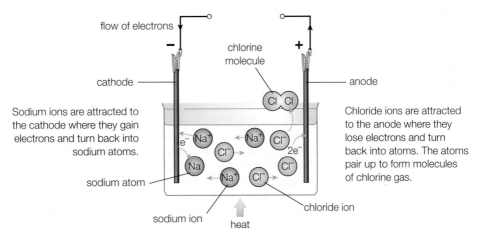

Figure 9.27 Electrolysis of molten sodium chloride

SUMMARY OF KNOWLEDGE

- Oxidation is:
  - **i** the gain of oxygen
  - **ii** the loss of hydrogen
  - **iii** the loss of electrons
  - **iv** increase in oxidation number
- Reduction is:
  - **i** the loss of oxygen
  - **ii** the gain of hydrogen
  - **iii** the gain of electrons
  - **iv** decrease in oxidation number

■ Half-equations represent the reduction and oxidation processes of a redox equation. A reduction half-equation uses electrons; an oxidation half-equation produces electrons.

■ A redox equation is constructed from two half-equations by combining them according to the principle that the number of electrons accepted by the reduction half-equation is equal to the number of electrons produced by the oxidation half-equation.

■ In aqueous solution hydrogen ions and water molecules may be involved. These species are manipulated into the equation, until all the oxygen and hydrogen atoms are balanced.

■ All full redox reactions must conform to the principle that the total oxidation number change in a chemical reaction is zero.

■ A disproportionation is a redox reaction in which atoms of the same element in the same oxidation state simultaneously undergo both oxidation and reduction.

■ Oxidizing agents bring about the oxidation of chemical species. Reducing agents bring about the reduction of chemical species.

■ The oxidation number of an atom in a chemical species expresses the number of electrons possessed by that atom, relative to the number possessed in the element. A positive oxidation number expresses a loss of that number of electrons and a negative number indicates a gain.

■ An uncombined element contains atoms with an oxidation number of zero.

■ A simple ion of an element has an oxidation number equal to its charge.

■ The sum of the oxidation numbers in a chemical species (molecule or ion) adds up to the overall charge on the species.

■ A change in oxidation number of one unit represents the transfer of one electron from an atom of lower electronegativity to an atom of higher electronegativity.

■ Oxidation numbers can be assigned according to rules that are based upon some elements having unchanging or fixed oxidation numbers: H (+1), O (–2), F (–1), group 1 (+1) and group 2 (+2).

■ The naming of inorganic compounds is based on oxidation numbers. Simple transition metal ions are named with the oxidation number as a Roman numeral. Related oxoanions have the suffix -*ate* together with the oxidation number as a Roman numeral.

■ An electric current is the movement of charged particles. In solution or a molten liquid, the positive and negative ions carry the current. For conduction to occur, ions must be mobile.

■ Decomposition caused by electricity is called electrolysis, and the liquid or solution that decomposes is an electrolyte.

■ During electrolysis positive ions (cations) are discharged at the cathode (negative electrode); negative ions (anions) at the anode (positive electrode). Metals are always discharged at the cathode; non-metals are always discharged at the anode.

■ Reduction always takes place at the negative electrode (cathode); oxidation always takes place at the positive electrode (anode).

■ Metals can be arranged in order of reactivity (reducing power) by comparing the reactions of metals with oxygen, water, dilute acid, metal oxides and aqueous solutions of metal ions.

■ A voltaic cell consists of two different metals and an electrolyte. Chemical energy is converted into electrical energy. The more reactive metal forms the negative electrode from which the electrons flow to the less reactive metal.

■ The further apart the two metals are in the reactivity series, the higher the voltage of the voltaic cell.

■ *Examination questions – a selection*

Paper 1 IB questions and IB style questions

Q1 When $MnO_4^-(aq)$ reacts in an acidic solution it produces:

A Mn^{2+} **B** Mn^{3+} **C** MnO_4^{2-} **D** MnO_2

Q2 In the reaction

$MnO_2(s) + 4HCl(aq) \rightarrow Cl_2(g) + MnCl_2(aq) + 2H_2O(l)$

A HCl is the oxidizing agent.
B Cl_2 is the oxidation product.
C MnO_2 is the reducing agent.
D H_2O is the reduction product.

Q3 The following reaction occurs in acid solution:

$\_H^+(aq) + \_NO_3^-(aq) + I^-(aq)$
$\rightarrow IO_3^-(aq) + \_NO_2(g) + \_H_2O(l)$

The equation is not balanced. What is the coefficient of NO_3^- in the balanced equation?
A 4   **B** 2   **C** 5   **D** 6

Q4 Which of the following reactions involves neither oxidation nor reduction?

A $Ag^+ + Br^- \rightarrow AgBr$
B $2H_2S + SO_2 \rightarrow 3S + 2H_2O$
C $2[Ag(NH_3)_2]^+ + Cu \rightarrow Cu[(NH_3)_4]^{2+} + 2Ag$
D $2Al + 2OH^- + 6H_2O \rightarrow 2Al(OH)_4^- + 3H_2$

Q5 Which one of the following equations represents the half-equation (or half-reaction) that occurs at the anode during the electrolysis of molten potassium iodide?

A $K^+ + e^- \rightarrow K$
B $2H_2O \rightarrow O_2 + 4H^+ + 4e^-$
C $2H_2O + 2e^- \rightarrow H_2 + 2OH^-$
D $2I^- \rightarrow I_2 + 2e^-$

Q6 All of the following equations represent oxidation–reduction reactions except:

A $2C_3H_7OH + 9O_2 \rightarrow 6CO_2 + 8H_2O$
B $Ni + 4CO \rightarrow Ni(CO)_4$
C $2Na + 2H_2O \rightarrow 2NaOH + H_2$
D $Cl_2 + 2NaBr \rightarrow Br_2 + 2NaCl$

Q7 The following information is given about reactions involving the metals X, Y and Z and solutions of their sulfates.

$X(s) + YSO_4(aq) \rightarrow$ no reaction
$Z(s) + YSO_4(aq) \rightarrow Y(s) + ZSO_4(aq)$

When the metals are listed in decreasing order of reactivity (most reactive first), what is the correct order?

A $Z > Y > X$ **C** $Y > X > Z$
B $X > Y > Z$ **D** $Y > Z > X$

Standard Level Paper 1, Nov 05, Q27

Q8 In the reaction

$2MnO_2 + 4KOH + O_2 + Cl_2 \rightarrow 2KMnO_4 + 2KCl + 2H_2O$

the oxidizing agent(s) is/are:

A $KMnO_4$ only   **C** MnO_2 and O_2
B MnO_2 only   **D** O_2 and Cl_2

Q9 Which one of the following represents an oxidation–reduction reaction?

A $I_2(s) + 2OH^-(aq) \rightarrow I^-(aq) + OI^-(aq) + H_2O(l)$
B $PO_4^{3-}(aq) + H_2O(l) \rightarrow HPO_4^{2-}(aq) + OH^-(aq)$
C $SO_3(g) + 2H_2O(l) \rightarrow HSO_4^-(aq) + H_3O^+(aq)$
D $Cu^{2+}(aq) + H_2S(aq) \rightarrow CuS(s) + 2H^+(aq)$

Q10 Which of the following reactions involves neither oxidation nor reduction?

A $Mg(s) + Cu^{2+}(aq) \rightarrow Mg^{2+}(aq) + Cu(s)$
B $2CrO_4^{2-}(aq) + 2H^+(aq) \rightarrow Cr_2O_7^{2-}(aq) + H_2O(l)$
C $C_3H_6(g) + H_2(g) \rightarrow C_3H_8(g)$
D $NH_4NO_2(s) \rightarrow N_2(g) + 2H_2O(g)$

Q11 All of the following would be expected to function as both oxidizing and reducing agents except:

A NO_2 **B** Cl^- **C** ClO^- **D** S

Q12 Magnesium is a more reactive metal than copper. Which is the strongest oxidizing agent?

A Mg   **B** Mg^{2+} **C** Cu   **D** Cu^{2+}

Standard Level Paper 1, Nov 03, Q26

Q13 Bromide ions are oxidized to bromine by all of the following except:

A $K_2Cr_2O_7/H^+$ **C** I_2
B Cl_2 **D** $KMnO_4$

Q14 When a direct current of electricity is conducted by an aqueous solution of an electrolyte, which one of the following statements is false?

A The movements of ions accounts for the current flow through the solution.
B During electrolysis, the solution remains electrically neutral.
C Electrons flow from the current source toward the solution at one electrode, and an equal number of electrons flows away from the solution at the other electrode.
D The number of positive ions moving toward one electrode is always equal to the number of negative ions moving toward the other electrode.

Q15 In acid solution, manganate(VII) ions, $MnO_4^-(aq)$, undergo reduction to manganese(II) ions, $Mn^{2+}(aq)$. What amount of $MnO_4^-(aq)$ is required to convert 5.36×10^{-3} moles of the ion $Y^{2+}(aq)$ to $YO_3^-(aq)$?

A 1.07×10^{-3} mol   **C** 5.36×10^{-3} mol
B 3.22×10^{-3} mol   **D** 8.93×10^{-3} mol

Q16 What is the oxidation number of chromium in $Cr_2O_7^{2-}$?

A +7 **B** +6 **C** –6 **D** –2

Q17 Which statement is correct for the electrolysis of molten sodium chloride?

A Sodium ions move toward the positive electrode.
B A gas is produced at the negative electrode.
C Only electrons move in the electrolyte.
D Both sodium and chloride ions move toward electrodes.

Q18 In which of the following does the metal undergo a change in oxidation state?

I $2MnO_4^{2-} + F_2 \rightarrow 2MnO_4^- + 2F^-$
II $2CrO_4^{2-} + 2H^+ \rightarrow Cr_2O_7^{2-} + H_2O$
III $[Fe(H_2O)_6]^{2+} + 6CN^- \rightarrow [Fe(CN)_6]^{3-} + 6H_2O$

A I only **C** I and II only
B II only **D** I and III only

Q19 Which one of the following does not represent a redox reaction?

A $Cu(NO_3)_2(aq) + Na_2S(aq) \rightarrow CuS(s) + 2NaNO_3(aq)$
B $2Na(s) + I_2(s) \rightarrow 2NaI(s)$
C $KH(s) + H_2O(l) \rightarrow KOH(aq) + H_2(g)$
D $H_2SO_4(aq) + 2HBr(g) \rightarrow SO_2(g) + 2H_2O(l) + Br_2(l)$

Q20 In which one of the following species does chlorine exhibit the highest oxidation number?

A Cl_2O **B** Cl_2 **C** $HClO_3$ **D** PCl_5

Paper 2 IB questions and IB style questions

Q1 a Use these equations, which refer to aqueous solutions, to answer the questions that follow:

$$Fe(s) + Cu^{2+}(aq) \rightarrow Fe^{2+}(aq) + Cu(s)$$

$$Cu(s) + 2Au^+(aq) \rightarrow Cu^{2+}(aq) + 2Au(s)$$

$$Mg(s) + Fe^{2+}(aq) \rightarrow Mg^{2+}(aq) + Fe(s)$$

(Au represents gold, which is below silver in the reactivity series.)

i List the metals above in order of **decreasing** reactivity. [1]
ii Define oxidation, in electronic terms, using **one** example from above. [2]
iii Define reduction, in terms of oxidation number, using **one** example from above. [2]
iv State and explain which is the **strongest reducing agent** in the examples above. [2]
v State and explain which is the **strongest oxidizing agent** in the examples above. [2]
vi Deduce whether a gold coin will react with aqueous magnesium nitrate. [2]

b Sketch a diagram of a cell used to electrolyse a molten salt. Label the essential components. [4]

c Describe how electrode reactions occur in an electrolytic cell and state the products at each electrode when molten copper(II) iodide is electrolysed. [4]

Q2 a Electrolysis can be used to obtain fluorine from molten potassium fluoride. Write an equation for the reaction occurring at each electrode and describe the two different ways in which electricity is conducted when the cell is in operation. [4]

b In one experiment involving the electrolysis of molten potassium fluoride, 0.1 mol of fluorine was formed. Deduce, giving a reason, the amount of potassium formed at the same time. [2]

c Sodium will displace aluminium from its chloride on heating:

$$3Na + AlCl_3 \rightarrow Al + 3NaCl$$

i Explain, by reference to electrons, why the reaction is referred to as a redox reaction. [2]
ii Deduce the oxidation numbers of sodium and aluminium in the reactants and products. [2]

Q3 A voltaic cell is set up with a silver reference electrode and a series of other metals immersed in an electrolyte. The cell voltages were recorded in the table below.

Metal	Cell voltage/V
Aluminium	2.47
Zinc	1.55
Iron	1.19
Copper	0.46
Silver	0.00

a What is the relationship between the voltage of the cell and the position of the metal in the reactivity series? [1]
b Is the metal acting as the negative or positive electrode?
Explain your answer. [2]
c Construct the half-cell equations for a voltaic cell in which the metal is zinc and the electrolyte is silver nitrate. [2]

10 Organic chemistry

- Organic chemistry is the chemistry of carbon-containing compounds (with the exception of the oxides and inorganic carbonates).
- Carbon is particularly versatile in terms of its bonding, and hence the number of compounds carbon can form is greater than for all the other elements combined.
- In order to study the vast range of organic compounds and make sense of the patterns involved, these compounds must be categorized into 'families' or homologous series.
- Hydrocarbons are organic compounds containing carbon and hydrogen only.
- The most basic homologous series of compounds is the saturated hydrocarbon series known as the alkanes. This series of compounds has the general formula C_nH_{2n+2}.
- The other homologous series consist of similar hydrocarbon chains but with the introduction of different functional groups; these give the compounds of the series their characteristic chemical properties.
- Another feature that develops from the complexity of carbon chemistry is the phenomenon of isomerism. It is possible for the same collection of atoms to arrange themselves in different structural ways to form isomers. Isomers are compounds with the same molecular formula but with different structural arrangements.
- In order to refer unambiguously to specific compounds when discussing their properties it is essential to have a clearly defined and universally agreed system of naming organic compounds. This system is the IUPAC system of nomenclature with its specific means of referring to chain length, functional groups and other aspects of organic molecules.
- There are different levels to discussing the chemical composition and arrangement of molecules. This gives rise to three types of chemical formula that are used in organic chemistry: the empirical formula, the molecular formula and the structural formula of a compound.
- There are a range of different functional groups that give the members of a homologous series the characteristic properties of that series; examples of such functional groups include the C=C double bond of the alkenes, the hydroxyl (–OH) group of the alcohols and the acid (–COOH) group of the carboxylic acids.
- The alkanes are an important but relatively unreactive homologous series. They are a 'family' of saturated hydrocarbons. They are most important as major fuels.
- The alkanes can undergo substitution reactions with chlorine when exposed to sunlight or ultraviolet radiation. The reaction is a chain reaction based on a mechanism involving free radicals.
- The alkenes are an unsaturated series of hydrocarbons whose molecules contain a C=C double bond in the hydrocarbon chain.
- The alkenes are unsaturated hydrocarbons and are significantly more reactive than the alkanes. They undergo a range of addition reactions with molecules such as the halogens. In these reactions atoms are added across the double bond and the unsaturation is removed.
- The alkenes can form a very useful set of addition polymers.
- The alcohols are a series of compounds containing the hydroxyl (–OH) functional group. They can be oxidized to aldehydes/carboxylic acids or ketones depending on the type of alcohol and the conditions used.
- Halogenoalkanes are a series of compounds containing halogen atom(s) as the functional group. They undergo nucleophilic substitution reactions with aqueous sodium hydroxide. There are two types of substitution mechanism (S_N1 and S_N2).
- The reactions of organic molecules are interlinked and reaction pathways involving more than one stage can be used to synthesize a compound from a known starting point.

10.1 Introduction

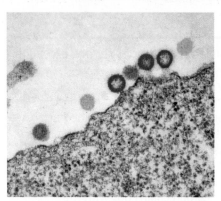

Figure 10.1 HIV viruses budding from a cell

Viruses straddle the junction between the inanimate world and life (Figure 10.1). They are not themselves living, but can reproduce themselves amazingly quickly when they infect a host. They manipulate the genetic 'machinery' of the host cells to use it for their own ends.

Viruses can be crystallized and their structures analysed. These structures are made up of carbon-containing molecules such as DNA or RNA and proteins. Molecules such as these are the basis of life and illustrate the versatility of carbon to form a range of complex molecules. All living things on Earth, from micro-organisms such as bacteria to the largest plants and animals, reproduce and grow using systems based on nucleic acids and proteins. These are macromolecules – molecules on a very large scale. Proteins are made by assembling amino acids into long chains. These chains then fold and organize themselves into complex structures. For example, a molecule of hemoglobin contains four protein chains. Each of these chains is made up of more than 100 amino acids. The molecules of life are based on the distinctive properties of one element – carbon. Carbon is unique in the variety of molecules it can form. The chemistry of these molecules forms a separate branch of chemistry known as organic chemistry.

Modern organic chemistry deals with both naturally occurring and synthetic compounds, including plastics, pharmaceuticals, petrochemicals, fuels and foods. As it provides a link between the properties of atoms and the functioning of living organisms, it is through organic chemistry that we come to biochemistry and hence to the chemical foundations of life itself.

History of Chemistry

'Vitalism' and the birth of organic chemistry

The nature of life and the features essential to its chemistry have long been the subject of speculation. Historically, organic molecules were believed to be a distinctive type of chemical substance unique to living things (part of a set of ideas known as 'vitalism'). It was thought that organic molecules could not be made outside a living organism. However, in 1828, the German chemist Friedrich Wöhler synthesized urea from inorganic substances without the presence of any biological tissue. He wrote to his mentor, Berzelius, saying: 'I must tell you that I can make urea without the use of kidneys, either man or dog. Ammonium cyanate is urea.' The synthesis of urea had not been the intention of his experiment, but in attempting to prepare ammonium cyanate from silver cyanide and ammonium chloride he had completed a revolutionary experiment. Upon analysis the white powder produced proved to have the composition and properties of urea, a compound that had previously been isolated from urine.

Wöhler pursued these experiments further and discovered that urea and ammonium cyanate had the same chemical formula, but very different chemical properties. This was an early discovery of isomerism, since urea has the formula $CO(NH_2)_2$ and ammonium cyanate has the formula NH_4CNO. Wöhler's results conclusively destroyed the belief that there was a distinction between the chemistry of life and general inorganic chemistry. It opened up the door to a whole branch of organic chemistry centred on the properties and reactions of carbon-containing compounds.

Friedrich Wöhler

Friedrich Wöhler (1800–1882) was a pioneer in organic chemistry as a result of his accidental synthesis of urea in 1828 (see above). This synthesis undermined the vitalism theory, by showing that organic compounds could be made from inorganic materials.

He was born in 1800 near Frankfurt in Germany. In 1823 Wöhler completed his medical studies in Heidelberg and moved to Stockholm to work under Jakob Berzelius. He returned to Germany in 1826 to teach chemistry, firstly at the Polytechnic School in Berlin and then at the Higher Polytechnic in Kassel. It was in Berlin that he carried out the experiments that most made his name. In 1836 he became Professor of Chemistry in the medical faculty at the University of Göttingen, where he remained until his death in 1882. He is noted for the isolation of aluminium, and important studies of the elements boron, silicon, beryllium and titanium.

Figure 10.2
Friedrich Wöhler

The nature of a homologous series

10.1.1 Describe the features of a homologous series.

Around *six million* compounds of carbon are already known! This versatility is made possible by certain unique properties of carbon. Carbon is a non-metal in group 4 of the periodic table and forms predominantly covalent compounds. There are three special features of covalent bonding involving carbon.

- Carbon atoms can join to each other to form long chains. Atoms of other elements can then attach to the chain.
- The carbon atoms in a chain can be linked by single, double or triple covalent bonds.
- Carbon atoms can also arrange themselves in rings.

Atoms of other elements can copy some of this versatility to a limited extent (e.g. silicon atoms can form short chains, while sulfur atoms can arrange themselves in rings). But only carbon can achieve all these different bonding arrangements, and do so to an amazing extent. The ability of carbon to form chains and rings is known as **catenation**.

■ Extension: Other distinctive features of carbon

There are other features of carbon which help to reinforce its unique position as the most versatile of the elements as regards compound formation.

- Carbon not only forms multiple bonds with itself, but it can also form double and triple bonds with other elements such as oxygen (the carbonyl group, $>C=O$, is an important feature of aldehydes, ketones and carboxylic acids), and with nitrogen (in nitriles, which contain the $-CN$ group).
- The C–C bond is particularly strong compared with the strength of similar bonds between other group 4 elements (Table 10.1). This contributes to the thermal stability of organic compounds.

Bond	Average bond enthalpy /kJ mol⁻¹
C–C	348
Si–Si	226
Ge–Ge	188
Sn–Sn	151

Table 10.1 The bond strengths of X–X bonds for the group 4 elements

Allied to the thermal stability of the C–C bond is the fact that the C–H bond is also significantly more stable than other comparable bonds, such as the Si–H bond (bond enthalpies: C–H = 412 kJ mol^{-1}; Si–H = 318 kJ mol^{-1}).

When a carbon atom is bonded to four other atoms it is kinetically stable as the outer shell of the carbon atom (electron shell $n = 2$) has a full octet of electrons which cannot be expanded. Thus tetrachloromethane (CCl_4) cannot be hydrolysed by water whereas silicon tetrachloride ($SiCl_4$) can be. Water molecules are thought to attack the $SiCl_4$ molecules through lone pairs on their oxygen atoms expanding the octet of electrons around the central silicon atom in each molecule.

These properties suggest a versatility and stability of carbon-containing compounds that cannot be matched by any other element. One highly speculative consequence of this is that, in considering the possibility of life on distant planets, it is difficult to imagine complex life forms based on any element other than carbon.

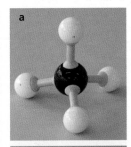

With so many possible molecules to describe and categorize, it is important to find ways of relating the different types of structure to the chemical properties of the compounds. The formation of chains or rings of carbon atoms provides the basis of a means of classifying organic compounds. When looking at the basic structure of any organic compound, therefore, it is important to see whether the structure is based on a chain or a ring. This forms the 'backbone' of the molecule.

The most obvious basic structure is that of a chain of carbon atoms with only hydrogen attached. It is easy then to think of a series of compounds that result from simply extending the chain progressively by one carbon atom at a time. Figure 10.3 shows models of the simplest compound of this series – one carbon atom with four hydrogen atoms covalently bonded to it. This is a model of the simplest hydrocarbon: methane, CH_4.

Thus a series of compounds exists in which the molecules get progressively extended by a carbon atom, or, more precisely, by a $-CH_2-$ group (see Table 10.2). Such a series of compounds is known as a **homologous series**; and this particular one is known as the alkanes.

Figure 10.3 Models of the structure of methane, CH_4. These models show the tetrahedral structure of the molecule. **a** A ball-and-stick model, showing the four single C–H bonds; **b** a space-filling model

TOK Link

One key to understanding the microscopic world of atoms and molecules is to build models that help us 'visualize' the unseeable. We can do this with words and with mathematical models but in chemistry there has been the very practical development of various types of model structures such as those shown in Figure 10.3. The most famous and significant model-building exercise in history must surely have been Watson and Crick's elucidation of the structure of DNA. Model-building kits have their virtues and limitations: the bond lengths in ball-and-stick models give a false impression of the space between the atoms, for instance. Recent developments in computer modelling make it easy to switch between various means of depicting structures and display the distances and angles in the structures with great accuracy. Figure 10.4 shows two computer-generated models of methane. The second is a space-filling model aimed at showing the inter-penetration of atoms as they bond to make a simple molecule.

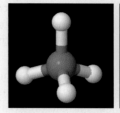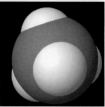

Figure 10.4 Computer-generated models of methane. These are generated using a computer program known as Rasmol. A similar program, Chime, allows the molecule to be rotated on-screen

Such models are useful in giving us an understanding of the crucial ideas of shape and the 3-dimensional arrangement of molecules. Figure 10.5 shows different types of model of ethanol. All of these have their uses, and you will encounter them in your reading.

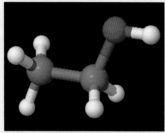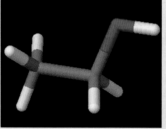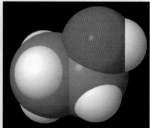

Figure 10.5 Three different styles of model for the structure of ethanol. Red is the code for oxygen

The merits of the computer modelling of molecules come into their own when depicting more complex molecules. Pharmaceutical companies invest a great deal in sophisticated programs for molecular design. Figure 10.6 shows the ring structure of glucose as an example of a slightly more complex molecule.

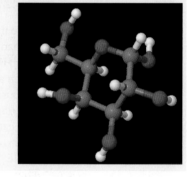

Figure 10.6 The ring form of D-glucose. Note the six-membered ring involving five carbon atoms and one oxygen atom

Language of Chemistry

Both in solid models and computer graphics, the following colour code is used to indicate which element is being represented:

black or grey = carbon   white = hydrogen

red = oxygen   yellow = sulfur

blue = nitrogen   green = chlorine ∎

Alkane	Molecular formula C_nH_{2n+2}	Number of carbon atoms	Melting point/K	Boiling point/K		Physical state at room temperature and pressure
Methane*	CH_4	1	91	109		gas
Ethane	C_2H_6	2	90	186	b.p. increasing	gas
Propane	C_3H_8	3	83	231		gas
Butane	C_4H_{10}	4	135	273		gas
Pentane	C_5H_{12}	5	144	309		liquid
Hexane	C_6H_{14}	6	178	342		liquid

*The naming of these compounds will be discussed shortly (page 268).

Table 10.2 Some details of the early members of a homologous series – the alkanes

This homologous series of compounds illustrates certain key features of all such series. These features are as follows.

- The names of the compounds all contain a consistent feature which denotes the series; in this case the names all have the same ending, *-ane*.
- The formulas show the increase in chain length of a $-CH_2-$ group between one member and the next.
- The molecules all have the same general formula, C_nH_{2n+2}, where *n* is the number of carbon atoms in the chain.
- There is a progressive and gradual change in basic physical properties as the chain length increases – illustrated here by the increasing boiling point of the compounds in the series. Figure 10.7 shows how the structures of these first six alkanes develop with the lengthening of the chain.

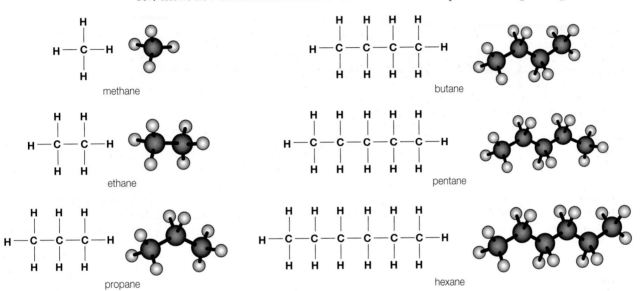

Figure 10.7 The structures of the first six alkanes

Worked examples

The homologous series of alkanes has the general formula C_nH_{2n+2}. The first four alkanes are gases at room temperature but the fifth member is a liquid, as are the next 11 members of the series.

a What is the molecular formula of octane given that its molecules each contain eight carbon atoms?

b The members of the series continue to be liquids up to carbon-16. What are the molecular formulas of this liquid alkane and the first of the alkanes that is a waxy solid at room temperature?

c Table 10.2 gives details of the melting and boiling points of the first six alkanes. Suggest a possible melting and boiling point for **i** heptane, which has seven carbon atoms and **ii** octane, which has eight carbon atoms.

a The number of carbon atoms (*n*) = 8; therefore $2n + 2 = 18$
The molecular formula of octane is C_8H_{18}.

b The final liquid alkane at room temperature has the molecular formula $C_{16}H_{34}$. While the first waxy solid in the series is $C_{17}H_{36}$.

c The prediction of melting points is the more difficult of the two as the trend is not as smooth.
  i Heptane would appear to have a melting point around 188 ± 5 K (actual value 182 K), and a boiling point of 372 ± 2 K (actual value 371 K).
  ii Octane would appear to have a melting point of 213 ± 5 K (actual value 222 K), and a boiling point of 147 ± 2 K (actual value 147 K).

Table 10.2 shows the trend in boiling points for the first six alkanes. They show a gradual, though not linear increase in value. Do other physical properties of the alkanes show a similar gradual progression?

The densities of the early members of the alkane series are given in Table 10.3. Plot a graph of the densities against the number of carbon atoms in each compound, and see if there is a similar regular trend in these values.

Alkane	Molecular formula	Density at 273 K/(g cm⁻³)
Methane	CH_4	0.466*
Ethane	C_2H_6	0.572*
Propane	C_3H_8	0.585*
Butane	C_4H_{10}	0.601*
Pentane	C_5H_{12}	0.626
Hexane	C_6H_{14}	0.659
Heptane	C_7H_{16}	0.684
Octane	C_8H_{18}	0.703
Nonane	C_9H_{20}	0.718
Decane	$C_{10}H_{22}$	0.730

*These four alkanes are gases at 273 K; the densities quoted here are values at a temperature just below the boiling point of each compound.

Table 10.3 The densities of members of the alkane series

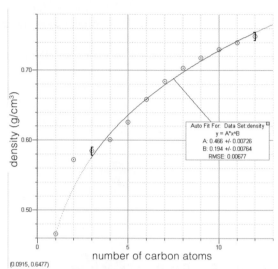

Figure 10.8 Graph of the densities of the early members of the alkane series. This graph is drawn, and the line fitted, by feeding the data into the computer program *Graphical Analysis* (Vernier)

From the table we can see that the values do increase with chain length. Plotting the graph shows that, for the alkanes that are liquid at 273 K, there is a regular smooth increase in the density values. Indeed, the values for propane and butane are not too far from fitting the pattern (Figure 10.8).

The alkanes are the simplest homologous series because the molecules contain only two elements, carbon and hydrogen, and all the carbon–carbon bonds are single bonds. However, it is easy to imagine a different series of molecules where one of the hydrogen atoms in the chain is replaced by a hydroxyl (−OH) group. This series of compounds is known as the **alcohols**. Table 10.4 gives some details of the early compounds of the alcohol homologous series.

Alcohol	Molecular formula $C_nH_{2n+2}O$ ($C_nH_{2n+1}OH$)	Boiling point/K (at atmospheric pressure)	
Methanol	CH_4O (CH_3OH)	338	
Ethanol	C_2H_6O (C_2H_5OH)	351	b.p. increasing
Propan-1-ol	C_3H_8O (C_3H_7OH)	370	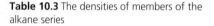
Butan-1-ol	$C_4H_{10}O$ (C_4H_9OH)	390	
Pentan-1-ol	$C_5H_{12}O$ ($C_5H_{11}OH$)	410	
Hexan-1-ol	$C_6H_{14}O$ ($C_6H_{13}OH$)	431	

Table 10.4 Some data on the early members of the alcohol homologous series

The properties of this series of compounds are dictated by the presence of the −OH group which gives the alcohols their distinctive properties. For this reason the hydroxyl (−OH) group is known as the **functional group** of the alcohols.

This adds the final distinctive feature to our definition of a homologous series. In summary, a homologous series is a group of organic compounds that:

- contain the same functional group
- have the same general formula, with successive members of the series having an additional −CH₂− group
- have similar chemical properties
- show a steady gradation in certain basic physical properties.

Table 10.5 shows some of the different functional groups present in several homologous series. These groups are attached to, or part of, the hydrocarbon 'backbone' or 'skeleton' of the molecules in the series. They give the molecules the distinctive properties of the particular chemical 'family' or homologous series.

Homologous series	Condensed structural formula	Structure of the functional group
Alkanes	$-CH_2-CH_2-$	*
Alkenes	$-CH=CH-$	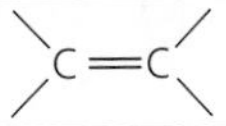 C=C
Halogenoalkanes	$-X$ (where X = F, Cl, Br, I)	$-X$ (where X = F, Cl, Br, I)
Alcohols	$-OH$	$-O-H$
Aldehydes	$-CHO$	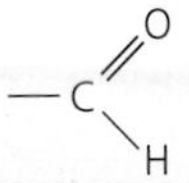
Ketones**	$-CO-$	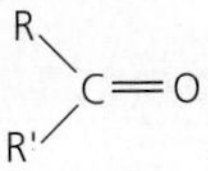
Carboxylic acids	$-COOH$ or $-CO_2H$	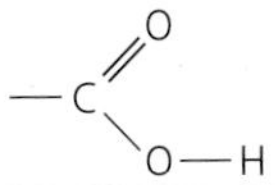

Table 10.5 Some functional groups

*The alkane structure is the basic backbone into which the functional groups are introduced.
**R and R' represent hydrocarbon chains (alkyl groups) attached to the group. These chains can be identical or different (as represented here).

■ Extension: The alkynes

Ethyne, C_2H_2, is the first member of a homologous series of hydrocarbons known as the alkynes. They have the general formula, C_nH_{2n-2}. Simple alkynes contain one carbon–carbon triple bond. The two carbon atoms forming the triple bond are sp hybridized (Chapter 14). Ethyne is unsaturated and undergoes addition reactions. For example, ethyne and bromine react together to form 1,1,2,2-tetrabromoethane:

$$HC\equiv CH + 2Br_2 \rightarrow CHBr_2CHBr_2$$

10.1.2 Predict and **explain** the trends in boiling points of members of a homologous series.

You will see from Tables 10.2 and 10.4 that the boiling points of a series of compounds in a homologous series increase steadily with the lengthening of the carbon chain. The alkanes show a smooth variation in boiling point, becoming less volatile with increasing molecular mass.

The alkanes are a useful starting point in considering the effect of increasing molecular size on such essential physical properties because, by their non-polar nature, the interactive forces between the molecules are solely van der Waals' forces (Chapter 4). These forces are based on interactions between temporary dipoles created by momentary shifts in electron distribution. The strength of the forces is related to the number of electrons involved in the structure and the surface area of the molecules over which the interactions can be spread. Increasing the chain length of the molecules increases both these features and so the strength of the van der Waals' forces increases with increasing molecular size. Physical properties dependent on these interactions, such as melting point, boiling point and enthalpy of vaporization (ΔH_{vap}), will also show an increase with chain length (Figure 10.9).

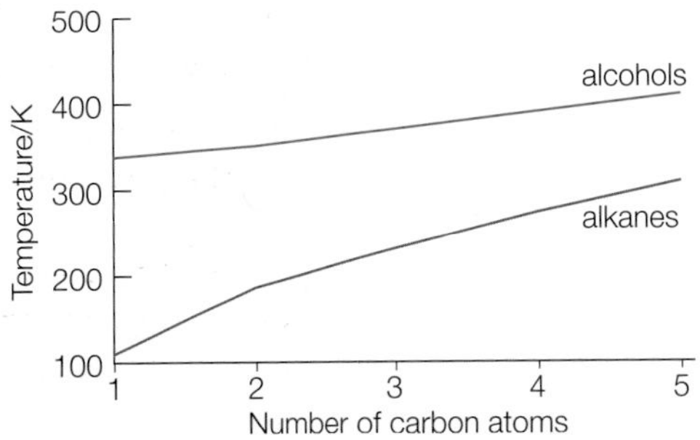

Figure 10.9 Graph of the increasing boiling points of the early alkanes and alcohols with increasing chain length

If the plot is extended to larger molecules, the effect of the addition of each $-CH_2-$ becomes less significant as the chain gets longer; the change is not linear, but it is regular and smooth. The gradient of the curves for melting point and boiling point for the alkanes decreases with increasing chain length.

Figure 10.10 An oil refinery at night

The transition in boiling point values for the early alcohols shows an almost linear increase with increasing molecular size (Figure 10.9). The values are all significantly higher than for the corresponding alkane, suggesting that there are stronger intermolecular forces acting in this case. This idea will be looked at in more detail later in the chapter (see page 275).

The regular increase in boiling point of the alkanes with increasing chain length is of immense commercial importance. It is the basis of the initial refinery process of fractional distillation whereby crude oil is separated into useful 'fractions' (Figures 10.10 and 10.11). Crude oil (or petroleum) is the major commercial source of hydrocarbons for a variety of uses as fuels and chemical feedstock. Figure 10.11 shows the components present in the major fractions obtained by distillation at an oil refinery.

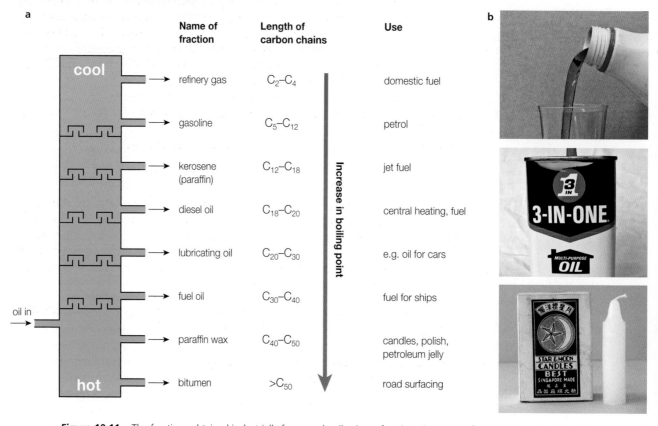

Name of fraction	Length of carbon chains	Use
refinery gas	C_2–C_4	domestic fuel
gasoline	C_5–C_{12}	petrol
kerosene (paraffin)	C_{12}–C_{18}	jet fuel
diesel oil	C_{18}–C_{20}	central heating, fuel
lubricating oil	C_{20}–C_{30}	e.g. oil for cars
fuel oil	C_{30}–C_{40}	fuel for ships
paraffin wax	C_{40}–C_{50}	candles, polish, petroleum jelly
bitumen	$>C_{50}$	road surfacing

Increase in boiling point

Figure 10.11 a The fractions obtained industrially from crude oil using a fractionating tower; **b** some examples of useful components

Formulas

10.1.3 **Distinguish** between empirical, molecular and structural formulas.

In the discussion so far we have referred to two different ways of writing the formulas of the organic molecules that have been used as examples. Tables 10.2 and 10.4 show the molecular formulas of the alkanes and alcohols, while Figure 10.7 gives the full structural formulas of the first six straight-chain alkanes. These different types of formula provide different levels of information and are each useful in their distinctive ways. Three different types of formulas are used for organic compounds: empirical, molecular and structural formulas.

The **empirical** formula of a compound is the simplest whole number ratio of the atoms it contains (Chapter 1). For some alkanes, methane and propane for instance, the empirical formula is the same as the actual molecular formula. However, for others this is not true. The empirical formula of ethane, whose molecular formula is C_2H_6, is CH_3. In practical terms the empirical formula is the formula that can be derived from the percentage composition data obtained from combustion analysis. In order to use this to establish the actual formula, data on the relative molecular mass (M_r) of the compound is required.

■ Extension: The practical determination of formulas

Historically, our knowledge of the chemical composition of compounds and their formulas came from practical work on elemental analysis. Subsequently, more sophisticated methods have been added to the armoury of techniques by which we can determine not just the composition of compounds, but also their structures.

- The empirical formula of an organic compound can be obtained from its percentage composition by mass. For compounds that contain only carbon, hydrogen and oxygen this can be found in a quantitative combustion experiment in an excess of oxygen. The number of carbon atoms in an organic molecule can be calculated if the volume of carbon dioxide produced by complete combustion of a known volume of pure gaseous compound is measured (Chapter 1).
- The presence of functional groups can be established using chemical tests, for example bromine water for alkenes, and infrared spectroscopy (Chapter 21).
- The molecular formula can be obtained from empirical formula if the relative molecular mass is known (Chapter 1). The relative molecular mass can be accurately measured by mass spectrometry (Chapter 2).
- The structural formulas of organic compounds can be determined by nuclear magnetic resonance (NMR) and mass spectrometry (Chapter 21).

The **molecular formula** of a compound is the actual number of atoms of each type present in a molecule. For example, the molecular formula of ethane is C_2H_6, of ethanol is C_2H_6O and of ethanoic acid is $C_2H_4O_2$. For many organic compounds the molecular formula and empirical formula are not the same. However, the molecular formula must be a whole number multiple of the empirical formula and therefore can be deduced if you know both the empirical formula and the relative molecular mass (M_r) of the compound.

Worked example

A halogenoalkane has a relative molecular mass of 99. Calculations based on elemental analysis of the compound show that it has an empirical formula of CH_2Cl. What is the molecular formula of this halogenoalkane?

The relative molecular mass of the empirical formula, $CH_2Cl = 12 + (2 \times 1) + 35.5 = 49.5$

The actual relative molecular mass = 99

Therefore the actual molecular formula = $2 \times (CH_2Cl) = C_2H_4Cl_2$

Molecular formulas such as those for ethanol (C_2H_6O) are of limited value in that they give no indication of the functional group(s) involved in a compound, and hence no clue as to the properties of the compound. Often the formula of ethanol is written as C_2H_5OH, which has the advantage of showing the presence of the alcohol group. Similarly the formula for ethanoic acid is written as CH_3COOH to indicate the presence of a carboxylic acid group in the structure.

These representations of the molecules begin to show exactly how the atoms are bonded to each other – the structure of the molecule. The **full structural formula** (also known as the graphic formula or displayed formula) shows *every* bond and atom (see Figure 10.7 for the full structural formulas of the first six alkanes). Note that all the carbon atoms and, importantly, all the hydrogens must be shown. It is easy to assume that the hydrogens will be understood but structures drawn without them will not be accepted in IB examinations. Sometimes it is sufficient to use a **condensed structural formula** that omits bonds where they can be assumed and groups atoms together. So, for example, propane can be written as $CH_3CH_2CH_3$, and butane can be written either as $CH_3CH_2CH_2CH_3$ or as $CH_3(CH_2)_2CH_3$. Figure 10.12 shows the full structural formulas, with models, for ethanol and ethanoic acid, both of which are important organic compounds.

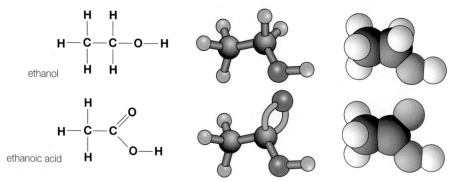

Figure 10.12 The structural formulas and models of ethanol and ethanoic acid

It is important that any structural formula is unambiguous and there is only one possible structure that could be described by the formula as it is written.

TOK Link

Chemistry is immersed in symbolism of all kinds. Amongst the most important of these are the symbols we use to represent and convey our ideas of the 'unseen' at a molecular and atomic level. Organic molecules are so complex that different ways of depicting the formulas of these compounds have been developed, depending on how much information is required.

Systems of symbols have their limitations, however. When representing organic molecules on the two-dimensional page we usually use structural formulas involving 90° and 180° angles when showing the bonds because this is the clearest way of showing them in this context. However, this does not show the true geometry of the molecule. When carbon forms four single bonds, as in methane or ethane, the arrangement is tetrahedral with the bonds at 109.5° to each other (Figure 10.13a and b). When it forms a double bond, as in ethene, the arrangement is trigonal planar, with bonds at 120° (Figure 10.13b). (Ethene is the first member of the homologous series called the alkenes, in which the molecules contain a C=C double bond – see Table 10.5.)

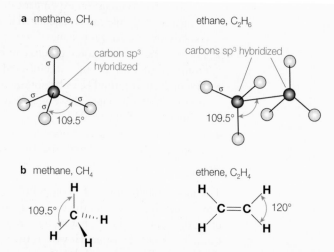

Figure 10.13 a The hybridization of each carbon atom in any alkane is sp³ hybridization. **b** Diagrams showing the bond angles in methane and ethene

These bond angles are consistent with the hybridization of the atomic orbitals of the carbon atom involved in the bonds: sp³ hybridization in methane, sp² hybridization in ethene (see Chapter 14). For some molecules it is particularly useful to show the relative three-dimensional positions of atoms or groups around a selected carbon atom – the stereochemistry of the molecule. To show this, the convention is that a bond sticking forwards from the page is shown as a solid, enlarging wedge, whereas a bond directed behind the page is shown as a broken line (Figure 10.13b). Figure 10.14 shows models illustrating the bonding in some simple hydrocarbon molecules.

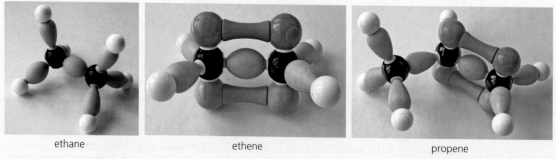

Figure 10.14 These models illustrate the shape of the molecule around each carbon atom and the sigma (σ) and pi (π) bonds in these molecules. The model of propene shows clearly the different orientations of the carbon atoms within the same molecule

Language of Chemistry

When drawing a structural formula, it is possible to miss off the hydrogen atoms. A structural formula with the hydrogen atoms omitted is known as a skeletal structure or skeletal formula (Figure 10.15). In certain cases of complex molecules this can be quite useful. The skeletal formula of the type in Figure 10.15a is very rarely used, and certainly should not be used in an examination answer. Skeletal formulas of the type in Figure 10.15b can be very useful for quickly writing out alternative structures and for depicting molecules of substantial complexity (Figure 10.16). We will use this type of skeletal structure later when working out the possible structural isomers for a given molecular formula.

Figure 10.15 The skeletal formula of pentane. In **a** the hydrogen atoms are simply missed off. In **b** there are no symbols for atoms at all; the carbon atoms (with the appropriate number of hydrogen atoms) are located at the joints

Figure 10.16 A representation of the structure of a prostaglandin in skeletal form. This structure involves some double bonds and also uses the spatial notation to give some idea of the three-dimensional arrangement

There are other ways in which an abbreviation to a formula can be made.

- Sometimes we do not need to show the exact details of the hydrocarbon, or alkyl, part of the molecule, so we can abbreviate this to R.
- For molecules which contain a benzene ring – aromatic compounds – we use ⬡ to show the ring. ■

Language of Chemistry

In discussing the alkane series of hydrocarbons we have described the molecules as chains which are extended by a $-CH_2-$ unit as we progress up the series. These molecules are often referred to as **straight-chain hydrocarbons** as there are no branches from the main chain. However, because each carbon atom is involved in a tetrahedral arrangement of bonds, the actual progression is zigzagged (Figure 10.17).

Figure 10.17 A model of the alkane, $C_{14}H_{30}$ showing the zigzag chain of the carbon atoms

There are hydrocarbon molecules which are genuinely **branched-chain** molecules, and these are an important group of compounds. One of the most significant is 2,2,4-trimethylpentane, whose ignition properties are the basis of the octane rating of gasoline (petrol) for cars. The system for numbering the positions where the branches attach to the main hydrocarbon chain is outlined on page 268.

2,2,4-trimethylpentane

In discussing the properties of hydrocarbons in general there are two other terms that it is important to understand clearly. These are the terms **saturated** and **unsaturated**. The alkanes are saturated hydrocarbons as all the C to C bonds in the chain are single bonds. The molecules have as much hydrogen as possible attached to the carbon chain. If there is a C=C bond in the chain then additional hydrogen atoms could be attached to the chain, and the molecule is said to be unsaturated. Thus the alkenes are a series of unsaturated hydrocarbons. ■

■ **Extension:** Cyclic alkanes

Molecules such as butane, $CH_3CH_2CH_2CH_3$, have their carbon atoms connected in a chain. Carbon atoms can also be joined together in rings, in which case a **cyclic** molecule is formed. The structures of these cyclic molecules are often represented by the appropriate polygon – the corners of the polygon represent a carbon atom together with the hydrogen atoms joined to it (this is similar to the skeletal structures for chain alkanes). The cycloalkanes have a general formula of C_nH_{2n}. Some examples of these cycloalkanes are shown in Figure 10.18.

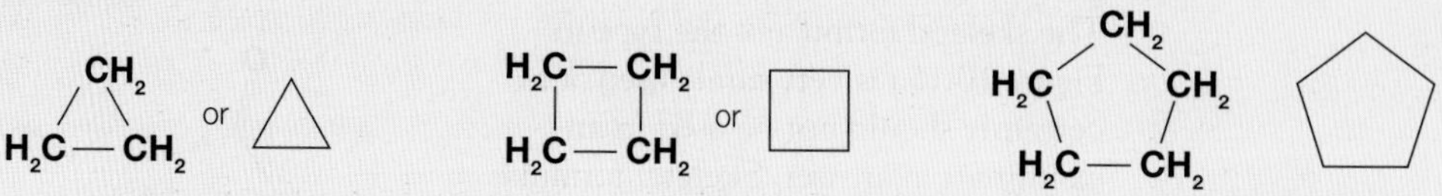

cyclopropane   cyclobutane   cyclopentane

Figure 10.18 The structure of the first three members of the cycloalkane homologous series

Despite their general formula, the cycloalkanes are saturated molecules and should not be confused with the alkenes. There is a series of unsaturated cycloalkenes that includes such compounds as cyclopentene and cyclohexene.

Structural isomerism

10.1.4 Describe structural isomers as compounds with the same molecular formula but with different arrangement of atoms.

The complexity and diversity of organic chemistry is increased by the fact that molecular formulas involving a reasonable number of carbon atoms can represent several different structures, i.e. different compounds. For the alkane series (general formula C_nH_{2n+2}) the first three formulas, CH_4, C_2H_6 and C_3H_8, are unambiguous; there is no other way in which the atoms can be arranged other than as shown in Figure 10.7.

However, looking at the condensed structural formula of butane, $CH_3CH_2CH_2CH_3$, there is an alternative way of arranging the atoms which involves a branched chain. This alternative form is $CH_3CH(CH_3)CH_3$, where there is a $-CH_3$ group branching off the middle carbon atom. Figure 10.19 shows the full structural formula of these two forms.

These two possible ways in which the carbon and hydrogen atoms can be bonded are both valid as each carbon atom has four bonds and each hydrogen atom has one bond. The two structures represent different compounds: one is butane and the other 2-methylpropane (formerly known as *iso*-butane). Their chemical properties are quite similar, but their physical properties show differences. For instance, the two compounds have different melting and boiling

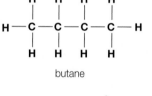

butane

methyl group

2-methylpropane

Figure 10.19 Two isomers of formula C_4H_{10}

points. When two or more compounds have the same molecular formula but a different structural formula they are known as **structural isomers**. In the alcohol series the first possible examples of isomerism occur with propanol and butanol. Here the isomerism depends on the fact that the alcohol group (−OH) can be attached on the terminal carbon atom, or to one in the middle of the chain (Figure 10.20).

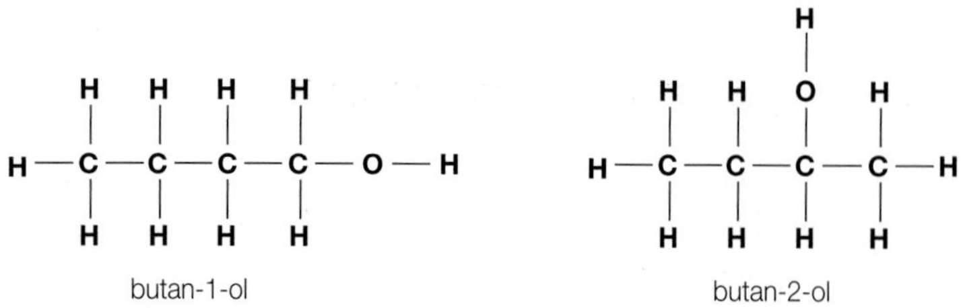

butan-1-ol   butan-2-ol

Figure 10.20 The two structural isomers of butanol

Note that it is important when working out possible isomers to remember the limitations of the two-dimensional representation of structures on paper. Thus the structures represented in Figures 10.21a and 10.21b are not isomers at all. In Figure 10.21a the chain appears to 'turn a corner' on paper, but remember that, in reality, the structure around each carbon atom is tetrahedral and that there is free rotation around each bond. In Figure 10.21b, one structure is just the other turned over on the paper. It is crucial to remember that *isomers are compounds with the same molecular formula but with different arrangements of atoms in the molecules.*

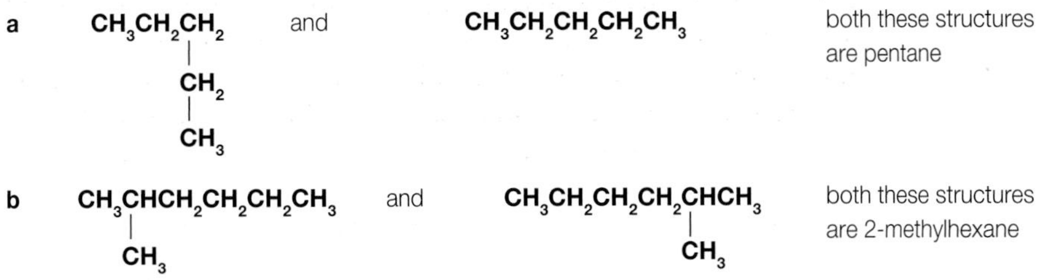

a   CH$_3$CH$_2$CH$_2$ and   CH$_3$CH$_2$CH$_2$CH$_2$CH$_3$ both these structures
             |
          CH$_2$ are pentane
             |
          CH$_3$

b   CH$_3$CHCH$_2$CH$_2$CH$_2$CH$_3$ and   CH$_3$CH$_2$CH$_2$CH$_2$CHCH$_3$ both these structures
         | | are 2-methylhexane
         CH$_3$ CH$_3$

Figure 10.21 Structural formulas that could be mistaken as isomers

Deducing structural formulas of alkanes

10.1.5 Deduce structural formulas for the isomers of the non-cyclic alkanes up to C$_6$.

After butane, the longer the carbon chain the more structural isomers are possible for a given molecular formula. For example, there are 75 isomers with the formula C$_{10}$H$_{22}$, and over 350 000 with the formula C$_{20}$H$_{42}$! When trying to work out the different straight-chain and branched isomers that fit a particular molecular formula it is important to remember the points mentioned above regarding the free rotation about a single C–C bond, and not to be fooled by the limitations of the two-dimensional representation of the molecular structures. Often just the simplicity of the skeletal formula can help clarify possibilities. Figure 10.22 shows the skeletal forms of two possible branched isomers of C$_5$H$_{12}$.

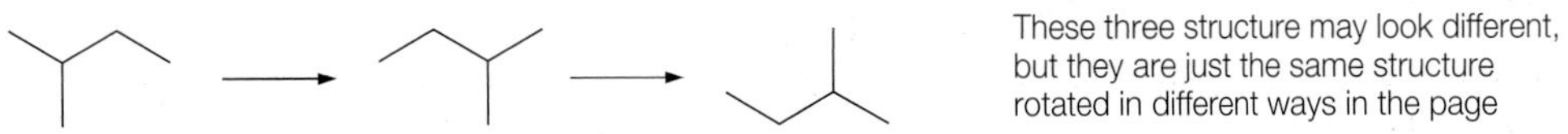

Figure 10.22 Using skeletal formulas to establish whether structures are isomers or not – these are not!

These three structure may look different, but they are just the same structure rotated in different ways in the page

Worked example

Making use of the clues in Figure 10.22, work out all the structural isomers, straight-chain and branched that have the following molecular formulas:

a   C$_5$H$_{12}$ b   C$_6$H$_{14}$

a   There are three isomers of C$_5$H$_{12}$. The skeletal formulas can be helpful in visualizing them.

1 straight-chain isomer   2 single-branched isomer   3 double-branched isomer

The condensed structural formulas are:

CH$_3$CH$_2$CH$_2$CH$_2$CH$_3$ CH$_3$CH(CH$_3$)CH$_2$CH$_3$ CH$_3$C(CH$_3$)$_2$CH$_3$

b   There are five isomers of C$_6$H$_{14}$.

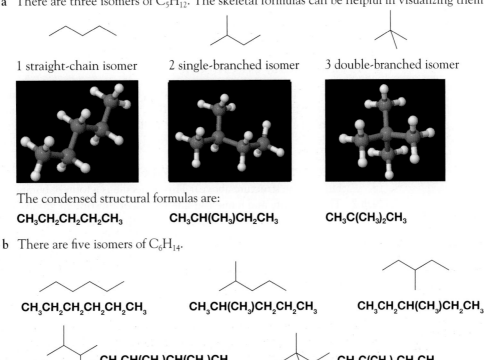

CH$_3$CH$_2$CH$_2$CH$_2$CH$_2$CH$_3$ CH$_3$CH(CH$_3$)CH$_2$CH$_2$CH$_3$ CH$_3$CH$_2$CH(CH$_3$)CH$_2$CH$_3$

CH$_3$CH(CH$_3$)CH(CH$_3$)CH$_3$ CH$_3$C(CH$_3$)$_2$CH$_2$CH$_3$

Naming alkanes

10.1.6 Apply IUPAC rules for naming the isomers of the non-cyclic alkanes up to C_6.

You have now seen how complicated the naming of even just one homologous series of organic compounds can become. In this context a systematic method of naming compounds becomes very important. The system must be unambiguous and universal. Such a system has been devised by IUPAC (International Union of Pure and Applied Chemistry).

Methane, ethane, propane and butane appear to have little logic in their names except that they all end in *-ane*, which signifies that they are all *alkanes* (see Figure 10.7). By looking at the next member of the series, you can begin to learn how the IUPAC system of naming works. For the next members of the alkane series, from $n = 5$ onwards, the prefix in the name follows the Greek prefixes for these numbers. Simply remember the names of the geometrical figures – *penta*gon, *hexa*gon and *hepta*gon, etc. – to help you remember these. Table 10.7 shows the names of the straight-chain alkanes.

Prefix of name	Number of carbon atoms in chain	Name of alkane	Condensed structure
Meth-	1	Methane	CH_4
Eth-	2	Ethane	CH_3CH_3
Prop-	3	Propane	$CH_3CH_2CH_3$
But-	4	Butane	$CH_3CH_2CH_2CH_3$
Pent-	5	Pentane	$CH_3CH_2CH_2CH_2CH_3$
Hex-	6	Hexane	$CH_3CH_2CH_2CH_2CH_2CH_3$

Table 10.7 Naming alkanes

For the branched isomers of the alkanes the system needs to give names that apply to the side-chain groups of atoms. Table 10.8 gives the names used for the hydrocarbon side-chains of differing lengths. You can see that the same prefix is used to designate the number of carbon atoms, followed by the ending *-yl*. These groups are known generally as alkyl groups.

Name of side-chain (R group)	Condensed structure
Methyl	$-CH_3$
Ethyl	$-CH_2CH_3$
Propyl	$-CH_2CH_2CH_3$
Butyl	$-CH_2CH_2CH_2CH_3$

Table 10.8 Names of some alkyl groups

Knowing the component parts of the names for alkanes, you can follow some simple steps to generate the name for both straight-chain and branched-chain molecules. The example here shows how to apply the steps to one of the isomers of C_6H_{14} identified earlier, namely the one with the condensed structure:

$$^1CH_3{}^2CH(CH_3)^3CH(CH_3)^4CH_3$$

Step 1 First, identify the longest continuous chain of carbon atoms; this gives the stem of the name using the prefixes in Table 10.7. When identifying the longest straight chain, do not be confused by the way the molecule is drawn on paper, as sometimes the same molecule can be represented differently owing to the free rotation around the C–C single bonds.

So, in the example above, the longest chain is four carbon atoms long (identified in blue on the structure above). This molecule is a form of butane.

Step 2 Then identify and name the side-chains or substituent groups as the first part or prefix of the name (see Table 10.8). In this case there are two different methyl groups.

Step 3 Where there is more than one side-chain of the same type, as in this case, use the prefixes *di-*, *tri-*, *tetra-* and so on, to indicate this. If there are several side-chains within a molecule, put them in alphabetical order, separated by dashes. Here there are two methyl groups – hence the prefix **dimethyl** in this case.

Step 4 The position of these side-chains is then identified. This is done using a number which refers to the number of the carbon atom in the stem. The carbon chain is numbered starting at the end which will give the substituent groups the smallest number.

In this case one methyl group is attached to carbon atom number 2; the other to carbon number 3. These numbers precede the name. This means that the name of this compound is **2,3-dimethylbutane**.

Worked example

Name the following hydrocarbons:
a $C(CH_3)_4$
b $CH_3CH(C_2H_5)CH_3$
c $CH_3CH_2CH(C_2H_5)CH_2CH_3$

a Look at this structure carefully. The longest chain is three carbons long. The central carbon atom of the three has two methyl groups attached. The name of this compound is **2,2-dimethylpropane**.
b Again this structure needs to be drawn out carefully. The longest chain is four carbons long, with a methyl group attached to the second carbon. The name of this compound is **2-methylbutane**.
c The longest chain in this molecule is five carbon atoms long. There is an ethyl group attached to the third carbon in the chain. The name of this structure is **3-ethylpentane**.

Deducing structural formulas of alkenes

10.1.7 Deduce structural formulas for the isomers of the non-cyclic alkenes up to C_6.

The second homologous series of hydrocarbons is the alkenes (see Table 10.5). These compounds are distinguished by the fact that they contain a C=C double bond at some point in the hydrocarbon chain. They have the general formula C_nH_{2n}.

The simplest alkene is ethene, C_2H_4, and the series develops by adding a $-CH_2-$ group to the chain. In the alkene molecules containing more than three carbon atoms the double bond can exist in different positions along the chain. The chain can also be branched. Thus there are two straight-chain isomers, and one branched-chain isomer, having the formula C_4H_8.

CH₃CH₂CH=CH₂   **CH₃CH=CHCH₃**   **CH₂=C(CH₃)₂**

Worked example

How many chain isomers are there with the molecular formula C_5H_{10}? Remember that the double bond can be moved, and the chain can be branched as well as straight.

There are five structural isomers of C_5H_{10}.

CH₂=CHCH₂CH₂CH₃   **CH₃CH=CHCH₂CH₃**

CH₂=C(CH₃)CH₂CH₃   **CH₃C(CH₃)=CHCH₃**   **CH₃CH(CH₃)CH=CH₂**

Naming isomers of non-cyclic alkenes

10.1.8 Apply IUPAC rules for naming the isomers of the non-cyclic alkenes up to C_6.

The IUPAC system for naming organic compounds extends to the naming of alkenes by stipulating how to indicate the position of the double bond in the chain. The basic names of the alkenes are assembled as for the alkanes, except that the names end in *-ene* rather than *-ane*. A fifth step needs to be added to those we identified before so that we can indicate the position in the chain of the C=C double bond.

Step 5   The position of the double bond is noted by inserting the number of the carbon atom at which the C=C bond 'starts'. Thus the two straight-chain isomers of C_4H_8 are known as but-1-ene and but-2-ene, while the branched-chain isomer is 2-methylprop-1-ene (named according to the rules described earlier).

$CH_3CH_2{}^2CH={}^1CH_2$ ${}^1CH_3{}^2CH={}^3CHCH_3$ ${}^1CH_2={}^2C(CH_3)_2$
but-1-ene   but-2-ene   2-methylprop-1-ene

| **Worked example** | What are the names of the following alkenes? |

a $CH_3CH=CHCH_2CH_2CH_3$ **b** $CH_3CH_2CH(CH_3)CH=CH_2$ **c** $CH_2=C(CH_3)CH_2CH=CH_2$

a   This is a straight-chain alkene of six carbon atoms. The double bond starts at carbon number 2. This compound is **hex-2-ene**.
b   There are five carbon atoms in the chain. The double bond is terminal and there is a methyl group on carbon number 3. This alkene is **3-methylpent-1-ene**.
c   There are two double bonds in this molecule, one at each end. There is a methyl group on carbon number 2. There are five carbons in the chain. This alkene is **2-methylpent-1,4-diene**.

In this example, **a** and **b** are isomers, but not **c**. **c** has the same number of carbon atoms but two fewer hydrogen atoms because of the presence of the second carbon–carbon double bond. **c** is called 2-methylpent-1,4-diene rather than 4-methylpent-1,4-diene because of the rule that the numbering must be kept to a minimum.

Introducing the diversity of organic compounds

10.1.9   **Deduce** structural formulas for the compounds containing up to six carbon atoms with one of the following functional groups: alcohol, aldehyde, ketone, carboxylic acid and halide.
10.1.10   **Apply** IUPAC rules for naming the compounds containing up to six carbon atoms with one of the following functional groups: alcohol, aldehyde, ketone, carboxylic acid and halide.

The series of functional groups listed in Table 10.5 give rise to a range of differing homologous series that begin to illustrate the diversity of organic compounds. There are four oxygen-containing functional groups with characteristic suffixes that appear at the end of the name for the organic compound. Halide groups have characteristic prefixes.

Alcohols – general formula R–OH or $C_nH_{2n+1}OH$

These compounds are characterized by the presence of the hydroxyl (–OH) group. The names of alcohols end in *-ol*. The position of the group is designated by the number of the carbon atom in the chain. Thus $CH_3CH_2CH_2OH$ and $CH_3CH(OH)CH_3$ are known as propan-1-ol and propan-2-ol, respectively (Figure 10.23).

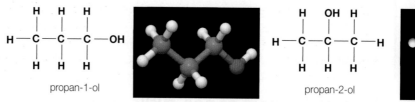

Figure 10.23 The structures of propan-1-ol and propan-2-ol

The alcohol group can be attached to any carbon atom in a straight chain or branched structure. The compound $CH_3C(CH_3)(OH)CH_3$ has the structure:

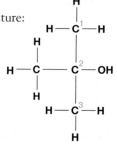

The longest chain here is three carbons long, with a methyl group and the hydroxyl group both attached to carbon-2 in the chain. The name of this compound then is 2-methylpropan-2-ol.

■ **Extension:** More complex structures

The condensed structural formula of 2-methylpropan-2-ol is given above as $CH_3C(CH_3)(OH)CH_3$. It is possible to condense this even further to $C(CH_3)_3OH$, but arguably this makes the formula more difficult to interpret.

Compounds exist that contain two and three hydroxyl (–OH) groups attached to a hydrocarbon stem. Anti-freeze for car engines is ethane-1,2-diol, $CH_2(OH)CH_2(OH)$. Glycerol, an important compound in biochemistry as the central component of triglyceride fats, has the structure $CH_2(OH)CH(OH)CH_2(OH)$. The IUPAC systematic name for glycerol is propane-1,2,3-triol (Figure 10.24). Do note the use of the prefixes *di-* and *tri-* in this context.

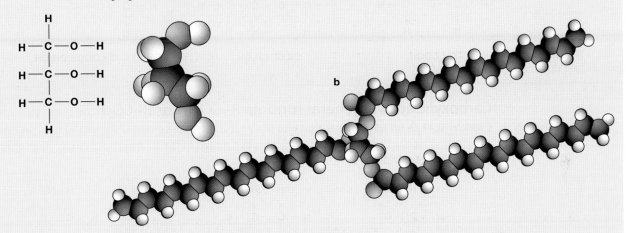

Figure 10.24 Glycerol (propane-1,2,3-triol) (**a**) and three stearic acid molecules combine together to form an ester (**b**) found in animal fat

When dealing with more complex structures that may contain more than one functional group, it is important to realize that there is a hierarchy to the functional groups when naming compounds. Most obviously, the carboxylic acid and the aldehyde group always take priority. Thus the compound lactic acid, that you may have heard of in connection with muscle cramps, has the condensed formula $CH_3CH(OH)COOH$. The systematic name for lactic acid is 2-hydroxypropanoic acid; here the acid group takes precedence in the naming, forcing the –OH group to be referred to as the hydroxyl group.

Aldehydes – general formula R–CHO or $C_nH_{2n+1}CHO$

The –CHO group always occurs at the end of a carbon chain – it is always terminal. Hence the carbon atom in the aldehyde group is always number 1 if any counting of atoms in the chain is required. Because of this, it is not usually necessary to number the position of the aldehyde group in the name of the compound (see the final example in Figure 10.25). The name of an aldehyde usually ends in *-al*.

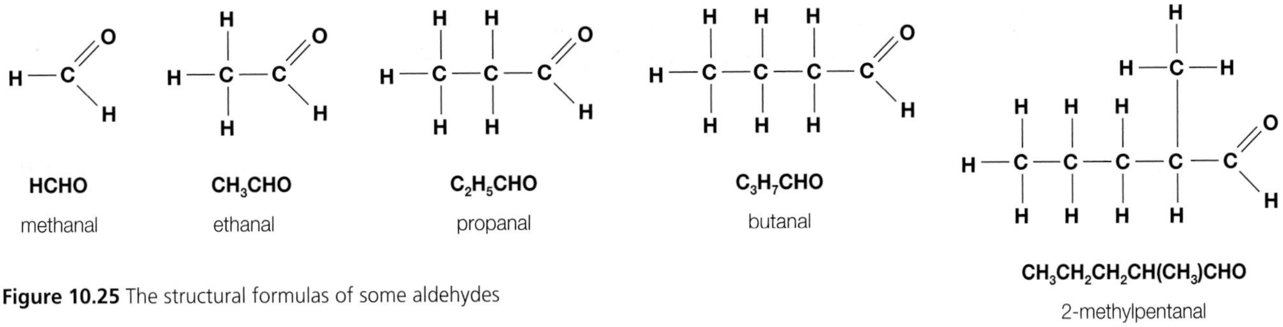

HCHO   **CH₃CHO**   **C₂H₅CHO**   **C₃H₇CHO**

methanal   ethanal   propanal   butanal

CH₃CH₂CH₂CH(CH₃)CHO

2-methylpentanal

Figure 10.25 The structural formulas of some aldehydes

Ketones – general formula R–CO–R' (where R' represents either the same alkyl group as R or a different alkyl group)

The name of a ketone always ends with the suffix *-one*. The group can be inserted anywhere in a hydrocarbon chain except at the end. Because of this, the carbon atom of the ketone group is counted when establishing the chain length and its position must be shown in the name, except for the first two members of the series (where there is no alternative position and a number is therefore unnecessary).

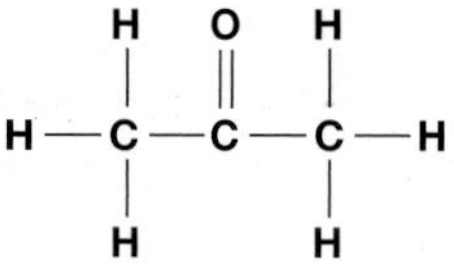

Figure 10.26 The formulas and structures of some ketones

CH_3COCH_3
propanone

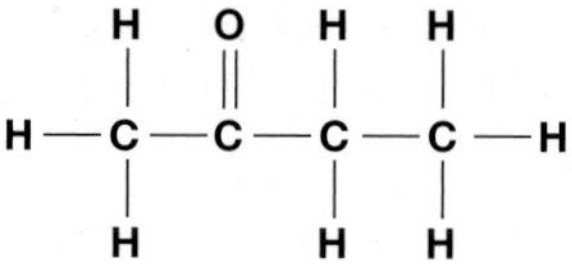

$CH_3COCH_2CH_3$
butanone

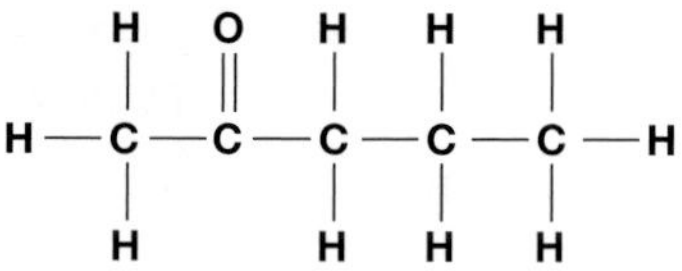

$CH_3COCH_2CH_2CH_3$
pentan-2-one

Carboxylic acids – general formula R–COOH or R–CO$_2$H

Carboxylic acids all have names that end in *-oic acid*. The acid group is always terminal and the carbon atom at the centre of the group is always counted as the first in the chain, no matter how complex the molecule.

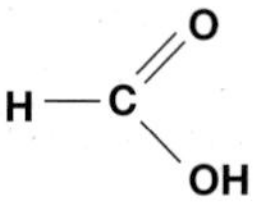

HCOOH
methanoic acid

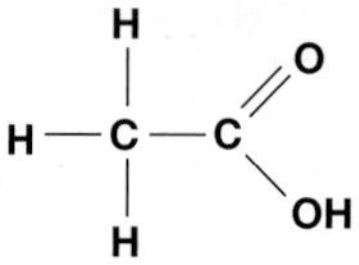

CH_3COOH
ethanoic acid

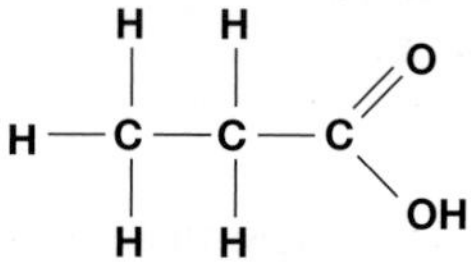

C_2H_5COOH
propanoic acid

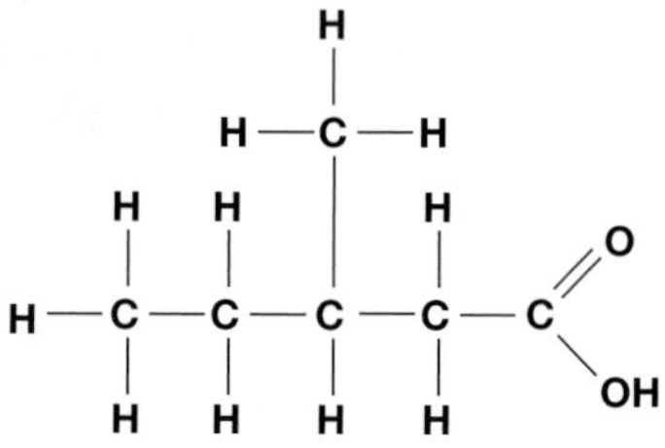

$CH_3CH_2CH(CH_3)CH_2COOH$
3-methylpentanoic acid

Figure 10.27 The structural formulas of some carboxylic acids

Halogenoalkanes – general formula R–X, where X = F, Cl, Br, I

Halogenoalkanes are an important and useful group of compounds. Their structure is straightforward in that the halogen atom simply replaces a hydrogen atom in the given hydrocarbon structure, whether that structure is a straight or branched chain.

Halogenoalkanes are named in a different way from the other homologous series examined so far. In the aldehydes, ketones and carboxylic acids, the part of the compound's name that indicates the functional group is placed as a suffix, at the end of the name. In halogenoalkanes the halogen is designated by a prefix: *fluoro-*, *chloro-*, *bromo-* or *iodo-*. The numbering of the carbon atom to which the halogen is attached follows the guidelines already established.

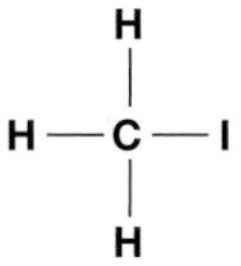

CH_3I
iodomethane

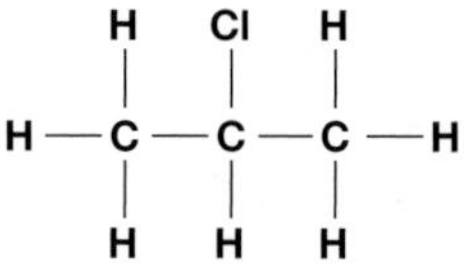

$CH_3CHClCH_3$
2-chloropropane

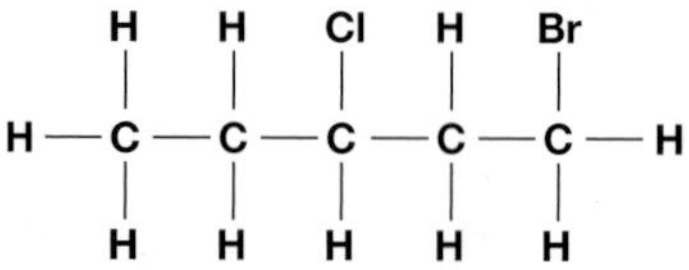

$CH_3CH_2CHClCH_2CH_2Br$
1-bromo-3-chloropentane

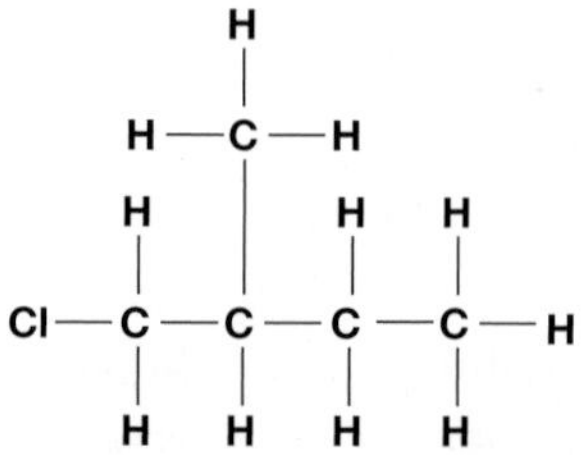

$CH_2ClCH(CH_3)CH_2CH_3$
1-chloro-2-methylbutane

Figure 10.28 The structural formulas of some halogenoalkanes

Further functional groups

10.1.11 **Identify** the following functional groups: amino (NH₂), benzene ring (), esters (RCOOR).

There are three other main functional groups. One of these groups is fundamental to a whole separate area of organic chemistry based on aromatic hydrocarbons. However, we will focus on two other types of compound first: amines and esters.

Amines – general formula R–NH₂

The amine group has the formula $-NH_2$ and can be found attached to a hydrocarbon backbone in a similar way to a halogen or an alcohol group. The presence of the group can be denoted by either the prefix *amino-* or the suffix *-amine* (Figure 10.29).

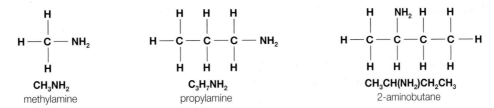

Figure 10.29 The structural formulas of some amines

Esters – general formula R–COOR′, where R′ is an alkyl group

Esters are derived from carboxylic acids by reaction with an alcohol. The second part of their name denotes the acid from which the compound is derived. Thus an 'ethanoate' is the product made from ethanoic acid, and so on. This is *preceded* by the name of the alkyl group (R′) that comes from the alcohol that reacted with the acid (Figure 10.30).

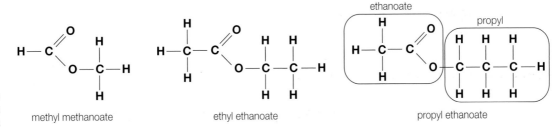

Figure 10.30 The structural formulas of some esters

Aromatic compounds

Aromatic compounds (**arenes**) represent a distinctive range of organic compounds where the features and properties of the benzene ring, or similar structures, produce a very different chemistry from that of the 'families' of compounds based on the straight and branched hydrocarbon chains examined so far.

Benzene has the formula C_6H_6. The carbon atoms are arranged in a ring structure and can be represented in several different ways (Figure 10.31).

Figure 10.31 Representations of the structure of the benzene molecule (C_6H_6). The three ring structures are all valid structural formulas. In the centre is a computer-generated electrostatic potential map

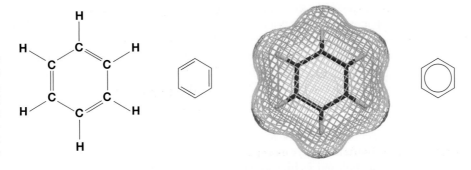

Instead of R representing an alkyl group, it can also mean an aryl group. Aryl groups are based on a benzene ring, with the simplest being the $-C_6H_5$ or phenyl group. This group can be attached to a wide range of other functional groups to produce a vast range of compounds (Figure 10.32).

Figure 10.32 The structure of some aromatic compounds

methylbenzene   chlorobenzene   phenol

History of Chemistry

Michael Faraday (Chapter 19) isolated benzene in 1825. He heated oil gas, a fuel derived from whale oil and used for lighting, and found a compound with the formula C_6H_6 in the distillate. In the past benzene was obtained from the distillation of coal, but it is now obtained by reforming hydrocarbons.

Language of Chemistry

Aromatic compounds are a group of hydrocarbons derived from benzene. They received this name because they were originally obtained from aromatic or sweet-smelling oils and resins. However, the term aromatic now refers to molecules (and ions) which have stabilizing pi electron delocalization. Compounds that do not contain a benzene ring are referred to as **aliphatic** compounds. ■

Primary, secondary and tertiary compounds

10.1.12 Identify primary, secondary and tertiary carbon atoms in alcohols and halogenoalkanes.

A functional group defines the chemistry of a particular homologous series, but its reactivity can be influenced by its position in the carbon chain. Consequently it is useful to be able to describe different positions in a structure exactly. This is done by applying the terms primary, secondary and tertiary to identify the location of the carbon atom to which the functional group (e.g. $-OH$) is attached in a molecule.

Figure 10.33 The structures of primary, secondary and tertiary alcohols. R, R', and R'' are alkyl groups (they may all be the same group, or different)

primary alcohol   secondary alcohol   tertiary alcohol

Note that from the examples in Figure 10.33 the following rules apply.

- A **primary carbon atom** is attached to the functional group ($-OH$ above) and also to *at least two hydrogen atoms*. Molecules with this arrangement are known as primary molecules. For example, ethanol is a primary alcohol, CH_3CH_2OH, while 1-chloropropane is a primary halogenoalkane, $CH_3CH_2CH_2Cl$.
- A **secondary carbon atom** is attached to the functional group and to *just one hydrogen atom* and two alkyl groups. These molecules are known as secondary molecules. For example, propan-2-ol is a secondary alcohol, $CH_3CH(OH)CH_3$, while 2-bromobutane is a secondary halogenoalkane, $CH_3CHBrCH_2CH_3$.
- A **tertiary carbon atom** is attached to the functional group and is also bonded to three alkyl groups. *There are no hydrogen atoms attached to a tertiary carbon atom.* These molecules are known as tertiary molecules. For example, 2-methylpropan-2-ol, $C(CH_3)_3OH$ is a tertiary alcohol, and 2-chloro-2-methylpropane, $CH_3C(CH_3)ClCH_3$, is a tertiary halogenoalkane.

| **Worked example** | Are the following molecules primary, secondary or tertiary? |

a 3-methylpentan-3-ol   **b** pentan-2-ol   **c** 1-chlorobutane

a Look at the structural formula: $CH_3CH_2C(CH_3)(OH)CH_2CH_3$. The marked carbon atom, to which the –OH is attached, has no hydrogen atoms bonded to it. The compound is a tertiary alcohol.

b The –OH group is bonded to the second carbon atom in the chain. This carbon will also have one hydrogen atom attached. The compound is a secondary alcohol.

c The chlorine atom is attached to the terminal carbon atom which will also have two hydrogen atoms bonded to it. The compound is a primary halogenoalkane.

Volatility

10.1.13 Discuss the volatility and solubility in water of compounds containing the functional groups listed in 10.1.9.

Members of each homologous series have the same functional group and therefore we expect them to have similar chemical properties. We also expect there to be a regular trend in their physical properties with increasing chain length – as shown for the alkanes and alcohols by the data presented on pages 258–260.

Volatility is a measure of how easily a substance evaporates – a highly volatile substance evaporates easily and has a low boiling point. How easily a substance evaporates depends on the molecules having sufficient kinetic energy to overcome the forces between the molecules. So substances with stronger intermolecular forces will evaporate less easily, and have higher boiling points. There are three factors which contribute to the observed pattern for the different homologous series.

First, volatility decreases and boiling point increase with increasing molecular size (chain length). As we go up a series the chain length increases by the addition of a $-CH_2-$ unit. This results in a longer molecule, stronger van der Waals' forces between the molecules and, therefore, an increase in the boiling point (see Figure 10.9). Thus, at room temperature, the early members of a series are generally gases or liquids, while the later members are more likely to be solids.

Second, a branched isomer of a compound is likely to have a lower boiling point than its straight-chain isomer. The branching of a chain results in a more spherical overall shape to the molecule. This means there is less surface contact between molecules other than for straight-chain isomers, so these branched isomers will have weaker intermolecular forces and hence lower boiling points (Figure 10.34).

Figure 10.34 a Elongated straight-chain molecules will have greater surface contact with each other and therefore stronger van der Waals' forces between molecules. **b** Branched molecules tend to be more spherical in shape, and therefore have a smaller contact area and weaker intermolecular forces

a

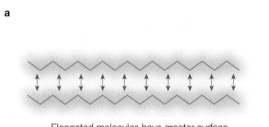

Elongated molecules have greater surface area for attraction

b

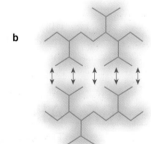

More oval or spherical molecules have less surface area for attraction

Finally, the nature of the functional group present in the molecules will influence volatility, depending on its effect on the intermolecular forces. Polar groups will lead to stronger dipole–dipole interactions between the molecules and hence higher boiling points. Groups that are capable of forming hydrogen bonds will result in even stronger forces between the molecules, giving rise to even higher boiling points (see Figure 10.9 for a comparison between the early alkanes and alcohols).

The factors that influence the physical properties of compounds containing the different functional groups are summarized in Figure 10.35.

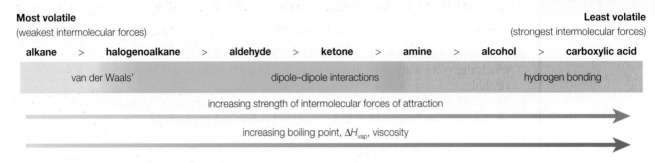

Most volatile
(weakest intermolecular forces)

Least volatile
(strongest intermolecular forces)

| alkane | > | halogenoalkane | > | aldehyde | > | ketone | > | amine | > | alcohol | > | carboxylic acid |

van der Waals'   dipole–dipole interactions   hydrogen bonding

increasing strength of intermolecular forces of attraction →

increasing boiling point, ΔH_{vap}, viscosity →

Figure 10.35 The influence of the functional group on intermolecular forces and certain physical properties

One crucial thing to remember when making comparisons of boiling points of compounds in different homologous series, or indeed other properties dependent on intermolecular forces, is to compare molecules that have similar M_r values. This may mean comparing molecules with different numbers of carbon atoms. For example, ethanol, C_2H_5OH (M_r = 46), has a boiling point of 78 °C (351 K) and can be usefully compared with propane, C_3H_8 (M_r = 44), whose boiling point is −42 °C (83 K). By comparing molecules of similar size, it becomes clear that the higher boiling point in ethanol is due to the presence of the alcohol (−OH) group which causes hydrogen bonding between the molecules, rather than being an effect of molecular size. In this context, some would argue that we should, in fact, count the total number of electrons in the molecules involved, as it is the distortions and shifting patterns of these that create the temporary dipoles that are the basis of the underlying van der Waals' forces.

Solubility in water

The solubility (which can be referred to as miscibility if dealing with a liquid) of an organic compound in water is largely determined by two factors that tend to have opposing effects. These factors relate to the two essential parts of the molecule – the functional group and the hydrocarbon 'backbone'.

- If the functional group in the compound is able to interact with water, for example by forming hydrogen bonds, then this will favour the compound being soluble.
- The hydrocarbon chain of the molecule is non-polar, and so does not help the solubility of the molecule in water. Indeed, a long hydrocarbon chain can counteract any solubility-favouring effect of the functional group. In general the solubility of organic molecules decreases with increasing chain length.

Applying these two factors to the types of homologous series met so far explains why the earlier members of the alcohols, aldehydes, ketones and carboxylic acids are quite soluble in water, but the solubility decreases as we progress up the series. Halogenoalkanes are not soluble in water as, despite their polarity, they are unable to form hydrogen bonds with water.

10.2 Alkanes

The alkanes are perhaps the simplest of the homologous series of organic compounds. The following are key points to remember:

- The alkanes are hydrocarbons, and therefore contain carbon and hydrogen *only*.
- The alkanes have the general formula C_nH_{2n+2}.
- Alkanes are **saturated hydrocarbons** (the term **saturated** means that all the carbon–carbon bonds are single bonds).
- Although they are relatively unreactive compounds, some of their reactions are highly significant and important.

In Section 10.1 we looked at the physical properties and structures of the alkanes. In this section we look at their chemical properties.

Figure 10.36 The burning of methane, a highly exothermic reaction, can provide a spectacular laboratory demonstration (note that the demonstrator should be wearing a lab coat)

Explaining the low reactivity of the alkanes

10.2.1 Explain the low reactivity of alkanes in terms of bond enthalpies and bond polarity.

Because of their chemical simplicity, alkanes contain only carbon–carbon single bonds and carbon–hydrogen bonds. Both these bonds are strong (C–C, 348 kJ mol^{-1} and C–H, 412 kJ mol^{-1}). Consequently alkane molecules will only react when a strong source of energy is being used, providing enough energy to break these bonds. Alkanes are generally stable under most conditions and can be stored, transported and even compressed safely. These latter points are particularly important in view of the uses these compounds are put to.

The C–C and C–H bonds are also characteristically non-polar as carbon and hydrogen have very similar electronegativities (the electronegativity difference is 0.4, $^{\delta-}$C–H$^{\delta+}$). This means that alkane molecules are not susceptible to attack by the most common attacking agents in organic chemistry: nucleophiles or electrophiles (electron pair donors or acceptors).

These two factors, together with those mentioned earlier in the chapter, are responsible for the very low reactivity of the alkanes. There are, however, two very significant reactions of alkanes to be considered here.

Language of Chemistry

The alkanes are chemically very unreactive. Indeed their old name was the *paraffins*, which came from the Latin and literally means 'little activity' (from the Latin *parum affinis*). This inertness may seem initially surprising given that the alkanes burn extremely well, resulting in their principal use as fuels. However, we have already discussed the reasons for this earlier in the chapter. The relatively strong covalent bonding in the molecules means that they are kinetically stable until sufficient activation energy is provided. ■

The combustion of alkanes

10.2.2 Describe, using equations, the complete and incomplete combustion of alkanes.

Because they release significant amounts of energy when they burn (Figure 10.36), the alkanes are widely used as fuels, for example in the internal combustion engines of cars and in aircraft engines and household heating systems (see also Figure 10.37). The combustion reactions of these molecules are highly exothermic. This is mainly a result of the high relative strength of the carbon–oxygen double bonds (C=O) in carbon dioxide and the oxygen–hydrogen (O–H) bonds in water molecules. These are the products formed in these combustion reactions and the large amount of heat energy released in making these bonds means that the reactions are strongly exothermic. Remember that bond formation is an exothermic, energy-releasing process (Chapter 5).

Alkanes burn in the presence of excess oxygen to produce carbon dioxide and water, for example:

$$CH_4(g) + 2O_2(g) \rightarrow CO_2(g) + 2H_2O(l) \qquad \Delta H_c^\ominus = -890 \text{ kJ mol}^{-1}$$
$$C_3H_8(g) + 5O_2(g) \rightarrow 3CO_2(g) + 4H_2O(l) \qquad \Delta H_c^\ominus = -2220 \text{ kJ mol}^{-1}$$

However, when the oxygen supply is limited, carbon monoxide and water can be produced, for example:

$$2CH_4(g) + 3O_2(g) \rightarrow 2CO(g) + 4H_2O(g)$$
$$2C_3H_8(g) + 7O_2(g) \rightarrow 6CO(g) + 8H_2O(g)$$

In conditions when oxygen is extremely limited, carbon itself can also be produced, for example:

$$C_3H_8(g) + 2O_2(g) \rightarrow 3C(s) + 4H_2O(g)$$

We can see that the incomplete combustion of an alkane can result in a mixture of carbon-containing products, including the element itself. The following equation is a possible reaction that may take place in a candle flame where the solid wax burns to give the characteristic flame we are familiar with (Figure 10.38a).

$$C_{20}H_{42}(s) + 15O_2(g) \rightarrow 11C(s) + 9CO(g) + 21H_2O(g)$$

Figure 10.37 Butane is used as a fuel in camping stoves. The butane is stored under pressure as a liquid in the canisters

It is the incandescent glow of the hot solid carbon particles (soot) that gives the flame its yellow colour. The same is true for the yellow (safety) flame of the Bunsen burner (Figure 10.38b).

Figure 10.38 a The yellow candle flame; **b** the safety flame of a Bunsen burner. Both these flames are characterized by the yellow glow of incandescent carbon particles

■ Extension: The burning of fossil fuels

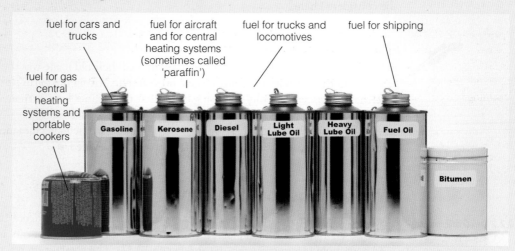

Figure 10.39 A representation of some of the useful products from the fractional distillation of crude oil. Emphasis here is on those uses involving combustion

Earlier in the chapter we looked at the range of products available from the distillation of crude oil (Figure 10.11). A significant number of these products are fuels for various forms of transport and heating systems (Figure 10.39). Methane is also the major component of natural gas.

The combustion reactions outlined earlier are thus amongst the most significant chemical reactions on the planet. They become even more important now that we have realized that products of all these reactions have a serious impact on the environment. This is why the burning of these and other 'fossil fuels' on a very large scale is now widely recognized as a global problem. Carbon dioxide and water are both 'greenhouse gases'. Such gases in the atmosphere absorb infrared radiation and so help retain heat from the Sun in the atmosphere, contributing to global warming. Rising levels of carbon dioxide caused by human activities – mainly the burning of fossil fuels – are being implicated in the significant increase in average world temperatures. The Intergovernmental Panel on Climate Change (IPCC) that met in Paris in January 2007 acknowledged that 11 of the preceding 12 years had been the warmest since 1850. Climate change has risen to the top of the political agenda in recent years. Growing awareness of this problem has raised the profile of the issues involved in fuel availability and consumption.

Applications of Chemistry

Methane is a 'greenhouse gas' released in large quantities from cattle, termite mounds, rice paddy fields and swamps. The methane produced is the product of bacteria living under anaerobic conditions. In recent years focus has been directed towards a potential source of methane that represents both an opportunity and a threat. Methane has been found stored in the sediments of the continental shelf beneath the deep ocean, underneath the permafrost of the

Arctic and in deep Antarctic ice cores (Figure 10.40). In these circumstances the methane is stored in the form of *methane clathrates*. Clathrates are structures formed by the inclusion of atoms or molecules of one kind, in this case methane, in cavities of the crystal lattice of another, in this case ice. The open, hydrogen-bonded structure of ice (see Chapter 4) lends itself to the formation of such caged structures.

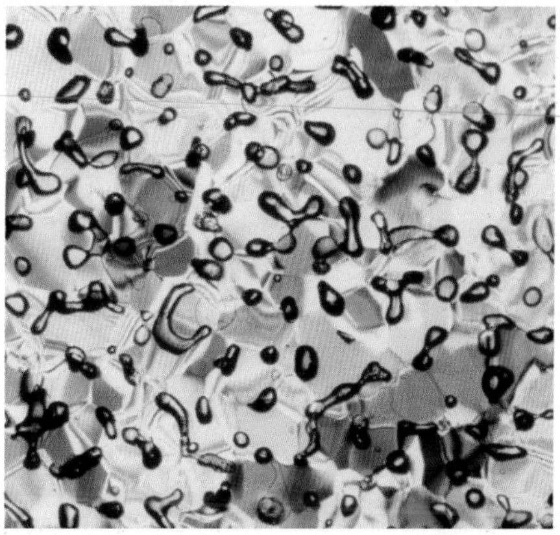

Many countries are investing in research into the extraction of methane from these clathrates. An exploration well drilled on land by Imperial Oil in 1971–72 discovered methane clathrates under the Canadian Arctic permafrost at the edge of the Mackenzie Delta and the Beaufort Sea. This site is known as the Mallik gas hydrate field. It has become a major research centre on methane clathrates with research groups from Canada, the USA, Japan and India working under the umbrella organization known as the International Continental Scientific Drilling Program (ICDP).

Figure 10.40 Bubbles of methane gas frozen in the polar ice. This is a polarized light micrograph of an ice sample extracted in Antarctica. The sample was from an ice core drilled to a depth of 234 m

However, there is potential for great concern regarding these methane deposits. Global warming is already causing the permafrost in the Arctic to melt, which in turn leads to release of methane gas from these terrestrial buried methane clathrates. Methane has a global warming potential greater than carbon dioxide. The potential contribution to global warming and climate change would be considerable. Thus increased methane emissions are of major environmental concern.

■ Extension: The role of the fuel in a petrol engine

The current use of hydrocarbon fuels represents a massive investment of research and development, while the issues surrounding the continued use of these fuels are a major focus of international discussion. It is worth considering how these fuels have developed to gain some insight into the demands placed on the fuel.

The gasoline (petrol) engine

Figure 10.41 The stages of the four-stroke cycle of a gasoline engine.

The car engine still in use in the majority of motor transport is based on the four-stroke cycle developed by Nikolaus Otto in the 19th century. This cycle places key requirements on the fuel at the various stages (Figure 10.41).

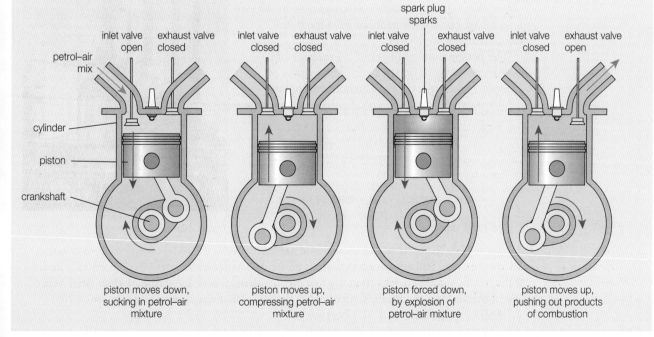

spark plug sparks

| inlet valve open | exhaust valve closed | inlet valve closed | exhaust valve closed | inlet valve closed | exhaust valve closed | inlet valve closed | exhaust valve open |

petrol–air mix

cylinder

piston

crankshaft

| piston moves down, sucking in petrol–air mixture | piston moves up, compressing petrol–air mixture | piston forced down, by explosion of petrol–air mixture | piston moves up, pushing out products of combustion |

The first downstroke of the piston

The fuel must be drawn into the cylinder from the carburettor through the inlet valve. For this to happen the fuel must be volatile enough to have formed a homogeneous fuel–air mixture in the carburettor. In a fuel-injection car, the first downstroke sucks in only air; the fuel is injected near the top of the first upstroke.

First upstroke

At this stage, the fuel–air mixture is compressed, being reduced to about one-tenth of its volume. This compression results in the temperature of the mixture being raised by several hundred degrees. This temperature rise is due to more van der Waals' interactions being formed between the gas molecules as they are pushed closer together. The fuel should not have an ignition temperature such that this self-heating causes pre-ignition of the fuel, before the piston has reached the top of the cylinder. This phenomenon is called 'knocking'.

Second downstroke

Following the sparking of the spark plug, the fuel–air mixture ignites and expands. The piston is pushed down the cylinder. The burning of the fuel must be smooth and completed in the fraction of a second that it takes the piston to travel down the cylinder.

Second upstroke

In the ideal situation, the fuel will have completely burnt to form gaseous carbon dioxide and water. However, if insufficient oxygen (air) has been drawn in with the fuel, or there has been poor mixing, or insufficient time for complete burning, then other substances will be produced.

Carbon particulates (soot) can be a problem, shorting the spark plug, and causing a smoky exhaust. Unburnt hydrocarbons and carbon monoxide can be components of the exhaust gases. These gaseous pollutants, along with nitrogen oxides produced at the high engine temperatures, can be removed at a later stage by using a catalytic converter fitted into the exhaust system.

The quality of petrol

Long-chain hydrocarbons tend to burn unevenly in car engines, tending to ignite too soon and cause a rattling noise ('knocking'). Branched-chain alkanes burn in a more controlled manner and so are added to the gasoline fraction when petrol is blended. Branched-chain alkanes are produced from straight-chain alkanes by catalytic cracking of fractions from fractional distillation (Figure 10.42).

The **octane number** of petrol is a measure of its quality, and is based on the ignition properties of an isomer of octane, C_8H_{18}. The branched alkane, 2,2,4-trimethylpentane, has good antiknock properties.

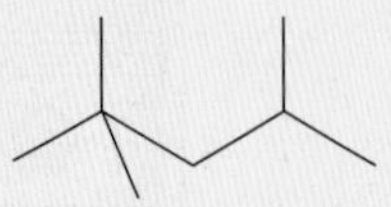

$$CH_3-C(CH_3)_2-CH_2-CH(CH_3)-CH_3$$

Figure 10.43 Structure of 2,2,4-trimethylpentane

Figure 10.42 A steam cracker at an oil refinery

2,2,4-Trimethylpentane is given an octane number value of 100. Heptane, a straight-chain alkane, has poor ignition properties and is given an octane number of 0. For a particular petrol, if the octane number is 100, then the fuel is equivalent to pure 2,2,4-trimethylpentane. Many petrol blends have an octane number of 70; they burn in a similar way to a mixture of 70% 2,2,4-trimethylpentane and 30% heptane.

Historically, tetraethyllead(IV) (Pb(C$_2$H$_5$)$_4$) has been used as a gasoline additive to improve the anti-knock properties of gasoline (known as leaded fuel). Leaded gasoline is now being phased out for antipollution reasons and also because it 'poisons' the platinum/rhodium catalyst in a catalytic converter. Catalytic converters change pollutants, such as carbon monoxide, unburnt hydrocarbons and nitrogen oxides, into more environmentally friendly compounds. Leaded petrol should therefore not be used in a car that contains a catalytic converter.

Reactions of alkanes with halogens

10.2.3 **Describe,** using equations, the reactions of methane and ethane with chlorine and bromine.

Figure 10.44 A representation of the substitution reaction between methane and chlorine

The alkanes are saturated molecules and as such the main type of reaction they can undergo is a **substitution** reaction in which a hydrogen atom is replaced by the atom of another element. Under appropriate conditions, a mixture of methane and chlorine gases reacts to form chloromethane and hydrogen chloride (Figure 10.44).

$$CH_4(g) + Cl_2(g) \rightarrow CH_3Cl(l) + HCl(g) \qquad \text{conditions: sunlight or ultraviolet radiation}$$

Further substitution can take place with excess chlorine to produce dichloromethane (CH$_2$Cl$_2$), trichloromethane (CHCl$_3$) and tetrachloromethane (CCl$_4$). For example:

$$CH_4(g) + 2Cl_2(g) \rightarrow CH_2Cl_2(l) + 2HCl(g) \quad \text{conditions: sunlight or ultraviolet radiation}$$

The degree of substitution achieved in these reactions cannot be easily controlled. The initial proportions of reacting gases can be varied but the result of the reaction is always likely to be a mixture of products.

Ethane and other alkanes will undergo similar substitution reactions with chlorine or bromine. For example:

$$C_2H_6(g) + Br_2(g) \rightarrow C_2H_5Br(l) + HBr(g) \qquad \text{conditions: sunlight or ultraviolet radiation}$$
$$\text{bromoethane}$$

$$C_2H_6(g) + 2Br_2(g) \rightarrow C_2H_4Br_2(l) + 2HBr(g)$$
$$\text{mixture of}$$
$$\text{dibromoethanes}$$

The product of this second reaction will be a mixture of dibromoethanes (1,1-dibromoethane and 1,2-dibromoethane) as there is no control over which hydrogen atom is substituted.

Worked example

Consider the reaction between ethane and bromine where the conditions are such as to produce dibromoethanes as products. Name the different products formed and estimate the proportions of the products.

The two possible products are 1,1-dibromoethane, CHBr$_2$CH$_3$, and 1,2-dibromoethane, CH$_2$BrCH$_2$Br.

There is nothing involved in the mechanism which directs the substitution to any particular carbon atom, so neither position is favoured. The product is approximately a 50:50 mixture of the two isomers.

Figure 10.45 Dichloromethane can be used to dissolve polystyrene in model making

The halogen molecules are able to act in this way because they are split into energized separate atoms which have an unpaired electron under these conditions. Such energized particles with an unpaired electron are known as **free radicals**. Once formed, these radicals initiate a chain reaction in which halogenoalkanes are produced. Studies on these reactions have shown that the reaction can be divided into a sequence of steps, known as the **reaction mechanism** for the substitution.

The halogenoalkane products of these substitution reactions have uses as solvents (dry cleaning for example), anaesthetics and fire retardants. Dichloromethane, for instance, can be used as a solvent glue for polystyrene in model making (Figure 10.45).

Language of Chemistry

A covalent bond between two atoms consists of a shared pair of electrons; for instance, the Cl–Cl bond in molecular chlorine. When this bond breaks there are two possible ways in which the electrons in the bond can distribute themselves.

In **homolytic fission**, the bond breaks so that one electron remains with each fragment.

$$Cl–Cl \rightarrow Cl\bullet + Cl\bullet$$

Here two chlorine free radicals are produced by the effect of ultraviolet radiation.

$$H–CH_3 + Cl\bullet \rightarrow \bullet CH_3 + HCl$$

The carbon–hydrogen bond is broken homolytically by the chlorine free radical ($Cl\bullet$) to produce a methyl free radical ($\bullet CH_3$). The chlorine free radical removes a hydrogen atom to form hydrogen chloride.

In **heterolytic fission**, the bond breaks so that one atom retains both electrons and ions are produced:

$$X–Y \rightarrow X^+ + Y^- \qquad \text{here atom Y retained the electrons from the bond}$$
or
$$X–Y \rightarrow X^- + Y^+ \qquad \text{here atom X retained the electrons}$$

Both these types of bond fission play their part in reaction mechanisms. In the case of the substitution of halogens into alkanes, it is homolytic fission that is involved in the reaction mechanism.

The prefix *homo* meaning 'the same' is from the Greek and refers to the fact that the two products have an equal assignment of electrons from the bond. The prefix *hetero* means 'different' and refers to the fact that the electrons are unequally shared between the fragments, producing oppositely charged ions. ■

The free-radical reaction mechanism

10.2.4 Explain the reactions of methane and ethane with chlorine and bromine in terms of a free-radical mechanism.

The formation of halogenoalkanes by substitution requires the presence of ultraviolet light. The reaction cannot take place in the dark. The reaction is a **photochemical reaction**.

The energy of a photon of ultraviolet radiation is of the order of $400\,kJ\,mol^{-1}$. This is enough energy to break a chlorine molecule into energized chlorine atoms, chlorine free radicals:

$$Cl–Cl \rightarrow Cl\bullet + Cl\bullet \qquad\qquad \Delta H = +242\,kJ\,mol^{-1}$$

This **homolytic fission** of the bond between the chlorine atoms is thought to be the initial step in this reaction.

The next stage is thought to involve a chlorine free radical reacting with a methane molecule. In this way a hydrogen chloride molecule is produced, along with a methyl free radical.

$$CH_4 + Cl\bullet \rightarrow \bullet CH_3 + \mathbf{HCl}$$

The methyl free radical reacts further:

$$\bullet CH_3 + Cl_2 \rightarrow \mathbf{CH_3Cl} + Cl\bullet$$

Do note that these two reactions produce one of the products of the overall reaction and, importantly, a further free radical. These free radicals can go on to produce further reactions. In this way the reaction propagates itself: it is a **chain reaction**.

As the reaction proceeds there is a build up of free radicals and this leads to the final stage of the sequence of reactions – the termination step. This step involves the recombination of two free radicals with each other. The three possible termination steps are shown below:

$$Cl\bullet + Cl\bullet \rightarrow Cl_2$$

$$\bullet CH_3 + Cl\bullet \rightarrow CH_3Cl$$

$$\bullet CH_3 + \bullet CH_3 \rightarrow C_2H_6$$

Note that in these reactions free radicals are being removed from the reaction mixture. The presence of small amounts of ethane in the final products of the reaction is an indication that this is indeed a plausible mechanism for the reaction. The experimental finding that for each original ultraviolet photon absorbed there are, on average, 10 000 molecules of chloromethane produced confirms the idea of this being a chain reaction.

Thus there are three main steps to this reaction mechanism: initiation, propagation and termination (Figure 10.46).

initiation	$Cl—Cl(g)$	$\xrightarrow{\text{ultraviolet}}$	$2Cl\bullet(g)$
propagation	$Cl\bullet(g) + CH_4(g)$	$\rightarrow$	$\bullet CH_3(g) + HCl(g)$
	$\bullet CH_3(g) + Cl_2(g)$	$\rightarrow$	$CH_3Cl(g) + Cl\bullet(g)$
possible termination steps	$Cl\bullet(g) + Cl\bullet(g)$	$\rightarrow$	$Cl_2(g)$
	$Cl\bullet(g) + \bullet CH_3(g)$	$\rightarrow$	$CH_3Cl(g)$
	$\bullet CH_3(g) + \bullet CH_3(g)$	$\rightarrow$	$C_2H_6(g)$
overall reaction	$Cl_2(g) + CH_4(g)$	$\xrightarrow{\text{ultraviolet}}$	$CH_3Cl(g) + HCl(g)$

Figure 10.46 The reaction mechanism for the free-radical substitution between methane and chlorine in ultraviolet light

Free-radical mechanisms are also thought to be important in other significant organic reactions such as the cracking of hydrocarbon chains and the formation of polymers such as poly(ethene).

10.3 Alkenes

The alkenes are also hydrocarbons, and therefore contain carbon and hydrogen *only*. They have the general formula C_nH_{2n}. The alkenes are **unsaturated hydrocarbons** containing a carbon–carbon double bond; this double bond is made up of a sigma (σ) bond and a pi (π) bond. The carbon atoms which form the double bond have an arrangement of groups around them which is trigonal planar with angles of 120°, as shown for ethene below. It is important to note that a carbon–carbon double bond is shorter than a carbon–carbon single bond.

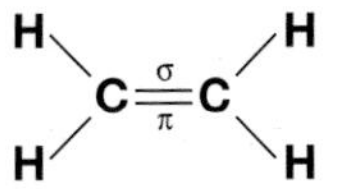

Alkenes are relatively more reactive compounds than alkanes because of the carbon–carbon double bond. They undergo a range of **addition** reactions.

The reactions of alkenes with hydrogen and halogens

10.3.1 **Describe,** using equations, the reactions of alkenes with hydrogen and halogens.

The carbon–carbon double bond is the functional group of the alkenes and is the site of chemical reactivity in the structure. The pi (π) bond is weaker than the sigma (σ) bond and is relatively easily broken without the molecule falling apart. Thus addition reactions can take place in which various molecules add to the carbon atoms originally participating in the double bond. The products of such reactions are all saturated molecules.

Hydrogenation
The simplest of these addition reactions is the addition of hydrogen across the carbon–carbon double bond to produce the alkane (Figure 10.47).

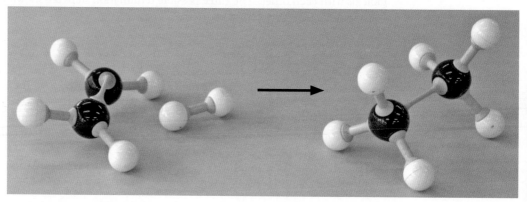

Figure 10.47
A representation of the hydrogenation of ethene to produce ethane

Hydrogen reacts with alkenes to form alkanes in the presence of a nickel catalyst at about 180 °C.

$$C_2H_4(g) + H_2(g) \longrightarrow C_2H_6(g)$$

ethene $\xrightarrow{\text{Ni, 180 °C}}$ ethane

$$H_2C = CHCH_2CH_3 + H_2 \xrightarrow{\text{Ni, 180 °C}} CH_3CH_2CH_2CH_3$$
but-1-ene   butane

$$CH_3CH = CHCH_3 + H_2 \xrightarrow{\text{Ni, 180 °C}} CH_3CH_2CH_2CH_3$$
but-2-ene   butane

Note that any isomerism due to the position of the double bond is lost once the hydrogen is added.

This process, known as **hydrogenation**, is used in the margarine industry to convert oils containing unsaturated hydrocarbon chains into more saturated compounds which have higher melting points. This is done so that margarine will be a solid at room temperature (Figure 10.48). However, there are now widespread concerns about the health effects of some of the fats produced in this way, known as *trans* fats (Chapter 25).

Figure 10.48 Margarine is solid but spreadable at room temperature.

Halogenation
Halogens react with alkenes to produce disubstituted compounds. These reactions take place readily at room temperature and are accompanied by the loss of colour of the reacting halogen. Note that the halogen atoms become bonded across the carbon–carbon double bond so the

structure of the product will have the halogen atoms on *adjacent* carbon atoms. The halogen is usually dissolved in a non-polar solvent such as hexane.

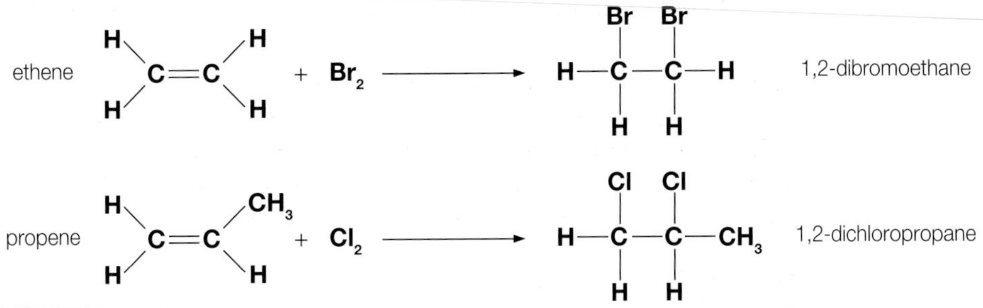

ethene + Br₂ ⟶ 1,2-dibromoethane

propene + Cl₂ ⟶ 1,2-dichloropropane

Further addition reactions

10.3.2 Describe, using equations, the reactions of symmetrical alkenes with hydrogen halides and water.

Hydrogen halides (HCl, HBr, etc.) react with alkenes to produce halogenoalkanes. These reactions take place rapidly in solution at room temperature. For example, ethene reacts with hydrogen chloride to form chloroethane, and but-2-ene reacts with hydrogen bromide to form 2-bromobutane.

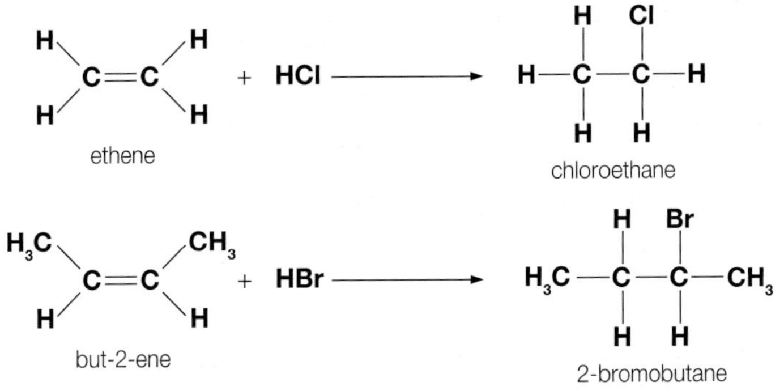

ethene + HCl ⟶ chloroethane

but-2-ene + HBr ⟶ 2-bromobutane

The hydrogen halides all react similarly, but the reactivity is in the order HI > HBr > HCl because of the decreasing strength of the hydrogen halide bond going down group 7. Thus hydrogen iodide, HI, having the weakest and longest H–X bond, reacts the most readily.

Hydration

The reaction with water is known as hydration and the alkene is converted into an alcohol. Water does not itself react directly with alkenes. However, in the laboratory, this reaction can be accomplished using concentrated sulfuric acid to form an addition product. The reaction involves an intermediate in which both H^+ and HSO_4^- ions are added across the double bond. Cold water is then added and hydrolysis takes place with replacement of the HSO_4^- by OH^- and re-formation of the sulfuric acid (H_2SO_4).

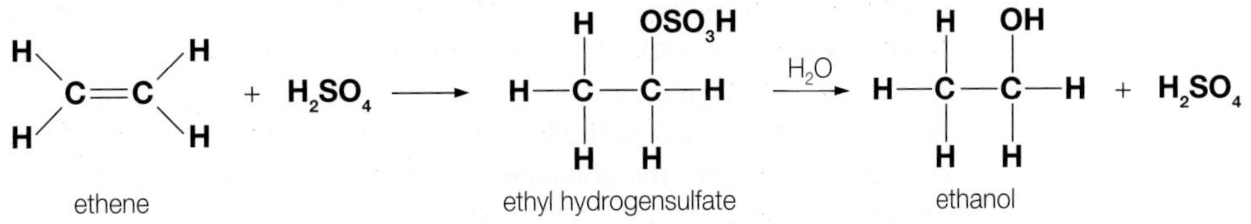

ethene + H₂SO₄ ⟶ ethyl hydrogensulfate —H₂O→ ethanol + H₂SO₄

There is an industrial process for synthesizing ethanol by hydration of ethene. This involves passing ethene and steam at high pressure (60 atmospheres) over a catalyst of immobilized phosphoric(V) acid at 300 °C. The phosphoric(V) acid is adsorbed on silicon dioxide pellets. An equilibrium is set up which achieves a conversion to ethanol of 5%. However, the unconverted ethene is recycled until it is all reacted (see Chapter 7). This method is of industrial significance because ethanol is a very important solvent and the product has a high degree of purity.

Testing for unsaturation

10.3.3 Distinguish between alkanes and alkenes using bromine water.

The fact that alkenes readily undergo addition reactions, whereas alkanes do not (they only undergo substitution reactions in the presence of ultraviolet light), can be used as the basis of chemical tests to distinguish between the two homologous series. If separate samples of an alkane (cyclohexane) and an alkene (cyclohexene) are shaken with bromine water at room temperature, the orange/yellow colour of the bromine water is immediately decolorized by the alkene. The alkane produces no reaction and so the colour remains unchanged.

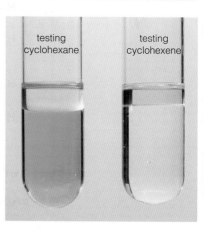

testing cyclohexane

testing cyclohexene

Figure 10.49 The bromine water test for unsaturated hydrocarbons. The unsaturated hydrocarbon decolorizes the bromine water; the saturated compound does not

It is worth noting that the test uses bromine water as the test reagent. In this case the product is not the dibromo- addition product because a hydroxyl (–OH) group replaces one of the bromine atoms.

$$Br_2(aq) \ + \ H_2O(l) \longrightarrow HBr(aq) \ + \ HBrO(aq)$$

orange/yellow

colourless

Extension: Using combustion to distinguish between hydrocarbons

Alkenes also differ from alkanes when they are burnt. Because they have a higher ratio of carbon to hydrogen, alkenes contain much more unburnt carbon than alkanes when they burn under similar conditions. This gives them a much dirtier, smokier flame. Aromatic compounds – which contain a benzene ring – have a higher carbon : hydrogen ratio still, and so burn with an even smokier flame.

Addition polymerization of alkenes

10.3.4 Outline the polymerization of alkenes.

Alkenes and substituted alkenes readily undergo addition reactions by breaking one of their double bonds (it is the pi (π) bond that breaks). Because of this they can be join together to produce long chains known as **polymers**. The alkene used in this type of reaction is known as the **monomer** and its chemical nature will determine the properties of the polymer. Polymers, typically containing thousands of molecules of the monomer, are among the major products of the organic chemical industry. Indeed, many of our most common and useful plastics are polymers of alkenes. Figure 10.50 shows an analogy for the polymerization process – the individual beads are the monomers, and they are able to join together in a long chain, which represents the polymer molecule.

For example, ethene polymerizes to form poly(ethene), commonly known as polythene (Figure 10.51). This molecule was first synthesized in 1935 at the Imperial Chemical Industries (ICI) in the United Kingdom. The process was discovered largely by accidental contamination of the reactants with oxygen, and the product was originally called 'alkathene'. It had excellent electrical insulating properties and was used in the development of radar during the Second World War. It is commonly used in household containers, carrier bags, water tanks and piping.

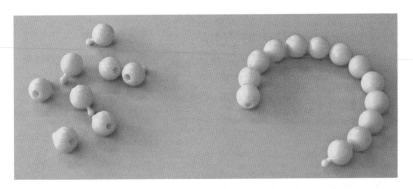

Figure 10.50 The assembly of beads into a chain represents a model of the polymerization process

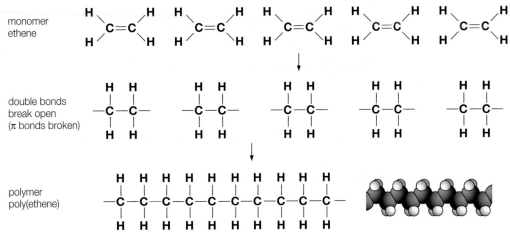

monomer ethene

double bonds break open (π bonds broken)

Figure 10.51 A diagrammatic representation of the polymerization of ethene to form poly(ethene)

polymer poly(ethene)

The process outlined in Figure 10.51 is summarized in an equation that is often written as shown below. Here n represents the number of repeating units and is a large number.

Language of Chemistry

Although the chains formed by the addition polymerization process are saturated, the names of the polymers formed still contain the suffix *-ene*. This is because the standard way of naming the polymer is to use the prefix *poly-* followed by the name of the monomer in brackets, for example poly(propene).

Do also note the standard way of representing a polymer in an equation – with the repeating unit in brackets followed by the letter n symbolizing a large number. ∎

Following the discovery of poly(ethene) and its usefulness, considerable research was carried out to produce other addition polymers with modified properties to suit many diverse practical uses. Propene polymerizes to form poly(propene), often called polypropylene. This polymer is used in the manufacture of clothing, especially thermal wear for outdoor activities.

propene   poly(propene)

■ Extension: Other addition polymers

Poly(chloroethene), also known as PVC (polyvinyl chloride), is very widely used in all forms of construction materials, packaging, electrical cable sheathing and so on. It is one of the world's most important plastics.

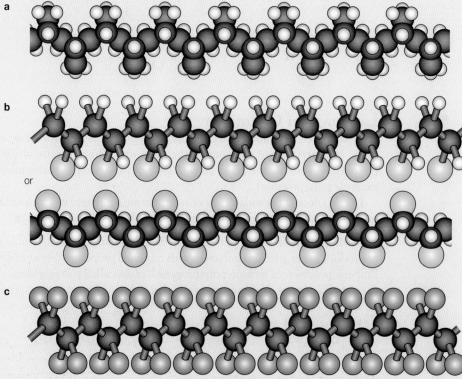

chloroethene
(vinyl chloride)

poly(chloroethene)
(PVC)

Its widespread use is, however, somewhat controversial as its synthesis is associated with some toxic byproducts known as dioxins which must be very carefully contained. Dioxins (Chapter 25) are linked to reproductive disorders and a variety of cancers.

Another interesting polymer is known as PTFE – poly(tetrafluoroethene). It has distinctive non-adhesive surface properties, and is widely used in non-stick pans under registered trademark names such as Teflon®. It also comprises one of the layers in the manufacture of waterproof and breathable fabrics such as Gore-tex®.

tetrafluoroethene

poly(tetrafluoroethene)
(PTFE)

Figure 10.52 shows model structures of these three highly useful polymers.

a

b

or

c

Figure 10.52
Representations of the chain structures of **a** poly(propene), **b** poly(chloroethene) and **c** poly(tetrafluoroethene). Do note that the –Cl or –CH$_3$ side-chains in **a** and **b** are attached to every *alternate* carbon atom in the chain

The uses of these manufactured polymers are varied and diverse and are summarized in Table 10.9.

Polymer (and trade-names(s))	Monomer	Properties	Examples of use
Poly(ethene) (polyethylene, polythene, PE)	Ethene $CH_2{=}CH_2$	Tough, durable	Plastic bags, bowls, bottles, packaging
Poly(propene) (polypropylene, PP)	Propene $CH_3CH{=}CH_2$	Tough, durable	Crates and boxes, plastic rope
Poly(chloroethene) (polyvinyl chloride, PVC)	Chloroethene $CH_2{=}CHCl$	Strong, hard (not as flexible as polythene)	Insulation, pipes and guttering
Poly(tetrafluoroethene) (polytetrafluoroethylene, Teflon, PTFE)	Tetrafluoroethene $CF_2{=}CF_2$	Non-stick surface, withstands high temperatures	Non-stick frying pans, non-stick taps and joints
Poly(phenylethene) (polystyrene, PS)	Phenylethene (styrene) $C_6H_5CH{=}CH_2$	Light, poor conductor of heat	Insulation, packaging (foam)

Table 10.9 Some uses of addition polymers

The economic importance of the reactions of alkenes

10.3.5 Outline the economic importance of the reactions of alkenes.

Alkenes readily undergo addition reactions and they are used as starting materials in the manufacture of many industrially important chemicals. Figure 10.53 shows some of the different major industrial addition reactions that involve ethene, which is obtained from the catalytic cracking of the hydrocarbon fractions in the distillation of crude oil.

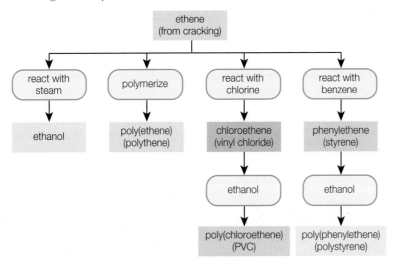

Figure 10.53 Some of the industrial products derived from ethene

---

History of Chemistry

Roy Plunkett (1910–1994) was an American industrial chemist who worked for DuPont. In 1938 he accidentally invented Teflon. He was working on tetrafluoroethene when he found that a sample had polymerized into a waxy solid with amazing properties such as low surface friction, high heat resistance and resistance to corrosion. He was also involved in the production of tetraethyllead(IV) and worked generally on halogenoalkanes.

10.4 Alcohols

Alcohols contain the hydroxyl (–OH) functional group, and have the general formula $C_nH_{2n+1}OH$. As the hydroxyl group is a polar group containing a hydrogen atom, it increases the solubility in water of the molecules relative to the corresponding alkanes. The most common alcohol, ethanol C_2H_5OH, is readily soluble in water. The alcohol molecules can form hydrogen bonds with water through the hydroxyl group.

The complete combustion of alcohols

10.4.1 **Describe,** using equations, the complete combustion of alcohols.

Alcohols burn in air or oxygen to form carbon dioxide and water. The reactions are strongly exothermic. Indeed, alcohols are important fuels and are used in alcohol burners and similar heaters. The amount of energy released per mole of alcohol increases as we go up the homologous series. This is mainly due to the fact that the amount of carbon dioxide produced per mole of the alcohol increases going up the series. We have seen before that it is the strength of the bonding in carbon dioxide that contributes greatly to the exothermic nature of these combustion reactions. The equations for the burning of a number of different alcohols are given below:

CO_2 : alcohol ratio

$$2CH_3OH(l) + 3O_2(g) \rightarrow 2CO_2(g) + 4H_2O(l) \qquad \Delta H_c^\ominus = -726 \, kJ \, mol^{-1} \qquad 1:1$$

$$C_2H_5OH(l) + 3\tfrac{1}{2}O_2(g) \rightarrow 2CO_2(g) + 3H_2O(l) \qquad \Delta H_c^\ominus = -1371 \, kJ \, mol^{-1} \qquad 2:1$$

$$2C_5H_{11}OH(l) + 15O_2(g) \rightarrow 10CO_2(g) + 12H_2O(l) \qquad \Delta H_c^\ominus = -3330 \, kJ \, mol^{-1} \qquad 5:1$$

Even though they contain oxygen, the alcohols will still behave as hydrocarbons do and produce carbon monoxide instead of carbon dioxide when burnt in a limited supply of oxygen.

For several years some countries such as Brazil have combined ethanol with gasoline to produce a fuel for cars known as gasohol. This makes a country less dependent on the supply of gasoline but weight for weight gasohol is not as efficient in terms of energy production.

Worked example

Working from the figures given for the burning of ethanol, and those for octane given in the equation below, show that one gram of octane produces over 60% more energy than the same mass of ethanol.

$$2C_8H_{18}(l) + 25O_2(g) \rightarrow 16CO_2(g) + 18H_2O(l) \qquad \Delta H_c^\ominus = -5512 \, kJ \, mol^{-1}$$

From the equation for ethanol:

1 mole (46 g) of ethanol produces 1371 kJ of energy

Therefore 1 g of ethanol produces 29.8 kJ

For octane:

1 mole (114 g) produces 5512 kJ of energy

Therefore 1 g produces 48.4 kJ

$$\frac{48.4}{29.8} \times 100\% = 162\%$$

So 1 g of octane produces 62% more energy than 1 g of ethanol when burnt.

Even though the alcohols are less energy efficient than hydrocarbon fuels, they still have the advantage that they can be produced form renewable sources. Ethanol can be produced by fermentation which initially suggested that ethanol was a 'carbon-neutral' fuel.

■ Extension: Biofuels

Following the early example of Brazil, which has used bioethanol as a transport fuel for many years (Figure 10.54), other countries have increased their production of fuels from carbohydrate-rich crops such as corn or sugar cane. These fuels were initially seen as 'carbon neutral' because

Figure 10.54 Both ethanol and methanol find their use as biofuels

re-growing the crops would absorb the same amount of carbon dioxide as was released when burning the fuel. In 2003 the European Union set a target of replacing 5.75% of all transport fossil fuels (petrol and diesel) with biofuels by 2010. The key advantage of these 'first-generation' biofuels was that they were easy to make with established technology. The methods used were not fundamentally different from those used to make vodka or cooking oil.

However, this increased production of such biofuels has quickly shown up some unforeseen economic shortcomings. Fuel crop production competes with food crops for land use and this situation is making food less affordable in developing countries. The green credentials of these early biofuels have come under question, too, as the intensive farming methods required for efficient mass production use significant amounts of energy. By the end of 2007, it had become clear that the potential solution had become part of the problem.

The oxidation reactions of alcohols

10.4.2 **Describe,** using equations, the oxidation reactions of alcohols.

The alcohol functional group is capable of being oxidized to other important organic molecules. Such reactions alter the functional group. The remaining part of the carbon skeleton is left unaffected. The products possible from oxidation depend on whether the alcohol concerned is primary, secondary or tertiary.

Various oxidizing agents can be used for these reactions. The most commonly used laboratory oxidizing agent is acidified potassium dichromate(VI). This is a bright orange solution. When the reaction mixture is heated a colour change takes place as the Cr(VI) is reduced to Cr(III), which is green, while the alcohol is oxidized (Figure 10.55).

Figure 10.55 The oxidation of alcohols with acidified potassium dichromate(VI) solution. The different alcohols react differently; tertiary alcohols do not react

In writing equations for these reactions it is often easier to show the oxidizing agent simply as [O]. The oxidation reactions of the different alcohols are described below.

Primary alcohols

Primary alcohols, such as ethanol, are oxidized in a two-stage process; firstly to an aldehyde, and then to a carboxylic acid. Thus ethanol is first oxidized to ethanal. This can be viewed as oxidation by removal of hydrogen.

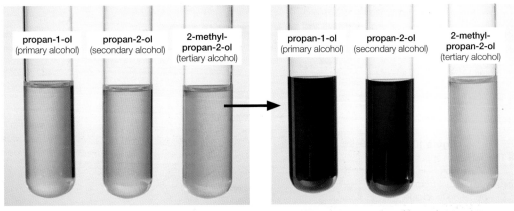

The second stage is the conversion of ethanal to ethanoic acid. The oxidation of ethanol to ethanoic acid is one of the oldest chemical oxidations practised by humans. It is the reaction used when wine is left exposed to air and bacterial action to produce vinegar.

$$CH_3CHO + [O] \longrightarrow CH_3COOH$$

In summary, any primary alcohol will undergo the following sequence of oxidation reactions:

Experimental conditions can be adjusted when carrying out these oxidations in order to prepare the different products. If the aldehyde is the desired product, then it is possible to remove it from the reaction mixture by distilling it off as it forms (Figure 10.56a). This is achievable because aldehydes have lower boiling points than either alcohols or carboxylic acids. Unlike the alcohols or the carboxylic acids, they do not have the capacity for hydrogen bonding between their molecules. An excess of the alcohol over the oxidizing agent can also favour the production of the aldehyde.

However, if we want to obtain the carboxylic acid as the product, we must leave the aldehyde in contact with the oxidizing agent for a prolonged period of time. In this case the apparatus is set up for reflux (Figure 10.56b), and an excess of the oxidizing agent is used to favour complete oxidation to the carboxylic acid.

Figure 10.56
a The distillation apparatus used to obtain the aldehyde product from the oxidation of ethanol; **b** the reflux apparatus used in the complete oxidation of ethanol to ethanoic acid

Secondary alcohols

Secondary alcohols have just a single hydrogen attached to the carbon atom that carries the functional group. This means that when secondary alcohols are oxidized there is just one product possible – a ketone. The oxidation of propan-2-ol produces propanone as the organic product.

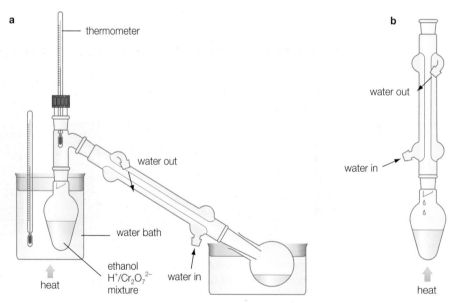

$$CH_3CH(OH)CH_3 + [O] \longrightarrow CH_3COCH_3 + H_2O$$

Tertiary alcohols

Tertiary alcohols are not readily oxidized under comparable mild conditions, as there is no hydrogen atom attached to the carbon atom to which the hydroxyl group is attached. Any oxidation of tertiary alcohols requires more drastic conditions as it is necessary to break the carbon skeleton of the molecule. Therefore we do not see a colour change in the acidified potassium dichromate(VI) oxidizing agent when it is heated with a tertiary alcohol (Figure 10.55).

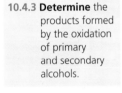

no oxidation possible as
no hydrogen atom on carbon
atom bonded to the alcohol group

■ **Extension:** ## Redox half-equations for the oxidation of alcohols

The oxidation of alcohols by acidified potassium dichromate(VI) can be represented by equations in which the oxidizing agent is written as [O] and the reactions are considered in terms of the removal of hydrogen or the gain of oxygen.

However, equations involving the loss of electrons can also be written and the overall reaction represented as the combination of two half-equations, one for the oxidation and the other for the associated reduction (see Chapter 9, for examples involving ethanol and ethanal).

Taking another example, the oxidation of propan-2-ol to propanone can be represented by the following half-equation:

$$CH_3CH(OH)CH_3(aq) \rightarrow CH_3COCH_3(aq) + 2H^+(aq) + 2e^-$$

Oxidation products of primary and secondary alcohols

10.4.3 **Determine** the products formed by the oxidation of primary and secondary alcohols.

The initial products of the oxidation of alcohols, whether from primary or secondary alcohols, all contain the $>C=O$ group. This group is present in both aldehydes and ketones.

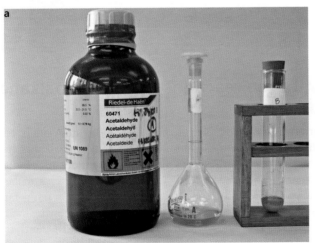

Figure 10.57 a The 2,4-dinitrophenylhydrazine test for aldehydes and ketones. An orange precipitate is formed in either case. **b** A commercial melting point apparatus

The standard test for an aldehyde or ketone is that they both form orange crystalline precipitates with 2,4-dinitrophenylhydrazine solution (Figure 10.57a). The precipitate can be recrystallized and its melting point determined (Figure 10.57b). Knowing the melting point of the crystals enables us to identify the particular aldehyde or ketone tested.

The 2,4-dinitrophenylhydrazine test does not distinguish between aldehydes and ketones. However, there are two simple tests which can do so, based on the fact that aldehydes can be oxidized whereas ketones cannot. Fehling's solution and Tollens' reagent are both mild oxidizing reagents that react with aldehydes to produce carboxylic acids (Figure 10.58).

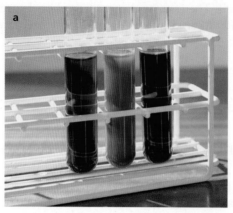

Figure 10.58 a Fehling's solution produces an orange-brown precipitate with aldehydes; **b** Tollens' reagent produces a 'silver mirror' on the inside of the test tube. The middle tube in each case contained ethanal, while the right-hand tube contained propanone. The first tube in each case is the unreacted starting reagent

Fehling's solution contains alkaline copper(II) sulfate and the precipitate is copper(I) oxide. Tollens' reagent is a solution of silver nitrate in ammonia and the precipitate of metallic silver coats the inside of the test tube producing a 'mirror'.

10.5 Halogenoalkanes

Halogenoalkanes contain an atom of fluorine, chlorine, bromine or iodine bonded to the carbon skeleton of the molecule. They have the general formula $C_nH_{2n+1}X$, where X = a halogen. They are generally oily liquids (Figure 10.59) that do not mix with water.

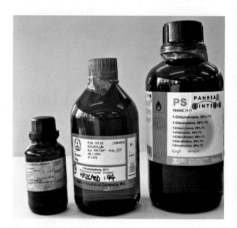

Figure 10.59
Halogenoalkanes are usually oily liquids

The substitution reactions of halogenoalkanes with sodium hydroxide

10.5.1 Describe, using equations, the substitution reactions of halogenoalkanes with sodium hydroxide

Halogenoalkanes are saturated molecules but the halogen atom can be replaced by other atoms or groups in **substitution reactions.** This means that halogenoalkanes are very useful in reaction pathways that enable us to synthesize a range of important organic products.

Halogenoalkanes are also used directly in many products. In particular, the group of compounds known as CFCs (chlorofluorocarbons) were used in refrigerants and aerosol propellants in many parts of the world from the 1930s. The growing awareness of their role in breaking down the stratospheric **ozone layer,** which protects the Earth from harmful ultraviolet radiation, has led to regulations for their distribution and use being introduced following the Montreal Protocol. Sadly, the stability of these molecules is such that even though they are no longer being released in large quantities they are likely to remain active and hence destructive in the atmosphere for generations.

The greater electronegativity of the halogen atom means that the carbon–halogen bond is polarized, resulting in a charge distribution as follows:

$$-\overset{|}{\underset{|}{C}}\overset{\delta+}{\longrightarrow}X^{\delta-}$$

The carbon atom attached to the halogen therefore has a partial positive charge and can be described as being electron deficient. This makes it susceptible to attack by a group of chemicals called **nucleophiles** – species which are themselves electron rich and hence are attracted to a region of electron deficiency. Nucleophiles have a lone pair of electrons and may be negatively charged.

The halogenoalkanes are saturated molecules (like alkanes), and as such they undergo substitution reactions. A good example of this is the substitution reaction involving the hydroxide ion (OH^-) from alkalis such as sodium hydroxide solution ($NaOH(aq)$). In this reaction, the OH^- ion is the nucleophile and will replace (substitute for) the halogen. In the process the halogenoalkane is converted into an alcohol. These reactions are commonly described as S_N reactions, standing for substitution nucleophilic.

The exact mechanism of these reactions depends on whether the halogenoalkane is primary, secondary or tertiary – as this influences the environment of the carbon–halogen bond. In organic reaction mechanisms, it is customary to use **curly arrows** to represent the movement of electron pairs. We will now look more closely at the different mechanisms possible for nucleophilic substitution.

Mechanisms for nucleophilic substitution

10.5.2 Explain the substitution reactions of halogenoalkanes with sodium hydroxide in terms of S_N1 and S_N2 mechanisms.

Primary halogenoalkanes are thought to undergo a substitution mechanism that involves a single reactive step. This one-stage reaction involves the simultaneous attack of the nucleophile and departure of the halide ion. We will use as an example the reaction between bromoethane and sodium hydroxide solution.

$$CH_3CH_2Br(aq) + OH^-(aq) \rightarrow CH_3CH_2OH(aq) + Br^-(aq)$$

Kinetic studies show that this reaction is a single-step reaction in which the halogenoalkane and hydroxide ion are both involved. The rate expression for the reaction is found experimentally to be:

$$\text{rate} = k[CH_3CH_2Br(aq)][OH^-(aq)]$$

Two species are involved in the rate-determining step, and the reaction is said to be bimolecular.

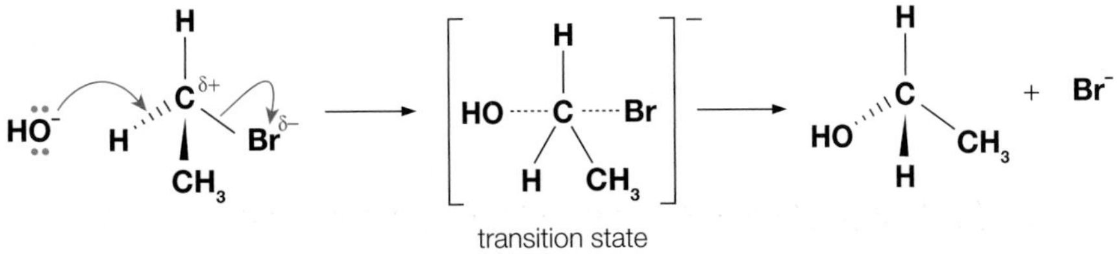

transition state

The nucleophile (OH^-) is attracted to the electron-deficient carbon atom and a transition state is formed in which the carbon–bromine bond is broken at the same time as a new carbon–oxygen bond is formed. The bromine atom then leaves as a bromide ion, and the alcohol (in this case ethanol) is formed. This mechanism is fully described as S_N2 (substitution nucleophilic bimolecular).

Tertiary halogenoalkanes also undergo a substitution reaction, but kinetic studies show that the mechanism is different from that occurring with primary halogenoalkanes. For example, consider the reaction between 2-bromo-2-methylpropane and hydroxide ions:

$$CH_3C(CH_3)_2Br(aq) + OH^-(aq) \rightarrow CH_3C(CH_3)_2OH(aq) + Br^-(aq)$$

$$\text{rate} = k[CH_3C(CH_3)_2Br(aq)]$$

Kinetic studies show that this reaction is unimolecular, with the rate-determining step involving just the halogenoalkane molecule.

This reaction has a different mechanism for several reasons. The presence of the three alkyl groups around the carbon of the carbon–halogen bond (see Figure 10.60b) causes what is called **steric hindrance**, meaning that these bulky groups make it difficult for an incoming group to attack this carbon atom. Note the difference when compared with the situation involving a primary molecule where hydrogen atoms are attached to the carbon atom (Figure 10.60a).

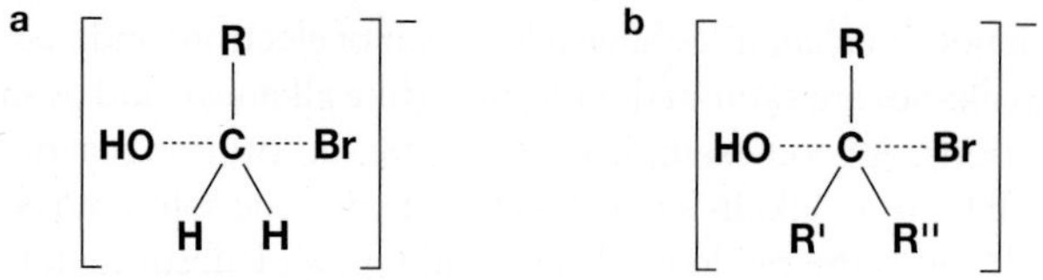

Figure 10.60 The structure of the intermediate in an SN2 reaction involving **a** a primary halogenoalkane, and **b** a tertiary halogenoalkane (the bulky R groups make this mechanism difficult)

Instead, the first step of the reaction with a tertiary halogenoalkane involves ionization of the halogenoalkane through the breaking of its carbon–bromine bond. This is an example of **heterolytic fission**. The pair of electrons in the bond both end up on the halogen, forming the bromide ion. This leaves a temporary positive charge on the electron-deficient carbon atom, which is known as a **carbocation**.

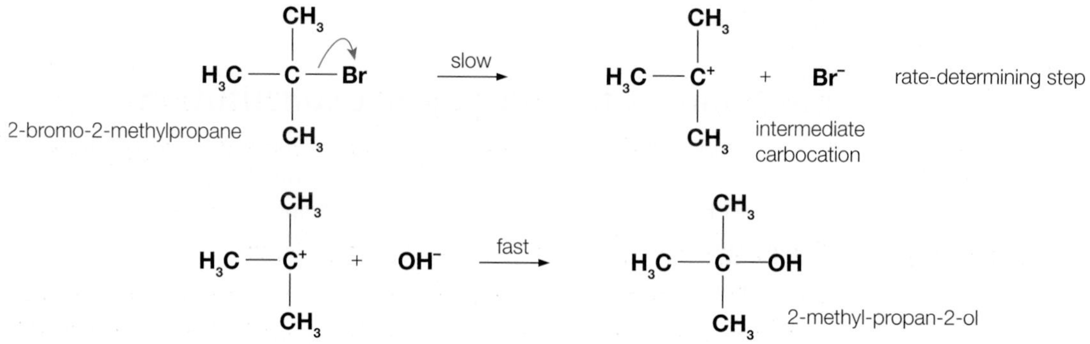

Another factor which favours this mechanism is that the carbocation is stabilized by the presence of the three alkyl groups, as each of these has an electron-donating effect (sometimes called a positive inductive effect), shown by the arrows in the structure on the right.

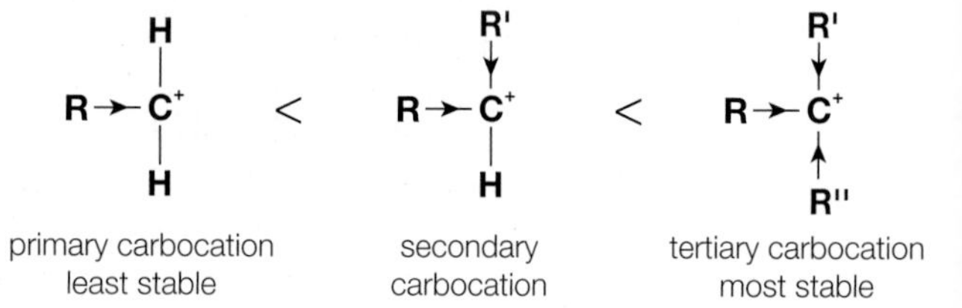

As the slow step of this reaction is determined by the concentration of only *one* reactant (the halogenoalkane), it is described as a **unimolecular** reaction. This reaction mechanism is therefore described as S_N1 (substitution nucleophilic unimolecular).

The mechanism of nucleophilic substitution in secondary halogenoalkanes is less easy to define as the data show that they usually undergo a mixture of both S_N1 and S_N2 mechanisms, depending on the reaction conditions, or, more likely, some mechanism in between the two.

The relative reactivity of the different halogens in these reactions depends on the strength of their bonds with carbon and this decreases as we go down the halogen group. So the iodoalkane with the longest, and hence weakest, carbon–halogen bond is the most reactive and the fluoroalkane is the least reactive. Kinetic studies on these reactions can be carried out by a variety of methods. One interesting method, which establishes the S_N1 mechanism for tertiary halogenoalkanes, is to follow the increase in conductivity of the reaction mixture when 2-chloro-2-methylpropane is hydrolysed by water.

As the reaction proceeds, the conductivity of the reaction mixture increases as the chloride ion (Cl^-) is released from the halogenoalkane molecule. The hydrogen ion (H^+) is also produced. Figure 10.61 shows one trace of the increase in conductivity obtained using a conductivity sensor linked to a datalogger. A series of such traces at different concentrations of halogenoalkane can demonstrate that the reaction is first order (there is just one molecule involved in the rate-determining step – see Chapter 16).

$$CH_3C(CH_3)(Cl)CH_3 + H_2O \rightarrow CH_3C(CH_3)(OH)CH_3 + HCl$$

$$\text{rate} = k[CH_3C(CH_3)(Cl)CH_3]$$

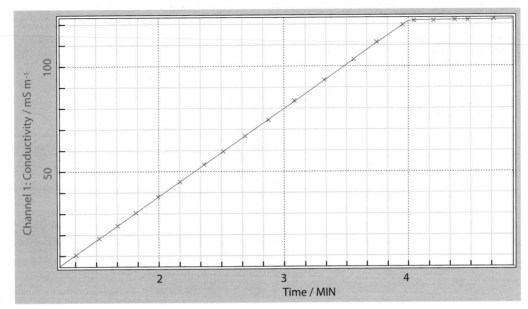

Figure 10.61 A datalogged trace of the increase in conductivity when 2-chloro-2-methylpropane is hydrolysed by water

■ Extension: Curly arrows and reaction mechanisms

In describing the detail of the transformations that take place in reaction mechanisms with the breaking of existing bonds and the making of new bonds we need a system that illustrates the movement of the electrons involved. Each covalent bond is made up of two electrons and so we are effectively having to denote the movement of *electron pairs*.

The generally accepted system is based on depicting the movement of pairs of electrons using 'curly arrows'. The blunt end of the arrow indicates the initial position of the electron pair – illustrated below by the arrow starting at the lone pair on the OH⁻ ion on the left. The arrow head is positioned to show where the electrons end up – thus the pair of electrons indicated will form a bond between the oxygen atom and the carbon atom.

$$HO^{\cdot}_{\cdot\cdot} \quad -\overset{\delta+}{\underset{|}{C}}-X^{\delta-} \longrightarrow HO-C \blacktriangleleft \; + \; X^-$$

It is worth noting the slightly different second 'curly arrow' on the diagram. This shows that the electrons involved in the C–X bond move from between those two atoms to a position on the more electronegative X atom. This shows the departure of the X atom as an X⁻ ion. You will be expected to use 'curly arrows' accurately in your IB examination papers. This system of indicating the movement of electrons is useful in depicting mechanisms such as S_N1 and S_N2.

There is a second aspect to the system of illustrating the movement of electrons which applies in the other reaction mechanism we have discussed in this chapter – namely the free-radical mechanism for substitution of halogens into alkanes. Here we indicate the movement of single electrons by the use of 'half arrows'. The diagrams below show the movement of the electrons in the two propagation steps involved in the chlorination of methane using these 'half arrows'.

$$Cl\cdot(g) \quad + \quad H-CH_3(g) \longrightarrow HCl(g) \quad + \quad \cdot CH_3(g)$$
chlorine atom + methane → hydrogen chloride + methyl radical

$$\cdot CH_3(g) \quad + \quad Cl-Cl(g) \longrightarrow CH_3Cl(g) \quad + \quad Cl\cdot(g)$$
methyl radical + chlorine → chloromethane + chlorine atom

Note that the IB syllabus does not require you to be able to use this 'half arrow' system in your examinations.

TOK Link

Organic reaction mechanisms are theories, and cannot be proved beyond doubt (Chapter 16). Chemists are led to accept a mechanism for a particular reaction because it provides the most satisfactory way of understanding all the data about that reaction. However, new facts may later be discovered which are not consistent with the current accepted theory and it must then be rejected or, more often, modified. So the field of organic chemistry is a continually developing one.

10.6 Reaction pathways

In reading through this chapter you may have noted that a number of very significant organic chemicals were discovered accidentally, suggesting that perhaps we should always investigate unexpected results carefully. However, many novel substances are developed intentionally as the result of purposeful research. Organic chemistry lends itself to this logic of synthesis since we are moving different combinations of defined 'building blocks' – the molecular skeletons and the different functional groups.

The development of new organic compounds – from pharmaceutical drugs to synthetic dyes, clothing fibres to new construction materials – represents a major part of modern industrial organic chemistry. The oil industry is the main source of organic compounds for starting these developmental processes, but it does not generally yield the required proportion of desired compounds.

As a result, organic chemists typically have to convert compounds from one form into another, often by linking reactions such as those met in this chapter into sequences of several steps, known as a **reaction pathway.** Deciding on a 'reaction route' between starting compound and desired product is a useful skill in modern organic chemistry.

10.6.1 Deduce reaction pathways given the starting materials and the product.

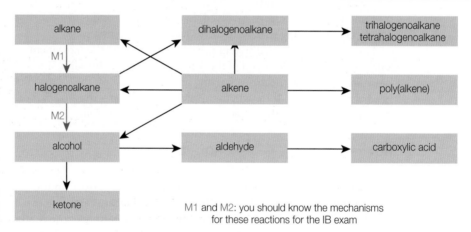

Figure 10.62 Reaction flowchart for organic synthesis

M1 and M2: you should know the mechanisms for these reactions for the IB exam

The flowchart in Figure 10.62 shows the interlinking of the reactions mentioned in this chapter. Using this chart you can see that it is possible, for instance, to convert 2-bromobutane into butanone using a two-step synthetic route:

$$\text{2-bromobutane} \xrightarrow[\text{reflux with NaOH(aq)}]{} \text{butan-2-ol} \xrightarrow[\text{reflux with H}^+/\text{ Cr}_2\text{O}_7{}^{2-}]{} \text{butanone}$$

Worked examples

Devise two-step syntheses of the following products from the stated starting material. Include any experimental conditions.

a Ethanoic acid from ethene. **b** Butan-1-ol from butane. **c** Propanal from 1-bromopropane.

a $\text{ethene} \xrightarrow[\text{conc. H}_2\text{SO}_4/\text{H}_2\text{O}]{} \text{ethanol} \xrightarrow[\text{reflux with H}^+/\text{Cr}_2\text{O}_7{}^{2-}]{} \text{ethanoic acid}$

b $\text{butane} \xrightarrow[\text{Br}_2/\text{UV light}]{} \text{1-bromobutane} \xrightarrow[\text{reflux with NaOH(aq)}]{} \text{butan-1-ol}$

c $\text{1-bromopropane} \xrightarrow[\text{reflux with NaOH(aq)}]{} \text{propan-1-ol} \xrightarrow[\text{reflux with H}^+/\text{Cr}_2\text{O}_7{}^{2-}]{} \text{propanal}$

SUMMARY OF KNOWLEDGE

- Carbon is a particularly versatile element in that its atoms can form single and multiple covalent bonds with other carbon atoms. Carbon atoms can also form themselves into long chains or rings. This versatility results in a vast range of organic compounds.

- The large variety of carbon-containing compounds can be categorized into homologous series defined firstly by the functional group they contain.

- The different members of a homologous series have the same general formula, show similar chemical properties and a regular gradation in certain physical properties as the chain length is extended.

- There is a set of functional groups which can be inserted into different hydrocarbon skeletons to produce the different homologous series.

- The IUPAC system is a means of systematically naming the compounds in the various series so that the properties of the homologous series can be rationally discussed and analysed.

- Linked to the systematic naming of the organic compounds is also an agreed method of representing the structures of these compounds. There are three levels of formula that begin to describe the structures involved in organic compounds: the empirical formula, the molecular formula and the structural formula.

- The alkanes are a series of organic compounds that are relatively unreactive. The alkanes are very important fuels as they burn well, releasing large amounts of heat energy.

- The alkanes undergo substitution reactions with chlorine and bromine. These reactions are photochemical chain reactions that require ultraviolet radiation. These substitution reactions proceed by a free-radical mechanism.

- The alkenes are a homologous series of unsaturated hydrocarbons. They are more reactive than the alkanes and undergo a range of important addition reactions. These addition reactions include hydrogenation (as used in the manufacture of margarine), hydration (as used in the industrial production of ethanol), halogenation and addition polymerization (as used in the production of poly(ethene)).

- The alcohols are an important homologous series of compounds that contain the hydroxyl (−OH) group as their functional group. They are increasingly important as fuels. They are able to be chemically oxidized to useful products, the nature of which depends on whether the starting alcohol is primary, secondary or tertiary. Primary alcohols yield aldehydes or carboxylic acids on oxidation, while secondary alcohols yield ketones. Tertiary alcohols cannot be oxidized under mild conditions.

- Halogenoalkanes are a homologous series of compounds that contain a halogen group attached to a hydrocarbon chain. They are relatively stable molecules used as solvents, fire retardants, aerosol propellants, etc.

- Halogenoalkanes undergo substitution reactions in the presence of nucleophiles such as hydroxide ions (OH^-). The reaction mechanism involved depends on whether the halogenoalkane has a primary, secondary or tertiary structure.

- Primary halogenoalkanes undergo S_N2 reactions, whereas tertiary halogenoalkanes take part in S_N1 reactions.

- Because of the nature of organic reactions, and the structure of the compounds involved, it is possible to devise organic synthesis pathways in order to make compounds of interest. For example, it is possible to synthesize a carboxylic acid from a halogenoalkane by the following synthetic route: halogenoalkane → alcohol → carboxylic acid

Examination questions – a selection

Paper 1 IB questions and IB style questions

Q1 How many structural isomers are possible with the molecular formula C_6H_{14}?

A 4   **B** 6   **C** 5   **D** 7

Q2 Which compound is a member of the aldehyde homologous series?

A $CH_3CH_2COCH_3$ **C** $CH_3CH_2CH_2OH$
B CH_3CH_2CHO **D** CH_3CH_2COOH

Q3 Which compound is a member of the same homologous series as 1-chloropropane?

A 1-chloropropene   **C** 1-bromopropane
B 1,2-dichloropropane   **D** 1-chlorobutane

Q4 The following is a three-dimensional representation of an organic molecule.

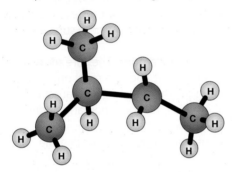

Which statement is correct?
A The correct IUPAC name of the molecule is 2-methylpentane.
B All the bond angles will be approximately 90°.
C One isomer of this molecule is pentane.
D The boiling point of this compound would be higher than that of pentane.

Standard Level Paper 1, Specimen 09, Q26

Q5 What is the organic product of the reaction between ethanol and ethanoic acid in the presence of concentrated sulfuric acid?

A CH_3CHO **C** $CH_3CH_2COOCH_3$
B $CH_3COOCH_2CH_3$ **D** CH_3COOCH_3

Q6 Which formulas represent butane or its isomer?

I $CH_3(CH_2)_2CH_3$
II $CH_3CH(CH_3)CH_3$
III $(CH_3)_3CH$

A I, II and III   **C** I and II only
B I and III only   **D** II and III only

Q7 A *gaseous* hydrocarbon, **X**, decolorizes aqueous bromine. Which one of the following molecular formulas could be **X**?

A C_2H_4 **B** C_6H_{14} **C** $C_{13}H_{28}$ **D** $C_{10}H_{20}$

Q8 What is the function of sunlight during the reaction between chlorine and methane?
A to dissociate the chlorine molecules into atoms
B to dissociate the chlorine molecules into ions
C to increase the temperature of the mixture
D to break C–H bonds in the methane molecules

Q9 Which one of the following is the best method of distinguishing between an alkane and an alkene?
A test with universal indicator paper
B burn the gases in excess oxygen
C test their solubility in water
D add bromine water

Q10 Which equation represents the combustion of methane in excess oxygen?
A $CH_4(g) + O_2(g) \rightarrow C(s) + 2H_2O(l)$
B $CH_4(g) + O_2(g) \rightarrow CO_2(g) + 2H_2(g)$
C $2CH_4(g) + 3O_2(g) \rightarrow 2CO(g) + 4H_2O(l)$
D $CH_4(g) + 2O_2(g) \rightarrow CO_2(g) + 2H_2O(l)$

Q11 Which of the following compounds is not formed by the reaction between excess chlorine and methane (in the presence of sunlight)?
A C_2H_4 **B** C_2H_6 **C** $CHCl_3$ **D** CCl_4

Q12 What type of alcohol is 2-methylpropan-2-ol?
A an unsaturated alcohol **C** a primary alcohol
B a tertiary alcohol   **D** a secondary alcohol

Q13 What is the IUPAC name for $(CH_3)_2C(OH)CH_2CH_3$?
A 2-methylbutan-2-ol   **C** 3-methylbutan-3-ol
B 4-methylbutan-3-ol   **D** pentan-3-ol

Q14 How do the bond angles and bond lengths in ethane and ethene compare?

	H–C–H bond angle in ethane	C–C bond length in ethane
A	larger	longer
B	smaller	longer
C	smaller	shorter
D	larger	shorter

Q15 Which compound is an ester?
A CH_3CH_2COOH **C** C_3H_7CHO
B $C_2H_5OC_2H_5$ **D** $HCOOCH_3$

Q16 The oxidation of propan-2-ol, $CH_3CH(OH)CH_3$, by sodium dichromate(VI) leads to the formation of:
A propanone (CH_3COCH_3)
B propan-l-ol ($CH_3CH_2CH_2OH$)
C propanal (CH_3CH_2CHO)
D propanoic acid ($CH_3CH_2CO_2H$)

Q17 When the compounds below are listed in order of **decreasing** boiling point (highest to lowest) what is the correct order?

1. ethane   2. choroethane
3. ethanol   4. ethanoic acid

A 2, 1, 3, 4   **C** 3, 4, 1, 2
B 4, 3, 2, 1   **D** 4, 3, 1, 2

Q18 What is the name of the compound whose condensed structural formula is $CH_3CH_2CH_2COOCH_3$?

A butyl methanoate   **C** methyl propanoate
B methyl butanoate   **D** pentanone

Q19 Which of the following descriptions can be correctly applied to the homologous series of alkanes?

  I Members of the series have the general formula C_nH_{2n+2}.
  II Members of the series have similar chemical properties.
  III Members of the series are isomers of each other.

A I only   **C** I and II only
B II only   **D** I and III only

Q20 The compound which is expected to have the lowest boiling point at a pressure of one atmosphere is:

A $CH_3CH_2CH_2F$ **C** CH_3CH_2COOH
B $CH_3CH_2CH_2OH$ **D** $CH_3CH_2CH_2NH_2$

Paper 2 IB questions and IB style questions

Q1 An alkane has the percentage composition 84.5% carbon and 15.5% hydrogen by mass.
  a Calculate the empirical formula of the alkane. [2]
  b The molecular mass of the alkane was found to be 142 using a mass spectrometer. What is the molecular formula? [2]
  c **i** The hydrocarbon can be used as a fuel. Write the balanced equation for the complete combustion of this alkane in oxygen. [2]
    ii Write a balanced equation for the incomplete combustion of this alkane in a limited supply of oxygen. [2]
  d When a hydrocarbon is cracked, it is broken into smaller molecules. Complete the following cracking reactions:
    i $C_8H_{18} \rightarrow C_4H_8 +$ _____ [1]
    ii $C_{13}H_{28} \rightarrow C_4H_{10} + C_4H_8 +$ _____ [1]

Q2 Ethene, propene and but-2-ene are members of the alkene homologous series.
  a Describe **three** features of members of a homologous series. [3]
  b State and explain which compound has the highest boiling point. [3]

  c Draw the structural formula and give the name of an alkene containing five carbon atoms. [2]
  d Write an equation for the reaction between but-2-ene and hydrogen bromide, showing the structure of the organic product. State the type of reaction occurring. [3]
  e Propene can be converted to propanoic acid in three steps:

         step1   step 2   step 3
propene $\rightarrow$ propan-1-ol $\rightarrow$ propanal $\rightarrow$ propanoic acid

    State the type of reaction occurring in steps 2 and 3 and the reagents needed. Describe how the conditions of the reaction can be altered to obtain the maximum amount of propanal, and in a separate experiment, to obtain the maximum amount of propanoic acid. [5]
  f Identify the strongest type of intermolecular force present in each of the compounds propan-1-ol, propanal and propanoic acid. List these compounds in decreasing order of boiling point. [4]
               Standard Level Paper 2, Nov 05, Q7

Q3 a An organic compound, **A**, containing only the elements carbon, hydrogen and oxygen was analysed.
    i **A** was found to contain 54.5% C and 9.1% H by mass, the remainder being oxygen. Determine the empirical formula of the compound. [3]
    ii The molecular mass of **A** is 88. What is the molecular formula of **A**? [2]
  b An organic compound **X** contains 40.00% carbon, 6.72% hydrogen and 53.28% oxygen by mass.
    i Determine the empirical formula of compound **X**. [2]
    ii Compound **X** has a relative molecular mass of 60.0. Deduce its molecular formula. [2]

Q4 a Give the structural formulas for the isomers of molecular formula C_4H_{10} and name each isomer. [4]
  b Several compounds have the molecular formula $C_3H_6O_2$. Three of them, **A**, **B** and **C**, have the following properties:
    A is soluble in water and is acidic.
    B and **C** are neutral and do not react with bromine or organic acids.
    Give a structural formula for each of these compounds and name them. [6]
  c **i** Explain the solubility and acidity of **A** in water. [2]
    ii Write an equation for the reaction of **A** with sodium hydroxide solution. [1]

iii Explain why **B** and **C** do not react with bromine. [1]

d State and explain which one of **A**, **B** or **C** has the highest boiling point. [2]

e i Name the class of compounds to which **B** and **C** belong and state a use of this class of compounds. [2]

   ii Name the **two** classes of compounds used to form **B** or **C**, and state the other product formed in this reaction. [3]

f Suggest the structural formula of an isomer of $C_3H_6O_2$ which does react rapidly with bromine. Name this type of reaction, and describe an observation that can be made during the reaction. [3]

Q5 a i List **three** characteristics of a homologous series. [3]

   ii Draw the **four** different structural isomers with the formula C_4H_9OH that are alcohols. [4]

b i Ethanoic acid reacts with ethanol in the presence of concentrated sulfuric acid and heat. Identify the type of reaction that takes place. Write an equation for the reaction, name the organic product formed and draw its structure. [4]

   ii State and explain the role of sulfuric acid in this reaction. [2]

   iii State **one** major commercial use of the organic product from this type of reaction. [1]

c Two compounds are shown below.

$$HCOOCH_2CH_3 \quad \text{and} \quad HCOOCHCH_2$$
$$\textbf{I} \qquad\qquad\qquad \textbf{II}$$

   i State and explain which of these two compounds can react readily with bromine. [2]

   ii Compound **II** can form polymers. State the type of polymerization compound **II** can take part in, and draw the structure of the repeating unit of the polymer. [2]

Q6 a The following is a computer-generated representation of the molecule methyl 2-hydroxyl benzoate, better known as oil of wintergreen.

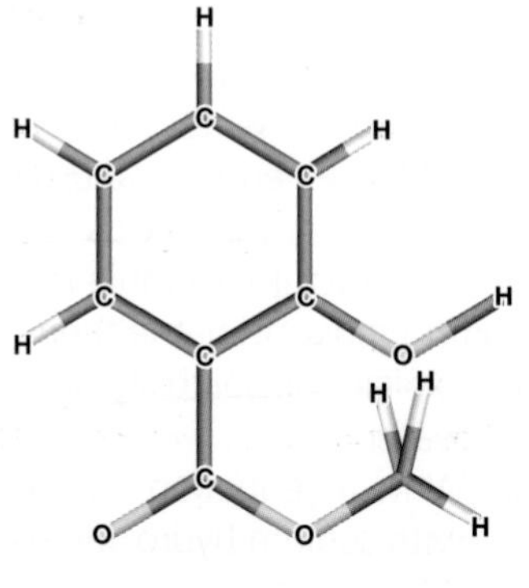

i Deduce the empirical formula of methyl 2-hydroxybenzoate and draw the full structural formula, including any multiple bonds that may be present. The computer-generated representation shown does not distinguish between single and multiple bonds. [2]

ii In this representation, two of the carbon–oxygen bond lengths shown are 0.1424 nm and 0.1373 nm, respectively. Explain why these are different and predict the carbon–oxygen bond length in carbon dioxide. [2]

iii Name two of the functional groups present in the molecule. [2]

b i State and explain the trend in the boiling points of the first six straight-chain alkanes. [2]

   ii Write an equation for the reaction between methane and chlorine to form chloromethane. Explain this reaction in terms of a free-radical mechanism. [5]

c i Identify the formulas of the organic products, **A–E**, formed in the reactions **I–IV**:

$$\textbf{I} \quad CH_3(CH_2)_8OH + K_2Cr_2O_7 \xrightarrow{H^+} \textbf{A} \xrightarrow{H^+} \textbf{B}$$
$$\textbf{II} \quad (CH_3)_3CBr + NaOH \rightarrow \textbf{C}$$
$$\textbf{III} \quad (CH_3)_2CHOH + K_2Cr_2O_7 \xrightarrow{H^+} \textbf{D}$$
$$\textbf{IV} \quad H_2C{=}CH_2 + Br_2 \rightarrow \textbf{E} \qquad [5]$$

ii $H_2C{=}CH_2$ can react to form a polymer. Name this **type** of polymer and draw the structural formula of a section of this polymer consisting of three repeating units. [2]

Standard Level Specimen Paper 2, 2009, Q8

Q7 a Name the following alcohols:

   i $CH_3CH(OH)CH_3$

   ii $CH_3CH_2CH_2OH$

   iii $CH_3CH_2C(OH)(CH_3)CH_3$

   iv $CH_2(OH)CH_2(OH)$ [4]

b For the four alcohols listed in **a**, state whether they are primary, secondary or tertiary alcohols. [4]

c If the alcohols in **a** are oxidized using acidified sodium dichromate(vi) under reflux, give the name and condensed structural formula of the organic product. [7]

11 Measurement and data processing

- No experimental measurement is completely accurate.
- All experimental measurements have a random uncertainty and lie within a range. This range may be expressed in absolute or percentage terms.
- Experimental errors are not mistakes.
- Experimental errors are of two types: random uncertainties and systematic errors.
- Random uncertainties give rise to a scatter of readings about the true value but may be reduced by averaging.
- Systematic errors give rise to bias (all readings are either too high or too low) and cannot be reduced by averaging.
- Accuracy is concerned with how close an experimental measurement is to the true value.
- Precise values are relatively close to each other.
- Precision and accuracy are not correlated.
- Precision increases with the number of significant figures to which the measurement is quoted.
- The precision of a calculation depends on the precision of the least precise measurement.
- Errors can be combined together to give an overall error from a calculation involving measurements with random uncertainties.
- Significant figures of a number are those digits that carry meaning contributing to its precision.
- There are a number of simple rules for determining the number of significant figures in a number.

11.1 Uncertainty and error in measurement

Quantitative chemistry involves the measurement of physical properties, for example mass, volume, temperature, voltage, pH, density and absorbance. A measurement involves comparing the property of a substance with a known **standard**.

Practical chemistry during your IB Chemistry Programme will involve recording many types of measurements – these will be assessed under the criterion Data Collection and Processing (DCP). Remember that when a measurement is recorded, there is *always* an experimental error or random uncertainty associated with the value. No experimental measurement can be exact.

■ Extension: Units

Base units

The SI base units relevant to the majority of IB chemical measurements and calculations are shown in Table 11.1.

Note the following points about the use of symbols:

- *Never* add 's' to indicate a plural form, for example 5 kg, *not* 5 kgs.
- A full stop is not written after symbols, except at the end of a sentence.
- Abbreviations of units named after a person have a capital letter for the first letter, for example Pa (named after Pascal).
- When the name of the unit is written in full it has a small letter, for example 5 newtons.

Measurement	Unit	Symbol
Length	metre	m
Mass	kilogram	kg
Time	second	s
Amount	mole	mol
Electric current	ampere	A
Temperature	kelvin	K

Table 11.1 A selection of SI base units

Derived units

A number of important derived SI units used in IB chemistry are shown in Table 11.2.

Measurement	Unit	Symbol
Frequency	hertz (reciprocal of seconds)	Hz (s^{-1})
Pressure	pascal (newtons per square metre)	Pa $(N\,m^{-2})$
Energy or enthalpy	joule	J
Electrical charge	coulomb	C
Potential difference	volt	V
Specific heat capacity	joules per kilogram per kelvin	$J\,kg^{-1}\,K^{-1}$
Heat capacity	joules per kelvin	$J\,K^{-1}$
Entropy	joules per kelvin per mole	$J\,K^{-1}\,mol^{-1}$
Enthalpy change or Gibbs free energy change	joules per mole	$J\,mol^{-1}$
Density	kilograms per cubic metre	$kg\,m^{-3}$

Table 11.2 A selection of SI derived units used in chemistry

Language of Chemistry

The word 'specific' in front of a quantity has the meaning 'per unit mass' and the word 'molar' in front of quantity means 'per mole'. Strictly speaking, entropy and enthalpy change are molar entropy and molar enthalpy change. ■

Multiple	Name
10^{-12}	pico (p)
10^{-9}	nano (n)
10^{-6}	micro (μ)
10^{-3}	milli (m)
10^{-1}	deci (d)
10^{3}	kilo (k)

Table 11.3 Common multiples of units

Multiples of units

The sizes of the units are not always the most suitable for certain measurements and decimal multiples are often used (Table 11.3).

Coherence

The SI system is a coherent system of units, that is, all the units for the derived physical quantities are obtained from the base units by multiplication or division without the introduction of numerical factors. This simplifies many calculations.

For example, the following calculation involving the ideal gas equation illustrates how numerical values of volume, amount, gas constant and absolute temperature in coherent SI units give a value for pressure in coherent SI units, namely pascals.

Suppose $0.250\,mol$ of gas occupies a volume of $6.34\,dm^3$ at a temperature of $300\,K$. The molar gas constant $R = 8.31\,J\,K^{-1}\,mol^{-1}$. Thus $n = 0.250\,mol$, $V = 6.34 \times 10^{-3}\,m^3$ and $T = 300\,K$. Substituting in the ideal gas equation:

$$P = \frac{nRT}{V} = \frac{0.250\,mol \times 8.31\,J\,K^{-1}\,mol^{-1} \times 300\,K}{6.34 \times 10^{-3}\,m^3}$$

$$P = 9.83 \times 10^4\,Pa$$

11.1.1 Describe and **give** examples of random uncertainties and systematic errors.

Random uncertainties and systematic errors

Errors or uncertainties can be caused by:

- imperfections in the apparatus used to record the measurement
- imperfections in the experimental method or procedure
- judgements made by the person operating the measuring apparatus.

There are two types of errors: **random uncertainties** and **systematic errors**.

Random uncertainties

A random uncertainty can make the measured value either smaller or larger than the true value. Chance alone determines if it is smaller or larger, and both are equally probable. Reading the scale of any instrument – balance, measuring cylinder, thermometer, pipette – produces random errors. Digital instruments, such as electronic balances and pH meters, also have random uncertainties. In other words, you can weigh a weighing bottle on a balance and get a slightly different answer each time simply due to random errors.

Random uncertainties *cannot* be avoided; they are part of the measuring process. Uncertainties are measures of random errors. These are errors incurred as a result of making measurements on imperfect apparatus which can only have a certain degree of accuracy. They are predictable, and the degree of error can be calculated. They can be reduced by repeating and *averaging* the measurement (page 309).

General examples of random uncertainties include:

- reading a scale (Figure 11.1)
- recording a digital readout
- reading a scale from the wrong position (parallax error) (Figure 11.2)
- taking a reading which changes with time.

Figure 11.1 Dual scale voltmeter showing two analogue scales

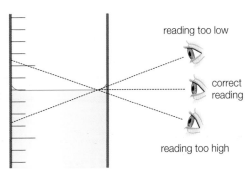

Figure 11.2 Parallax error with a measuring cylinder

Language of Chemistry

Random uncertainties are also known as random errors. However, the term error has the everyday meaning of mistake. Random uncertainties are *not* due to mistakes and cannot be avoided. ■

Systematic error

A systematic error makes the measured value always smaller or larger than the true value, but not both. In other words, a systematic error causes a **bias** in an experimental measurement in one direction, but always in the same direction. For example, all volumetric glassware is usually calibrated at 20 °C. Thus, when this equipment is used at any other temperature, a small systematic error is introduced. An experiment may involve more than one systematic error and these errors may cancel one another, but each alters the true value in one way only.

Accuracy (or validity) is a measure of the systematic error. If an experiment is accurate or valid then the systematic error is very small. Accuracy is a measure of how well an experiment measures what it was trying to measure. This is difficult to evaluate unless you have an idea of the expected value (e.g. a textbook value or a calculated value from a data book). Compare your experimental value to the literature value. If it is within the margin of error for the random errors, then it is most likely that the systematic errors are smaller than the random errors. If it is larger, then you need to determine where the systematic errors have occurred.

General examples of systematic errors include:

- non-zero reading on a meter (a **zero error**) (Figures 11.3 and 11.4)
- incorrectly calibrated scale
- reaction time of experimenter.

Zero errors can be avoided by checking for a 'zero reading' before starting the investigation or recording the measurement with two separate pieces of apparatus and checking that the readings agree (within experimental error). To correct for zero error the value should be subtracted from every reading. For example, a balance with a zero error of −0.2 g reports a mass of 100.0 g. The true mass (ignoring the random uncertainty) is 100.2 g.

Specific examples of systematic chemical errors:

- leaking gas syringes
- calibration errors in pH meters and balances
- use of equipment outside appropriate operating range
- changes in external influences, such as temperature and atmospheric pressure, which affect the measurement of gas volumes
- volatile liquids evaporating
- slow chemical reactions that make it difficult to judge end-points accurately; interfering reactions where a chemical species reacts with the titrant
- retention or loss of chemicals
- poor or no insulation during experiments involving calorimeters
- loss of enzyme activity.

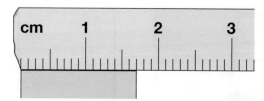

Figure 11.3 Zero error with a metre rule

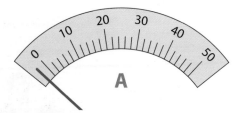

Figure 11.4 An ammeter reading with a zero error of about −2 A

Specific examples of chemical systematic and random errors

Consider a simple titration where a solution of sodium hydroxide is prepared from its solid. A sample of the solution is then titrated against hydrochloric acid in the presence of a suitable acid−base indicator (Chapter 1).

Systematic errors
- Sodium hydroxide is not a primary standard and absorbs water vapour and carbon dioxide from the atmosphere.
- Sodium hydroxide is left behind in the weighing bottle.

Random errors
- Judgements about whether the indicator has changed colour.
- Judgements about whether the bottom of the meniscus is touching the calibration line on a pipette.
- Temperature variations in the glassware and solutions.
- Random uncertainties in the measurement of the mass of sodium hydroxide.
- Random uncertainties in the measurement of the volumes of sodium hydroxide and hydrochloric acid concentrations.

Consider a simple investigation where small known masses of alcohols are burnt in a spirit burner placed underneath a copper can acting as a calorimeter (Figure 11.5). The masses of the alcohols and the water in the calorimeter are both determined using an electronic balance. The temperature of the water is measured before and after the combustion.

Systematic errors
- A large proportion of the heat released by the burning alcohol will be lost to the surrounding air.
- Some heat will be lost from the water; some will be used to heat up the thermometer.
- Some alcohol and water may evaporate.

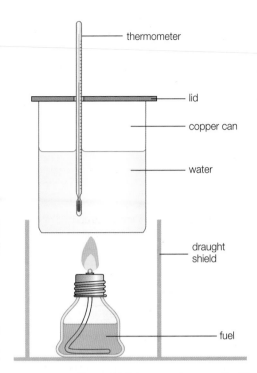

Figure 11.5 An approximate method for determining the enthalpy of combustion of a liquid hydrocarbon or alcohol

Random errors

- Random uncertainties in the measurement of the masses of the alcohols and water.
- Random uncertainties in the measurements of the temperatures of the water before and after combustion of the alcohol.

Systematic and random uncertainty errors can often be recognized from a graph of the results (Figure 11.6).

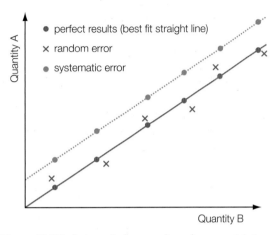

Figure 11.6 Perfect results (no errors), random uncertainties and systematic errors (positive bias) of two proportional quantities

■ **Extension:** Evaluating systematic errors

Systematic errors are difficult to evaluate unless you have an idea of the expected or true value (for example a textbook value or a calculated value from a data book). Compare your experimental value to the literature value. If it is within the range of error for the random uncertainties then it is most likely that the systematic errors are smaller than the random uncertainties. If it is larger then you need to determine where the systematic errors have occurred.

For example, consider a student who has determined via a back titration using sodium hydroxide (Chapter 1) the molar mass of an organic acid (molar mass $126 \, g \, mol^{-1}$). The total experimental uncertainty (see page 314) is calculated to be 1.5%. The student's experimentally determined value for the organic acid is $130 \, g \, mol^{-1} \pm 2 \, g \, mol^{-1}$. This means that the student's result lies between 132 and $128 \, g \, mol^{-1}$.

$$\text{The percentage error} = \frac{(130 - 126)}{126} \times 100 = 3.2\%$$

The percentage error is greater than the sum of the all random uncertainties present in the measurements recorded during the titration. Hence, there are systematic errors present in the investigation, for example the sodium hydroxide solution may have absorbed carbon dioxide from the air, thus reducing the concentration of hydroxide ions.

Precision and accuracy

11.1.2 **Distinguish** between precision and accuracy.

If a series of measurement is repeated and values are obtained which are close together, then the results are said to be **precise**. If the same student obtained these results then the method or procedure is said to be **repeatable**. If the same method or procedure was carried out by a number of different students, then the method or procedure can be said to be **reproducible**. If the results are close to the true value, then the results are described as **accurate**. The differences between accuracy and precision are summarized in Figures 11.7 and 11.8.

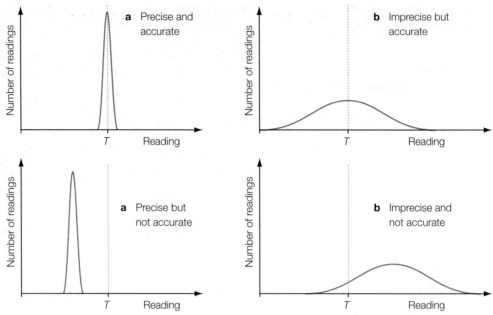

Figure 11.7 a Precise and accurate readings; **b** imprecise and accurate readings (where *T* represents the true value)

Figure 11.8 a Precise but not accurate; **b** imprecise and not accurate (where *T* represents the true value)

For example, a mercury thermometer could measure the normal boiling point of pure water as 99.5 °C (±0.5 °C) whereas a datalogging probe recorded it as 98.25 °C (±0.05 °C). In this example the mercury thermometer is more accurate whereas the datalogging probe is more precise.

Precision is related to how closely you can read the divisions on the scale of an instrument or measuring device. For example, a metre rule is commonly divided into centimetre measurements, but a 30 cm ruler is commonly divided into millimetre measurements (Figure 11.9). This means that the 30 cm ruler is more precise (× 10) than the metre ruler – it has smaller divisions.

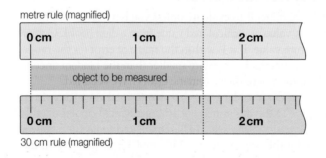

Figure 11.9 Rulers for measurement

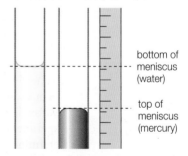

Figure 11.10 The correct approach to accurately reading the meniscuses of water (left) and mercury (right)

When measuring liquids that have a curve at the surface, you must measure from the bottom or top of the meniscus (Figure 11.10), depending on the liquid. The meniscus is the curve formed at the surface of a liquid due to attraction of the liquid for the sides of the container. The curve will be convex for water and aqueous solutions; concave for mercury.

Incorrect technique can lead to the recording of measurements which are precise but inaccurate. For example, suppose a student consistently reads the top of the meniscus in a pipette, burette or measuring cylinder when recording volumes used. These measurements may be precise but are inaccurate. However, if measurements are taken of the initial volume and the volume after adding or removing liquid and their *difference* recorded, then the two systematic errors will cancel.

The measurement of an object on an electronic balance will fluctuate. If this occurs, start with the numbers that are not fluctuating and then make your best guess as to what the next digit would be. For example, consider the following electronic balance readings:

13.345 g   13.320 g   13.349 g   13.357 g   13.327 g

This measurement could then be reported as 13.34 g.

Reducing the effect of random uncertainty

11.1.3 **Describe** how the effects of random uncertainties may be reduced.

Small random errors occur during all practical investigations and are beyond the control of the person recording the measurements. However, the effect of random errors can be reduced by carrying out repeated measurements. The average value from a set of repeated measurements should give a better estimate of the true value of the measurement.

■ **Extension:** ## Averages of non-linear measurements

If the measurement scale is *not* linear, simple averages may give a false value. For example, if three solutions have pH values of 7, 8 and 9, the mean pH is *not* 8 because the pH scale is logarithmic: $pH = -\log_{10}[H^+(aq)]$ (Chapter 18). To obtain the true mean the pH values should be converted to hydrogen ion concentrations ($[H^+(aq)] = 10^{-pH}$), the mean calculated and then converted back to pH.

Thus the average $[H^+(aq)] = \dfrac{(10^{-7} + 10^{-8} + 10^{-9})}{3} = 3.7 \times 10^{-8}$; average pH = 7.4

Random uncertainty as an uncertainty range

11.1.4 **State** random uncertainty as an uncertainty range (±).

Random uncertainties of measured quantities are reported as an uncertainty range. For example, a length may be reported as 5.2 ± 0.5 cm, which means that actual length is located between 4.7 and 5.7 cm. The last digit in the measurement is effectively an estimate. Generally, random uncertainties are expressed to one significant figure only.

A reading is the single determination of a value at one point on a measuring scale. *Generally*, a reading can be estimated to one half of the smallest division (**least count**) on a measuring scale. In the case of the ruler in Figure 11.11, half of the smallest division (least count) would be 0.5 mm or 0.05 cm. Hence the value of the reading is 2.45 cm. The maximum range within which the reading will lie is between 24.0 mm and 25.0 mm or 2.40 cm to 2.50 cm.

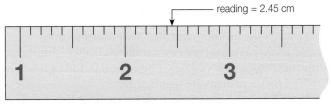

reading = 2.45 cm

Figure 11.11 A single reading obtained from a metre ruler

The *least count* is the smallest division that is marked on the scale of the apparatus. For example, a 50 cm³ burette will have a least count of 0.1 cm³, and an electronic chemical balance giving up to three decimal places of a gram (e.g. 1 g reads as 1.000 g) will have a least count of 0.001 g, that is, 1 mg. For a digital reading such as an electronic balance the last digit is rounded up or down by the instrument and so will also have a random error of plus or minus half the last digit. Hence, the random uncertainties in this burette and this balance may be reported as ±0.05 cm³ and ±0.0005 g.

Significant figures

11.1.5 **State** the results of calculations to the appropriate number of significant figures.

Experimental measurements always have some uncertainty associated with them. One method of expressing the uncertainty in a measurement is to express it in terms of significant figures. In this method, it is assumed that all the digits are known with certainty except the last digit, which is uncertain. Hence, a measurement is expressed in terms of a number which includes all digits which are certain and a last digit which is uncertain. The total number of digits in the number is called the number of **significant figures**.

The concept of significant figures is illustrated in Figure 11.12, which shows a magnified part of a thermometer scale. The temperature is obviously between 18.5 °C and 19.0 °C but three significant figures are justified. Reporting the temperature to three significant figures as 18.7 °C indicates that there is uncertainty in the final figure.

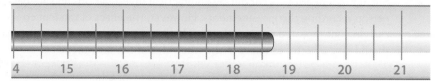

Figure 11.12 A magnified thermometer scale

Table 11.4 shows how the precision of a measurement increases with the number of significant figures. Random uncertainties (page 305) are also included and decrease as the precision of the measurement increases.

Measured value	Precision of measurement	Random uncertainty in the measurement	Significant figures of measured value
3 g	1 g	±0.5 g	1
3.1 g	0.1 g	±0.05 g	2
2.53 g	0.01 g	±0.005 g	3
2.531 g	0.001 g	±0.0005 g	4

Table 11.4 Masses with associated random uncertainties and significant figures

The number of significant figures in a measurement is the number of figures that are known with certainty plus one that is uncertain, beginning with the first non-zero digit. In order to determine the significant figures in a measurement the following rules should be applied:

- All non-zero digits are significant.
  For example, 549 g has three significant figures and 1.892 g has four significant figures.
- Zeros to the left of the first non-zero digit are not significant.
  For example, 0.000 034 g has only two significant figures (this is more easily seen if it is written in scientific notation as 3.4×10^{-5} g). The value 0.001 111 g has four significant figures.
- Zeros between non-zero digits are significant.
  For example, 4023 g has four significant figures and 50 014 g has five significant figures.
- Zeros to the right of the decimal point are significant.
  For example, 2.50 g has three significant figures and 5.500 g has four significant figures.
- Exact numbers, for example 2, and irrational numbers, for example π and 4/3, have an infinite number of significant figures.
- If a number ends in zeros that are not to the right of a decimal, the zeros may or may not be significant.

For example, 1500 g may have two, three or four significant figures. Numbers like this with trailing zeros are best written in scientific notation, where the number is written in the standard exponential form as $N \times 10^n$, where N represents a number with a single non-zero digit to the left of the decimal point and n represents some integer.

The mass above can be expressed in scientific notation in the following forms depending upon the number of significant figures:

$$1.5 \times 10^3 \, g \quad \text{(2 significant figures)}$$
$$1.50 \times 10^3 \, g \quad \text{(3 significant figures)}$$
$$1.500 \times 10^3 \, g \quad \text{(4 significant figures)}$$

In these expressions all the zeros to the right of the decimal point are significant. Scientific notation is an excellent way of expressing the significant figures in very large or very small measurements or physical constants, such as Avogadro's constant ($6.02 \times 10^{23}\,mol^{-1}$) and Planck's constant ($6.63 \times 10^{-34}\,J\,s$).

The measured value of any quantity can be characterized by two important terms:

- the maximum uncertainty
- the number of significant figures.

The maximum uncertainty is an indication of the scale sensitivity or the accuracy of the instrument used. Table 11.5 shows the sensitivity of some commonly used measuring instruments.

Instrument or apparatus	Tolerance	Example
Metre rule	0.001 m	0.544 m
Digital stopwatch	0.01 s	10.85 s*
Thermometer	0.5 °C	68.5 °C
Electronic balance	0.1 g	4.3 g
Electronic balance	0.01 g	6.03 g
Electronic balance	0.001 g	1.689 g
Voltmeter	0.05 V	1.35 V

Figure 11.13 Digital stopwatch

Table 11.5 Typical sensitivities of commonly used apparatus and measuring instruments

*Since the average human reaction time is about 0.2 seconds, the times obtained from a manually operated stopwatch (Figure 11.13) must be rounded to 1 decimal place. Hence the time in Table 11.5 should be reported as 10.9 s. The human reaction time is an example of a systematic error (see page 305).

For volumetric glassware, the manufacturers often print the random uncertainty (or tolerance) on the glass. Some typical tolerance values of apparatus in a school chemistry laboratory are shown in Table 11.6.

Apparatus	Manufacturer's tolerance
Pipette (Class B) (25.0 cm³)	± 0.06 cm³
Burette (Class B) (25.0 cm³)	± 0.1 cm³
Volumetric flask (100 cm³)	± 0.1 cm³
Volumetric flask (250 cm³)	± 0.3 cm³
Measuring cylinder (100 cm³)	± 0.1 cm³

Table 11.6 Tolerance values of apparatus in the laboratory

As general rules the uncertainty ranges due to readability from analogue scales and digital readings are summarized below in Table 11.7.

Instrument or apparatus	Example	Random uncertainty
Analogue scale	Rulers, voltmeters, colorimeters, volumetric glassware	± (half the smallest scale division (least count))*
Digital reading	Top pan balances, spectrophotometers, stop watches, pH meters	± (1 in the least significant digit)

Table 11.7 Estimating uncertainties from analogue scales and digital readings

*If the least count is relatively wide, then it can be mentally divided into fifths or tenths. A magnifying glass may help interpolate the scale in this way.

In Figure 11.14 the random uncertainty of the digital thermometer is ±0.1 °C, hence the temperature shown should be reported as 22.1 °C ± 0.1 °C.

Figure 11.14 Digital thermometer

Calculations with significant figures

When performing calculations with measured quantities the rule is that the accuracy of the final result is limited to the accuracy of the least accurate measurement. In other words, the final result cannot be more accurate than the least accurate number involved in that calculation.

Rounding off

The final result of a calculation often contains figures that are not significant. When this occurs the final result is rounded off. The following rules are used to round off a number to the required number of significant figures:

- If the digit following the last digit to be kept is less than five, the last digit is left unchanged. For example, 46.32 rounded to two significant figures is 46.
- If the digit following the last digit to be kept is five or more, the last digit to be kept is increased by one. For example, 52.87 rounded to three significant figures is 52.9.

Calculations involving addition and subtraction

In addition and subtraction, the final result should be reported to the same number of decimal places as the number with the least number of decimal places. For example:

$$35.52 + 10.3 = 45.82 \qquad \text{which is rounded to 45.8}$$

In this sum, the number 10.3 has digits to the least number of decimal places – one. The final result is therefore rounded to only one decimal place. The digit 2 is dropped and the sum is expressed as 45.8.

Here is an example involving subtraction:

$$3.56 - 0.021 = 3.539 \qquad \text{which is rounded to 3.54}$$

In this subtraction, the number 3.56 has digits to the least number of decimal places – two. The final result is therefore limited to two decimal places. The result will be rounded to 3.54.

Calculations involving multiplication and division

In multiplication and division, the final result should be reported as having the same number of significant figures as the number with the least number of significant digits. This rule is illustrated in the following example:

$$6.26 \times 5.8 = 36.308 \qquad \text{which is rounded to 36}$$

The number with the least significant figures is 5.8 – two significant figures. The final result is therefore limited to two significant digits.

Here is an example involving division:

$$\frac{5.27}{12} = 0.439 \qquad \text{which is rounded off to 0.44}$$

In this division, the number 12 has the least number of significant figures – two. The final result of the calculation is therefore rounded off to two significant figures.

A calculator can give a misleading impression of precision. For example:

$$3.02 \times 11.11 = 33.5522$$

This appears to be a very precise value, but the answer must be given as 33.5, as 3.02 has only three significant figures.

■ Extension: Logarithms and antilogarithms

When calculating the logarithm of a number, retain in the mantissa (the number to the right of the decimal point in the logarithm) the same number of significant figures as there are in the number whose logarithm is being found.

For example:

$$\log_{10}(3.000 \times 10^4) = 4.477\,121 \quad \text{which should be rounded to } 4.4771$$
$$\log_{10}(3.0 \times 10^4) = 4.477\,121 \quad \text{which should be rounded to } 4.5$$

When calculating the antilogarithm of a number, the resulting value should have the same number of significant figures as the mantissa in the logarithm. For example:

$$\text{antilog}(0.301) = 1.9998 \quad \text{which should be rounded to } 2.00$$
$$\text{antilog}(0.30) = 1.9953 \quad \text{which should be rounded to } 2.0$$

Multiple mathematical operations

If a calculation involves a combination of mathematical operations, then perform the calculation using more figures than will be significant to arrive at a final value. Then, go back and look at the individual steps of the calculation and determine how many significant figures would carry through to the final result based on the above rules.

For example:

$$\frac{(5.254 + 0.0016)}{34.6} - 2.231 \times 10^{-3}$$

Calculate the value of the expression using more digits than will be significant. In this example, $0.149\,664\,953\,8$ (depending on the calculator used).

Then, examine each part of the equation to determine the number of significant figures.

$$5.254 + 0.0016 = 5.256 \quad \text{(since the sum is limited to the thousandths place by 5.254)}$$

$$\frac{5.256}{34.6} = 0.152 \quad \text{(since the quotient is limited to three significant figures by 34.6)}$$

$$0.152 - 0.002231 = 0.150 \quad \text{(since the difference is limited to the thousandths place by 0.152)}$$

The value $0.149\,664\,953\,8$ initially obtained should be rounded to have three significant digits. Therefore, the final answer is 0.150 or 1.50×10^{-1}.

11.2 Uncertainties in calculated results

Absolute and percentage uncertainties

11.2.1 State uncertainties as absolute and percentage uncertainties.

For IB chemistry investigations that assess Data Collection and Processing (DCP), estimated uncertainties should be indicated for all measurements.

These uncertainties may be estimated in different ways:

- from the smallest division from a scale
- from the last significant figure in a digital measurement
- from data provided by the manufacturer.

The amount of uncertainty attached to a reading is usually expressed in the same units as the reading. This is the absolute uncertainty, for example, 25.4 ± 0.1 s.

(The mathematical symbol for absolute uncertainty is δx, where x represents the measurement: in the example: $x = 25.4$ and $\delta x = 0.1$.)

The absolute uncertainty is often converted to a **percentage uncertainty**.

For the example, this would be: $25.4\,\text{s} \pm 0.4\%$ ($\frac{0.1\,\text{s}}{25.4\,\text{s}} \times 100 = 0.4\%$).

(The mathematical symbol for fractional uncertainty is: $\delta x/x$).

Note that uncertainties are themselves approximate and are generally not reported to more than one significant figure (see page 312), so the percentage uncertainty reported is 0.4%, not 0.39370%.

The last significant figure in a measurement should be in the same place as the uncertainty. For example: 1261.29 mA ± 200 mA is incorrect, but 1300 mA ± 200 mA is correct.

Since the uncertainly is stated to the hundreds place, we also state the answer to the hundreds place. Note that the uncertainty determines the number of significant figures in the answer.

Uncertainties in results

11.2.2 **Determine** the uncertainties in results.

- When adding or subtracting uncertain values, add the absolute uncertainties:
    initial temperature = 34.50 °C (± 0.05 °C)
    final temperature = 45.21 °C (± 0.05 °C)
    change in temperature, ΔT = 45.21 − 34.50 = 10.71 °C (± 0.05 + 0.05 = ± 0.1 °C)
    Hence, the change in temperature, ΔT, should be reported as 10.7 °C ± 0.1 °C.
- When multiplying or dividing, add the percentage uncertainties:
    mass = 9.24 g ± 0.005 g and volume = 14.1 cm³ ± 0.05 cm³
    Perform the calculation:
    $$\text{density} = \frac{9.24\,\text{g}}{14.1\,\text{cm}^3} = 0.655\,\text{g cm}^{-3}$$
    Convert the absolute uncertainties to percentage uncertainties:
    $$\text{mass} = \frac{0.005}{9.24} \times 100 = 0.054\% \text{ and volume} = \frac{0.05}{14.1} \times 100 = 0.35\%$$
    Add the percentage uncertainties:
    0.054% + 0.35% = 0.40%; density = 0.655 g cm⁻³ (± 0.40%)
    Convert the total uncertainty back to an absolute uncertainty:
    $$0.655 \times \frac{0.4}{100} = 0.002\,62; \text{density} = 0.655 \pm 0.003\,\text{g cm}^{-3}$$
- Multiplying or dividing by a pure (whole) number: multiply or divide the uncertainty by that number.

    (4.95 ± 0.05) × 10 = 49.5 ± 0.5

- Powers: When raising to the nth power, multiply the percentage uncertainty by n. When extracting the nth root, divide the percentage uncertainty by n.
    $$(4.3 \pm 0.5\,\text{cm})^3 = 4.3^3 \pm (\frac{0.5}{4.3}) \times 3$$
    $$= 79.5\,\text{cm}^3\,(\pm 0.349\%)$$
    $$= 79.5 \pm 0.3\,\text{cm}^3$$
- Averaging: repeated measurements can lead to an average value for a calculated quantity. The final answer could be given to the propagated error of the component values in the average. For example:
    average $\Delta H_c^{\ominus}$ = [+100 kJ mol⁻¹ (± 10%) + 110 kJ mol⁻¹ (± 10%) + 108 kJ mol⁻¹ (± 10%)] ÷ 3
    average $\Delta H_c^{\ominus}$ = 106 kJ mol⁻¹ (± 10%)
    This is more appropriate than adding the percentage errors to generate 30%, since that would grossly exaggerate the error and be contrary to the purpose of repeating measurements.

(A more rigorous method for treating repeated measurements from a large sample is to calculate standard deviations and standard errors, but these statistical techniques are more appropriate to large-scale biological studies with many calculated results to average. This is not common in IB chemistry and is therefore not a requirement in chemistry internal assessment.)

■ Extension: Graphing

Graphing is an excellent way to average a range of values. When a range of experimental values (Table 11.8) is plotted each point could have error bars (Figure 11.15) drawn on it (but this is not a requirement of the current IB Chemistry Programme). The size of the bar is calculated from the uncertainty due to random errors. Any line that is drawn should be within the error bars of each point. If it is not possible to draw a line of best fit within the error bars, then the systematic errors are greater than the random errors.

Absolute temperature /K ±5 K	Volume of air/cm³ ±0.3 cm³
274	10.2
301	11.2
316	11.8
342	12.7
369	13.7

Table 11.8 Experimental data with random uncertainties

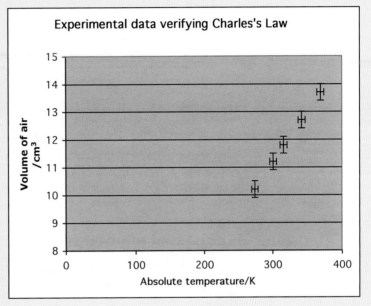

Figure 11.15 Excel-generated graph (with error bars) of experimental measurements of the volume of a fixed mass of air and absolute temperature (at constant pressure)

11.3 Graphical techniques

11.3.1 **Sketch** graphs to represent dependences and interpret graph behaviour.

TOK Link

Graphs are often used in chemistry because on a simple level they provide an instant visual representation of data. 'Visual learners' make up about 65% of the population and they absorb and recall information most effectively by seeing. 'Visual learners' relate best to written information, notes, diagrams, maps and graphs. However, graphs are not just visual representations of the physical world – they are powerful tools that can be used to establish numerical relationships and make predictions.

When drawing graphs from data the y-axis is often used to show values of a **dependent variable** and the x-axis shows the values of the **independent variable**. The dependent variable is the variable that is measured after the independent variable is changed.

For example, in showing how concentration changes with time, concentration is regarded as the dependent variable because its value depends upon time (the independent variable).

If an investigation involved changing the volume of a gas and recording the resulting changes in pressure, then pressure would be regarded as the dependent variable and be plotted along the y-axis.

A graph of pressure against volume for a fixed mass of ideal gas at constant temperature takes the form of a hyperbolic curve (Figure 11.16). The graph clearly illustrates the inverse dependence between the pressure and volume.

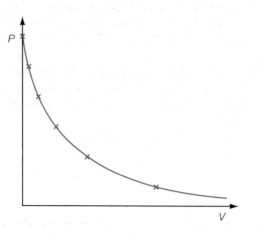

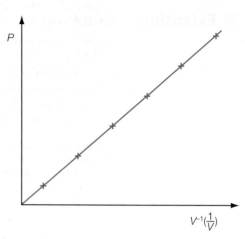

Figure 11.16 Sketch of gas pressure against gas volume (for a fixed mass of ideal gas at constant temperature)

Figure 11.17 Sketch of gas pressure against reciprocal of gas volume (for a fixed mass of ideal gas at constant temperature)

A linear graph can be obtained for this dependence by plotting a graph of pressure against the reciprocal of volume (Figure 11.17). This implies that the product of the pressure and volume will be a constant: $PV = k$. This is known as Boyle's law (Chapter 1). The equation for a straight line, $y = mx + c$, yields the expression $P = m/V$ which is equivalent to $PV = $ constant.

■ Extension: Exponential relationships

Exponential relationships are often found in physical chemistry. The function $y = e^x$ or $y = \exp(x)$ is referred to as the exponential function. The form e^x is referred to as 'e to the power of x' and $\exp(x)$ is referred to as 'exponential x'. General forms of the exponential function that appear in chemistry are $y = Ae^x$ and $y = Ae^{-x}$. These functions depend upon x as shown in Figure 11.18.

a

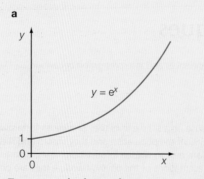

b

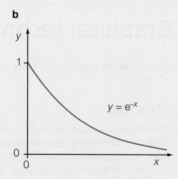

Figure 11.18 Sketches of positive and negative exponential functions

Exponential relationships encountered during the IB Chemistry Programme include the relationship between temperature and rate and concentration versus time for a first-order reaction (Chapter 16).

Constructing graphs

11.3.2 **Construct graphs from experimental data.**

Remember the following points when plotting graphs from experimental data:

- Plot the independent variable on the x-axis and the dependent variable on the y-axis.
- Choose a scale which makes full use of the graph paper. A useful rule is that if you can double the scale (either in the x- or y-direction) and still fit all the points on to the paper, you should do so.

■ Choose a convenient scale, such as 1 cm = 1 unit, or 2 units, or 5 units or 10 units.
■ Label each axis and include (where appropriate) the units (usually SI units).
■ Plot the points as accurately as possible with small crosses. A minimum of five readings is required.
■ For most graphs you will need to draw a straight line, the line of best fit, or a curve of best fit.
■ The graph needs a title and if there are two or more lines or curves, then it needs to have a key.
■ A best trend line is added. This line or curve *never* 'joins the dots' – it is added to show the overall trend (see page 318).
■ Any anomalous data points that do not agree with the line or curve of best fit must be identified.

(It is acceptable to use software, such as Excel, to plot a graph. Similar considerations to those above apply.)

Table 11.9 and Figure 11.19 summarize some idealized experimental data from burning known masses of magnesium and weighing the mass of magnesium oxide (the product). (For simplicity random uncertainties have not been included; graphs do not need to include error bars.)

Mass of magnesium/g	Mass of magnesium oxide/g
0.10	0.16
0.20	0.32
0.30	0.48
0.40	0.64
0.50	0.80

Table 11.9 Experimental data for the combustion of known masses of magnesium

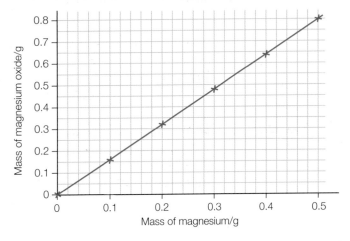

Figure 11.19 A graph of the mass of magnesium oxide versus the mass of magnesium oxide (residue)

Table 11.10 and Figure 11.20 summarize some idealized experimental data from the reaction between excess powdered calcium carbonate and hydrochloric acid.

Time/s	Volume of carbon dioxide released/cm³
10	25
20	45
30	60
40	70
50	75
60	78
70	80
80	80

Table 11.10

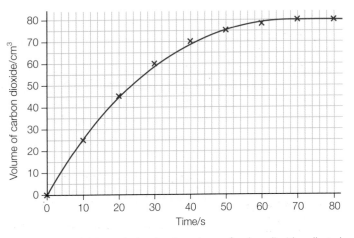

Figure 11.20 A graph showing the volume of carbon dioxide collected against time for the reaction between excess powdered calcium carbonate and hydrochloric acid

Extension: Empirical equations

Some chemical relationships take the form of an empirical equation of the form:

$$y = ax^n$$

where a represents a constant and n represents an unknown exponent.

An empirical relationship is one that is derived from experimental data, rather than from theory.

One method of determining the value of n is to take logarithms of the equation and to plot the logarithm of y against the logarithm of x. Logarithms to the base 10 or natural logarithms may be used.

For example, taking logarithms to base 10 gives the following equation:

$$\log_{10} y = \log_{10} a + n \log_{10} x$$

This equation is of the form $y = mx + c$, if you identify $\log_{10} y$ with y and $\log_{10} x$ with x. The straight line obtained by plotting $\log_{10} y$ against $\log_{10} x$ would have a gradient equal to n, which allows n to be determined.

This method can be used to determine the order of a chemical reaction (Chapter 16).

Fitting a line to a graph

11.3.3 Draw best-fit lines through data points on a graph.

When a graph is plotted from experimental data, it is often found that the data points do not fall on a smooth line or curve, but instead display a degree of random scatter. The scatter occurs from random uncertainties present in the data being plotted.

It is assumed that the variable being measured would vary in a regular way without scatter if the measurements were totally accurate. Hence, it is *incorrect* to join adjacent points together with straight lines ('dot-to-dot').

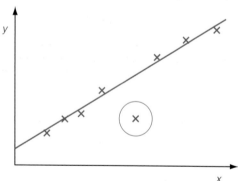

The data points marked on the graph shown in Figure 11.21 are assumed to indicate that there exists a linear relationship between the variable, y, and the variable x. The straight line assumes a linear relationship between the two variables and is known as the **line of best fit**.

It can be drawn by hand (there should roughly be the same number of points above the line as below the line) or using a mathematical approach known as the method of least squares. Spreadsheets, such as Excel, can be programmed to fit a line of best fit through the data with an assumed linear relationship.

The data point circled in Figure 11.21 is **anomalous** data since it is clear that it does not fit in with the trend shown by the other data points and hence suggests a mistake was made during the recording. Anomalous data is not included in the analysis of the data.

Figure 11.21 A graph of experimental data showing linear behaviour and anomalous data

One of the advantages of graphing raw data from an investigation is that it allows anomalous data to be easily recognized. If the graph is plotted (either manually or via a datalogger) when measurements are being taken, an anomalous point indicates that the measurement may be wrong and may need to be repeated.

Measuring the intercept, gradient and area under a graph

11.3.4 Determine the values of physical quantities from graphs.

Graphs can be used to analyse data. This is relatively easy and accurate for straight-line graphs, but the principles can also be applied to curved graphs. Determining the gradient and the intercept is often helpful. For a small number of graphs in chemistry, the area under the graph may be a useful quantity.

Intercept

A straight-line graph will intercept (cut) the axis once and often it is the y-intercept that has a physical significance (Figure 11.22). For example, the intercept in an Arrhenius plot (Chapter 16) gives the value of the Arrhenius constant, A.

If a graph has an intercept of zero it passes through the origin. Two quantities are said to be proportional if the graph is a straight line that passes through the origin.

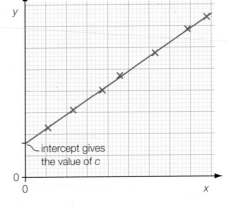

Figure 11.22 A straight-line graph with an intercept, c, on the y-axis

Gradient

The gradient of a straight-line graph is the increase in the y-axis value divided by the increase in the x-axis value. Note the following points:

- A straight-line graph has a constant gradient.
- The triangle used to calculate the gradient should be as large as possible to maximize accuracy.
- The gradient has units. It is the units on the y-axis divided by units on the x-axis.
- If the x-axis is a measurement of time then the gradient represents the rate at which the quantity on the y-axis changes.

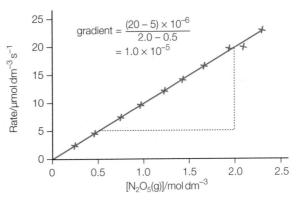

Figure 11.23 A graph of initial rate against concentration for a first-order reaction (decomposition of dinitrogen pentoxide): $2N_2O_5(g) \rightarrow 4NO_2(g) + O_2(g)$

A graph of initial rate against concentration for a first-order reaction (Chapter 16) gives a straight-line graph that passes through the origin (Figure 11.23). The gradient of the graph gives you the value of the rate constant, k. This is characteristic for the reaction under standard conditions.

Since the reaction is first order the rate equation is:

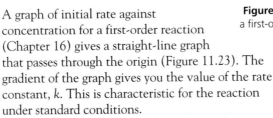

rate $= k\,[N_2O_5(g)]$

so $k = \text{rate}/[N_2O_5(g)]$

$\qquad = \text{gradient of graph}$

$\qquad = 1.0 \times 10^{-5}\,s^{-1}$

A rate versus concentration graph (Chapter 6) is obtained by drawing tangents to the curves of concentration versus time graphs (Figure 11.24) for a reactant or a product.

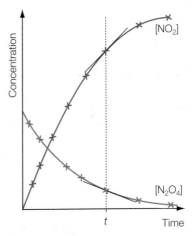

Figure 11.24 Concentration–time graphs for the decomposition of dinitrogen tetroxide, N_2O_4: $N_2O_4 \rightarrow 2NO_2$

Interpolation

Interpolation is a technique where a graph is used to determine data points between those at which you have taken measurements. Figure 11.25 is a graph of concentration of hydrogen peroxide against time. It is an exponential graph and the dotted construction lines are interpolation lines to 'prove' that it is a first-order reaction (Chapter 16). The half-life of the reaction is approximately 25 seconds.

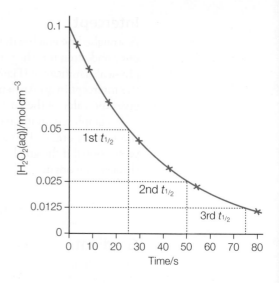

Figure 11.25 A graph showing the concentration of decomposing hydrogen peroxide against time

Extrapolation

Extrapolation is a technique used to find values outside the range for which measurements are made. The straight line or smooth curve is simply extended.

When the volume of an ideal gas is plotted against its temperature using the Celsius temperature scale a straight line with a positive gradient is obtained. This relationship (at constant pressure) is known as Charles's law (Chapter 1). If the line is extrapolated back to the intercept on the x-axis, it gives the value of the temperature at which the volume of gas would be zero (Figure 11.26). Accurate measurements give the value of $-273.15\,°C$. The same temperature is obtained regardless of the volume of gas used, the pressure at which the investigation is carried out or the nature of the gas. Absolute zero is the basis of the thermodynamic temperature scale which uses units of kelvin.

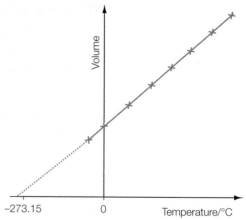

Figure 11.26 A graph of the volume of an ideal gas against temperature (in degrees Celsius) showing the theoretical derivation of absolute zero

Area under a graph

The area (Figure 11.27) under a straight-line graph can be easily calculated using simple arithmetic. If the graph is a curve the area can be estimated by dividing the shape into a number of squares (of known dimensions) and counting the squares. If the equation of the line is known, the area under the graph can be calculated using integration.

Figure 11.27 Areas under a straight-line graph and a curve graph

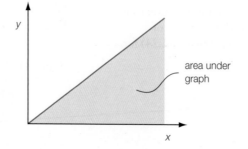

area under graph

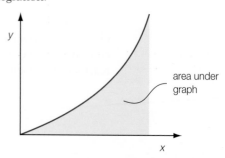

area under graph

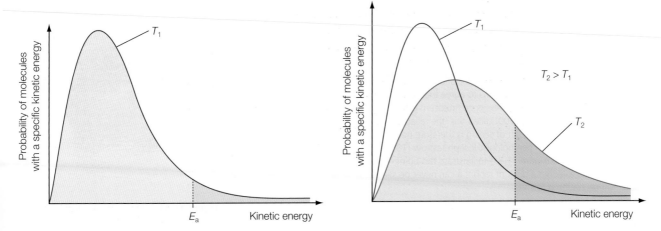

Figure 11.28 The Maxwell–Boltzmann distribution curve at temperature T_1

Figure 11.29 The Maxwell–Boltzmann distribution curves for temperatures T_1 and T_2

For the majority of the graphs in chemistry the area under the graph does not represent a useful physical quantity. However, the area under the curve is relevant to the Maxwell–Boltzmann distribution curve (Chapter 6), which is useful in accounting for the rate of reaction at different temperatures. It is a frequency distribution curve which shows the distribution of kinetic energies amongst reacting gas particles at a particular absolute temperature, T (in kelvin) (Figure 11.28).

The area under the graph in Figure 11.28 is proportional to the number of gas particles. The graph shows that a certain number of particles with kinetic energies equal to or greater than the activation energy E_a, are able to undergo reaction. At a higher temperature (T_2), a greater proportion or percentage of the gas particles have energies equal to or greater than the activation energy and hence more reactions occur, which increases the rate of reaction (Figure 11.29).

Displaying discontinuous data

Dot-to-dot graphs such as that shown in Figure 11.30 are useful for showing patterns. Strictly speaking, this is not a graph and is an incorrect approach to presenting this type of data. This is because type of hydrogen halide is not a type of continuous data. A histogram is an appropriate approach to presenting this type of data.

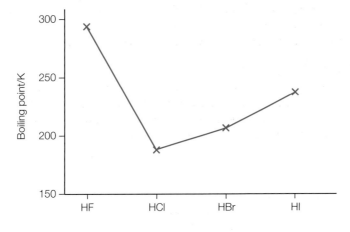

Figure 11.30 Boiling points of hydrogen halides

TOK Link

Seeing or perception is an active process, not a passive receiving of information. The English painter John Constable (1776–1837) remarked, 'the art of seeing nature is almost as much to be acquired as the art of reading Egyptian hieroglyphs'. What enters the eye is not really seen until it is organized by the brain. Making sense of the visual image requires context, inference, concepts, experience and interpretation. In Shakespeare's *Hamlet, Prince of Demark*, Hamlet gets Polonius to agree that a cloud looked like a camel, a weasel, and a whale. There is no 'innocent eye'. The German philosopher Nietzsche (1844–1900) called this 'the fallacy of the immaculate perception'. Hence, no single act of seeing is therefore necessarily the correct one. The American psychologist Joseph Jastrow (1863–1944) used a well-known drawing (Figure 11.31) to illustrate this point – it is a figure that can be seen as either a duck or rabbit; it can never be seen as both and neither interpretation is 'correct'.

Figure 11.31 Jastrow's duck–rabbit illusion

SUMMARY OF KNOWLEDGE

▥ Every experimental measurement possesses an error or uncertainty.

▥ Experimental errors result in a difference between the experimentally recorded value and the real value.

▥ Experimental errors can be classified as random uncertainties and systematic errors.

▥ Random uncertainties are errors which fluctuate or change from one measurement to the next. They produce results distributed or spread around a mean value.

▥ Random uncertainties may occur due to lack of sensitivity. An instrument may not be able to respond to a small change or to indicate it, or the observer may not be able to observe it. Uncertainties due to a lack of sensitivity can be reduced by using more precise instruments and apparatus.

▥ A common source of systematic errors is instruments with a zero error, which can be corrected by subtracting the zero error from every reading. Systematic errors may also arise if an instrument is incorrectly calibrated, if the instrument is not used correctly or if assumptions about the experiment are wrong.

▥ Random errors displace measurements in a random direction (unbiased) whereas systematic errors displace measurements in a single direction (biased).

▥ The effects of random uncertainties on accuracy can be reduced by averaging experimental results. Averaging does not reduce the effect of a systematic error on accuracy.

▥ An accurate measurement is one that has a relatively small systematic error; a precise measurement is a measurement that has a small random uncertainty.

▥ Random errors or uncertainties in experimental raw data are processed through a calculation to give an overall error or uncertainty in the final calculated result.

▥ When adding or subtracting quantities, the absolute uncertainties are added. When multiplying or dividing quantities, or raising them to a power, the percentage uncertainties are added.

▥ Percentage error is smaller in values with more significant figures.

▥ A larger measurement has a smaller percentage error on an instrument with a constant absolute uncertainty.

▥ Rules for determining degrees of precision in a measurement (significant figures) are:
 – all non-zero digits are significant
 – zeros after a decimal point and after a non-zero digit are significant
 – zeros between non-zero digits are significant
 – zeros at the end of numbers punctuated by a decimal point are significant.

▥ When adding and subtracting, your answer must have the same number of decimal places as the item of data with the fewest decimal places.

▥ When multiplying or dividing, your answer must have the same number of significant figures as the item of data with fewest significant figures.

■■■ Exact numbers can be treated as if they have an infinite number of significant figures.
■■■ When doing more than one calculation, do not round numbers until the end.
■■■ Graphs can be used to display or interpret raw or processed data.
■■■ A straight-line graph can be represented by $y = mx + c$, where m represents the gradient and c is the y-intercept (the value where the graph cuts the y-axis).
■■■ Line graphs are appropriate for continuous data; histograms are appropriate for discontinuous data.
■■■ Pie charts can be used to show the relative importance of differing components.
■■■ Graphs can be used to obtain information from the gradient; a value between points on the graph can be measured by interpolation and a value outside the measured range by extrapolation.

■ *Examination questions – a selection*

Paper 1 IB questions and IB style questions

Q1 Which one of the following numbers is given to four significant figures?
A 0.000 40 **C** 4.000
B 0.0040 **D** 4000

Q2 How many significant figures are there in this measured quantity?
0.040 930
A 7 **B** 5 **C** 4 **D** 3

Q3 How many significant figures are there in this measured quantity?
5.010×10^3
A 4 **B** 2 **C** 5 **D** 6 **E** 3

Q4 Perform the indicated operation and give the answer to the appropriate accuracy.
$48.2\,m + 3.87\,m + 48.4394\,m$
A 100.5094 m **C** 100.51 m
B 100.5 m **D** 101 m

Q5 Perform the indicated operation and give the answer to the appropriate accuracy.
$451\,g - 15.46\,g$
A 436 g **C** 435.5 g
B 435.54 g **D** $4.4 \times 10^2\,g$

Q6 The dimensions of a cube are measured. The measured length of each side is 40 mm ± 0.1 mm. What is the approximate uncertainty in the value of its volume?

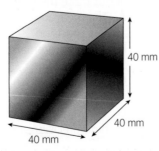
40 mm
40 mm
40 mm

A 1/64% **B** 1/4% **C** 4/10% **D** 0.75%

Q7 What is the percentage random uncertainty in a mass of pure water measured to the nearest microgram (µg) in a kilogram (kg)? ($1\,000\,000\,µg = 1\,g$)
A 10^{-4} **B** 10^{-6} **C** 10^{-7} **D** 10^{-12}

Q8 An object of mass 2.000 kg is placed on four different balances (**A**, **B**, **C** and **D**) and for each balance the reading is recorded five times. The table shows the values obtained with the averages.
Which balance has the smallest systematic error *but* is imprecise?

Balance	1	2	3	4	5	Average/kg
A	2.000	2.000	2.002	2.001	2.002	2.001
B	2.011	1.999	2.001	1.989	1.995	1.999
C	2.012	2.013	2.012	2.014	2.014	2.013
D	1.993	1.987	2.002	2.000	1.983	1.993

Q9 An IB chemistry student records a series of precise measurements from which the student calculates the enthalpy of combustion of a hydrocarbon as $327.66\,kJ\,mol^{-1}$. The student estimates that the result is accurate to ±3%.

Which of the following gives the student's result expressed to the appropriate number of significant figures?
A $300\,kJ\,mol^{-1}$ **C** $330\,kJ\,mol^{-1}$
B $328\,kJ\,mol^{-1}$ **D** $327.7\,kJ\,mol^{-1}$

Q10 Which experimental technique reduces the systematic error in the investigation?
A Adjusting an electronic balance to remove its zero error before weighing a chemical.
B Repeating a titration a number of times and calculating an average titre.
C Using larger amounts of an indicator during a titration.
D Using a magnifying glass to enlarge the meniscus in a burette.

Q11 A titration is carried out by a large number of students in class and the number N of measurements giving a titre volume x is plotted against x. The true value of the titre volume is x_0.

Which graph best represents precise measurements with poor accuracy?

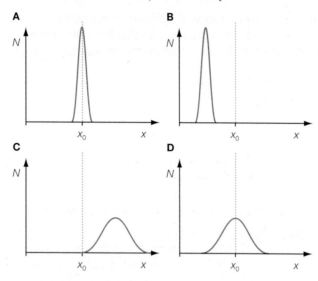

Q12 Which of the following four recorded measurements has the smallest percentage error?
  A 9.99 cm ± 0.005 cm
  B 4.44 cm ± 0.005 cm
  C 1.11 cm ± 0.005 cm
  D 5.55 cm ± 0.005 cm

Q13 The first mass reading is a weighing bottle and sodium hydroxide. The second mass reading is the empty weighing bottle.

What is the mass of the sodium hydroxide, and what is the random uncertainty in the value?

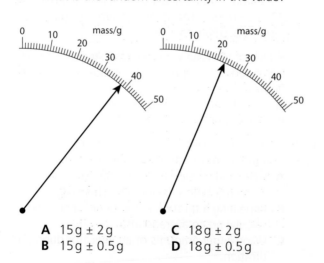

  A 15 g ± 2 g   **C** 18 g ± 2 g
  B 15 g ± 0.5 g   **D** 18 g ± 0.5 g

Q14 What are the greatest and smallest possible values for the following length:

34.3 cm?

  A 34.35 cm and 34.25 cm
  B 34.2 cm and 34.4 cm
  C 34.6 cm and 34.0 cm
  D 35.3 cm and 33.3 cm

Q15 What is the following measurement with the possible random uncertainty shown as a percentage instead of an absolute error:

$6.25 \text{ cm}^3 \pm 0.005 \text{ cm}^3$?

  A $6.25 \text{ cm}^3 \pm 0.4\%$
  B $6.25 \text{ cm}^3 \pm 0.8\%$
  C $6.25 \text{ cm}^3 \pm 0.04\%$
  D $6.25 \text{ cm}^3 \pm 0.08\%$

Q16 When comparing systematic errors and random uncertainties during an investigation, the following pairs of properties of errors in an experimental measurement may be considered:
  I error can possibly be removed
  II error cannot possibly be removed
  III error is of constant sign and size
  IV error is of varying sign and size
  V error will be reduced by averaging repeated measurements
  VI error will not be reduced by averaging repeated measurements

Which properties apply to random uncertainties?
  A I, II, III   **C** II, IV, V
  B I, IV, VI   **D** II, III, V

Q17 The diagram shows a set of experimental data points, X, determined when one experimental measurement was repeated four times. The centre of the diagram represents the ideal value calculated from theory. What statement is correct about these measurements?
  A The measurements involve low accuracy and low precision.
  B The measurements involve low accuracy and high precision.
  C The measurements involve high accuracy and low precision.
  D The measurements involve high accuracy and high precision.

Q18 In a school laboratory, which of the pieces of apparatus listed below has the greatest random uncertainty in a measurement?
  A A 50 cm^3 burette when used to measure 25 cm^3 of ethanol.
  B A 25 cm^3 pipette when used to measure 25 cm^3 of ethanol.
  C A 50 cm^3 measuring cylinder when used to measure 25 cm^3 of ethanol.
  D An analytical balance (4 decimal places) when used to weigh 25 cm^3 of ethanol.

Q19 Perform the following density calculation to the correct number of significant figures:
$$\frac{1.00\,g}{3.00\,cm^3}$$

  A 0.333 g cm^{-3} **C** 0.3 g cm^{-3}
  B 0.3333 g cm^{-3} **D** 0.33 g cm^{-3}

Q20 An experiment to determine the molar mass of solid hydrated iron(II) sulfate, $FeSO_4.5H_2O$ gave a result of 258 g mol^{-1}.

  What is the experimental error?
  A 0.07%   **C** 0.7%
  B 7%   **D** 77%

Paper 2 IB questions and IB style questions

Q1 One method a chemist can use to investigate acid-base reactions is a titration. A pH titration is performed by adding small, accurate amounts of sodium hydroxide solution to hydrochloric acid of unknown concentration. The pH is recorded and is plotted versus the volume of base added to the acid solution.

State how the following would affect the calculated concentration of acid:
  a i The burette is dirty and drops of sodium hydroxide cling to the side walls of the burette as it is drained. [2]
  ii The burette is not rinsed with sodium hydroxide prior to filling. [2]
  iii The burette tip is not filled at the start of the titration. [1]
  iv The sodium hydroxide solution is added too rapidly in the region of rapid pH change. [1]
  b i It is suspected that the pH meter consistently gives a reading 0.5 units above the actual value. What type of error is this? [1]
  ii How would you verify this error? [1]

Q2 The graph below shows how the mass of copper deposited during electrolysis varies with time.

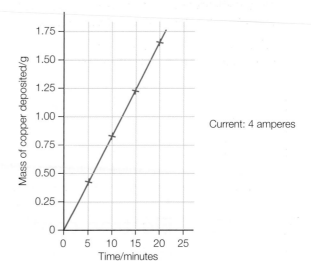

Current: 4 amperes

  a Identify the dependent, independent and controlled variables. [3]
  b State the relationship between the independent and dependent variables. [1]
  c Calculate the rate (in g min^{-1}) of copper deposition (to two decimal places). [2]
  d State **two** useful mathematical operations that could be performed on the graph. [2]

Q3 The length of a piece of paper was measured as 298 mm ± 1 mm. Its width was measured as 210 mm ± 1 mm.
  a Calculate the percentage random uncertainty in its length. [1]
  b Calculate the percentage random uncertainty in its width. [1]
  c Calculate the area of the piece of paper and its random uncertainty. [2]

Q4 A rectangular block with a density of 2.50 g cm^{-3} has the following dimensions and random uncertainties. Calculate the uncertainty in the density. [2]
  mass = 25.0 g ± 0.1 g
  length = 5.00 cm ± 0.01 cm
  width = 2.00 cm ± 0.01 cm
  height = 1.00 cm ± 0.01 cm

Q5 A weighing bottle plus a sample of a pure liquid has a combined mass of 120.2 g. The weighing bottle has a mass of 119.0 g. The density of the liquid is 2.05 g cm^{-3}. Calculate the volume of the liquid to the correct number of significant digits. [2]

Atomic structure

- Shells (energy levels) can be divided into sub-shells (sub-levels).
- Electrons occupy regions of space known as orbitals.
- Orbitals have different shapes and energies.
- Orbitals are filled according to certain rules.
- Electrons can be removed from atoms and the energy changes (ionization energies) measured.
- Ionization energies provide evidence for shells (energy levels), sub-shells (sub-levels) and electron pairing.
- A simple electrostatic model of the atom can account for many of its properties.

12.1 Electron configuration

Ionization energy

12.1.1 **Explain** how evidence from first ionization energies across periods accounts for the existence of main energy levels and sub-levels in atoms.

The first **ionization energy** is the minimum energy per mole required to remove electrons from one mole of isolated gaseous atoms to form one mole of gaseous unipositive ions under standard thermodynamic conditions. For example, the first ionization energy of chlorine is the energy required to bring about the reaction:

$$Cl(g) \rightarrow Cl^+(g) + e^-$$

The electron is removed from the outer sub-shell (energy sub-level) of the chlorine atom (that is, a 3p electron). Some examples of ionization energy are the enthalpy (energy) changes for the reactions shown in Table 12.1. Ionization energies are listed on page 8 of the IB *Chemistry data booklet*.

Element	Ionization equation	First ionization energy/ kJ mol^{-1}
Oxygen	$O(g) \rightarrow O^+(g) + e^-$	1310
Sulfur	$S(g) \rightarrow S^+(g) + e^-$	1000
Copper	$Cu(g) \rightarrow Cu^+(g) + e^-$	745

Table 12.1 Selected ionization energies

Factors that affect ionization energy

Values of ionization energies depend on the following factors:

- the size of the atom (or ion)
- the nuclear charge
- the **shielding** effect.

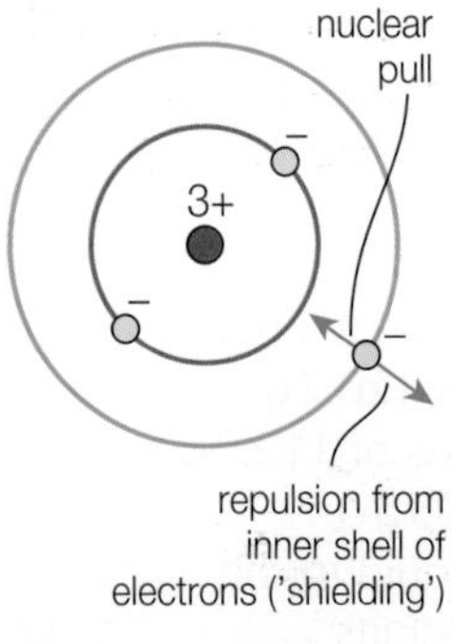

nuclear pull

3+

repulsion from inner shell of electrons ('shielding')

Figure 12.1 Electrostatic forces operating on the outer or valence electron in a lithium atom

Atomic radius

As the distance of the outer electrons from the nucleus increases, the attraction of the positive nucleus for the negatively charged electrons falls. This causes the ionization energy to decrease. Hence, ionization energy decreases as the atomic or ionic radius increases.

Nuclear charge

When the nuclear charge becomes more positive (due to the presence of additional protons), its attraction on all the electrons increases. This causes the ionization energy to increase.

Shielding effect

The outer or valence electrons are repelled by all the other electrons in the atom in addition to being attracted by the positively charged nucleus. The outer electrons are shielded from the attraction of the nucleus by the **shielding effect** (an effect of electron–electron repulsion) (Figure 12.1).

In general, the shielding effect is most effective if the electrons are close to the nucleus. Consequently, electrons in the first shell (energy level), where there is high electron density, have a stronger shielding effect than electrons in the second shell, which in turn have a stronger shielding effect than electrons in the third shell. Electrons in the same shell exert a relatively small shielding effect on each other.

Figure 12.2 shows the first ionization energies for the chemical elements of periods 1, 2 and 3. The general increase in ionization energy across each period is due to the increase in nuclear charge. This occurs because across the period each chemical element has one additional proton, which increases the nuclear charge by +1.

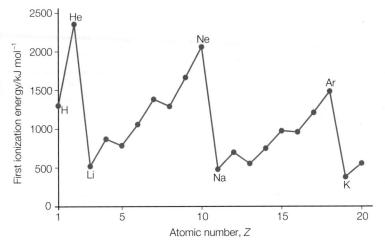

Figure 12.2 First ionization energies for periods 1, 2 and 3

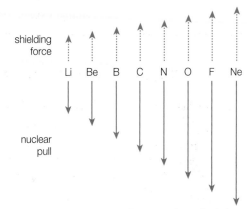

Figure 12.3 A diagram illustrating how the balance between shielding and nuclear charge changes across period 2

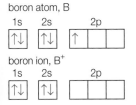

Figure 12.4 Orbital notations for boron and beryllium atoms and their unipositive ions

The increase in nuclear charge increases the force of attraction on all the electrons, so they are held closer and hence more strongly. Each additional electron across a period enters the same shell (energy level) and hence the increase in shielding is minimal (Figure 12.3).

Although the general trend is for the ionization energy to increase across the period, there are two distinct dips in ionization energy across periods 2 and 3 (Figure 12.2). These dips can only be explained using an *orbital model of electronic structure* (see pages 330–340).

The first decrease in each period is the result of a change in the sub-shell (sub-level) from which the electron is lost and a change in electron shielding. These have a greater effect than the increase in nuclear charge and decrease in atomic radius. In period 2, this first decrease occurs between the elements beryllium and boron. When it is ionized, the beryllium atom ($1s^2 2s^2$) loses a 2s electron, whereas a boron atom ($1s^2 2s^2 2p^1$) loses a 2p electron (Figure 12.4). More energy is required to remove an electron from the lower energy 2s orbital in beryllium than from the higher energy 2p orbital in boron. Although the 2s and 2p sub-shells are in the same shell, the energy difference is relatively large. Recall (Chapter 2) that the energy gap between shells and sub-shells becomes smaller with an increase in shell number. In addition, a single electron in the 2p sub-shell is more effectively shielded by the inner electrons than the $2s^2$ electrons (Figure 12.5).

Figure 12.5 Electron density clouds of the 2s and 2p orbitals (only one lobe shown). The dotted line shows the extent of the 1s orbital; the 2s electron can partially penetrate the 1s orbital, increasing its stability

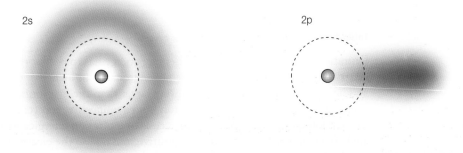

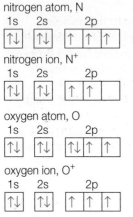

nitrogen atom, N

nitrogen ion, N⁺

oxygen atom, O

oxygen ion, O⁺

Figure 12.6 Orbital notation for nitrogen and oxygen atoms and their unipositive ions

A similar explanation also accounts for the first decrease observed in period 3 for the elements magnesium and aluminum. The decrease in first ionization energy from magnesium ($1s^22s^22p^63s^2$) to aluminium ($1s^22s^22p^63s^23p^1$) arises largely because the electrons in the filled 3s orbital are more effective at shielding the electron in the 3p orbital than they are at shielding each other.

The second decrease in first ionization energy in period 2 occurs between nitrogen ($1s^22s^22p_x^{1}2p_y^{1}2p_z^{1}$) and oxygen ($1s^22s^22p_x^{2}2p_y^{1}2p_z^{1}$). The three valence (outer) electrons in the nitrogen atom are in three separate orbitals. This is in accordance with **Hund's rule**, which states that every orbital in a sub-shell is singly occupied with one electron before any one orbital is doubly occupied (page 336). However, in the oxygen atom two electrons are in the same 2p orbital. The two electrons in the same orbital experience severe repulsion. This electron–electron repulsion makes it easier to remove one of these $2p_x$ electrons than an unpaired electron from a half-filled $2p_z$ orbital. Hence, the decrease in first ionization energy between nitrogen and oxygen is due to the additional repulsion present in the $2p^4$ configuration of the oxygen atom (Figure 12.6).

A similar explanation accounts for the decrease in first ionization energy observed between phosphorus and sulfur in period 3. The first ionization energy of sulfur ($1s^22s^22p^63s^23p_x^{2}3p_y^{1}3p_z^{1}$) is less than that of phosphorus ($1s^22s^22p^63s^23p_x^{1}3p_y^{1}3p_z^{1}$) because less energy is required to remove an electron from the $3p^4$ orbitals of sulfur than from the half-filled 3p orbitals of phosphorus.

The pattern of first ionization energy across periods 2 and 3 is identical except that all the corresponding ionization energies for period 2 are higher. This is because the electrons being removed are in a second shell closer to the nucleus, compared to the third shell for the period 3 elements. The outer electrons in period 2 experience a higher nuclear charge than those in period 3.

■ **Extension:**   Electron affinity

The **electron affinity** is the *reverse* of ionization energy. It is the energy change per mole for the reaction:

$$X(g) + e^- \rightarrow X^-(g)$$

The electron affinity is important when deciding what type of bonding non-metals should show and is used in calculating lattice enthalpies for ionic compounds (Chapter 15). Lattice enthalpies are a measure of the energetic stability of ionic substances.

Relating successive ionization energy data to electron configuration

12.1.2 Explain how successive ionization energy data is related to the electron configuration of an atom.

The first ionization energy is the minimum energy required to remove a mole of electrons from a mole of gaseous atoms to form a mole of unipositive ions. The second ionization energy is the minimum energy required to remove a mole of electrons from a mole of gaseous unipositive ions to form a mole of dipositive ions. The third ionization energy is the minimum energy required to remove a mole of electrons from a mole of gaseous dipositive ions to form a mole of tripositive ions (Table 12.2).

Ionization energy	Ionization equation	Ionization energy/ kJ mol⁻¹
First	$O(g) \rightarrow O^+(g) + e^-$	+1310
Second	$O^+(g) \rightarrow O^{2+}(g) + e^-$	+3388
Third	$O^{2+}(g) \rightarrow O^{3+}(g) + e^-$	+5301

Table 12.2 Successive ionization energies for oxygen

Ionization energies are always endothermic: energy has to be absorbed and **work** done so that the negatively charged electron can be removed from the influence of the positively charged nucleus.

The second ionization energy is always larger than the first ionization energy because more energy is required to remove an electron from a unipositive ion (compared with a neutral atom for the first ionization energy). Further **successive ionization energies** increase because the electrons are being removed from increasingly positive ions and so the electrostatic forces are greater. There are also fewer remaining electrons to provide shielding.

Figure 12.7 shows the successive ionization energies for potassium atoms. It is possible in a mass spectrometer to progressively raise the kinetic energy of the bombarding beam of electrons to remove the electrons one at a time from gaseous atoms and to measure the amount of energy required to carry out each individual removal.

Figure 12.7 provides strong experimental evidence that the electron configuration of a potassium atom is 2,8,8,1 (or 2.8.8.1), as the greatest increases in ionization energy required to remove an electron occur after 1, 9 and 17 electrons are removed (1 + 8 = 9, 1 + 8 + 8 = 17). These large increases in ionization energy correspond to removing electrons from inner shells whose electrons are progressively located nearer to the nucleus. A logarithmic scale is used to reduce the size of these 'jumps'.

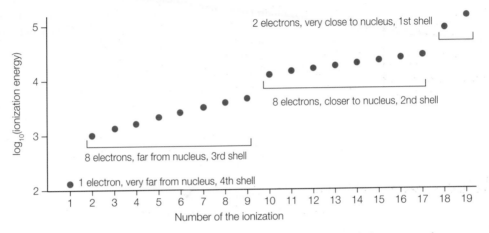

Figure 12.7 Successive ionization energies for a potassium atom

Understanding the relationship between successive ionization data and electron configuration allows you to make deductions about the shell structure of elements. You can also use your knowledge to sketch a graph of successive ionization values for a given element. Examples of the sorts of question you may be asked are given below.

Worked example

Sketch a graph of the successive ionization energies of silicon against the number of electrons removed.

Figure 12.8 shows the answer for this question.

Notes on the answer

The trend for 14 successive ionization energies of the silicon atom should be shown. Large increases in ionization energy must be shown between electrons 4 and 5, and between electrons 12 and 13. These correspond to changes in shells (energy levels), since the electronic arrangement of a silicon atom

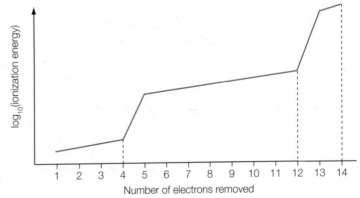

Figure 12.8 Sketch of the successive ionization energies of silicon against the number of electrons removed

is 2,8,4. (Unless otherwise specified, small variations due to changes in sub-shell (sub-level) are not required.)

Always remember to start on the *outside* of the atom and work your way towards the nucleus. It may therefore be helpful to write the electron configuration of the atom back to front, for example, 4,8,2.

You should also be able to deduce to which group in the periodic table a chemical element belongs from successive ionization data listed as numbers.

Worked example

The first eight experimentally determined ionization energies for a chemical element are as follows (in kJ mol⁻¹):

| 580 | 1800 | 2750 | 11 580 | 14 850 | 18 400 | 23 300 | 27 500 |

Deduce the following:

- the number of electrons in the outer shell of an atom of the element
- the group in the periodic table to which the element belongs
- the outer detailed electron configuration of the atom.

There is a relatively large increase in the ionization energy when the fourth electron is removed. The increase is much larger than the increase for the first four electrons. Hence, the first three electrons must be located in the outer or valence shell. The element must be located in group 3 since it has three outer electrons. All group 3 elements have three valence electrons.

The detailed outer electron configuration is:

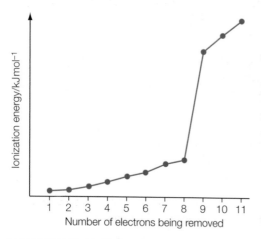

Successive ionization energy data can also be used to provide evidence for sub-shells (sub-levels) and electron pairing in atoms. Consider the successive ionization energies of the argon atom (2,8,8 or 2.8.8) (Figure 12.9). The reason for the relatively large increase in ionization energy between electrons 8 and 9 is that up to electron number 8, the electrons were removed from the third shell, but electron 9 is removed from the second shell. The slight increase in ionization energy between electrons 6 and 7 is due to a change in sub-shell from 3p to 3s. The slight increase in ionization energy for electrons 4, 5 and 6 relative to electrons 1, 2 and 3 can be accounted for by enhanced electron–electron repulsion.

Figure 12.9 Eleven successive ionization energies of an argon atom

■ Extension: Effective nuclear charge

Effective nuclear charge is the charge experienced by the valence electrons after the number of shielding electrons that surround the nucleus has been taken into account. Consider a fluorine atom (2,7). The two electrons in the first energy level (shell) experience a +9 charge because that is the charge on the nucleus. However, the electrons that are in the valence energy level (second shell) would be shielded from the nucleus by the two shielding electrons. The +9 nuclear charge is shielded by 2 electrons to give an effective nuclear charge of +7; this is the charge experienced by the valence electrons. Beyond the valence electrons, the effective nuclear charge is zero because the +9 charge of the nucleus is cancelled out by the 9 electrons.

Orbitals and energy levels

12.1.3 State the relative energies of s, p, d and f orbitals in a single energy level.
12.1.4 State the maximum number of orbitals in a given energy level.

Electrons are arranged in shells (energy levels) around the nucleus of an atom. The shells are usually numbered 1 (first), 2 (second), 3 (third) etc., starting from the nucleus (Chapter 2).

Each shell consists of a number of **sub-shells** (sub-levels), labelled s, p, d or f. The existence of sub-shells is confirmed experimentally by the fine structure present in successive ionization energy data (page 327).

The number of sub-shells is equal to the number of shells. Hence, the first shell is composed of one sub-shell, the second shell is composed of two sub-shells, and so on. The sub-shell composition of the first four shells is summarized in Table 12.3.

Table 12.3 Structure of the first four electron shells

Shell (energy level) number	Number of sub-shells in the shell	Sub-shells (sub-levels)
1	1	1s
2	2	2s and 2p
3	3	3s, 3p and 3d
4	4	4s, 4p, 4d and 4f

Each sub-shell contain a number of **orbitals** in which the electrons are placed (Figure 12.10). The number of orbitals in each sub-shell depends on the type of sub-shell. Table 12.4 summarizes the number of orbitals and electrons in the four common types of sub-shell. Each orbital can be represented by a 'square box'. Each orbital can hold a maximum of two electrons. An electron is represented by an arrow: ↑.

Figure 12.10 Representation of a 1s orbital, 2p orbital and 3d orbital, each containing one electron

Table 12.4 Number of orbitals in the four types of sub-shells

Type of sub-shell (sub-level)	Number of orbitals	Maximum number of electrons in the sub-shell (sub-level)
s	1	2
p	3	6
d	5	10
f	7	14

Figure 12.11 shows how the energy shells are split into sub-shells (not applicable to hydrogen). The structure of these shells is summarized in Table 12.5.

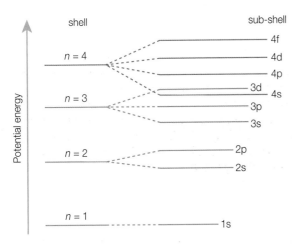

Figure 12.11 Energy levels of electron sub-shells

Table 12.5 Structure of sub-shells (not hydrogen)

Shell (energy level)	Sub-shell	Number of electrons in sub-shell	Maximum number of electrons in the shell (energy level)
$n = 1$	s	2	2
$n = 2$	s	2	
	p	6	8
$n = 3$	s	2	
	p	6	
	d	10	18
$n = 4$	s	2	
	p	6	
	d	10	
	f	14	32

Figure 12.12 shows the energy levels of the atomic orbitals (except hydrogen). Note that the 4s sub-shell (sub-level) has a lower energy than the 3d sub-shell and hence electrons fill the 4s sub-shell before they occupy the 3d sub-shell (see page 335). This sub-shell overlap (Figure 12.13) first occurs with the first row of the d-block metals (Chapter 13). However, the 3d sub-shell is then stabilized across the first row of the d-block metals.

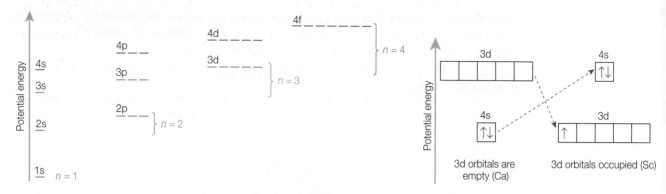

Figure 12.12 Orbital structure of atoms

Figure 12.13 The 3d–4s sub-shell overlap

■ **Extension:** **Energy levels in hydrogen**

In a hydrogen atom the sub-shells all have the same energy. For example, the 2s and 2p sub-shells have the same energy. In helium the two sub-shells have different energies. This is because of electron–electron repulsion in the helium atom between the two electrons. (This type of repulsion occurs in all atoms except hydrogen.)

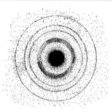

Figure 12.14 Drawing of an electron diffraction photograph obtained by G.P. Thomson

History of Chemistry

Louis de Broglie (1892–1987) was a French physicist who was awarded the Nobel Prize in Physics in 1929. The de Broglie hypothesis, formulated in 1925, suggests that sub-atomic particles exhibit wave behaviour as well as behaving as particles. Three years later G.P. Thomson (the son of J.J. Thomson) confirmed that electrons do undergo diffraction (spreading out) when passing through a thin metal film (Figure 12.14). De Broglie's new way of viewing the behaviour of matter helped the development of a new field in physics, wave mechanics, which treated the physics of light and matter as similar phenomena. Electrons were now considered to have the properties of both waves and particles. This is termed **wave–particle duality**. De Broglie described the electron in a hydrogen atom as a standing wave. However, only electrons whose orbits allowed a *whole number* of wavelengths would be able to exist (Figure 12.15).

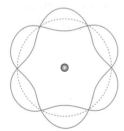

Figure 12.15 Allowed standing electron wave with a whole number of wavelengths

Shapes of orbitals

12.1.5 Draw the shape of an s orbital and the shapes of the p_x, p_y and p_z orbitals.

Electrons do not occupy fixed positions within an atom, nor do they follow orbits in the shells. Electrons occupy volumes or regions of space called **orbitals** (Figure 12.16). The four types of orbitals, s, p, d and f, all have different shapes. (The shapes and energies of atomic orbitals are obtained by solving the Schrödinger wave equation.)

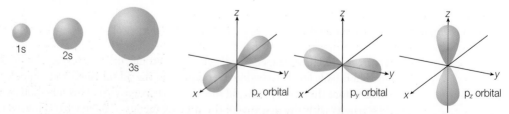

Figure 12.16 Shapes of s and p orbitals

All s orbitals can be represented as spheres. They differ only in size and energy. The 3s orbital is larger than the 2s orbital, which is larger than the 1s orbital. The larger orbitals are described as being more *diffuse* since the electron density is less.

The p orbitals have two **lobes** forming a 'dumb-bell' shape and have different orientations in space. They are arranged at right angles to each other and are labelled p_x, p_y and p_z to reflect their orientation. The three p orbitals all have the same energy – the orbitals are said to be **degenerate**. The 3p orbitals have the same shape as the 2p orbitals but are larger.

(Knowledge of the shapes of the 3d orbitals is not required by the IB Chemistry Syllabus, but they are given in Chapter 13.)

The most important orbitals are those in the outer shells which are involved in the formation of chemical bonds. Covalent bonds are formed when atomic orbitals overlap and merge to form molecular orbitals (Chapter 14).

■ Extension: Quantum mechanical model

The quantum mechanical model is a *probability* model (which uses a wave function to describe mathematically the location of the electron). The orbitals described previously are drawn as volumes of space where electrons spend 95% of their time. A more accurate description of the 1s orbital in a hydrogen atom is shown in Figure 12.17. It is a computer-generated image showing the positions of the electron in a hydrogen atom over a very short interval of time. The boundary of the 1s orbital is clearly visible, but it is seen to be 'fuzzy' owing to the existence of some electron density outside the boundary surface of the orbital.

It is also helpful to compare and contrast Bohr's concept of an *orbit* (Chapter 2) with Schrödinger's concept of an *orbital*. The differences are summarized in Table 12.6.

boundary within which there is a 95% chance of finding an electron

Figure 12.17 Quantum mechanical 'electron cloud' model of a 1s orbital in a hydrogen atom

Orbit	Orbital
A well-defined circular orbit followed by a revolving electron around the nucleus.	A region of space where an electron is likely to be located.
It represents planar motion.	It represents three-dimensional motion.
Orbits are non-directional and hence cannot account for the shape of molecules.	Orbitals have different shapes.
The maximum number of electrons in an orbit is $2n^2$, where n represents the number of the orbit.	An orbital cannot accommodate more than two electrons.
The electron is viewed as a localized particle.	The electron is viewed as mathematical wave function, indicating the probability of finding an electron in a particular region of space.

Table 12.6 Summary of the differences between an orbit and an orbital

Language of Chemistry

The word *quantum* is originally Latin, meaning 'how much'. Its plural form is *quanta*. ■

Applications of Chemistry

Computational chemistry (Figure 12.18) is a branch of chemistry that uses computer programs to solve chemical problems. It uses quantum mechanical models to calculate the properties of molecules and solids. It can be used to predict the properties of previously unprepared molecules and is widely used in drug design (Chapter 24). Software can be used to predict the most stable conformation or shape of a molecule, together with its spectroscopic properties (Chapter 21). Two methods are used in computational chemistry: *ab initio*, from the fundamental principles of physics, and semi-empirical, using approximations.

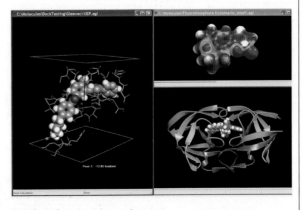

Figure 12.18 A screenshot from Argus Lab, a molecular modelling, graphics and drug design program

History of Chemistry

Erwin Schrödinger (1887–1961) was an Austrian physicist awarded the Nobel Prize in Physics in 1933. He used de Broglie's ideas to show how the properties of waves could be used to explain the behaviour of the electron in a hydrogen atom. Schrödinger published his ideas in 1926 and his quantum mechanical description is known as the **Schrödinger wave equation**. The solutions of Schrödinger's equation give the energy levels of the hydrogen atom (identical to that of Bohr) and the shapes and energies of the orbitals as probability densities. The Schrödinger equation cannot be solved exactly for other atoms, but a variety of approximate methods give similar descriptions of orbitals.

The German physicist **Werner Heisenberg** (1901–1976), using a different but related theory to Schrödinger, proposed that it was impossible to measure with complete accuracy both the position and momentum (the product of mass and velocity) of an electron. This is known as **Heisenberg's uncertainty principle**. A scientist can measure the position of an electron to some accuracy, but then its momentum will be inside a very large range of values. Likewise, a scientist can measure the momentum accurately, but then its position is unknown. In trying to record a measurement on an electron, the act of measuring would change the value. Locating an electron in space requires light to hit it and return to the detector. However, the interaction between the light used in locating the electron and the electron itself causes its momentum to change drastically. Hence the measurement of both quantities simultaneously would never be accurate. Thus the uncertainty principle puts a definite limit to the knowledge accessible to scientists. One comment attributed to Heisenberg is: 'atoms form a world of potentialities or possibilities rather than one of things or facts'. He remained in Germany during the Second World War and was involved in the Nazi's attempts to build atomic weapons. Historians of science continue to debate his role, with some suggesting he deliberately derailed and stalled work.

■ Extension: Quantum numbers

Each electron in an atom can be uniquely described by a set of four **quantum numbers**: principal quantum number (n), angular momentum quantum number (l), magnetic quantum number (m) and spin quantum number (s) (Table 12.7). The magnetic quantum number can be any whole number from $+l$ to $-l$, which accounts for the number of orbitals in each sub-shell. Pauli's exclusion principle states that no electron in the same atom can have the same four quantum numbers. n defines the energy of the orbital, l defines the shape of the orbital, m defines the orientation of the p, d or f orbital and s defines the spin of the electron.

n	l (0 to $n-1$)	Sub-shell notation	m	Number of orbitals in sub-shell (sub-level)	Total number of orbitals in the main shell (energy level)
1	0	1s	0	1	1
2	0	2s	0	1	
	1	2p	1, 0, −1	3	4
3	0	3s	0	1	
	1	3p	1, 0, −1	3	
	2	3d	2, 1, 0, −1, −2	5	9
4	0	4s	0	1	
	1	4p	1, 0, −1	3	
	2	4d	2, 1, 0, −1, −2	5	
	3	4f	3, 2, 1, 0, −1, −2, −3	7	16

Table 12.7 The relationship between the values of principal, angular momentum and magnetic quantum numbers

12.1.6 Apply the Aufbau principle, Hund's rule and the Pauli exclusion principle to write electron configurations for atoms and ions up to $Z = 54$.

Filling atomic orbitals

The electrons are arranged in atomic orbitals according to certain principles:

- Each orbital can hold up to a maximum of two electrons. This, in simplified form, is the **Pauli exclusion principle**.
- Electrons enter and occupy an empty atomic orbital with the lowest energy. This is known as the **Aufbau principle** (see Figures 12.19 and 12.20).

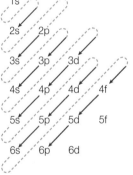

Figure 12.19 The Aufbau principle for filling atomic orbitals with electrons

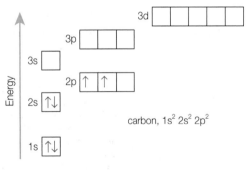

hydrogen, $1s^1$

carbon, $1s^2\,2s^2\,2p^2$

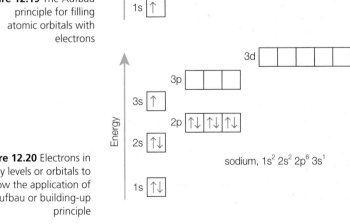

Figure 12.20 Electrons in energy levels or orbitals to show the application of the Aufbau or building-up principle

sodium, $1s^2\,2s^2\,2p^6\,3s^1$

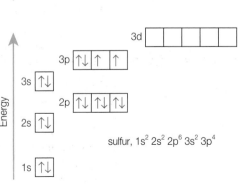

sulfur, $1s^2\,2s^2\,2p^6\,3s^2\,3p^4$

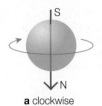

a clockwise

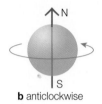

b anticlockwise

Figure 12.21 Electron spin

- Within a sub-shell electrons experience repulsion and hence enter two different orbitals of the same energy. This is known as **Hund's rule**.
- Electrons behave as particles and hence possess a classical property known as spin. An electron can spin in two different directions: clockwise and anticlockwise, shown by the symbols ↑ or ↓. Two electrons in the same orbital must have *opposite* spins,

i.e. ↑↓ and not ↑↑.

Single electrons in the same sub-shell must have the *same* (parallel) spin,

i.e. ↑ ↑ and not ↑ ↓ .

When a charged particle spins on its axis, a magnetic field is produced (Figure 12.21). Thus, electrons have magnetic properties. Protons show similar behaviour to electrons and also produce a magnetic field. This property is exploited in the technique of nuclear magnetic resonance (NMR) (Chapter 21).

Electron configurations of atoms

The detailed electron configuration of the hydrogen atom (atomic number 1) is:

1s
↑

It can be written as $1s^1$. The large number represents the shell number (principal quantum number), the letter represents the sub-shell and the superscript number represents the number of electrons in the sub-shell.

The detailed electron configuration of the helium atom (atomic number 2) is:

1s
↑↓

It can be written as $1s^2$. The two electrons must form a **spin pair**.

The lithium atom (atomic number 3) has three electrons. Two electrons enter the 1s orbital (as a spin pair). The 1s orbital is now full; so the third electron enters the 2s orbital (the next orbital with the lowest energy). This is in accordance with Hund's rule. The detailed electron configuration is:

1s 2s
↑↓ ↑

It can be written as $1s^2 2s^1$.

The beryllium atom (atomic number 4) has the detailed electron configuration shown below.

1s 2s
↑↓ ↑↓

It can be written as $1s^2 2s^2$.

The boron atom (atomic number 5) has five electrons. The first four electrons occupy the 1s and 2s orbitals. The fifth electron occupies the 2p orbital. The correct detailed electron configuration is:

1s 2s 2p
↑↓ ↑↓ ↑

It can be written as $1s^2 2s^2 2p^1$.

The carbon atom (atomic number 6) has six electrons and the correct detailed electron configuration is:

1s 2s 2p
↑↓ ↑↓ ↑ ↑

It can be written as $1s^2 2s^2 2p^2$.

Note that the following detailed electron configurations are *not allowed* (forbidden) for a carbon atom in its ground state.

Reason for error (principle violated)	Detailed electron configuration
The 2p electrons should occupy different orbitals. Hund's rule has been violated.	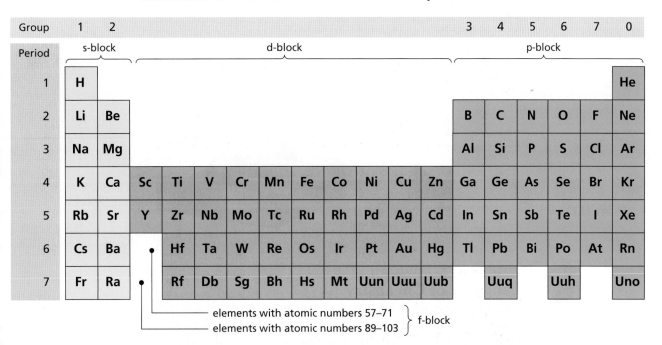
The single electrons in the same sub-shell should have the same spin. The Pauli exclusion principle has been violated.	
The 2s orbital can accept one more electron, so it should contain two electrons. The Aufbau principle has been violated.	

Table 12.8 Forbidden electron configurations

Language of Chemistry

Chemists often refer to empty orbitals when talking about atomic structure or chemical bonding. This is, of course, a linguistic convenience, since the orbital is the electron – there is no such thing as an empty orbital. ∎

Division of the periodic table into blocks

The long form of the periodic table is divided into four blocks: the s-, p-, d- and f-blocks (Chapter 3). This division reflects the filling of the outermost orbitals with electrons (Figure 12.22).

The detailed electron configurations of the first 54 elements are shown in Table 12.9. These are the ground state (lowest energy) configurations of the atoms.

Two elements in the first row of the d-block have unexpected electron configurations (highlighted in Table 12.9) that do not obey (they violate) the Aufbau principle. The outer electron configuration of the chromium atom is $4s^1 3d^5$ and not $4s^2 3d^4$ as expected. The outer configuration of the copper atom is $4s^1 3d^{10}$ and not $4s^2 3d^9$. A *simplified explanation* for these observations is that a half-filled or completely filled 3d sub-shell is a particularly stable electron configuration. The outer electron configurations for copper and chromium atoms can also be written as $3d^5 4s^1$ and $3d^{10} 4s^1$ so that the 3d sub-shell is placed into the third shell.

Group	1	2											3	4	5	6	7	0
Period	s-block						d-block								p-block			
1	H																	He
2	Li	Be											B	C	N	O	F	Ne
3	Na	Mg											Al	Si	P	S	Cl	Ar
4	K	Ca	Sc	Ti	V	Cr	Mn	Fe	Co	Ni	Cu	Zn	Ga	Ge	As	Se	Br	Kr
5	Rb	Sr	Y	Zr	Nb	Mo	Tc	Ru	Rh	Pd	Ag	Cd	In	Sn	Sb	Te	I	Xe
6	Cs	Ba		Hf	Ta	W	Re	Os	Ir	Pt	Au	Hg	Tl	Pb	Bi	Po	At	Rn
7	Fr	Ra		Rf	Db	Sg	Bh	Hs	Mt	Uun	Uuu	Uub	Uuq		Uuh		Uno	

elements with atomic numbers 57–71 } f-block
elements with atomic numbers 89–103

Figure 12.22 Long form of the periodic table marked into s-, p-, d- and f-blocks

You will *not* be expected to know the electron configurations of elements 39 to 48 (the second row of the d-block). However, their atoms behave like the first row – they ionize via loss of the 5s and then the 4d electrons. A number of other elements, in addition to copper and chromium, have anomalous configurations. These elements are highlighted in Table 12.9.

Atomic number	Chemical symbol of element	Electron configuration	Atomic number	Chemical symbol of element	Electron configuration	Atomic number	Chemical symbol of element	Electron configuration
1	H	$1s^1$	19	K	$[Ar]4s^1$	37	Rb	$[Kr]5s^1$
2	He	$1s^2$	20	Ca	$[Ar]4s^2$	38	Sr	$[Kr]5s^2$
3	Li	$[He]2s^1$	21	Sc	$[Ar]4s^23d^1$	39	Y	$[Kr]5s^24d^1$
4	Be	$[He]2s^2$	22	Ti	$[Ar]4s^23d^2$	40	Zr	$[Kr]5s^24d^2$
5	B	$[He]2s^22p^1$	23	V	$[Ar]4s^23d^3$	41	Nb	$[Kr]5s^14d^4$
6	C	$[He]2s^22p^2$	24	Cr	$[Ar]4s^13d^5$	42	Mo	$[Kr]5s^14d^5$
7	N	$[He]2s^22p^3$	25	Mn	$[Ar]4s^23d^5$	43	Tc	$[Kr]5s^14d^6$
8	O	$[He]2s^22p^4$	26	Fe	$[Ar]4s^23d^6$	44	Ru	$[Kr]5s^14d^7$
9	F	$[He]2s^22p^5$	27	Co	$[Ar]4s^23d^7$	45	Rh	$[Kr]5s^14d^8$
10	Ne	$[He]2s^22p^6$	28	Ni	$[Ar]4s^23d^8$	46	Pd	$[Kr]4d^{10}$
11	Na	$[Ne]3s^1$	29	Cu	$[Ar]4s^13d^{10}$	47	Ag	$[Kr]5s^14d^{10}$
12	Mg	$[Ne]3s^2$	30	Zn	$[Ar]4s^23d^{10}$	48	Cd	$[Kr]5s^24d^{10}$
13	Al	$[Ne]3s^23p^1$	31	Ga	$[Ar]4s^23d^{10}4p^1$	49	In	$[Kr]5s^24d^{10}5p^1$
14	Si	$[Ne]3s^23p^2$	32	Ge	$[Ar]4s^23d^{10}4p^2$	50	Sn	$[Kr]5s^24d^{10}5p^2$
15	P	$[Ne]3s^23p^3$	33	As	$[Ar]4s^23d^{10}4p^3$	51	Sb	$[Kr]5s^24d^{10}5p^3$
16	S	$[Ne]3s^23p^4$	34	Se	$[Ar]4s^23d^{10}4p^4$	52	Te	$[Kr]5s^24d^{10}5p^4$
17	Cl	$[Ne]3s^23p^5$	35	Br	$[Ar]4s^23d^{10}4p^5$	53	I	$[Kr]5s^24d^{10}5p^5$
18	Ar	$[Ne]3s^23p^6$	36	Kr	$[Ar]4s^23d^{10}4p^6$	54	Xe	$[Kr]5s^24d^{10}5p^6$

Table 12.9 Detailed electron configurations of gaseous isolated atoms in the ground state

■ Extension: The Stern–Gerlach experiment

The Stern–Gerlach experiment (Figure 12.23) was performed in 1921 by two German physicists, Otto Stern and Walther Gerlach. It provides strong experimental evidence for electron spin and hence supports quantum mechanics. Stern and Gerlach passed a beam of silver atoms through a non-uniform magnetic field and found that the beam was split into two bands. The interpretation of this result is based on the model in which each silver atom contains one unpaired electron. This electron can have a clockwise or anticlockwise spin. The two bands correspond to the two spin orientations.

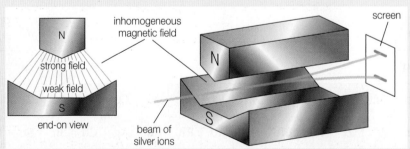

Figure 12.23 Stern–Gerlach apparatus

Language of Chemistry

The Aufbau principle takes its name from the German *Aufbauprinzip*, 'building-up principle', rather than being named for a scientist. In fact, it was formulated by the Danish physicist Niels Bohr around 1920. The principle postulates a hypothetical process in which an atom is 'built up' by progressively adding electrons. ■

■ Extension: Quantization

The two diagrams in Figure 12.24 provide a useful analogy for understanding the concept of quantization. A staircase is a quantized system. A person walking down the staircase must place their feet on the stairs. Hence they will possess certain definite values of potential energy corresponding to the energies of the steps. The potential energy of the person is quantized. In contrast, if a person walks down a ramp, then their potential energy changes continuously and they can have any value of potential energy corresponding to any specific point on the ramp. Potential energy in this case is not quantized.

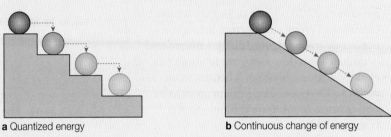

a Quantized energy **b** Continuous change of energy

Figure 12.24 Analogies of quantized and non-quantized systems

■ Extension: Electron configurations of excited species

When one or more electrons absorb thermal or electrical energy, they are promoted into higher energy orbitals. The atoms and electrons are in an excited state (Chapter 2).

A specific example of an excited sodium atom is shown below in Figure 12.25. The return of the excited electron to the ground state will give rise to emission of electromagnetic radiation corresponding to a specific line in the emission spectrum of sodium atoms.

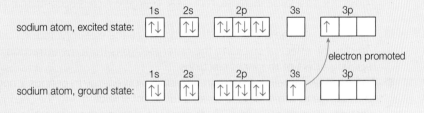

Figure 12.25 Orbital notation for sodium atoms in ground and excited states

Electronic configuration of ions

Hund's rule, the Pauli exclusion principle and the Aufbau principle also apply when extra electrons are added to form negative ions (anions). The fluoride ion (Figure 12.26) is formed when a fluorine atom ($1s^2 2s^2 2p^5$) gains an additional electron.

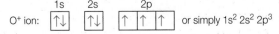

Figure 12.26 Orbital notation and detailed electron configuration for a fluoride ion, F^-

To deduce the electron configuration of positive ions (cations), electrons are removed in reverse order (that is, the last electron is removed first). (An exception to this 'rule' occurs with the transition metals – see Chapter 13.)

For example, the $O^+(g)$ ion is formed by the removal of one electron from an oxygen atom ($1s^2 2s^2 2p^4$) Figure 12.27. This ionization process can be made to occur inside a mass spectrometer (Chapter 2). The electron removed is the last electron from the 2p sub-shell.

O^+ ion: [↑↓] [↑↓] [↑ | ↑ | ↑]   or simply $1s^2 2s^2 2p^3$

1s 2s 2p

Figure 12.27 Orbital notation and detailed electron configuration for an $O^+(g)$ ion

The octet rule

The electron arrangements of noble gases are relatively stable and their atoms do not lose or gain electrons to form ions. Atoms of noble gases, with the exception of helium, have eight electrons in their outer shells. This arrangement is known as an **octet**.

According to the **octet rule**, atoms usually form stable ions by losing or gaining electrons to attain an octet (Chapter 4). For example, the nitrogen atom gains three electrons to attain the stable electron arrangement of neon, the nearest noble gas (Figure 12.28). The calcium atom loses two electrons to attain the electron arrangement of the noble gas, argon (Figure 12.29). Lithium and beryllium atoms lose electrons to attain the electronic arrangement of a helium atom, with two electrons. The lithium atom loses one electron to form the lithium ion, Li⁺ (Figure 12.30).

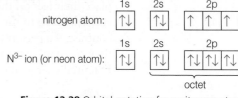

Figure 12.28 Orbital notation for a nitrogen atom and the nitride ion, N^{3-}

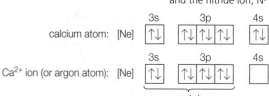

Figure 12.29 Orbital notation for a calcium atom and the calcium ion, Ca^{2+}

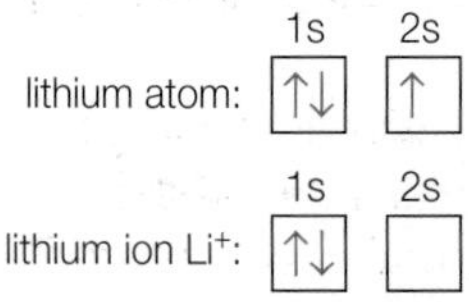

lithium atom:

lithium ion Li⁺:

Figure 12.30 Orbital notation for a lithium atom and the lithium ion, Li⁺

SUMMARY OF
KNOWLEDGE

- The first ionization energy is the minimum energy required to remove a mole of electrons from a mole of gaseous atoms to form a mole of gaseous unipositive ions (under standard thermodynamic conditions).
- The ionization energy of an atom or ion is determined by the following factors: nuclear charge, shielding effect and atomic radius (distance between nucleus and outer electrons).
- On moving across periods 1, 2 and 3 there is a general increase in first ionization energy. This is due to a large increase in nuclear charge which is accompanied by a small increase in shielding.
- The nuclear charge is the force that attracts all the electrons to the nucleus. The shielding effect is the effect of shielding the outer electrons from the attraction of the nucleus by the repelling effect of the inner electrons.
- There are two small decreases in first ionization energy in periods 2 and 3. The first is due to a change in sub-shell (sub-level) and loss of shielding and the second is due to enhanced electron–electron repulsion. These factors outweigh the increase in nuclear charge with increasing atomic number.
- Electrons can be progressively removed from gaseous atoms and successive ionization energies measured and plotted against number of electrons removed. The graphs offer strong experimental evidence for shells (main energy levels) and sub-shells (sub-levels). The relatively large increases in successive ionization energies correspond to changes in shells; smaller changes correspond to changes in sub-shells or electron pairing.
- Each shell can hold a maximum number of electrons: first shell (two electrons), second shell (eight electrons), third shell (18 electrons) and fourth shell (32 electrons). (The number of electrons in each shell is given by $2n^2$, where n represents the shell number.)
- Each shell (energy level) is composed of one or more sub-shells (sub-levels). The first shell is composed of one sub-shell, the second shell is composed of two sub-shells, the third shell is composed of three sub-shells and the fourth shell is composed of four sub-shells.
- Electrons occupy regions or volumes of space called orbitals. Each atomic orbital can accommodate a maximum of two electrons.
- Electrons can occupy four types of orbitals: s, p, d and f. s orbitals are spherical in shape; p orbitals are 'dumb-bell' shaped and arranged mutually perpendicularly to each other. p, d and f orbitals all have the same energy within the same shell or sub-shell.
- Orbitals (of the same type) retain the same shape but become larger (and their electron density more diffuse) with an increase in shell number.

- The first shell has an s sub-shell (one orbital); the second shell has an s and a p sub-shell (three orbitals); the third shell has an s sub-shell, a p sub-shell and a d sub-shell (five orbitals); and the fourth shell has an s sub-shell, a p sub-shell, a d sub-shell and an f sub-shell (seven orbitals).
- Electrons are assigned to atomic orbitals of atoms (in their ground state) according to the Aufbau principle, the Pauli exclusion principle and Hund's rule.
- The Aufbau principle states that electrons occupy atomic orbitals in the order of the energy levels of the orbitals.
- Copper and chromium atoms have anomalous electron configurations which violate the Aufbau principle. The chromium atom is $3d^54s^1$ and the copper atom is $3d^{10}4s^1$ because d sub-shells that are half-filled or fully filled are particularly stable.
- The Pauli exclusion principle states that only two electrons may occupy the same orbital and that these two electrons must have opposite spins.
- Hund's rule states when electrons are placed in a set of atomic orbitals with equal energies, the electrons must occupy them singly with parallel spins before they occupy the orbitals in pairs. As a consequence of Hund's rule atoms tend to maximize the number of unpaired electrons.
- The electron configuration of an atom describes how electrons are distributed among the various orbitals in the various shells and sub-shells and is denoted by an orbital diagram or spdf notation.
- In orbital notation each orbital is represented by a box and each electron by an arrow. The arrow head indicates the direction of electron spin.
- In spdf notation the electron configuration is denoted by writing the symbol for the occupied sub-shell and adding a superscript to indicate the number of electrons in that sub-shell. This can be condensed by describing the core electrons with the nearest noble gas electron configuration.
- Simple positive ions are formed by removing the appropriate number of outer electrons; simple negative ions are formed by adding the appropriate number of electrons to the outer sub-shell. Excited atoms have one or more electrons promoted to higher energy levels.
- Many ions formed by non-transition metals obey the octet rule. These ions have a full outer shell of eight electrons.
- Excited atoms are formed when one or more electrons are promoted to higher energy orbitals.

■ *Examination questions – a selection*

Paper 1 IB questions and IB style questions

Q1 What is the electron configuration for an atom with $Z = 22$?
A $1s^22s^22p^63s^23p^63d^4$
B $1s^22s^22p^63s^23p^64s^24p^2$
C $1s^22s^22p^63s^23p^63d^24p^2$
D $1s^22s^22p^63s^23p^64s^23d^2$
Higher Level Paper 1, May 03, Q5

Q2 For the species below, which one would require the most energy for the removal of an electron?
A Na⁺   B F   C F⁻   D Ar

Q3 How many unpaired electrons would the Fe^{2+} ion be expected to have?
A 1   B 3   C 4   D 6

Q4 An atom of chlorine has the electron configuration $[Ne]3s^23p^5$. What is the number of orbitals occupied by at least one electron?
A 7   B 9   C 13   D 17

Q5 For which of the following pairs of species are the chemical properties most similar?
A ^2_1H and $^2_1H^+$ C $^{23}_{11}Na$ and $^{39}_{19}K$
B $^{14}_6C$ and $^{14}_7N$ D $^7_3Li^+$ and ^9_4Be

Q6 Which of the following elements has the smallest first ionization energy?
A $_{19}K$ B $_6C$ C $_{12}Mg$ D $_4He$

Q7 What is the total number of electrons in d orbitals in a tin atom $_{50}Sn$?
A 5   B 10   C 20   D 0

Q8 Which of the following atoms has the smallest first ionization energy?
A Na   B F   C Be   D Cl

Q9 What is the electron configuration of Co^{3+}?
A [Ar]
C $[Ar]3d^5$
B $[Ar]4s^23d^4$
D $[Ar]3d^6$

Q10 What is the ground state electron configuration of the Fe^{3+} ion?
A $[Ar]4s^13d^5$
C $[Ar]3d^5$
B $[Ar]4s^23d^3$
D $[Ar]3d^3$

Q11 Which set of chemical species is arranged in order of decreasing ionic radius?
A $Ca^{2+} > Sr^{2+} > Ba^{2+}$
C $K^+ > Cl^- > S^{2-}$
B $Al^{3+} > Mg^{2+} > Na^+$
D $O^{2-} > F^- > Na^+$

Q12 Which ground state configuration for the following atoms is incorrect?
A $_{12}Mg$ $1s^22s^22p^63s^2$
B $_{25}Mn$ $[Ar]4s^23d^4$
C $_{35}Br$ $[Ar]3d^{10}4s^24p^5$
D $_{36}Kr$ $[Ar]4s^23d^{10}4p^6$

Q13 Which chemical species has the smallest radius?
A Na^+ **B** Na   **C** Mg^{2+} **D** Al^{3+}

Q14 Which of the following species would you expect to have the largest radius?
A Na^+ **B** N^{3-} **C** Ne   **D** F^-

Q15 How many unpaired electrons would the cobalt atom (element number 27) in its ground state electronic configuration be expected to have?
A 5   **B** 2   **C** 4   **D** 3

Q16 Which property changes along with increasing ionization energy?
A increasing non-metallic properties
B increasing atomic radii
C decreasing electron affinity
D decreasing nuclear charge

Q17 For which element would neutral isolated atoms in the ground state have two half-filled orbitals?
A $_{15}P$ **B** $_4Be$ **C** $_6C$ **D** $_7N$

Q18 What is the maximum number of electrons that can occupy the 5d sub energy level?
A 20   **B** 10   **C** 32   **D** 25

Q19 Which of the following elements has the largest first ionization energy?
A $_{18}Ar$ **B** $_{15}P$ **C** $_6C$ **D** $_{55}Cs$

Q20 The first three ionization energies for two elements, X and Y, are given below.

Element	First ionization energy/kJ mol^{-1}	Second ionization energy/kJ mol^{-1}	Third ionization energy/kJ mol^{-1}
X	520	7300	11800
Y	1086	2350	4620

Which pair of elements could represent X and Y?
A $_3Li$ and $_6C$
C $_8O$ and $_{16}S$
B $_4Be$ and $_{15}P$
D $_2He$ and $_4Be$

Q21 Which of the following chemical species has the largest radius?
A Al^{3+} **B** Mg^{2+} **C** F^- **D** S^{2-}

Q22 Which atom or ion could have a $3d^8$ electronic configuration?
A Mn   **B** Ni   **C** Cu^{2+} **D** Ni^{2+}

Q23 Elements in the first transition series differ from each other mainly in the number of which type of electrons?
A p electrons
B s and p electrons
C p and d and f electrons
D d electrons

Q24 Which is the best periodic correlation between atomic number and another atomic property?
A atomic masses
B ionic and atomic radii
C first ionization energies
D electron configuration

Q25 In which of the following series are the atoms arranged in order of increasing first ionization energy?
A Be, Mg, Ca
C Be, B, C
B O, F, Ne
D Ne, O, F

Q26 What increases **in equal steps of one** from left to right in the periodic table for the elements lithium to neon?
A the number of occupied electron energy levels
B the number of neutrons in the most common isotope
C the number of electrons in the atom
D the atomic mass

Higher Level Paper 1, May 05, Q6

Q27 In which of the following ground-state electron configurations are unpaired electrons present?
I $[He]2s^22p^2$
II $[He]2s^22p^3$
III $[He]2s^2p^4$
A I only
C I, II and III
B I and II only
D II and III only

Q28 What is the number of unpaired electrons in the Cr^{3+} ion?

A 0   **B** 2   **C** 3   **D** 5

Q29 A transition metal ion has the electronic configuration $X^{3+} = [Ar]3d^4$. What is the atomic number of element X?

A 24   **B** 22   **C** 25   **D** 26

Q30 What is the total number of p orbitals containing one or more electrons in germanium (atomic number 32)?

A 2   **B** 3   **C** 5   **D** 8

Higher Level Paper 1, May 04, Q5

Paper 2 IB questions and IB style questions

Q1 The graph below shows the variation in first ionization energy of some chemical elements. Figure 1 refers to the chemical elements and Figure 2 refers to chemical elements in the same group as element C.

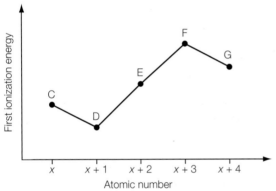

Figure 1

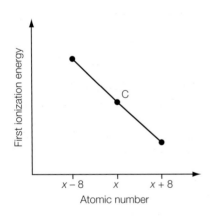

Figure 2

a Define the term *first ionization energy* of an element. [2]

b Element C, with atomic number x, is in group 2 of the periodic table. Justify this using all the information from Figure 1. [4]

c Explain the trend in the first ionization energy as shown in Figure 2. [3]

d State which period element C is in and explain your reasoning. [3]

e Why is the first ionization energy of element G lower than that of element F? [2]

Q2 The successive ionization energies of germanium are shown in the following table:

	1st	2nd	3rd	4th	5th
Ionization energy/ kJ mol⁻¹	760	1540	3300	4390	8950

a Identify the sub-level from which the electron is removed when the first ionization energy of germanium is measured. [1]

b Write an equation, including state symbols, for the process occurring when measuring the second ionization energy of germanium. [1]

c Explain why the difference between the 4th and 5th ionization energies is much greater than the difference between any two other successive values. [2]

Higher Level Paper 2, Nov 05, Q2

Q3 The graph shows the variation in second ionization energy of the elements silicon to argon.

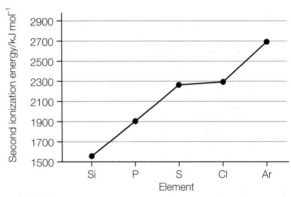

a Explain the general increase in second ionization energy from silicon to argon. [3]

b Give the detailed electron configurations of the $S^+(g)$ and $Cl^+(g)$ ions. [2]

c Explain why the increase from sulfur to chlorine is significantly less than the increase to sulfur from phosphorus. [1]

STARTING POINTS

- There are clear trends in the empirical formulas of chlorides and oxides in period 3.
- Compounds on the left of period 3 tend to be ionic; compounds on the right of period 3 tend to be covalent.
- Ionic compounds are good conductors when molten; covalent compounds are usually poor conductors.
- Ionic chlorides dissolve in water; many ionic oxides react with water.
- Covalent chlorides usually react with water; covalent oxides may react with water.
- Some period 3 non-metallic elements can form more than one stable chloride and oxide.
- The d-block elements are metals whose atoms are filling up a 3d sub-shell with electrons.
- The d-block metals have characteristic physical and chemical properties.
- Vertical trends in the d-block are less significant than in groups 1 to 7 in the periodic table.
- The d-block metals have physical properties that allow them to be used as structural materials.
- Scandium and zinc are not transition elements.
- Many transition elements are used in industry as heterogeneous catalysts.
- Many enzymes contain transition elements or ions in their active sites.

13.1 Trends across period 3

13.1.1 Explain the physical states (under standard conditions) and electrical conductivity (in the molten state) of the chlorides and oxides of the elements in period 3 in terms of their bonding and structure.

13.1.2 Describe the reactions of chlorine and the chlorides referred to in 13.1.1 with water.

Oxides of period 3

Reactions of the elements in period 3 with oxygen

The reactions between oxygen with the elements of period 3 are summarized in Table 13.1.

Name of element	Description of reaction of element with oxygen	Equation for reaction of element and oxygen
Sodium	Burns in air with an orange flame to give a white solid.	$4Na(s) + O_2(g) \rightarrow 2Na_2O(s)$
Magnesium	Burns with a brilliant white flame to form a white solid.	$2Mg(s) + O_2(g) \rightarrow 2MgO(s)$
Aluminium	Burns on heating in oxygen to give a white solid.	$4Al(s) + 3O_2(g) \rightarrow 2Al_2O_3(s)$
Silicon	Burns on heating in oxygen to give a white solid.	$Si(s) + O_2(g) \rightarrow SiO_2(s)$
Phosphorus	White phosphorus catches fire spontaneously in air to give a white solid.	Limited air (lower oxide): $P_4(s) + 3O_2(g) \rightarrow P_4O_6(s)$ Excess air (higher oxide): $P_4(s) + 5O_2(g) \rightarrow P_4O_{10}(s)$
Sulfur	Burns with a blue flame giving a colourless gaseous oxide.	$S(s) + O_2(g) \rightarrow SO_2(g)$
Chlorine	Does not directly react with oxygen.	

Table 13.1 Reactions of the elements of period 3 with oxygen

Sodium oxide

Sodium oxide is a **basic oxide**, which means that it will react with acids to form a salt and water, for example:

$$Na_2O(s) + 2HCl(aq) \rightarrow 2NaCl(aq) + H_2O(l)$$

The net ionic equation is

$$O^{2-}(aq) + 2H^+(aq) \rightarrow H_2O(l)$$

where the oxide ion is again acting as a base (hydrogen ion acceptor) (Chapter 8).

Magnesium oxide

Magnesium oxide (Figure 13.1) reacts reversibly with water and some dissolves to form a weakly alkaline solution of magnesium hydroxide in an equilibrium reaction:

$$MgO(s) + H_2O(l) \rightleftharpoons Mg(OH)_2(aq)$$

$$O^{2-}(aq) + H_2O(l) \rightarrow 2OH^-(aq)$$

Figure 13.1 Magnesium oxide

Magnesium oxide, like sodium oxide, is a basic oxide and will react with dilute aqueous solutions of acids, for example:

$$MgO(s) + 2HCl(aq) \rightarrow MgCl_2(aq) + H_2O(l)$$

$$O^{2-}(aq) + 2H^+(aq) \rightarrow H_2O(l)$$

Applications of Chemistry

Magnesium hydroxide is slightly alkaline. A suspension of it in water is a milky white liquid, known as Milk of Magnesia. It is used as an antacid (Chapter 24) to neutralize excess stomach acid. It also acts as a laxative and helps relieve constipation.

■ Extension: Side reactions

Sodium and magnesium both undergo side reactions when burnt in air. Some yellowish sodium peroxide (Na_2O_2 [$2Na^+$ $2O^{2-}$]) is also formed, especially when the air is in plentiful supply:

$$2Na(s) + O_2(g) \rightarrow Na_2O_2(s)$$

In the case of magnesium some magnesium nitride (Mg_3N_2 [$3Mg^{2+}$ $2N^{3-}$]) is formed as a result of combination with nitrogen in the air:

$$3Mg(s) + N_2(g) \rightarrow Mg_3N_2(s)$$

Aluminium oxide

Unlike sodium and magnesium oxides, aluminium oxide does not react with water, although it does react slowly with warm, aqueous solutions of dilute acids to form salts, for example:

$$Al_2O_3(s) + 6HCl(aq) \rightarrow 2AlCl_3(aq) + 3H_2O(l)$$
$$Al_2O_3(s) + 6H^+(aq) \rightarrow 2Al^{3+}(aq) + 3H_2O(l)$$

Aluminium oxide also reacts with warm concentrated solutions of strong alkalis to form aluminates, for example:

$$Al_2O_3(s) + 2NaOH(aq) + 3H_2O(l) \rightarrow 2NaAl(OH)_4(aq)$$
$$Al_2O_3(s) + 2OH^-(aq) + 3H_2O(l) \rightarrow 2Al(OH)_4^-(aq)$$

Aluminium oxide is described as **amphoteric** since it reacts with both acids and bases. Aluminium is found in large quantities as the impure hydrated form, $Al_2O_3.2H_2O$, known as bauxite. Aluminium is extracted from this ore on an industrial scale by electrolysis (Chapter 23). Its insolubility in water is largely due to its high lattice energy (Chapter 15).

Figure 13.2 Aluminium oxide powder

Applications of Chemistry

Aluminium oxide (Figure 13.2) is a soft white powder with a very high melting point and, like magnesium oxide, is used as a refractory. The term refractory refers to the quality of a material to retain its strength at high temperatures. Refractory materials are used to make crucibles and linings for furnaces, kilns and incinerators.

Silicon dioxide

Figure 13.3 Quartz powder

Silicon dioxide is found in nature as quartz (Figure 13.3) and upon weathering forms sand (often coloured yellow due to the presence of iron(III) oxide).

The non-metal oxides are all **acidic oxides**. Silicon dioxide will react with molten sodium hydroxide to form sodium silicate:

$$SiO_2(s) + 2NaOH(l) \rightarrow Na_2SiO_3(l) + H_2O(g)$$

It does not react with aqueous sodium hydroxide except at high pressure and temperature. The need for these extreme conditions is due to the strength of the giant covalent lattice of silicon dioxide (Chapter 4).

Applications of Chemistry

A thick solution of sodium silicate is known as water glass and was used during the Second World War in the UK as a preservative for eggs. Sodium silicate is also used in laboratories to grow 'crystal gardens'. These are formed when crystals of highly soluble salts such as copper(II) sulfate and iron(II) chloride are added. Highly insoluble precipitates of copper and iron(II) silicate are formed.

Phosphorus oxides

Phosphorus(III) oxide is a white solid that reacts with cold water to form phosphoric(III) acid:

$$P_4O_6(s) + 6H_2O(l) \rightarrow 4H_3PO_3(aq)$$

Phosphorus(V) oxide is a white solid that reacts violently with cold water to give a solution of phosphoric(V) acid:

$$P_4O_{10}(s) + 6H_2O(l) \rightarrow 4H_3PO_4(aq)$$

■ **Extension:** ## Structures of phosphorus oxides

The two oxides of phosphorus have interesting 'cage structures' (Figure 13.4) and exist as units of double their empirical formulas: P_2O_3 and P_2O_5. The cages are formed from tetrahedral arrangements of phosphorus molecules, P_4, 'bridged' by oxygen.

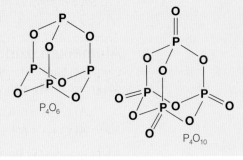

Figure 13.4 The oxides of phosphorus

Sulfur oxides

$O \overset{\cdot\cdot}{\underset{\underset{\cdot\cdot}{S}}{}} O$

sulfur dioxide, SiO_2

$O = S \overset{O}{\underset{O}{}}$

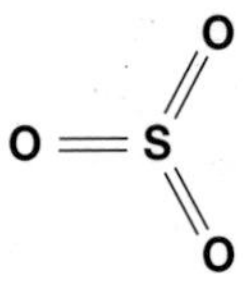

sulfur trioxide, SiO_3

Figure 13.5 Structures of sulfur dioxide and sulfur trioxide molecules

Sulfur dioxide is a colourless gas with a pungent smell. It forms sulfurous acid in water. It acts as a reducing agent. The presence of sulfur dioxide in the air contributes to the formation of acid rain (Chapter 25).

$$SO_2(g) + H_2O(l) \rightarrow H_2SO_3(aq)$$

Sulfur trioxide reacts violently with water to form sulfuric acid. (It can act as an oxidizing or dehydrating agent when concentrated.)

$$SO_3(g) + H_2O(l) \rightarrow H_2SO_4(aq)$$

Sulfur dioxide can be oxidized in the presence of a catalyst to form sulfur trioxide. This is the basis of the industrial manufacture of sulfuric acid by the Contact process (Chapter 7).

The Lewis structures for the oxides of sulfur are usually drawn with the sulfur atom exceeding the octet rule (Chapter 4), but still with an even number of electrons (Figure 13.5).

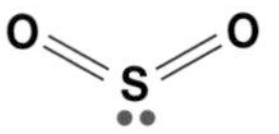

Applications of Chemistry

Sulfurous acid or aqueous sulfur dioxide is used in the food industry to inhibit bacterial growth in wine and tinned food. It also acts as an antioxidant in wine, preventing any browning and keeping the wine fresher longer (Chapter 26).

Oxides of chlorine

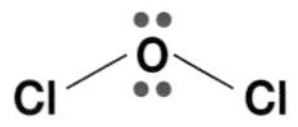

Figure 13.6 Structure of the dichlorine monoxide molecule

Dichlorine(I) oxide or dichlorine monoxide (Figure 13.6) is prepared by passing dry chlorine over mercury(II) oxide.

It is a brownish-yellow gas which condenses to a golden-brown liquid. It is a highly unstable compound that is liable to explode to form its elements:

$$2Cl_2O(g) \rightarrow 2Cl_2(g) + O_2(g)$$

It dissolves in water forming chloric(I) acid, a weak acid and a moderately strong oxidizing agent:

$$Cl_2O(g) + H_2O(l) \rightleftharpoons 2HClO(aq) \rightleftharpoons 2H^+(aq) + 2ClO^-(aq)$$

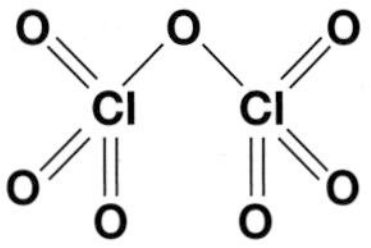

Figure 13.7 Structure of the chlorine heptoxide molecule

Chlorine(VII) oxide or chorine heptoxide (Figure 13.7) is a colourless and oily liquid that, like dichlorine monoxide (and other oxides of chlorine), is liable to explode violently. It dissolves and reacts with water slowly to form a solution of chloric(VII) acid (perchloric acid):

$$Cl_2O_7(l) + H_2O(l) \rightarrow 2HClO_4(aq)$$

■ Extension: Electronegativity and acid–base character

An element represented by M bonded to an —OH group can ionize in two different ways:

As an acid $MOH \rightarrow M–O^- + H^+$
As a base: $MOH \rightarrow M^+ + OH^-$

The electronegativity of M largely controls which type of ionization occurs (Table 13.2). Low values of electronegativity favour ionic bonding between M and OH, whereas high values of electronegativity favour covalent bonding between M and OH. Intermediate values will lead to polar covalent bonding (Chapter 4).

Electronegativity of M	Acid–base character
<1.5	Basic
1.5 to 2.5	Amphoteric
> 2.5	Acidic

Table 13.2 Acid–base behaviour of MOH (if water is used as the solvent)

For example, consider sodium hydroxide, NaOH. The electronegativity of sodium is 0.9 and hence it is predicted to be basic. Sodium hydroxide is a strong alkali:

$$NaOH(aq) \rightarrow Na^+(aq) + OH^-(aq)$$

Aluminium has an electronegativity value of 1.5 and its hydroxide is predicted to show amphoteric behaviour, which is confirmed experimentally. Sulfuric acid, when anhydrous, is composed of molecules with the structure $HOSO_2OH$. Sulfur has an electronegativity value of 3.5 and hence is predicted to form an acidic 'hydroxide'.

The properties of the oxides of the elements in period 3 are summarized in Table 13.3.

Element	Sodium	Magnesium	Aluminium	Silicon	Phosphorus	Sulfur	Chlorine
Formula of oxide(s)	Na_2O	MgO	Al_2O_3	SiO_2	P_4O_6 P_4O_{10}	SO_2 SO_3	Cl_2O Cl_2O_7
State at RTP*	Solid	Solid	Solid	Solid	Solid Solid	Gas Liquid	Gas Liquid
Oxidation number	+1	+2	+3	+4	+3 +5	+4 +6	+1 +7
Structure	Ionic	Ionic	Ionic	Giant covalent	Molecular covalent	Molecular covalent	Molecular covalent
Melting point/K	1405	3173	2313	1883	297 853 (under pressure)	198 290	
Boiling point/K	Decomposes at 2223	3873	3253	2503	448 P_4O_{10} sublimes at 573	263 318	Decomposes
Acid–base nature	Basic	Basic	Amphoteric	Acidic	Acidic	Acidic	Acidic
Effect of water on oxide; approximate pH of saturated solution	Strongly alkaline solution pH ≅ 13	Slightly alkaline solution pH ≅ 9	Insoluble	Insoluble	Strongly acidic solution pH ≅ 2	Strongly acidic solution pH ≅ 2	Strongly acidic solution pH ≅ 2
Electrical conductivity in molten state	High	High	High	Very low	Nil	Nil	Nil

* RTP is 298 K (25 °C) and 1 atm pressure.

Table 13.3 Summary of the oxides of the elements in period 3

Summary of trends

Physical states

The oxides which are ionic or have a giant covalent structure are solids at room temperature and pressure. Those oxides with a simple covalent structure tend to be liquids or gases. If they are solids, such as the phosphorus oxides, they will have a relatively low melting point.

Formula

There is a clear trend in the formulas of the oxides. For each of the oxides and higher oxides there is an increase of 0.5 mole of oxygen per mole of the element.

$$Na_2O \rightarrow MgO \rightarrow Al_2O_3 \rightarrow SiO_2 \rightarrow P_4O_{10} \rightarrow SO_3 \rightarrow Cl_2O_7$$
$$0.5 \quad 1.0 \quad 1.5 \quad 2.0 \quad 2.5 \quad 3.0 \quad 3.5$$

The maximum or higher oxidation number (Chapter 9) of the element in the oxide corresponds to the number of electrons used for bonding and accounts for the trend in oxide empirical formula (Chapter 9). It also corresponds to the group number.

Acid–base nature of the oxides

The oxides in period 3 change gradually from being ionic and basic on the left to being covalent and acidic on the right-hand side of the period. This change correlates with a change in structure from ionic via giant covalent to simple molecular.

The acid–base nature of oxides also depends on the electronegativity of the period 3 element to which the oxygen is chemically bonded. The less electronegative this chemical element is, the more basic the oxide. The oxide ion of ionic oxides accepts hydrogen ions to form water. The covalent oxides react with water, forming products where the covalent bonding between the original element and oxygen persists.

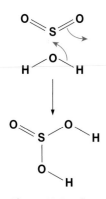

Figure 13.8 A diagram showing a simple mechanism for the reaction between sulfur dioxide and water molecules

This is illustrated in Figure 13.8 for sulfur dioxide reacting with water to form sulfurous acid. The 'curly arrows' represent the movement of electron pairs. This notation is widely used in organic chemistry (Chapters 10, 20 and 27).

Conductivity

The change in electrical conductivity correlates with the change in bonding from ionic to covalent. Ionic compounds when melted release ions which allow an electric current to flow when a voltage is applied through the liquid. Simple covalent compounds are composed of uncharged molecules and hence the molten liquids are non-conductors.

However, note that covalent compounds that react with water, for example the oxides of phosphorus, will form solutions containing ions. These aqueous solutions will be excellent conductors when a voltage is applied due to the presence of a high concentration of ions.

Chlorides of period 3

Reactions of the elements in period 3 with chlorine

The reactions between chlorine with the elements of period 3 are summarized in Table 13.4. (Chlorides of sulfur can also be formed by heating sulfur and chlorine, but knowledge of them is not a requirement of the IB Chemistry syllabus.)

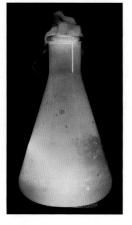

Figure 13.9 Sodium burning in chlorine

Name of chemical element	Description of reaction	Equation for reaction of element with chlorine
Sodium	Hot sodium reacts in chlorine (Figure 13.9) to give a white solid	$2Na(s) + Cl_2(g) \rightarrow 2NaCl(s)$
Magnesium	Hot magnesium reacts in chlorine to give a white solid	$Mg(s) + Cl_2(g) \rightarrow MgCl_2(s)$
Aluminium	Reacts on heating in chlorine to give a pale yellow solid	$2Al(s) + 3Cl_2(g) \rightarrow 2AlCl_3(s)$
Silicon	Reacts on heating in chlorine to give a colourless liquid	$Si(s) + 2Cl_2(g) \rightarrow SiCl_4(l)$
Phosphorus	White phosphorus reacts with limited chlorine to give a colourless liquid	Limited chlorine: $2P(s) + 3Cl_2(g) \rightarrow 2PCl_3(l)$ Excess chlorine: $2P(s) + 5Cl_2(g) \rightarrow 2PCl_5(s)$

Table 13.4 Reactions of the elements of period 3 with chlorine

Sodium chloride

Sodium chloride (Figure 13.10) is a colourless solid that dissolves in water to form a neutral solution.

$$NaCl(s) + (aq) \rightarrow NaCl(aq)$$

Figure 13.10 Crystals of sodium chloride

Magnesium chloride

Magnesium chloride (Figure 13.11) is a colourless solid that dissolves in water to form a slightly acidic solution. A small proportion of the magnesium chloride reacts reversibly with water to form hydrochloric acid.

Main reaction: $MgCl_2(s) + (aq) \rightarrow Mg^{2+}(aq) + 2Cl^-(aq)$

Side reaction: $MgCl_2(s) + 2H_2O(l) \rightleftharpoons Mg(OH)_2(s) + 2HCl(aq)$

Figure 13.11 Crystals of hydrated magnesium chloride, $MgCl_2.6H_2O$

Aluminium chloride

Anhydrous aluminium chloride is a volatile white solid that sublimes at a relatively low temperature. Anhydrous aluminium chloride undergoes rapid **hydrolysis** with a *small* amount of water to form aluminium hydroxide and hydrochloric acid.

$$AlCl_3(s) + 3H_2O(l) \rightarrow Al(OH)_3(s) + 3HCl(aq)$$

If aluminium chloride is dissolved in a *large* excess of water then the solution will be acidic due to the reaction between the hydrated aluminium ions and water molecules.

$$AlCl_3(s) + (aq) \rightarrow [Al(H_2O)_6]^{3+}(aq) + 3Cl^-(aq)$$

The hydrated aluminium ion behaves exactly like some transition metal ions: the small, highly charged metal ion polarizes (withdraws electron density from) the water molecules that are attached to the aluminium ion through dative covalent bonds (Chapter 8). This makes the hydrogen atoms positive and susceptible to attack from solvent water, which is acting as a base. The complex ion then loses hydrogen ions, causing the solution to be acidic from the formation of hydrogen ions, $H^+(aq)$:

$$[Al(H_2O)_6]^{3+}(aq) \rightarrow [Al(H_2O)_5OH]^{2+}(aq) + H^+(aq)$$

■ Extension: The dimerization of aluminium chloride

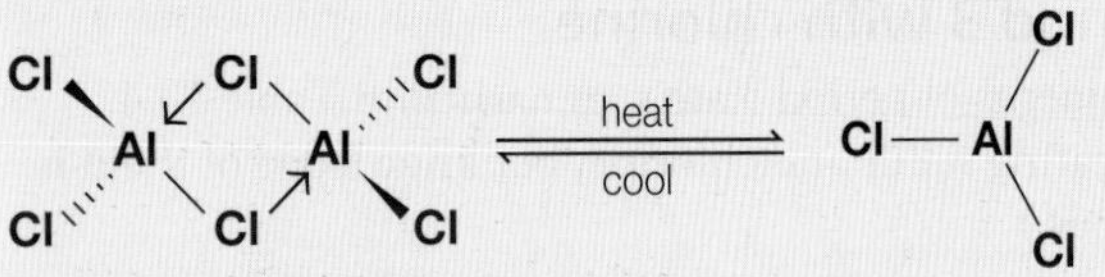

Figure 13.12 The dimerization of aluminium chloride at low temperatures

Analysis has revealed that aluminium chloride gas just above its sublimation temperature exists as molecules with the formula Al_2Cl_6. On raising the temperature further these **dimers** dissociate into $AlCl_3$ molecules. The structure of the dimer is shown in Figure 13.12.

Each arrow indicates the donation and sharing of a lone pair of electrons from a chlorine atom. This type of covalent bond is known as a dative or coordinate covalent bond (Chapter 4). In organic chemistry anhydrous aluminium chloride is used as a Lewis acid in the Friedel–Crafts reaction (Chapter 27).

Silicon tetrachloride

Silicon tetrachloride is a colourless, volatile covalent liquid that reacts vigorously with water (Figure 13.13) to form a strongly acidic solution of hydrochloric acid:

$$SiCl_4(l) + 2H_2O(l) \rightarrow SiO_2(s) + 4HCl(g)$$

This reaction is another example of hydrolysis. The structure and shape of the silicon tetrachloride molecule is shown in Figure 13.14. The molecular shape can be deduced from VSEPR theory (Chapter 4).

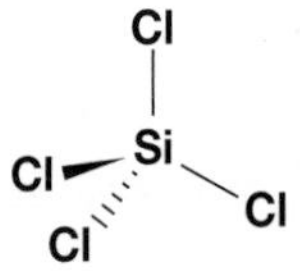

Figure 13.14 Structure and shape of silicon tetrachloride

Figure 13.13 The reaction between silicon tetrachloride and water vapour to form acidic fumes of hydrochloric acid that turn blue litmus red

Chlorides of phosphorus

Phosphorus(III) chloride (phosphorus trichloride) is a colourless, covalent liquid that fumes in moist air due to the action of hydrolysis. It is hydrolysed in ice cold water to form phosphoric(III) (phosphorous) acid:

$$PCl_3(l) + 3H_2O(l) \rightarrow H_3PO_3(aq) + 3HCl(aq)$$

Phosphorus(v) chloride (phosphorus pentachloride) is a pale yellow covalent solid that sublimes, but at high temperatures it undergoes dissociation:

$$PCl_5(s) \rightarrow PCl_3(l) + Cl_2(g)$$

Phosphorus pentachloride fumes in moist air and reacts violently with water to eventually form phosphoric(v) acid:

$$PCl_5(s) + 4H_2O(l) \rightarrow H_3PO_4(aq) + 5HCl(aq)$$

Phosphorus trichloride obeys the octet rule, but phosphorus pentachloride exceeds the octet rule (Figure 13.15).

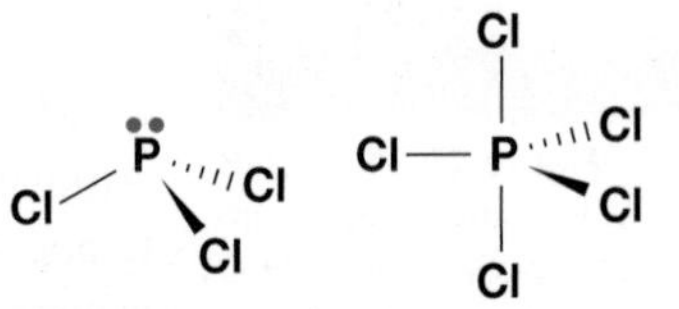

Figure 13.15 Structures and shapes of phosphorus(III) and (v) chloride molecules

■ Extension: Phosphorus(v) chloride

In the gaseous state phosphorus(v) chloride exists as molecules, but on cooling it forms an ionic solid containing a 1 : 1 molar ratio of $[PCl_4]^+$ and $[PCl_6]^-$ ions (Figure 13.16).

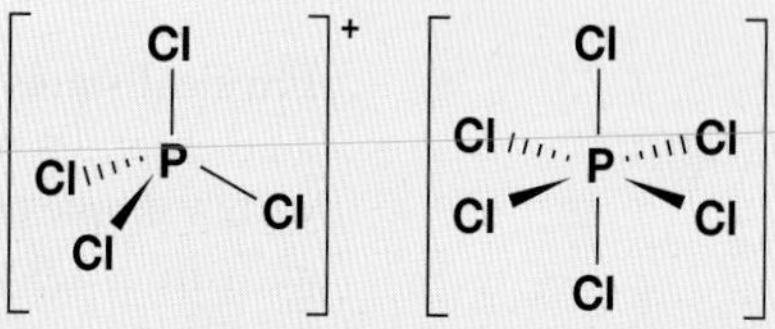

Figure 13.16 Structures and shapes of the ions present in solid phosphorus(v) chloride

Figure 13.17 Structure and shape of sulfur hexafluoride

Chlorides of sulfur

The highest chloride of sulfur is SCl_4 and not the expected SCl_6. However, there is a stable sulfur hexafluoride molecule, SF_6 (Figure 13.17). Sulfur hexachloride, SCl_6, is predicted to be unstable because six large chlorine atoms cannot be packed around a central sulfur atom. Fluorine atoms are smaller and can be packed around a sulfur atom without severe repulsion occurring.

Chlorine

Chlorine can be regarded as 'chlorine chloride'. It is a poisonous pale green gas with a pungent odour. It is moderately soluble in water, forming a yellowish green solution called 'chlorine water'. Some of the chlorine reacts with the water to form a mixture of hydrochloric and chloric(I) ('hypochlorous') acids:

$$Cl_2(g) + H_2O(l) \rightleftharpoons HCl(aq) + HClO(aq)$$

The formation of chlorine water is an example of disproportionation (Chapter 9). The chlorine undergoes both oxidation and reduction.

Applications of Chemistry

Hydrochloric acid is a strong acid and is responsible for the strongly acidic nature of chlorine water. The chloric(I) acid is responsible for the bleaching and disinfecting properties of chlorine water:

$$HClO(aq) + dyestuff \rightarrow HCl(aq) + oxidized\ dyestuff$$

The bleaching process is an oxidation process that involves the addition of oxygen to the dye.

Element	Sodium	Magnesium	Aluminium	Silicon	Phosphorus	Chlorine
Formula of chloride(s)	NaCl	MgCl$_2$	AlCl$_3$	SiCl$_4$	PCl$_3$ PCl$_5$	Cl$_2$
State at RTP	Solid	Solid	Solid	Liquid	Liquid Solid	Gas
Oxidation number	+1	+2	+3	+4	+3 +5	0
Structure	Ionic	Ionic	Covalent (two-dimensional polymer)	Simple covalent	Simple covalent (gas phase); ionic (solid)	Simple covalent
Melting point/K	1081	987	Sublimes at 453	203	179.4 (PCl$_3$) PCl$_5$ sublimes at 435	172
Boiling point/K	1738	1691		330	349 (PCl$_3$)	238
Effect of water on chloride; approximate pH of saturated solution	Neutral solution	Slightly acidic solution	Fumes of hydrochloric acid	Fumes of hydrochloric acid	Fumes of hydrochloric acid	Hydrochloric and chloric(I) acids formed
	pH ≅ 7	pH ≅ 6.5	pH ≅ 2	pH ≅ 1	pH ≅ 1	pH ≅ 2
Electrical conductivity in molten state	High	High	Very low	Nil	Nil	Nil

Table 13.5 Summary of the chlorides of the element in period 3

Summary of trends

Physical states

The chlorides which are ionic will be solids at room temperature and pressure. The chlorides with a simple covalent structure will tend to be volatile liquids or gases. If they are solids, such as anhydrous aluminium chloride, they have a relatively low melting point. This physical property is indicative of significant covalent character in aluminium chloride and a reflection of the polarizing power of the aluminium ion (Chapter 4).

Formulas

There is a clear trend in the empirical formulas of the chlorides. For each of the chlorides and higher chlorides there is an increase of one atom of chlorine per mole of the element.

$$NaCl \rightarrow MgCl_2 \rightarrow AlCl_3 \rightarrow SiCl_4 \rightarrow PCl_5$$
$$1 \qquad 2 \qquad 3 \qquad 4 \qquad 5$$

The oxidation number (Chapter 9) of the element in the chlorides, like the oxides, corresponds to the number of electrons used for bonding and accounts for the trend in chloride empirical formula (Chapter 9). Again, it also corresponds to the group number.

Acid–base nature of the chlorides

The chlorides in period 3 change gradually from being ionic and neutral on the left to being covalent and 'acidic' on the right-hand side of the period. The term 'acidic' refers to substances that can be hydrolysed by water to form acidic solutions. This change correlates with a change in structure from ionic to simple molecular.

The ionic chlorides dissolve in water to release hydrated ions and form neutral solutions. The covalent chlorides have polar covalent bonds that attract water molecules. The lone pair of electrons on the oxygen atom is attracted to the central non-metal atom.

Conductivity

The change in electrical conductivity correlates with the change in bonding from ionic to covalent. Ionic chlorides when melted release ions which allow an electric current to flow when a voltage is applied through the liquid. Simple covalent chlorides are composed of uncharged molecules and hence the molten liquids are non-conductors. (PCl_5 is a conductor because the liquid contains PCl_4^+ and PCl_6^- ions.)

However, note that covalent chlorides that *react* with water, for example the chlorides of phosphorus, will form solutions containing ions. These aqueous solutions will therefore be excellent conductors.

■ Extension: Noble gas fluorides

The noble gases have stable electronic structures, but in 1962 British chemist Neil Bartlett (1932–) prepared the first noble gas compound, $Xe^+[PtF_6]^-$. He later produced several other compounds of xenon: XeF_2, XeF_4 and XeF_6 (Chapter 14). These are stable covalent compounds. Argon, in period 3, is a smaller and more electronegative atom than xenon: its electrons are closer to the nucleus and hence less likely to enter into bond formation. However, in August 2000, the first argon compounds were created by researchers at the University of Helsinki, Finland. When they shone ultraviolet light onto frozen argon containing a small amount of hydrogen fluoride, argon hydrofluoride (HArF) was formed.

■ Extension: Group 4 chlorides

Silicon, germanium, tin and lead tetrachlorides all undergo hydrolysis when placed in water. However, carbon tetrachloride is inert towards water (Figure 13.18). Carbon ($1s^22s^22p^2$), at the top of group 4, *cannot* form more than four covalent bonds. When bonded to four other atoms its valence shell is filled.

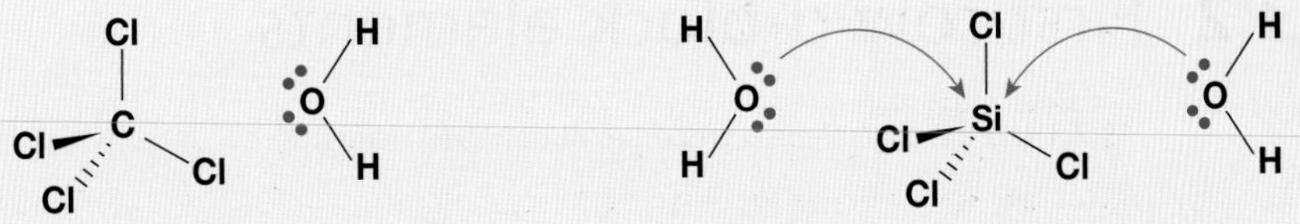

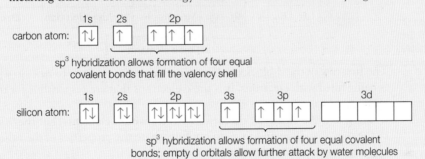

No possible attack by water molecules   Attack by water molecules on central atom

Figure 13.18 Water molecules with carbon and silicon tetrachloride molecules

However, silicon ($1s^22s^22p^63s^23p^2$) and the other members of group 4 have empty 3d orbitals in their valency shell (Figure 13.19) which can be used to accommodate the lone pairs of electrons from the attacking water molecules. These empty d orbitals are of relatively low energy since they are in the same shell as the outer 3p sub-shell. In contrast, carbon's outer sub-shell is in the second shell. There are no d orbitals in the second shell and the empty 3d orbitals are unavailable due to their high energy.

Another factor is that the CCl_4 molecule is so small that there is no space for nucleophilic attack (Chapter 10) on the carbon centre. The water acts as a nucleophile in the hydrolysis of silicon tetrachloride by donating a lone pair of electrons. A nucleophile is an electron pair donor. The silicon atom is much larger and is more accessible to water molecules. Consequently, the hydrolysis of carbon tetrachloride is thermodynamically favoured *but* kinetically hindered, meaning that the activation energy barrier to the reaction is very high.

carbon atom:

1s	2s		2p	

sp³ hybridization allows formation of four equal covalent bonds that fill the valency shell

silicon atom:

| 1s | 2s | | 2p | | 3s | | 3p | | | 3d | | | | |

sp³ hybridization allows formation of four equal covalent bonds; empty d orbitals allow further attack by water molecules

Figure 13.19 Diagrams of orbitals in the carbon and silicon atoms of the tetrachlorides

■ Extension: Period 3 hydrides

The period 3 elements form a number of hydrides that show clear trends in properties (Table 13.6). Sodium and magnesium form ionic hydrides containing the hydride ion, H^-. This behaves as a strong base in water, releasing hydrogen gas.

$$H^-(s) + H_2O(l) \rightarrow OH^-(aq) + H_2(g)$$

The hydrides of phosphorus and silicon catch fire in air spontaneously due to weak bonds. Phosphine (PH_3) is insoluble and unreactive towards water because it cannot form hydrogen bonds. Hydrogen sulfide is a very poisonous gas that acts as a weak acid in water.

Table 13.6 Properties of the hydrides of period 3

	NaH	MgH₂	AlH₃	SiH₄	PH₃	H₂S	HCl
Type of hydride	←——— Alkaline ———→			←—— Neutral ——→		←—— Acidic ——→	
Type of bonding	←——— Ionic ———→		Giant covalent ←	←——————— Covalent ———————→			
Reaction in dry air	←——————— Stable ———————→			←—— Catches fire ——→		←— Stable —→	

13.2 First-row d-block elements

13.2.1 **List** the characteristic properties of transition elements.

d-block metals

The **d-block metals** are a group of metals that occur in a large block between group 2 (s-block) and group 3 (p-block) of the periodic table (Figure 13.20). These elements have *similar* physical and chemical properties.

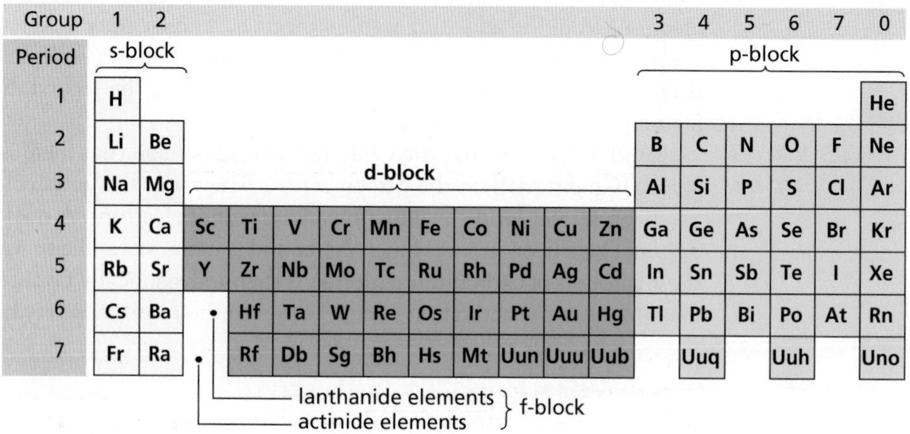

Figure 13.20 Position of the d-block metals in the periodic table

The first row of the d-block contains ten elements, scandium to zinc, in which the 3d sub-shell is being filled with electrons. It is these electrons that are responsible for their characteristic properties.

Eight of these elements are classified as **transition elements**, but the first and last members of the row, scandium and zinc, do not fully share the properties of the other eight and are not classified as transition elements.

The characteristic properties of the transition elements are:

- high densities, melting and boiling points
- the ability to exist in a variety of stable oxidation states; that is, they can form a variety of ions, both simple ions, for example manganese(II), Mn^{2+} (in solid compounds) and **oxoanions**, for example manganate(VII), MnO_4^-
- the formation of coloured ions (Figure 13.21)
- the ability to form a variety of **complex ions** (page 358), where the transition metal ion becomes datively bonded to molecules or ions
- the ability to act as catalysts (Chapter 6) and increase the rates of chemical reactions.

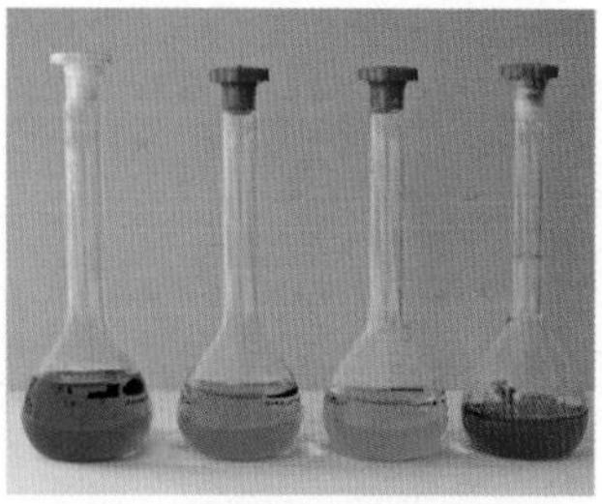

Figure 13.21 Aqueous solutions of copper(II), dichromate(VI), chromate(VI) and colbalt(II) ions

Zinc and scandium do not share these properties: they have relatively low melting points, boiling points and densities compared to the transition metals. Zinc and scandium have only one stable oxidation state: two and three, respectively. The ions they form are colourless. Zinc and scandium show some catalytic properties. However, zinc and scandium do both form complex ions, although this property is not unique to transition metals.

13.2.2 **Explain** why Sc and Zn are not considered to be transition elements.

Electron configurations of the d-block elements

The last element before the first member of the d-block is calcium, whose atom has the detailed electron configuration $1s^22s^22p^63p^64s^2$. However, with the next element, scandium, the additional electron is placed in an empty 3d sub-shell which was unoccupied (empty) in the calcium atom.

For calcium the 3d sub-shell was too high in energy for electrons to enter, so the extra 3d electron enters the first sub-shell of the fourth shell. But for scandium the extra proton has lowered the energy of the d orbitals so they can now be filled (Figure 13.22).

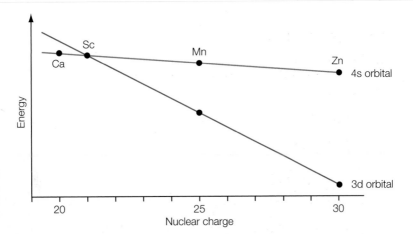

Figure 13.22 The 3d and 4s energy levels on crossing the first row of the d block

The 3d sub-shell has five orbitals into which successive electrons are placed according to the **Aufbau** or building up principle (Chapter 12), in particular:

- Electrons are, if possible, placed in 3d orbitals without being paired up, unless there are no more empty orbitals.
- If electrons are paired up in the same 3d orbitals, then a spin pair results.

The electron configurations of the first row d-block metals are given in Table 13.7.

Element	Atomic number	Electron configuration		3d					4s
Sc	21	[Ar]3d¹4s²	[Ar]	↑					↑↓
Ti	22	[Ar]3d²4s²	[Ar]	↑	↑				↑↓
V	23	[Ar]3d³4s²	[Ar]	↑	↑	↑			↑↓
Cr	24	[Ar]3d⁵4s¹	[Ar]	↑	↑	↑	↑	↑	↑
Mn	25	[Ar]3d⁵4s²	[Ar]	↑	↑	↑	↑	↑	↑↓
Fe	26	[Ar]3d⁶4s²	[Ar]	↑↓	↑	↑	↑	↑	↑↓
Co	27	[Ar]3d⁷4s²	[Ar]	↑↓	↑↓	↑	↑	↑	↑↓
Ni	28	[Ar]3d⁸4s²	[Ar]	↑↓	↑↓	↑↓	↑	↑	↑↓
Cu	29	[Ar]3d¹⁰4s¹	[Ar]	↑↓	↑↓	↑↓	↑↓	↑↓	↑
Zn	30	[Ar]3d¹⁰4s²	[Ar]	↑↓	↑↓	↑↓	↑↓	↑↓	↑↓

Table 13.7 Outer electron configurations of the first row d-block metals, where [Ar] represents the electron configuration of the noble gas argon

There are, however, two unexpected or anomalous electron configurations that break the Aufbau principle, namely, those of chromium and copper. A *simple explanation* to explain the existence of these electronic arrangements is to suggest that a half-filled and filled 3d sub-shell are particularly stable electron configurations.

For all the d-block metals the 3d and 4s sub-shells, despite being from different shells, are relatively close in energy. The low energy difference means that the 3d and 4s electrons can *both* be regarded as valence electrons and involved in bonding.

Ions of the d-block elements

When a d-block metal ionizes to form a simple positive ion the first electrons to be lost are the 4s electrons, followed by the 3d electrons. In other words, when a d-block metal ionizes, positive ions are formed which possess $4s^03d^n$ electron configurations.

For example, if the iron(II) ion is formed, only the two 4s electrons are lost, but if the iron(III) ion is formed an additional electron is lost from the spin pair of the 3d sub-shell. Some examples of common d-block simple ions are shown in Table 13.8.

d-block metal	Simple ion	Detailed outer electron configuration
Scandium	Sc^{3+}	$3d^04s^0$
Titanium	Ti^{3+}	$3d^14s^0$
	Ti^{4+}	$3d^04s^0$
Vanadium	V^{2+}	$3d^34s^0$
	V^{3+}	$3d^24s^0$
Chromium	Cr^{3+}	$3d^34s^0$
Manganese	Mn^{2+}	$3d^54s^0$
	Mn^{4+}	$3d^34s^0$
Iron	Fe^{2+}	$3d^64s^0$
	Fe^{3+}	$3d^54s^0$
Cobalt	Co^{2+}	$3d^74s^0$
Nickel	Ni^{2+}	$3d^84s^0$
Copper	Cu^+	$3d^{10}4s^0$
	Cu^{2+}	$3d^94s^0$
Zinc	Zn^{2+}	$3d^{10}4s^0$

Table 13.8 Selected examples of common simple ions from the first row of the d-block

A **transition element** is defined as a d-block metal that forms at least one stable cation with an incomplete 3d sub-shell. All the elements in Table 13.8 conform to this *except* zinc and scandium. Zinc and scandium are therefore not transition elements. Copper is regarded as a transition element since it forms the stable copper(II) ion, which has an incomplete d sub-shell.

Ions with a half-filled 3d sub-shell ($3d^5$) or a filled 3d sub-shell ($3d^{10}$) are usually relatively stable, but a number of factors are involved in determining the stability of transition metal compounds in the solid state.

Extension: Trends in physical properties

Melting and boiling points

The d-block metals typically have relatively high melting and boiling points (Figure 13.23) compared to the non-d-block metals (except mercury and cadmium). This is a consequence of strong metallic bonding (Chapter 4) because the first row d-block metals have valence electrons from the 3d and 4d sub-shells.

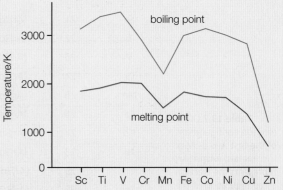

Figure 13.23 Melting and boiling points of the first row of the 3d-block metals

Atomic radii

The atoms become smaller in passing across the first row of the d-block metals from titanium to nickel. The extra electrons are entering an inner 3d sub-shell and it is the increase in the nuclear charge that causes the atoms to shrink and their radii to decrease. However, the additional electron enters the *inner* 3d sub-shell, providing an effective shield between the nucleus and the outer 4s electron. Hence the decrease in radius is small.

Ionization energies and electrode potentials

In general ionization energies rise slightly in passing from scandium to nickel, which follows the trend in increasing nuclear charge holding the electrons more strongly and closer to the nucleus (Figure 13.24).

Electrode potentials (Chapter 19) of M^{2+} tend to increase (either becoming more positive or less negative) across the first row of the d-block. This means that M^{2+} ions become weaker reducing agents and less willing to release electrons and be converted to M^{3+}.

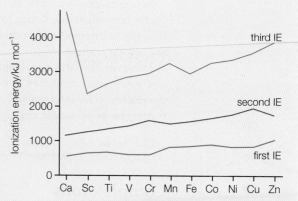

Figure 13.24 Ionization energies of the first row of the d-block metals and calcium (s-block)

The values of *third* ionization energies gradually increase from scandium to zinc, but drop from manganese to iron. The formation of the iron(III) ion, Fe^{3+}, involves the removal of an electron from a doubly occupied 3d orbital; this electron is more easily removed because it is repelled by the other electron in the orbital (Figure 13.25).

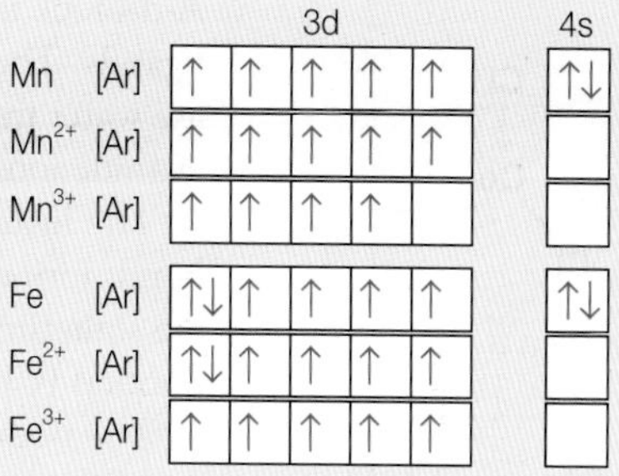

Figure 13.25 The electronic configurations of manganese and iron atoms and their simple ions

Chemical properties

Oxidation states

The common oxidation states for the first row of the d-block are shown in Figure 13.26. The most common oxidation state is +2 resulting from the loss of two 4s electrons to form an M^{2+} ion. The maximum stable oxidation state frequently corresponds to the maximum number of electrons (3d and 4s) available for bonding. For example, manganese ($3d^5 4s^2$) has a maximum oxidation state of +7.

Sc	Ti	V	Cr	Mn	Fe	Co	Ni	Cu	Zn
								+1	
	+2	+2	+2	+2	+2	+2	+2	+2	+2
+3		+3	+3		+3	+3			
	+4	+4	+4		+4				
		+5							
			+6	+6					
				+7					

Figure 13.26 Common oxidation states for the first row of the d-block

The variable oxidation states are due, in part, to the relatively small energy difference between the 3d and 4s sub-shells. This can be illustrated by examining the successive ionization energies of iron and magnesium (Table 13.9).

Element	First ionization energy/kJ mol⁻¹	Second ionization energy/kJ mol⁻¹	Third ionization energy/kJ mol⁻¹	Fourth ionization energy/kJ mol⁻¹
Magnesium	736	1450	7740	10500
Iron	762	1560	2960	5400

Table 13.9 Successive ionization energies of iron and magnesium

When iron and magnesium form ions, energy must be supplied to remove electrons from the atoms of the two elements. Ionization is an endothermic process.

When magnesium forms an ionic compound, such as magnesium oxide, MgO [Mg^{2+} O^{2-}], the energy released during lattice formation (Chapter 15) is much greater than the energy required to form the magnesium and oxide ions (the sum of the first and second ionization energies and first and second electron affinities).

However, the lattice enthalpy is *not* able to supply the energy required to remove the third electron from a magnesium atom. This is because the electron is from an inner shell and close to the nucleus. This electron and others in the inner shells are known as 'core' electrons and never participate in ion formation.

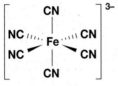

octahedral

tetrahedral

square planar

linear

Figure 13.27 Common shapes of complex ions

> **13.2.4 Define** the term *ligand*.
> **13.2.5 Describe** and **explain** the formation of complexes of d-block elements.

Complex ions

A d-block metal **complex ion** consists of a d-block metal ion surrounded by a definite number of **ligands**. These are molecules or negative ions that have a lone pair of electrons. Common ligands are water molecules, H_2O, ammonia molecules, NH_3, chloride ions, Cl^-, hydroxide ions, OH^-, and cyanide ions, CN^-.

The ligands share their lone pair with empty orbitals in the central d-block metal ion. The bonds formed between the d-block metal ion and the ligands are dative bonds (Chapter 4). They are sometimes called coordinate bonds. The ligands are behaving as Lewis bases (electron pair donors) (Chapter 8).

The number of dative bonds formed by the ligands with the d-block metal ion is known as the **coordination number**. Common coordination numbers are four and six. Two is less common.

Complexes with a coordination number two will be linear, those with four are usually tetrahedral (occasionally square planar) and those with six are octahedral (Figure 13.27). (Some octahedral complexes may have distorted shapes.)

The net charge on a complex ion is the sum of the charge on the d-block metal ion and the charge on the ligands (if they are ions). The net charge may be positive, negative or zero.

Ions of d-block metals have a strong tendency to form complex ions because they are relatively small and highly charged. They are highly polarizing due to a high charge density, which favours covalent bond formation. Ions from groups 1 and 2 are less polarizing and form ion–dipole bonds (Chapter 4) with ligands.

■ Extension: Polydentate ligands

Water, ammonia, chloride and cyanide ions usually behave as **monodentate** ligands, meaning that they form only one dative bond with the central d-block metal ion. A number of larger ligands are able to form two or more dative bonds with the central metal ion and are said to be **polydentate**. The resulting complexes can be very stable and are known as chelates or chelating complexes. Polydentate ligands are often known as chelating agents. Ethanedioate ions and 1,2-diaminoethane molecules are bidentate ligands, while the negative ion derived from EDTA (**ethan**e**diaminetetraethanoic acid**) is hexadentate, forming up to six dative bonds per ion (Figure 13.28).

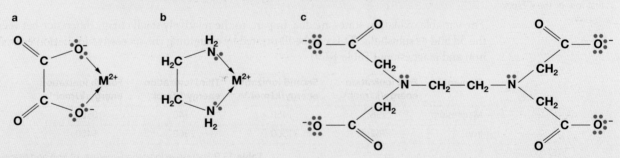

Figure 13.28
a The ethanedioate ion acts as a bidentate ligand via the lone pairs on its charged oxygens

b The 1,2-diaminoethane molecule acts as a bidentate ligand via the lone pairs on its nitrogen atoms in the amine groups

c The EDTA⁴⁻ ion acts as a hexadentate ligand, using lone pairs on both its nitrogen atoms and charged oxygens

Applications of Chemistry

Ethanediaminetetraethanoic acid (Figure 13.29), usually just called EDTA and abbreviated to H₄EDTA, is a widely used chelating agent (Figure 13.30). It is present in shampoos (Figure 13.31), fertilizers, cosmetics and soft drinks containing ascorbic acid and sodium benzoate to prevent formation of benzene. It is also used as a food preservative (Chapter 26).

EDTA and other polydentate ligands are also used in chelation therapy, where chelating agents are used to remove heavy metals, such as mercury, lead and uranium, from the body. One famous example of successful chelation therapy was Harold McCluskey (nicknamed the 'Atomic Man'), a nuclear worker who became badly contaminated with the radioactive isotopes of americium. He was treated with a chelating agent over many years and he did not develop cancer, but died from a heart attack.

Figure 13.29 The sodium salt of EDTA

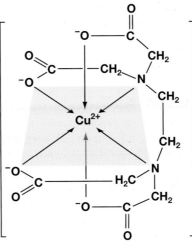

Figure 13.30 The [Cu(EDTA)]²⁻ complex

Figure 13.31 Shampoos contain the EDTA⁴⁻ ion which helps soften the water by complexing calcium ions present in hard water

Language of Chemistry

The term chelation is derived fom the Greek word *chelè*, meaning claw. ■

Ligand replacement reactions

All the stable, simple d-block metal ions exist in water in the hydrated form where the d-block metal ion is surrounded by six octahedrally arranged water molecules. For example, the hydrated copper(II) ion, $Cu^{2+}(aq)$, is more accurately represented by $[Cu(H_2O)_6]^{2+}$.

If aqueous ammonia solution is added to a solution of a copper(II) salt a process of **ligand replacement** (Figure 13.32) occurs; four of the water molecules are replaced by ammonia molecules and the colour changes from light blue to dark blue.

$$[Cu(H_2O)_6]^{2+}(aq) + 4NH_3(aq) \rightarrow [Cu(H_2O)_2(NH_3)_4]^{2+}(aq) + 4H_2O(l)$$

blue — dark blue

If a solution of a copper(II) salt is treated with an excess of concentrated hydrochloric acid, the water molecules are replaced in a similar ligand replacement reaction. The six water molecules are replaced by four chloride ions to form a yellow complex ion.

$$[Cu(H_2O)_6]^{2+}(aq) + 4Cl^-(aq) \rightarrow [CuCl_4]^{2-}(aq) + 6H_2O(l)$$

blue — yellow

A ligand replacement reaction also occurs when concentrated hydrochloric acid is added to a solution of cobalt(II) chloride. The colour changes from pink to blue .

$$[Co(H_2O)_6]^{2+}(aq) + 4Cl^-(aq) \rightarrow [CoCl_4]^{2-}(aq) + 6H_2O(l)$$

pink — blue

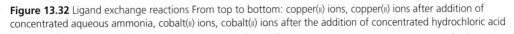

Figure 13.32 Ligand exchange reactions From top to bottom: copper(II) ions, copper(II) ions after addition of concentrated aqueous ammonia, cobalt(II) ions, cobalt(II) ions after the addition of concentrated hydrochloric acid

■ **Extension:** **Standard electrode potentials**

Changing the ligands not only changes the colour of a complex ion, but can also affect the standard electrode potential (Chapter 19) of a complexed d-block metal. This is because different ligands provide different 'environments' for the ion.

For example, consider the following reduction potentials for iron(III) ions complexed with water and cyanide ions.

$$[Fe(H_2O)_6]^{3+}(aq) + e^- \rightarrow [Fe(H_2O)_6]^{2+}(aq) \qquad E^\ominus = +0.77\,V$$
$$[Fe(CN)_6]^{3-}(aq) + e^- \rightarrow [Fe(CN)_6]^{4-}(aq) \qquad E^\ominus = +0.36\,V$$

The decrease in electrode potential means that the iron(III) ion in the more stable cyano complex has less tendency to accept electrons and hence is a weaker oxidizing agent than the aqua complex.

Many enzymes involved in redox reactions contain d-block metal ions and evolution has used this principle to produce related enzymes, such as the cytochromes, each with a slightly different redox strength (Chapter 22), even though the metal ion is identical.

History of Chemistry

Alfred Werner (Figure 13.33) (1866–1919) was a French born Swiss chemist and the first to propose correct structures for coordination compounds containing complex ions. For example, it was known that cobalt formed a 'complex' with the formula $CoCl_3.6NH_3$, but the nature of the bonding specified by the dot was not known. Werner proposed the familiar structure $[Co(NH_3)_6]^{3+}[Cl^-]_3$, implying a cobalt(III) ion, Co^{3+}, surrounded by six ammonia molecules at the corners of an octahedron. He also prepared a number of isomers of complex ions and correctly accounted for their observed properties.

Figure 13.33 Alfred Werner

Applications of Chemistry

The phthalocyanines are intense blue or green coloured pigments. Copper phthalocyanine or Monastral blue (Figure 13.34) is used in paint, coloured plastics, printing inks and enamels. The plastic discs that are designed to become CDs (Figure 13.35) are coated with a very thin layer of phthalocyanines, and a laser beam writes the information on the disc by melting pits into the plastic. The dye helps this heating process by absorbing the laser light.

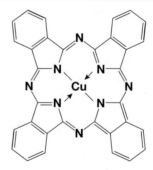

Figure 13.34 The planar structure of copper phthalocyanine

Figure 13.35 A CD-ROM

■ **Extension:** **Naming complex ions**

Complex ions are named according to the following rules:
- The prefixes, *di* (2), *tetra* (4) and *hexa* (6) are used to indicate the number of ligands.
- The type of ligand(s) present is indicated by names: ammonia (*ammine*), chloride (*chloro*), water (*aqua*) and cyanide (*cyano*).
- If the complex ion has a positive charge then the name of the complex ion ends with the d-block metal ion and its oxidation state as a Roman numeral.
  For example, $[Fe(H_2O)_6]^{3+}$ is the hexaaquairon(III) ion.
- If the complex has a negative charge then the name ends with a shortened name, or the Latin name, of the element followed by *-ate*.
  For example, $[CoCl_4]^{2-}$ is the tetrachlorocobaltate(II) ion.

Other complex ions

The formation of complex ions is *not* unique to d-block metals; for example, tin and lead can form complex ions such as $[PbCl_4]^{2-}$, $[PbCl_6]^{2-}$ and $[Sn(H_2O)_6]^{2+}$. The formation of these complex ions is due to the presence of empty low energy d orbitals that can accept lone pairs of electrons.

Some ligands are known as ambidentate ligands. An ambidentate ligand has two lone pairs of electrons, for example the thiocyanate ion (Figure 13.36). Hence it can form a dative bond via the sulfur or nitrogen. Other examples of ambidentate ligands include the nitrite ion, NO_2^-, and the cyanide ion, CN^-.

$$^-\!:\!\ddot{S}\!\!-\!\!C\!\!\equiv\!\!N\!:$$

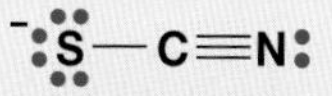

Figure 13.36 One resonance structure for the thiocyanate ion

■ Extension: Isomerism in complex ions

Geometric isomerism

Where a complex ion contains two different ligands, it can exist as two isomers in which there is a different arrangement of the ligands. This is often termed *cis–trans* isomerism (Figure 13.37), referring to the relative positions of the two ligands. *Cis* means same side, *trans* means opposite side. This form of isomerism is only possible for complex ions with coordination numbers greater than or equal to four. The *cis* complex $[PtCl_2(NH_3)_2]$, systematic name *cis*-diamminedichloroplatinum(II), is known as cisplatin (Figure 13.38) and is used in chemotherapy for treating certain types of cancer (Chapter 24). The *trans* isomer, known as transplatin, has no anti-cancer properties.

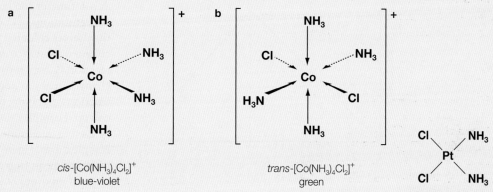

a *cis*-[Co(NH₃)₄Cl₂]⁺
blue-violet

b *trans*-[Co(NH₃)₄Cl₂]⁺
green

Figure 13.37 Geometric isomerism in a transition metal complex ion

Figure 13.38 Structure of cisplatin

Optical isomerism

Octahedral isomers containing a bidentate ligand can form a pair of complexes that are mirror images of each other (Figure 13.39). These molecules are known as enantiomers. This form of isomerism is called optical isomerism (Chapter 20) because the two forms will rotate plane-polarized light in equal but opposite directions.

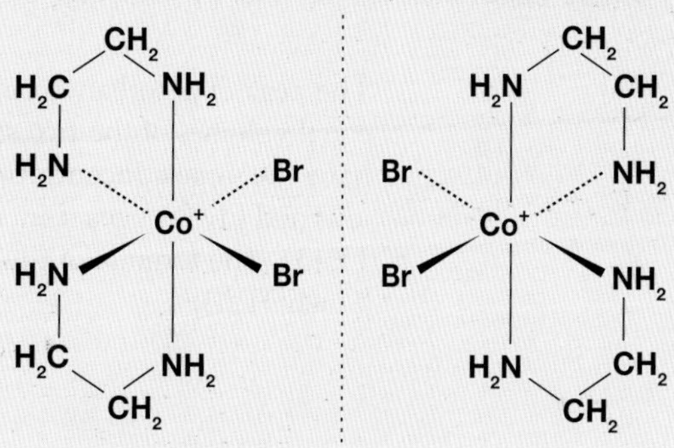

Figure 13.39 The enantiomers of *cis* [Co(en)₂Br₂]⁺, where en represents 1,2-diaminoethane ($H_2NCH_2CH_2NH_2$)

■ Extension: Bonding in complex ions

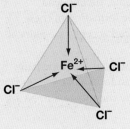

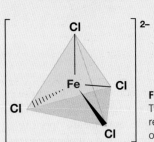

Figure 13.40
Two equivalent representations of the tetrahedral tetrachloroferrate(II) ion

Ions of the d-block metals can use the empty orbitals of the 3d, 4s and 4p sub-shells to form dative bonds with ligands. An example of this type of bonding is exhibited by the tetrachloroferrate(II) ion, $[FeCl_4]^{2-}$ (Figure 13.40).

Detailed outer electron configurations for an iron atom, an iron(II) ion and an iron(II) ion surrounded by four chloride ligands are shown in Figure 13.41. (This is a very *simple* model of bonding in complex ions. Hybridization and molecular orbital theories (Chapter 14) are also used to describe their bonding.)

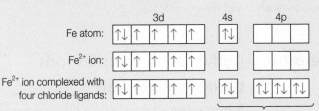

orbitals used to form four coordinate bonds with lone pairs from four Cl⁻ ions

Figure 13.41 Detailed outer electron configurations for an iron atom, an iron(II) ion and an iron(II) ion complexed with four chloride ligands

■ Extension: Finding the formula of a complex ion

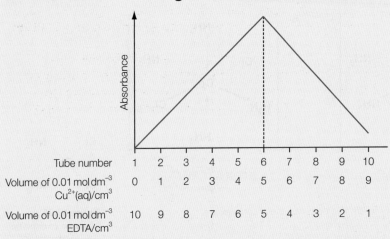

Tube number	1	2	3	4	5	6	7	8	9	10
Volume of 0.01 mol dm⁻³ Cu²⁺(aq)/cm³	0	1	2	3	4	5	6	7	8	9
Volume of 0.01 mol dm⁻³ EDTA/cm³	10	9	8	7	6	5	4	3	2	1

Figure 13.42 Plot of absorbance of ten mixtures of 0.01 mol dm⁻³ Cu²⁺(aq) ions and 0.01 mol dm⁻³ EDTA⁴⁻(aq)

A colorimeter can be used to measure the concentrations of chemicals that are themselves coloured or which produce a coloured substance during a chemical reaction (Chapter 6). By following the changes in absorbance that take place in reactions involving the formation of coloured complex ions, a colorimeter can be used to determine the formulas of these complex ions. Figure 13.42 shows the results of measuring the absorbance of a series of mixtures of 0.01 mol dm⁻³ copper(II) ions, Cu²⁺(aq), and 0.01 mol dm⁻³ EDTA⁴⁻(aq). EDTA⁴⁻ is a multidentate ligand and forms very stable complex ions with d-block metal ions.

The peak of absorbance corresponds to mixing equal volumes of the two solutions which both have the same concentration. Hence, copper(II) ions and EDTA⁴⁻ react in a 1:1 molar ratio (Figure 13.43). The formula of the complex is $[Cu(EDTA)]^{2-}$.

Figure 13.43 Structure of the complex ion formed between copper(II) ions and EDTA⁴⁻ ions

The colours of complex ions

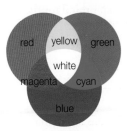

Figure 13.44 Primary and secondary colours

White light can result from the combination of only red, green and blue light. When equal brightnesses of these are combined and projected on a screen white light is produced (Figure 13.44). The screen appears yellow when red and green light alone overlap. The combination of red and blue light produces the bluish-red colour of magenta. Green and blue produce the greenish-blue colour called cyan. Almost any colour can be made by overlapping light of three colours and adjusting the brightness of each colour. Red, green and blue light are known as primary colours since all other colours can be formed from them.

When white light is shone on a chemical substance, either as a solid or in solution, some light is absorbed and some is reflected.

- If all the light is absorbed then the substance appears 'black'.
- If only certain wavelengths are absorbed then the compound will appear coloured.
- If all the light is reflected then the colour of the substance will appear 'white'.

Most d-block metal compounds are coloured, both in solution and in the solid state. The colours of many (but not all) of these compounds are due to the presence of incompletely filled 3d sub-shells.

Language of Chemistry

Black and white are not regarded as colours. Black is the absence of colour; white is the presence of all colours (with equal brightness or intensity). ∎

In an isolated gaseous d-block metal element atom, the five 3d sub-shells all have different orientations in space (shapes), but *identical* energies. However, in a complex ion the 3d sub-shells are orientated differently relative to the ligands. The 3d electrons close to a ligand will experience repulsion and be raised in energy. The 3d electrons located further away from the ligand will be reduced in energy.

The 3d sub-shell has now been 'split' into two energy levels. Octahedral complexes are very common and the splitting of the d sub-shells is shown in Figure 13.45 for the hydrated titanium(III) ion. Two d orbitals are raised in energy; three d orbitals are lowered in energy. (Different d–d splitting patterns are observed in tetrahedral, square planar and linear complexes.)

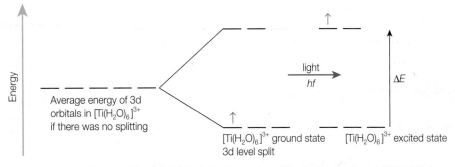

Figure 13.45 The splitting of the 3d sub-shells in the titanium(III) ion in the octahedral $[Ti(H_2O)_6]^{3+}$(aq) complex

Figure 13.46 Colour wheel of complementary colours

The energy difference or energy gap ΔE between the two sets of energy levels is related to the frequency of light necessary to cause an electron to be excited from the lower energy level to the higher energy level.

$$\Delta E = hf$$

where h is Planck's constant and f is the frequency (Chapter 2).

The colour of hydrated titanium(III) ions, $[Ti(H_2O)_6]^{3+}$(aq), is violet because yellow-green light (of a particular frequency) is absorbed and the colour of the complex ions will be *complementary* to that (Figure 13.46).

When white light is absorbed by a solution of titanium(III) ions then some of the light waves will have energies that correspond to the energy gap or energy difference, ΔE, between the two groups of 'split' 3d energy levels. This light will be absorbed and a single 3d electron will become excited and be promoted to the higher energy level. This is known as a d–d transition.

The energy difference, and hence the colour of d-block metal complex ions, depends on three factors:

- *The nuclear charge*
  For example, vanadium(II) ions ($3d^3$) are lavender in colour, while chromium(III) ions (also $3d^3$) are green.
- *The number of d electrons*
  d–d transitions occur only if there is an incomplete 3d sub-shell, so there is an orbital for the 3d electron to be promoted into.
  Scandium(III), copper(I) and zinc(II) compounds are all colourless due to the absence of 3d electrons (Sc^{3+}, $3d^0$) or the presence of a filled 3d sub-shell (Cu^+ and Zn^{2+}, $3d^{10}$).
- *The nature of the ligand*
  Different ligands, because of their different sizes and charges, produce different energy gaps between the two groups of 3d sub-shell energy levels. For example, ammonia ligands produce a larger energy gap than water molecules, causing a colour change when aqueous ammonia solution is added to the solution of a copper(II) salt. This means that light of higher energy and frequency, and hence lower wavelength, is absorbed.

Absorption spectra of selected complex ions can be found in Chapter 21. The chapter also outlines how the concentration of complex ions can be determined by means of colorimetry and describes the effect of different ligands on d–d splitting.

Table 13.10 displays the elements of the first row of the d-block elements, giving an example of one coloured simple ion for each, where there is one.

Transition metal ion	Sc^{3+}	Ti^{4+} Ti^{3+}	V^{2+}	Cr^{3+}	Mn^{2+}	Fe^{3+} Fe^{2+}	Co^{2+}	Ni^{2+}	Cu^{2+}	Zn^{2+}
Colour in aqueous solution	Colourless	Colourless Violet	Violet	Green	Very pale pink	Yellow-brown Pale green	Pink	Green	Pale blue	Colourless

Table 13.10 Simple ions of the first row of the d-block elements

■ Extension: Determining the value of the energy gap, ΔE

Planck's equation and the wave equation (Chapter 2) can be used to determine an *approximate* value for the energy gap of a transition metal ion complex from the frequency or wavelength of light *most strongly absorbed* by the ions of a complex ion. (This is an approximate value because the absorption spectrum is broad – it is not a single sharp peak.)

Worked example

The wavelength of light most strongly absorbed by nickel(II) ions, Ni^{2+}(aq), is 410 nm. Calculate the value of the energy gap or field splitting in hydrated nickel ions in joules and in kilojoules per mole (to the nearest integer).

$$c = f\lambda \qquad \text{and} \qquad \Delta E = hf$$

Therefore

$$\Delta E = \frac{hc}{\lambda} = \frac{6.63 \times 10^{-34}\,\text{J s} \times 3 \times 10^8\,\text{m s}^{-1}}{410 \times 10^{-9}\,\text{m}}$$

$$= 4.85 \times 10^{-19}\,\text{J}$$

When multiplied by the Avogadro constant ($6.02 \times 10^{23}\,\text{mol}^{-1}$), this becomes 292 kJ mol^{-1}.

■ Extension: The shapes of d orbitals

The shapes of the 3d atomic orbitals that make up the 3d sub-shell are shown in Figure 13.47. They all have the same energy (in the gas phase ions), but different orientations in space relative to one another. The + and − symbols in the lobes do *not* represent electrical charge. They represent the phases of the waves (Chapter 14).

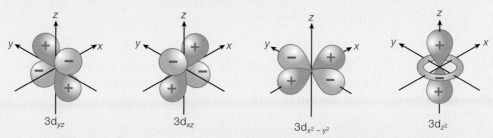

$3d_{xy}$ $3d_{yz}$ $3d_{xz}$ $3d_{x^2-y^2}$ $3d_{z^2}$

Figure 13.47 The shapes of the 3d orbitals

The model described here to account for colours of transition metal ions is a simplified version of crystal field theory. The theory is based on the idea that the bonding in complex ions is purely electrostatic and that the ligands behave as point negative charges. The most common type of complex ion is octahedral, where six ligands form an octahedron around the metal ion. In octahedral symmetry the d orbitals split into two sets with an energy difference ΔE. The d_{xy}, d_{xz} and d_{yz} orbitals will be lower in energy than the d_{z^2} and $d_{x^2-y^2}$, because they are further from the ligands than the latter and therefore electrons in these orbitals experience *less repulsion*.

Chemistry of selected d-block metals

Vanadium

Vanadium can exist in four different oxidation states in compounds. In aqueous solution, ions containing vanadium in the different oxidation states have different colours (Table 13.11).

A common vanadium compound is ammonium vanadate(v), NH_4VO_3. In acidic conditions the trioxovanadate(v), $VO_3^-(aq)$ ions are converted to dioxovanadium(v), $VO_2^+(aq)$ ions:

$$2H^+(aq) + VO_3^-(aq) \rightarrow VO_2^+(aq) + H_2O(l)$$

These dioxovanadium(v) ions will react with zinc and dilute sulfuric acid (a powerful reducing agent) and undergo stepwise reduction to vanadium(II) ions, $V^{2+}(aq)$, displaying the colours of the ions as reaction proceeds (Figure 13.48).

Vanadium(v) oxide, V_2O_5, is used as a catalyst in the Contact process (Chapter 7).

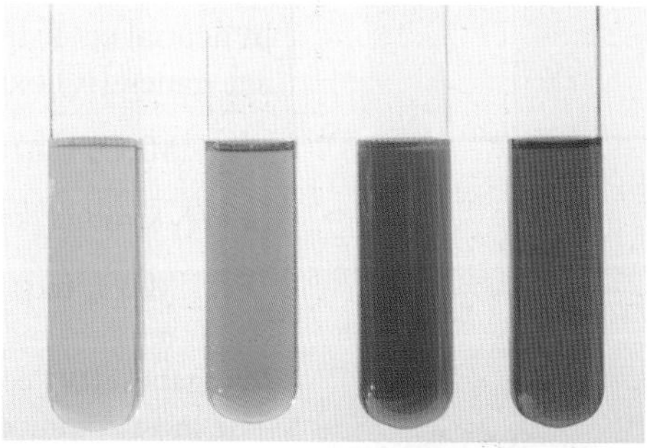

Oxidation number	Formula of aqueous ion	Colour of aqueous ion
+2	V^{2+}	Lavender
+3	V^{3+}	Green
+4	VO^{2+}	Blue
+5	VO_2^+, VO_3^-	Yellow

Table 13.11 Formulas, colours and oxidation numbers of some vanadium ions

Figure 13.48 The colour changes that can be seen here are due to reduction of the acidified solution of VO_2^+ (yellow), through VO^{2+} (blue), V^{3+} (green) to V^{2+} (lavender)

Figure 13.49 Aqueous solutions of chromate(vi) and dichromate(vi) ions

Chromium

Chromium has the oxidation number +6 in both chromate(vi), CrO_4^{2-}, and in dichromate(vi), $Cr_2O_7^{2-}$ (Figure 13.49). Yellow chromate(vi) ions can be converted to orange dichromate(vi) ions by the addition of dilute acid. This is an example of a condensation reaction (Chapter 10).

$$2CrO_4^{2-}(aq) + 2H^+(aq) \rightarrow Cr_2O_7^{2-}(aq) + H_2O(l)$$

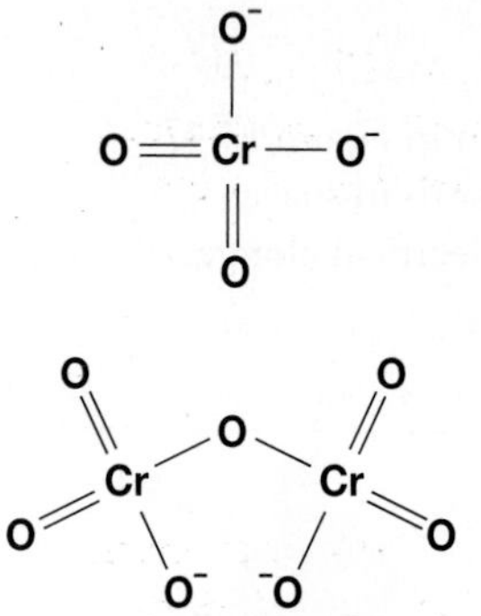

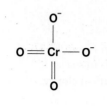

Figure 13.50 Structures of chromate(VI) and dichromate(VI) ions

Dichromate(VI) ions can be converted to chromate(VI) ions by the addition of dilute alkali:

$$Cr_2O_7^{2-}(aq) + 2OH^-(aq) \rightarrow 2CrO_4^{2-}(aq) + H_2O(l)$$

The structures of the chromate(VI) and dichromate(VI) ions are shown in Figure 13.50. The ions have resonance or delocalized bonding (Chapter 14), not shown in the diagram.

The dichromate(VI) ion is a powerful oxidizing agent. The half-equation below shows how it reacts with reducing agents in acidified aqueous solution. The reduction results in the formation of the green chromium(III) ion.

$$Cr_2O_7^{2-}(aq) + 14H^+(aq) + 6e^- \rightarrow 2Cr^{3+}(aq) + 7H_2O(l) \qquad E^{\ominus} = +1.33\,V$$

The high positive value for the standard electrode potential ($E^{\ominus}$) indicates that the dichromate(VI) ion has a strong affinity for electrons under these conditions.

The reaction is used in redox titrations (Chapter 1 and Chapter 9) where it can be used to find the concentrations of reducing agents. Acidified aqueous solutions of potassium dichromate(VI), $K_2Cr_2O_7$, are used to oxidize alcohols (Chapter 10).

In dilute acidified solutions, dichromate(VI) ions react with hydrogen peroxide solution (acting here as a reducing agent) to form green chromium(III) ions. The two half-equations are:

$$Cr_2O_7^{2-}(aq) + 14H^+(aq) + 6e^- \rightarrow 2Cr^{3+}(aq) + 7H_2O(l)$$

$$H_2O_2(aq) \rightarrow 2H^+(aq) + O_2(g) + 2e^-$$

Another important chromium compound is chromium(VI) oxide, CrO_3, a covalent compound which is precipitated when concentrated sulfuric acid is added to solutions of chromate(VI) or dichromate(VI) ions. It is a red crystalline solid that reacts with water to form a solution of dichromate(VI) ions. Chromium(III) oxide, Cr_2O_3, is a green solid (insoluble in water) that can be prepared by heating ammonium dichromate(VI):

$$(NH_4)_2Cr_2O_7(s) \rightarrow N_2(g) + Cr_2O_3(s) + 4H_2O(g)$$

Manganese

Manganese can exist in four oxidation states: +2, +4, +6 and +7.

Manganese(VII) compounds

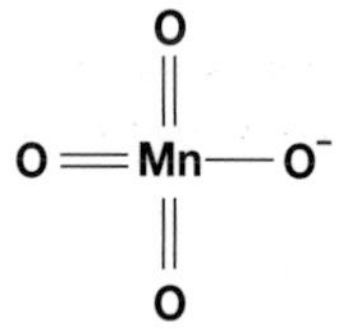

The most familiar manganese(VII) compound is potassium manganate(VII), which dissolves in water to give a deep purple solution that contains MnO_4^- ions (Figure 13.51).

Potassium manganate(VII) is commonly used as an oxidizing agent in the presence of excess dilute sulfuric acid and under these conditions is usually reduced to the nearly colourless manganese(II) ions, and the purple solution decolorised:

$$MnO_4^-(aq) + 8H^+(aq) + 5e^- \rightarrow Mn^{2+}(aq) + 4H_2O(l) \qquad E^{\ominus} = +1.52\,V$$

In neutral or slightly alkaline conditions the manganate(VII) ions are reduced to brown manganese(IV) oxide:

$$MnO_4^-(aq) + 4H^+(aq) + 3e^- \rightarrow MnO_2(s) + 2H_2O(l) \qquad E^{\ominus} = +1.67\,V$$

In very strong alkaline solutions, manganate(VII) ions are reduced to green manganate(VI) ions:

$$MnO_4^-(aq) + e^- \rightarrow MnO_4^{2-}(aq) \qquad E^{\ominus} = +0.56\,V$$

Figure 13.51 Structure of the manganate(VII) ion

Manganese(IV) compounds

The most common manganese(IV) compound is manganese(IV) oxide (Figure 13.52). It is a strong oxidizing agent and is reduced to manganese(II) ions:

$$MnO_2(s) + 4H^+(aq) + 2e^- \rightarrow Mn^{2+}(aq) + 2H_2O(l) \qquad E^{\ominus} = +1.23\,V$$

Manganese (IV) oxide is a catalyst for the decomposition of hydrogen peroxide (Chapter 6).

Figure 13.52 Manganese(IV) oxide powder

Figure 13.53
Manganese(ɪɪ) carbonate

Figure 13.54 Hydrated
iron(ɪɪ) sulfate crystals,
FeSO$_4$.7H$_2$O

Figure 13.55 Copper(ɪ)
oxide

Manganese(ɪɪ) compounds

Manganese(ɪɪ) salts are generally pale pink in colour. Manganese(ɪɪ) chloride, MnCl$_2$, and manganese(ɪɪ) sulfate, MnSO$_4$, are prepared by reacting manganese with dilute hydrochloric and sulfuric acids, respectively. Other manganese(ɪɪ) salts are made via precipitation reactions, for example, manganese(ɪɪ) carbonate (Figure 13.53).

Iron

The two most common oxidation states of iron are +2 and +3.

Iron(ɪɪ) compounds

Soluble iron(ɪɪ) salts, for example hydrated iron(ɪɪ) sulfate (Figure 13.54), FeSO$_4$.7H$_2$O, are green crystalline solids that release the hexaaquairon(ɪɪ) ion, [Fe(H$_2$O)$_6$]$^{2+}$, in solution. The solutions of soluble iron(ɪɪ) salts give (in the absence of air) a pale green precipitate of iron(ɪɪ) hydroxide when treated with a base.

$$Fe^{2+}(aq) + 2OH^-(aq) \rightarrow Fe(OH)_2(s)$$

Acidified solutions of iron(ɪɪ) salts are readily oxidized to iron(ɪɪɪ) by oxidizing agents, for example chlorine water, and are oxidized slowly by oxygen in the air.

$$[Fe(H_2O)_6]^{2+}(aq) \rightarrow [Fe(H_2O)_6]^{3+}(aq) + e^- \qquad E^\ominus = +0.77\,V$$

Iron(ɪɪɪ) compounds

Soluble iron(ɪɪɪ) salts, for example iron(ɪɪɪ) chloride, FeCl$_3$, are usually yellow or brown crystalline solids that release the violet hexaaquairon(ɪɪɪ) ion, [Fe(H$_2$O)$_6$]$^{3+}$, in strongly acidic solution.

Iron(ɪɪɪ) salts are acidic in aqueous solution (see Section 13.1). The solutions of soluble iron(ɪɪɪ) salts give (in the presence of air) a rusty brown precipitate of iron(ɪɪɪ) hydroxide on reaction with base.

$$Fe^{3+}(aq) + 3OH^-(aq) \rightarrow Fe(OH)_3(s)$$

Iron(ɪɪɪ) salts are readily reduced to iron(ɪɪ) salts by a variety of reducing agents, for example iodide ions, copper metal and copper(ɪ) ions.

Copper

The common oxidation states of copper are +1 and +2. The hydrated copper(ɪɪ) ion is stable in water, but the hydrated copper(ɪ) ion is unstable in water and is immediately converted into copper(ɪɪ) ions and copper metal.

$$2Cu^+(aq) \rightarrow Cu^{2+}(aq) + Cu(s)$$

The reaction between copper(ɪ) ions and water is an example of disproportionation (Chapter 9). Some copper(ɪ) compounds are insoluble, for example copper(ɪ) oxide, Cu$_2$O (Figure 13.55) and copper(ɪ) chloride, CuCl.

Copper(ɪ) chloride can be prepared by boiling copper(ɪɪ) chloride with copper (a reducing agent) and pouring the complex ion mixture into air-free water to precipitate white copper(ɪ) chloride. The overall equation for the reaction is:

$$CuCl_2(aq) + Cu(s) \rightarrow 2CuCl(s)$$

Copper(ɪ) chloride is not stable in aqueous solution, but the reaction is slow. Copper(ɪɪ) ions can be reduced to copper(ɪ) ions by powerful reducing agents. For example, a solution of aqueous copper(ɪɪ) ions reacts with potassium iodide solution to give a white precipitate of copper(ɪ) iodide and iodine:

$$2Cu^{2+}(aq) + 4I^-(aq) \rightarrow 2CuI(s) + I_2(aq)$$

However, the yellow or brown colour of the iodine solution obscures the white precipitate.

Figure 13.56 Zinc oxide powder

Zinc

The only stable oxidation state of zinc is +2. The compounds of zinc are white (Figure 13.56). Zinc complexes usually have a coordination number of four and these are tetrahedral. Addition of alkali to an aqueous solution of a zinc salt initially produces a white precipitate of the hydroxide but this is amphoteric and dissolves in excess to give a colourless solution of zincate ions:

$$Zn^{2+}(aq) + 2OH^-(aq) \rightarrow Zn(OH)_2(s)$$

$$Zn(OH)_2(s) + 2OH^-(aq) \rightarrow [Zn(OH)_4]^{2-}(aq)$$

Transition metals as catalysts

13.2.7 State examples of the catalytic action of transition elements and their compounds.

Transition metals and their compounds are often used as catalysts to increase the rates of industrial processes (Chapter 23). Their catalytic properties are due to their ability to exist in a number of stable oxidation states and the presence of empty orbitals for temporary bond formation.

The transition metal catalysts used in industry are heterogeneous catalysts, where the catalyst is in a different physical state from the reactants. Typically, the catalyst is a powdered solid (often on the surface of an inert support) and the reactants are a mixture of gases. Some examples of transition metal based catalysts are given in Table 13.12.

Process	Reaction catalysed	Products	Catalyst
Haber	$N_2(g) + 3H_2(g) \rightleftharpoons 2NH_3(g)$	Ammonia	Iron, Fe
Contact	$2SO_2(g) + O_2(g) \rightleftharpoons 2SO_3(g)$	Sulfuric acid	Vanadium(v) oxide, V_2O_5
Hydrogenation of unsaturated oils to harden them	$RCH{=}CHR' \rightarrow RCH_2CH_2R'$	Semi-solid saturated fat	Nickel, Ni
Hydrogenation	Alkene to alkane	Alkane	Nickel, Ni
Ziegler–Natta polymerization of alkenes	$nCH_2{=}CHR \rightarrow -[CH_2-CHR]_n-$	Stereoregular polymer	Complex of $TiCl_3$ and $Al(C_2H_5)_3$

Table 13.12 Some transition metal based industrial catalysts

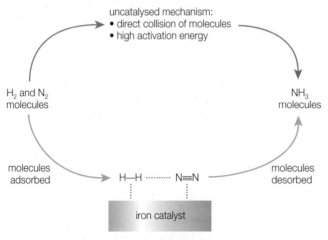

uncatalysed mechanism:
• direct collision of molecules
• high activation energy

H_2 and N_2 molecules

NH_3 molecules

molecules adsorbed

molecules desorbed

H—H ·········· N≡N

iron catalyst

catalysed mechanism:
• molecules adsorbed and react on surface of catalyst
• lower activation energy

Figure 13.57 A simplified model of heterogeneous catalysis, using the Haber process as an example

Heated nickel can also be used in the laboratory to catalyse the reduction of nitriles to amines using hydrogen gas (Chapter 20). The Haber and Contact processes are both examples of chemical equilibria (Chapter 7) and the resulting mixture will contain both reactants and products. Heterogeneous catalysis can be demonstrated in the laboratory by adding a small amount of manganese(IV) oxide, MnO_2, to a dilute solution of hydrogen peroxide (Chapter 6).

During heterogeneous catalysis the liquid or gas molecules are adsorbed on the surface of the solid catalyst (Figure 13.57), where their bonds are weakened via complex formation so the products are produced more rapidly (Chapter 6).

The strength of the adsorption helps to determine the activity of a catalyst. Some transition metals adsorb very strongly, so products are released slowly. Others adsorb weakly so the concentration of reactants on the surface is low.

Transition metal ions also exhibit homogeneous catalysis where they and the reactant(s) are in the same physical state; usually all are in solution. For example, iron(III) ions act as a homogeneous catalyst for the reaction between peroxodisulfate and iodide ions to form sulfate ions and iodine molecules (Chapter 16). Iron catalyses the reaction by interconverting between its two common oxidation states, iron(II) and iron(III). This facilitates the electron transfer processes that occur. Many metal-containing enzymes, especially those in the electron transport chain, act in a similar way inside cells (Chapter 22).

Homogeneous catalysis is also exhibited in the reaction between manganate(VII) ions and ethanedioate ions. This is an example of autocatalysis, as manganese(II) ions, one of the products, is a catalyst for the reaction (Chapter 16).

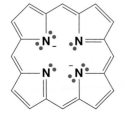

Figure 13.58 Basic porphyrin structure

Biomolecules containing a metal

A number of molecules in living organisms contain one or more strongly bonded metal ions as part of their molecular structure. The metal ion will be attached to the biological molecule via donor atoms, usually nitrogen, oxygen or sulfur, present in the molecules. They act as ligands to the metal ion. The coordination number of the metal ion is usually six and the coordination geometry often octahedral. The ligands may be the side-chains of amino acids or a porphyrin ring (Figure 13.58), a planar tetradentate ligand (Chapter 26).

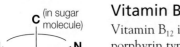

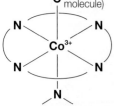

Figure 13.59 Simplified structure of vitamin B_{12}

Vitamin B_{12}

Vitamin B_{12} is a large molecule that contains a single cobalt(III) ion complexed in the middle of a porphyrin type ring. The cobalt is bonded to five nitrogen atoms and one carbon atom giving it a coordination number of six. A simplified structure of vitamin B_{12} is given in Figure 13.59.

Vitamin B_{12} is essential for the production of red blood cells in the body. The vitamin is a co-factor and attaches itself to an enzyme. The enzyme only works in the presence of vitamin B_{12}. The mechanism of action of vitamin B_{12} is a complicated free-radical mechanism, but one key feature is that Co^{3+} is converted to Co^{2+} and then back to Co^{3+}.

Hemoglobin

A hemoglobin molecule is made up a heme unit covalently attached to a protein chain (globin). Each hemoglobin molecule contains four heme units and four globin chains. In the hemoglobin molecule the iron(II) ions are bonded to five nitrogen atoms and one oxygen atom (in the water molecule). A simplified structure of hemoglobin is shown in Figure 13.60.

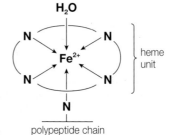

Figure 13.60 Simplified structure of hemoglobin

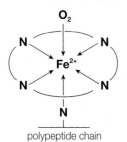

Figure 13.61 Simplified structure of oxyhemoglobin

Hemoglobin absorbs oxygen from the lungs by forming oxyhemoglobin (Figure 13.61). The oxygen molecule displaces the water molecule and forms a dative bond with the iron. This reaction is reversible and at low pressures the oxygen leaves, reforming hemoglobin.

Catalysts in industry

13.2.8 Outline the economic significance of catalysts in the Contact and Haber processes.

Catalysts play a critical role in the chemical industry by increasing the rates of reaction and decreasing the time needed for a reaction to reach equilibrium. They increase the efficiency of industrial processes and help to reduce costs and so increase profits. The ammonia and sulfuric acid produced by the Haber and Contact processes are vital feedstocks for many other parts of the chemical industry. Considerable research is conducted to find cheaper, 'greener' and more effective catalysts for industrial reactions.

One reason for recent increases in the prices of metals such as platinum, rhodium and palladium is that they are used in catalytic converters in vehicles (Chapter 25) and as catalysts in some industrial processes. For example, a platinum–rhodium alloy can be used in the Contact process (Chapter 7).

■ **Extension:** Identifying d-block metal ions

The hydroxides of d-block metal ions can be precipitated from their aqueous solution by the addition of sodium hydroxide solution. The colour of the precipitate (Figure 13.62) can be used to identify the d-block metal present (Table 13.13).

d-block metal ion present	Formula of precipitate	Colour of precipitate
Chromium(III)	$Cr(OH)_3$	Green
Manganese(II)	$Mn(OH)_2$	Pale brown
Iron(II)	$Fe(OH)_2$	Pale green (surface turns rusty brown)
Iron(III)	$Fe(OH)_3$	Rusty brown
Cobalt(II)	$Co(OH)_2$	Pink
Nickel	$Ni(OH)_2$	Green
Copper(II)	$Cu(OH)_2$	Blue
Zinc	$Zn(OH)_2$	White

Table 13.13 Colours and formulas of selected d-block metal hydroxides

Figure 13.62 Precipitates of hydrated d-block metal hydroxides: from left to right, cobalt(II), zinc, iron(III), iron(II) and copper(II)

SUMMARY OF KNOWLEDGE

■ All the elements in period 3, except chlorine and argon, combine directly with oxygen to form oxides.

■ Sodium oxide (Na_2O) and magnesium oxide (MgO) are ionic oxides; aluminium oxide (Al_2O_3) is ionic with some covalent character; silicon dioxide (SiO_2) has a giant covalent structure; the oxides of phosphorus (P_4O_6 and P_4O_{10}), sulfur (SO_2 and SO_3) and chlorine (Cl_2O and Cl_2O_7) have simple molecular structures.

■ The metal oxides are ionic and hence have a high melting point. MgO and Al_2O_3 have a higher melting point than Na_2O since the charges are higher and the ions smaller, resulting in a stronger electrostatic attraction between the ions. They are electrolytes when molten due to the release of mobile ions at the melting point.

■ Silicon dioxide has a giant covalent structure and hence a high melting point. There are strong covalent bonds between all the atoms and thus a large amount of energy is required to break them.

■ The oxides of the other non-metallic elements are simple molecular and hence only weak intermolecular forces exist between the molecules. The melting points are thus much lower and increase with molecular mass.

■ The non-metallic oxides are non-electrolytes when molten since molecules are released at the melting point.

■ Sodium and magnesium oxides contain the oxide ion. This is a strongly basic ion which reacts with water to produce hydroxide ions:
$O^{2-}(aq) + H_2O(l) \rightarrow 2OH^-(aq)$
The oxides therefore become more basic on moving from right to left in the periodic table.

■ The covalent oxides do not contain ions, but have a strongly positive dipole on the atom which is not oxygen. This attracts the lone pair on water molecules, releasing hydrogen ions:

$MO(s) + H_2O(l) \rightarrow MO(OH)^-(aq) + H^+(aq)$

The oxides therefore become more acidic on moving from left to right in the periodic table.

■ Aluminium oxide is amphoteric and reacts with both acids and alkalis.

■ All the elements of period 3 except argon combine directly with chlorine to give chlorides. Chlorine may be regarded as chlorine chloride.

■ Sodium chloride (NaCl) and magnesium chloride ($MgCl_2$) are ionic chlorides; anhydrous aluminium chloride ($AlCl_3$) behaves like a simple molecular substance; silicon tetrachloride ($SiCl_4$) has a simple molecular structure; the chlorides of phosphorus (PCl_3 and PCl_5) and chlorine (Cl_2) all have simple molecular structures.

■ Sodium and magnesium chlorides are ionic and hence have high melting points. They are electrolytes when molten due to the release of mobile ions at the melting point. They dissolve in water to release ions and act as electrolytes.

■ The other chlorides are simple molecular substances, and so only intermolecular forces exist between the molecules. The melting points are thus much lower than those of the ionic chlorides.

■ Magnesium and sodium chlorides dissolve in water to give neutral solutions. The ions released do not react with water.

■ The covalent chlorides react readily with water at room temperature to form the oxide or hydroxide and fumes of hydrogen chloride. Covalent chlorides thus react with water in a hydrolysis reaction to give acidic solutions. The acidity is due to the formation of dissolved hydrochloric acid.

■ d-block elements have electron structures that feature an outer pair of 4s electrons, within which there is a partially or totally filled 3d sub-shell. Copper and chromium adopt $3d^54s^1$ and $3d^{10}4s^1$ configurations to reflect the stability of half-full and filled 3d sub-shells.

■ The d-block metals have high densities and moderate to low reactivities compared with the metals in the s-block. They form compounds that show variable oxidation number (with the exception of zinc and scandium). Their simple ions have a strong tendency to form complex ions with ions or molecules (known as ligands) that can form dative bonds.

■ They give rise to coloured compounds (with the exception of zinc and scandium), in which the colours also depend upon the oxidation number and the identity of the ligand.

■ d-block metals and their compounds can act as a catalysts. This occurs because of their variable oxidation states and ability to bond with ligands.

■ All the d-block metals can form +2 simple ions; the majority can also form +3 ions. Ion formation may involve the 3d and 4s electrons. Higher oxidation states by d-block metals in the middle of the row involve the formation of polar covalent bonds. They are often oxidizing agents.

■ The coordination number of most complex ions is 2, 4 or 6 and the geometries associated with these numbers are linear, tetrahedral or square planar (less common), and octahedral.

■ Ligand replacement reactions occur when one ligand displaces another ligand in a complex. This may occur because the competing ligand is a more effective ligand and/or the competing ligand is present in large excess.

■ Colour in d-block elements arises because of a split in the energy levels of the 3d sub-shell caused by the approach of ligands to a transition metal ion. The colours are caused by the absorption of light, which is used to promote transitions of electrons from lower energy to higher energy d orbitals.

■ A transition metal is a metal in the d-block of the periodic table which has at least one of its ions with a partly filled d sub-shell. (This definition excludes scandium and zinc in the first row of the d-block; these metals form colourless ions and their compounds show less catalytic properties than other d-block metals.)

■ *Examination questions – a selection*

Paper 1 IB questions and IB style questions

Q1 Which is an essential feature of a ligand?
  A a negative charge
  B an odd number of electrons
  C the presence of two or more atoms
  D the presence of a non-bonding pair of electrons
  Higher Level Paper 1, May 05, Q8

Q2 The electron configuration of atoms of a transition metal atom is $1s^2 2s^2 2p^6 3s^2 3p^6 3d^3 4s^2$. What is the maximum oxidation state?
  A +5 **B** +4 **C** +3 **D** +2

Q3 In which one of the series listed below would the elements have most nearly the same atomic radius?
  A B, Si, As, Te **C** F, Cl, Br, I, At
  B Sc, Ti, V, Cr **D** Na, Mg, Al

Q4 Transition metals differ from group 1 and group 2 metals in all of the following respects except for:
  A being more chemically reactive
  B having higher melting points
  C being more likely to form coloured compounds
  D being more likely to form complex ions in aqueous solution

Q5 Oxides of sulfur and magnesium are dissolved in water and the resulting solutions are tested with litmus, an acid–base indicator. What can be deduced from the observations?
  A Magnesium forms an acidic oxide and sulfur forms an acidic oxide.
  B Magnesium forms a basic oxide and sulfur forms an acidic oxide.
  C Magnesium forms an acidic oxide and sulfur forms a basic oxide.
  D Magnesium forms a neutral oxide and sulfur forms a basic oxide.

Q6 The cyanide ion, CN^-, can form two complex ions with iron ions. The formulas of these ions are $[Fe(CN)_6]^{4-}$ and $[Fe(CN)_6]^{3-}$. What is the oxidation state of iron in the two complex ions?

	$[Fe(CN)_6]^{4-}$	$[Fe(CN)_6]^{3-}$
A	−4	−3
B	+2	+3
C	+3	+2
D	−3	−4

  Higher Level Paper 1, Nov 04, Q8

Q7 Which one of the oxides below, when added to water, will produce the acid indicated?
  A $NO + H_2O \rightarrow HNO_2$ **C** $Cl_2O_7 + H_2O \rightarrow HClO_4$
  B $P_4O_{10} + H_2O \rightarrow H_3PO_3$ **D** $SO_2 + H_2O \rightarrow H_2SO_4$

Q8 Which of the following species will not be able to act as a ligand in the formation of complex ions with transition metals?
  A $C_6H_5NH_2$ **B** $C_2H_5NH_2$ **C** NH_3 **D** NH_4^+

Q9 Which formula is incorrect for the period 3 oxides?
  A Na_2O **B** MgO_2 **C** SiO_2 **D** Cl_2O

Q10 Which one of the following is not a characteristic property of a d-block metal or its compounds?
  A catalytic activity
  B formation of coloured ions
  C high standard electrode potential
  D formation of complex ions

Q11 Which period 3 oxide has a simple molecular structure?
  A Na_2O **B** Al_2O_3 **C** SiO_2 **D** P_4O_{10}

Q12 Which one of the following is most likely to be coloured?
  A LiBr **C** $Mg(NO_3)_2$
  B $NiSO_4$ **D** $Pb(CH_3COO)_2$

Q13 0.02 mole samples of the following oxides were added to separate $1\,dm^3$ portions of water. Which will produce the most acidic solution?
  A $Al_2O_3(s)$ **B** $SiO_2(s)$ **C** $Na_2O(s)$ **D** $SO_2(g)$

Q14 Which oxidation number is the most common among first row transition elements?
  A +1 **B** +2 **C** +3 **D** +5

Q15 The colours of the compounds of d-block elements are due to electron transitions:
  A between different d orbitals (within the same shell).
  B between d orbitals and p orbitals.
  C among the attached ligands.
  D from the metal ion to the attached ligands.

Q16 Based on melting points the dividing line between ionic and covalent chlorides of the elements Mg to S lies between:
  A Al and Si **C** Si and P
  B Mg and Al **D** P and S

Q17 Which of the following oxides is unlikely to dissolve in hot aqueous sodium hydroxide?
  A Al_2O_3 **B** Cl_2O_7 **C** MgO **D** SiO_2

Q18 For which transition metal does its atom (in the ground state) have an unpaired electron in a 4s orbital?
  A iron **C** manganese
  B zinc **D** chromium

Q19 In which one of the following reactions does the transition metal or compound not behave as a catalyst?
- **A** The formation of ethanal from ethanol, using acidified potassium dichromate(VI).
- **B** The formation of oxygen from hydrogen peroxide using manganese(IV) oxide.
- **C** The formation of ammonia from its elements in the presence of iron.
- **D** The use of vanadium(V) oxide in the Contact process.

Q20 Transition metals can be distinguished from non-transition metals by the fact that:
- **A** non-transition metals have higher relative atomic masses than transition metals.
- **B** non-transition metals only have +1 or +2 oxidation states.
- **C** only the transition metals can form complex ions.
- **D** transition metals have a greater tendency to form coloured compounds than non-transition metals.

Q21 Titanium has the electronic structure $1s^2 2s^2 2p^6 3s^2 3p^6 3d^2 4s^2$.

Which one of the following compounds is unlikely to exist?
- **A** Na_2TiO_4 **B** $TiCl_4$ **C** TiO **D** TiO_2

Q22 Which of the following sets contains a basic, an acidic and an amphoteric oxide?
- **A** Al_2O_3 SiO_2 P_4O_{10} **C** MgO P_4O_6 SO_3
- **B** MgO Al_2O_3 P_4O_{10} **D** MgO Na_2O SO_2

Q23 The conversion of chromate ions, $CrO_4^{2-}(aq)$ to dichromate ions, $Cr_2O_7^{2-}(aq)$ is represented by the following equation:

$$2CrO_4^{2-}(aq) + 2H^+(aq) \rightleftharpoons Cr_2O_7^{2-}(aq) + H_2O(l)$$

Which statement is not true?
- **A** The $CrO_4^{2-}(aq)$ ion acts as a base.
- **B** There is a colour change.
- **C** None of the ions are complex ions.
- **D** The conversion of $CrO_4^{2-}(aq)$ to $Cr_2O_7^{2-}(aq)$ involves a change of oxidation state.

Q24 Which of the following has the greatest number of unpaired electrons?
- **A** Fe^{3+} **B** Fe^{2+} **C** Cr^{3+} **D** Co^{2+}

Q25 Which one of the following characteristics of transition metals is associated with their catalytic behaviour?
- **A** coloured ions
- **B** low standard electrode potentials
- **C** variable oxidation states
- **D** high melting and boiling points

Paper 2 IB questions and IB style questions

Q1 Give the formula of the chlorides (except sulfur) of the elements in period 3 of the periodic table. Describe and account for the bonding in these chlorides and their reaction (if any) with water. [12]

Q2 State and explain how the acid–base behaviour of the oxides varies across period 3 of the periodic table. [8]

Q3 Two characteristics of the d-block (transition) elements are that they exhibit variable oxidation states and form coloured compounds.
- **a** State two possible oxidation states for iron; explain these in terms of electron arrangements. [2]
- **b** Explain why many compounds of d-block (transition) elements are coloured. [3]
  Higher Level Paper 1, May 05, Q8

Q4 **a** State and explain what would be observed when aqueous ammonia is added to aqueous copper(II) nitrate, dropwise, until the aqueous ammonia is in excess. [4]
- **b** The green MnO_4^{2-} ion, when treated with acid, forms a brown precipitate of MnO_2 in a purple solution of MnO_4^-.
  - **i** Write a balanced equation for this redox reaction. [2]
  - **ii** Deduce the oxidation numbers of manganese in all three manganese compounds. [3]
  - **iii** What type of behaviour does MnO_2 exhibit when placed in hydrogen peroxide solution? [1]
- **c** Explain why an aqueous and concentrated solution of zinc ions is colourless. [2]

Q5 **a** State the electronic configuration of the chromium atom and predict **two** of the likely oxidation states of chromium. [2]
- **b** Explain why transition elements often show variable oxidation states in their compounds, whereas other metals, such as calcium and aluminium, do not. [2]
- **c** For the element manganese state the formula of a compound in each of the most common oxidation states +2, +4 and +7. [3]
- **d** Describe the bonding between the ligand and the ion of a transition metal in a complex. [2]

Q6 Prussian blue is made by mixing together aqueous solutions of $FeCl_3$ and $K_4Fe(CN)_6$, which contains $[Fe(CN)_6]^{4-}$ ions.
- **a** State the oxidation states of the iron ions in each of these solutions. [2]
- **b** Describe the shape of the $[Fe(CN)_6]^{4-}$ ion. [2]
- **c** Iron is a transition metal. Define the term *transition metal*. [1]
- **d** Explain the activity of iron in hemoglobin. [2]

14 Bonding

STARTING POINTS
■ The principles of VSEPR theory can be applied to molecules and ions with five and six centres of negative charge.
■ Resonance occurs when more than one valid Lewis structure can be written for a molecule or polyatomic ion.
■ The molecular orbital theory is an alternative chemical bonding theory to the Lewis (electron dot) model.
■ Covalent bonds are formed when atomic orbitals overlap and merge to form molecular orbitals.
■ The greater the degree of overlap the stronger the covalent bond formed.
■ There are two types of molecular orbitals: sigma (σ) and pi (π).
■ Single bonds are σ bonds.
■ Sigma and pi bonds are present in double and triple bonds.
■ Hybridization is a process atoms are postulated to undergo prior to covalent bond formation.
■ Common types of hybridization are sp, sp^2 and sp^3.

14.1 Shapes of molecules and ions

14.1.1 **Predict** the shape and bond angles for species with five and six negative charge centres using VSEPR theory.

We have seen earlier (Chapter 4) that the valence shell electron pair repulsion theory (VSEPR theory) can be very usefully applied to explain the shapes of simple covalent molecules and polyatomic ions built around a central atom.

This theory can also be applied to molecules and ions that have five and six centres of negative charge. The basic shapes (Figure 14.1) adopted by molecules with five or six electron pairs are trigonal bipyramidal and octahedral, respectively. These shapes minimize the repulsion between electron pairs in the valence shell. The trigonal bipyramid has three equatorial bonds and two axial bonds; the octahedron has four equatorial bonds and two axial bonds.

Table 14.1 summarizes how the numbers of bonding electrons and lone pairs of electrons determine the geometries of molecules with five and six centres of negative charge.

Figure 14.1 The basic shapes for molecules with five and six electron pairs

Total number of electron pairs	Number of negative charge centres		Molecular shape	Examples
	Bonding pairs	Lone pairs		
5	2	3	Linear	ICl_2^-, XeF_2 and I_3^-
5	3	2	T-shaped	ClF_3 and BrF_3
5	4	1	See-saw (distorted tetrahedral)	SF_4
5	5	0	Trigonal bipyramidal	PCl_5
6	6	0	Octahedral	SF_6 and PF_6^-
6	5	1	Square pyramidal	BrF_5 and ClF_5
6	4	2	Square planar	XeF_4 and ICl_4^-

Table 14.1 Summary of molecular shapes for species with five and six centres of negative charge

For species with five and six centres of negative charge there are alternative positions for lone pairs of electrons. The favoured positions will be those where the lone pairs are located furthest apart, thus minimizing the repulsive forces in the molecule. Consequently, lone pairs will *usually* occupy equatorial positions.

A multiple bond is still treated as if it is a single electron pair and the two or three electron pairs of a multiple bond are treated as a single pair. For example, the xenon trioxide molecule has a pyramidal shape. The valence shell of the xenon atom in xenon trioxide contains 14 electrons: eight from the xenon and two each from the three oxygen atoms.

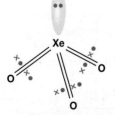

Figure 14.2 Structure and shape of the XeO_3 molecule

Worked examples

Deduce the shape of PF_5.

The valence shell of the phosphorus atom in phosphorus(V) fluoride contains ten electrons: five from the phosphorus and one each from the five fluorine atoms. The shape will be a trigonal bipyramid with bond angles of 120°, 180° and 90° (Figure 14.3).

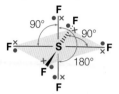

Figure 14.3 Structure and shape of the PF_5 molecule

Deduce the shape of SF_6.

The valence shell of the sulfur atom in the sulfur(VI) fluoride molecule contains 12 electrons: six from the sulfur and one each from the six fluorine atoms. The shape will be an octahedron with bond angles of 90° and 180° (Figure 14.4).

Figure 14.4 Structure and shape of the SF_6 molecule

Deduce the shape of SF_4.

The valence shell of the sulfur atom contains ten electrons: six from the sulfur and one each from the four fluorine atoms. There are four bonding pairs and one lone pair. The basic shape adopted by the electron pairs in the molecule is trigonal bipyramidal. In this arrangement, the electron pairs at the equatorial positions experience less repulsion compared to axial electron pairs. Hence the lone pair occupies an equatorial position and thus the shape of the molecule itself resembles a see-saw (Figure 14.5). As a general rule, for a molecule where the negative charge centres adopt a trigonal bipyramid structure, any lone pairs will occupy equatorial positions.

Figure 14.5 Structure and shape of the SF_4 molecule

Deduce the shape of ClF_3.

The valence shell of the chlorine atom contains ten electrons: seven from the chlorine and one each from the three fluorine atoms. There are three bonding pairs and two lone pairs. The basic shape adopted by the electron pairs in the molecule is a trigonal bipyramid. To minimize the repulsion between bonding pairs and lone pairs of electrons, the two lone pairs of electrons occupy the equatorial positions. Hence, the molecule has a T-shape (Figure 14.6).

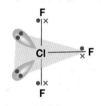

Figure 14.6 Structure and shape of the ClF_3 molecule

Deduce the shape of $XeCl_4$.

The valence shell of the xenon atom contains 12 electrons: eight from the xenon and one each from the four chlorine atoms. There are four bonding pairs and two lone pairs. The basic shape adopted by the molecule is octahedral. However, there are two possible arrangements for the lone pairs. The first structure, square planar, minimizes the strongest repulsion (the lone pairs are at 180° to each other)

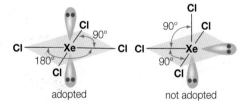

Figure 14.7 Possible structures of $XeCl_4$

and is hence adopted as the molecular shape (Figure 14.7). As a general rule, for a molecule where the negative charge centres adopt an octahedral structure, any lone pairs will occupy axial positions.

Deduce the shape of ICl_2^-.

The valence shell of the central iodine atom contains ten electrons: seven from the iodine atom, two from the two chlorine atoms and an additional electron responsible for the negative charge. There are two bonding pairs and three lone pairs. The basic shape adopted by the molecule is a trigonal bipyramid (Figure 14.8). The three lone pairs occupy the equatorial positions to minimize the repulsion between the bonding pairs and the lone pairs of electrons.

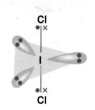

Figure 14.8 Structure and shape of ICl_2^-

■ **Extension:** ## The shape of the XeF$_6$ molecule

Xenon forms a number of compounds with oxygen and fluorine. The highest fluoride of xenon is xenon hexafluoride, XeF$_6$. It undergoes rapid hydrolysis with water to form xenon trioxide:

$$XeF_6(s) + 3H_2O(l) \rightarrow XeO_3(s) + 6HF(aq)$$

The shape of xenon hexafluoride is based upon a *distorted* octahedron (Figure 14.9). However, the lone pair is *not* in a fixed position and moves around the molecule. XeF$_6$ is an example of a fluxional molecule.

Figure 14.9 The molecular shape of XeF$_6$

History of Chemistry

The valence shell electron pair repulsion (VSEPR) theory is a simple extension of G.N. Lewis's ideas (Chapter 4) and it is very successful for predicting the shapes of polyatomic molecules. Although many people have been involved in its development, the theory stems from suggestions made by **Nevil Sidgwick** (1873– 1952) and Herbert Powell in 1940. Their ideas were extended and put into a more modern context by Ronald Gillespie and Ronald Nyholm. **Nyholm** (1917–1971) taught chemistry in Sydney, and was also a Professor of Chemistry at University College London (UCL). **Gillespie** (1924–) also taught chemistry at UCL before joining McMaster University, Canada, in 1958. He continues to carry out research into molecular geometry.

■ **Extension:** ## Expansion of the octet

Nitrogen forms one chloride, nitrogen trichloride, NCl$_3$. Phosphorus, however, forms two chlorides: phosphorus trichloride, PCl$_3$, and phosphorus pentachloride, PCl$_5$ (Chapter 13). Phosphorus can form five bonds since it has low energy d orbitals (in the third shell) to accommodate the extra electrons, while nitrogen has no such orbitals available (in the second shell). There are no 2d orbitals.

The energy required to promote a 3s electron in the phosphorus atom (Figure 14.10) to the empty 3d orbital to form five unpaired electrons is more than offset by an even larger amount of energy released when two extra P–Cl bonds are formed in PCl$_5$. Compounds with more than eight electrons in their outer or valence shell, such as PCl$_5$ and SF$_6$ (Figure 14.11), are termed **hypervalent** and are said to have 'expanded their octet'.

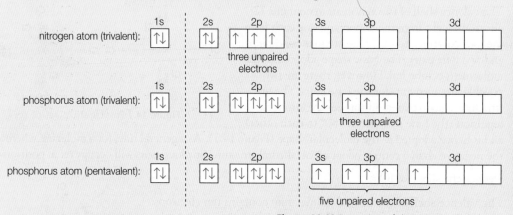

Figure 14.10 Expansion of the octet in phosphorus

Figure 14.11 Molymod models of phosphorus pentachloride, PCl$_5$, and sulfur hexafluoride, SF$_6$

14.2 Hybridization

14.2.1 **Describe** σ and π bonds.

Molecular orbitals

Sigma bonding

Covalent bonding was previously described (Chapter 4) by means of 'dot-and-cross' diagrams where the electron pairs making the covalent bonds are represented by dots and crosses. A single bond is a shared pair of electrons, a double bond is two shared pairs of electrons and a triple bond is three shared pairs. A simple electrostatic model was used to describe molecules.

A *better* description of the nature of covalent bonding is derived from examining covalent bonding in terms of interactions between atomic orbitals (Chapter 12). A covalent bond is formed by the overlap and merging of two atomic orbitals on different atoms, each containing an unpaired electron. A molecular orbital is formed and is the region where the electron density is concentrated. This theory of bonding is known as **molecular orbital theory (MO theory)**. The strength of a bond is directly related to the match in energy levels for the two atomic orbitals and to the degree of the overlap.

A dative covalent bond (Chapter 4) is a formed by the similar merging of two atomic orbitals, but in this case one orbital initially contains both the electrons that will form the bond. This orbital merges with a vacant atomic orbital of appropriate orientation and energy level on the other atom.

The simplest molecule formed is that produced by the overlap of the 1s orbitals of two hydrogen atoms (Figure 14.12).

Figure 14.12 Molecular orbital for the hydrogen molecule, H_2

The electron density distribution for a hydrogen molecule, H_2, can also be illustrated by means of an electron density map. Figure 14.13 shows a contour map of the charge distribution for the hydrogen molecule. Imagine a hydrogen molecule cut in half by a plane which contains the nuclei. The amount of charge at every point in space is determined, and all points having the same value for the electron density in the plane are joined by a contour line. The important feature of a σ bond is that the electron density occupies the space between and around the two nuclei.

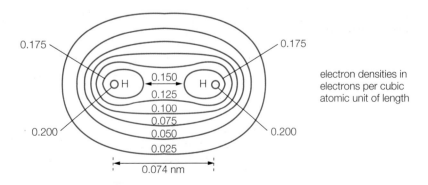

Figure 14.13 Electron density map of a hydrogen molecule

0.175   0.175
0.150
0.125
0.100
0.075
0.200   0.050   0.200
0.025

H   H

0.074 nm

electron densities in electrons per cubic atomic unit of length

Language of Chemistry

The single bond in the hydrogen molecule, and all other single covalent bonds in other molecules, are known as σ bonds. They are so named because, if you imagine looking down the bond, the electron density would appear to be circular (Figure 14.14). By analogy with the spherically symmetrical s orbitals, these bonds are termed sigma, σ being the Greek letter corresponding to s. The distinguishing feature of a σ bond (or σ bonding orbital) is that the overlap region lies directly between the two nuclei. ■

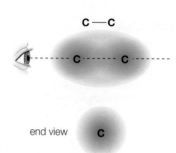

end view

Figure 14.14 The carbon–carbon single bond (σ bond)

History of Chemistry

Molecular orbital theory was developed in the 1930s by Robert Mulliken, John C. Slater, John Lennard-Jones and Friedrich Hund. **Hund** (1896–1997) was a German physicist who worked with Dirac, Heisenberg and Schrödinger (Chapter 12). Hund's rule (Chapter 12) is named after him. **Mulliken** (1896–1986) was an American chemist and physicist who was awarded the Nobel Prize in Chemistry in 1966. **Slater** (1900–1976) was a theoretical chemist who introduced exponential functions to describe atomic orbitals. **Lennard-Jones** (1894–1954) was a mathematician and theoretical physicist who was particularly interested in atomic and molecular forces. The Lennard-Jones potential describes how potential energy varies with distance for interacting atoms (Chapter 4).

The development of MO theory by this group illustrates how scientists of different backgrounds and disciplines can work together to make a conceptual advance.

■ Extension: Molecular orbitals in hydrogen

Consider the hydrogen molecule, H_2. Two hydrogen atoms, each with an electron in the 1s orbital, approach each other. The two 1s orbitals overlap and merge to form two new molecular orbitals: σ (sigma) and σ* (sigma star) (Figure 14.15).

The σ orbital is of lower energy than the original atomic orbital, and is known as a **bonding orbital**. The σ* orbital is of higher energy than the original atomic orbital and is known as the **antibonding orbital**. In a manner similar to the filling of electrons into the orbitals in atoms, the two 1s electrons enter the lower energy σ bonding orbital as a *spin pair* (Chapter 12). The potential energy of the hydrogen molecule is *lower* than the potential energy of the uncombined atoms and hence bond formation is exothermic (Figure 14.16).

(MO theory is based upon the quantum mechanical model which regards electrons as standing waves (Chapter 12). The bonding and antibonding σ orbitals correspond to in-phase and out-of-phase wave combinations. The electron density is a *minimum* in the centre of a σ* orbital because that is where the two electron waves *cancel*.)

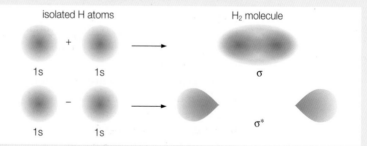

isolated H atoms H_2 molecule

1s   1s   σ

1s   1s   σ*

Figure 14.15 Overlap of two 1s orbitals to form two molecular orbitals, σ and σ*

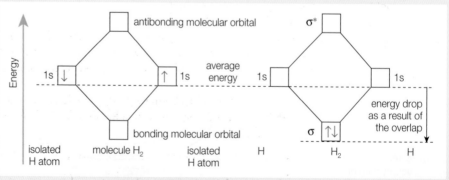

Figure 14.16 The energy levels of the molecular orbitals σ and σ*

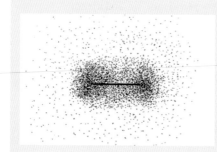

The two electrons in the bonding molecular orbital of H_2 are equivalent to a shared pair of electrons (single covalent bond). The bonding molecular orbital corresponds to an increase in electron density concentrated around the nuclei and on an imaginary axis, known as the **inter-nuclear axis**, connecting the two hydrogen nuclei (Figure 14.17). The electron density reduces the repulsion between the two nuclei and is the source of attractive forces that hold the two hydrogen atoms together (Chapter 4).

Figure 14.17 An electron dot density plot for the bonding σ orbital of the hydrogen molecule

Pi bonding

Sigma bonds are formed in all the examples described by the **axial** (head-on) overlap of orbitals (Figure 14.18). Such σ bonds are relatively unreactive, but are important in determining the shape or 'skeleton' of a molecule or ion (Chapter 4). However, there is another way in which p orbitals can overlap if they are brought together in the correct orientation.

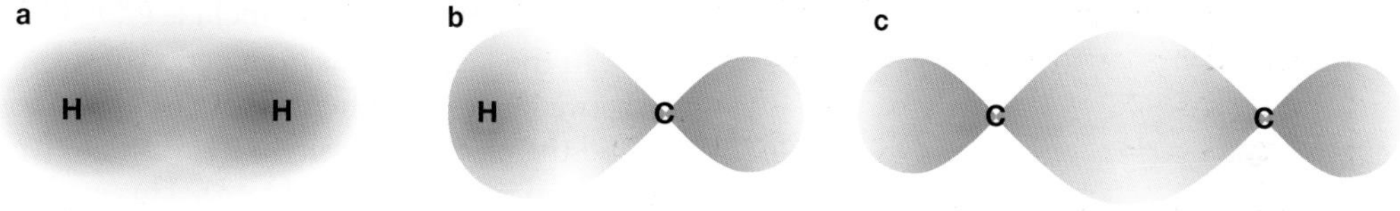

Figure 14.18 Sigma molecular orbitals formed between **a** hydrogen atoms, **b** carbon and hydrogen atoms, **c** carbon atoms

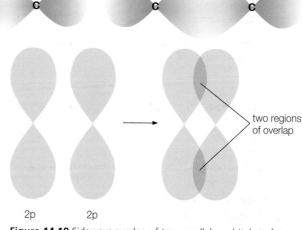

Two parallel p orbitals can also undergo sideways overlap to produce a π bond (Figures 14.19). Pi bonds (Figure 14.20) occur in molecules containing double or triple covalent bonds. In practice the formation of a π bond requires the prior formation of a σ bond between the atoms concerned; the formation of the σ bond brings the p orbitals together so that they can overlap sideways. A double bond consists of one σ and one π bond; a triple bond consists of one σ bond and two π bonds.

two regions of overlap

2p   2p

Figure 14.19 Sideways overlap of two parallel p orbitals to form a π bond

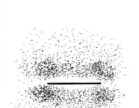

Figure 14.20 An electron dot density plot for a bonding π orbital

In a π bond the two nuclei are poorly shielded from each other and the electrons are further away from the nuclei than the electrons in the σ bond. As a consequence the π electrons are more *polarizable* (Figure 14.21) and hence chemically reactive. They are often involved in initiating chemical reactions, for example addition reactions of alkenes (Chapter 27).

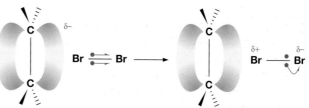

Figure 14.21 Polarization of the bromine molecule by the π bond of an ethene molecule

Bond	Bond enthalpy /kJ mol⁻¹
C—C	348
C=C	612
C≡C	837

Table 14.2 Bond enthalpies of carbon

When a double bond forms between two atoms, the two bonds are different. There is a strong σ bond and a weaker π bond. Evidence supporting this model is derived from bond enthalpies (Table 14.2).

A double bond, although stronger than a single bond, is *not* twice as strong. If a carbon–carbon double bond consisted of two identical bonds, then its bond enthalpy would be $2 \times 348\,\text{kJ mol}^{-1}$ ($= 696\,\text{kJ mol}^{-1}$).

strength of carbon–carbon single bond (σ bond) = $348\,\text{kJ mol}^{-1}$

strength of carbon–carbon double bond (σ bond + π bond) = $612\,\text{kJ mol}^{-1}$

Hence, the extra strength due to the π bond = $612\,\text{kJ}\,\text{mol}^{-1} - 348\,\text{kJ}\,\text{mol}^{-1} = 264\,\text{kJ}\,\text{mol}^{-1}$.

This is based on the assumption that the σ bond has the same strength in double and single carbon–carbon bonds.

Language of Chemistry

In a π bond the electron density is concentrated in two regions, one above and one below the inter-nuclear axis (Figure 14.22). The end-on view has the same symmetry as, and looks like, an atomic p orbital. The letter π is the Greek equivalent of p. ∎

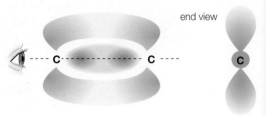

Figure 14.22 A carbon–carbon double bond showing the σ and π bond electron density

∎ Extension: π bond formation and the periodic table

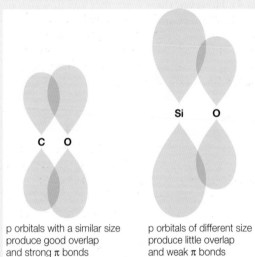

p orbitals with a similar size produce good overlap and strong π bonds

p orbitals of different size produce little overlap and weak π bonds

Figure 14.23 Effective and ineffective overlap of p orbitals

The formation of strong π bonds only occurs in molecules formed by the elements carbon, nitrogen and oxygen from period 2 of the periodic table. Examples include CO_2, CO, N_2, O_2, HCN and NO_2.

In the larger atoms from period 3, strong π bonds are not formed between atoms of the *same element* because the p orbitals are larger and the electron density is more diffuse (spread out), hence the overlap is poor (Figure 14.23). This explains why there are no stable P_2 or S_2 molecules. (However, sulfur and phosphorus can form π bonds with oxygen. These double bonds are greatly strengthened by their polar character.)

Silicon, also in period 3, has p orbitals significantly larger than those of oxygen. Their differences in size and energy mean that extensive overlap does not occur. Consequently silicon dioxide, SiO_2, has a giant covalent structure (see Chapter 4) where the bonds are –O–Si–O– bonds. No stable SiO_2 (O=Si=O) molecules are possible in the solid state.

Sigma (σ) bond	Pi (π) bond
This bond is formed by the axial overlap of atomic orbitals.	This bond is formed by the sideways overlap of atomic orbitals.
This bond can be formed by the axial overlap of s–s or s with a hybridized orbital.	It involves the sideways overlap of parallel p orbitals only.
The bond is stronger because overlapping can take place to a larger extent.	The bond is weaker because the overlapping occurs to a smaller extent.
The electron cloud formed by axial overlap is symmetrical about the inter-nuclear axis and consists of a single electron cloud.	The electron cloud of the π bond is discontinuous and consists of two charged electron clouds above and below the plane of atoms.
There can be a free rotation of atoms around the σ bond.	Free rotation of atoms around the π bond is not possible because it involves the breaking of the π bond.
The σ bond may be present between the two atoms either alone or along with the π bond.	The π bond is always present between the two atoms with the σ bond, i.e. it is always superimposed on the σ bond.
The shape of the molecule or oxoanion is determined by the σ framework around the central atom.	The π bonds do not contribute to the shape of the molecule.

Table 14.3 A summary of the differences between σ and π bonds

The different features of σ and π bonding are compared in Table 14.3. However, consideration of these two bonding types does not offer a complete explanation of covalent bonding even in simple molecules such as methane, ethane and ethene (Figure 14.24). The explanation of the known spatial arrangements and shapes of these molecules has led to chemists developing the concept of hybridization (see below).

Figure 14.24 Molymod models of methane, ethene and ethane

The C–H bonds in hydrocarbons do *not* arise through the overlap of the 1s orbital on the hydrogen atom and a 2p orbital on the carbon atom. *Nor* is the carbon–carbon double bond formed via axial ('end-on' or 'head-on') overlap of two 2p orbitals of adjacent carbon atoms. Chemists postulate s–p mixing in a process called hybridization and the orbital overlaps described occur with hybridized carbon atoms (see page 383 for a discussion of the hybridization of carbon atoms).

Hybridization

14.2.2 Explain hybridization in terms of the mixing of atomic orbitals to form new orbitals for bonding.

Atomic orbitals can overlap on adjacent atoms to form σ and π molecular orbitals. In addition, orbitals *of the same atom* can overlap, merge and undergo a process of **hybridization** to form a new set of *hybrid atomic orbitals*. The hybrid or hybridized orbitals formed have a specific shape and relative orientation depending on the number and type of atomic orbitals that have hybridized. The number of hybrid orbitals formed is equal to the number of atomic orbitals involved in the hybridization process. The three common types of hybridization are sp, sp^2 and sp^3. All atoms (except hydrogen) are *postulated* to undergo a process of hybridization prior to bonding.

The hybrid orbital has electron density concentrated on one side of the nucleus, i.e. it has one lobe relatively larger than the other. Hence, the hybrid orbitals can form stronger bonds compared to unhybridized atomic orbitals because they can undergo more effective overlap. The hybrid orbitals repel each other and adopt a configuration that minimizes the electron repulsion. Hybridization is simply a mathematical model that is convenient for describing localized bonds. It is *not* a phenomenon that can be studied or measured.

sp hybridization

A single s and a single p orbital will overlap and merge to form two identical sp hybridized orbitals (Figure 14.25). The orbitals will be orientated at 180° to each other and will have identical energies and shapes. The potential energy of the hybrid orbital will be intermediate between the energies of the s and p orbitals.

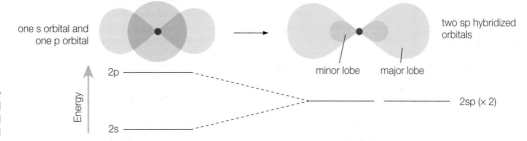

Figure 14.25 The formation of sp hybridized orbitals from an s orbital and a p orbital

Beryllium in the beryllium fluoride molecule, BeF$_2$(g), provides an example of sp hybridization (Figure 14.26). The beryllium atom has the electron configuration 1s^{2}2s^2. After hybridization the electrons in the outer shell of the beryllium atom are in two sp hybrid orbitals. Single bonds between the beryllium and fluorine atoms involve the sharing of the unpaired electrons in the sp hybrid orbital of beryllium and the sp³ hybrid orbitals of the fluorine atom (Figure 14.27).

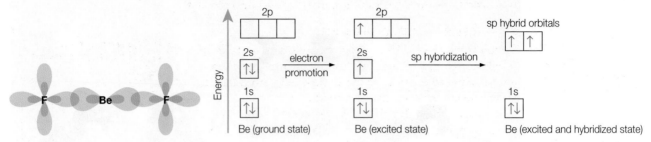

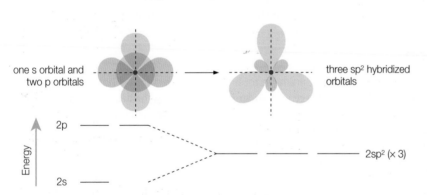

Figure 14.26 sp hybrid orbitals in the beryllium fluoride molecule

Figure 14.27 Atomic orbitals in beryllium (ground state), beryllium (excited state) and beryllium fluoride (hybridized state)

sp² hybridization

sp² hybridization involves the combination of *one* s and *two* p orbitals. These three hybrid orbitals have identical shapes and orientations. They point towards the corners of an equilateral triangle (Figure 14.28).

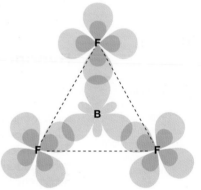

Figure 14.28 The formation of sp² hybridized orbitals from an s orbital and two p orbitals

Figure 14.29 sp² hybrid orbitals in the boron trifluoride molecule

Boron in the boron trifluoride molecule, BF_3, provides an example of sp² hybridization (Figure 14.29). The boron atom has the electron configuration $1s^2 2s^2 2p^1$. After hybridization the electrons in the outer shell of the boron atom are in three sp² hybrid orbitals. Single bonds between the boron and fluorine atoms involve the sharing of the unpaired electrons in the sp² hybrid orbitals of boron and the sp³ orbitals of the fluorine atoms (Figure 14.30).

Figure 14.30 Atomic orbitals in boron (ground state), boron (excited state) and boron trifluoride (hybridized state)

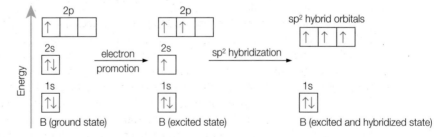

sp³ hybridization

sp³ hybridization involves the combination of one s orbital and three p orbitals (Figure 14.31). The sp³ hybridized orbitals formed are arranged tetrahedrally (to minimize repulsion).

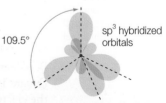

Figure 14.31 The formation of sp³ hybridized orbitals from an s orbital and three p orbitals

Hybridization in carbon

Carbon provides examples of all three types of hybridization: sp, sp² and sp³. A detailed discussion follows on these three types in the context of organic molecules.

sp³ hybridization in alkanes

Figure 14.32 Carbon atom in the ground state

Carbon atoms (Figure 14.32) have the configuration $1s^2 2s^2 2p^2$ in their ground state. Since the 2s sub-shell is full and there are two unpaired electrons in the 2p sub-shell, carbon might be expected to form *two* covalent bonds.

Figure 14.33 Carbon atom in an excited state

However, carbon always forms *four* bonds in familiar and stable compounds, such as methane, CH_4, and carbon dioxide, CO_2. To account for the ability of a carbon atom to form four bonds a 2s electron has to be unpaired and promoted into an empty orbital of the 2p sub-shell (Figure 14.33).

This is an endothermic process, but the excess energy is regained when the two extra bonds are formed (Figure 14.34).

The 2s orbital (now containing just an unpaired electron) and the three 2p orbitals undergo hybridization which leads to the production of four identical sp³ hybrid orbitals, each containing a single electron (Figure 14.35).

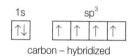

carbon – hybridized

Figure 14.35 Orbitals in an sp³ hybridized carbon atom

The use of the idea of hybridization also explains the observed shape of the methane molecule. The four sp³ hybridized carbon atoms can then overlap with the 1s electrons of four hydrogen atoms (Figure 14.36) to form four σ bonds arranged tetrahedrally around the central carbon atom. Atoms joined by single covalent bonds can usually rotate freely about the bond. The linear shape of BeF_2 and the trigonal planar shape of BF_3 can be understood similarly.

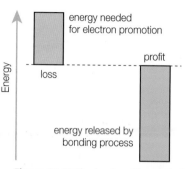

Figure 14.34 The 'cost and benefit' of hybridization

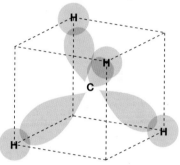

Figure 14.36 sp³ hybrid orbitals in a methane molecule

sp² hybridization in alkenes

In many organic compounds, for example the alkenes and the carbonyl compounds (Chapter 10), carbon forms bonds with three other atoms: two single bonds and a double bond. An electron from the 2s sub-shell is again promoted into the 2p sub-shell. The 2s electron hybridizes with *two* of the 2p² hybrid orbitals to give *three* sp² hybrid orbitals (Figure 14.37). The 2p_z orbital does *not* participate in the hybridization process.

Figure 14.37 Excited carbon atom and an sp² hybridized carbon atom

The hypothetical formation of ethene, C_2H_4, is summarized in Figure 14.38. One sp² hybrid orbital on each carbon atom overlaps and merges with its neighbour to form a carbon–carbon σ bond. The other four sp² hybrid orbitals overlap with the 1s orbitals on four hydrogen atoms to form four carbon–hydrogen bonds. The *unhybridized* 2p orbital on each carbon atom overlaps sideways with its neighbouring carbon atom to form the carbon–carbon π bond.

The two areas of orbital overlap in the π bond (above and below the plane of the molecule) cause ethene to be a rigid molecule: there is no rotation around the carbon–carbon bond. This is in contrast to ethane, and other non-cyclic alkanes, and leads to the existence of geometric (*cis–trans*) isomerism in alkenes (Chapter 20).

Any rotation about the σ bond of ethene joining the two carbon atoms would result in the reduction of p overlap and weaken the π bond (Figure 14.39). The breaking of the π bond will only occur at relatively high temperatures. There is a high activation energy barrier to π bond rotation. Geometric isomers will only interconvert at high temperatures when a sufficient proportion of molecules have enough energy to overcome the π bond.

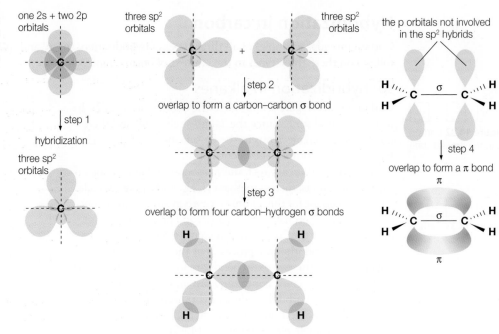

Figure 14.38 The formation of the ethene molecule from hybridized orbitals

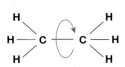

rotation possible around the single bond

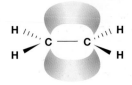

no rotation possible

Figure 14.39 The π bond in ethene prevents free rotation around the carbon–carbon double bond

sp hybridization in alkynes

In the ethyne molecule, C_2H_2, the two carbon atoms form sp hybrid orbitals. An electron from the 2s sub-shell is again promoted into the 2p sub-shell. The 2s electron hybridizes with *one* of the 2p orbitals to give *two* sp hybridized orbitals. The $2p_z$ and $2p_y$ orbitals do *not* participate in the hybridization process (Figure 14.40).

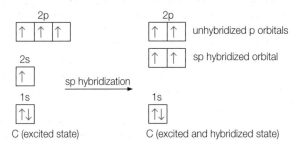

Figure 14.40 Excited carbon atom and an sp hybridized carbon atom

Figure 14.41 Pi bond formation in ethyne

Triple bonds are formed in ethyne, C_2H_2, when *two unhybridized* p orbitals overlap sideways, forming two π bonds (Figure 14.41), in addition to the σ bond already present. The total electron density of the two combined π orbitals takes the shape of a cylinder that encircles the molecule.

Type of hybridized orbital	sp³	sp²	sp
Atomic orbitals used	s, p, p, p	s, p, p	s, p
Number of hybridized orbitals	4	3	2
Number of atoms bonded to the carbon atom	4	3	2
Number of σ bonds	4	3	2
Number of unhybridized p orbitals	0	1	2
Number of π bonds	0	1	2
Bonding arrangement(s)	Tetrahedral; four single bonds only	Trigonal planar; two single bonds and a double bond	Linear; one single bond and a triple bond or two double bonds
Examples:	CH_4, C_2H_6 and CCl_4	$H_2C=CH_2$ and H_2CO	$H–C≡C–H$, $H_2C=C=CH_2$

Table 14.4 Summary of hybridization in carbon

Language of Chemistry

The concept of hybridization can be understood by the analogy of mixing paint. If red and white paints are mixed together in equal proportions, the colour of the resulting mixture is pink. Two quantities of pink paint have been produced. The pink paint is a hybrid of red and white paint because its shade is intermediate between red and white.

In biology the term hybridization may refer to the result of interbreeding between two animals or plants of different species. For example, the tigron is a hybrid cross between a female lion and a male tiger. Tigrons have a blend of lion and tiger features, for example, they may show spots and stripes.

The term sp^3 hybridization indicates that the hybrid orbitals are derived from one s and three p orbitals. Each orbital is said to have 25% s character and 75% p character. The term sp hybridization indicates that the hybrid orbitals are derived from one s and one p orbital. Each orbital is said to have 50% s character and 50% p character. Consequently, sp hybrid orbitals look more like s orbitals and sp^3 hybridized orbitals look more like p orbitals. ■

TOK Link

Students studying the IB Chemistry Programme are exposed to a number of different chemical models: the Bohr model of atomic structure (Chapter 2) and then later VSEPR theory (Chapter 4) followed by hybridization. The quantum mechanical model is briefly introduced in Chapter 12, but knowledge of it is not a syllabus requirement.

You may ask which of these models is 'correct' or 'more correct'? You should be aware that chemistry is not really a search for absolute truth. Chemical models are merely the simplest possible ideas that fit the facts (experimental data).

To quote the late Michael Dewar (1918–1997): 'There is of course no question of a model being true or false'. The same rule applies to scientific theories, which are simply definitions of scientific models. The question 'Is it true?' is meaningless in science. The correct questions to ask are 'Does it work?' and 'Is it useful?'.

At the IB level you should be satisfied with simple models that rationalize and predict the properties of the majority of examples you encounter. You should 'learn to live with exceptions', instead of 'living by exceptions'. In other words, you should be content with a simple model that account for 95% of molecules and their properties, rather than a very complicated model that accounts for 100%.

The simple Lewis model with its emphasis on the idea that electrons form pairs in molecules is capable of explaining a vast amount of chemistry. However, it cannot explain why liquid oxygen is magnetic or the existence of molecules with unpaired electrons, such as NO. These observations can only be rationalized with quantitative MO theory.

You should be aware that hybridization is not a phenomenon that can be studied experimentally or observed. It is just a model for describing localized bonds. Essentially it is a mathematical convenience of reorganizing and reshaping atomic orbitals to make them consistent with the experimentally observed geometries of molecules.

Lewis structures and hybridization

14.2.3 **Identify** and **explain** the relationships between Lewis structure, molecular shapes and types of hybridization (sp, sp^2 and sp^3).

The application of VSEPR theory (Chapter 4) to determine the shape of a simple molecule or polyatomic ion from its Lewis structure (Table 14.5) can also be used to quickly identify the hybridization state of the central atom in the structure.

- sp^3 hybridization can be used to describe the bonding in any structure where there are four negative charge centres around the centrally bonded atom. A negative charge centre is a lone pair or a covalent bond (whether single, double or triple).
- sp^2 hybridization can be used to describe the bonding in any structure where there are three negative charge centres around the centrally bonded atom.
- sp hybridization can be used to describe the bonding in any structure where there are two negative charge centres around the centrally bonded atom.

Hybridization state of central atom	Number of negative charge centres	Number of covalent bonds	Number of lone pairs	Shape	Examples
sp	2	2	0	Linear	BeF_2, CO_2
sp^2	3	3	0	Trigonal planar	BF_3, graphite, fullerenes, SO_3 and CO_3^{2-}
sp^2	3	2	1	V-shaped or bent	SO_2 and NO_2^-
sp^3	4	4	0	Tetrahedral	CH_4, diamond, ClO_4^- and SO_4^{2-}
sp^3	4	3	1	Pyramidal	NH_3, NF_3, PCl_3 and H_3O^+
sp^3	4	2	2	V-shaped or bent	H_2O, H_2S and NH_2^-

Table 14.5 The relationship between the Lewis structure and the hybridization of the central atom

(Hybridization models involving *d orbitals*, such as sp^3d and sp^3d^2, have been used to explain molecules with an expanded octet of electrons, for example PCl_5 and SF_6 (see page 376). This is one model used to explain the observed shapes of certain molecules. However, serious questions are currently being posed in university research as to whether this is the best model regarding these structures.)

Worked example

Deduce the hybridization state of the central atoms of the following ions and molecules: CS_2, $AlCl_3$, PH_3 and NH_4^+.

CS_2

The carbon atom forms two double bonds; there are no lone pairs of electrons on the carbon atom, hence the hybridization is sp.

$AlCl_3$

The aluminium atom forms three single bonds; there are no lone pairs of electrons on the aluminium atom, hence the hybridization is sp^2.

PH_3

The phosphorus atom forms three single bonds; there is one lone pair of electrons on the phosphorus atom, hence the hybridization is sp^3.

NH_4^+

The nitrogen atom forms four single bonds; there are no lone pairs of electrons on the nitrogen atom, hence the hybridization is sp^3.

14.3 Delocalization of electrons

14.3.1 Describe the delocalization of π electrons and **explain** how this can account for the structures of some species.

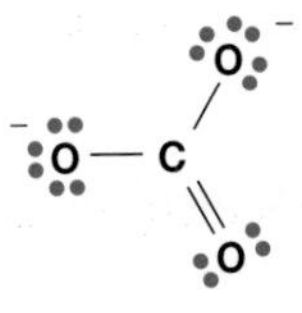

Figure 14.42 A single Lewis structure for the carbonate ion

Resonance

For some molecules and compound ions, it is possible to draw several Lewis structures that differ in the positions of the π-bonding electrons and lone (non-bonded) pairs. For example, we could draw the carbonate ion as shown in Figure 14.42.

However, we could also draw the carbonate ion as either of the two structures shown in Figure 14.43 (assuming that none of the atoms have changed position).

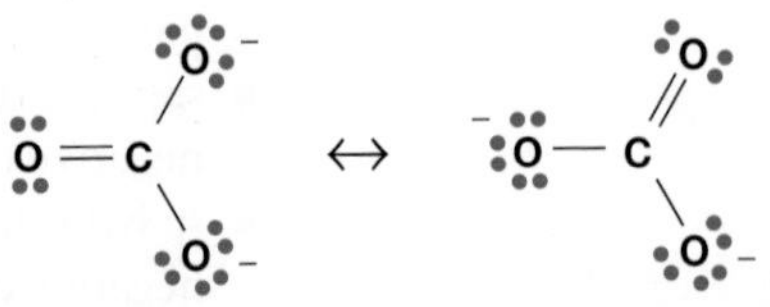

Figure 14.43 Two additional Lewis structures for the carbonate ion

Language of Chemistry

The double-headed single arrow often drawn between the different forms represents resonance and is not to be confused with the double arrow symbol for a reversible reaction at equilibrium ($\rightleftharpoons$, see Chapter 7). ■

Figure 14.44
Resonance hybrid for the carbonate ion

The carbonate ion is *not* correctly described by any of these structures, but exists as a form, known as a **resonance hybrid**, which is a ' blend' of all three structures (Figure 14.44). Each of the three structures, known as **resonance structures**, contributes to the resonance hybrid depending on its energy: the lower the energy of the resonance structure, the greater its contribution to the hybrid. In the case of the carbonate ion all three resonance structures are of equal energy (due to their symmetry) and make an equal contribution to the resonance hybrid.

The dotted lines indicate that all three carbon–oxygen bonds are identical or equivalent to each other and each has partial double bond character. Similarly, each of the three oxygen atoms has a partial negative charge. The real hybrid is thus a 'blend' with equal weighting from the three imaginary resonance structures that have the familiar single and double bonds with two of the oxygen atoms carrying a complete formal negative charge.

Experimental evidence supports the resonance description of the carbonate ion. X-ray diffraction data (Chapter 4) of metal carbonate crystals reveals that all the carbon–oxygen bond lengths in this ion are equivalent and are shorter than a single C–O bond, but slightly longer than a C=O bond.

When resonance occurs the resonance hybrid is more stable than any of the resonance structures. The difference between the energy of the most stable form and the hybrid is known as the resonance energy. This concept is illustrated in Figure 14.45 for the carbonate ion.

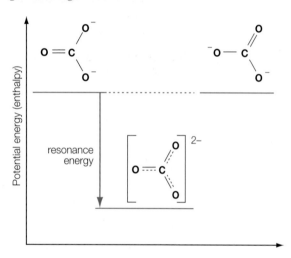

Figure 14.45 A diagram illustrating the concept of resonance energy for the carbonate ion

For significant resonance stabilization to occur within a molecule or ion the suggested resonance structures must meet all of the following requirements, as exemplified by the carbonate ion.

- All resonance forms must have the same distribution of atoms or nuclei in space, that is the same molecular shape.
  All three resonance structures of the carbonate ion have the three oxygen atoms distributed around a central carbon atom in a trigonal planar arrangement.
- No resonance form may have a very high energy. In particular no resonance form containing carbon, oxygen and nitrogen (or other atom from the second row of the periodic table) may have more than eight valence electrons (or for hydrogen, two valence electrons). In other words, the octet (and duplet) rules must be obeyed.
  The three oxygen atoms and one carbon atom of the carbonate ion each obey the octet rule – in all three resonance forms they have eight valence electrons.
- All resonance forms must contain the same number of electron pairs.
  All the resonance structures of the carbonate ion have three σ pairs, one π pair and eight lone (non-bonded) pairs.
- All resonance forms must carry the same total or net charge.
  All the resonance structures of the carbonate ion have a total charge of minus two.

To understand the importance of resonance in the stabilization of a structure, and the contribution of a particular resonance form, the following points must be considered.

- In the case of ions, the most stable resonance structures will have negative charges on the electronegative atoms, usually oxygen and sulfur, and positive charges on the less electronegative atoms, usually carbon and nitrogen.
- The most important resonance structures are those where as many atoms as possible obey the octet rule, have the maximum number of bonded electrons and which have the fewest charges.
- Resonance effects are always stabilizing and *generally* the greater the number of resonance structures that can be drawn, the greater the stability of the resonance hybrid.

Examples of resonance-stabilized ions and simple molecules

Two other inorganic ions that are stabilized by resonance and described by a series of symmetrical resonance structures are the nitrate(III) ('nitrite') ion, NO_2^-, and the nitrate(V) ('nitrate') ion, NO_3^- (Figure 14.46). Curly arrows may be used to show the movement of electron pairs to generate resonance structures.

nitrite ion

nitrate ion

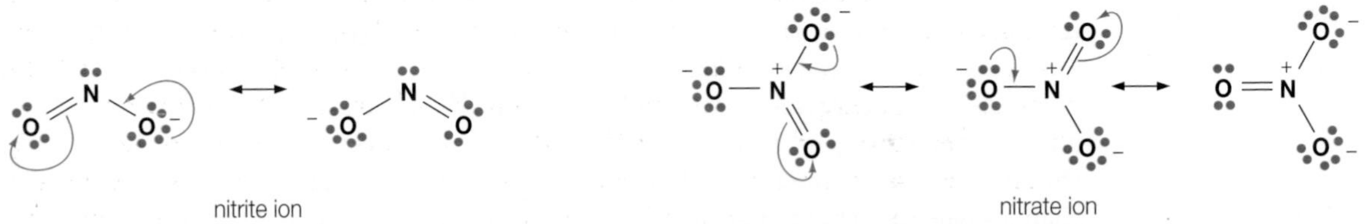

Figure 14.46 Resonance structures and hybrids for the nitrite and nitrate ions

Another familiar ion extensively stabilized by resonance is the sulfate(VI) ('sulfate') ion, SO_4^{2-} (Figure 14.47).

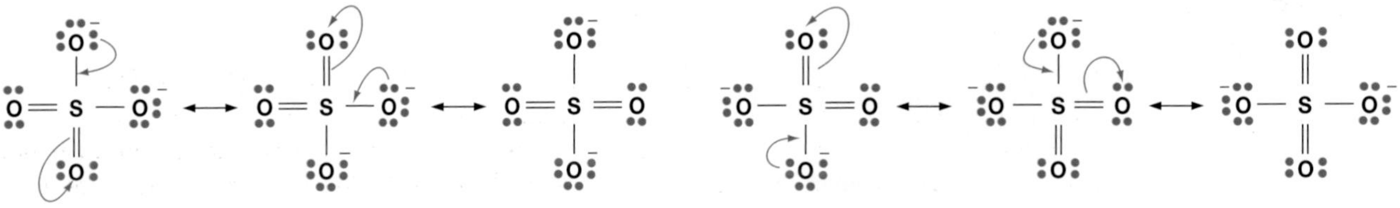

Figure 14.47 Resonance structures for the sulfate(VI) ion

Ozone (trioxygen, O_3) is an allotrope of oxygen that plays an important role in the ozone layer, protecting the Earth's surface from the harmful effects of ultraviolet radiation (Chapter 25). The ozone molecule is V-shaped and is a resonance hybrid described by two major resonance forms (Figure 14.48).

Figure 14.48 Resonance structures for ozone

Many transition metal oxoanions (Chapter 13) are resonance stabilized, for example manganate(VII) ('permanganate'), MnO_4^-, and chromate(VI) ('chromate'), CrO_4^{2-} (cf. sulfate).

Language of Chemistry

In physics, resonance is the tendency of a system, for example a simple pendulum, to oscillate at maximum amplitude at certain frequencies, known as the system's resonant frequencies. At these frequencies, even a small driving force can produce large amplitude vibrations, because the system stores vibrational energy.

Owing to confusion with the physical meaning of the word resonance, as no elements actually appear to be resonating, it has been suggested by some chemists that the term *resonance* be abandoned in favour of *delocalization*. Resonance energy would become delocalization energy and a resonance structure becomes a contributing structure. The double-headed arrows would be replaced by commas. ■

Examples of resonance in organic chemistry

A number of organic molecules and ions exist as resonance hybrids, for example, the ethanoate ion, CH_3COO^-, formed during the dissociation of ethanoic acid, CH_3COOH (Chapter 9). The ethanoate ion, like the carbonate ion, has symmetrical resonance involving equivalent resonance structures making equal contributions to the resonance hybrid (Figure 14.49). As with the carbonate ion the two oxygens cannot be distinguished and the two carbon–oxygen bond lengths are identical.

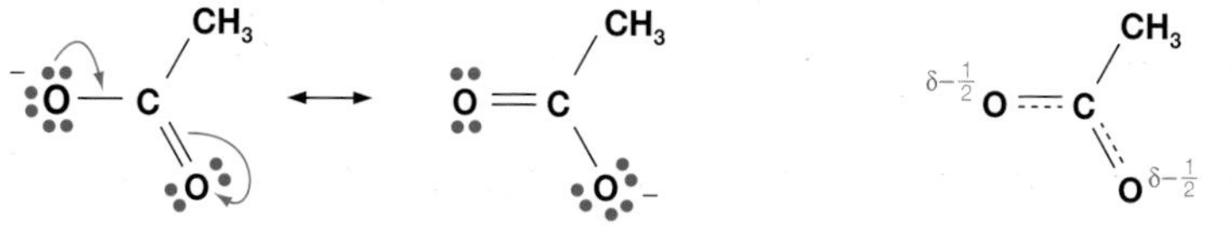

Figure 14.49 Resonance structures and resonance hybrid of the ethanoate ion

Simple resonance theory can be used to explain why ethanoic acid, CH_3COOH, is a stronger acid than ethanol, C_2H_5OH. Neither the ethanol molecule nor the ethoxide ion ($C_2H_5O^-$) are stabilized by resonance, hence the low dissociation of ethanol into ions.

Although ethanoic acid itself is stabilized by resonance, this stabilization is not very effective compared with that in the ethanoate ion. In the latter both the negative charge and electrons are delocalized, and the resonance is symmetrical. However, in ethanoic acid (Figure 14.50) the charged form has unlike charges separated and is of relatively high energy. (We assume for the purposes of this argument that entropy (Chapter 15) and hydration effects are similar enough to be ignored.)

The concept of resonance and another electronic effect called the inductive effect can be used to explain and predict the differences in strengths between other organic acids and bases (Chapter 27).

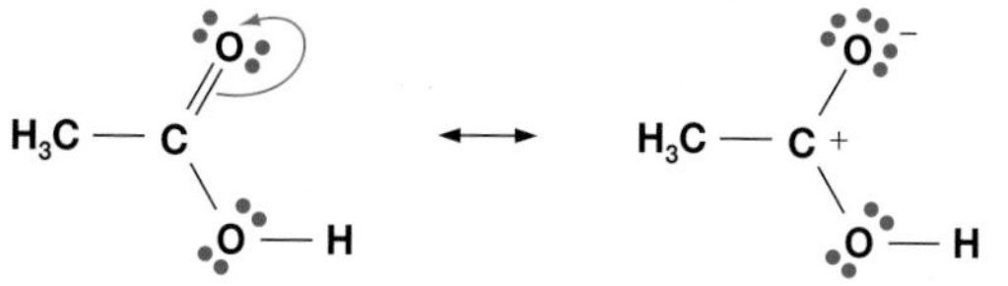

Figure 14.50 Resonance in ethanoic acid

Propanone, $(CH_3)_2CO$, the simplest ketone, can be described as a resonance hybrid of two resonance structures (Figure 14.51). The first resonance structure is known as the major resonance form and makes the largest contribution to the resonance hybrid; the second resonance structure, with the separation of charge, makes a smaller contribution to the resonance hybrid.

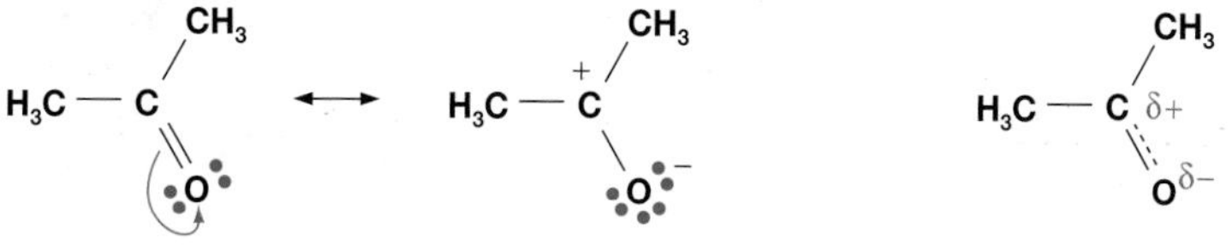

Figure 14.51 Resonance structures and hybrid for propanone

The second resonance structure in propanone makes a significant contribution because oxygen is appreciably more electronegative than carbon and is often found bearing a complete or formal negative charge and forming a single covalent bond, for example, the hydroxide ion, OH^-.

This description for propanone is consistent with the chemical reactions of ketones, which frequently involve nucleophilic attack on the carbon atom of the carbonyl group (Chapter 27).

The best known example of resonance in organic chemistry concerns the cyclic hydrocarbon benzene, C_6H_6, which can be described by two major resonance structures called Kekulé structures (Figure 14.52).

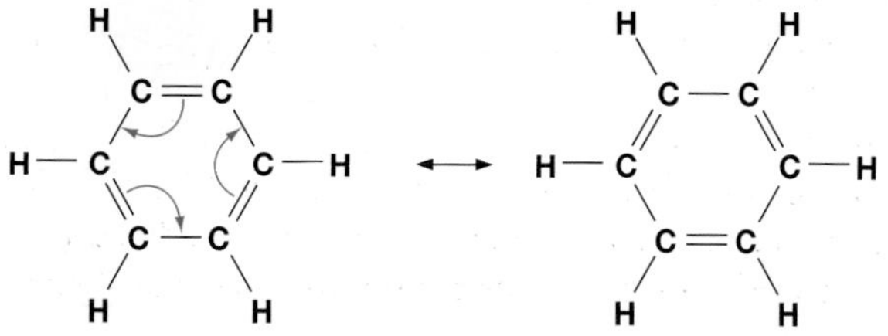

Figure 14.52 Major resonance structures for benzene

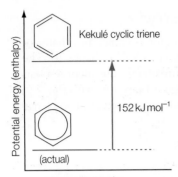

Figure 14.53 The relative energetic stabilities of benzene and cylcohexatriene (the Kekulé structure)

The two resonance structures are symmetrical and make equal contributions to the resonance hybrid where all the carbon–carbon bonds are equivalent, being intermediate between single and double carbon–carbon bonds, both in strength and length (0.139 nm and 518 kJ mol⁻¹).

The differences in energies between the resonance hybrid (the 'real' benzene) and the resonance structures (Kekulé structures) is known as the resonance energy (Figure 14.53). An approximate value can be calculated from enthalpies of hydrogenation or combustion data (Chapter 27).

Molecular orbital theory

An alternative but equivalent model for describing benzene (and other resonance stabilized structures) is molecular orbital theory. We have already seen how this theory can explain the formation of molecular structures such as methane, ethene and others. In 'localized' molecules like ethene, C_2H_4, two unhybridized p_z orbitals overlap to form a π molecular orbital in which a pair of electrons is shared between the nuclei of two carbon atoms. In molecular orbital theory resonance-stabilized structures are described in terms of 'delocalized' π orbitals where the π electron clouds extend over three or more atoms.

Delocalization in benzene

The benzene molecule has a distinctive structure in which the six carbon atoms (each with one hydrogen atom attached) are arranged in a planar hexagonal 'skeleton'. Each of the six carbon atoms is bonded to three other atoms (two carbons and one hydrogen) and so sp^2 hybridization can be used to describe the atomic orbitals of each carbon. Each of the three sp^2 hybridized carbon atoms forms three σ bonds with neighbouring atoms, forming the σ framework of the benzene ring (Figure 14.54).

Each carbon atom has an unhybridized $2p_z$ orbital (with lobes above and below the ring) which overlap with two carbon atoms on either side to give a delocalized cyclic π orbital above and below the plane of the benzene ring (Figure 14.55).

The six π electrons are therefore delocalized over the six carbon atom in a 'doughnut'-shaped π molecular orbital (a symmetrical torus – Figure 14.56). It is these π electrons which are responsible for the kinetic stability of benzene and its tendency to undergo substitution reactions, rather than addition reactions (Chapter 27).

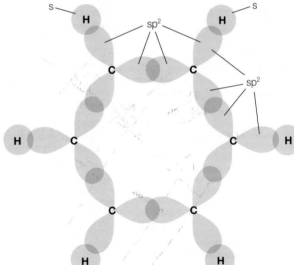

Figure 14.54 Sigma framework of the benzene molecule

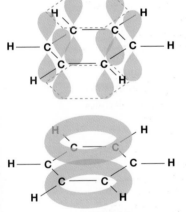

Figure 14.55 Overlapping of p_z orbitals to form a delocalized cyclic π orbital

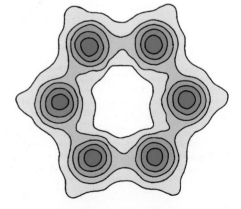

Figure 14.56 Pi electron density map for benzene (in the plane of the molecular π orbital)

Delocalization in ions

Another example of π delocalization occurs in the carboxylate ion, RCOO⁻ (e.g. the ethanoate ion). A lone pair of electrons in the $2p_z$ orbital of the oxygen atom overlaps and merges with the π orbital of the adjacent carbonyl group ($>C=O$). This is an alternative but equivalent description to the resonance model described earlier. Four π electrons (two from the carbonyl group and two from the lone pair) are delocalized over three atoms. The resulting molecular orbital is known as a *three-centre* delocalized π orbital (Figure 14.57).

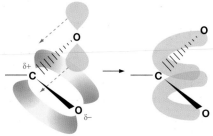

Figure 14.57 Delocalization in a carboxylate ion

All molecules and ions in which the bonding can be represented by a series of resonance structures can also be described in terms of delocalized π orbitals. For example, in the carbonate ion, each of the lone pairs in $2p_z$ orbitals can overlap with the π orbital of the $>C=O$ group, creating a four-centre delocalized π orbital (Figure 14.58).

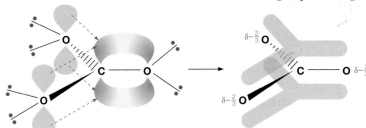

If there is to be delocalization of π electrons across a series of σ-bonded atoms, then the p_z orbitals must be parallel for maximum overlap. Hence, the ion or molecule must be *flat* (planar). The system of overlapping p orbitals corresponds to the extent of resonance or delocalization within the molecule or ion.

Figure 14.58 Delocalization in the carbonate ion

TOK Link

During the middle of the 19th century an elementary understanding of bonds between atoms began to develop. The term bond, the concept of valency (combining power) and the use of structural formulas were introduced by the English chemist Edward Frankland in 1852. In 1857, a German chemist, August Kekulé (Figure 14.59), proposed that carbon was tetravalent and formed four covalent bonds.

The chemists at that time grouped organic molecules into two large groups: unreactive saturated compounds, where each carbon atom is bonded to four other atoms, and relatively reactive unsaturated compounds, where there are one or more double and/or triple carbon–carbon bonds. All of the structures they proposed were linear or branched chains.

In 1860 a number of chemists applied their simple bonding theories to benzene and proposed a variety of linear structures, for example $CH_2=CH–C\equiv C–CH=CH_2$. However, these linear unsaturated structures were incompatible with the unreactivity of benzene and its general property of forming only one isomer of its mono-substitution products. In 1869 Kekulé proposed the familiar hexagonal ring of carbon atoms that are bonded to each other by alternating single and double bonds, with one hydrogen atom attached to each carbon atom.

Kekulé claimed that the structure he proposed for benzene came to him in a dream, as described below in his own words:

> I was sitting writing at my textbook, but the work did not progress; my thoughts were elsewhere. I turned my chair to the fire [after having worked on the problem for some time] and dozed. Again the atoms were gamboling before my eyes. This time the smaller groups kept modestly to the background. My mental eye, rendered more acute by repeated vision of this kind, could not distinguish larger structures, of manifold conformation; long rows, sometimes more closely fitted together; all twining and twisting in snakelike motion. But look! What was that? One of the snakes had seized hold of its own tail, and the form whirled mockingly before my eyes. As if by a flash of lighting I awoke ... Let us learn to dream, gentlemen.

The ancient image of a snake eating its own tail is an archetypal symbol of 'life living on life', and of wholeness, and it has appeared in art and dreams all around the world. However, his early training in architecture may have also played a critical role in helping him conceive his structural theories about chemical bonding. Kekulé may have therefore brought outside abilities to the field of chemistry. His architect's spatial and structural sense may have given him another approach to the problem of benzene, which his contemporaries did not possess. In addition, he was also able to place a problem in his subconscious mind and turn his dreams loose on it.

When Kekulé published his suggestion that benzene had a cyclic structure with an alternating series of double and bonds he did not mention that he had been inspired by a dream. This was presumably to avoid ridicule from his scientific peers, since his model was not derived from experimental observations. Kekulé only made public reference to his 'dream' in 1890 during a chemical conference held in Berlin, when he was an established figure in the German chemical community. However, a number of chemists have argued that Kekulé invented the dream as a cover up for plagiarizing, or at least being strongly influenced by, the work of an Austrian contemporary called Loschmidt (Chapter 1).

Figure 14.59 August Kekulé

- The shapes of molecules are determined by the number of electron pairs (bonding and non-bonding) in the outer shell of the central atom. These pairs arrange themselves in space so as to minimize electrostatic repulsion between the pairs.
- Repulsion decreases in the order: lone pair/lone pair > lone pair/bonding pair > bonding pair/bonding pair. This is due to the fact that lone pairs are located further from the nucleus.
- In order to minimize electrostatic repulsion, whenever alternative structures are possible, the one that is adopted is either that in which the lone pairs are as far apart as possible.
- Covalent bonds are formed when atomic orbitals overlap and merge to form molecular orbitals.
- Each atomic orbital forming a covalent bond normally contains an unpaired electron, so the resulting molecular orbital contains two electrons (in a spin pair). (The exception is the formation of a co-ordinate bond (dative covalent bond)).
- Sigma bonds are formed when two atomic orbitals overlap in an axial manner. Sigma bonds are single bonds and undergo free rotation (unless part of a cyclic system).
- In σ bonds the electron density is located around the nuclei and along the internuclear axis.
- Sigma bond formation may involve the following combinations of atomic orbitals: s–s, s and hybridized orbitals (sp, sp^2 and sp^3).
- Pi bond formation involves the side-on overlap of parallel p orbitals.
- Pi bonds are weaker than σ bonds (for the same chemical element).
- Pi bonds have two regions of high electron density located above and below the internuclear axis.
- A double bond has one σ bond and one π bond. A triple bond consists of a σ bond and two π bonds with the π bonds in different planes.
- Hybridization is a model that proposes atoms must mix or hybridize their s and p orbitals so that they match the observed bond angles.
- Hybridized atomic orbitals are formed by the combination and rearrangement of orbitals from the same atom. The hybridized orbitals are all of equal energy and electron density, and the number of hybridized orbitals is equal to the number of pure atomic orbitals that combine.
- sp^3 hybridization results from the combination of the s orbital and three p orbitals in the same shell. sp^2 hybridization uses the s orbital and two of the p orbitals from the same shell to form three hybrid orbitals – there is a single unhybridized p orbital. sp hybridization uses the s orbital and one of the p orbitals from the same shell to form two hybrid orbitals – there are two unhybridized p orbitals.
- In sp hybridization, the two hybrid orbitals lie in a straight line; in sp^2 hybridization, the three hybrid orbitals are directed toward the corners of a triangle; in sp^3 hybridization, the four hybrid orbitals are directed toward the corners of a tetrahedron.
- Summary of hybrid orbitals, centres of negative charge and arrangement of centres of negative charge:
  – sp, two centres of negative charge and linear
  – sp^2, three centres of negative charge and trigonal planar
  – sp^3, four centres of negative charge and tetrahedral.
- Resonance structures exist when the bonding in an ion or molecule can be represented by two or more different Lewis structures (electron dot diagrams). Such structures arise from different arrangements of π electrons.
- Resonance is a model that was introduced to attempt to describe a real molecule or ion in which electrons are not fully localized in terms of Lewis structures.
- Resonance leads to charge being delocalized over an ion or molecule and leads to increased energetic stability.
- Delocalized molecular orbitals, in which electrons are free to move around a whole molecule or group of atoms, are formed by electrons in p orbitals of adjacent atoms. Delocalized molecular orbitals are an alternative to resonance structures in explaining observed molecular properties.

Examination questions – a selection

Paper 1 IB questions and IB style questions

Q1 What is the hybridization state of carbon in ethyne (C_2H_2), graphite and diamond?
  A sp, sp², sp³   C sp³, sp², sp
  B sp, sp³, sp²   D sp, sp³, sp³

Q2 What types of orbitals are used by the hydrogen atoms in the formation of the hydrogen molecule, H_2?
  A s and sp   C s and p
  B s and sp²   D s and s

Q3 Which of the following species does not contain an sp³ hybridized carbon atom?
  A CH_3OH C $CH_3(CH_2)_2CH_3$
  B CH_2CHF D diamond

Q4 Which of the following species does *not* contain an sp³ hybridized oxygen atom?
  A H_2O_2 B H_3O^+ C H_2O D CH_3CHO

Q5 Which of the following molecules does *not* have a π bond?
  A CO_2 B CO   C H_2O_2 D SO_3

Q6 Which type of bond is formed by the sideways or lateral overlap of p orbitals?
  A π bond   C ionic bond
  B dative bond   D σ bond

Q7 What is the number of σ and π bonds in $(NC)_2C=C(CN)_2$?
  A 5, 4   B 6, 6   C 9, 4   D 9, 9

Q8 Which one of the following species is octahedral?
  A SF_6 B PF_5 C BF_4^- D BO_3^{3-}

Q9 Which one of the following types of hybridization leads to three-dimensional geometry of bonds around a carbon atom?
  A sp²   B sp³   C sp   D none of these

Q10 Which one of the following pairs of molecules has identical shapes for both species?
  A CCl_4, SF_4 C BCl_3, PF_3
  B BeF_2, CO_2 D PF_5, IF_5

Q11 Which one of the following contains the largest number of lone pairs on the central atom?
  A ClO_3^- B XeF_4 C I_3^- D SF_4

Q12 The length of a carbon–carbon single bond is 0.154 nm and the length of a carbon–carbon double bond is 0.134 nm. What is the likely carbon–carbon bond length within the benzene molecule, C_6H_6?
  A 0.164 nm
  B 0.124 nm
  C 0.139 nm
  D 0.154 nm for three bonds and 0.134 nm for the other three bonds

Q13 Sulfur hexafluoride is non-polar (zero dipole moment) while ammonia is polar (non-zero dipole moment). How can these observations be explained?
  A Ammonia is able to hydrogen bond in the liquid and solid states.
  B Nitrogen is more electronegative than sulfur.
  C Fluorine is the most electronegative element.
  D Sulfur hexafluoride is octahedral; ammonia is pyramidal with a lone pair.

Q14 Which one of the following chemical species is expected to show resonance or delocalization?
  A NO_3^- B HNO_3 C N_2F_4 D NH_4^+

Q15 Which one of the following statements is incorrect about the sulfur hexafluoride molecule?
  A The oxidation number of the sulfur in the molecule is the same as the number of its electrons it uses in the covalent bonding.
  B All the S–F bonds are equivalent.
  C The molecule is octahedral.
  D The sulfur atom has the electronic structure of argon.

Q16 Which statements correctly describe the NO_2^- ion ?
  I It can be represented by resonance structures.
  II It has two lone pairs of electrons on the N atom.
  III The N atom is sp² hybridized.
  A I and II only   C II and III only
  B I and III only   D I, II and III
  Higher Level Paper 1, Nov 05, Q12

Q17 What is the hybridization of the carbon atoms in 1,3-butadiene, $CH_2CHCHCH_2$?
  A sp³   B sp²   C sp   D s²p

Q18 Which one of the following statements concerning resonance is not true?

A The contributing resonance structures differ only in the arrangement of the π electrons.

B The resonance hybrid is intermediate between the contributing resonance structures.

C Resonance describes the chemical equilibrium between two or more Lewis structures.

D A single Lewis structure does not provide an adequate representation of bonding.

Q19 In which one of the following does π bonding not occur?

A carbon monoxide, CO

B hydrazine, N_2H_4

C methylbenzene, $C_6H_5CH_3$

D ozone (trioxygen), O_3

Q20 When the carbon–carbon bonds in the compounds C_2F_2, C_2F_4 and C_2F_6 are arranged in order of increasing length, the correct order is:

A $C_2F_6 < C_2F_4 < C_2F_2$

B $C_2F_4 < C_2F_2 < C_2F_6$

C $C_2F_4 < C_2F_6 < C_2F_2$

D $C_2F_2 < C_2F_4 < C_2F_6$

Q21 Difluoroethyne molecules, C_2F_2, can be prepared at low temperatures. What is the correct description of the bonding?

A two σ bonds and one π bond

B two σ bonds and three π bonds

C three σ bonds and two π bonds

D six σ bonds

Q22 The hybridization of the carbon atom in the carbonate ion, CO_3^{2-}, is best described as:

A sp   C sp^2

B sp^3 D unhybridized

Q23 In the Lewis structure for the molecule ClF_3, the number of lone pairs around the central chlorine atom is:

A 4   B 3   C 1   D 2

Q24 For which one of the following species would one usually draw resonance structures?

A O_2 C CH_3COOH

B $(CH_3)_2CO$ D $CH_3CH_2COO^-$

Q25 In the Lewis structure for the ion, ClF_2^+, what is the number of lone pairs around the central atom?

A 3   B 2   C 1   D 0

Paper 2 IB questions and IB style questions

Q1 a For each of the molecules C_2H_2, C_2Cl_4 and SF_4, draw their Lewis (electron dot) structures, and use the valence shell electron pair repulsion (VSEPR) theory to predict their shape and bond angles. [10]

b State the type of hybridization in C_2H_2 and C_2Cl_4. [2]

c Draw two resonance structures for each of the ethanoate ion (CH_3COO^-) and the benzene molecule. [4]

Higher Level Paper 2, May 01, Q8

Q2 a Explain what is meant by a σ bond and a π bond. Describe a double and a triple bond in terms of σ and π bonds. [4]

b Define the term *delocalization*. [2]

Higher Level Paper 2, Nov 01, Q6

Q3 Phosphorus(III) bromide is a colourless liquid. Phosphorus(V) bromide is a yellow solid which sublimes when heated.

The solid is ionic and consists of $[PBr_4]^+$ and Br^- ions, but in the gas phase PBr_5 molecules are present.

Phosphorus(V) bromide is hydrolysed by water producing hydrogen bromide gas:

$$PBr_5(s) + 4H_2O(l) \rightarrow H_3PO_4(aq) + 5HBr(aq)$$

a Write Lewis electron dot structures for PBr_3 and PBr_5 and sketch their molecular shapes. [4]

b State the bond angles present in PBr_5 and account for the bond angle in PBr_3 being approximately 107°. [4]

c Deduce the shape of the $[PBr_4]^+$ ion. [1]

d Give the hybridization of the phosphorus in phosphorus(III) bromide and the oxygen in water. [2]

15 Energetics

STARTING POINTS
- Thermochemical equations are balanced equations which include enthalpy data.
- Hess's law states that the enthalpy change for any chemical reaction is independent of the route or path taken, provided that the initial and final conditions are identical.
- The standard enthalpy change of reaction is the enthalpy change which takes place when the reactants in the balanced chemical equation react together under standard thermodynamic conditions to give the products. It is usually measured per mole of reactant or product, depending on the change taking place.
- The standard enthalpy change of combustion is the enthalpy change which occurs when one mole of a substance is completely burned in excess oxygen, under standard conditions.
- The standard enthalpy change of solution is the enthalpy change which takes place when one mole of a solute dissolves to form an infinitely dilute solution.
- The bond enthalpy is the enthalpy change which occurs when one mole of a particular covalent bond is broken in the gaseous state.
- The bond enthalpy (bond energy) is the average value of several individual bond enthalpies.
- The ionization energy is the enthalpy change when a mole of electrons is removed from a mole of gaseous ions or atoms, under standard conditions.
- The electron affinity is the enthalpy change which takes place when one mole of gaseous atoms or ions gains one mole of electrons, under standard conditions.
- The standard enthalpy change of atomization is the enthalpy change which takes place when one mole of gaseous atoms is formed from the element in its standard state under standard conditions.
- Ionic bonding arises from the attraction between oppositely charged ions in a lattice.
- The lattice enthalpy is a measure of the strength of bonding in an ionic lattice, and is affected by the relative sizes and charges of the cations and anions. It can be calculated experimentally or theoretically.
- Entropy is a measure of the degree of disorder and is involved in determining whether reactions or changes occur at a specific temperature.

15.1 Standard enthalpy changes of reaction

Enthalpy change of formation

15.1.1 Define and apply the terms standard state, standard enthalpy change of formation ($\Delta H_f^{\ominus}$) and standard enthalpy change of combustion ($\Delta H_c^{\ominus}$).

The **enthalpy change of formation**, $\Delta H_f^{\ominus}$, of a substance is the heat change (at constant pressure) on production of one mole of the pure substance from its elements in their standard states under standard thermodynamic conditions (298 K and 1 atm pressure).

The **standard state** is *generally* the most thermodynamically stable form of the pure element that exists under standard thermodynamic conditions. For carbon it is graphite and for phosphorus it is white phosphorus, $P_4(s)$. (However, red phosphorus is more stable than white phosphorus.)

The enthalpy change of formation of silver bromide, AgBr, is the enthalpy change for the reaction:

$$Ag(s) + \tfrac{1}{2}Br_2(l) \rightarrow AgBr(s) \qquad \Delta H_f^{\ominus} = -99.5 \, \text{kJ mol}^{-1}$$

The following balanced equations do *not* represent enthalpy changes of formation:

$$2Ag(s) + Br_2(l) \rightarrow 2AgBr(s) \qquad \text{two moles of silver bromide are formed}$$

$$Ag(s) + \tfrac{1}{2}Br_2(g) \rightarrow AgBr(s) \qquad \text{bromine is } not \text{ in its standard state}$$

Enthalpy changes of formation are often difficult to measure in practice due to competing side reactions and slow rates of reaction. For example, methane and potassium manganate(VII) *cannot* be prepared from their elements via the following thermochemical equations:

$$C(s) + 2H_2(g) \rightarrow CH_4(g) \qquad\qquad \Delta H_f^\ominus = -75\,kJ\,mol^{-1}$$

$$K(s) + Mn(s) + 2O_2(g) \rightarrow KMnO_4(s) \qquad\qquad \Delta H_f^\ominus = -813\,kJ\,mol^{-1}$$

The IB *Chemistry data booklet* tabulates enthalpies of formation for selected organic compounds on pages 12 and 13. Note that enthalpy changes of formation for elements (in their standard states) are zero since the thermochemical equation representing the formation of an element is a *null* reaction: no reaction is involved in their formation.

For example, the standard enthalpy change of formation of oxygen is represented by:

$$O_2(g) \rightarrow O_2(g) \qquad\qquad \Delta H_f^\ominus = 0\,kJ\,mol^{-1}$$

However, the enthalpy changes of formation for ozone ($O_3(g)$) and diamond (C(s, diamond) are *not* zero since these are *not* the standard states of the elements oxygen and carbon.

A number of enthalpies of formation can be directly determined, for example the enthalpy of formation of water:

$$H_2(g) + \tfrac{1}{2}O_2(g) \rightarrow H_2O(l) \qquad\qquad \Delta H_f^\ominus = -286\,kJ\,mol^{-1}$$

This enthalpy change is also equivalent to the enthalpy of combustion of hydrogen. The **enthalpy of combustion** is the enthalpy change (at constant pressure) when one mole of a pure substance undergoes complete combustion under standard thermodynamic conditions.

For example, the standard enthalpies of combustion of hydrogen, methane and ethanol are represented by:

$$H_2(g) + \tfrac{1}{2}O_2(g) \rightarrow H_2O(l) \qquad\qquad \Delta H_c^\ominus = -286\,kJ\,mol^{-1}$$

$$CH_4(g) + 2O_2(g) \rightarrow CO_2(g) + 2H_2O(l) \qquad\qquad \Delta H_c^\ominus = -890\,kJ\,mol^{-1}$$

$$CH_3CH_2OH(l) + 3O_2(g) \rightarrow 2CO_2(g) + 3H_2O(l) \qquad \Delta H_c^\ominus = -1370\,kJ\,mol^{-1}$$

Enthalpies of formation are usually calculated indirectly from other enthalpy changes of reaction including bond enthalpies. Enthalpy changes of formation are commonly used to calculate enthalpy changes of reaction, using Hess's law (Chapter 5).

Enthalpy changes of formation are usually negative, that is, the corresponding reactions are exothermic. However, some compounds have positive enthalpies of formation, for example benzene and nitrogen monoxide.

$$6C(s) + 3H_2(g) \rightarrow C_6H_6(l) \qquad\qquad \Delta H_f^\ominus = +49\,kJ\,mol^{-1}$$

$$\tfrac{1}{2}N_2(g) + \tfrac{1}{2}O_2(g) \rightarrow NO(g) \qquad\qquad \Delta H_f^\ominus = +90\,kJ\,mol^{-1}$$

These compounds are energetically unstable relative to their elements. However, the Gibbs free energy change of formation, $\Delta G_f^\ominus$, is the criterion that determines the *thermodynamic* stability of a compound relative to its elements (see Section 15.4).

Both benzene and nitrogen monoxide are *kinetically* stable – there is a large activation energy barrier to decomposition at room temperature. However, nitrogen monoxide does undergo significant decomposition in the presence of a platinum catalyst, as the catalyst lowers the activation energy barrier to the reaction (Chapter 6).

The normal 'rules' for manipulating enthalpy changes apply to enthalpies of formation (Chapter 5). If a thermochemical equation is reversed, then the sign of the enthalpy change is reversed. If the balanced thermochemical equation is multiplied (or divided) by a constant, the enthalpy change is multiplied (or divided) by the same constant.

Enthalpy changes of reaction from enthalpy changes of formation

15.1.2 Determine the enthalpy change of a reaction using standard enthalpy changes of formation and combustion.

The enthalpy change of any reaction can be determined by calculation, from the enthalpy changes of formation of all the substances in the chemical equation, using Hess's law.

In words:

$$\text{enthalpy change of a reaction} = \text{sum of enthalpies of formation of products} - \text{sum of enthalpies of formation of reactants}$$

In symbols:

$$\Delta H = \Sigma \Delta H_f^{\ominus} \text{[products]} - \Sigma \Delta H_f^{\ominus} \text{[reactants]}$$

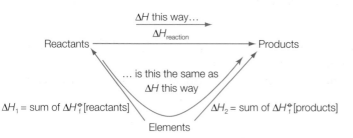

As we saw in Chapter 5, the calculation can be carried out using an energy level diagram, a Hess's law cycle (Figure 15.1) or with algebra.

Figure 15.1 Hess's law cycle for calculating the standard enthalpy of a reaction from standard enthalpies of formation

Worked example

Calculate the enthalpy change of the following reaction:

$$3CuO(s) + 2Al(s) \rightarrow 3Cu(s) + Al_2O_3(s)$$

$$\Delta H_f^{\ominus} \text{[CuO]} = -155 \, \text{kJ mol}^{-1}$$

$$\Delta H_f^{\ominus} \text{[Al}_2\text{O}_3\text{]} = -1669 \, \text{kJ mol}^{-1}$$

This is a redox reaction (Chapter 9) and involves a more reactive metal displacing a less reactive metal from its oxide.

Figure 15.2 shows there are two routes for forming the products of the reaction from their elements. Route 1 is the direct route and route 2 is the indirect route where the reactants are formed from their elements and then the reactants are converted to products.

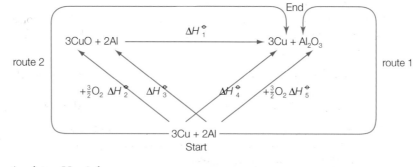

Figure 15.2 Hess's law energy cycle for the reaction between copper(II) oxide and aluminium

Applying Hess's law:

enthalpy change in route 1 = enthalpy change in route 2

Hence:

$$\Delta H_2^{\ominus} + \Delta H_3^{\ominus} + \Delta H_1^{\ominus} = \Delta H_4^{\ominus} + \Delta H_5^{\ominus}$$

$$\Delta H_1^{\ominus} = \Delta H_4^{\ominus} + \Delta H_5^{\ominus} - \Delta H_2^{\ominus} - \Delta H_3^{\ominus}$$

$$\Delta H_1^{\ominus} = (0 \, \text{kJ mol}^{-1}) + (-1669 \, \text{kJ mol}^{-1}) - 3 \times (-155 \, \text{kJ mol}^{-1}) - (0 \, \text{kJ mol}^{-1})$$

$$= -1204 \, \text{kJ mol}^{-1}$$

(i.e. $\Delta H = \Sigma \Delta H_f^{\ominus}$ [products] $- \Sigma \Delta H_f^{\ominus}$ [reactants]

Enthalpy changes of reaction from enthalpy changes of combustion

An enthalpy change can also be calculated from enthalpy changes of combustion. However, this can only be done for reactions in which the substances on both sides of the equation can be burnt in oxygen (Figure 15.3). This method is widely used for organic compounds.

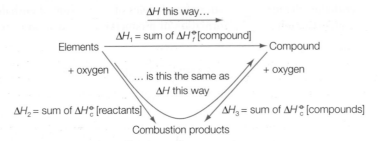

Figure 15.3 Hess's law cycle for calculating the standard enthalpies of formation from standard enthalpies of combustion

Worked example

Calculate the enthalpy change of reaction for the hydrogenation of propene to form propane.

$$CH_3{-}CH{=}CH_2(g) + H_2(g) \rightarrow CH_3{-}CH_2{-}CH_3(g)$$

$$\Delta H_c^{\ominus} \, [C_3H_6(g)] = -2509 \, kJ \, mol^{-1}$$

$$\Delta H_c^{\ominus} \, [H_2(g)] = -286 \, kJ \, mol^{-1}$$

$$\Delta H_c^{\ominus} \, [C_3H_8(g)] = -2220 \, kJ \, mol^{-1}$$

The diagram in Figure 15.4 shows two routes for the combustion of propene and hydrogen into carbon dioxide and water. In route 1 the reactants are burnt directly in oxygen. In route 2 the reactants are first converted to propane, which is then burnt completely.

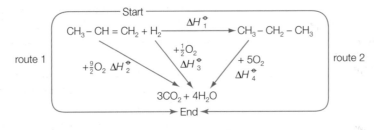

Figure 15.4 Hess's law energy cycle for calculating the enthalpy of hydrogenation of propene from the enthalpies of combustion of propene, hydrogen and propane

Applying Hess's law:

enthalpy change in route 1 = enthalpy change in route 2

$$\Delta H_2 + \Delta H_3 = \Delta H_1 + \Delta H_4$$
$$\Delta H_1 = \Delta H_2 + \Delta H_3 - \Delta H_4$$
$$\Delta H_1 = (-2509 \, kJ \, mol^{-1}) + (-286 \, kJ \, mol^{-1}) - (-2220 \, kJ \, mol^{-1})$$
$$= -575 \, kJ \, mol^{-1}$$

Enthalpy of atomization

The **enthalpy change of atomization** of an element is the heat change (at constant pressure) when one mole of separate gaseous atoms of the element is formed from the element (in its standard state) under standard conditions.

The following thermochemical equations each describe the enthalpy of atomization of an element. In each case the enthalpy change of the reaction is equivalent to the enthalpy change of atomization.

$$Fe(s) \rightarrow Fe(g) \qquad \Delta H_{at}^{\ominus} = +418 \, kJ \, mol^{-1}$$
$$\tfrac{1}{2}Br_2(l) \rightarrow Br(g) \qquad \Delta H_{at}^{\ominus} = +96.5 \, kJ \, mol^{-1}$$

Note that the enthalpy change for the atomization of a halogen or other diatomic gaseous molecule is equivalent to *half* the bond enthalpy (Chapter 5).

All the enthalpy changes of atomization of the noble gases are zero. This is because the elements are already in the form of separate gaseous atoms under standard conditions.

Enthalpy changes of atomization are always positive, because energy must be absorbed to pull the atoms apart and overcome the chemical bonds. Enthalpy changes of atomization are usually found indirectly by calculation from other enthalpy changes, using Hess's law. Enthalpy changes of atomization are used in Born–Haber cycle calculations (see Section 15.2).

■ Extension: Enthalpy of atomization and bonding

The enthalpy of atomization is an indication of the strength of bonding in a substance. For a solid to break up into atoms, it may first melt (accompanied by the standard enthalpy change of fusion – see Chapter 5), then evaporate (accompanied by the standard enthalpy change of vaporization – see Chapter 16) and finally, in the gas phase, any remaining bonds break to give individual atoms.

The size of the enthalpy changes associated with the different processes depends on the type of substance. For a metal, the enthalpy changes associated with melting and evaporation are the most important, whilst for a covalent substance, the bond enthalpies are most important.

The concept of enthalpy of atomization can also be applied to compounds. For example, the following equation describes the enthalpy of atomization of methane:

$$CH_4(g) \rightarrow C(g) + 4H(g) \qquad \Delta H^\ominus = \Delta H^\ominus_{at}[CH_4(g)]$$

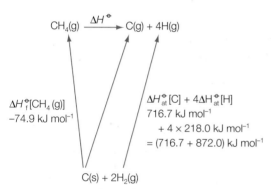

Figure 15.5 Hess's law cycle to calculate the average bond enthalpy in methane

Dividing the standard enthalpy of atomization between four bonds gives an average value for the C–H bond enthalpy (Chapter 5). The sum of all the bond enthalpies for a compound is the standard enthalpy of atomization of that compound in the gaseous state. The standard enthalpy of formation of a compound is composed of two terms: the bond enthalpies and the standard enthalpies of atomization of all the atoms, which are themselves derived from the bond enthalpy of the element.

The Hess's law cycle in Figure 15.5 can be used to calculate the average bond enthalpy in methane using the enthalpy of formation of methane and the enthalpies of atomization of carbon and hydrogen.

$$\Delta H^\ominus_f[CH_4(g)] + \Delta H^\ominus = \Delta H^\ominus_{at}[C] + 4\Delta H^\ominus_{at}[H]$$
$$-74.9\,kJ\,mol^{-1} + \Delta H^\ominus = 716.7\,kJ\,mol^{-1} + 872.0\,kJ\,mol^{-1}$$
$$\Delta H^\ominus = 1588.7\,kJ\,mol^{-1} - (-74.9\,kJ\,mol^{-1}) = +1663.6\,kJ\,mol^{-1}$$

Hence, the average C–H bond enthalpy for methane is $416\,kJ\,mol^{-1}$.

If similar calculations are performed on molecular liquids, for example tetrachloromethane, $CCl_4(l)$, then the cycle must show the bonds breaking in the gaseous state, rather than the liquid (Figure 15.6). It is essential to include an enthalpy of vaporization term in the bond energy calculation (Chapter 17).

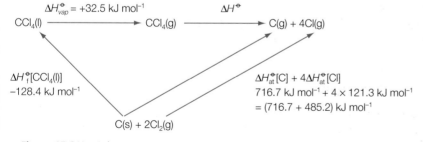

Figure 15.6 Hess's law cycle to calculate the average bond enthalpy in tetrachloromethane

$$\Delta H^\ominus_f[CCl_4(l)] + \Delta H^\ominus_{vap} + \Delta H^\ominus = \Delta H^\ominus_{at}[C] + 4\Delta H^\ominus_{at}[Cl]$$

$$-128.4\,kJ\,mol^{-1} + 32.5\,kJ\,mol^{-1} + \Delta H^\ominus = 716.7\,kJ\,mol^{-1} + 485.2\,kJ\,mol^{-1}$$

$$\Delta H^\ominus = 1201.9\,kJ\,mol^{-1} - (-95.9\,kJ\,mol^{-1}) = +1297.8\,kJ\,mol^{-1}$$

Hence, the average C–Cl bond enthalpy for tetrachloromethane is a quarter of this value, i.e. $324\,kJ\,mol^{-1}$.

Enthalpies of physical change

The standard enthalpy change that accompanies a change in physical state is called the standard enthalpy of transition. The standard enthalpies of fusion and vaporization (Chapter 17) are two examples of enthalpies of transition.

The standard enthalpy change of vaporization is the enthalpy change which occurs (at constant pressure) when one mole of a pure liquid is completely vaporized under standard conditions. The standard enthalpy change of fusion is the enthalpy change which occurs (at constant pressure) when one mole of a pure solid is completely melted under standard conditions.

The two thermochemical equations represent the enthalpies of fusion and vaporization of benzene:

$$C_6H_6(s) \rightarrow C_6H_6(l) \qquad \Delta H^\ominus_{fus} = 11\,kJ\,mol^{-1}$$

$$C_6H_6(l) \rightarrow C_6H_6(g) \qquad \Delta H^\ominus_{vap} = 353\,kJ\,mol^{-1}$$

Hess's law (see Section 5.3) can be applied to enthalpies of transition. For example, the conversion of ice to water vapour can be regarded as occurring by sublimation:

$$H_2O(s) \rightarrow H_2O(g) \qquad \Delta H^\ominus_{sub}$$

or occurring in two steps, first melting (fusion) and then vaporization of the liquid (Figure 15.7):

$$H_2O(s) \rightarrow H_2O(l) \qquad \Delta H^\ominus_{fus}$$

$$H_2O(l) \rightarrow H_2O(g) \qquad \Delta H^\ominus_{vap}$$

Overall:

$$H_2O(s) \rightarrow H_2O(g) \qquad \Delta H^\ominus_{fus} + \Delta H^\ominus_{vap}$$

Figure 15.7 shows that, because all enthalpies of fusion are positive, the enthalpy of sublimation of a substance is greater than its enthalpy of vaporization. Another consequence of Hess's law is that the changes of a forward process and its reverse process differ in sign (Figure 15.8).

For example, the enthalpy of vaporization of water is $+44\,kJ\,mol^{-1}$ (at 298 K), but the enthalpy of condensation is $-44\,kJ\,mol^{-1}$ (at 298 K).

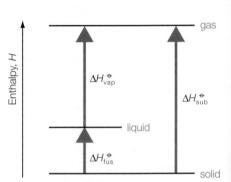

Figure 15.7 The relationship between the enthalpies of fusion, vaporization and sublimation

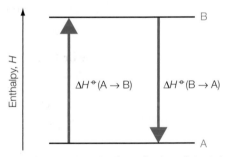

Figure 15.8 A simple application of Hess's law

Language of Chemistry

IUPAC notation is to attach the name of the transition to the symbol Δ. Hence $\Delta_{vap}H$, is the modern convention for the standard enthalpy change of vaporization. However, the older convention, ΔH_{vap} is still widely used by the IBO and other organizations. The new convention is more logical because the subscript identifies the type of change, not the physical observable. ∎

15.2 Born–Haber cycle

Lattice enthalpy

15.2.1 Define and **apply** the terms *lattice enthalpy* and *electron affinity*.

The lattice enthalpy ($\Delta H^\ominus_{lattice}$) of an ionic crystal is the heat energy absorbed (at constant pressure) when one mole of solid ionic compound is decomposed to form gaseous ions separated to an infinite distance from each other (under standard thermodynamic conditions). (This is the definition used in the IB *Chemistry data booklet*.)

Note that the sign of the lattice enthalpy must always be included in the thermochemical equation; the *reverse* of lattice enthalpy is the heat energy released (at constant pressure) when one mole of ionic solid is formed from gaseous ions (Figure 15.9).

For example, the lattice enthalpy of potassium chloride is the enthalpy change for the reaction:

$$KCl(s) \rightarrow K^+(g) + Cl^-(g) \qquad \Delta H^{\ominus}_{lattice} = +701 \, kJ \, mol^{-1}$$

The reaction is illustrated in Figure 15.10. Lattice energies are a measure of the stability of a crystal. The greater the value of the lattice energy, the more energetically stable the lattice. This results in a higher melting and boiling point. The size of the lattice energy has an effect on the solubility (if any) of an ionic salt. The size of the lattice energy is controlled by the charges on the ions, their ionic radii and the packing arrangement of the ions (type of lattice).

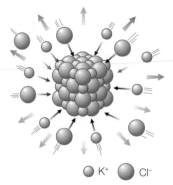

● K⁺   ● Cl⁻

Figure 15.9 The *reverse* of lattice enthalpy is the energy which would be released to the surroundings (short red arrows) if one mole of an ionic compound could form directly from infinitely free gaseous ions rushing together (black arrows) and forming a lattice

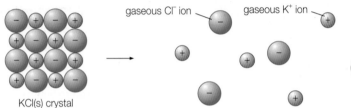

gaseous Cl⁻ ion   gaseous K⁺ ion

Figure 15.10 The lattice energy of potassium chloride, $\Delta H^{\ominus}_{lattice}$

KCl(s) crystal

Electron affinity

When an electron is acquired by an atom, energy is released, for example:

$$Cl(g) + e^- \rightarrow Cl^-(g) \qquad \Delta H^{\ominus}_{EA(1)} = -364 \, kJ \, mol^{-1}$$

The **first electron affinity**, $\Delta H^{\ominus}_{EA(1)}$, is the energy released when one mole of gaseous atoms accepts one mole of electrons to form singly charged negative ions.

The **second electron affinity**, $\Delta H^{\ominus}_{EA(2)}$, is the energy absorbed when one mole of gaseous ions with a single negative charge accept one mole of electrons, for example:

$$O^-(g) + e^- \rightarrow O^{2-}(g) \qquad \Delta H^{\ominus}_{EA(2)} = +844 \, kJ \, mol^{-1}$$

It is always endothermic because energy is required to overcome the mutual repulsion between the negatively charged O⁻ ion and the electron.

First electron affinities *generally* correlate with electronegativity (Chapter 3). The halogens show clear trends in electronegativity and electron affinity (with the exception of fluorine, which has an unexpectedly low electron affinity) (Table 15.1).

Halogen	Electronegativity (Pauling scale)	First electron affinity/kJ mol⁻¹
Fluorine	4.0	**−348**
Chlorine	3.0	−364
Bromine	2.8	−342
Iodine	2.5	−314

Table 15.1 Values of electronegativity and first electron affinity for the halogens

---

■ Extension: Explaining the reactivity of fluorine

Fluorine is much *more reactive* than chlorine (despite the *lower* electron affinity) because the energy released in other steps in its reactions more than makes up for the lower amount of energy released as electron affinity. Fluorine's first electron affinity is less exothermic than expected because the fluorine atom is relatively small and the additional electron experiences enhanced repulsion when it enters the second shell. The filled first shell has a very compact electron density.

Lattice enthalpies of ionic compounds

15.2.2 Explain how the relative sizes and the charges of ions affect the lattice enthalpies of different ionic compounds.

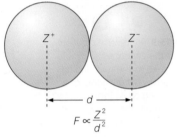

Figure 15.11 Ions with charges of Z^+ and Z^- separated by a distance of d

In an ionic compound, the lattice enthalpy depends on the attractive electrostatic forces operating between oppositely charged ions in the crystal. The larger the electrostatic forces of attraction, the larger the size of the lattice enthalpy.

According to Coulomb's law, the attractive force F operating between two adjacent oppositely charged ions in contact (Figure 15.11) can be expressed as follows:

$$F \propto \frac{\text{charge on positive ion} \times \text{charge on negative ion}}{d^2}$$

where d represents the distance between the nuclei of the two ions.

Ionic charge has a large effect on the size of the lattice enthalpy. For example, the lattice enthalpy of magnesium oxide, MgO, is about four times greater than the lattice enthalpy of sodium fluoride, NaF.

$$\Delta H^{\ominus}_{\text{lattice}}[\text{MgO(s)}] = +3889 \, \text{kJ mol}^{-1}$$

$$\Delta H^{\ominus}_{\text{lattice}}[\text{NaF(s)}] = +902 \, \text{kJ mol}^{-1}$$

This is largely due to the doubling of the charges of the ions. The sum of ionic radii and the lattice structures of magnesium oxide and sodium fluoride are similar.

The effect of ionic radius can be clearly seen in the sodium halides (Table 15.2). The smaller the distance between the centres of ions (sum of ionic radii), the larger the value of lattice enthalpy.

Sodium halide	Radius of halide ion/nm	Distance between centres of ions/nm	Lattice enthalpy/kJ mol⁻¹
Na⁺ F⁻	0.133	0.231	+1902
Na⁺ Cl⁻	0.181	0.276	+771
Na⁺ Br⁻	0.196	0.291	+733
Na⁺ I⁻	0.219	0.311	+684

Table 15.2 The effect of ionic radius on lattice enthalpy

■ **Extension:** ## Lattice type

The relative sizes of the cation and anion determine the type of lattice an ionic compound adopts. For example, although caesium and sodium are both in the same group of the periodic table, the chlorides crystallize with different types of lattice. Sodium chloride adopts the simple cubic structure (Chapter 4), whereas caesium chloride adopts the lattice shown in Figure 15.12. In caesium chloride, the caesium ions cannot get as close to the chloride ions as the smaller sodium ions. Eight caesium ions can pack around a chloride ion if they are positioned at the corners of a cube.

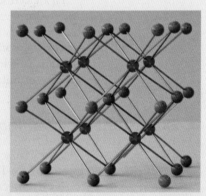

Figure 15.12 A model of the caesium chloride lattice: red balls represent caesium ions; green balls represent chloride ions

Using Born–Haber cycles to calculate enthalpy changes

15.2.3 Construct a Born–Haber cycle for group 1 and 2 oxides and chlorides and use it to calculate an enthalpy change.

Lattice enthalpies *cannot* be found directly from experiments (partly because ionic crystals form ion pairs when heated, not free gaseous ions). Therefore lattice enthalpies must be calculated *indirectly* from other known enthalpy changes of reaction using a **Born–Haber cycle**. This is an enthalpy level diagram derived from Hess's law used to follow the enthalpy changes which occur when an ionic compound is formed from its chemical elements and gaseous ions (Figure 15.13).

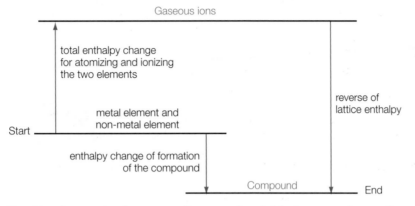

Figure 15.13 The main features of a generalized Born–Haber cycle

Consider the reaction between sodium metal and chlorine gas to form sodium chloride:

$$Na(s) + \tfrac{1}{2}Cl_2(g) \rightarrow NaCl(s) \qquad \Delta H_f^\ominus = -411\,kJ\,mol^{-1}$$

This reaction can be described by an equivalent pathway that involves a number of steps each with its own individual enthalpy change.

The atoms in the solid sodium must be converted into gaseous sodium atoms:

$$Na(s) \rightarrow Na(g) \qquad \Delta H_{at}^\ominus = +108\,kJ\,mol^{-1}$$

This is the enthalpy change of atomization of sodium.

The gaseous chlorine molecules are then dissociated into gaseous chlorine atoms:

$$\tfrac{1}{2}Cl_2(g) \rightarrow Cl(g) \qquad \Delta H_{at}^\ominus = +122\,kJ\,mol^{-1}$$

This is the enthalpy change of atomization of chlorine. It is equivalent to *half* the bond enthalpy of the chlorine molecule.

Once gaseous sodium and chlorine atoms are formed, electron transfer can take place. The sodium atom loses its outer electron and donates it to the chlorine atom, which forms the chloride ion.

$$Na(g) \rightarrow Na^+(g) + e^- \qquad \Delta H_{IE(1)}^\ominus = +494\,kJ\,mol^{-1}$$

$$Cl(g) + e^- \rightarrow Cl^-(g) \qquad \Delta H_{EA(1)}^\ominus = -364\,kJ\,mol^{-1}$$

These energy changes are the first ionization energy ($\Delta H_{IE(1)}^\ominus$) of sodium and the first electron affinity ($\Delta H_{EA(1)}^\ominus$) of chlorine.

The oppositely charged ions exert powerful attractive electrostatic forces and form an ionic lattice of solid sodium chloride.

$$Na^+(g) + Cl^-(g) \rightarrow NaCl(s) \qquad -\Delta H_{lattice}^\ominus = -771\,kJ\,mol^{-1}$$

or

$$NaCl(s) \rightarrow Na^+(g) + Cl^-(g) \qquad \Delta H_{lattice}^\ominus = +771\,kJ\,mol^{-1}$$

The reactions leading to the formation of sodium chloride from its elements are summarized in graphic form in Figure 15.14.

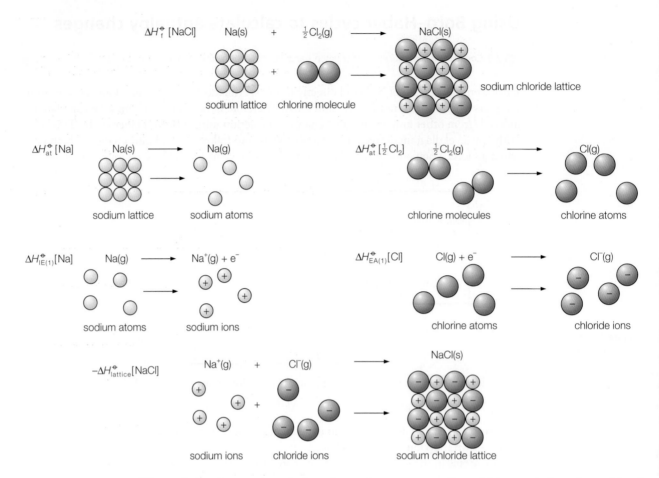

Figure 15.14 A summary of the steps occurring during the formation of sodium chloride from sodium and chlorine in their standard states

The enthalpy changes described can be used to construct a Born–Haber cycle for sodium chloride (Figure 15.15). The Born–Haber cycle can be used to calculate an individual enthalpy change if all the others are known. The unknown value is usually the lattice enthalpy.

The Born–Haber cycle clearly shows that the principal reason why stable ionic compounds form from their elements is because the reverse of the lattice enthalpy is very exothermic. Large amounts of energy are released during lattice formation which more than compensates for a number of endothermic steps.

Note that polyatomic ions, such as NH_4^+, NO_3^-, CO_3^{2-}, OH^- and SO_4^{2-}, cannot be used in Born–Haber cycle calculations as they do *not* have atomization energies, ionization energies or electron affinities. Lattice energies of these compounds have to be calculated theoretically.

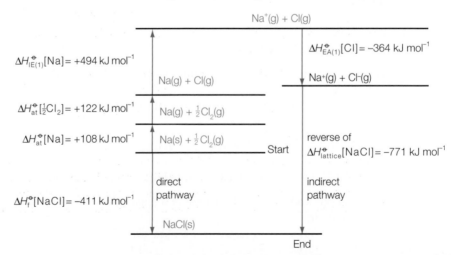

Figure 15.15 The Born–Haber cycle for sodium chloride

History of Chemistry

The Born–Haber thermochemical cycle is named after the two German physical chemists, Max Born and Fritz Haber (Chapter 7), who first used it in 1919. **Max Born** (1882–1970) was born in Poland and had a distinguished career as a mathematician in Germany before he was forced to leave in 1933 to escape the Nazi regime. He became a lecturer in Cambridge before being appointed Professor of Applied Mathematics in Edinburgh, where he assembled a distinguished research group mainly of European refugees. He received the Nobel Prize in Physics in 1954 for his work on quantum mechanics (Chapter 12).

Worked example

Use the Born–Haber cycle shown in Figure 15.16 to calculate the value of the lattice enthalpy for magnesium chloride.

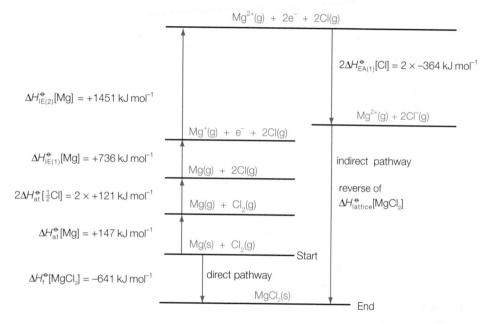

Figure 15.16 The Born–Haber cycle for magnesium chloride

Applying Hess's law:

indirect pathway = direct pathway

$$\Delta H_{at}^{\ominus}[Mg] + 2 \times \Delta H_{at}^{\ominus}[\tfrac{1}{2}Cl_2] + \Delta H_{IE(1)}^{\ominus}[Mg] + \Delta H_{IE(2)}^{\ominus}[Mg]$$
$$+ 2 \times \Delta H_{EA(1)}^{\ominus}[Cl] + -\Delta H_{lattice}^{\ominus}[MgCl_2] = \Delta H_f^{\ominus}[MgCl_2]$$

where subscripts IE(1) and IE(2) represent the first and second ionization energies and EA(1) represents the first electron affinity.

$$+147\,kJ\,mol^{-1} + (2 \times 121\,kJ\,mol^{-1}) + 736\,kJ\,mol^{-1} + 1451\,kJ\,mol^{-1}$$
$$+ (2 \times -364\,kJ\,mol^{-1}) + -\Delta H_{lattice}^{\ominus}[MgCl_2] = -641\,kJ\,mol^{-1}$$

$$-\Delta H_{lattice}^{\ominus}[MgCl_2] = -641\,kJ\,mol^{-1} - (1848\,kJ\,mol^{-1}) = -2489\,kJ\,mol^{-1}$$

$$\Delta H_{lattice}^{\ominus} = +2489\,kJ\,mol^{-1}$$

With ions possessing charges greater than one (+ or −), second ionization energies (or higher) are needed for the metal and second electron affinities (or higher) are needed for the non-metal. The need for more than one particular ion also multiplies the relevant enthalpy value by that number. This is illustrated in Figure 15.16 for magnesium chloride and in Figure 15.17 for calcium oxide.

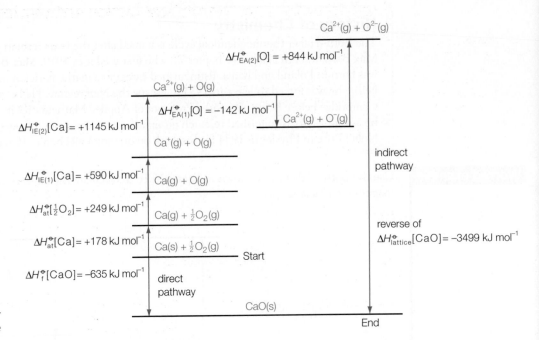

Figure 15.17 Born–Haber
cycle for calcium oxide

Large negative values of enthalpy of formation are typical of stable ionic compounds. This can be illustrated by examining the Born–Haber cycle for the hypothetical ionic compound, CaCl [Ca⁺ Cl⁻] (Figure 15.18). Using a theoretically calculated lattice enthalpy (see page 407) in the Born–Haber cycle, CaCl has an *estimated* enthalpy of formation of −69 kJ mol⁻¹. In contrast, CaCl₂ has an enthalpy of formation of −795 kJ mol⁻¹.

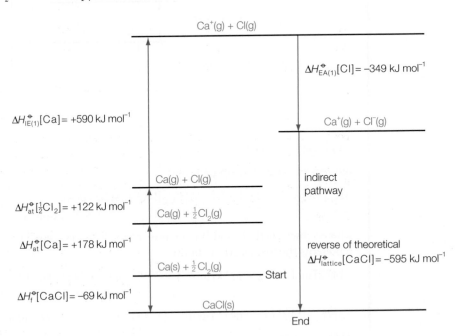

Figure 15.18 Born–Haber
cycle for CaCl

Experimental and theoretical lattice enthalpies

15.2.4 Discuss the difference between theoretical and experimental lattice enthalpy values of ionic compounds in terms of their covalent character.

The lattice enthalpy of an ionic compound can be found experimentally by means of a Born–Haber cycle (see page 400).

It is also possible to calculate the lattice enthalpy of an ionic compound from a theoretical model (a number of which exist). A simple model of an ionic bond is two point charges separated by an interionic distance. The potential energy between two ions changes as their interionic distance changes.

The model can be extended to a three-dimensional lattice (Figure 15.19) to give a theoretical value for the lattice enthalpy. Table 15.3 compares theoretical values of lattice enthalpies with experimentally determined values from a Born–Haber cycle calculation.

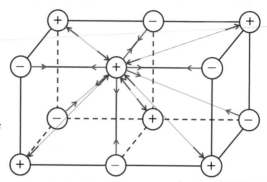

Figure 15.19 A selection of the many attractions and repulsions between ions which have to be taken into account when using the laws of electrostatics to calculate a theoretical value for the lattice enthalpy of an ionic crystal (with a simple cubic lattice)

Compound	Theoretical lattice enthalpy /kJ mol⁻¹	Experimental value (from a Born–Haber cycle calculation)/kJ mol⁻¹	Percentage difference
NaCl	766	771	0.06
NaBr	732	733	0.01
NaI	686	684	0.29
KCl	690	701	0.15
KBr	665	670	0.75
KI	632	629	0.48
AgCl	770	905	14.9
AgBr	758	890	14.8
AgI	736	876	16.0

Table 15.3 Theoretical and experimental values of lattice enthalpies

For the alkali metal halides, there is very *good agreement* between the theoretical values calculated from an ionic model using point charges and experimental values derived from a Born–Haber cycle calculation. A purely ionic model (Chapter 4) is therefore a good description of the bonding within the alkali metal halides. This is because there is a large difference in electronegativity and hence there is almost complete electron transfer and little electron density between ions. There is little or no covalent bonding.

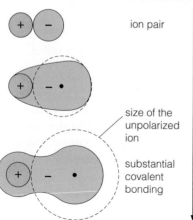

ion pair

size of the unpolarized ion

increasing polarization of the negative ion by the positive ion

substantial covalent bonding

However, for the silver halides there is *poor agreement* between the theoretical and experimental values. A purely ionic model is therefore not a good description of the bonding for the silver halides. The bonding in the silver halides is stronger than that predicted by a purely ionic model. The bonding is not purely ionic because of incomplete transfer of valence electrons between the atoms. As a result, there is electron density between the cation and anion. There is some degree of covalent bonding. The partial covalent bond strengthens the ionic bond (Figure 15.20).

Figure 15.20 Ionic bonding with increasing degrees of electron sharing because the positive ion has polarized the negative ion. Dotted circles show the electron cloud of the unpolarized ions

■ Extension: Dissolving ionic solids in water

Many ionic compounds, such as sodium chloride, dissolve well in water to form solutions. Others, such as silver chloride, are virtually insoluble. When an ionic substance dissolves the lattice of the ionic crystal needs to be broken up (Figure 15.21). Lattice enthalpies are always endothermic, so if the compound is to dissolve the energy needed to achieve this must be supplied from enthalpy changes that occur during the dissolving process.

Figure 15.21 The formation of a sodium chloride solution from a sodium chloride lattice

solid sodium chloride – a regular ionic lattice

○ Na⁺   ● Cl⁻

sodium chloride dissolved in water

○ Na⁺(aq)   ● Cl⁻(aq)

The ions in the crystal lattice become separated as they are hydrated by water molecules (Figure 15.22). Water molecules are polar (Chapter 4) and are attracted to both positive and negative ions. The hydration process involves the cations and anions being surrounded by a number of water molecules. Ion–dipole bonds are formed and hence the hydration process is an exothermic process.

The separation of the ions in the lattice is strongly endothermic and the hydration of the ions is an exothermic process. The enthalpy change of solution is the difference between these two enthalpy changes.

The **enthalpy change of solution**, $\Delta H^{\ominus}_{sol}$, is the enthalpy change when one mole of solute dissolves in an infinite volume of water. The value cannot be determined directly by experiment and must be found by a process of extrapolation. In practice, there comes a point when further dilution has no measurable effect on the value of the enthalpy change of solution, and this may be taken as infinite dilution.

The **enthalpy change of hydration**, $\Delta H^{\ominus}_{hyd}$, is the enthalpy change when a mole of gaseous ions becomes hydrated by a large excess of water and refers to the process:

$$A^{n+}(g) \rightarrow A^{n+}(aq)$$

where the concentration of A^{n+} in the aqueous solution approaches zero. Enthalpies of hydration for ions are always negative because strong ion–dipole bonds are formed when the gas-phase ion is surrounded by water. Small, highly charged ions have the most negative values. (As with lattice energies, it is possible to calculate theoretical values of enthalpy changes of hydration.)

The overall process of dissolving an ionic solid can be represented by a Hess's law cycle. Figure 15.23 shows the energy cycle for sodium chloride. Ionic substances that are soluble *generally* have a large negative value for the enthalpy change of solution. Conversely, insoluble ionic substances *generally* have a large positive value for the enthalpy of solution.

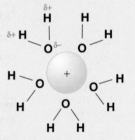

Figure 15.22 Hydrated anion and cation

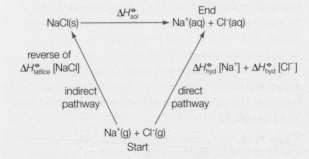

Figure 15.23 Hess's law energy cycle summarizing the dissolving process for sodium chloride

However, these are 'rules of thumb' and a number of exceptions occur. This is because free energy changes determine solubility and they include entropy (see Section 15.3). Figure 15.24 is an enthalpy level diagram that shows the processes which occur when silver chloride, a virtually insoluble salt, is placed in water. Figure 15.25 is an energy level diagram that shows the processes which occur when silver fluoride, a very soluble salt, is placed in water. Note the larger positive enthalpy of solution for silver chloride compared to silver fluoride.

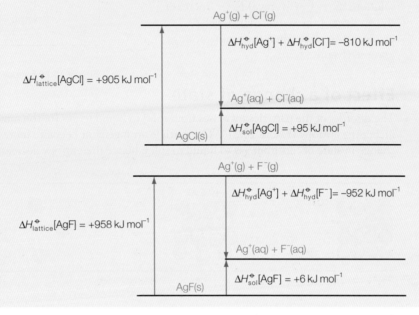

Figure 15.24 Enthalpy level diagram for the dissolving of silver chloride

Figure 15.25 Enthalpy level diagram for the dissolving of silver fluoride

15.3 Entropy

15.3.1 State and **explain** the factors that increase the entropy in a system.

Entropy can be regarded (in a crude and simplified way) as a measure of the disorder or dispersal of energy in a system. Entropy is given the symbol S. The disorder refers to the arrangement of particles (atoms, ions or molecules) and the kinetic energies of the particles in a system.

For example, comparing an ionic solid and a gas (maintained at the same temperature), the gas has a significantly greater entropy because its particles are moving rapidly in all directions. In an ionic solid the particles are in fixed positions within a lattice.

A collection of particles with a larger range of kinetic energies has a greater entropy than a sample of the same amount of particles which have a smaller range of kinetic energies. This corresponds to high temperature and low temperature.

Language of Chemistry

The term entropy was introduced in about 1865 by the German physicist Clausius, who derived it from the Greek word *entrope*, meaning 'transformation or turning.' The symbol S for entropy indicates that it is a state function. This is a function whose value depends solely upon the current state of the system, and is totally independent of how that state was reached. Enthalpy (Chapter 5) is another state function. ∎

■ Extension: The Third Law of Thermodynamics

Theoretically, at absolute zero (Chapter 1) all matter would be in a crystalline solid state. At this temperature the particles have a perfectly ordered arrangement and do not vibrate. The substance will have the maximum order and no disorder and hence has zero entropy. This is a way of stating the Third Law of Thermodynamics.

Effect of a change in temperature

When the temperature is increased, disorder and hence entropy increases. For example, the particles of solids vibrate more, which makes the arrangement of the their particles slightly less orderly. The particles in liquids and solutions move (on average) faster, increasing the disorder of the system. The reverse changes take place when the temperature is lowered (Figure 15.26).

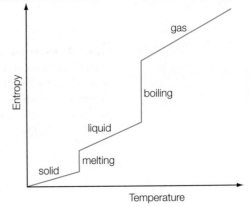

Figure 15.26 Entropy changes with temperature

Effect of a change of state

The disorder of the particles increases from solid to liquid to gas (of the same substance), increasing entropy (Figure 15.27). The particle arrangement becomes more orderly when a change in state occurs from gas to liquid to solid, hence the entropy decreases.

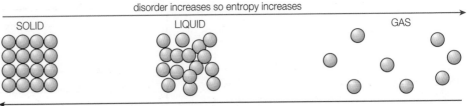

Figure 15.27 Effect of change in state on entropy

Effect of a change in the number of particles

If the number of particles increases, disorder and hence entropy increases. This is especially significant in reactions involving gases, for example the thermal decomposition of dinitrogen tetroxide:

$$N_2O_4(g) \rightarrow 2NO_2(g)$$

Conversely, the entropy decreases when the number of particles decreases, for example the hydrogenation of ethene:

$$CH_2{=}CH_2(g) + H_2(g) \rightarrow C_2H_6(g)$$

Effect of mixing of particles

The mixing of particles increases disorder, resulting in an increase in entropy (Figures 15.28 and 15.29). If the two sets of particles have different average kinetic energies (that is, different temperatures), then the kinetic energy is randomly dispersed in collisions until a new equilibrium temperature is attained.

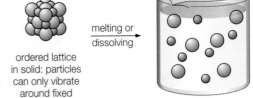

ordered lattice in solid: particles can only vibrate around fixed positions

melting or dissolving

particles free to move around in a liquid or solution

Figure 15.28 A generalized illustration showing that entropy increases when particles are mixed during melting or dissolving

Figure 15.29 Two gas jars separated by a removable partition: the gas particles spread out between the two gas jars when the partition is removed, increasing disorder and hence increasing entropy.

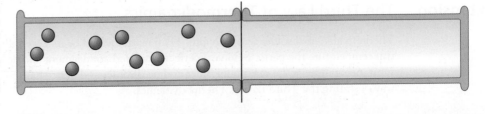

■ **Extension:** The Second Law of Thermodynamics

The Second Law of Thermodynamics states that the total entropy of the universe tends to increase. This means that for a chemical reaction or a physical change to take place, the entropy of the system (the reactants) and their surroundings must increase. Some chemical reactions and physical changes *appear* to have a decrease in entropy but this is because the entropy of the surroundings increases by a greater amount, giving an overall increase in entropy (Figure 15.30).

Figure 15.30 A reaction involving the formation of a solid from a solid and a gas. The small decrease in entropy in the system is accompanied by an even larger increase in entropy in the surroundings

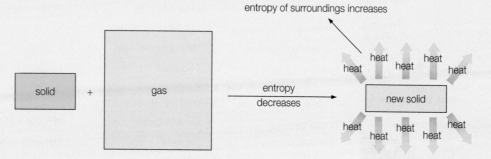

History of Chemistry

Josiah Willard Gibbs (1839–1903) was an American mathematical physicist best known for his work in chemical thermodynamics. He described thermodynamic equilibrium and developed the science of statistical mechanics. He spent his entire career at Yale University. One quote attributed to Gibbs (Figure 15.31) is: 'A mathematician may say anything he pleases, but a physicist must be at least partially sane'. He remained a bachelor and lived in his childhood home close to the campus.

Figure 15.31 Josiah Willard Gibbs

Language of Chemistry

The system is the sample or reaction mixture. Outside the system are the surroundings which include the apparatus. The universe is the system plus the surroundings (Figure 15.32). ■

Figure 15.32 A diagram illustrating the thermodynamic concepts of system and surroundings

Predicting the sign of a change in entropy

15.3.2 Predict whether the entropy change (ΔS) for a given reaction or process is positive or negative.

In many chemical reactions and physical processes it is possible to predict the sign of the entropy change (of the system), ΔS, by examining the reactants and products in the balanced equation. If the products are more disordered than the reactants, then the entropy change (of the system), ΔS, is positive. If the products are less disordered than the reactants, then the entropy change (of the system), ΔS, is negative.

Some of the more common examples are summarized in Table 15.4. The largest entropy changes occurs when a gas is produced from a liquid or solid (or vice versa) and when there is a change in the number of gas molecules.

Figure 15.33 Ice cubes melting (a positive entropy change)

Figure 15.34 Condensation of steam (a negative entropy change)

Chemical reaction or physical change	Entropy change	Example
Melting (Figure 15.33)	Increase	$H_2O(s) \rightarrow H_2O(l)$
Boiling	Large increase	$H_2O(l) \rightarrow H_2O(g)$
Condensing (Figure 15.34)	Large decrease	$H_2O(g) \rightarrow H_2O(l)$
Sublimation	Very large increase	$I_2(s) \rightarrow I_2(g)$
Vapour deposition	Very large decrease	$I_2(g) \rightarrow I_2(s)$
Freezing	Decrease	$H_2O(l) \rightarrow H_2O(s)$
Dissolving a solute to form a solution	Generally an increase (except with highly charged ions)	$NaCl(s) + (aq) \rightarrow NaCl(aq)$
Precipitation	Large decrease	$Pb^{2+}(aq) + 2Cl^-(aq) \rightarrow PbCl_2(s)$
Crystallization from a solution	Decrease	$NaCl(aq) \rightarrow NaCl(s)$
Chemical reaction: solid or liquid forming a gas	Large increase	$CaCO_3(s) \rightarrow CaO(s) + CO_2(g)$
Chemical reaction: gases forming a solid or liquid	Large decrease	$2H_2S(g) + SO_2(g) \rightarrow 3S(s) + 2H_2O(l)$
Increase in number of moles of gas	Large increase	$2NH_3(g) \rightarrow N_2(g) + 3H_2(g)$

Table 15.4 Qualitative entropy changes for the system of common reactions and physical changes

Calculating entropy changes

15.3.3 Calculate the standard entropy change for a reaction (ΔS) using standard entropy values(s).

A change in entropy is represented by ΔS. The units of entropy, S, and of entropy change, ΔS, are both joules per kelvin per mole, $J\,K^{-1}\,mol^{-1}$. Entropy values are absolute values and can be measured experimentally (via specific heat capacities). (It is also possible to calculate entropy values theoretically, as for lattice enthalpy.)

The value of an entropy change can be calculated from absolute values of entropies using the following expression:

standard entropy change = sum of entropies of products − sum of entropies of reactants

In symbols this can be expressed as:

$$\Delta S^{\ominus} = \sum S^{\ominus}_{[products]} - \sum S^{\ominus}_{[reactants]}$$

Entropy values (under standard thermodynamic conditions) for selected organic compounds are listed on pages 12 and 13 of the IB *Chemistry data booklet*.

Worked example

Calculate the entropy change that occurs during the complete combustion of ethane:

$$C_2H_6(g) + 3\tfrac{1}{2}O_2(g) \rightarrow 2CO_2(g) + 3H_2O(l)$$

$$S^{\ominus}_{[C_2H_6(g)]} = 230\,J\,K^{-1}\,mol^{-1}$$

$$S^{\ominus}_{[O_2(g)]} = 205\,J\,K^{-1}\,mol^{-1}$$

$$S^{\ominus}_{[CO_2(g)]} = 214\,J\,K^{-1}\,mol^{-1}$$

$$S^{\ominus}_{[H_2O(l)]} = 70\,J\,K^{-1}\,mol^{-1}$$

$$\Delta S^{\ominus} = \sum S^{\ominus}_{[\text{products}]} - \sum S^{\ominus}_{[\text{reactants}]}$$

$$\Delta S^{\ominus} = [(2 \times 214) + (3 \times 70)] - [230 + (3.5 \times 205)]$$

$$= -310 \, \text{J K}^{-1} \, \text{mol}^{-1}$$

As expected, because of the decrease in the amount of gas (from 4.5 moles to 2 moles), there is an increase in the order of the system, hence the entropy change is negative.

Note that, in contrast to standard enthalpy changes of formation, the absolute entropy values of elements, such as oxygen in this example, are *not* zero (under standard thermodynamic conditions).

■ Extension: Entropy and time

According to the Second Law of Thermodynamics, all spontaneous processes lead to an increase in the overall entropy of the universe. Hence, the passage of time is accompanied by an increase in the entropy of the universe. You can tell that time has passed by observing a process, for example the melting of ice cubes at room temperature, which involves an increase in entropy.

Entropy is the only quantity in the physical sciences that 'picks' a particular direction for time, sometimes called an arrow of time. As we go 'forward' in time, the Second Law of Thermodynamics tells us that the entropy of an isolated system can only increase or remain the same; it cannot decrease. Hence, from one perspective, entropy measurement is thought of as a kind of clock (Figure 15.35). For example, if we see a broken egg becoming whole again while watching a video, we immediately know that the video is running in reverse. With the egg becoming whole again, what is seen is entropy decreasing, which does not happen spontaneously. The time arrow always points in the direction of increasing entropy.

Figure 15.35 Analogue clock (an entropy meter)

The Second Law predicts that the universe (if it is a closed system) will reach a completely disordered state with maximum entropy. In this condition, known as heat death, time would have no meaning because there would be no overall entropy change to observe. The temperature would be close to absolute zero and no life would be possible.

The increase in entropy and hence the beginning of time may be associated with the expansion caused by the 'Big Bang' at the beginning of the universe about 14 billion years ago. Physicists previously argued there is enough matter in the universe for gravitational forces to overcome this expansion and cause the universe to contract resulting in a 'Big Crunch'. One highly controversial but subsequently retracted theory, which originated with Stephen Hawking, is that the 'arrow of time' will run backward during the contraction of the universe. If this occurs, then natural processes would result in an overall increase in order and 'effects' would precede their 'causes'. However, the current thinking is that the universe will experience accelerating expansion, and the Second Law of Thermodynamics cannot be violated.

15.4 Spontaneity

15.4.1 Predict whether a reaction or process will be spontaneous by using the sign of $\Delta G^{\ominus}$.

Chemists want to know whether a physical change or chemical reaction is spontaneous under standard conditions (1 atm pressure and 298 K). A spontaneous process has a natural tendency to occur. A spontaneous process involves an increase in the entropy of the universe. Spontaneous processes may need no initiation, for example:

- the evaporation of water
- the dissolving of sucrose in water to form a solution
- the diffusion of gases.

Some spontaneous chemical processes may need initiation, for example:

- hydrogen reacts rapidly with oxygen to form water if a small brief spark is applied
- carbon reacts with the oxygen in the air when ignited. The reaction then produces its own heat.

Spontaneous processes may occur very quickly or very slowly.

Some reactions and processes are non-spontaneous, for example copper does not react with dilute hydrochloric acid and water does not freeze (under standard thermodynamic conditions). A non-spontaneous process would result in a decrease in the entropy of the universe.

However, some processes and reactions which are non-spontaneous under standard conditions may become spontaneous when the temperature is increased (see page 418) or when energy is supplied continuously from an external source. For example, the decomposition of water into hydrogen and oxygen is a non-spontaneous process under standard thermodynamic conditions. However, water will undergo decomposition into hydrogen and oxygen when an electric current is passed through it (Chapter 9). The decomposition stops immediately when the electrical energy is no longer supplied.

The **spontaneity** of a process or chemical reaction is determined by the sign of the Gibbs free energy change, $\Delta G^{\ominus}$. It is calculated from the following relationship (Gibbs equation):

$$\Delta G^{\ominus} = \Delta H^{\ominus} - T\Delta S^{\ominus}$$

where T represents the absolute temperature, $\Delta S^{\ominus}$ represents the entropy change occurring in the system (chemicals) and $\Delta H^{\ominus}$ represents the enthalpy change (Chapter 5).

The Gibbs free energy change as a criterion of spontaneity is summarized as follows:

- If the Gibbs free energy change, $\Delta G^{\ominus}$, is negative then the reaction or process will be spontaneous.
- If the Gibbs free energy change, $\Delta G^{\ominus}$, is positive then the reaction or process will be non-spontaneous.
- If the Gibbs free energy change, $\Delta G^{\ominus}$, is zero, then the reaction or process will be at equilibrium: the rate of the forward reaction will equal the rate of the backward reaction (see Chapter 7). ($\Delta G = \Delta G^{\ominus}$ when the concentrations of all reactants and products are $1 \, mol \, dm^{-3}$.)

Worked example

Calculate the Gibbs free energy change for the following reaction under standard conditions:

$$N_2(g) + 3H_2(g) \rightarrow 2NH_3(g) \qquad \Delta H^{\ominus} = -95.4 \, kJ \, mol^{-1}; \; \Delta S^{\ominus} = -198.3 \, J \, K^{-1} \, mol^{-1}$$

$$\Delta G^{\ominus} = \Delta H^{\ominus} - T\Delta S^{\ominus}$$

$$\Delta G^{\ominus} = -95.4 \, kJ \, mol^{-1} - 298 \, K \times -0.1983 \, kJ \, K^{-1} \, mol^{-1}$$

(Note the conversion from $J \, K^{-1} \, mol^{-1}$ to $kJ \, K^{-1} \, mol^{-1}$ for the entropy change – this is because the units of Gibbs free energy changes are $kJ \, mol^{-1}$).

$$\Delta G^{\ominus} = -36.3 \, kJ \, mol^{-1}$$

The negative sign indicates that the reaction is spontaneous at this temperature.

Calculate the temperature above which the reaction ceases to occur spontaneously.

$$\Delta G^{\ominus} = 0 = \Delta H^{\ominus} - T\Delta S^{\ominus}$$

$$\Delta H^{\ominus} = T\Delta S^{\ominus}$$

$$T = \frac{\Delta H}{\Delta S} = \frac{-95\,400 \, J \, mol^{-1}}{198.3 \, J \, K^{-1} \, mol^{-1}} = 481 \, K = 208 \, ^{\circ}C$$

Language of Chemistry

The Gibbs energy change, ΔG, is often referred to as 'free energy'. Gibbs free energy changes refer to chemical reactions and physical processes which take place at constant temperature and *constant pressure*.

The Gibbs free energy, G, refers to the maximum amount of non-expansion work that can be done by the system. The non-expansion work is termed useful work (Figure 15.36). $T\Delta S$ (which has units of energy) is *not* available to do work and can be regarded as 'un-free energy', that is, energy that is dispersed into the random motion of particles. A useful analogy is a game of squash or tennis. The heat energy lost through perspiration (sweat) is not useful energy, rather it is the energy to hit the ball which is 'free' to do useful work. Work is done when a force moves an object through a distance (in the direction of the force). ∎

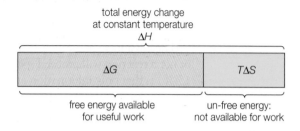

Figure 15.36 A diagram showing how an enthalpy change can be split into two parts: $\Delta H = \Delta G + T\Delta S$

An example of expansion work is shown in Figure 15.37. When the calcium carbonate decomposes, the carbon dioxide produced must push back the surrounding atmosphere (represented by the mass resting on the piston), and hence must do work on its surroundings.

Non-expansion work refers to any work done by the system other than that due to expansion. For redox reactions the maximum amount of work they can do can be easily measured by constructing an electrochemical cell and measuring the potential difference (voltage) generated (Chapter 19). The voltage can be converted to a Gibbs free energy change, ΔG (see page 420).

Figure 15.37 The decomposition of calcium carbonate (within a closed system) illustrating expansion work

■ Extension: Deriving the Gibbs equation

The total entropy change (in the system and surroundings) in a chemical reaction or physical process is given by the following expression:

$$\Delta S_{universe} = \Delta S_{system} + \Delta S_{surroundings}$$

The entropy change in the surroundings is given by the following relationship:

$$\Delta S_{surroundings} = -\frac{\Delta H}{T}$$

where ΔH represents the enthalpy change of the chemical reaction or physical process and T represents the absolute temperature (in kelvin). (Note the change of sign: the heat that leaves the system enters the surroundings, so a decrease in the enthalpy of the system corresponds to an addition of heat to the surroundings.)

Hence:

$$\Delta S_{universe} = \Delta S_{system} - \frac{\Delta H}{T}$$

Multiplying both sides of the equation by $-T$ gives the expression:

$$-T\Delta S_{universe} = \Delta H - T\Delta S_{system}$$

This means that, at constant temperature, the total entropy change is proportional to $\Delta H - T\Delta S$. $-T\Delta S_{universe}$ is defined as the Gibbs free energy change, ΔG. It is the free energy change for a chemical reaction or physical process.

Hence:

$$\Delta G = \Delta H - T\Delta S$$

Under standard thermodynamic conditions:

$$\Delta G^{\ominus} = \Delta H^{\ominus} - T\Delta S^{\ominus}$$

The advantage of using $\Delta G^{\ominus}$ to predict whether or not a reaction will proceed spontaneously (Figure 15.38) is that only information about the chemical system is required and the entropy change in the surroundings need not be calculated.

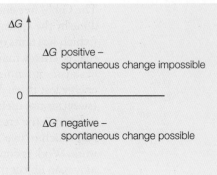

Figure 15.38 A diagram showing the possibility of change related to the value of $\Delta G^{\ominus}$

Worked example

Calculate the entropy change of the surroundings when water condenses on a window at 25 °C.

$$H_2O(g) \rightarrow H_2O(l) \qquad \Delta H^{\ominus} = -44.0\,\text{kJ mol}^{-1} \qquad \Delta S^{\ominus} = -118\,\text{J K}^{-1}\,\text{mol}^{-1}$$

(where $\Delta S^{\ominus}$ represents the entropy change of the system)

$$\begin{aligned} \Delta S_{\text{surroundings}} &= -\frac{\Delta H}{T} \\ &= -\frac{(-44\,000\,\text{J mol}^{-1})}{298\,\text{K}} \\ &= +147.7\,\text{J K}^{-1}\,\text{mol}^{-1} \end{aligned}$$

(Note the conversion from kJ mol^{-1} to J mol^{-1} for the enthalpy change – this is because the units of entropy and entropy changes are $\text{J K}^{-1}\,\text{mol}^{-1}$.)

The entropy change in the surroundings is positive (favourable), but the entropy change in the system is negative (unfavourable) (Figure 15.39).

The overall entropy change in the universe can then be calculated:

$$\Delta S_{\text{universe}} = \Delta S_{\text{system}} + \Delta S_{\text{surroundings}}$$

$$\Delta S_{\text{universe}} = -118\,\text{J K}^{-1}\,\text{mol}^{-1} + 147.7\,\text{J K}^{-1}\,\text{mol}^{-1}$$

$$= +29.7\,\text{J K}^{-1}\,\text{mol}^{-1}$$

Hence, the overall entropy change in the universe is positive (favourable) and the process is spontaneous at this temperature.

Spontaneity can also be predicted by calculating the Gibbs free energy change, $\Delta G^{\ominus}$:

$$\Delta G^{\ominus} = \Delta H^{\ominus} - T\Delta S^{\ominus}$$

$$\Delta G = -44.0\,\text{kJ mol}^{-1} - (298\,\text{K} \times 0.118\,\text{kJ K}^{-1}\,\text{mol}^{-1})$$

$$= -79.2\,\text{kJ mol}^{-1}$$

A negative value for the Gibbs free energy change, $\Delta G^{\ominus}$, and a positive value for the entropy change of the universe, $\Delta S_{\text{universe}}$, are *equivalent* criteria for spontaneity.

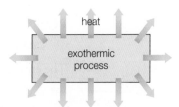

energy released to surroundings, the entropy of which therefore increases

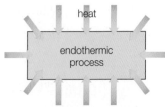

energy absorbed from surroundings, the entropy of which therefore decreases

Figure 15.39 Changes in the entropy of the surroundings during exothermic and endothermic reactions

■ Extension: Evolution, life and entropy

One expression of the Second Law of Thermodynamics (due to Clausius) is his famous statement: 'The energy of the Universe is constant; the entropy of the Universe tends towards a maximum.' There would appear to be a conflict between the evolution of life and the continued presence of living organisms (Figure 15.40) and the Second Law. Simple chemical calculations suggest that if only equilibrium processes were operating then the most complex molecules to evolve would be small peptides. The spontaneous formation of DNA and other complex molecules, let alone simple organisms such as bacteria, is extremely improbable.

Animals are maintained by energy transferred from nutrients, especially glucose. This is broken down to release energy that can be used in polymerization reactions that build up complex molecules, such as protein and DNA, from smaller, simpler monomers (resulting in a decrease in entropy). Animals, however, release heat to their surroundings. The entropy increase of the surroundings more than compensates for the entropy decrease in the organism.

Figure 15.40 A domestic cat: an example of a highly ordered (low entropy) entity maintained by a constant input of nutrients which release useful energy for endothermic syntheses

Gibbs free energy change of formation

Each compound has a **Gibbs free energy change of formation**, $\Delta G_f^\ominus$. This is the free energy change that occurs when one mole of a compound is formed from its elements in their standard states under standard conditions. (This is analogous to expressing the standard enthalpy of the reaction in terms of the standard enthalpies of formation of each chemical species taking part in the reaction – see page 397.)

For example, the Gibbs free energy change of formation of water and benzene:

$$H_2(g) + \tfrac{1}{2}O_2(g) \rightarrow H_2O(l) \qquad \Delta G_f^\ominus = -51\,\text{kJ}\,\text{mol}^{-1}$$
$$6C(s) + 3H_2(g) \rightarrow C_6H_6(l) \qquad \Delta G_f^\ominus = +49\,\text{kJ}\,\text{mol}^{-1}$$

The Gibbs free energy changes of formation (under standard thermodynamic conditions) for a selection of organic compounds are listed on pages 12 and 13 of the IB *Chemistry data booklet*. Note that the Gibbs free energy change of formation of elements (in their standard state) is zero.

The Gibbs free energy change of formation can be regarded as a measure of a compound's thermodynamic stability relative to its elements. Compounds that have a negative Gibbs free energy change of formation are more thermodynamically stable than their elements. The compound can be synthesized from its elements and there is no tendency for it to decompose back into the elements (Figure 15.41).

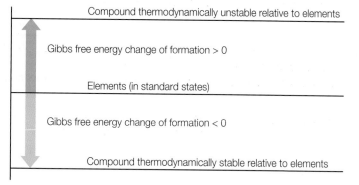

Figure 15.41 A diagram illustrating compounds with negative and positive Gibbs free energy changes of formation

Compounds that have a positive Gibbs free energy change of formation are less thermodynamically stable than their elements. The compound cannot be synthesized from its elements and there is a tendency for it to decompose back into its constituent elements (Figure 15.41).

However, compounds such as this, for example benzene (Chapters 14 and 27), are often kinetically stable because there is a large activation energy barrier to decomposition (see Chapter 6).

Gibbs free energy changes of formation can be used to calculate the Gibbs free energy change, $\Delta G^\ominus$, for a reaction using the following relationship:

| Gibbs free energy change | = | sum of Gibbs free energy of formation of products | − | sum of Gibbs free energy of formation of reactants |

In symbols:
$$\Delta G^\ominus \quad = \quad \Sigma \Delta G_f^\ominus(products) \quad - \quad \Sigma \Delta G_f^\ominus(reactants)$$

Worked example

Use the following Gibbs free energy changes of formation, $\Delta G_f^\ominus$, to calculate the free energy change, $\Delta G^\ominus$, for the decomposition of magnesium carbonate:

$$MgCO_3(s) \rightarrow MgO(s) + CO_2(g)$$

$\Delta G_f^\ominus(MgCO_3(s)) = -1012\,kJ\,mol^{-1}$

$\Delta G_f^\ominus(MgO(s)) = -569\,kJ\,mol^{-1}$

$\Delta G_f^\ominus(CO_2(g)) = -394\,kJ\,mol^{-1}$

$\Delta G^\ominus = [(-569 + (-394)] - (-1012) = +49\,kJ\,mol^{-1}$

The positive sign of the Gibbs free energy change, ΔG, indicates that the reaction is non-spontaneous under standard conditions.

Spontaneity and temperature

15.4.3 Predict the effect of a change in temperature on the spontaneity of a reaction, using standard entropy and enthalpy changes and the equation $\Delta G^\ominus = \Delta H^\ominus - T\Delta S^\ominus$

The Gibbs equation, $\Delta G^\ominus = \Delta H^\ominus - T\Delta S^\ominus$, gives rise to four different types of reaction depending on whether the enthalpy change, $\Delta H^\ominus$, and the entropy change (of the system), $\Delta S^\ominus$, are positive or negative.

Positive entropy changes are favourable; negative enthalpy changes are favourable and hence help to 'drive' the reaction forward. Negative entropy changes are unfavourable; positive enthalpy changes are unfavourable and hence help to 'drive' the reaction backward.

The results for these four combinations on the sign of the Gibbs free energy change, ΔG, are shown in Table 15.5. The results are also summarized graphically in Figure 15.42. A reaction is only spontaneous if the sign of the Gibbs free energy change is *negative*.

Enthalpy change, ΔH	Entropy change, ΔS	Gibbs free energy change, ΔG	Spontaneity
Positive (endothermic)	Positive (products more disordered than reactants)	Depends on the temperature	Spontaneous at high temperatures, when $T\Delta S > \Delta H$
Positive (endothermic)	Negative (products less disordered than reactants)	Always positive	Never spontaneous
Negative (exothermic)	Positive (products more disordered than reactants)	Always negative	Always spontaneous
Negative (exothermic)	Negative (products less disordered than reactants)	Depends on the temperature	Spontaneous at low temperatures, when $T\Delta S < \Delta H$

Table 15.5 The four types of thermodynamic reactions

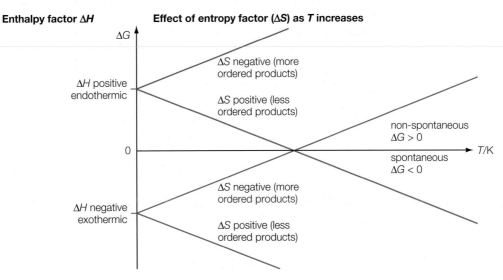

Figure 15.42 A graphical summary of the conditions for spontaneity (values of entropy and enthalpy remain constant with a change in temperature)

The relationship between free energy and temperature can also be modelled using a spreadsheet. Ian Bridgwood has written an interactive Excel model (Figure 15.43) where the user can alter the temperature using a set of buttons, which in turn dynamically updates a graph. The spreadsheet can be downloaded at www.excellencegateway.org.uk/page.aspx?o=ferl.aclearn.resource.id1789.

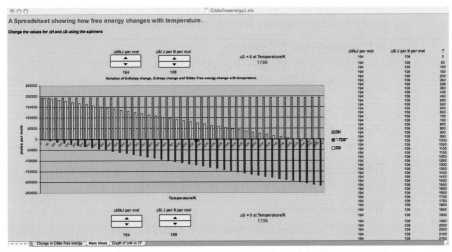

Figure 15.43 Free energy changes with temperature spreadsheet written by Ian Bridgwood, Wyggeston and Queen Elizabeth I College, Leicester

A change in temperature can alter the thermodynamics of the reaction. This dependence on temperature is shown by the presence of absolute temperature, T, in the Gibbs equation. However, temperature affects the spontaneity of exothermic and endothermic reactions in different ways.

For endothermic processes with a positive entropy change, the enthalpy change, ΔH, is positive (unfavourable and opposing the forward reaction) and hence, for the process to be spontaneous $T\Delta S$ must be positive and greater than the enthalpy change, ΔH. With an increase in temperature $T\Delta S$ increases but ΔH remains constant. However, with a decrease in temperature $T\Delta S$ decreases and at a specific temperature will become less than ΔH. The Gibbs free energy change, ΔG, would become positive and the process would become non-spontaneous. So for an *endothermic process* or reaction a *higher temperature* favours the spontaneity of the process.

For exothermic processes with a negative entropy change, the enthalpy change, ΔH, is negative (favourable and favouring the forward reaction). For such reactions to be spontaneous, the enthalpy change, ΔH, must be greater than $T\Delta S$. With an increase in temperature, the opposing factor $T\Delta S$ increases but ΔH remains constant. As the temperature increases the opposing factor $T\Delta S$ will become greater than ΔH. The Gibbs free energy change, ΔG, would become positive and the process would become non-spontaneous. So for an *exothermic process* or reaction a *lower temperature* favours the spontaneity of the process.

The effect of temperature on spontaneity

Using example values for $\Delta H^\ominus$ and $\Delta S^\ominus$ we can calculate the Gibbs free energy change for reactions at different temperatures (10 K and 10 000 K) for positive and negative values of $\Delta H^\ominus$ and $\Delta S^\ominus$.

$\Delta H^\ominus$ and $\Delta S^\ominus$ both positive

$\Delta H^\ominus = 200 \, \text{kJ mol}^{-1}$ and $\Delta S^\ominus = 200 \, \text{J K}^{-1} \text{mol}^{-1}$

At $T = 10 \, \text{K}$:

$\Delta G^\ominus = 200 \, \text{kJ mol}^{-1} - (10 \, \text{K} \times 0.2 \, \text{kJ K}^{-1} \text{mol}^{-1}) = +198 \, \text{kJ mol}^{-1}$

At $T = 10 000 \, \text{K}$:

$\Delta G^\ominus = 200 \, \text{kJ mol}^{-1} - (10 000 \, \text{K} \times 0.2 \, \text{kJ K}^{-1} \text{mol}^{-1}) = -1800 \, \text{kJ mol}^{-1}$

This corresponds to the first row in Table 15.5: the sign of the Gibbs free energy change depends on the temperature and the reaction is spontaneous at high temperatures, when $T\Delta S > \Delta H$.

$\Delta H^\ominus$ positive and $\Delta S^\ominus$ negative

$\Delta H^\ominus = 200 \, \text{kJ mol}^{-1}$ and $\Delta S^\ominus = -200 \, \text{J K}^{-1} \text{mol}^{-1}$

At $T = 10 \, \text{K}$:

$\Delta G^\ominus = 200 \, \text{kJ mol}^{-1} - (10 \, \text{K} \times -0.2 \, \text{kJ K}^{-1} \text{mol}^{-1}) = +202 \, \text{kJ mol}^{-1}$

At $T = 10 000 \, \text{K}$:

$\Delta G^\ominus = 200 \, \text{kJ mol}^{-1} - (10 000 \, \text{K} \times -0.2 \, \text{kJ K}^{-1} \text{mol}^{-1}) = +2200 \, \text{kJ mol}^{-1}$

This corresponds to the second row in Table 15.5: the sign of the Gibbs free energy is always positive and the reaction is never spontaneous.

$\Delta H^\ominus$ negative and $\Delta S^\ominus$ positive

$\Delta H^\ominus = -200 \, \text{kJ mol}^{-1}$ and $\Delta S^\ominus = 200 \, \text{J K}^{-1} \text{mol}^{-1}$

At $T = 10 \, \text{K}$:

$\Delta G^\ominus = -200 \, \text{kJ mol}^{-1} - (10 \, \text{K} \times 0.2 \, \text{kJ K}^{-1} \text{mol}^{-1}) = -202 \, \text{kJ mol}^{-1}$

At $T = 10 000 \, \text{K}$:

$\Delta G^\ominus = -200 \, \text{kJ mol}^{-1} - (10 000 \, \text{K} \times 0.2 \, \text{kJ K}^{-1} \text{mol}^{-1}) = -2200 \, \text{kJ mol}^{-1}$

This corresponds to the third row in Table 15.5: the sign of the Gibbs free energy is always negative and the reaction is always spontaneous.

$\Delta H^\ominus$ and $\Delta S^\ominus$ both negative

$\Delta H^\ominus = -200 \, \text{kJ mol}^{-1}$ and $\Delta S^\ominus = -200 \, \text{J K}^{-1} \text{mol}^{-1}$

At $T = 10 \, \text{K}$:

$\Delta G^\ominus = -200 \, \text{kJ mol}^{-1} - (10 \, \text{K} \times -0.2 \, \text{kJ K}^{-1} \text{mol}^{-1}) = -198 \, \text{kJ mol}^{-1}$

At $T = 10 000 \, \text{K}$:

$\Delta G^\ominus = -200 \, \text{kJ mol}^{-1} - (10 000 \, \text{K} \times -0.2 \, \text{kJ K}^{-1} \text{mol}^{-1}) = +1800 \, \text{kJ mol}^{-1}$

This corresponds to the fourth row in Table 15.5: the sign of the Gibbs free energy depends on the temperature and the reaction is spontaneous at low temperatures, when $T\Delta S < \Delta H$.

Voltaic cells

In a working voltaic cell (Chapter 19) a spontaneous change is occurring and hence the Gibbs free energy change, ΔG, must be negative. The relationship between the free energy change and the cell's potential is given by the following relationship:

$$\Delta G^\ominus = -nFE^\ominus_{\text{cell}}$$

where n represents the number of electrons transferred and F represents the Faraday constant ($96 485 \, \text{C mol}^{-1}$). A *negative* free energy change will only result from a *positive* cell potential. The measurement of cell potential provides one approach to the measurement of Gibbs free energy changes for redox reactions.

■ **Extension:** Ellingham diagrams

On page 420 the effect of temperature on the values and signs of Gibbs free energy changes was illustrated using a series of calculations. These calculations can be simplified by plotting changes in free energy against temperature.

Consider the following reaction which involves the formation of a mixture of carbon monoxide and hydrogen (known as water gas synthesis) from the reaction between carbon and steam:

$$C(s) + H_2O(g) \rightarrow CO(g) + H_2(g) \qquad \Delta H^{\ominus} = +131.3 \, kJ \, mol^{-1}; \quad \Delta S^{\ominus} = +133.8 \, J \, K^{-1} \, mol^{-1}$$

The enthalpy change for this reaction can be calculated from the standard enthalpies of formation of water vapour and carbon monoxide. Enthalpy change = sum of enthalpies of formation of products − sum of enthalpies of formation of reactants.

The standard enthalpy and entropy changes for these reactions are:

$$C(s) + \tfrac{1}{2}O_2(g) \rightarrow CO(g) \qquad \Delta H^{\ominus} = -110.5 \, kJ \, mol^{-1}; \qquad \Delta S^{\ominus} = +89.4 \, J \, K^{-1} \, mol^{-1}$$

$$H_2(g) + \tfrac{1}{2}O_2(g) \rightarrow H_2O(g) \qquad \Delta H^{\ominus} = -241.8 \, kJ \, mol^{-1}; \qquad \Delta S^{\ominus} = -44.4 \, J \, K^{-1} \, mol^{-1}$$

The variation of the Gibbs free energy change with temperature for the overall process and for each of these steps is shown graphically in the form of an **Ellingham diagram** in Figure 15.44.

The temperature at which the reaction of carbon with steam becomes spontaneous, 981 K, is also the point at which the lines for the formation of carbon monoxide and water meet. In these reactions, hydrogen and carbon are both competing to bond with the oxygen. The reaction with the *more negative* free energy change will be favoured and so steam will not oxidize carbon below a temperature of 981 K.

Ellingham diagrams can be used to show that iron oxide may be reduced by carbon at the temperatures reached in a blast furnace. Carbon is also able to reduce aluminium oxide, but the very high temperature required cannot be achieved in a blast furnace. Hence, aluminium is extracted from aluminium oxide via the electrolysis of the molten oxide (Chapter 23).

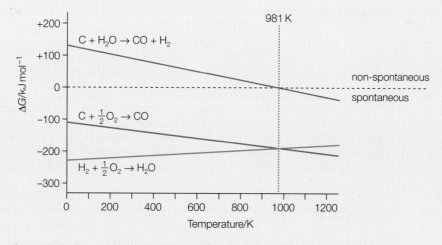

Figure 15.44 The variation in the Gibbs free energy change with absolute temperature

■ Extension: Gibbs free energy and equilibrium

For a chemical reaction or physical process to be spontaneous at a given temperature, the Gibbs free energy change, $\Delta G^{\ominus}$, must be negative. In the case of a reversible reaction (Chapter 7) there is a decrease in Gibbs free energy during the course of reaction until equilibrium is reached. This decrease occurs whether the reaction begins with reactants or products.

Consider a simple reaction $A \rightleftharpoons B$ performed in a closed vessel. As the reaction progresses, the Gibbs free energy decreases and the composition of the reaction mixture changes. When equilibrium is reached the composition of the reaction mixture remains constant, and therefore the ratio [B]/[A] remains constant.

If the Gibbs free energy is plotted against the composition of mixture as the reaction progresses, then a curve is obtained. The *minimum* of the curve corresponds to the equilibrium composition.

It does not matter whether the equilibrium state is reached from A or B as the reactant, the Gibbs free energy change, $\Delta G^{\ominus}$, is negative. Once the system reaches equilibrium then no further change in composition occurs because a change would lead to an increase in Gibbs free energy, an unfavourable process.

The graphs in Figures 15.45 and 15.46 show curves where the equilibrium favours products ($K_c > 1$) and where the equilibrium favours reactants ($K_c < 1$). If the minimum is midway between A and B, then K_c would be equal to one.

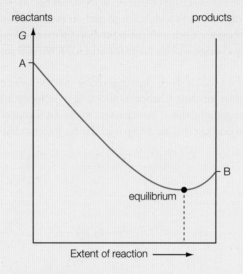

Figure 15.45 Plot of Gibbs free energy versus change in composition ($K_c > 1$)

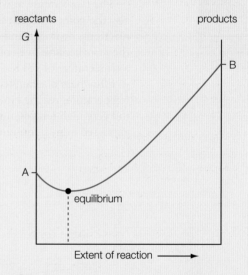

Figure 15.46 Plot of Gibbs free energy versus change in composition ($K_c < 1$)

Consider the reaction

$$A \rightleftharpoons B$$

There are three possible ways in which the free energy can change for reactions of this type (Figure 15.47).

- The reaction goes virtually to completion, the minimum value for free energy being when there is almost 100% B.
- The reaction hardly goes at all, the free energy being at a minimum when hardly any of A has reacted.
- This is an equilibrium reaction, the minimum value of free energy being when significant amounts of A and B are both present. Such reactions have values of $\Delta G^{\ominus}$ between approximately +30 and −30 kJ mol⁻¹.

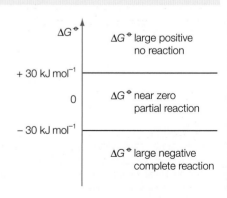

Figure 15.47 A diagram showing how the possibility of change is related to values of $\Delta G^{\ominus}$

History of Chemistry

Ludwig Boltzmann (1844–1906) was an Austrian physicist famous for developing statistical thermodynamics. He also made important contributions to kinetic theory including the Maxwell–Boltzmann distribution for molecular speeds in a gas (Chapter 6). He was also a strong supporter of atomic theory when the model was still very controversial. Boltzmann studied Physics at the University of Vienna and among his teachers were Josef Loschmidt (Chapter 1). He suffered from mental illness and committed suicide during an attack of depression. He is buried in Vienna and his tombstone (Figure 15.48) reads '$S = k \log W$', where k represents the Boltzmann constant and W represents the number of ways of arranging the particles and energy in a system (this equation was first written down by Planck). This is the cornerstone of statistical thermodynamics and is a logarithmic connection between entropy and probability. Statistical thermodynamics gives identical results to the simple thermodynamic approach in this chapter, but provides a deeper and more subtle description of entropy.

Figure 15.48
Boltzmann's tomb in the Zentralfriedhof (Central Cemetery)

Application of Chemistry

Information theory is a branch of applied mathematics and engineering based on earlier work by Maxwell, Boltzmann and Gibbs which involves quantifying information. It was developed by the American engineer and mathematician Claude Shannon (1916–2001) and is widely used in the study of electronic communication. Information theory uses a form of statistical entropy known as Shannon or information entropy to quantify the information in a message, usually in bits (digital 1s and 0s). Shannon is regarded as the founder of the digital revolution and his theories underpin our computers, networks and digital devices, such as MP3 players. Lossless data compression used in zip files and Apple's iPod MP3 files (Figure 15.49) can be traced back to work published by Shannon in 1948.

Figure 15.49 An iPod player from Apple Inc.

The short passage below is from William Shakespeare's *Julius Caesar*, Act IV, Scene 3. It is spoken by Brutus when he and Cassius are discussing the final phase of their civil war with the forces of Octavian and Marc Antony.

> *There is a tide in the affairs of men,*
> *Which, taken at the flood, leads on to fortune…*

In addition to the literal meaning of the words, the passage has many hidden and implied meanings. It reflects a complex series of events in the play and echoes the play's ideas on conflict, ambition, and the demands of leadership. In simple terms Brutus is saying that they must act while the ratio of forces is most advantageous.

In effect the passage is low in entropy, but rich in information. However, if the letters making up this passage were allowed to fall into a completely random pattern, as shown in Figure 15.50, they would have no meaning. In this form the letters would contain little or no information, but be high in entropy. Hence, entropy can be regarded as a measure of unavailable information.

Figure 15.50 Random arrangement of letters from passage

- The standard enthalpy change of formation, $\Delta H_f^{\ominus}$, is the enthalpy change when one mole of a pure substance is formed from its elements in their standard state.
- Enthalpies of formation for elements (in their standard states) are zero.
- The majority of enthalpies of formation are negative, corresponding to exothermic reactions.
- The standard enthalpy change of combustion, $\Delta H_c^{\ominus}$, is the enthalpy change when one mole of a pure substance is completely burnt in oxygen under standard conditions.
- Enthalpy changes may be calculated from enthalpies of formation using the following expression: $\Delta H = \sum \Delta H_f^{\ominus}\ \text{products} - \sum \Delta H_f^{\ominus}\ \text{reactants}$.
- Enthalpy changes may also be calculated for reactions in which all the reactants and products can be made to undergo combustion: $\Delta H^{\ominus} = \sum \Delta H_c^{\ominus}\ \text{reactants} - \sum \Delta H_c^{\ominus}\ \text{products}$.
- The standard enthalpy change of atomization, $\Delta H_{at}^{\ominus}$, is the energy change when one mole of atoms is formed from the element in its standard state.

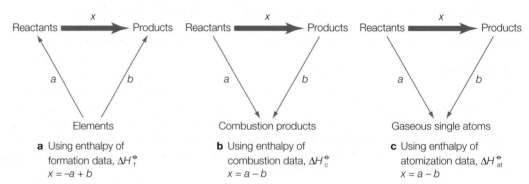

Figure 15.52 Generalized Hess's law cycle showing three common types of enthalpy data

- Electron affinity is the enthalpy change when one mole of electrons is added to a mole of gaseous atoms (under standard thermodynamic conditions).
- The lattice enthalpy is the enthalpy change when one mole of solid ionic compound forms gaseous ions (separated to an infinite distance from each other).
- The values of lattice enthalpy are determined by the charges and sizes of the ions. Large values are favoured by small, highly charged ions.
- Experimental values of lattice enthalpies can be found using a Born–Haber cycle. Theoretical values of lattice enthalpies can also be calculated from the crystal structure, assuming the substance is completely ionic in nature.
- A significant difference between the experimental and theoretical values of lattice enthalpy indicates that the lattice contains significant covalent character.
- In thermodynamics, the term 'system' describes the reactants and products of a reaction, together with any solvent. The term 'surroundings' describes the rest of the universe.
- Entropy (S) is a measure of the disorder or dispersal of energy in a system or its surroundings. In a spontaneous process, the sum of the entropy of the system and the entropy of the surroundings increases ($\Delta S_{universe} > 0$).
- The entropy of the surroundings can be calculated from the enthalpy change in the system:

$$\Delta S_{surroundings} = -\frac{\Delta H}{T}.$$

- The entropy change in the system is calculated from standard entropies of reactants and products. It is calculated from: $\Delta S^{\ominus} = \sum S_{[products]}^{\ominus} - \sum S_{[products]}^{\ominus}$.
- The entropy of the surroundings increases in an exothermic reaction, when ΔH is negative. This means that, at low temperatures, the reaction is usually spontaneous. At high temperatures, $\Delta S^{\ominus}$ becomes increasingly important and reactions in which $\Delta H^{\ominus}$ and $\Delta S^{\ominus}$ are positive become spontaneous.
- $\Delta G^{\ominus}$ gives no information about the rate of a reaction (which is controlled by the activation energy).

- The spontaneity of changes is governed by the Gibbs equation: $\Delta G^{\ominus} = \Delta H^{\ominus} - T\Delta S^{\ominus}$, where $\Delta G^{\ominus}$ represents the Gibbs free energy change and $\Delta S^{\ominus}$ the entropy change of the system.
- $\Delta G^{\ominus}$ is negative for spontaneous reactions; this is equivalent to a positive change in the entropy of the universe, $\Delta S_{universe}$. The value of the Gibbs free energy change of a chemical reaction changes in sign when the direction of the reaction is reversed.
- For reactions at chemical equilibrium, ΔG is zero.
- Values of $\Delta G^{\ominus}$ can also be calculated from Gibbs free energy changes of formation: $\Delta G^{\ominus} = \Sigma \Delta G_f^{\ominus}(\text{products}) - \Sigma \Delta G_f^{\ominus}(\text{reactants})$.
- The Gibbs free energy change of formation is the free energy change associated with the formation of one mole of a pure compound from its elements (in their standard states).

■ *Examination questions – a selection*

Paper 1 IB questions and IB style questions

Q1 Which reaction has the highest positive value of ΔS?
- **A** $2Al(s) + 3S(s) \rightarrow Al_2S_3(s)$
- **B** $CO_2(g) + 3H_2(g) \rightarrow CH_3OH(g) + H_2O(g)$
- **C** $CH_4(g) + H_2O(g) \rightarrow 3H_2(g) + CO(g)$
- **D** $S(s) + O_2(g) \rightarrow SO_2(g)$

Q2 The standard enthalpy change of formation values of two oxides of phosphorus are:

$$P_4(s) + 3O_2(g) \rightarrow P_4O_6(s) \qquad \Delta H_f^{\ominus} = -1600\,kJ\,mol^{-1}$$

$$P_4(s) + 5O_2(g) \rightarrow P_4O_{10}(s) \qquad \Delta H_f^{\ominus} = -3000\,kJ\,mol^{-1}$$

What is the enthalpy change, in $kJ\,mol^{-1}$, for the reaction below?

$$P_4O_6(s) + 2O_2(g) \rightarrow P_4O_{10}(s)$$

A +4600 **B** +1400 **C** −1400 **D** −4600
Higher Level Paper 1, May 06, Q20

Q3 Which is a correct equation to represent the lattice enthalpy of magnesium oxide?
- **A** $MgO(s) \rightarrow Mg(g) + O(g)$
- **B** $MgO(s) \rightarrow Mg(s) + O(s)$
- **C** $MgO(s) \rightarrow Mg^+(g) + O^-(g)$
- **D** $MgO(s) \rightarrow Mg^{2+}(g) + O^{2-}(g)$

Q4 For the reaction

$$2MgO(s) \rightarrow 2Mg(s) + O_2(g)$$

at 1 atmosphere pressure the values of $\Delta H^{\ominus}$ and $\Delta S^{\ominus}$ are positive. Which statement is correct?
- **A** $\Delta G^{\ominus}$ is temperature dependent.
- **B** The decrease in entropy is the main driving force.
- **C** At high temperatures $\Delta G^{\ominus}$ is positive.
- **D** The forward reaction is exothermic.

Q5 The average bond enthalpy for the C–H bond is $412\,kJ\,mol^{-1}$. Which process has an enthalpy value closest to this value?
- **A** $CH_4(g) \rightarrow CH_3^-(g) + H^+(g)$
- **B** $CH_4(g) \rightarrow C(g) + 2H_2(g)$
- **C** $CH_4(g) \rightarrow C^{4-}(g) + 4H^+(g)$
- **D** $CH_4(g) \rightarrow CH_3(g) + H(g)$

Q6 Some hydrogen gas is placed in a flask of fixed volume at room temperature. Which change will cause a decrease in the entropy of the system?
- **A** adding a small amount of hydrogen
- **B** adding a small amount of iodine
- **C** cooling the flask
- **D** exposing the flask to ultraviolet radiation

Q7 Which reaction has the most negative $\Delta H^{\ominus}$ value?
- **A** $LiF(s) \rightarrow Li^+(g) + F^-(g)$
- **B** $Li^+(g) + F^-(g) \rightarrow LiF(s)$
- **C** $NaBr(s) \rightarrow Na^+(g) + Br^-(g)$
- **D** $Na^+(g) + Br^-(g) \rightarrow NaBr(s)$

Q8 Which reaction occurs with the largest increase in entropy?
- **A** $Pb(NO_3)_2(s) + 2KI(s) \rightarrow PbI_2(s) + 2KNO_3(aq)$
- **B** $BaCO_3(s) \rightarrow BaO(s) + CO_2(g)$
- **C** $3H_2(g) + N_2(g) \rightarrow 2NH_3(g)$
- **D** $H_2(g) + Cl_2(g) \rightarrow 2HCl(g)$

Q9 Which of the following statements help(s) to explain why the value for the lattice enthalpy of lithium fluoride is less than that for strontium fluoride?
- **I** The ionic radius of lithium is less than that of strontium.
- **II** The ionic charge of lithium is less than that of strontium.

- **A** I only
- **B** II only
- **C** neither I nor II
- **D** I and II

Q10 When $\Delta G^{\ominus}$ for a reaction is negative, the reaction must be:
- **A** slow
- **B** exothermic
- **C** reversible
- **D** spontaneous

Q11 Consider the reaction

$$NH_4Br(s) \rightarrow NH_3(g) + HBr(g)$$

Which row gives the correct signs for the values of ΔH and ΔS?
- **A** ΔH +ve $\quad$ ΔS +ve
- **B** ΔH −ve $\quad$ ΔS −ve
- **C** ΔH −ve $\quad$ ΔS +ve
- **D** ΔH +ve $\quad$ ΔS −ve

Q12 The Born–Haber cycle for the formation of potassium fluoride includes the steps below:
- **I** $K(g) \rightarrow K^+(g) + e^-$
- **II** $\frac{1}{2}F_2(g) \rightarrow F(g)$
- **III** $F(g) + e^- \rightarrow F^-(g)$
- **IV** $K^+(g) + F^-(g) \rightarrow KF(s)$

Which of these steps are exothermic?
- **A** II and III only
- **B** III and IV only
- **C** I, II and III only
- **D** I, III and IV only

Q13 When the substances $H_2(g)$, $O_2(g)$ and $H_2O_2(l)$ are arranged in order of increasing entropy values at 25°C, what is the correct order?
- **A** $H_2(g)$, $O_2(g)$, $H_2O_2(l)$
- **B** $H_2(g)$, $H_2O_2(l)$, $O_2(g)$
- **C** $O_2(g)$, $H_2(g)$, $H_2O_2(l)$
- **D** $H_2O_2(l)$, $H_2(g)$, $O_2(g)$

Q14 Which one of the following would have a standard enthalpy change of formation that is not zero?
- **A** $O_2(l)$
- **B** $Br_2(l)$
- **C** $F_2(g)$
- **D** $Hg(l)$

Q15 Which of the following must have a negative value for an exothermic reaction?
- **A** enthalpy change, $\Delta H^{\ominus}$
- **B** entropy change, $\Delta S^{\ominus}$
- **C** equilibrium constant, K_c
- **D** standard electrode potential, $E^{\ominus}$

Q16 The bond enthalpies for $H_2(g)$ and $HF(g)$ are 435 kJ mol⁻¹ and 565 kJ mol⁻¹, respectively. For the reaction $\frac{1}{2}H_2(g) + \frac{1}{2}F_2(g) \rightarrow HF(g)$, the enthalpy of reaction is −268 kJ mol⁻¹ of HF produced.

What is the bond energy of F_2 in kJ mol⁻¹?
- **A** 464
- **B** 138
- **C** 243
- **D** 159

Q17 In the reaction, $PbI_2(s) \rightarrow Pb^{2+}(aq) + 2I^-(aq)$, the solubility of PbI_2 is determined by the tendency of the reaction system to attain which of the following?
- **A** maximum entropy and maximum enthalpy
- **B** maximum entropy and minimum enthalpy
- **C** minimum entropy and maximum enthalpy
- **D** minimum entropy and minimum enthalpy

Q18 Which equation shows a reaction in which there is an increase in entropy?
- **A** $Ag^+(aq) + Br^-(aq) \rightarrow AgBr(s)$
- **B** $H_2(g) + \frac{1}{2}O_2(g) \rightarrow H_2O(l)$
- **C** $MgO(s) + CO_2(g) \rightarrow MgCO_3(s)$
- **D** $Na(s) + H_2O(l) \rightarrow Na^+(aq) + OH^-(aq) + \frac{1}{2}H_2(g)$

Q19 The standard enthalpy of formation of sodium fluoride corresponds to which reaction?
- **A** $Na(g) + F(l) \rightarrow NaF(s)$
- **B** $Na(g) + F(g) \rightarrow NaF(g)$
- **C** $Na(s) + F(l) \rightarrow NaF(s)$
- **D** $Na(s) + \frac{1}{2}F_2(g) \rightarrow NaF(s)$

Q20 For the reaction, $C(s) + CO_2(g) \rightarrow 2CO(g)$, which is spontaneous only at temperatures higher than 1100 K, one would conclude which of the following?
- **A** ΔH is negative and ΔS is negative
- **B** ΔH is positive and ΔS is negative
- **C** ΔH is positive and ΔS is positive
- **D** ΔH is negative and ΔS is positive

Q21 Which of the following is the correct equation for the standard enthalpy change of formation of carbon monoxide?
- **A** $C(s) + \frac{1}{2}O_2(g) \rightarrow CO(g)$
- **B** $C(g) + \frac{1}{2}O_2(g) \rightarrow CO(g)$
- **C** $C(g) + O(g) \rightarrow CO(g)$
- **D** $2C(s) + O_2(g) \rightarrow 2CO(g)$

Q22 For the reaction $2NO(g) + O_2(g) \rightarrow 2NO_2(g)$ at one atmosphere pressure, the values of ΔH and ΔS are both negative and the process is spontaneous at room temperature. Which of the following is also true?
- **A** ΔG is temperature dependent.
- **B** The change in entropy is the driving force of the reaction.
- **C** At high temperatures, ΔH becomes positive.
- **D** The reaction is endothermic.

Q23 Which one of the following processes is not exothermic?
- **A** $2CH_3(g) \rightarrow C_2H_6(g)$
- **B** $Cl_2(g) \rightarrow 2Cl(g)$
- **C** $Br(g) + e^- \rightarrow Br^-(g)$
- **D** $4Fe(s) + 3O_2(g) \rightarrow 2Fe_2O_3(s)$

Q24 Which of the following processes would be expected to have an entropy change value close to zero?
- **A** $2H_2(g) + O_2(g) \rightarrow 2H_2O(g)$
- **B** $H_2O(s) \rightarrow H_2O(g)$
- **C** $Br_2(g) + Cl_2(g) \rightarrow 2BrCl(g)$
- **D** $CO_2(g) + (aq) \rightarrow CO_2(aq)$

Q25 For reaction systems at equilibrium, which of the following must always be true?

  A $\Delta G = 0$ **B** $\Delta H = 0$ **C** $\Delta S = 0$ **D** $K_c = 0$

Paper 2 IB questions and IB style questions

Q1 When solid blue copper(II) sulfate pentahydrate, $CuSO_4.5H_2O$, loses water the white solid, copper(II) sulfate monohydrate, $CuSO_4.H_2O$ is produced as represented by the following equation:

$$CuSO_4.5H_2O(s) \rightleftharpoons CuSO_4.H_2O(s) + 4H_2O(g)$$

The thermodynamic data for the substances involved in the reversible process are:

Substance	$\Delta H_f^\ominus$/kJ mol^{-1}	$S^\ominus$/J K^{-1} mol^{-1}
$CuSO_4.5H_2O(s)$	-2278	305
$CuSO_4.H_2O(s)$	-1084	150
$H_2O(g)$	-242	189

a i Name and define the terms $\Delta H_f^\ominus$ and $S^\ominus$ and explain the standard symbol $^\ominus$. [5]

 ii Explain why, in the case of $S^\ominus$, the symbol 'Δ' is not included. [1]

 iii What is the $\Delta H_f^\ominus$ value of elemental copper? [1]

b i Calculate the value of $\Delta H^\ominus$ for the above reaction and state what information the sign of $\Delta H^\ominus$ provides about this reaction. [4]

 ii Calculate $\Delta S^\ominus$ for the reaction and state the meaning of the sign of $\Delta S^\ominus$ obtained. [4]

 iii Identify a thermodynamic function that can be used to predict reaction spontaneity and state its units. [2]

c i Use the values obtained in **b** above to determine if the following reaction is spontaneous or non-spontaneous at 25 °C.

$$CuSO_4.5H_2O(s) \rightleftharpoons CuSO_4.H_2O(s) + 4H_2O(g)$$

 Identify which compound is more stable at 25 °C, $CuSO_4.5H_2O(s)$ or $CuSO_4.H_2O(s)$. [5]

 ii Use the values obtained in **b** to determine the Celsius temperature above which the other compound in **c i** is more stable. [3]

 Higher Level Paper 2, Nov 00, Q6

Q2 a Define the term *lattice enthalpy*. [2]

 b State and explain the relationship between ionic radius and ionic charge and the size of the lattice enthalpy. [2]

 c Draw a Born–Haber cycle for rubidium oxide, Rb_2O, and use the following data to calculate the lattice enthalpy.

$$\Delta H_{atl}^\ominus[Rb] = 80.9 \text{ kJ mol}^{-1}$$
$$\Delta H_{IE(1)}^\ominus[Rb] = 403.0 \text{ kJ mol}^{-1}$$
$$\Delta H_{atl}^\ominus[\tfrac{1}{2}O_2] = 249.2 \text{ kJ mol}^{-1}$$
$$\Delta H_{EA(1)}^\ominus[O] = -146.1 \text{ kJ mol}^{-1}$$
$$\Delta H_{EA(2)}^\ominus[O] = 795.5 \text{ kJ mol}^{-1}$$
$$\Delta H_f^\ominus[Rb_2O] = -339.0 \text{ kJ mol}^{-1} \quad [4]$$

 d The experimental value of lattice enthalpy for rubidium oxide is very similar to its theoretical value. However, there is a significant difference between the theoretical and calculated values of lattice enthalpies of silver bromide. Explain these observations. [2]

 e Explain why the lattice enthalpy of sodium oxide would be expected to be higher than that of rubidium oxide. (Assume both compounds have the same lattice structure.) [2]

Q3 a Define the term *standard enthalpy change of formation*. [2]

 b Define the term *standard enthalpy change of combustion*. [2]

 c State Hess's law. [1]

 d Calculate the standard enthalpy change of formation of propane, C_3H_8, given the following standard enthalpies of combustion:

$$\Delta H_c^\ominus[C_3H_8(g)] = -2220 \text{ kJ mol}^{-1}$$

$$\Delta H_c^\ominus[C_{graphite}(s)] = -393 \text{ kJ mol}^{-1}$$

$$\Delta H_c^\ominus[H_2(g)] = -286 \text{ kJ mol}^{-1}$$

 Draw an energy cycle and use Hess's law to produce an equation for the enthalpy change of formation. [4]

16 Kinetics

- The rates of many reactions are affected by changes in the concentrations of their reactants.
- Order describes how rate is affected by a change in a concentration.
- Simple collision theory suggests that doubling the concentration of a reactant should double the rate of reaction.
- Many, but not all, reactions exhibit this behaviour, which is known as first-order kinetics.
- For some reactions changing the concentration of a reactant has no effect on the rate.
- A rate expression is a mathematical summary of how changes in concentrations of reactants affect the rate of a reaction.
- A rate constant is a specific value with associated units that is specific and characteristic for a reaction (at a fixed temperature).
- Many chemical reactions occur in two or more elementary steps.
- A reaction mechanism is a set of elementary steps at the molecular level.
- Reaction rates increase exponentially with temperature.
- The activation energy can be determined from repeated measurements of rate or rate constant and temperature.

16.1 Rate expression

The rate expression and order of reaction

16.1.1 **Distinguish** between the terms *rate constant*, *overall order of reaction* and *order of reaction* with respect to a particular reactant.
16.1.2 **Deduce** the rate expression for a reaction from experimental data.
16.1.3 **Solve** problems involving the rate expression.
16.1.4 **Sketch, identify** and **analyse** graphical representations for zero-, first- and second-order reactions.

Many reactions that take place in solution have rates (Chapter 6) that are affected by changes in the concentrations of their reactants. Similar behaviour is shown by reactions involving gases. The way in which the concentration of a reactant affects the rate of a chemical reaction, known as its **order**, can only be found by carrying out experiments, or by knowing the mechanism of the reaction.

You *cannot* deduce the order of a reaction from looking at a balanced equation for the reaction: any similarity is purely coincidental. Experiments will show the relationship between the rate of a reaction and the concentrations of the reactants. The **rate expression** is a precise mathematical way of summarizing this information about concentration changes.

A very common rate expression is rate $\propto$ [A]. This means that if the concentration of the reactant A is *doubled*, the rate is *doubled*. Conversely, if the concentration of A is *halved*, the rate is *halved*. The equation specifies a *directly proportional* relationship between rate and concentration of the reactant A.

In a general reaction such as:

$$A + B \rightarrow C + D$$

where A and B represent reactants and C and D represent products, the generalized rate expression is:

$$\text{rate} \propto [A]^a[B]^b \qquad \text{or} \qquad \text{rate} = k[A]^a[B]^b$$

Square brackets indicate concentrations (as they do in an equilibrium expression – Chapter 7), the exponents a and b are the **individual orders** of the reaction with respect to the reactants A and B, and k is the rate constant. The values of a, b and k have to be determined experimentally. The sum of the individual orders, a and b, is known as the **overall order** of the reaction.

Rate expressions may not include *all* the reactants and they may include substances, such as acid and alkali, which although present in the reaction mixture do not appear in the equation because they act as catalysts. Rate expressions may contain products but *never* intermediates – chemicals that appear temporarily during the reaction.

Rate expressions have two main uses:

■ The rate expression (together with the rate constant) can be used to predict the rate of a reaction from a mixture of reactants of known concentrations.
■ A rate expression will help to formulate a mechanism for the reaction: this is a description of the intermediates and the simple reactions (known as elementary steps) by which many reactions occur.

Examples of rate expressions

■ $H_2(g) + I_2(g) \rightarrow 2HI(g)$

The rate expression is: rate = $k[H_2(g)][I_2(g)]$

■ $2N_2O_5(g) \rightarrow 4NO_2(g) + O_2(g)$

The rate expression is: rate = $k[N_2O_5(g)]$

This emphasizes that the order of reaction is *not* obtained from the stoichiometric coefficient in the chemical equation: it is obtained experimentally.

■ $CH_3COCH_3(aq) + I_2(aq) \rightarrow CH_2ICOCH_3(aq) + HI(aq)$

The reaction between propanone and iodine is catalysed by hydrogen ions.
The rate expression is: rate = $k[CH_3COCH_3(aq)][H^+(aq)]$

The rate expression does *not* include iodine, $I_2(aq)$, the other reactant in the equation. That is, the concentration of iodine does not affect the rate of reaction. This reaction does not occur via a one-step reaction involving iodine directly reacting with propanone. It consists of a number of individual reactions known as elementary steps.

Some rate expressions are shown in Table 16.1 for reactions commonly encountered during the IB Chemistry course.

Reaction	Rate expression
$H_2(g) + I_2(g) \rightarrow 2HI(g)$	Rate = $k[H_2(g)][I_2(g)]$
$2NO(g) + O_2(g) \rightarrow 2NO_2(g)$	Rate = $k[NO(g)]^2[O_2(g)]$
$H_2O_2(aq) + 2H^+(aq) + 2I^-(aq) \rightarrow 2H_2O(l) + I_2(aq)$	Rate = $k[H_2O_2(aq)][I^-(aq)]$
$S_2O_8{}^{2-}(aq) + 2I^-(aq) \rightarrow 2SO_4{}^{2-}(aq) + I_2(aq)$	Rate = $k[S_2O_8{}^{2-}(aq)][I^-(aq)]$
$BrO_3^-(aq) + 5Br^-(aq) + 6H^+(aq) \rightarrow 3Br_2(aq) + 3H_2O(l)$	Rate = $k[BrO_3^-(aq)][Br^-(aq)][H^+(aq)]$
$CH_3Br(aq) + OH^-(aq) \rightarrow CH_3OH(aq) + Br^-(aq)$	Rate = $k[CH_3Br(aq)][OH^-(aq)]$
$CH_3COOC_2H_5(aq) + OH^-(aq) \rightarrow CH_3COO^-(aq) + C_2H_5OH(aq)$	Rate = $k[CH_3COOC_2H_5(aq)][OH^-(aq)]$
$I_2(aq) + CH_3COCH_3(aq) \rightarrow CH_2ICOCH_3(aq) + HI(aq)$	Rate = $k[CH_3COCH_3(aq)][H^+(aq)]$
$CH_3COOH(aq) + C_2H_5OH(aq) \rightarrow CH_3COOC_2H_5(aq) + H_2O(l)$	Rate = $k[CH_3COOH(aq)][C_2H_5OH(aq)]$
$(CH_3)_3CCl(aq) + H_2O(l) \rightarrow (CH_3)_3COH(aq) + Cl^-(aq) + H^+(aq)$	Rate = $k[(CH_3)_3CCl(aq)]$

Table 16.1 Selected chemical reactions and their experimentally determined rate expressions

Worked example

The rate expression for the reaction:

$2A + B + C \rightarrow 2D + E$

is: rate = $k[A][B]^2$

Deduce the overall order and the individual orders for the reactants.

The overall order is three (1 + 2).
The reaction is first order with respect to A.
The reaction is second order with respect to B.
The reaction is zero order with respect to C.

Iodine reacts with thiosulfate ions as shown below:

$$I_2(aq) + 2S_2O_3^{2-}(aq) \rightarrow S_4O_6^{2-}(aq) + 2I^-(aq)$$

Experiments have shown that the reaction is first order with respect to the concentration of iodine and first order with respect to the concentration of thiosulfate ions. The overall order is therefore two (1 + 1).

First order with respect to the iodine concentration means that doubling the concentration of iodine would double the initial rate of the overall reaction. First order with respect to the thiosulfate ion concentration similarly means that doubling the concentration of thiosulfate ions would double the initial rate of the overall reaction. The initial rate of a reaction is the rate of reaction measured just after the reaction has started.

Halving the concentration of each reactant separately would reduce the rate of reaction by half. In a first-order reaction the initial rate of reaction is *directly proportional* to the concentration of that reactant. The overall order of two means that doubling the concentrations of both reactants, iodine and thiosulfate, would increase the overall rate of reaction four-fold: its initial rate would be four times faster. Halving the concentrations of both reactants would decrease the rate of reaction by a factor of four: its initial rate of reaction would be four times slower.

The rate expression for this reaction is therefore:

$$\text{rate} = k[I_2(aq)][S_2O_3^{2-}(aq)]$$

In general a second-order rate expression will be:

$$\text{rate} = k[A][B] \qquad \text{or more simply} \qquad \text{rate} = k[A]^2$$

For the simpler rate expression, if the concentration of A is doubled then the initial rate of reaction would quadruple (2 × 2). Conversely, if the concentration of A were halved, then the initial rate of reaction would decrease by a factor of four.

Some reactions have an overall order of three and are usually first order with respect to one reactant and second order with respect to another reactant:

$$\text{rate} = k[A]^2[B]$$

This means that if the concentration of A is doubled, then the initial rate will be quadrupled (2 × 2). If the concentrations of A and B are both doubled, then the initial rate will increase by a factor of eight (2 × 2 × 2). For example:

$$\text{rate} = k[NO(g)]^2[O_2(g)]$$

However, another possibility is:

$$\text{rate} = k[A][B][C]$$

where the rate is first order with respect to A, B and C.

This means that if the concentration of any one of A, B or C is doubled, then the initial rate will be doubled. If the concentrations of two of the reactants are both doubled, then the initial rate will increase by a factor of four (2 × 2). If the concentrations of all three reactants are doubled, then the initial rate will be increased by a factor of eight (2 × 2 × 2). For example:

$$\text{rate} = k[BrO_3^-(aq)][Br^-(aq)][H^+(aq)]$$

A few reactions are zero order with respect to a particular reactant, which means that changing the concentration of that reactant has no effect on the initial rate of reaction. This occurs when the reactant does not participate until after the slowest step, or **rate-determining step**, of the mechanism.

The rate expression will then be:

$$\text{rate} = k[A]^0 \qquad \text{or} \qquad \text{rate} = k \qquad (\text{since } [A]^0 = 1)$$

For example, for the reaction:

$$(CH_3)_3CCl(aq) + H_2O(l) \rightarrow (CH_3)_3COH(aq) + Cl^-(aq)$$

the rate expression is:

$$\text{rate} = k[(CH_3)_3CCl(aq)]$$

Therefore for water:

$$\text{rate} = k[H_2O(l)]^0$$

Worked example

The following equation represents the oxidation of bromide ions in acidic solution.

$$BrO_3^-(aq) + 5Br^-(aq) + 6H^+(aq) \rightarrow 3Br_2 + 3H_2O$$

The rate expression is:

rate = $k[BrO_3^-(aq)][Br^-(aq)][H^+(aq)]$

Deduce the effect on the rate if the concentration of bromate is *halved*, but the concentration of hydrogen ions is *quadrupled* (at constant temperature and bromide concentration).

The quadrupling of the hydrogen ion concentration results in a quadrupling of the rate (first-order kinetic behaviour); the halving of the bromate concentration results in the rate being reduced by a factor of two (first-order kinetic behaviour). This is equivalent to the overall rate being doubled.

You may be asked to deduce the effect on the relative rate if two reactant concentrations are changed at the same time.

The rate constant

The rate of a reaction usually decreases during the reaction and is dependent on the concentrations of reactants. Hence, a reaction can exhibit many different rates depending on the concentrations and the extent of the reaction. Rate is proportional to an expression involving reactant concentrations and this is transformed into an equation by introducing a constant of proportionality, k.

The **rate constant, k**, is a numerical value included in the rate expression. It is characteristic of a particular reaction: each reaction has its own unique rate constant in terms of a value and associated units (at a particular temperature and pressure). The rate constant does *not* depend on the extent of the reaction or vary with the concentrations of the reactants.

Rate constants will also vary with a change in solvent if the reaction is performed in solution. Low values of rate constants are associated with slow reactions, while high values are associated with fast reactions. Since rate constants vary with temperature they are usually quoted at a specific temperature and pressure.

Worked example

The rate expression for a reaction at 800 K is:

rate = $k[A]^2[B]$

The initial rate of reaction was $55.0 \times 10^{-5}\,mol\,dm^{-3}\,s^{-1}$ when the concentrations of A and B were $3.00 \times 10^{-2}\,mol\,dm^{-3}$ and $6.00 \times 10^{-2}\,mol\,dm^{-3}$. Calculate the rate constant (to 1 decimal place).

$$k = \frac{rate}{[A]^2[B]}$$

$$= \frac{55.00 \times 10^{-5}\,mol\,dm^{-3}\,s^{-1}}{(3.00 \times 10^{-2})^2\,(mol\,dm^{-3})^2 \times 6.00 \times 10^{-2}\,mol\,dm^{-3}}$$

$$= 10.2\,dm^6\,mol^{-2}\,s^{-1}$$

The units for the rate constant of a reaction depend on whether the reaction is first, second or third order overall. For example, in a first-order reaction:

rate = $k[A]$ and hence $k = \dfrac{rate}{[A]}$

Substituting the units for rate and concentration gives:

$$k = \frac{mol\,dm^{-3}\,s^{-1}}{mol\,dm^{-3}}$$

which cancels to s^{-1}.

However, for an equation with an overall order of two:

rate = $k[A][B]$ and hence $k = \dfrac{rate}{[A] \times [B]}$

Substituting the units for rate and concentration gives $\dfrac{mol\,dm^{-3}\,s^{-1}}{(mol\,dm^{-3})^2}$ which simplifies to $\dfrac{mol\,dm^{-3}\,s^{-1}}{mol^2\,dm^{-6}}$ before cancelling to $mol^{-1}\,dm^3\,s^{-1}$. These units are sometimes written as $dm^3\,mol^{-1}\,s^{-1}$.

For a third-order reaction:

$$\text{rate} = k[A][B]^2 \qquad \text{and hence} \qquad k = \dfrac{\text{rate}}{[A] \times [B]^2}$$

Substituting the units for rate and concentration gives:

$$k = \dfrac{mol\,dm^{-3}\,s^{-1}}{(mol\,dm^{-3})(mol\,dm^{-3})^2}$$

$$= \dfrac{mol\,dm^{-3}\,s^{-1}}{mol^3\,dm^{-9}}$$

$$= mol^{-2}\,dm^6\,s^{-1}$$

These units are sometimes written as $dm^6\,mol^{-2}\,s^{-1}$.

In general, for a reaction of nth order, the units of the rate constant are:

$$k = (mol\,dm^{-3})^{1-n}\,s^{-1}$$

Worked examples

Deduce the overall order of the reaction and the units of the rate constant, k.

$$3NO(g) \rightarrow N_2O(g) + NO_2(g) \qquad \text{rate} = k[NO(g)]^2$$

The overall order is two.

Rate constant, $k = \dfrac{\text{rate}}{[NO(g)]^2}$

$$\text{Units of } k = \dfrac{mol\,dm^{-3}\,s^{-1}}{(mol\,dm^{-3})^2}$$

$$= \dfrac{mol\,dm^{-3}\,s^{-1}}{mol^2\,dm^{-6}}$$

$$= mol^{-1}\,dm^3\,s^{-1}$$

If the rate constant and the rate expression are known for a reaction, then the initial rates of reaction can be calculated.

Worked example

Propene reacts with bromine to form 1,2-dibromopropane.

propene + bromine → 1,2-dibromopropane

$$CH_3CH{=}CH_2(g) + Br_2(g) \rightarrow CH_3CHBr{-}CH_2Br(l)$$

The rate equation for this reaction is:

$$\text{rate} = k[CH_3CH{=}CH_2(g)][Br_2(g)]$$

and the rate constant is $30.0\,dm^3\,mol^{-1}\,s^{-1}$.

Calculate the initial rate of reaction when the concentrations of propene and bromine are both $0.040\,mol\,dm^{-3}$.

Substituting into the rate expression:

$$\text{rate} = 30.0\,dm^3\,mol^{-1}\,s^{-1} \times 0.040\,mol\,dm^{-3} \times 0.040\,mol\,dm^{-3} = 0.048\,mol\,dm^{-3}\,s^{-1}$$

$$2HI(g) \rightarrow H_2(g) + I_2(g)$$

The rate expression for the reaction is:

$$\text{rate} = k[HI(g)]^2$$

At a temperature of $700\,K$ and a concentration of $2.00\,mol\,dm^{-3}$, the rate of decomposition of hydrogen iodide is $25.0 \times 10^{-5}\,mol\,dm^{-3}\,s^{-1}$.

Calculate the rate constant, k, at this temperature and calculate the number of hydrogen iodide molecules that decompose per second in $1.00\,dm^3$ of gaseous hydrogen iodide under these conditions. (Avogadro constant = $6.02 \times 10^{23}\,mol^{-1}$.)

$$k = \frac{rate}{[HI(g)]^2} = \frac{25.0 \times 10^{-5}\,mol\,dm^{-3}\,s^{-1}}{4.00\,(mol\,dm^{-3})^2} = 6.25 \times 10^{-5}\,dm^3\,mol^{-1}\,s^{-1}$$

The number of molecules decomposing in one cubic decimetre every second
$= 25.0 \times 10^{-5} \times 6.02 \times 10^{23} = 1.51 \times 10^{20}$ molecules.

The important differences between the rate and the rate constant are summarized in Table 16.2.

Rate of reaction	Rate constant, k
It is the speed at which the reactants are converted into products at a specific time during the reaction.	It is a constant of proportionality in the rate expression.
It depends upon the concentration of reactant species at a specific time.	It refers to the rate of reaction when the concentration of every reacting species is unity (one).
It generally decreases with time.	It is constant and does not vary during the reaction.

Table 16.2 Differences between the rate and rate constant of a reaction

Experimental determination of the rate expression

For a reaction such as:

$$A + B \rightarrow C + D$$

where A and B represent reactants and C and D represent products, the following procedure is used to establish the rate expression.

- Carry out the reaction with known concentrations of A and B and measure the initial rate of reaction.
- Repeat the experiment using double the concentration of A, but keeping the concentration of B the same as in the first experiment. Therefore any change in the rate of reaction can only be caused by a change in the concentration of A.

If the rate is doubled in the second experiment then the reaction is first order with respect to A, if the initial rate is increased four times, the reaction is second order with respect to A and if there is no change in the initial rate then the reaction is zero order with respect to A. Similar experiments can be carried out to establish the order with respect to B. This method of finding orders is known as the **initial rates method**.

Worked example

Iodine reacts with propanone according to the following equation:

$$I_2(aq) + CH_3COCH_3(l) \rightarrow ICH_2COCH_3(aq) + HI(aq)$$

The kinetics of this reaction were investigated in four experiments carried out at constant temperature. The initial rate of reaction was measured at different concentrations of propanone, iodine and hydrogen ions (Table 16.3). Use these data to determine the individual orders and overall order of the reaction.

Experiment number	Propanone concentration/mol dm^{-3}	Hydrogen ion concentration/mol dm^{-3}	Iodine concentration /mol dm^{-3}	Initial rate /mol dm^{-3} s^{-1}
1	6.0	0.4	0.04	18×10^{-6}
2	6.0	0.8	0.04	36×10^{-6}
3	8.0	0.8	0.04	48×10^{-6}
4	8.0	0.4	0.08	24×10^{-6}

Table 16.3 Initial rate data for the iodination of propanone

Comparing experiments 1 and 2 in Table 16.3, there is a doubling in the concentration of hydrogen ions, while the other two reactant concentrations are kept constant. There is a doubling in rate between experiments 1 and 2, indicating that the reaction is first order with respect to hydrogen ions. (Rate is proportional to concentration of hydrogen ions.)

Comparing experiments 3 and 4, there is a doubling in the concentration of iodine, while the propanone concentration is kept constant and the hydrogen ion concentration is halved. The initial rate between experiments 3 and 4 is also halved, which means the reaction is zero order with respect to iodine, since the halving of the initial rate is due to the change in concentration of hydrogen ions. (Rate is not affected by iodine concentration.)

Comparing experiments 2 and 3, there is an increase in the concentration of propanone of $\frac{4}{3}$. There is also an increase in rate between experiments 2 and 3 of $\frac{4}{3}$. Therefore, the rate is proportional to the propanone concentration, that is, it is first order with respect to propanone.

$$\text{rate} = k[H^+(aq)]^1[I_2(aq)]^0[\text{propanone}]^1$$

or, since $[I_2(aq)]^0 = 1$, $\quad\quad \text{rate} = k[H^+(aq)][\text{propanone}]$

The overall order of the reaction is therefore two.

The order of a reaction with respect to a reactant can also be determined from a concentration–time graph. If the reaction is zero order with respect to a reagent, then the graph produced is a straight line (Figure 16.1).

In the case of first-order (Figure 16.2) and second-order reactions (Figure 16.3), the graph obtained is a curve. If the reaction is second order then a 'deeper' curve is obtained for the graph of concentration versus time. The first-order curve is an exponential curve and the second-order curve is a quadratic curve. It can be hard to distinguish between these two types of kinetic behaviour with experimental data with random uncertainties.

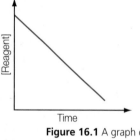

Figure 16.1 A graph of concentration against time for a zero-order reaction

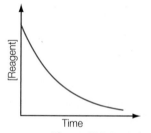

Figure 16.2 A graph of concentration against time for a first-order reaction

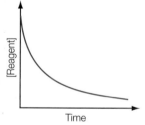

Figure 16.3 A graph of concentration against time for a second-order reaction

The order of a reaction can also be determined by plotting the rate against different initial concentrations of a reactant (Chapter 6), or the 'rate' for reactions like the iodine clock and that between thiosulfate ions and dilute acid (Chapter 6).

For a first-order reaction, the initial rate of reaction is directly proportional to the concentration of the reactant and the resulting graph is a sloping straight line. For a second-order reaction, the initial rate of reaction increases with concentration in a quadratic manner and the resulting curve is known as a parabola. For a zero-order reaction, the rate remains constant as the reactant is depleted, leading to a horizontal line for the rate, but one with a definite end.

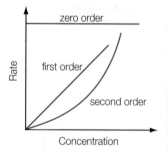

Figure 16.4 Rate–concentration graphs for zero-order, first-order and second-order reactions

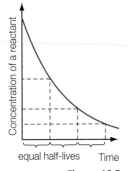

Figure 16.5
Concentration–time curve
for a first-order reaction

Concept of half-life for a first-order reaction

The time taken for the concentration of reactant to be halved during a chemical reaction is called the **half-life**. Identical kinetic behaviour is exhibited by substances undergoing radioactive decay (Chapter 2), with the exception that this physical process is unaffected by changes in temperature. For a reaction that has an overall order of one, the half-life is constant and is independent of the initial concentration of the reactants (see Figure 16.5).

For any other order the half-life is *not* constant: it will constantly change during the experiment. In first-order reactions, the half-life, $t_{1/2}$, is related to the rate constant k by the following equation:

$$t_{1/2} = \frac{\ln 2}{k} \text{ (where } \ln 2 = 0.693)$$

This expression is obtained by integrating the first-order differential equation:

$$\frac{dx}{dt} = k(a - x)$$

where a represents the amount of reactants and x represents the amount of products (see page 449).

Worked examples

The half-life of a first-order reaction is 100 seconds. Calculate its rate constant.

$$k = \frac{\ln 2}{t_{1/2}} = \frac{0.693}{100\,\text{s}} = 6.93 \times 10^{-3}\,\text{s}^{-1}$$

A first-order reaction has a rate constant of $0.100\,\text{s}^{-1}$. Calculate the half-life for this reaction.

$$t_{1/2} = \frac{\ln 2}{k} = \frac{0.693}{0.100\,\text{s}^{-1}} = 6.93\,\text{s}$$

The greater the value of the rate constant, k, the more rapid the exponential decrease in the concentration of the reactant (Figure 16.6). Hence, a first-order rate constant is a measure of the rate: the greater the value of the rate constant, the faster the reaction.

Another way of calculating the rate constant for a first-order reaction is to plot a graph of the natural logarithm of the concentration against the time. For a first-order reaction the graph (Figure 16.7) will take the form of a straight line with a slope (gradient) with a value of $-k$.

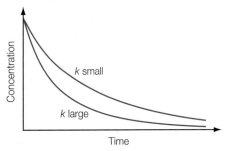

Figure 16.6 First-order concentration–time graphs with different rate constant values

Overall reaction order	Zero order A → products	First order A → products	Simple second order* A → products A + B → products
Rate expression	Rate = $k[A]_0$ **	Rate = $k[A]$	Rate = $k[A]^2$ Rate = $k[A][B]$
Data to plot for a straight-line graph	[A] versus t	ln [A] versus t	$\frac{1}{[A]}$ versus t
Slope or gradient equals	$-k$	$-k$	$+k$
Changes in the half-life as the reactant is consumed	$\frac{[A]_0}{2k}$ becomes shorter	$\frac{\ln 2}{k}$ is a constant	$\frac{1}{k\,[A]_0}$ becomes longer
Units of k	mol dm^{-3} s^{-1}	s^{-1}	dm^3 mol^{-1} s^{-1}

Table 16.4 Summary of graphical methods of finding orders and half-lives
* A simple second-order reaction is a reaction which is second order with respect to one reactant, that is rate = $k[A]^2$.
** $[A]_0$ is the initial concentration (at time $t = 0$).

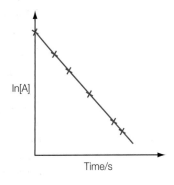

Figure 16.7 Determining the rate constant for a first-order reaction

■ Extension: Pseudo first-order reactions

Chemists often change the concentration of reactants so that they can study the effect the change has on the rate of the reaction. For example, consider the reaction of manganate(VII) ions in an acidic solution with ethanol to form ethanoate ions. The balanced equation for this reaction is:

$$5C_2H_5OH(aq) + 4MnO_4^-(aq) + 17H^+(aq) \rightarrow 5C_2H_5O_2^-(aq) + 4Mn^{2+}(aq) + 11H_2O(l)$$

Five moles of ethanol molecules are needed to react with four moles of manganate(VII) ions to form five moles of ethanoate ions and four moles of the manganese(II) ions. If the concentrations of ethanol and acid are raised to a high level relative to the manganate(VII) concentration, the kinetics of the appearance of the manganese(II) ion can be studied. In a similar manner, if the concentration of the hydrogen ion is raised above the stoichiometric requirement of the reaction, then the interaction of the other two reactants can be studied.

Each participant in the reaction can be studied in turn using this technique. The method is a pseudo first-order reaction because the kinetics of the single reactant can be studied as if the concentration were first order while the other reactants are held almost constant because their concentration is so large relative to the species being studied.

The reaction is first order with respect to all the reactants:

$$\text{rate} = k[C_2H_5OH(aq)][MnO_4^-(aq)][H^+(aq)]$$

However, if this reaction is carried out with manganate(VII) and hydrogen ions present in large excess, only a small amount of these reactants will be consumed during the hydrolysis. The concentration of these two species is then effectively constant, and the rate of reaction will depend only upon the concentration of the ethanol:

$$\text{rate} = k'[C_2H_5OH(aq)]$$

where $k' = k[MnO_4^-(aq)][H^+(aq)]$ and is termed an apparent first-order constant. The reaction thus exhibits under these experimental conditions 'pseudo' first-order kinetics with respect to the ethanol.

Figure 16.8 illustrates how two reactant concentrations would vary in a reaction if B were present in a large excess (greater than ×10). The reaction between A and B occurs with a 1:1 stoichiometry and the concentration of B at the start of the reaction is ten times greater than the concentration of A at the start of the reaction. The initial concentrations are represented by $[A]_0$ and $[B]_0$. The concentration of A rapidly approaches zero, but the concentration of B remains relatively constant during the reaction (it decreases by a much smaller proportion of its initial value).

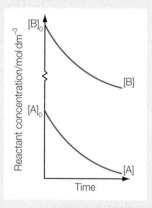

Figure 16.8 The changes in concentrations for two reactants, A and B, where B is present in large excess

A change from first-order to zero-order kinetics is observed during the decomposition of gases on the surface of a homogeneous catalyst. A simplified explanation is that at low concentrations the gas molecules are adsorbed on to the surface of the catalyst and the reaction exhibits first-order kinetics.

However, at high concentrations of gas molecules all the catalytic surface sites are occupied. The rate then becomes independent of the concentration of the reactant molecules and so the reaction exhibits zero-order kinetics. Similar 'saturation behaviour' (Figure 16.9) is shown by enzymes (Chapter 22).

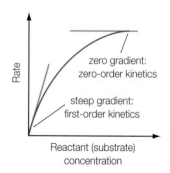

Figure 16.9 Graph of the rate of an enzyme-catalysed reaction against substrate concentration.

16.2 Reaction mechanism

Elementary steps

All chemical reactions, except the very simplest, actually take place via a number of simple reactions called **elementary steps**, or simply steps, which are collectively termed the **mechanism** of the reaction. It is a theoretical model of what chemists believe occurs during a chemical reaction at the molecular level.

Elementary steps are described as unimolecular if only one chemical species (atom, ion, radical or molecule) is involved and bimolecular if two chemical species are involved. True termolecular steps in the gas phase involving the simultaneous collision of three chemical species are virtually *impossible*, since the statistical chance of three species colliding is considerably less than that of two colliding.

Unimolecular steps involve either the decomposition or dissociation of a molecule into two or more smaller molecules or ions, or the rearrangement of a molecule:

$$A \rightleftharpoons B \text{ (rearrangement)}$$

or

$$A \rightleftharpoons B + C \text{ (decomposition or dissociation)}$$

Bimolecular steps involve two species colliding and reacting with each other:

$$A + B \rightleftharpoons C \qquad \text{or} \qquad A + A \rightleftharpoons D$$

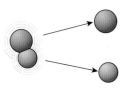

Figure 16.10
Unimolecular step

Unimolecular and bimolecular steps are illustrated in Figures 16.10 and 16.11.

It is important to distinguish between the elementary steps of a mechanism, indicated in this text by $\Rightarrow$, and the overall chemical reaction, indicated by $\rightarrow$. This is illustrated in the two specific examples of well established mechanisms outlined below.

The thermal decomposition of dinitrogen oxide is given by:

$$2N_2O(g) \rightarrow 2N_2(g) + O_2(g)$$

This reaction is believed to occur via the following steps:

$$N_2O(g) \Rightarrow N_2(g) + O(g)$$

$$N_2O(g) + O(g) \Rightarrow N_2(g) + O_2(g)$$

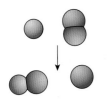

Figure 16.11
Bimolecular step

The oxygen atoms are **intermediates** and are produced and consumed during the reaction. However, the reaction:

$$2HI(aq) + H_2O_2(aq) \rightarrow I_2(aq) + 2H_2O(l)$$

has a more complex mechanism and is believed to occur via the following four bimolecular elementary steps:

$$I^-(aq) + H_2O_2(aq) \Rightarrow IO^-(aq) + H_2O(l)$$

$$H^+(aq) + IO^-(aq) \Rightarrow HIO(aq)$$

$$I^-(aq) + HIO(aq) \Rightarrow I_2(aq) + OH^-(aq)$$

$$H^+(aq) + OH^-(aq) \Rightarrow H_2O(l)$$

In these steps iodic(I) acid, HOI, and iodate(I) ions, IO⁻, are both intermediates.

Note that when the equations for the elementary steps are summed together, they give the overall equation.

Much of the evidence for proposed mechanisms of reactions come from kinetic studies that identify the products and intermediates formed during a reaction, as well as determining the individual orders of reactants and overall order of the reaction. Other evidence for reaction mechanisms comes from stereochemical studies which follow the changes in shape that occur as the reactants are converted into intermediates and products. To find out which bonds are broken and formed, chemists frequently use isotopically labelled reactants that incorporate heavy oxygen, ^{18}O, or heavy hydrogen, ^{2}H, which can be easily detected in the products and intermediates (Figure 16.12).

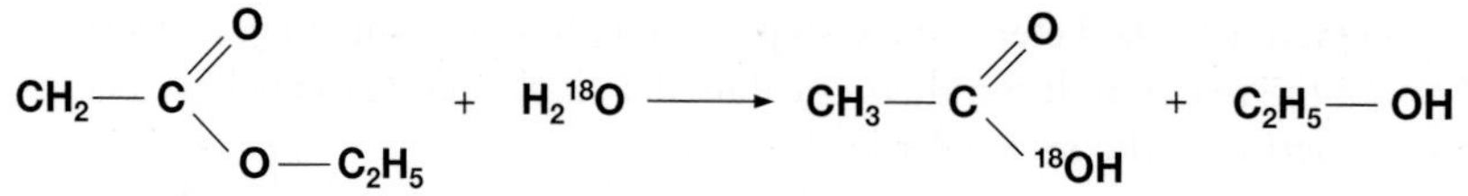

Figure 16.12 Use of labelling to investigate bond breaking during ester hydrolysis

$$CH_2-C\begin{smallmatrix}O\\\\O-C_2H_5\end{smallmatrix} \quad + \quad H_2{}^{18}O \quad \longrightarrow \quad CH_3-C\begin{smallmatrix}O\\\\{}^{18}OH\end{smallmatrix} \quad + \quad C_2H_5-OH$$

Kinetic studies only *suggest* a mechanism: they do not 'prove' that a particular mechanism is operating. They can perhaps disprove another candidate mechanism. Like any other chemical theory, a reaction mechanism may be revised or replaced as a result of later and more thorough investigations or more accurate calculations.

The elementary steps proposed for a mechanism must meet both of the following requirements:

■ The sum of the elementary steps must give the overall balanced equation for the reaction. For example, the reaction between sulfur dioxide and oxygen is:

$$2SO_2(g) + O_2(g) \rightarrow 2SO_3(g)$$

This reaction is catalysed by nitrogen monoxide via the following two elementary steps:

$$2NO(g) + O_2(g) \rightleftharpoons 2NO_2(g)$$

$$NO_2(g) + SO_2(g) \rightleftharpoons SO_3(g) + NO(g)$$

The sum of the first and second (multiplied through by two) elementary steps is:

$$2NO(g) + O_2(g) + 2NO_2(g) + 2SO_2(g) \rightleftharpoons 2NO_2(g) + 2SO_3(g) + 2NO(g)$$

Cancelling the nitrogen dioxide intermediate and the nitrogen monoxide catalyst gives the overall reaction:

$$\cancel{2NO(g)} + O_2(g) + \cancel{2NO_2(g)} + 2SO_2(g) \rightarrow \cancel{2NO_2(g)} + 2SO_3(g) + \cancel{2NO(g)}$$

This and similar catalysed reactions can be represented by cyclical diagrams as shown in Figure 16.13. The catalyst remains 'rotating' within the middle, and the reactants and products enter and leave the cycle.

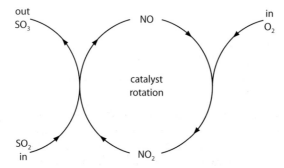

Figure 16.13 Catalytic cycle for the oxidation of sulfur dioxide by oxygen

■ The mechanism must agree with the experimentally determined rate expression.

Each elementary step determines the order of the reaction with respect to the reactant or catalyst. Specifically, a bimolecular step will have second-order kinetics and a unimolecular reaction will have first-order kinetics.

However, these orders will *only* appear in the rate expression *if* they occur during the rate-determining step of the mechanism, *or* in an equilibrium directly before the rate-determining step. The rate-determining step is the slowest step of the mechanism and determines the rate of the overall reaction.

Outlined opposite are mechanisms of five reactions with their corresponding rate expressions. Intermediates do *not* appear in a rate expression.

Overall reaction:	$A + B \rightarrow D$		
Mechanism:	$A + B \Rightarrow D$		rate = $k[A][B]$

Overall reaction:	$P + Q \rightarrow R$		
Mechanism:	$P \Rightarrow S$	slow	
	$S + Q \Rightarrow R$	fast	rate = $k[P]$

Overall reaction:	$K + 2M \rightarrow N$		
Mechanism:	$2M \rightleftharpoons M_2$	fast	
	$M_2 + K \Rightarrow N$	slow	rate = $k[M]^2[K]$

Overall reaction:	$A + B \rightarrow C + D$		
Mechanism:	$A \Rightarrow M + D$	slow	
	$M + B \Rightarrow C$	fast	rate = $k[A]$

Overall reaction:	$T + R \rightarrow P$		
Mechanism:	$T + H^+ \rightleftharpoons TH^+$	fast	
	$TH^+ + R \Rightarrow P + H^+$	slow	rate = $k[T][R][H^+]$

(Hydrogen ions are acting as a catalyst here.)

Molecularity

The number of particles participating in the rate-determining step is known as the **molecularity**. (The term molecularity should be reserved for describing the elementary steps of a complex reaction. It should not be used to describe an overall reaction of a complex mechanism.) It may or may not be the same value as the overall order of the reaction and, unlike the order, will always be a positive integer.

So, for example, a reaction may involve several elementary steps and yet still exhibit second-order kinetics. For example, consider the reaction:

$$2A + B \rightarrow C + D$$

This occurs via the following elementary steps:

$A + B \Rightarrow X$	slow
$X + A \Rightarrow C + D$	fast

where X is an intermediate in the reaction. The first and second steps are both bimolecular, but only the kinetics of the first step, the rate-determining step, will appear in the rate expression. The reaction will be first order with respect to A and first order with respect to B. The overall order of the reaction is two. During the reaction all of the intermediate X formed reacts with A as fast as it formed. The rate of formation of C and D under these conditions equals the rate of formation of X.

Only reactants that take part in the rate-determining step will appear in the final rate expression for a reaction (after intermediates have been removed). A reactant that does not appear in the rate-determining step will not appear in the rate expression. This means that changing the concentration of that reactant will have no effect on the rate of that reaction. It is said to have zero-order kinetics.

The reaction between iodine and propanone shows zero-order kinetics with respect to iodine. Changing the concentration of iodine has *no* effect on the rate of the reaction. The chemical equation for the reaction is:

$$CH_3COCH_3(aq) + I_2(aq) \rightarrow CH_3COCH_2I(aq) + HI(aq)$$

The rate expression is:

$$\text{rate} = k[\text{CH}_3\text{COCH}_3(\text{aq})][\text{H}^+(\text{aq})]$$

The accepted mechanism that explains this rate equation involves a mechanism with four elementary steps.

Step 1 Rate-determining step: rapid protonation

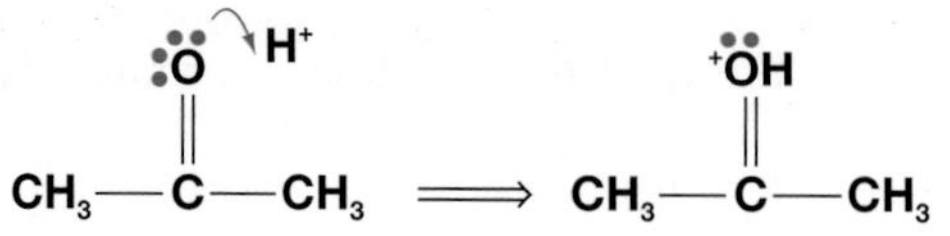

Step 2 Deprotonation and formation of propene-1,2-diol (an intermediate)

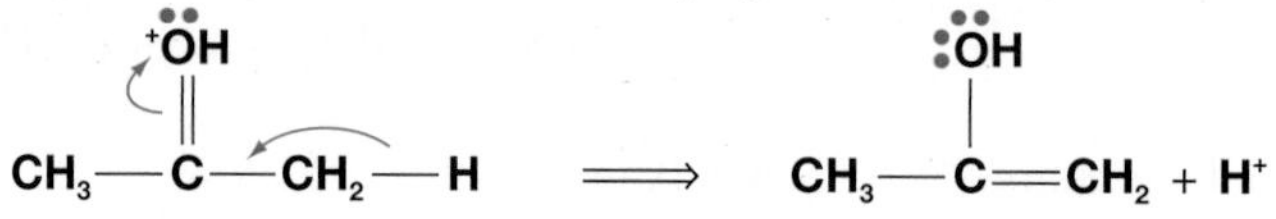

Step 3 Reaction of propene-1,2-diol with iodine to form a carbocation intermediate

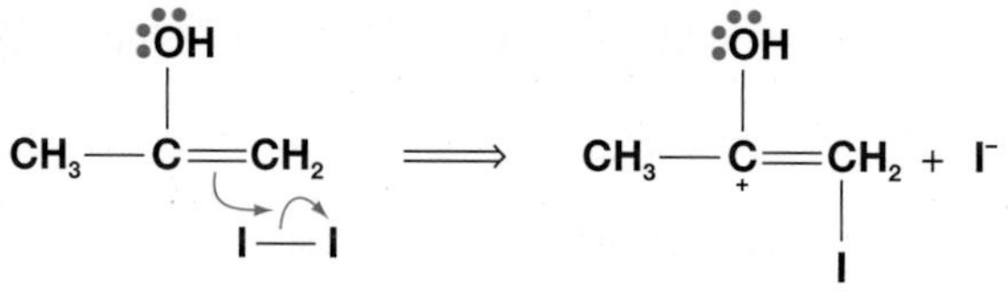

Step 4 Deprotonation and formation of product, iodopropanone

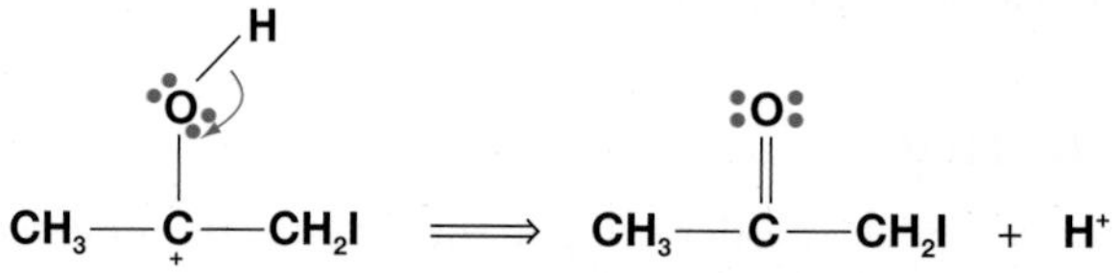

The first slow step of the mechanism is the rate-determining step. It has a molecularity of two since only propanone molecules and hydrogen ions are involved. The molecularity of two in the rate-determining step gives the reaction an order of one with respect to both the propanone and acid. The iodine is *not* involved in the rate-determining step and therefore changing its concentration has *no* effect on the overall rate.

Kinetic studies provide much of the data required for proposing reaction mechanisms. Two organic examples (Chapter 20) are outlined below. Bromoethane, $\text{C}_2\text{H}_5\text{Br}$, undergoes rapid hydrolysis in the presence of aqueous alkali. The products of the reaction are ethanol, $\text{C}_2\text{H}_5\text{OH}$, and bromide ions:

$$\text{C}_2\text{H}_5\text{Br}(\text{aq}) + \text{OH}^-(\text{aq}) \rightarrow \text{C}_2\text{H}_5\text{OH}(\text{aq}) + \text{Br}^-(\text{aq})$$

Experimental investigations reveal that the reaction is first order with respect to the concentration of hydroxide ions and first order with respect to the amount of bromoethane:

$$\text{rate} = k[\text{C}_2\text{H}_5\text{Br}(\text{aq}][\text{OH}^-(\text{aq})]$$

This implies that bromoethane and hydroxide ions are both involved in the rate-determining step. The mechanism is thought to involve the transition state or activated complex shown below:

$$\text{OH}^- + \text{C}_2\text{H}_5\text{Br} \Rightarrow [\text{HO} \cdots \text{C}_2\text{H}_5 \cdots \text{Br}]^- \Rightarrow \text{C}_2\text{H}_5\text{OH} + \text{Br}^-$$

The reaction is therefore a bimolecular second-order reaction termed $\text{S}_\text{N}2$, where S indicates substitution (the replacement of one atom (or group) by another atom (or group), in this example bromo by hydroxyl); N indicates that the organic species is attacked by a nucleophile (an electron pair donor), in this example, the hydroxide ion; and 2 indicates a molecularity of two. (The two *may* also indicate the order *but* if the reaction is carried out with a large excess of hydroxide ions then pseudo first-order kinetics are observed.)

The dotted lines in the transition state indicate partial bonds and the negative charge is delocalized or 'spread out' over both the partial bonds. The transition state is preceded and followed by the two related structures shown below, which show the HO–C bond getting shorter and stronger as the C–Br bond gets longer and weaker:

$$[HO\cdots\cdots\cdots\cdots C_2H_5\cdots Br]^- \qquad \text{and} \qquad [HO\cdots\cdots C_2H_5\cdots\cdots\cdots\cdots\cdots Br]^-$$

In contrast, the hydrolysis of 2-bromo-2-methylpropane (tertiary-butyl bromide), $(CH_3)_3CBr$, exhibits first-order kinetics:

$$\text{rate} = k[(CH_3)_3CBr]$$

and so the initial rate is *independent* of the concentration of the alkali. In other words, it exhibits zero-order kinetics with respect to hydroxide ions. The kinetic data suggests a unimolecular mechanism that involves only the 2-bromo-2-methylpropane in a rate-determining step:

Step 1 $(CH_3)_3Br \rightarrow (CH_3)_3C^+ + Br^-$

Step 2 $(CH_3)_3C^+ + H_2O \rightarrow (CH_3)_3COH + H^+$

This hydrolysis is a unimolecular first-order reaction termed S_N1, where S indicates substitution, N indicates that the organic species is attacked by a nucleophile (electron pair donor, in this example the hydroxide ion), and 1 indicates a molecularity of one.

(Note: a more detailed description of this mechanism using curly arrows to show electron pair movement is given in Chapter 20.)

■ Extension: Chain reactions

Reactions with very high rates often involve chemical species called radicals which contain one or more unpaired electrons. These are formed when a covalent bond is split homolytically. During this process each of the atoms or groups involved in the covalent bond accepts one of the bonding or shared pair of electrons:

$$A : B \rightarrow A + B\bullet$$

Homolytic cleavage of bonds frequently takes place at high temperatures, in the presence of ultraviolet light or when reactions are carried out in non-polar solvents. Homolytic cleavage in a molecule usually occurs in the weakest bond, which is often a single bond.

The explosive reaction between hydrogen and chlorine in the presence of sunlight is an example of a so-called **chain reaction** that involves free radicals as intermediates. The first step of the mechanism is an endothermic step involving the absorption of the ultraviolet light present in sunlight. The chlorine molecules dissociate into chlorine atoms, which contain seven valence electrons, one of which is unpaired:

$$Cl_2 \rightarrow 2Cl\bullet$$

(Note: for simplicity only the single unpaired electron is drawn to emphasize the radical nature of the chlorine atom.) This step of the mechanism is known as the **initiation** step.

The chlorine atoms or radicals then react with hydrogen molecules to release hydrogen atoms or radicals. This and the subsequent step in the mechanism are termed **propagation** steps, since in these steps the product, HCl, is formed along with another free radical:

$$Cl\bullet + H_2 \rightarrow HCl + H\bullet$$

The hydrogen atoms or radicals formed react with chlorine molecules to regenerate the chlorine atoms or radicals:

$$H\bullet + Cl_2 \rightarrow HCl + Cl\bullet$$

The two propagation steps often occur many times, constituting a chain reaction. The reaction rates of the two propagation steps are very high because of the relatively low activation energies for these two steps of the mechanism.

The chain reaction stops when two free radicals combine to form molecules. These steps are called chain **termination** steps and three operate in the mechanism:

$$2Cl\bullet \to Cl_2; \qquad 2H\bullet \to H_2; \qquad H\bullet + Cl\bullet \to HCl$$

Free-radical chain reactions also occur during the chlorination of methane (Chapter 10) and the methyl group of methylbenzene. Ozone depletion by chlorofluorocarbons (CFCs), acid rain formation and formation of photochemical smog (Chapter 25) also involve free-radical reactions. (Free-radical reactions are also operating in unpolluted atmospheres and play an important role in all chemical reactions that occur in the gas phase.) The combustion of hydrocarbons, such as petrol, also proceeds via a free-radical mechanism, which has important consequences for the smooth running and performance of combustion engines. Chain reactions may also have ions as intermediates, as opposed to free radicals.

Transition-state theory

This theory suggests that as molecules collide and bond breaking and bond formation take place, the interacting molecules are temporarily in a high-energy and unstable state. This state is known as a **transition state** or **activated complex**. It is invariably of higher enthalpy or potential energy than either reactants or products and is inherently unstable. The transition state can either decompose to re-form the reactants, or it can undergo further changes to form the product molecules (or an intermediate). This is illustrated in Figure 16.14 by the simple bimolecular reaction between hydrogen and iodine to form hydrogen iodide:

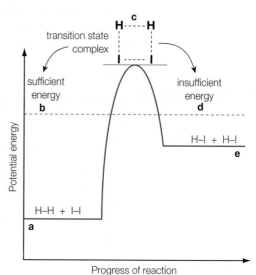

Figure 16.14 Enthalpy or potential energy level diagram for the formation of hydrogen iodide

a   represents the enthalpy of the individual reactant hydrogen and iodine molecules;
b   weak covalent bonds start to form between the hydrogen and iodine atoms of the hydrogen and iodine molecules. Simultaneously, the hydrogen–hydrogen and iodine–iodine bonds start to lengthen and weaken;
c   represents the formation of the transition state or activated complex;
d   the two hydrogen–iodine bonds continue to shorten and strengthen, while the hydrogen–hydrogen and iodine–iodine bonds continue to lengthen and weaken. In other words, electron density steadily increases between the hydrogen and iodine atoms;
e   represents the formation of the two hydrogen iodide molecules.

Hydrogen iodide is only formed if the colliding hydrogen and iodine molecules have sufficient kinetic energy to overcome the energy barrier, which corresponds here to the activation energy. A successful reaction also requires the hydrogen and iodine molecules to collide in a 'sideways' fashion: a so-called steric factor. Transition-state theory (TST) can be used to calculate reaction rates and transition states from a knowledge of the molecular structures and shapes of reactants.

TOK Link

The concept of a transition state originates from **transition-state theory**. Before transition-state theory chemists had explained rates of reactions in terms of collision theory, which is based on the kinetic theory of gases. It treats collisions by regarding the reacting molecules as hard spheres colliding with one another.

Transition-state theory does not conflict with collision theory. It assumes that reactions involve collisions but takes into account some of the details of the collision, such as how the reacting molecules must approach one another for reaction to be possible and what the effect of a solvent might be. None of these factors are accounted for by simple collision theory.

One of the assumptions of transition-state theory is that the transition state is, in a certain sense, at equilibrium with the reacting molecules. This special kind of equilibrium is termed a quasi-equilibrium. Transition states do not exist except as the state corresponding to the highest energy value on a reaction coordinate plot: they cannot be captured or directly observed.

However, use of a technique called femtochemical IR spectroscopy allows chemists to probe molecular structure extremely close to the transition point. The Egyptian-born chemist Ahmed H. Zewail was awarded the 1993 Nobel Prize in Chemistry for his work in this area.

Transition-state theory was first proposed in a paper published in 1933 by an American chemist called Henry Eyring. The theory has withstood the test of time – so far – but it has not been successful in predicting, from first principles, the rates of chemical reactions.

16.3 Activation energy

16.3.1 **Describe** qualitatively the relationship between the rate constant (k) and temperature (T).

The effect of temperature

When the temperature increases, the rate of a chemical reaction increases very rapidly. It has been found that for many reactions, the initial rate and the rate constant, k, vary with temperature in an exponential manner (Figure 16.15). This relationship between absolute temperature and the rate constant can be approximately described or modelled by the **Arrhenius equation**:

$$k = A e^{-E_a/RT}$$

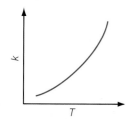

Figure 16.15 Plot of rate constant, k, against absolute temperature, T

The Arrhenius equation describes reactions involving gases, as well as those occurring in solution or on the surface of a catalyst. E_a and A are both constants characteristic of a particular reaction, R is a fundamental physical constant for all reactions and T and k are variables. None of the three constants change significantly with temperature. The expression ($e^{-E_a/RT}$) is known as the **exponential factor** and allows for the large effect of an increase in temperature in the Arrhenius equation.

A represents the **Arrhenius constant** (which has the same units as the rate constant, k, usually $dm^3 mol^{-1} s^{-1}$), E_a represents the **activation energy** (units of $kJ mol^{-1}$), T represents the absolute temperature (K) and R represents the **gas constant** ($8.31 J K^{-1} mol^{-1}$). The Arrhenius constant is a measure of the proportion of molecules that collide with enough kinetic energy to react and which also have the correct orientation to react.

The activation energy, E_a, is commonly interpreted as being a measure of the 'energy barrier' that a reaction has to overcome before it can proceed. Its value controls the 'sensitivity' of the reaction to changes in temperature. Low activation energies give rise to fast rates of reaction and a low sensitivity to changes in temperature. Large activation energies give rise to slow reactions at low temperatures and a high sensitivity to changes in temperature.

The Arrhenius equation is used to calculate the activation energy and the Arrhenius constant of a reaction. First you need to experimentally measure the rate constants of the reaction at several different temperatures. The modified form of the Arrhenius equation shown below is used to transform the data so that an Arrhenius plot can be produced (see page 445).

$$\ln k = \ln A - \frac{E_a}{RT}$$

Extension: Arrhenius temperature dependence

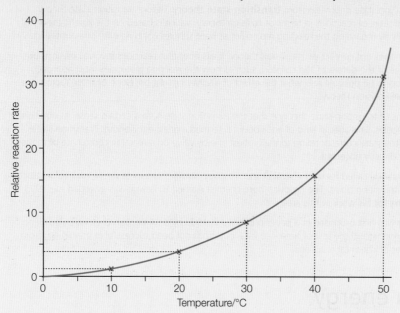

Some reactions have an activation energy of approximately $50\,\mathrm{kJ\,mol^{-1}}$ which means they exhibit so-called **Arrhenius temperature dependence**: a rise in temperature of $10\,°C$ will approximately double the initial rate and rate constant of the reaction over a range of temperatures (Figure 16.16). However, since values of activation energies vary considerably, reactions may be either much faster or much slower.

Figure 16.16 Arrhenius temperature dependence

History of Chemistry

Svante Arrhenius (1859–1927) was a Swedish physical chemist who was awarded the Nobel Prize in Chemistry in 1903 for his work on ionic solutions. The Arrhenius equation was first proposed by the Dutch chemist J.H. van't Hoff in 1884, but Arrhenius provided a physical justification and interpretation for it. He also proposed the idea of panspermia: that life might have been carried between planets by spores. His theory of acids and bases defined acids as substances that produce hydrogen ions in solution (as the only positive ions) and bases as substances that produce hydroxide ions (as the only negative ions). He also developed a theory to explain ice ages and the greenhouse effect (Chapter 25).

Figure 16.17 Svante Arrhenius

Calculating activation energies

16.3.2 Determine activation energy (E_a) values from the Arrhenius equation by a graphical method.

If you take natural logarithms of both sides of the original form of the Arrhenius equation and rearrange it, you get:

$$k = A\,e^{-E_a/RT}$$

$$\ln k = \ln A + \ln e^{-E_a/RT}$$

$$\ln k = \ln A - \frac{E_a}{RT}$$

$$\ln k = \ln A - \left(\frac{E_a}{R}\right)\frac{1}{T}$$

This form of the Arrhenius equation fits the general formula for a straight line, that is, $y = mx + c$. Here $\ln k$ is analogous to y, m to $-E_a/R$, x to T^{-1} and $\ln A$ to c. Essentially, the original Arrhenius equation describing a curve has been transformed into a more useful linear or 'straight line' form (Figure 16.18).

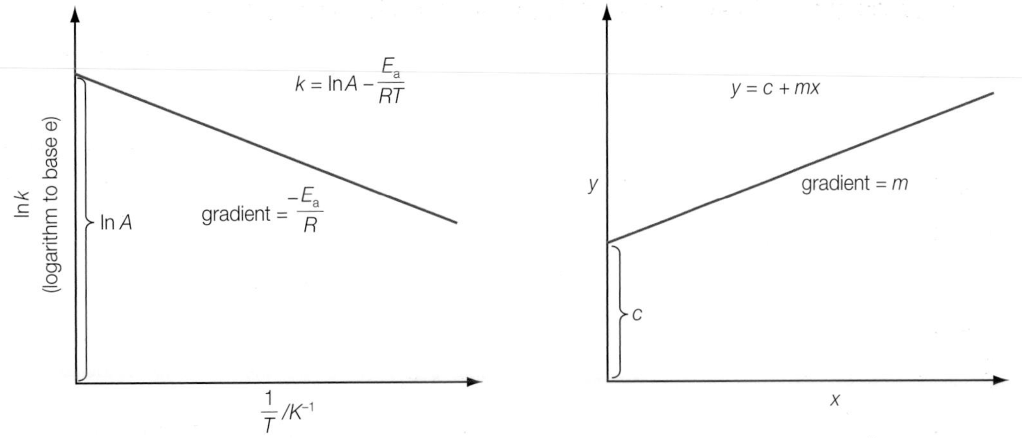

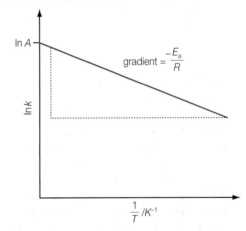

$$k = \ln A - \frac{E_a}{RT}$$

gradient $= \frac{-E_a}{R}$

$\ln A$

$\ln k$ (logarithm to base e)

$\frac{1}{T}$ /K^{-1}

$y = c + mx$

gradient $= m$

c

y

x

Figure 16.18 Comparing the equation $\ln k = \ln A - \frac{E_a}{RT}$ with $y = c + mx$

An **Arrhenius plot** is a graph of the natural logarithm of the rate constants, k, against the reciprocal of the corresponding absolute temperatures (T^{-1}). As the temperature, T, increases $\frac{-E_a}{RT}$ becomes less negative, and $\ln k$, and therefore the rate constant k, increases.

A sloping straight-line graph (Figure 16.19) is obtained, which can be used to calculate the experimental activation energy and Arrhenius factor. The slope or gradient has a value of $-E_a/R$ and the intercept on the rate constant axis is $\ln A$. Alternatively, once the activation energy, E_a, has been determined, the Arrhenius constant, A, can be calculated by substituting into the Arrhenius equation. The Arrhenius plot will give an initial value for the activation energy in J mol^{-1} provided temperatures are measured in kelvin and the gas constant is expressed in J mol^{-1} K^{-1}.

$\ln A$

gradient $= \frac{-E_a}{R}$

$\ln k$

$\frac{1}{T}$ /K^{-1}

Figure 16.19 An Arrhenius plot of $\ln k$ against T^{-1}

Worked example

The rate constant, k, was determined for a reaction at various temperatures. The results are given on the right.

a Plot a graph of $\ln k$ against T^{-1} where T must be expressed as an absolute temperature. (Take the $\ln k$ axis from -12 to -8 and the T^{-1} axis from 0.0030 to 0.0038.)

Temperature/°C	Second-order rate constant, k /mol^{-1} dm^3 s^{-1}
5	6.81×10^{-6}
15	1.40×10^{-5}
25	2.93×10^{-5}
35	6.11×10^{-5}

b Calculate the gradient (slope) of your Arrhenius plot and use it to determine a value for the activation energy, E_a, in kJ mol^{-1}.

c Calculate an approximate value for the Arrhenius constant, A, using the Arrhenius plot.

a First, draw up a table showing T^{-1} and $\ln k$ values.

T^{-1}/K^{-1}	$\ln k$
3.597×10^{-3}	-11.90
3.472×10^{-3}	-11.18
3.356×10^{-3}	-10.44
3.247×10^{-3}	-9.70

Use these values to draw the Arrhenius plot (Figure 16.20).

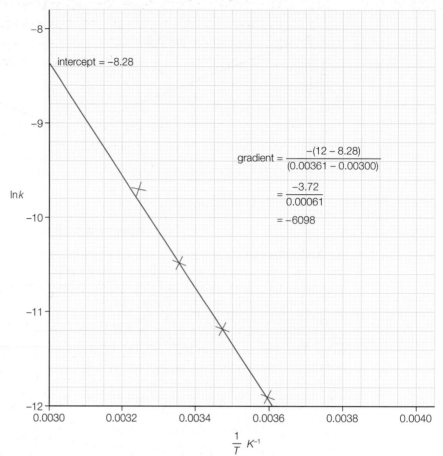

intercept = −8.28

$$\text{gradient} = \frac{-(12 - 8.28)}{(0.00361 - 0.00300)}$$

$$= \frac{-3.72}{0.00061}$$

$$= -6098$$

ln k

$\frac{1}{T}$ K^{-1}

Figure 16.20 The Arrhenius plot for the worked example

b From the graph,

y-intercept = −8.28

gradient = −6098

So $\qquad y = -6098x - 8.28$

The straight line gives:

$$\text{slope} = -\frac{E_a}{R} = -6098\,\text{K}$$

$$-E_a = -6098\,\text{K} \times 8.31\,\text{J}\,\text{mol}^{-1}\,\text{K}^{-1}$$

$$E_a = 50.7\,\text{kJ}\,\text{mol}^{-1}$$

c Intercept $\quad \ln A = -8.28$

$$A \cong 2.5 \times 10^{-4}\,\text{mol}^{-1}\,\text{dm}^3\,\text{s}^{-1}$$

If Arrhenius plots are drawn on the same axes for two reactions with different activation energies (Figure 16.21) you can see that the reaction with the higher activation energy has a steeper gradient. This indicates that the rate constant, and hence the initial rate, will change with temperature much more quickly than the reaction with the lower activation energy. This is because the value of the activation energy is given by the expression: $-R \times$ gradient.

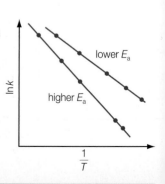

Figure 16.21 Arrhenius plots for two reactions with different activation energies

Using a formula to calculate the activation energy (from a pair of rate constant and temperature measurements)

The activation energy can also be calculated from two values of the rate constant, k_1 and k_2, at only two temperatures, T_1 and T_2, by using the following formula:

$$\ln k_2 - \ln k_1 = \frac{-E_a}{R}\left(\frac{1}{T_2} - \frac{1}{T_1}\right)$$

Worked example

The rate constant for a reaction increases by a factor of 1.65 when the temperature is increased from 20 °C to 40 °C. Calculate the activation energy.

$$\ln(1.65) = \frac{-E_a}{8.31 \times 10^{-3}}\left(\frac{1}{313} - \frac{1}{293}\right)$$

$$E_a = 19.1\,\text{kJ}\,\text{mol}^{-1}$$

Catalysis

Catalysts are substances that increase the rate constant of a particular chemical reaction but remain chemically unchanged. There are three types of catalysts: **homogeneous catalysts**, **heterogeneous catalysts** and enzymes.

Homogeneous catalysts are in the same physical state as the reactants. Often both the catalyst and the reactants are in solution.

Heterogeneous catalysts are in a different physical state or phase from the reactants. Often the reactants are gases and the catalyst is a solid, frequently a transition metal or transition metal compound. Many industrial processes use heterogeneous catalysts (Chapter 13). The action of catalysts may be modified by the presence of low concentrations of certain substances, which may be classified as either promoters, inhibitors or catalyst poisons.

Promoters increase rates of reactions. For example, in the Haber process (Chapter 7) traces of the metal molybdenum act as a promoter for the iron catalyst. **Inhibitors** slow down the rates of catalysed reactions by reacting and removing intermediates. **Catalyst poisons** greatly reduce the rates of catalysed reactions by binding to catalytic sites on the surface of the heterogeneous catalyst. Examples include arsenic, carbon monoxide and hydrogen cyanide.

The rate of catalysed reactions depends on the 'amount' of catalyst present. For a homogeneous catalyst the reaction rate depends on the concentration of the catalyst. The rate of a heterogeneous catalysed reaction depends on the surface area of the catalyst.

Enzymes are biological catalysts present in living cells (Chapter 22). They are large globular protein molecules, consisting of a large number of amino acid molecules polymerized together (Chapter 20). They frequently contain a metal ion in their **active site** where the catalysis occurs.

Unlike other catalysts, enzymes only increase reaction rates over a narrow pH range (typically about 5 to 8) and a narrow temperature range (typically about 20–40 °C). They are also very sensitive to the presence of various inhibitors which affect their kinetic behaviour (Chapter 22).

Language of Chemistry

The term catalysis was coined by Jakob Berzelius, who first noted in 1835 that certain chemicals speed up chemical reactions. The word catalyst is derived from the Greek word *katalyein* meaning 'to dissolve'. ■

Homogeneous catalysis

Homogeneous catalysis usually involves the formation of an intermediate during the reaction, which then decomposes to form the product and the unchanged catalyst (Figure 16.22). The presence of a catalyst provides an alternative pathway that is more energetically favourable. Generally, the rate is directly proportional to the concentration of the catalyst.

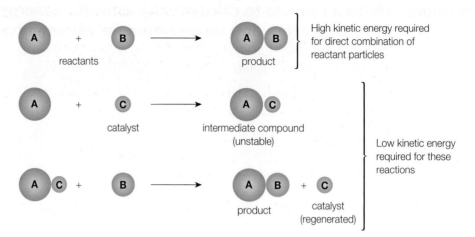

Figure 16.22 The principle of homogeneous catalysis

The reaction between iodide and peroxodisulfate ions requires homogeneous catalysis by iron(II) ions (Figure 16.23). The uncatalysed reaction involving direct reaction between negatively charged iodide and peroxodisulfate ions has a high activation energy and hence a slow reaction rate:

$$S_2O_8^{2-}(aq) + 2I^-(aq) \rightarrow 2SO_4^{2-}(aq) + I_2(aq)$$

The presence of iron(II) ions provides an alternative mechanism that involves two elementary steps involving reactions between *oppositely charged* ions which have lower activation energies and hence higher rate constants:

$$S_2O_8^{2-}(aq) + 2Fe^{2+}(aq) \rightleftharpoons 2SO_4^{2-}(aq) + 2Fe^{3+}(aq)$$

$$2Fe^{3+}(aq) + 2I^-(aq) \rightleftharpoons 2Fe^{2+}(aq) + I_2(aq)$$

The iron(II) ion catalyst is consumed during the first elementary step, but is 'regenerated' in the second elementary step. The iron(III) ion intermediate is readily detected and studied by spectroscopic techniques (Chapter 21).

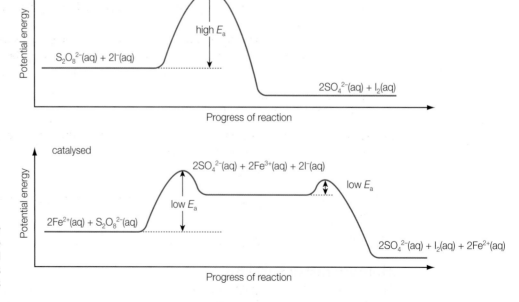

Figure 16.23 Energy level diagrams for the uncatalysed and catalysed oxidation of iodide ions and peroxodisulfate ions

Heterogeneous catalysis

Usually the process involves the reaction of two gases on the surface of a solid catalyst, which is often in the form of a powder. Such reactions are not only important in many industrial processes, but are also involved in acid rain formation and ozone depletion.

All heterogeneous catalysis occurs at a phase boundary, often involving the reaction of gas molecules on the surface of a solid. Figures 16.24 and 16.25 show the five steps that are thought to occur during heterogeneous catalysis, using the hydrogenation of ethene by hydrogen in the presence of nickel catalyst as an example.

The **first** step in heterogeneous catalysis involves the diffusion of the reacting gas molecules onto the surface of the catalyst. The **second** step involves the **adsorption** (Figure 16.24) of the reacting gas molecules onto the surface of the metal where they are temporarily bonded to the surface by weak intermolecular forces and/or dative or coordinate covalent bonds.

The **third** step involves the breaking of and formation of chemical bonds to bring about the formation of the product molecules. The activation energy for the reaction is lowered (relative to the uncatalysed reaction) because the reaction follows a different pathway, one with lower activation energy barriers. During the **fourth** step, **desorption**, the product molecules break free from the surface of the catalyst; this is the reverse of the adsorption step (Figure 16.25).

Finally, in the **fifth** step the molecules of the gaseous product diffuse away from the surface of the catalyst. Their places on the catalytic surface are then occupied by unreacted gas molecules.

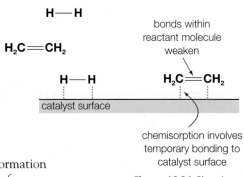

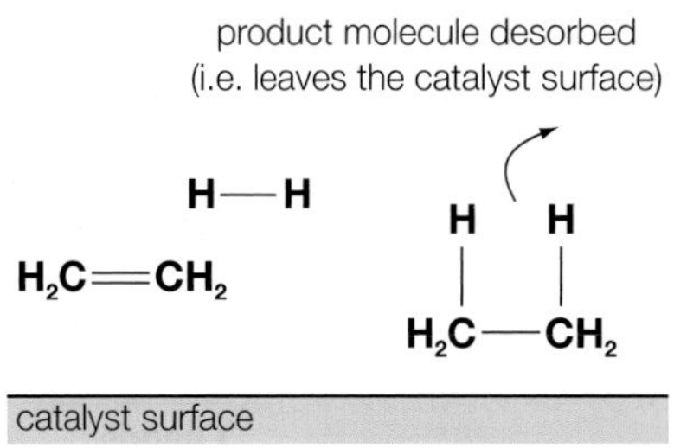

Figure 16.24 Chemisorption

Figure 16.25 Desorption

■ Extension: Integration of a first-order differential equation

The half-life of a first-order rate expression can be obtained by integrating the first-order differential rate equation.

Let a represent the initial concentration of a reactant A which is converted into products. After a time interval, t, a concentration of product represented by x has been formed.

	[A]	[Products]
At time = 0	a	0
At time = t	$a - x$	x

For a first-order reaction, the rate of formation of products is the product of the rate constant and the concentration of the reactant. Thus:

$$\text{rate} = k[\text{A}]$$

Using calculus notation, an expression can be written for the rate of formation of products at time t:

$$\frac{\mathrm{d}x}{\mathrm{d}t} = k(a - x)$$

Rearranging and integrating:

$$\int k\,\mathrm{d}t = \int \frac{\mathrm{d}x}{(a - x)}$$

$$kt = \ln\!\left(\frac{a}{a - x}\right)$$

Rearranging the equation using the exponential function yields:

$$\frac{a}{(a-x)} = \exp(kt) \qquad \text{or} \qquad a - x = a\exp(-kt)$$

The equation is often written as:

$$[A] = [A]_0(1 - e^{-kt})$$

where $[A]_0$ represents the concentration at time zero. This is known as an integrated rate expression.

If the equation $kt = \ln\left(\dfrac{a}{a-x}\right)$ is rearranged, then:

$$\ln a - \ln(a-x) = kt$$

and $\quad \ln(a-x) = -kt + \ln a$

A plot of $\ln(a-x)$ against t will generate a straight line whose slope is $-k$.

The half-life of a first-order reaction occurs when $x = \dfrac{a}{2}$.

Substituting this into the equation demonstrates that $k \times t_{1/2} = \ln 2$, or $t_{1/2} = \dfrac{0.693}{k}$, which does not depend on the initial concentration, a.

Table 16.3 summarizes the changes that affect the rate of reaction and the rate constant. Rate constants are unaffected by changes in concentration and are only affected by temperature (as described by the Arrhenius equation) or the presence of a catalyst, which provides a new pathway or reaction mechanism. Rates increase with concentration and pressure (if gaseous reactants are involved), which can be accounted for by simple collision theory (Chapter 6).

Change	Effect on rate of reaction	Effect on rate constant, k	Notes
Increase in concentration	Increased	No change	
Increase in pressure	Increased	No change	Only applies to gaseous reactants
Increase in temperature	Increased	Increased	
Use of a catalyst	Increased	Increased	A catalyst changes the rate expression

Table 16.3 Summary of the changes that affect the rate of reaction and the rate constant

▦ For the reaction A + B → products, the reaction rate can be expressed as:
reaction rate = $k[A]^a [B]^b$
This is known as the rate expression. k represents the rate constant and its value is constant for a given reaction at a particular temperature. The rate constant is independent of the concentration of the reactants. The units of a rate constant depend on the order of the reaction.

▦ The rate expression summarizes the relationship between the rate of a reaction and its reactants (and any catalysts). The rate expression can only be obtained experimentally. It *cannot* be deduced from the stoichiometric equation. A rate expression may contain all or some of the reactants. It may also include a homogeneous catalyst.

▦ The individual order of a reaction with respect to a reactant is the power of that reactant's concentration in the rate expression. The overall order is the sum of the powers of the concentration terms in the rate expression.

▦ Zero-order reaction: rate = k; the rate is independent of concentration.
First-order reaction: rate = $k[A]$; the rate is directly proportional to the concentration of A.
Second-order reaction: rate = $k[A][B]$, or $k[A]^2$ or $k[B]^2$. If the concentration of B is doubled in the last rate expression then the rate will be quadrupled.

▤ Graphs of reactant concentration against time elapsed since the start of the reaction have different shapes for zero-, first- and second-order reactions. Zero order is a straight line with a negative slope, first order is exponential and second order is a curve similar to $y = \dfrac{1}{x}$ (not exponential).

▤ The half-life is the time taken for the concentration of reactant to be halved during a chemical reaction. The half-life, $t_{1/2}$, of a first-order reaction is constant.
$t_{1/2} = \dfrac{\ln 2}{k}$.
Reactions with orders greater or less than one do *not* exhibit a constant half-life.

▤ One common approach used during kinetic analysis is to isolate one reactant and keep all the other reactants in large excess. Effectively, one reactant is consumed and the concentrations of other reactants may be considered as constant. (The concentrations of catalysts remain constant whether or not they are present in excess because catalysts are not consumed.) Only one reactant undergoes a significant change during the reaction and the data will show how the concentration (and hence rate) varies with the concentration of the selected reactant.

▤ An alternative approach in kinetic analysis to the 'isolation method' is the initial rates method. The rate is measured by recording the concentration change very early in the reaction. The investigation is repeated with one reactant concentration changed and the other reactant concentrations unchanged. The rate is then measured again. The changes in the values of the initial rates will indicate the individual orders for the rate expression.

▤ Many chemical reactions proceed by a sequence of elementary steps, each of which involves one or two particles of a chemical species. This series of elementary steps is collectively termed the mechanism of the reaction. The sum of the elementary steps yields the overall equation of the reaction.

▤ The molecularity of an elementary step in a reaction mechanism is the number of particles of a chemical species taking part in that step. Elementary steps that involve one particle of a chemical species are known as unimolecular; those that involve two are known as bimolecular. Mechanisms do not contain termolecular steps.

▤ The slowest step in the reaction mechanism determines the overall reaction rate. The slowest elementary step is known as the rate-determining step. Only reactants that participate in the rate-determining step or in an equilibrium immediately preceding the rate-determining step will appear in the rate expression.

▤ The mechanism for a chemical reaction is related to the rate expression. It *cannot* be deduced from the stoichiometric equation.

▤ Before changing into products, reactants form a transition state or activated complex which lies at the maximum in potential energy on an enthalpy level diagram.

▤ For chemical reactions there is often an exponential relationship between temperature and rate or rate constant. It cannot be accounted for by an increase in collision frequencies.

▤ This exponential relationship is described by the Arrhenius equation: $\ln k = \ln A - \dfrac{E_a}{RT}$, where E_a represents the activation energy and A the Arrhenius constant.

▤ The activation energy can be obtained from the gradient of an Arrhenius plot where the natural logarithm of the rate constant, k, is plotted against the reciprocal of the absolute temperature (T^{-1}).

▤ A catalyst does *not* change the position of equilibrium; it increases the rate of the forward reaction to the same extent as the reverse reaction. Catalysts do *not* affect the energetics (ΔH) and thermodynamics (ΔG) of the reaction.

▤ A homogeneous catalyst is in the same physical state as the reactants. A heterogeneous catalyst is in a different physical state to the reactants.

▤ Homogeneous catalysts often function by interconverting between two stable oxidation states.

▤ Heterogeneous catalysts adsorb reactant molecules onto their surfaces and help to increase their concentration and to weaken and break bonds.

Examination questions – a selection

Paper 1 IB questions and IB style questions

Q1 For the reaction, $2N_2O_5(g) \rightarrow 4NO_2(g) + O_2(g)$, the rate is expressed as $\dfrac{\Delta[O_2(g)]}{\Delta t}$. An equivalent expression would be:

A $-\dfrac{1}{2}\dfrac{\Delta[N_2O_5(g)]}{\Delta t}$

B $-2\dfrac{\Delta[N_2O_5(g)]}{\Delta t}$

C $8\dfrac{\Delta[NO_2(g)]}{\Delta t}$

D $-\dfrac{1}{8}\dfrac{\Delta[NO_2(g)]}{\Delta t}$

Q2 For the reaction

$(CH_3)_3CBr + OH^- \rightarrow (CH_3)_3COH + Br^-$

it is experimentally found that doubling the concentration of $(CH_3)_3CBr$ causes the reaction rate to be increased by a factor of two, but doubling the concentration of OH^- has no effect on the rate. What is the rate expression?

A rate = $k[(CH_3)_3CBr]^2[OH^-]$
B rate = $k[(CH_3)_3CBr][OH^-]$
C rate = $k[(CH_3)_3CBr]$
D rate = $k[(CH_3)_3COH][Br^-]$

Q3 In a chemical reaction at constant temperature, the addition of a catalyst:

A increases the fraction of reacting particles with more than a given kinetic energy.
B increases the equilibrium constant.
C increases the concentration of products at equilibrium.
D provides an alternative energy pathway with a different activation energy.

Q4 In aqueous solution, iodine reacts with propanone as represented by the following stoichiometric equation:

$I_2 + CH_3COCH_3 \rightarrow CH_3COCH_2I + H^+ + I^-$

The experimental rate expression is:

rate = $k[H^+][CH_3COCH_3]$

From this information it can be concluded that increasing the iodine concentration will:

A decrease the value of the equilibrium constant.
B increase the value of the equilibrium constant.
C decrease the rate of the reaction.
D not affect the rate of reaction.

Q5 The rate constant for the *first-order* decomposition of N_2O_5 to give NO_2 and O_2 is $0.166\,s^{-1}$ at $150\,^\circ C$. If two containers at $150\,^\circ C$ and $1\,atm$ pressure contain respectively $40\,g$ of N_2O_5 (container 1) and $20\,g$ of N_2O_5 (container 2), which of the following is true about the amount of time required for $\frac{3}{4}$ of the N_2O_5 to decompose in each container?

A Container 1 requires twice as much time as container 2.
B Container 1 requires 1.5 as much time as container 2.
C Container 1 requires half as much time as container 2.
D Container 1 requires the same amount of time as container 2.

For questions 6–8 refer to the gas phase reaction:

$2A + B_2 \rightarrow C + D$

whose experimental rate expression has been found to be:

rate = $k[A][B_2]^2$

Q6 The overall order of the reaction is:

A zero   **B** first   **C** second   **D** third

Q7 If the concentration of A is tripled and the concentration of B is doubled, the reaction rate would increase by the factor:

A 9   **B** 6   **C** 12   **D** 16

Q8 Which one of the following would increase the value of the rate constant, k?

A increasing the temperature
B increasing the concentration of A
C increasing the concentration of B
D adding an inert gas

Q9 For the reaction, $A + 2B \rightarrow 2C + D$

A rate = $k[A][B]^2$
B rate = $k[A][B]$
C rate = $\dfrac{k[A][B]^2}{2}$
D The rate expression is impossible to determine without experimental data.

Q10 Which of the following questions are answered by kinetic principles instead of thermodynamic principles?

    I How fast will a reaction be at a specific temperature?
    II Will a reaction be spontaneous at a specific temperature?
    III What are the energy changes that occur during a reaction?
    IV What is the reaction mechanism?

    A II and III **B** I and II **C** I and IV **D** III and IV

Q11 For a certain second-order decomposition reaction, the rate is $0.30\,mol\,dm^{-3}\,s^{-1}$ when the concentration of the reactant is $0.20\,mol\,dm^{-3}$. What is the rate constant ($dm^3\,mol^{-1}\,s^{-1}$) for this reaction?
    A 2.2   **B** 1.5   **C** 0.06   **D** 7.5

Q12 Two reactants A and B are mixed and the reaction is timed until a cloudy precipitate is formed. The data are:

[A]	[B]	Time/s
0.100	0.140	25
0.050	0.140	50
0.100	0.070	100

What is the order of the reaction with respect to A?
    A 0   **B** 1   **C** 3   **D** 2

Q13 For the following reaction:

$$H_2(g) + I_2(g) \rightarrow 2HI(g)$$

the experimental rate expression is:

$$rate = k\,[H_2(g)]\,[I_2(g)]$$

When time is given in seconds and the concentration is in $mol\,dm^{-3}$, the units for the rate constant are:
    A $mol\,dm^{-3}\,s^{-1}$ **C** $mol^{-1}\,dm^1\,s^{-1}$
    B $mol^{-1}\,dm^3\,s^{-1}$ **D** s^{-1}

Q14 For the reaction:

$$I(g) + I(g) \rightarrow I_2(g)$$

the reaction must be:
    A first order and exothermic.
    B first order and endothermic.
    C second order and endothermic.
    D second order and exothermic.

Q15 To what does A refer in the Arrhenius equation $k = Ae^{-E_a/RT}$?
    A activation energy   **C** gas constant
    B rate constant   **D** collision geometry
    Higher Level Paper 1, Nov 05, Q21

Q16 The rate expression for a reaction is shown below:

$$rate = k[A]^2[B]^2$$

Which statements are correct for this reaction?

    I The reaction is second order with respect to both A and B.
    II The overall order of the reaction is 4.
    III Doubling the concentration of A would have the same effect on the rate of reaction as doubling the concentration of B.

    A I and II only   **C** II and III only
    B I and III only   **D** I, II and III
    Higher Level Paper 1, Nov 03, Q20

Q17 In a certain reaction, in which the concentration of X was monitored over time, the following kinetic results are obtained:

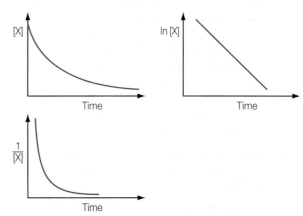

Which type of kinetic behaviour is being exhibited?
    A zero order   **C** third order
    B first order   **D** second and zero order

Q18 The rate of a gaseous reaction is given by the expression, $rate = k\,[P][Q]$. If the volume of the reaction vessel is reduced to $\frac{1}{4}$ of the initial volume, what will be the ratio of the new rate to the original rate?
    A $1:4$ **B** $1:16$ **C** $4:1$ **D** $16:1$
    Higher Level Paper 1, Nov 02, Q23

Q19 $2NO(g) + 2H_2(g) \rightarrow N_2(g) + 2H_2O(g)$

The following data were obtained for the reaction above. Use these data to determine the orders for the reactants nitrogen monoxide, NO, and hydrogen, H_2.

[NO(g)]/mol dm^{-3}	[H$_2$(g)]/mol dm^{-3}	Rate/mol dm^{-3} s^{-1}
0.040	0.240	0.066
0.080	0.240	0.264
0.240	0.040	0.400
0.240	0.080	0.800

Order of reaction

	NO	H$_2$
A	2	1
B	1	1
C	1	0
D	2	2

Q20 A certain chemical reaction can be represented by the overall equation:

$2A(g) + B(g) \rightleftharpoons C(g)$

At a particular temperature the initial rate of this reaction was measured for various initial concentrations of A and B, as shown below:

Experiment	Initial concentration of A/mol dm^{-3}	Initial concentration of B/mol dm^{-3}	Initial rate /mol C h^{-1}
1	0.5	0.5	1.0×10^{-3}
2	1.0	0.5	4.0×10^{-3}
3	1.0	1.0	4.0×10^{-3}
4	1.5	1.0	9.0×10^{-3}

On the basis of the evidence provided, it appears that the mechanism of this reaction would involve two or more steps. Given this experimental data, a possible rate-determining step might be:

A $A + B \Rightarrow$ intermediate
B $A + A \Rightarrow$ intermediate
C $A + B \Rightarrow C$
D $A + AB \Rightarrow C$

Paper 2 IB questions and IB style questions

Q1 Dinitrogen oxide decomposes to give nitrogen and oxygen according to the following equation:

$2N_2O(g) \rightarrow 2N_2(g) + O_2(g)$ $\qquad \Delta H = -82\,kJ\,mol^{-1}$

a The decomposition is a first-order reaction in the presence of gold as a catalyst. The half-life of the catalysed reaction at 834 °C is 1.62×10^4 s.

i Calculate the rate constant (velocity constant), k, for the reaction at this temperature and give the units of k. [1]

ii Calculate the activation energy of the reaction at this temperature, given the Arrhenius constant, $A = 25\,s^{-1}$. [2]

Higher Level Paper 2, Nov 99, Q3

Q2 Evidence suggests that the reaction between the gases nitrogen dioxide and fluorine is a two-step process:

$2NO_2(g) + F_2(g) \rightarrow 2NO_2F(g)$

Step 1 $NO_2 + F_2 \rightarrow NO_2F + F$ (slow)

Step 2 $F + NO_2 \rightarrow NO_2F$ (fast)

a State and explain which step is the rate-determining step. [1]
b State and explain which of the two steps is expected to have the higher activation energy. [2]
c Give the rate expression of the reaction based on your answer to **a**. [1]

Higher Level Paper 2, Nov 00, Q5

Q3 The following data were obtained for the reaction between A and B:

$A(aq) + 2B(aq) \rightarrow 3C(aq) + D(aq)$

Experiment	Initial concentration of reactant A/ mol dm^{-3}	Initial concentration of reactant B/ mol dm^{-3}	Initial rate of reaction/ mol dm^{-3} h^{-1}
1	0.400	0.400	1.00
2	0.800	0.400	4.00
3	0.800	1.600	16.00

a Give the order with respect to A. [1]
b Give the order with respect to B. [1]
c Write the rate expression for this reaction. [1]
d Using the first experiment, calculate the value of the rate constant. [1]

17 Equilibrium

STARTING POINTS
- The kinetic theory postulates that a gas is composed of tiny particles in rapid and random motion, moving in straight lines between collisions.
- Gases exert pressure due to collisions between their particles and the walls of their container. This is known as vapour pressure and increases with temperature.
- Gases near their boiling points are often referred to as vapours.
- Evaporation is the movement of particles from the liquid state to the gas state at temperatures below the boiling point.
- Volatile liquids are those that evaporate at a high rate and have a relatively high vapour pressure at a given temperature.
- Attractive intermolecular forces operate between molecules in a liquid.
- A dynamic equilibrium is established in a closed system when the rate of the forward reaction equals the rate of the backward reaction.
- At equilibrium, the measurable and observable properties of the system become constant.
- A state of equilibrium can be approached from either direction and, at a particular temperature, this equilibrium position is the same whichever direction it is approached from.
- Le Châtelier's principle states that when a system at equilibrium is subjected to a change, the system shifts its position to partially counteract the change and restore a state of equilibrium.
- Factors that may affect the position of equilibrium are: the concentration of a reactant or product, the pressure exerted on the system and the temperature of the system.
- The presence of a catalyst results in equilibrium being achieved faster, but has no effect on the position of equilibrium.
- Reactions at equilibrium obey the equilibrium law: the ratio of the multiple of the equilibrium concentrations of the products to the multiple of the equilibrium concentrations of the reactants is a constant (K_c). In this expression each concentration term is raised to the power equal to the coefficient of the substance in the balanced equation. Thus, for the reaction:

$$aA + bB \rightleftharpoons cC + dD$$

$$K_c = \frac{[C]^c[D]^d}{[A]^a[B]^b}$$

- The position of equilibrium is measured by the equilibrium constant, K_c.
- Reactions that favour products have large values of K_c; reactions that favour reactants have low values of K_c.
- Temperature is the only factor that influences the value of K_c. It is independent of the initial concentrations of reacting species.

17.1 Liquid–vapour equilibrium

Introduction

Figure 17.1 An oil painting of the *SS Dunedin*, the first ship to complete the successful transport of refrigerated meat on a journey from New Zealand to Britain

It is difficult to over-estimate the importance of the invention of the modern refrigerator in the context of food transportation and storage. The invention of refrigerated transport for food led to a revolution in the globalization of markets and the availability of important commodities across, and between, continents. Commercial organizations experimented with refrigerated shipping in the mid-1870s. The first commercial success came when William Davidson fitted a compression refrigeration unit to the New Zealand sailing vessel *SS Dunedin* in 1882 (Figure 17.1). These developments led to a meat and dairy boom in Australia, New Zealand and South America.

A refrigerator takes advantage of the energy transfers when a volatile liquid evaporates and condenses. The key stage of the system depends on the fact that evaporation is an endothermic process, withdrawing heat from the surroundings (Chapter 5). Within the body of a refrigerator (Figure 17.2) a pump circulates a liquid with a low boiling point around a circuit of pipes. This volatile liquid vaporizes in the pipes inside the refrigerator, taking in heat energy from the air inside the refrigerator and keeping the food inside cool.

Continuing round the circuit, the vapour is compressed by the pump as it flows out at the bottom of the refrigerator. The compressed vapour is hot. As it flows through the pipes at the back of the refrigerator the fluid cools and condenses back to a liquid, giving out energy and heating up the air around the back of the cabinet. Overall, the circulating fluid transfers energy from inside the refrigerator to the air in the room. The use of the reversible evaporation–condensation cycle of volatile liquids in refrigeration and air conditioning (Figure 17.3) is one of the features of modern living.

the coolant uses heat energy from the air in the cabinet to vaporize in the coils around the ice box

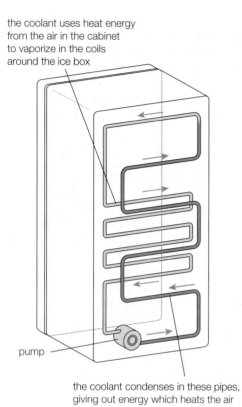

pump

the coolant condenses in these pipes, giving out energy which heats the air

Figure 17.2 The coolant system of a refrigerator

Evaporation takes place when particles at the liquid surface have sufficient kinetic energy to overcome the forces holding them within the liquid. These particles have kinetic energy greater than the attractive intermolecular forces of attraction between the particles. Thus some of the particles at the surface enter the space above the liquid and become a gas.

Figure 17.3 A domestic air conditioning unit

Evaporation takes place at the surface of the liquid at any temperature. As we saw earlier (Chapter 7), an equilibrium can be established between a liquid and its vapour if the system is contained in a closed vessel so that the vapour cannot escape.

History of Chemistry

The refrigerated storage and transport of food

The first gas absorption refrigeration system was developed by Ferdinand Carré of France in 1859 and patented in 1860. He used gaseous ammonia dissolved in water (referred to then as '*aqua ammonia*'). Such systems were not developed for use in homes because of the toxicity of ammonia, however they were used to manufacture ice for sale. In the United States, the consumer public at that time still used the ice box with ice brought in from commercial suppliers, many of whom were still harvesting ice in winter (from frozen lakes, for instance) and storing it in icehouses.

An original slant on the significance of the ice-making is highlighted in Paul Theroux's novel *Mosquito Coast* (and the subsequent film starring Harrison Ford (Figure 17.4)). In his escape to nature from the trappings of affluence in the USA, the one item that Allie Fox, the central character of the book, takes with him is the engineering know-how to build an ice house. The idiosyncratic inventor has frequent battles with his ice-making machine as he tries to establish his family in the inhospitable surroundings of the Honduran coast. The ice he produces is clearly central to how he sees his family surviving and relating to the local people.

Figure 17.4 Harrison Ford in the film *Mosquito Coast*

Language of Chemistry

'Dynamic' is a key word in our understanding of what is happening in liquid–vapour equilibria. It implies continuous activity. Except at the beginning and end of the day, a busy store is often at dynamic equilibrium, with the number of customers arriving matching the number leaving. Dynamic equilibrium is quite unlike the 'static equilibrium' of a ball at rest at the foot of a hill (where 'static' indicates an absence of activity). A store is at static equilibrium before it opens for business.

A further example of a dynamic equilibrium is a fish swimming upstream at the same speed as the stream is flowing down. The fish appears to be static (not moving), but it is in dynamic equilibrium with the stream. ■

Figure 17.5 Dynamic equilibrium: the fish appears to be still. However, it is swimming upstream at the same velocity as the stream is flowing in the opposite direction

The equilibrium between a liquid and its own vapour

17.1.1 Describe the equilibrium established between a liquid and its own vapour and how it is affected by temperature changes.

We met the phenomenon of liquid–vapour equilibria in Chapter 7. There we looked at the dynamic equilibrium established in a sealed flask of bromine.

$$Br_2(l) \rightleftharpoons Br_2(g)$$

This equilibrium situation can be generalized for any liquid in a sealed container, and becomes:

liquid $\rightleftharpoons$ vapour

The position of the equilibrium will depend on the liquid being used and the temperature.

This phenomenon can be explored more precisely if the apparatus is designed so that no other substance is present in the space occupied by the liquid/vapour system being studied. A volatile liquid (water or ethanol, for instance) can be injected into the evacuated space above a column of mercury in a barometer tube (Figure 17.6). The liquid will float on top of the mercury column as it has a much lower density.

The volatile liquid evaporates and the vapour fills the space above the mercury. This creates a vapour pressure which lowers the level of the mercury a short distance. If too little liquid is injected then it will all evaporate. However, provided some liquid remains above the mercury, an equilibrium will be set up. While some liquid remains the vapour pressure rises and finally reaches a maximum value. This value will be constant at a given temperature and is known as the **saturated vapour pressure**. The rate of evaporation equals the rate of condensation, and liquid and vapour are in equilibrium.

$$\text{liquid} \underset{\text{condensation}}{\overset{\text{evaporation}}{\rightleftharpoons}} \text{vapour}$$

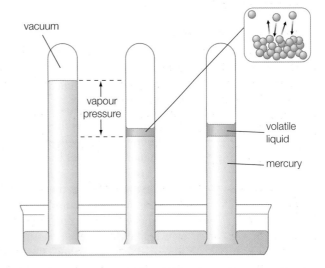

Figure 17.6 An apparatus for studying the pressure exerted by the vapour of a liquid. The vapour pressure is the same however much liquid is present and whatever the surface area of the liquid. When the liquid and vapour are in equilibrium, the rate at which molecules leave the surface is equal to the rate at which they return

Liquid	Saturated vapour pressure*/Pa
Water	3170
Methanol	16 350
Ethanol	7850
Benzene	12 600
Mercury	0.226

*Saturated vapour pressures measured in this way used to be stated in mmHg, but the SI unit of pressure is the pascal (Pa).

Table 17.1 The saturated vapour pressures of certain liquids at 298 K

Table 17.1 gives some values of saturated vapour pressure for certain liquids at 298 K. You will note that the value for mercury is very much lower than for the other liquids because of the strong metallic bonding in the liquid. This is crucial to the validity of the method.

Examine the figures given in Table 17.1. We will be looking at the factors that affect the magnitude of the vapour pressure of a liquid later in the chapter. However, from your previous knowledge you should be able to think about the following questions.

Worked example

a Can you identify which factors you think are the cause of the differences you see in Table 17.1?

b Why is it crucial that the value for mercury is so much lower than the others?

a The vapour pressure above a liquid will depend on how readily molecules can escape from the surface of the liquid. This will depend on the relative strength of the intermolecular forces in the liquid. In benzene there will just be van der Waals' forces between the molecules, although these will be relatively strong as benzene is quite a large molecule. In methanol and ethanol there is also the capacity for hydrogen bonding between molecules, with each molecule involved in one hydrogen bond per molecule. Methanol is a relatively small molecule, so the van der Waals' forces will be weaker here than in ethanol and benzene. Water is a molecule capable of hydrogen bonding, with each molecule being involved in two hydrogen bonds per molecule.

b If the vapour pressure of mercury was not so much lower than the values for the other liquids then it would interfere with our estimation of the values for these other liquids.

The value for the vapour pressure is a reflection of a dynamic process. When the liquid, water for instance, is first introduced into the vacuum above the mercury, some of its molecules leave the liquid phase and form the vapour. When molecules that have escaped from the liquid water strike the water's surface, they may be recaptured by the attractive intermolecular forces. As the number of molecules in the vapour increases, more of them will strike the surface until eventually a point is reached at which the number of molecules returning to the liquid exactly matches the number escaping from it. At this stage, the vapour is condensing as fast as the liquid is evaporating (Figure 17.7). That is, the rate of evaporation, in moles of H_2O per second, is equal to the rate of condensation. At this point the concentration of molecules in the vapour, and hence its pressure, remains constant, and the liquid and vapour are in 'dynamic equilibrium'.

In a closed container, a liquid and its vapour reach dynamic equilibrium when the pressure of the vapour has risen to a particular value that depends on two factors:

- the liquid being studied
- the temperature.

Figure 17.7a shows the alternative situations possible in a closed container under different conditions. As the liquid is injected into the container then the situation will resemble that in **i**, with more molecules leaving the liquid to eventually establish the equilibrium described in **iii**. If, for instance, situation **iii** – an equilibrium – had been established at a particular temperature and the container was then cooled, we would move into situation **ii**. Here more molecules are entering the liquid state then leaving it, and so condensation is taking place. This will continue until the equilibrium conditions appropriate to the new, lower temperature are achieved, i.e. a new equilibrium vapour pressure is attained. This new equilibrium position will involve a lower saturated vapour pressure.

The change just described is consistent with Le Châtelier's principle as lowering the temperature always favours the exothermic process in an equilibrium system. Condensation is the exothermic process in this case.

$$\text{liquid} \underset{\text{exothermic}}{\overset{\text{endothermic}}{\rightleftharpoons}} \text{vapour}$$

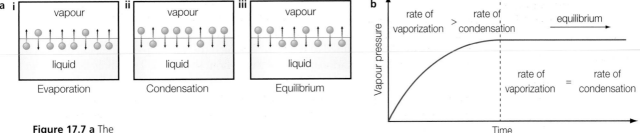

Figure 17.7 a The alternative situations possible for a liquid in contact with its vapour in a closed vessel **b** A graphical illustration of what happens when a liquid is introduced into a closed container

Figure 17.7b illustrates graphically what happens after a liquid is introduced into a closed container. Molecules will escape the surface of the liquid and the rate of vaporization will be greater than the rate of condensation until equilibrium is reached. The vapour pressure at this point will be the saturated vapour pressure of the liquid at that temperature.

It is possible to define saturated vapour pressure as follows:

The **saturated vapour pressure** of a liquid is the pressure exerted by its vapour when the two phases are in dynamic equilibrium in a closed system at a given temperature.

Under conditions where evaporation is very slow, the pressure of the vapour does not need to be very high for the condensation rate, which is proportional to the pressure, to match it. Hence a low saturated vapour pressure is a sign that molecules are leaving the surface at a relatively low rate. On the other hand, if evaporation is rapid the vapour pressure will reach a relatively high value before condensation occurs at a matching rate. Hence a high saturated vapour pressure is a sign that molecules are leaving the liquid surface at a greater rate. The nature of the liquid being studied is important as different intermolecular forces are involved for different liquids. Thus the value of the saturated vapour pressure of a liquid at 298 K gives us an indication of the intermolecular forces at work in that liquid.

Vapour pressure, temperature and kinetic theory

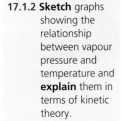

17.1.2 Sketch graphs showing the relationship between vapour pressure and temperature and **explain** them in terms of kinetic theory.

Applying the ideas of the kinetic theory (Chapter 1) to this situation helps us understand the effect of temperature on the system. Remember, evaporation takes place from the surface of the liquid because the molecules there are less strongly bonded and can escape to the vapour more easily than those in the body of the liquid (Figure 17.8a).

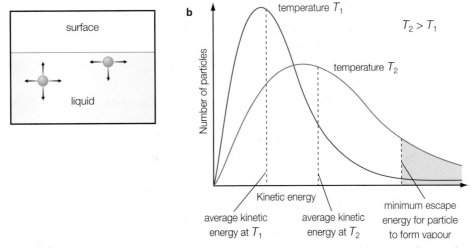

Figure 17.8 a A diagram illustrating how the forces acting on a molecule at the surface of a liquid are less than those acting on a molecule in the body of the liquid **b** The Maxwell–Boltzmann distribution of molecular energies in a liquid. At a higher temperature (T_2) more particles possess the necessary kinetic energy to escape the liquid

With increasing temperature the molecules in the warm liquid have (on average) more kinetic energy; they move at higher (average) speed and more will have sufficient kinetic energy to escape from the surface (Figure 17.8b). This minimum 'escape energy' needed to leave the surface of the liquid is analogous to the activation energy (E_a) required for molecules to participate in a chemical reaction. Molecules with high kinetic energy will overcome the attraction of the neighbouring molecules at the surface and are released into the vapour above the liquid. The vapour pressure increases with temperature (Figure 17.9).

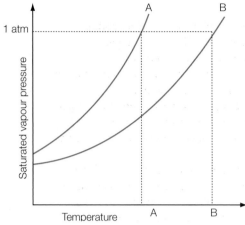

Figure 17.9 Saturated vapour pressure curves for two liquids A and B, where A (e.g. ethanol, b.p. 78°C) is more volatile than B (e.g. water, b.p. 100°C)

Figure 17.9 shows how the saturated vapour pressure of a liquid increases with temperature. It also shows why two different liquids do not have the same boiling point. As a liquid is heated, its vapour pressure increases. When the temperature of the liquid is raised to the point at which the vapour pressure *equals* the external pressure, vaporization can occur throughout the liquid. Thus, bubbles of vapour form in the liquid and rise to the surface. Since vaporization no longer occurs only at the surface, it can proceed very rapidly. This situation we call 'boiling'; the temperature at which it occurs is the 'boiling point'. If the container is not sealed then the liquid will boil when its vapour pressure is equal to that of the atmosphere.

Liquid A is more volatile than B. We can see on the graph that, at any temperature, liquid A (ethanol, for instance) has a higher vapour pressure than liquid B (say, water). Consequently, liquid A boils at a lower temperature than liquid B (Figure 17.9). So, at 1 atmosphere, ethanol boils at 78°C, while water boils at 100°C.

Looking more closely at these curves it is possible to see why water, for instance, boils at a lower temperature at a higher altitude. In Denver, Colorado – the mile-high city – water boils at 95°C; while it would boil at 69°C at the summit of Everest. As we rise above sea level, atmospheric pressure is reduced. As water is heated at altitude its vapour pressure reaches equality with this reduced atmospheric pressure at a lower temperature. And so the water boils at this lower temperature.

Applications of Chemistry

Scientists make use of the lowering of boiling point under reduced pressure in the laboratory when using a rotary evaporator (Figure 17.10a). This apparatus is particularly useful when trying to purify organic substances that may be unstable at temperatures near their normal boiling point. The compound can be distilled over at a lower boiling temperature, protecting it from thermal decomposition. This type of apparatus is a particular example of the more general process of distillation under reduced pressure.

It is sometimes desirable to raise the boiling point of water above 100°C to achieve a desired outcome. Pressure cookers are useful for cooking food faster than would normally be possible. Weights on a valve in the lid of a cooker designed to withstand pressure (Figure 17.10b) raise the pressure inside. The raised pressure means that the boiling temperature of the water in the cooker is increased above 100°C and the food cooks faster. A laboratory use of this phenomenon is the autoclave, which is used to sterilize instruments and apparatus for certain biological experiments.

Figure 17.10
a Rotary evaporator
b pressure cooker

Language of Chemistry

It is important to realize the difference in meaning of the terms 'evaporation' and 'boiling'. Both are related to the change in state between the liquid and gaseous phases, but evaporation can take place at any temperature, whereas boiling happens at a particular temperature determined by the external pressure.

As we have seen, evaporation is a surface phenomenon, with molecules escaping from the surface of the liquid. Boiling, on the other hand, takes place throughout the whole body of the liquid, with bubbles of gas forming anywhere in the liquid (Figure 17.11).

The **normal boiling point** of a liquid is the temperature at which its vapour pressure is equal to the atmospheric pressure. The stronger the forces of intermolecular attraction in a liquid, the higher the boiling point. ■

Figure 17.11 Boiling a liquid involves bubbles of gas forming at any point in the liquid, not simply at the surface

The role of intermolecular forces

17.1.3 State and explain the relationship between enthalpy of vaporization, boiling point and intermolecular forces.

In Chapter 7 we referred very briefly to the phenomenon that it is possible to feel a cooling effect on the palm of your hand as a result of the evaporation of a volatile liquid such as ether (ethoxyethane) or propanone. This simple observation shows that evaporation is an endothermic process. The heat energy from the surroundings (your hand) is used to overcome the intermolecular forces and allow molecules to escape from the surface of the liquid.

The energy involved in evaporation is referred to as the **enthalpy of vaporization** of the liquid (ΔH_{vap}). The enthalpy of vaporization is the amount of energy required to convert one mole of pure liquid to one mole of the gas at its normal boiling point. The energy is used to overcome the bonds and/or intermolecular forces operating in the liquid.

A volatile liquid has a high vapour pressure at room temperature as its intermolecular forces are weak. In contrast, liquids containing molecules capable of forming hydrogen bonds are much less volatile than others. Water is a covalent substance with low molar mass, but it has strong hydrogen bonds between its molecules. This explains why water has a relatively low vapour pressure and a relatively high enthalpy of vaporization. The following worked example should remind you of some of the factors involved here.

Worked example

The following question illustrates the background to the ideas we are discussing in this section and draws on the concepts relating to the forces between molecules covered in Chapter 4.

The physical properties of a simple molecular compound, such as its melting point, boiling point, vapour pressure or solubility, are related to the strength of attractive forces between the molecules of that compound. These relatively weak attractive forces are called intermolecular forces. They differ in their strength and include the following:

A van der Waals' forces (interactions involving temporary induced dipoles)

B dipole–dipole interactions (forces between permanent dipoles)

C hydrogen bonds.

a By using the letters **A**, **B** or **C**, state the *strongest* intermolecular force present in *each* of the following compounds and explain your answers.

i ethanal	CH_3CHO	**iii** 2-methylpropane	$(CH_3)_2CHCH_3$
ii ethanol	CH_3CH_2OH	**iv** methoxymethane	CH_3OCH_3

b On the basis of the answers in **a**, put the compounds **i**, **ii** and **iii** in order of increasing boiling point.

a i B, interactions between permanent dipoles. The $>C=O$ bond is polarized, creating a permanent dipole in each molecule of ethanal. There is no capacity for hydrogen bonding as there are no hydrogen atoms attached to an electronegative atom in the molecule.

ii C, the C–O and O–H bonds are polarized and there is capacity for hydrogen bonding between the –OH groups on neighbouring molecules.

iii A, this molecule contains no atoms that would set up any permanent dipoles. The interactions between molecules are those between temporary dipoles (van der Waals' forces).

iv B, again the C–O bond in each molecule is polarized to create a permanent dipole. However, there is no capacity for hydrogen bonding.

b The order of increasing boiling point is

2-methylpropane < ethanal < ethanol

weakest		strongest
intermolecular		intermolecular
forces		forces

(It is always important to read questions like this carefully to make sure you write the order in the correct direction.)

This section discusses how the enthalpy (heat) of vaporization, ΔH_{vap}, of a compound also fits into this pattern. Evaporation separates the particles in a liquid so the values for enthalpy of vaporization are an approximate measure of the strength of the forces between particles in liquids. Figure 17.12 shows apparatus that can be used to find the experimental value of ΔH_{vap} for ethanol. The immersion heater is connected to the electricity supply via a joulemeter to measure the energy transferred to the liquid as it boils. Knowing the mass of ethanol distilled over and the heat energy supplied, a value for ΔH_{vap} can be calculated.

Substances with strong ionic or metallic bonding have much higher boiling points and enthalpies of vaporization than substances consisting of molecules with weak intermolecular forces. Figure 17.13 shows the correlation between boiling point and the enthalpy of vaporization for a wide range of substances of varying bonding types.

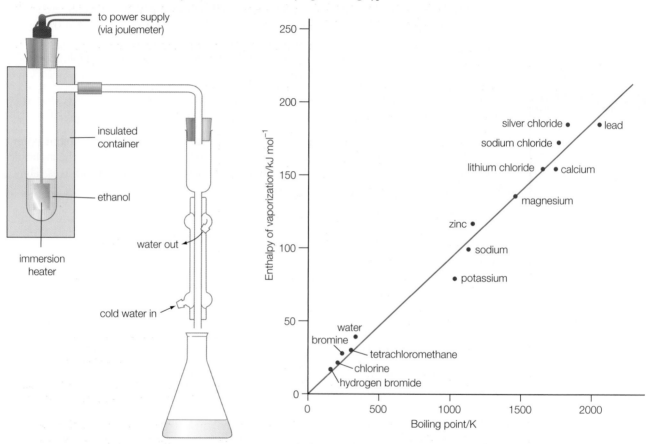

Figure 17.12 Apparatus for measuring the enthalpy of vaporization of a liquid

Figure 17.13 Graph showing the correlation between the boiling points and ΔH_{vap} for a wide range of substances

History of Chemistry

In 1884 the Irish physical chemist **Frederick Trouton (1863–1922)** discovered that, for many liquids, the enthalpy of vaporization divided by the normal boiling point is a constant.

$$\Delta H_{vap} / T = 90 \, \text{J} \, \text{K}^{-1} \, \text{mol}^{-1}$$

where ΔH_{vap} represents the enthalpy of vaporization ($\text{kJ} \, \text{mol}^{-1}$) and T represents the normal boiling point (K).

This expression is known as Trouton's rule and implies a linear relationship between the enthalpy of vaporization and the normal boiling point. Figure 17.13 shows an approximate relationship between these two properties.

Worked example

Table 17.3 shows the enthalpy (heat) of vaporization, ΔH_{vap}, of the hydrides of elements in groups 5 and 6.

a For the hydrides of group 6 elements:

Plot a graph of ΔH_{vap} on the vertical axis against relative molecular mass (M_r) on the horizontal axis. Then use it to estimate the value of ΔH_{vap} for water if there were no hydrogen bonding present.

Hydrides of group 5	ΔH_{vap}/kJ mol^{-1}	Hydrides of group 6	ΔH_{vap}/kJ mol^{-1}
NH_3	23.4	H_2O	40.7
PH_3	14.6	H_2S	18.7
AsH_3	17.5	H_2Se	19.3
		H_2Te	23.2

Table 17.3 The values of ΔH_{vap} for the hydrides of the elements of groups 5 and 6

Then, by subtracting this estimated value from the actual ΔH_{vap} value, obtain a measure of the hydrogen bonding contribution in water.

Use this value to get an estimate of the strength of 1 mole of hydrogen bonds by dividing it by two.

b Repeat a similar estimation for the strength of the hydrogen bonding in ammonia.

a The data here is limited, and so the extrapolation and estimation will not be that accurate. Values in the range 18.0–18.5 kJ mol^{-1} for the estimate of the ΔH_{vap} of water if there is no hydrogen bonding are justifiable. This leads to a value of about 11 kJ mol^{-1} for the strength of 1 mole of hydrogen bonds in water. (Taking 18.5 kJ mol^{-1} as the extrapolated value for water without hydrogen bonding; we get 40.7 – 18.5 = 22.2 kJ mol^{-1}. Each water molecule participates in two hydrogen bonds per molecule. So the estimated strength of the hydrogen bonding in water is approximately 11 kJ mol^{-1}.)

The graphing and extrapolation shown in Figure 17.14 has been carried out using the computer program 'Graphical Analysis 3.0'.

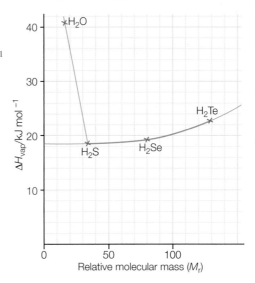

Figure 17.14 A plot of the boiling point ΔH_{vap} values for the hydrides of group 6 elements

b Repeat the exercise for the group 5 hydrides. There is less data here so the estimation for ammonia without hydrogen bonding can just be done by inspection. Values in the range 9–12 kJ mol^{-1} can be justified for ammonia.

■ Extension: The vapour pressure of solids

Figure 17.15 A block of frozen carbon dioxide ('dry ice') subliming

Figure 17.13 shows values of ΔH_{vap} for some solids. It is important to realize that apparently involatile solids do have a measurable vapour pressure associated with them. A solid in a closed container will fill the space above it with its vapour, thus exerting a characteristic sublimation vapour pressure. However, the vapour pressures of most solids at room temperature are much smaller than those of liquids, and the rate at which a solid evaporates is often so low that its vapour never reaches its final pressure.

There are some solids that we are aware of that do have a discernible vapour pressure. The smell of traditional mothballs is created because naphthalene sublimes. Solid carbon dioxide sublimes, which is why it is referred to as 'dry ice' (Figure 17.15).

Language of Chemistry

The three states of matter (Chapter 1) can also be referred to as phases. A **phase** is defined as a homogenous part of a system which is chemically and physically uniform throughout. Phases are separated from one another by physical boundaries known as phase boundaries. A number of foods consist of one phase dispersed through another phase (Chapter 26). These are known as colloids or dispersed phases.

A single gas in contact with a liquid, for example a carbonated drink (carbonic acid) or liquid bromine in a sealed container (Chapter 7), is a two-phase system. Two liquids that do not mix, for example hexane and water, also form a two-phase system. However, two liquids that completely mix, for example ethanol and water, or any mixture of gases, form single-phase systems.

Equilibria that involve changes from one phase to another are known as **phase equilibria**. **Phase changes** occur when heat energy is added to or removed from a system. Examples of phase changes include melting, boiling, condensing and sublimation (Chapter 1).

Some solids undergo two sharp phase changes when heated. They first melt sharply giving a cloudy liquid, and then at a higher temperature they form a clear liquid. The cloudy state is known as a liquid crystal (Chapter 23). ∎

Applications of Chemistry

Naphthalene is an aromatic molecule that consists of two fused benzene rings. It is a volatile crystalline solid used to make mothballs (Figure 17.16). Mothballs are small balls of naphthalene used to stop moth larvae from eating stored clothes. They can also be used as an insecticide. The mothballs can be left in a closed cupboard and the vapour pressure will slowly increase as the naphthalene sublimes.

Figure 17.16 'Mothballs' made from naphthalene

Language of Chemistry

Foul smells!

Putrescine originates in putrefying and rotting flesh. It is one of the breakdown products of some of the amino acids found in animals. Although the molecule is a poisonous solid, as flesh decays the vapour pressure of the putrescine it contains becomes sufficiently large to allow its disgusting smell to be detected. It is usually accompanied by cadaverine (named after the cadavers that give rise to it), a poisonous syrupy liquid with an equally disgusting smell. Putrescine (1,4-diaminopentane, $H_2N(CH_2)_4NH_2$) and cadaverine (1,5-diaminopentane, $H_2N(CH_2)_5NH_2$) are both amines.

Vapour trails

Snow sublimes when the surrounding temperature is below 0 °C. The white cloud-like trail (Figure 17.17) that is observed coming out of the high-flying jet is water vapour from the aircraft exhaust, being converted directly into ice which is slowly converted back into water vapour without passing through the liquid state (sublimation). The water vapour from the combustion of kerosene inside the aircraft's engines contributes to global warming (Chapter 25). ∎

Figure 17.17 Vapour trail or contrail from a commercial airliner

■ Extension: Phase diagrams

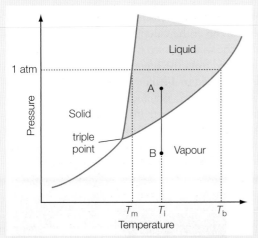

Figure 17.18 A typical phase diagram for a pure substance

Figure 17.9 showed how the saturated vapour pressure of a liquid varies with temperature. If we cool the liquid, it freezes to a solid. We have also seen how solids have their own vapour pressure (page 463). As with liquids, the vapour pressure of a solid will also be temperature dependent and this relationship can be plotted on a graph. The solid/vapour curve is similar in shape to that of the liquid/vapour curve but the values are lower. By combining these two curves representing how the saturated vapour pressures of a solid and its liquid vary with temperature we have the beginning of what is known as a phase diagram for a substance (Figure 17.18).

A **phase diagram** is a graphical plot of pressure versus temperature. A typical phase diagram has three regions: solid, liquid and gas. Each of these regions is separated by a phase boundary line. These lines show how the equilibria between the three phases vary with temperature and pressure. In the areas to each side of the lines only one phase exists. A phase diagram allows us to predict the state of matter which is stable at any give temperature and pressure.

A gas at temperature T_1 can be liquefied by pressurizing it. This can be represented by the vertical line A–B on Figure 17.18. If a horizontal line at a pressure of one atmosphere is drawn on Figure 17.18, the temperature at which it crosses the solid–liquid boundary is the melting pointing of the substance, T_m. The temperature at which the one-atmosphere line crosses the liquid–gas boundary is the boiling point of the substance, T_b. The point where the three boundary lines meet (intersect) is known as the **triple point** (Chapter 1).

The value for the triple point of carbon dioxide involves a pressure more than five times atmospheric pressure. This means that, at atmospheric pressure, carbon dioxide changes straight from solid to gas, that is, it sublimes. You can get a sense of liquid carbon dioxide, however, by swirling a cylinder of carbon dioxide. Here it is liquid because the pressure in the cylinder is much higher, about 70 times atmospheric pressure.

■ Extension: Distillation

Simple distillation

Distillation is used in the purification of liquids, especially organic liquids. It involves heating a molecular liquid to its boiling point and then cooling the vapour in a condenser to form the purified liquid. In the distillation process, differences in the boiling points (volatilities) are used to achieve a separation. A common application of distillation is to separate water from dissolved salts. This is done on a large scale with sea water in some parts of the world and is known as desalination (Chapter 25). The liquid obtained by the condensation of the vapour in a distillation is called the distillate. Any involatile solid remaining in the flask is known as the residue.

Fractional distillation

Fractional distillation is used to separate a mixture of volatile molecular liquids into their component liquids, based on their different boiling points. Figure 17.19 shows the apparatus for fractional distillation. The **fractionating column** is packed with inert glass beads with a large surface area. They provide a surface on which vapour can condense. Vapour, produced by boiling the liquid in the flask, rises inside the column and condenses in the cooler portions of the column. This process of condensation–evaporation is repeated many times as the vapour slowly rises through the fractionating column. The temperature in the column takes the form of a gradient: hottest at the bottom and becoming progressively cooler further up the column. The composition of the condensing liquid changes as it moves up the column, with the liquid having an increasingly higher concentration of the more volatile (lower boiling point) component as it gets higher up the column.

Numerous cycles of boiling and condensing occur in the column before the vapour leaves the top of the column. The vapour is condensed into a liquid in the water-cooled condenser. As the more volatile component is removed from the liquid mixture in the flask, the temperature must be increased to distil over the less volatile components. The first sample (known as a fraction) contains a high concentration of the most volatile component; later fractions will contain a high concentration of the less volatile components.

Fractional distillation is used in the petroleum or oil industry to separate petroleum or crude oil into its various fractions, for example, petrol (gasoline), kerosene and lubricating oil (Chapters 10 and 23). It can also be used to separate the gases in liquid air (Figure 17.20).

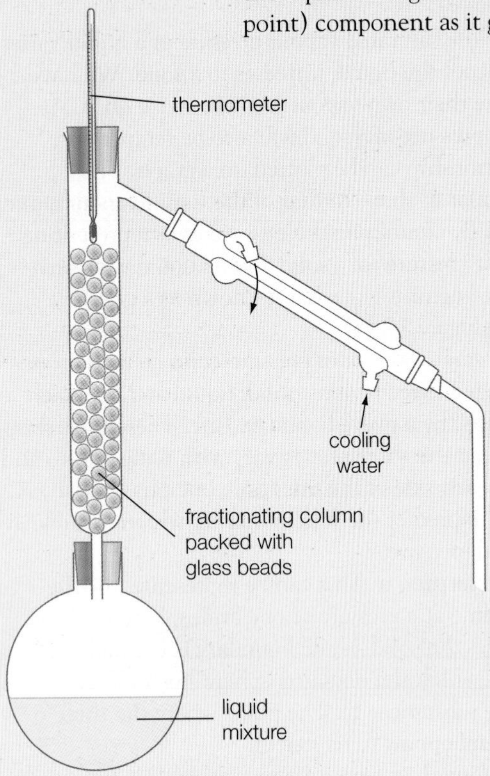

Figure 17.20 Pouring liquid nitrogen that has been separated by fractional distillation of liquid air

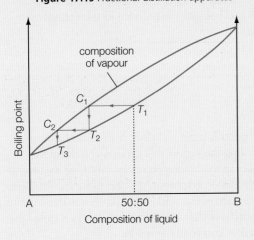

Figure 17.19 Fractional distillation apparatus

Phase equilibria and fractional distillation

A boiling-point–composition graph can be used to explain fractional distillation with reference to the principles of phase equilibria. Let us start with a 50:50 mixture of two volatile liquids, A and B, A having the lower boiling point (greater volatility). The vapour will contain a greater proportion of A than B. The composition of this vapour is indicated by point C_1 in Figure 17.21. The vapour of composition C_1 that is formed at temperature T_1 condenses at a lower temperature, T_2, than the 50:50 equimolar mixture that we started with. Further distillation results in a mixture of composition C_2 and boiling point T_3. The process may be repeated until almost pure A is obtained. This repeated boiling followed by condensation of the enriched vapour is what takes place in the fractionating column.

Figure 17.21 Fractional distillation of a mixture of ideal liquids A and B. The 50:50 mixture boils at temperature T_1. The vapour has composition C_1 and condenses at temperature T_2. The process is continued until nearly pure A is obtained as the distillate.

17.2 The equilibrium law

In Chapter 7 we saw how it is possible to derive an equilibrium expression for K_c for a homogeneous equilibrium mixture produced by a reversible reaction in a closed system. The value of K_c is constant for a given reaction at a particular temperature. For example, K_c for the reaction

$$PCl_5(g) \rightleftharpoons PCl_3(g) + Cl_2(g)$$

is given by the expression

$$K_c = \frac{[PCl_3][Cl_2]}{[PCl_5]}$$

The value of K_c is a very useful parameter as it gives an indication of the position of the equilibrium and the proportion of products in the equilibrium mixture. The next section of this chapter gives examples of some calculations based on K_c. Later we will develop ideas about different equilibrium situations.

Calculations on homogeneous equilibria

17.2.1 Solve homogeneous equilibrium problems using the expression for K_c.

The fact that such a highly significant relationship as that of the equilibrium constant can be established for *any* reversible reaction occurring under conditions where equilibrium can be reached is very useful in quantitative chemistry. Whether dealing with an industrial process (Chapter 7), a biochemical interaction or a wide range of other important areas of chemistry including electrochemistry (Chapter 19), it is important that we can 'put numbers' to the process and calculate the shifting proportions of reactants and products during a reaction. The calculations involved at IB level are entirely confined to those relating to homogeneous equilibrium systems. Such calculations require knowledge of the relevant chemical equation and the ability to write equilibrium expressions confidently. The following discussion and series of worked examples illustrate the different levels of sophistication needed to solve problems in this area.

The most straightforward type of question in this area would require the writing of an equilibrium expression, followed by the 'feeding in' of some equilibrium concentration values to find K_c.

Worked example

Nitrogen(II) oxide, NO, is a pollutant released into the atmosphere from car exhausts. It is also formed when nitrosyl chloride, NOCl, dissociates according to the following equation:

$$2NOCl(g) \rightleftharpoons 2NO(g) + Cl_2(g)$$

To study this reaction, different amounts of the three gases were placed in a closed container and allowed to come to equilibrium at 503 K and at 738 K.

The equilibrium concentrations of the three gases at each temperature are given Table 17.4.

Temperature/K	Concentration/mol dm⁻³		
	NOCl	NO	Cl₂
503	2.33×10^{-3}	1.46×10^{-3}	1.15×10^{-2}
738	3.68×10^{-4}	7.63×10^{-3}	2.14×10^{-4}

Table 17.4 Equilibrium concentrations for the reaction $2NOCl(g) \rightleftharpoons 2NO(g) + Cl_2(g)$ at two different temperatures

a Write the expression for the equilibrium constant, K_c, for this reaction.

b Calculate the value of K_c at each of the two temperatures given.

c Is the forward reaction endothermic or exothermic? (Explain your answer based on ideas covered in Chapter 7.)

a $K_c = \dfrac{[NO]^2[Cl_2]}{[NOCl]^2}$

b at 503 K $\quad K_c = \dfrac{(1.46 \times 10^{-3})^2 \times 1.15 \times 10^{-2}}{(2.33 \times 10^{-3})^2}$

$\qquad\qquad\quad = 4.5 \times 10^{-3}$

$\quad$ at 738 K $\quad K_c = \dfrac{(7.63 \times 10^{-3})^2 \times 2.14 \times 10^{-4}}{(3.68 \times 10^{-4})^2}$

$\qquad\qquad\quad = 9.2 \times 10^{-2}$

c The value of K_c is greater at 738 K; K_c increases with temperature, with the forward reaction being favoured to increase the proportion of products in the equilibrium mixture. This suggests that the forward reaction is endothermic.

A slightly more difficult question depends on you being able to use the equation for the reaction to work out the concentrations of the various substances in the equilibrium mixture.

Worked example

The acid-catalysed hydrolysis of ethyl ethanoate can be achieved by mixing the ester with dilute hydrochloric acid.

$$CH_3COOC_2H_5(l) + H_2O(l) \overset{H^+}{\rightleftharpoons} CH_3COOH(l) + C_2H_5OH(l)$$

If 1.00 mole of ethyl ethanoate is mixed with 1.00 mole of water and the reaction allowed to reach equilibrium at a particular temperature then 0.30 moles of ethanoic acid is found in the equilibrium mixture.

Calculate the value of K_c at this temperature.

It is useful to set out the first stage of the calculation as follows, focusing on the information that can be worked out from the equation:

	$CH_3COOC_2H_5(l)$ +	$H_2O(l)$ $\rightleftharpoons$	$CH_3COOH(l)$ +	$C_2H_5OH(l)$
Starting amount (moles)	1.00	1.00	0.00	0.00
Equilibrium amount (moles)			0.30	

We need to fill in the gaps in the second line by applying the stoichiometry built into the equation. It is important to note that the coefficients in the equation are all '1'. The reaction has proceeded to produce 0.30 moles of ethanoic acid, and the molar ratio of ethanoic acid and ethanol is 1 : 1. This means that the amount of ethanol in the equilibrium mixture is also 0.30 moles.

If 0.30 moles of each product are present in the equilibrium mixture, then they must have been produced from the reaction of 0.30 moles of the ester and water. This means that the amount of ester and water remaining at equilibrium must be (1.00 – 0.30) moles of each.

This means we can complete the table above as follows:

	$CH_3COOC_2H_5(l)$ +	$H_2O(l)$ $\rightleftharpoons$	$CH_3COOH(l)$ +	$C_2H_5OH(l)$
Starting amount(moles)	1.00	1.00	0	0
Equilibrium amount (moles)	(1.00 – 0.30) = 0.70	(1.00 – 0.30) = 0.70	0.30	0.30

Many questions on equilibria will require the drawing up of a similar table to this. It is important to get this line of equilibrium amounts correct.

However, there is still one more line to put in. To calculate K_c we need the equilibrium concentration values to put into the equilibrium expression. But we have not been given the volume of the reaction mixture. So let us say that the volume is V dm³. Thus we have an additional, and final, line to our table.

Equilibrium concentration (mol dm⁻³)   0.70/V   0.70/V   0.30/V   0.30/V

We are now in a position to calculate K_c:

$$K_c = \frac{[CH_3COOH][C_2H_5OH]}{[CH_3COOC_2H_5][H_2O]}$$

$$K_c = \frac{(0.30/V)(0.30/V)}{(0.70/V)(0.70/V)} \quad \text{All the '}V\text{' terms cancel out}$$

$$= \frac{0.30 \times 0.30}{0.70 \times 0.70}$$

$$= 0.18$$

In other examples you may be given the total volume of the reaction mixture (if the reaction is in the liquid phase or solution) or the volume of the container (if the reaction is gaseous). In these cases you would need to use the numerical values provided in your calculations.

In the examples so far you have been asked to calculate K_c. Obviously it is possible that a question may be posed that provides you with that value and asks you to calculate the equilibrium concentrations of the reactants and/or products.

An organic compound X exists in equilibrium with its isomer, Y, in the liquid state at a particular temperature.

$X(l) \rightleftharpoons Y(l)$

Calculate how many moles of Y are formed at equilibrium if 1 mole of X is allowed to reach equilibrium at this temperature, if K_c has a value of 0.02.

Let the number of moles of Y at equilibrium = y moles

	X(l) $\rightleftharpoons$	Y(l)
Starting amount (moles)	1.00	0.00
Equilibrium amount (moles)		y

The process now is similar to that we have used previously. From the equation, if y moles of the isomer Y are present then y moles of X must have reacted. Therefore, $(1.00 - y)$ moles of X must remain at equilibrium. Also, if we call the volume of liquid $V\,dm^3$, then we can complete the table as follows:

	X(l) $\rightleftharpoons$	Y(l)
Starting amount (moles)	1.00	0.00
Equilibrium amount (moles)	$(1.00 - y)$	y
Equilibrium concentration ($mol\,dm^{-3}$)	$(1.00 - y)/V$	y/V

$$K_c = \frac{[Y]}{[X]}$$

$$0.02 = \frac{y/V}{(1.00 - y)/V} \quad \text{the 'V' terms cancel}$$

$$0.02 = \frac{y}{(1.00 - y)}$$

$$0.02(1.00 - y) = y$$

$$0.02 - 0.02y = y$$

and so $\qquad 1.02y = 0.02$

therefore $\qquad y = \dfrac{0.02}{1.02} \qquad = 0.0196\ \text{moles}$

This type of calculation involves using some basic algebra. The chemical equation here was as simple as possible, involving just one reactant and one product. Most chemical reactions are more complicated than this! Equilibrium calculations similar to the above can be solved for these more complex reactions, provided sufficient numerical information is given.

Phosphorus(v) chloride undergoes thermal decomposition as follows:

$PCl_5(g) \rightleftharpoons PCl_3(g) + Cl_2(g)$

and therefore

$$K_c = \frac{[PCl_3][Cl_2]}{[PCl_5]}$$

Some PCl_5 was placed in an evacuated flask of volume $1.0\,dm^3$ at $500\,K$. An equilibrium was then established in which the concentration of PCl_5 was $4.0 \times 10^{-2}\,mol\,dm^{-3}$. The value of K_c for this reaction at $500\,K$ is 1.00×10^{-2}. Calculate the concentration of chlorine in the equilibrium mixture.

Let the concentration of Cl_2 at equilibrium = $x \, mol \, dm^{-3}$

$$PCl_5(g) \rightleftharpoons PCl_3(g) + Cl_2(g)$$

Equilibrium concentrations ($mol \, dm^{-3}$) 4.0×10^{-2} x x

$$K_c = 1.0 \times 10^{-2} = \frac{x^2}{4.0 \times 10^{-2}}$$

therefore $4 \times 10^{-4} = x^2$

so $x = 2 \times 10^{-2} \, mol \, dm^{-3}$

This example is straightforward but there are problems where the solution generates a quadratic equation for working out the unknown concentration. The IB syllabus specifically states that calculations that would require the use of the formula for solving quadratic equations will not be asked (an example is given as extension work here).

■ Extension: Calculations using the formula for quadratic solutions

For the esterification reaction

$$CH_3COOH(l) + C_2H_5OH(l) \rightleftharpoons CH_3COOC_2H_5(l) + H_2O(l)$$

What amount of ethyl ethanoate will be formed at equilibrium when 1.0 mole of ethanol is reacted with 2.0 moles of ethanoic acid at 373 K, given that the value of K_c is 4.0 at this temperature?

Let the number of moles of ethyl ethanoate at equilibrium = x moles, and the volume of the reacting mixture = $V \, dm^3$

	$CH_3COOH(l)$	$+ \; C_2H_5OH(l)$	$\rightleftharpoons CH_3COOC_2H_5(l)$	$+ \; H_2O(l)$
Starting amount (moles)	2.00	1.00	0.00	0.00
Equilibrium amount (moles)	$(2.0 - x)$	$(1.0 - x)$	x	x
Equilibrium concentration ($mol \, dm^{-3}$)	$(2.0 - x)/V$	$(1.0 - x)/V$	x/V	x/V

$$K_c = \frac{[CH_3COOC_2H_5][H_2O]}{[CH_3COOH][C_2H_5OH]}$$

$$= \frac{(x/V)(x/V)}{((2.0 - x)/V)((1.0 - x)/V)} \qquad \text{note that the 'V' term cancels}$$

$$= \frac{x^2}{(2.0 - x)(1.0 - x)}$$

therefore:

$$4.0 = \frac{x^2}{(2.0 - x)(1.0 - x)} = \frac{x^2}{x^2 - 3x + 2}$$

This rearranges to:

$$3x^2 - 12x + 8 = 0$$

The solution of this quadratic equation requires the use of the general expression:

$$x = \frac{-b \pm \sqrt{b^2 - 4ac}}{2a} \qquad \text{for the general quadratic } ax^2 + bx + c = 0$$

Using this expression gives possible values for x of 0.85 or 3.15 moles. The second of these solutions is impossible as we only started with 1.0 moles of ethanol. Therefore the number of moles of ethyl ethanoate at equilibrium is 0.85 moles.

■ Extension: Gaseous equilibria

For reactions involving gases, the equilibrium constant is often expressed in terms of the partial pressures of the gases in the equilibrium mixture rather than their concentrations. The equilibrium constant is then given the symbol K_p. The reason this approach is feasible is that it can be shown that the partial pressure of a gas in a mixture is directly proportional to its concentration.

The partial pressure of a gas is related to the amount of gas by the ideal gas equation (Chapter 1).

$$PV = nRT$$

Hence:

$$P = \frac{n}{V} RT$$

However, $\frac{n}{V} = c$, where c represents the concentration (in $mol\,dm^{-3}$) of the gas.

Therefore:

$$P = cRT$$

Since R is the gas constant and T is a specific temperature, the multiple RT is numerically constant at a given temperature. This shows that the pressure of a gas is *directly proportional* to its concentration at a specific temperature and it is valid therefore to express an equilibrium constant in terms of partial pressures for reactions involving gases. The structure of the expression for the equilibrium constant is identical to that for K_c (see the following worked example).

Worked example

In an equilibrium mixture, the partial pressures of N_2, H_2 and NH_3 are as follows:

$P_{N_2} = 149\,atm$, $P_{H_2} = 40\,atm$ and $P_{NH_3} = 11\,atm$

Calculate K_p for the following equilibrium reaction.

$$N_2(g) + 3H_2(g) \rightleftharpoons 2NH_3(g)$$

$$K_p = \frac{(P_{NH_3})^2}{(P_{N_2})(P_{H_2})^3} = \frac{(11)^2}{(149)(40)^3} = 1.3 \times 10^{-5}$$

Language of Chemistry

The partial pressure of a gas, in a mixture of ideal gases (Chapter 1), is the pressure that gas would exert if it alone occupied the container. The concept was introduced by John Dalton (Chapter 2), who stated that the total pressure of a mixture of gases is the sum of the partial pressures of the individual gases in the mixture (Dalton's law of partial pressures). ■

■ Extension: Relationship between the equilibrium constant and Gibbs free energy

The value of the equilibrium constant K_c (or K_p) does *not* give any information about the rate of reaction. Equilibrium constants are *independent* of the kinetics of the reaction. *However*, the chemical equilibrium constant, K_c, is directly related to the Gibbs free energy change, $\Delta G^{\ominus}$ (Chapter 15) by the following equation (van't Hoff's equation):

$$\Delta G^{\ominus} = -RT \ln K_c$$

where R represents the molar gas constant and T the absolute temperature in kelvin. The relationship between K_c and $\Delta G^{\ominus}$ obtained from this expression is summarized in Table 17.5.

$\Delta G^{\ominus}$	$\ln K_c$	K_c
Negative	Positive	>1
Zero	Zero	= 1
Positive	Negative	<1

Table 17.5 A summary of the relationship between $\Delta G^{\ominus}$ and K_c

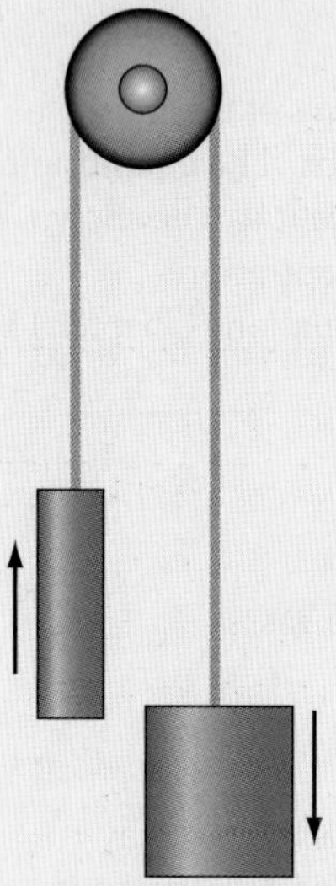

Figure 17.22 A mechanical analogy to illustrate the concept of a coupled reaction

Table 17.5 indicates the following broad relationship between $\Delta G^{\ominus}$ and the equilibrium constant, K_c. If $\Delta G^{\ominus}$ is negative, K_c is greater than 1 and the products predominate in the equilibrium mixture. Alternatively, if $\Delta G^{\ominus}$ is positive, K_c will be less than 1 and the reactants will predominate in the equilibrium mixture.

Coupled reactions

An equilibrium where the reactants are favoured over the products ($K_c < 1$) may be 'driven forward' by a reaction that is more spontaneous, that is, has a more negative value for the Gibbs free energy change, ΔG, (Chapter 15) and hence strongly favours the products over the reactants ($K_c > 1$).

A simple mechanical analogy is a pair of masses joined by a rope passing over a pulley (Figure 17.22). The smaller of the two masses will be pulled up as the heavier mass falls: its coupling to the heavier mass results in it being raised. The thermodynamic equivalent is a reaction with a small positive value of ΔG being forced to occur by its coupling to a reaction with a very large negative value of ΔG. The sum of the two values of ΔG is negative.

The role of adenosine triphosphate (ATP) in cells is to act as the short-term energy source. The function of the hydrolysis of ATP is to couple with non-spontaneous reactions, for example, polymerization, and provide sufficient free energy to make them spontaneous (Chapter 22).

Figure 17.22 A mechanical analogy to illustrate the concept of a coupled reaction

The relation of equilibrium composition to reaction rate

The concept of the equilibrium constant for a reversible reaction in a closed system was discovered by analysing experimental data and has since been theoretically justified in terms of a thermodynamic approach involving consideration of $\Delta G^{\ominus}$ values. However, there is another approach to understanding the nature of K_c which, provided it is viewed carefully, gives an understanding of the dynamics of how equilibrium is achieved.

This approach is based on the fact that equilibrium is achieved when the forward and reverse reactions taking place in a reaction mixture have the same rates. Since reaction rates depend on (and change with) concentration, then, at a particular temperature, there will be a unique set of reactant and product concentrations that correspond to these forward and reverse rates of reaction. The equilibrium constant expresses the relationship between the concentrations that guarantee this equality of rates. To look at this more closely, consider a general reaction. Here the forward and reverse reactions are both single-step, bimolecular reactions with 1:1 stoichiometry.

$$A + B \rightleftharpoons C + D$$

Forward reaction:	$A + B \rightarrow C + D$	rate = $k_f[A][B]$
Reverse reaction:	$C + D \rightarrow A + B$	rate = $k_r[C][D]$

At equilibrium these two rates are equal. Therefore:

$$k_f[A][B] = k_r[C][D]$$

This can be re-arranged to give:

$$\frac{[C][D]}{[A][B]} = \frac{k_f}{k_r} = \text{constant}$$

This is the form of the equilibrium expression for K_c, and implies that the equilibrium constant is the ratio of the forward and reverse rate constants.

$$K_c = \frac{k_f}{k_r}$$

If the rate constant for the forward reaction is large relative to that of the reverse reaction, then the equilibrium constant is large and the production of products favoured (Figure 17.23a). On the other hand, if the rate constant for the reverse reaction is relatively large compared to that of the forward reaction, then the reactants will be favoured and the equilibrium constant will be small (Figure 17.23b).

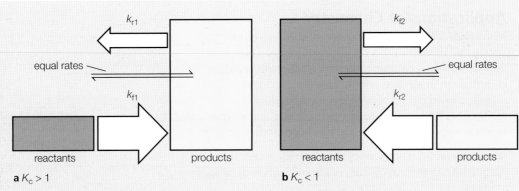

Figure 17.23
The relationship between the values of K_c, the position of the equilibrium and the rate constants of the forward and reverse reactions

a $K_c > 1$ 　　　　　　　**b** $K_c < 1$

One apparent problem with this argument relating K_c to k_f and k_r is that it is only easily derived for a single-step reaction. In practice, many reactions have complex rate expressions (obtained by experiment) where reaction orders are not related to reaction stoichiometry. Mechanisms may not be known, and where they are, they are often multi-step. Nevertheless, even in these cases, when the system is at equilibrium, we have a distinctive situation in which all the individual elementary steps of a reaction mechanism must be in equilibrium too (see Chapter 16 for the use of the idea of elementary steps in the context of reaction mechanisms). As a result each elementary step of the reaction sequence can be treated as an equilibrium in itself. Equilibrium constants for each step can then be expressed in terms of the forward and reverse rate constants found for each elementary step. Using the relationship we discussed in Chapter 7 for finding the overall equilibrium constant of a sequence of connected equilibria, it can be demonstrated that it is valid to evaluate the expression for K_c directly from the overall stoichiometric equation, even though the reaction may take place in several steps.

Language of Chemistry

Homogeneous and heterogeneous equilibria

The IB syllabus specifies that questions will deal only with homogeneous equilibria. But it is still worthwhile understanding what this means, and indeed what type of reaction represents the alternative, heterogeneous, equilibria.

An equilibrium in which all the substances are present in the same phase is known as a **homogeneous equilibrium**. For example, the Haber process reaction,

$$N_2(g) + 3H_2(g) \rightleftharpoons 2NH_3(g)$$

is an example of a homogeneous equilibrium (Chapter 7).

An equilibrium in which the substances involved are present in different phases is known as a **heterogeneous equilibrium**. For example

$$H_2O(l) \rightleftharpoons H_2O(g) \quad \text{and} \quad AgCl(s) + (aq) \rightleftharpoons Ag^+(aq) + Cl^-(aq)$$

are examples of heterogeneous equilibria. It is worth noting that pure solids and pure liquids do *not* appear in the equilibrium expression for a heterogeneous reaction. For example, in the case of the thermal decomposition of limestone in a closed system,

$$CaCO_3(s) \rightleftharpoons CaO(s) + CO_2(g)$$

it is found that $K_c = [CO_2(g)]$. This situation occurs because the 'activity' of a pure liquid or solid is 1. The terms homogeneous and heterogeneous are also applied to catalysts (Chapter 6). ∎

Applications of Chemistry

Many chemical reactions are *not* performed in a closed system and therefore do *not* reach equilibrium. The thermal decomposition of calcium carbonate reaches an equilibrium if it is carried out in a closed system. However, industrially it is most usefully carried out in an open system to produce lime (calcium oxide) from limestone (calcium carbonate). Lime is often added to acidic soils to increase their pH and optimize crop yields.

When calcium carbonate is strongly heated, it undergoes decomposition to form calcium oxide and carbon dioxide gas:

$$CaCO_3(s) \rightarrow CaO(s) + CO_2(g)$$

Lime is manufactured in large gas-fuelled lime kilns (Figure 17.24). Pieces of limestone are added to the kiln and a draught of air is allowed to enter the kiln. Gaseous fuel is the passed into the kiln and burned to produce heat. The limestone is heated in the kiln for several hours at about 1000 °C. The lime is removed from the bottom and the carbon dioxide is allowed to escape into the atmosphere.

The draught of air which flows through the kiln is important because it helps to remove the carbon dioxide as soon as it is formed. This helps to prevent the reaction between calcium oxide and carbon dioxide to reform calcium carbonate. The reaction is *not* allowed to reach equilibrium.

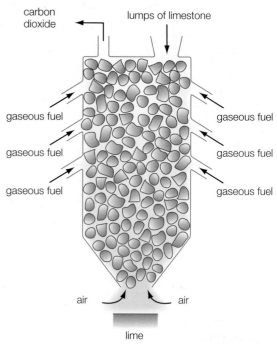

Figure 17.24 A cross section of a lime kiln

■ Extension: Examples of heterogeneous equilibria

Solubility equilibria

When a solute is added to a given amount of solvent at a given temperature, a point is finally reached when no more solute dissolves in the solvent. At that point, the solution is described as being saturated.

$$\text{solute + solvent} \underset{\substack{\text{crystallization or} \\ \text{precipitation}}}{\overset{\text{dissolving}}{\rightleftharpoons}} \text{saturated solution}$$

In a saturated solution, a dynamic equilibrium is established between the dissolved solute in the solution and the undissolved solute. At equilibrium, the rate of the forward reaction (dissolving) equals the rate of the reverse reaction (crystallization or precipitation). The solubility of a substance is usually expressed as the mass or amount of solute present in $1\,dm^3$ of solution.

Compounds are often regarded as being soluble *or* insoluble. However, many ionic compounds, such as silver chloride (Chapter 3), are *sparingly soluble* in water. When increasing quantities of a sparingly soluble ionic solid are added to water, a saturated solution is eventually formed. There is a dynamic equilibrium between the undissolved salt and its dissolved ions (Figure 17.25).

$$AgCl(s) \rightleftharpoons Ag^+(aq) + Cl^-(aq)$$

The product of the concentrations of ions in a saturated solution of silver chloride is an equilibrium constant termed the **solubility product** (K_{sp}):

$$K_{sp} = [Ag^+(aq)] \times [Cl^-(aq)]$$

Calculating a solubility product from the solubility and vice versa is discussed in Chapter 25. The common ion effect, a consequence of Le Châtelier's principle, is also discussed in this chapter in the context of removing unwanted ions in the environment.

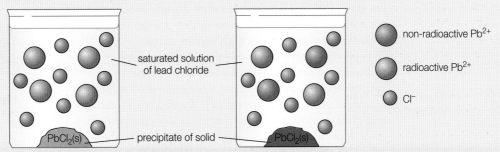

Figure 17.25 If solid lead chloride is placed in a saturated solution of lead chloride labelled with radioactive lead ions, Pb²⁺ (shown in red), the solid becomes radioactive

Studies on such saturated solutions of sparingly soluble salts have provided evidence for our ideas on dynamic equilibria. Radioactive labelling experiments with lead(II) chloride solution have provided evidence for the exchange of ions in an equilibrium situation. Solid lead(II) chloride, $PbCl_2$, is only slightly soluble in cold water. Some solid lead(II) chloride is placed in a saturated solution of radioactive lead(II) chloride. The solution contains radioactive $Pb^{2+}(aq)$ ions. Although the solution is saturated and no more lead chloride can dissolve overall, the solid takes up some of the radioactivity. This shows that some of the radioactive lead ions in the solution have been precipitated into the solid, and an equal number of non-radioactive lead ions from the solid must have dissolved to keep the solution saturated.

Applications of Chemistry

A 'barium meal' before an X-ray is used to diagnose cancers and ulcers in the intestine or stomach. The patient drinks a barium meal (barium sulfate and water) before having the X-ray taken. Aqueous barium ions, $Ba^{2+}(aq)$, are highly toxic. However, the 'barium meal' is not poisonous because barium sulfate is highly insoluble – this is indicated by its very low solubility product.

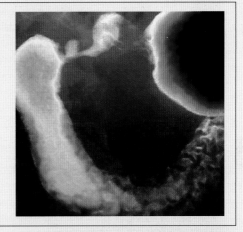

Figure 17.26 A coloured photograph of the entrance to the duodenum showing the blocking of the passage of the barium meal by an ulcer in the centre

Partitioning

When a small quantity of iodine is shaken with a mixture of water and hexane (Figure 17.27), some iodine will dissolve in *both* liquids. The water and hexane do not mix and are said to be a pair of immiscible liquids.

If the mixture is allowed to separate into two layers and left until equilibrium is established (Figure 17.28) the iodine concentration in each layer can be determined by titration with sodium thiosulfate (Chapter 9).

$$I_2(aq) \rightleftharpoons I_2(hexane)$$

Experiments using different masses of iodine, water and hexane show that at a fixed temperature the ratio of the concentrations of the iodine in the two layers is constant:

$$K_D = \frac{\text{concentration of iodine in water}}{\text{concentration of iodine in hexane}}$$

Figure 17.27 Iodine dissolved in a mixture of hexane (upper layer) and water (lower layer)

The constant K_D is called the partition coefficient or distribution coefficient for the solute distributed between two solvents at a given temperature. The partition or distribution law states that at a fixed temperature a solute distributes itself between two immiscible solvents so that the ratio of the concentrations of solute in each layer is *constant*.

Paper chromatography and other chromatographic techniques (Chapter 21) depend on the principle of partitioning. It is also the basis for the separation technique of solvent extraction (Chapter 20). We also need to know particular partition coefficients when developing pesticides and insecticides. Pesticides and insecticides need to be soluble in the fatty tissues of the animals that they are designed to kill, but much less soluble in water so that the chemicals are not simply washed away by rain.

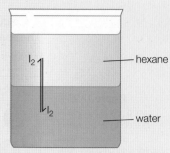

Figure 17.28 Iodine molecules between the two layers – a dynamic equilibrium is set up

■ Extension: Further examples of homogeneous equilibria in solution

The effect of a non-volatile solute on vapour pressure

If a non-volatile solute, such as glucose or sodium chloride, is added to a volatile liquid the vapour pressure is lowered compared to the pure solvent (Figure 17.29). This effect can be explained by considering the surface area of the liquid from which evaporation occurs. In the case of the solution, parts of the liquid's surface are occupied by solute particles, which are non-volatile. Hence, evaporation takes place over a smaller surface area and so is slower. Consequently, the vapour pressure is reduced.

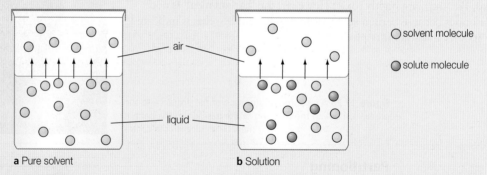

Figure 17.29 The lowering of vapour pressure by adding a non-volatile solute to a solvent

a Pure solvent   **b** Solution

Henry's law

The vapour pressure has a major effect on gas solubility, since gases are more compressible than liquids or solids. When, at a given pressure, the same numbers of gas molecules above a saturated solution enter and leave the solution per unit time, then a dynamic equilibrium is established.

gas + solvent ⇌ saturated solution

When the pressure is increased, gas molecules collide more often with the surface of the solvent, leading to more gas molecules entering the solvent per unit time. To re-establish the equilibrium, more gas molecules dissolve to reduce this change. Henry's law is discussed on a quantitative basis in Chapter 25.

■ Extension: Ionic equilibrium in solutions

In 1824 Michael Faraday (Chapter 19) classified substances into electrolytes and non-electrolytes on the basis of the conductivity behaviour of their aqueous solutions. Substances that were good conductors were termed electrolytes; non-conductors were termed non-electrolytes. Arrhenius (Chapter 16) suggested that electrolytes dissolve in water to release ions. Reactions such as acid–base reactions (Chapters 8 and 18) can be described as chemical equilibria involving ions in aqueous solution.

Figure 17.30 A bottle of antifreeze for adding to the water in a car radiator in cold winters

Language of Chemistry

The properties of solutions that depend on the *number* of dissolved solute particles in the solution are called colligative properties. One colligative property familiar to those living in cold countries is freezing point depression: salt (either rock salt or calcium chloride) is spread over icy roads and this helps to melt the ice as long as the temperature outdoors is above the lowest freezing point of the salt–water mixture. Sodium chloride can melt ice at temperatures as low as –21 °C.

A further example of this is the use of antifreeze (ethane-1,2-diol) which is added to the water in a car radiator in the cold winter months to prevent the radiator freezing up (Figure 17.30). This is crucial, as water expands on freezing and would crack the radiator, causing leaks in the cooling system that would be dangerous. Fish that live in the cold waters of the northern or southern oceans have their own natural antifreeze in their blood.

Another colligative property is the raising of the boiling point of a liquid when a non-volatile solute is added. This phenomenon is known as boiling point elevation. Note that the overall effect of adding an involatile solute to a liquid is to increase the 'distance' between the melting and boiling points of the liquid. ■

■ Extension: Osmosis

Osmosis (Figure 17.31) is the movement of water molecules through a semi-permeable membrane from a dilute solution into a more concentrated one. The water molecules can pass through the membrane, but the large glucose molecules cannot pass through. The process is largely driven by an increase in entropy (Chapter 15).

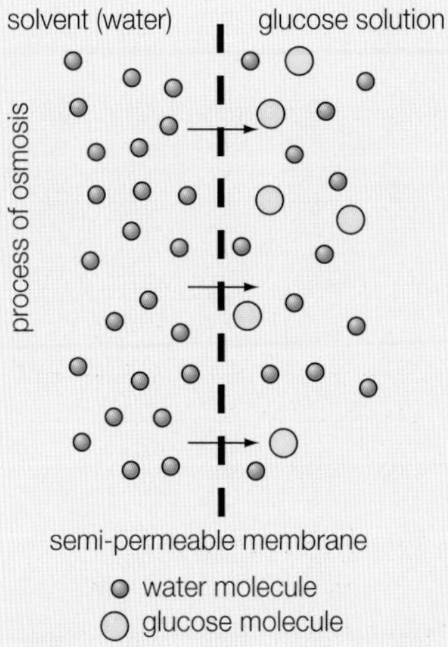

Osmosis can be demonstrated in the apparatus shown in Figure 17.32. A concentrated glucose solution is placed in a thistle funnel sealed with a semi-permeable membrane. The water from the beaker will diffuse through the membrane into the funnel. The glucose solution rises up the funnel stem until the downward pressure exerted by the solution above the membrane stops the upward flow of water. The *minimum* pressure required to prevent the osmosis of pure water into a solution is known as its osmotic pressure (π). The osmotic pressure is directly proportional to the concentration of the solute particles.

If the apparatus is set up so that a pressure *greater* than the osmotic pressure is applied on the solution, then water starts passing from the solution across the semi-permeable membrane in the opposite direction to normal, producing more pure water. This is termed reverse osmosis and is used in the purification of sea water and waste water (Chapter 25). It is of particular importance as a method of desalination in countries where there is not a large supply of fresh water and in submarines.

solvent (water) glucose solution

process of osmosis

semi-permeable membrane

○ water molecule
◯ glucose molecule

Figure 17.31 The process of osmosis

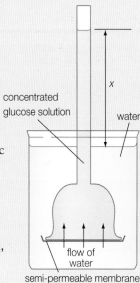

x

concentrated glucose solution

water

flow of water

semi-permeable membrane

Figure 17.32 An illustration of osmosis

SUMMARY OF
KNOWLEDGE

- The particles in a liquid are in constant motion. As a result of random collisions, some particles at the surface gain sufficient kinetic energy to overcome the intermolecular forces holding them within the liquid.
- The process by which liquid particles with sufficient kinetic energy escape from the liquid surface and become a gas is called vaporization or evaporation.
- Evaporation occurs at the surface of a liquid at any temperature below the boiling point of the liquid.
- Vaporization (or evaporation) is an endothermic process; the reverse process of condensation is exothermic.
- The saturated vapour pressure of a liquid is the pressure exerted by vapour in equilibrium with liquid in a closed container. Different liquids have different saturated vapour pressures at the same temperature.
- The vapour pressure of a liquid increases exponentially with temperature since an increasing proportion of the molecules have kinetic energies equal to or greater than the energy required to escape from the liquid surface.
- Liquids with weak intermolecular forces are volatile and have high values of saturated vapour pressure. Liquids with stronger intermolecular forces are less volatile and have lower values of saturated vapour pressure.
- The boiling point of a liquid is defined as the temperature at which its vapour pressure equals the external pressure.
- The normal boiling point is defined as the temperature at which the vapour pressure reaches 1 atmosphere (1.01×10^5 Pa).
- At the same temperature, a molecular liquid which is more volatile will have a higher vapour pressure but a lower boiling point.
- Boiling points vary with the external pressure: a decrease in pressure decreases the boiling point; an increase in pressure raises the boiling point.
- The enthalpy of vaporization is the amount of energy required to convert one mole of a pure liquid to one mole of the gas at its normal boiling point. The energy is used to overcome the bonds and/or intermolecular forces operating in the liquid.
- The enthalpy of vaporization reflects the strength of the attraction between molecules. The higher the value of the enthalpy of vaporization the stronger the bonds and/or intermolecular forces in the liquid.
- If the amounts of the reactants at the start of a reaction are known, analysis of one of the substances at equilibrium is sufficient to calculate the amounts of all the others using the stoichiometric (balanced) equation.

■ *Examination questions – a selection*

Paper 1 IB questions and IB style questions

Q1 The sequence of diagrams shown represents the system as time passes for a gas phase reaction in which reactant **X** is converted to product **Y**.

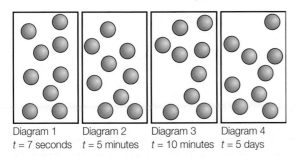

Diagram 1   Diagram 2   Diagram 3   Diagram 4
$t = 7$ seconds $t = 5$ minutes $t = 10$ minutes $t = 5$ days

Time, t

Which statement is correct?

A At $t = 5$ days the rate of the forward reaction is greater than the rate of the backward reaction.

B At $t = 7$ seconds the reaction has reached completion.

C At $t = 10$ minutes the system has reached a state of equilibrium.

D At $t = 5$ days the rate of the forward reaction is less than the rate of the backward reaction.

Higher Level Paper 1, Specimen 09, Q2

Q2 For the reaction below:

$H_2(g) + I_2(g) \rightleftharpoons 2HI(g)$

at a certain temperature, the equilibrium concentrations, in $mol\,dm^{-3}$, are

$[H_2(g)] = 0.50, \quad [I_2(g)] = 0.50, \quad [HI(g)] = 5.0$

What is the value of K_c?

A 1.0×10^{-2} **C** 55
B 10   **D** 1.0×10^2

Q3 A volatile liquid and its vapour are at equilibrium inside a sealed container. Which change will alter the equilibrium vapour pressure of the liquid in the container?

A adding more volatile liquid
B adding more vapour
C decreasing the volume of the container
D decreasing the temperature

Q4 The diagram below represents the results of introducing three different volatile liquids into mercury columns (initially they are all at the same height) in evacuated glass tubes, in such a way that some liquid remains in each tube.

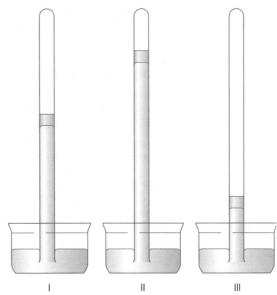

I   II   III

Which of the following statements is true about the vapour pressures (VP_I etc.) of the liquids?

A $VP_I \gg VP_{II} \gg VP_{III}$
B $VP_{III} \gg VP_I \gg VP_{II}$
C $VP_{II} \gg VP_I \gg VP_{III}$
D $VP_I = VP_{II} = VP_{III}$

Q5 The vapour pressure curves for several substances (1, 2, 3 and 4) are given below.

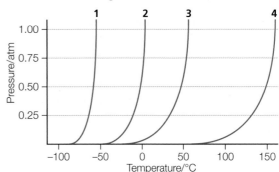

For vapour pressures near 0.5 atmospheres, a 2 °C change in temperature will affect the vapour pressure of which substance the most?

A 3   **B** 2   **C** 1   **D** 4

Q6 Hydrogen and carbon dioxide react as shown in the equation below:

$$H_2(g) + CO_2(g) \rightleftharpoons H_2O(g) + CO(g)$$

For this reaction the values of K_c at different temperatures are:

Temperature/K	K_c
500	7.76×10^{-3}
700	1.23×10^{-1}
900	6.01×10^{-1}

Which statement for the reaction is correct?
A The forward reaction is endothermic.
B The products ($H_2O(g)$ and $CO(g)$) are more stable than the reactants ($H_2(g)$ and $CO_2(g)$).
C The reaction does not proceed at high temperatures.
D The reverse reaction is favoured by high temperatures.

Q7 Which statement is correct about the behaviour of a catalyst in a reversible reaction?
A It decreases the enthalpy change of the forward reaction.
B It increases the enthalpy change of the reverse reaction.
C It decreases the activation energy of the forward reaction.
D It increases the activation energy of the reverse reaction.

Q8 The manufacture of sulfur trioxide can be represented by the equation below:

$$2SO_2(g) + O_2(g) \rightleftharpoons 2SO_3(g) \qquad \Delta H^\ominus = -197\,kJ\,mol^{-1}$$

What happens when a catalyst is added to an equilibrium mixture from this reaction?
A The rate of the forward reaction increases and that of the reverse reaction decreases.
B The rates of both forward and reverse reactions increase.
C The value of $\Delta H^\ominus$ decreases.
D The yield of sulfur trioxide increases.

Q9 A sealed container at room temperature is half full of water. The temperature of the container is increased and time allowed for equilibrium to be re-established at the higher temperature.

Which statement is correct when the equilibrium is re-established at the higher temperature?
A The rate of vaporization is greater than the rate of condensation.
B The amount of water vapour is greater than the amount of liquid water.
C The amount of water vapour is greater than it is at the lower temperature.
D The rate of condensation is greater than the rate of vaporization.

Q10 The position of equilibrium in a reversible reaction is shifted to the right until it reaches equilibrium again. Which statement must be true for the reaction when the new position of equilibrium is reached?
A The rate of the forward reaction is greater than the rate of the reverse reaction.
B Once the new equilibrium is established, concentrations of reactants and products do not change.
C The equilibrium concentrations of reactants and products are equal.
D The value of K_c is greater than 1.

Q11 Which change will shift the position of equilibrium to the right in this reaction?

$$N_2(g) + 3H_2(g) \rightleftharpoons 2NH_3(g) \qquad \Delta H^\ominus = -92\,kJ\,mol^{-1}$$

A increasing the temperature
B decreasing the pressure
C adding a catalyst
D removing ammonia from the equilibrium mixture

Q12 Which potential energy value(s) will change when a catalyst is added?

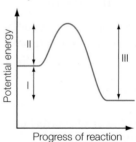

A I, II and III
B I only
C II only
D II and III only

Q13 For a reaction which almost goes to completion, the equilibrium constant, K_c, is:
A $\gg 1$ B $= 1$ C $= 0$ D $\ll 1$

Q14 The value of the equilibrium constant for the reaction:

$$2HI(g) \rightleftharpoons H_2(g) + I_2(g)$$

is 0.25 at 440 °C. What would the value of the equilibrium constant be for the following reaction at the same temperature?

$$H_2(g) + I_2(g) \rightleftharpoons 2HI(g)$$

A 0.25   **B** 0.50   **C** 2.0   **D** 4.0
Higher Level Paper 1, Nov 04, Q23

Q15 $N_2O_4(g) \rightleftharpoons 2NO_2(g)$ $K_c = 5.0 \times 10^{-3}$

In an equilibrium mixture of these two gases $[N_2O_4] = 5 \times 10^{-1}\,mol\,dm^{-3}$. What is the equilibrium concentration of NO_2 in $mol\,dm^3$?

A 5.0×10^{-1} **C** 5.0×10^{-3}
B 5.0×10^{-2} **D** 2.5×10^{-4}
Higher Level Paper 1, Nov 01, Q25

Q16 $10.0\,cm^3$ of liquid bromine is placed in an empty $100\,cm^3$ flask which is then sealed and left to reach equilibrium at room temperature. What happens first?

A The rate of evaporation is greater than the rate of condensation.
B The rate of condensation is greater than the rate of evaporation.
C The rate of evaporation is equal to the rate of condensation.
D There is no evaporation or condensation.

Q17 In which of the following combinations are the terms matched correctly for a given molecular substance?

	Boiling point	ΔH_{vap}	Intermolecular forces
A	high	small	strong
B	high	large	strong
C	low	large	weak
D	low	small	strong

Q18 Which of the following factors indicates the presence of strong intermolecular forces of attraction between the molecules of a liquid?

A very low surface tension
B very low vapour pressure
C very low heat of vaporization
D very low viscosity

Q19 $2H_2O(l) \rightleftharpoons H_3O^+(aq) + OH^-(aq)$

The equilibrium constant for the reaction above is 1.0×10^{-14} at 25 °C and 2.1×10^{-14} at 35 °C. What can be concluded from this information?

A $[H_3O^+]$ decreases as the temperature is raised.
B $[H_3O^+]$ is greater than $[OH^-]$ at 35 °C.
C Water is a stronger electrolyte at 25 °C.
D The ionization of water is endothermic.
Standard Level Paper 1, May 00, Q22

Q20 For the reaction,

$$Br_2(aq) + 2H_2O(l) \rightleftharpoons HBrO(aq) + H_3O^+(aq) + Br^-(aq)$$

the concentration of HBrO(aq) can be decreased by adding:

A Ag^+, which forms a precipitate of insoluble AgBr with Br^-
B OH^- (i.e. an alkali)
C sodium chloride
D hydrochloric acid (i.e. addition of H^+/H_3O^+)

Q21 Equal quantities of a volatile liquid, propan-1-ol, are placed in two glass flasks, one with a volume of $250\,cm^3$, the other of $1000\,cm^3$. Both flasks are sealed and maintained at the same temperature. The propan-1-ol evaporates, but some liquid remains in both flasks. If the vapour pressure of propan-1-ol in the smaller flask is 5.3 kPa, what would it be in the larger one?

A 1.3 kPa   **B** 5.3 kPa   **C** 4.0 kPa   **D** 10.6 kPa

Q22 Which of the following changes will *increase* the vapour pressure of a sample of the organic solvent methylbenzene, $C_6H_5CH_3$?

   I increasing the temperature of the solvent
   II increasing the volume of the container
   III adding sodium chloride to the solvent

A I only   **B** III only   **C** I and III   **D** II and IIII

Q23 A sealed flask containing methylbenzene (a hydrocarbon solvent), iodine (soluble in the solvent), and air is at equilibrium at 25 °C. If the temperature of the system were raised to 35 °C, which one of the following would **not** be expected to increase?

A the entropy of the system
B the pressure of the vapour phase
C the average kinetic energy of the gas particles
D the amount of air dissolved in the solvent

Q24 The thermal decomposition of hydrogen iodide may be represented by the equation:

$$2HI(g) \rightleftharpoons H_2(g) + I_2(g)$$

When 1.0 mol of hydrogen iodide was placed in a 1.0 dm³ reaction vessel and allowed to reach equilibrium at 547 °C, 0.76 moles of hydrogen iodide remained.

What is the value of the equilibrium constant for this system at this temperature?
A 0.025　**B** 0.360　**C** 0.160　**D** 0.100

Q25 For the reaction:

$$PCl_5(g) \rightleftharpoons PCl_3(g) + Cl_2(g)$$

at 290 °C, $K_c = 5 \times 10^{-12}$. At equilibrium, what is the concentration of chlorine molecules if the concentration of gaseous PCl_5 is 0.050 mol dm⁻³?
A 5.0×10^{-7} mol dm⁻³　**C** 2.5×10^{-5} mol dm⁻³
B 0.22 mol dm⁻³　**D** 2.5×10^{-4} mol dm⁻³

Paper 2 IB questions and IB style questions

Q1 Methanol is an important industrial solvent and fuel. It can be produced from carbon monoxide and hydrogen according to the following equation:

$$CO(g) + 2H_2(g) \rightleftharpoons CH_3OH(g) \qquad \Delta H^{\ominus} = -91 \text{ kJ mol}^{-1}$$

The effect of different catalysts on this reaction is investigated using the following apparatus:

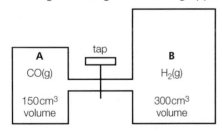

A contains 1 mole of carbon monoxide and **B** contains 2 moles of hydrogen. The gases in both containers are at the same temperature and pressure. The tap is closed at the start of the experiment.
a What pressure change will occur, if any, in the containers when the tap is opened
　i and the gases are allowed to mix (but before they start to react)? [1]
　ii as the reaction takes place? [1]

b **i** What will happen to the temperature as the gases begin to react? [1]
　ii What will happen to the concentration of methanol if the system is allowed to reach equilibrium at a lower temperature? [1]
c **i** Write the equilibrium expression for the above reaction. [1]
　ii Calculate a value for K_c if the maximum yield of methanol is 85%. [3]
　iii When this reaction is carried out on an industrial scale, the yield is about 60%. Suggest a reason for this. [1]
　iv Copper is a good catalyst for this reaction. What effect, if any, will the addition of copper have on the value of K_c? [1]
　　　　　　Higher Level Paper 2, Nov 00, Q4

Q2 When 1.0 mole of ethanoic acid is mixed with 1.0 mole ethanol, and the mixture allowed to reach equilibrium, the following reaction occurs:

$$CH_3COOH(l) + C_2H_5OH(l) \rightleftharpoons CH_3COOC_2H_5(l) + H_2O(l)$$

The amounts of ethyl ethanoate and water at equilibrium are both 0.67 moles.
a **i** What is meant by the term *equilibrium*? [2]
　ii Write an expression for K_c for this reaction. [1]
　iii Calculate the value of K_c for this reaction. [2]
b For the dissociation:

$$H_2O(l) \rightleftharpoons H^+(aq) + OH^-(aq)$$

the ionic product is given by:

$$K_w = [H^+(aq)][OH^-(aq)].$$

The value of K_w is 1.0×10^{-14} at 298 K and 2.4×10^{-14} at 310 K.

Using Le Châtelier's principle, deduce whether the dissociation of water is exothermic or endothermic. [3]

Q3 **a** An industrial gas mixture is produced by the catalytic reforming of methane using steam.

$$CH_4(g) + H_2O(g) \rightleftharpoons CO(g) + 3H_2(g)$$
$$\Delta H^{\ominus} = +206 \text{ kJ mol}^{-1}$$

By choosing from the appropriate letter(s) below, identify the change(s) that would shift the position of equilibrium to the right.
A increasing the temperature
B decreasing the temperature
C increasing the pressure
D adding a catalyst
E decreasing the pressure
F increasing the concentration of H_2 [2]

b The following graph represents the change of concentration of reactant and product during the reaction.

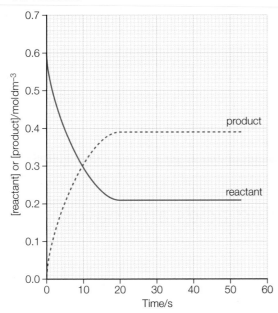

i Calculate the average rate of reaction over the first 15 s, stating the units. [3]

ii After 19 s the concentrations of the reactant and product do not change. State what this indicates about the reaction. [1]

Higher Level Paper 2, May 03, Q3

Q4 a When a solution of iodine in a hydrocarbon solvent is shaken with water the following equilibrium occurs:

$I_2(hydrocarbon) \rightleftharpoons I_2(water)$

Write down an expression for K_c for this equilibrium. [1]

b How would you expect the value of K_c to change when the concentration of iodine molecules, I_2, in the hydrocarbon solvent at the start is doubled? [1]

c The experiment was repeated using potassium iodide solution instead of water. The following equilibrium was set up in the aqueous layer:

$I_2(aq) + I^-(aq) \rightleftharpoons I_3^-(aq)$

How would you expect this to affect the concentration of iodine in the hydrocarbon layer at equilibrium? Justify your answer. [3]

Q5 Pent-1-ene (C_5H_{10}) reacts with ethanoic acid to produce pentyl ethanoate. The following equilibrium is established:

$C_5H_{10} + CH_3COOH \rightleftharpoons CH_3CO_2C_5H_{11}$

When a solution of 0.02 mol of pent-1-ene and 0.01 mol of ethanoic acid in 600 cm³ of an inert solvent was allowed to reach equilibrium at 15 °C, 0.009 mol of pentyl ethanoate was formed.

a Calculate how many moles of pent-1-ene and ethanoic acid were present in the equilibrium mixture. [2]

b Write down the equilibrium expression for this reaction. [1]

c Calculate the concentrations, in mol dm⁻³, of the three substances at equilibrium. [3]

d Use your answers in **c** to calculate the value of K_c. [2]

18 Acids and bases

STARTING POINTS
- Water is slightly dissociated into hydroxide and hydrogen ions.
- Acid and base strength are measured with equilibrium constants.
- pH and pOH are measures of hydrogen and hydroxide ion concentrations.
- Buffer solutions resist changes in pH and behave according to Le Châtelier's principle.
- Indicators are usually weak acids.
- One or both ions from a salt may undergo hydrolysis with water.
- pH can change abruptly during the titration of an acid by an alkali.
- The shape of a titration curve depends on the strength of the acid and alkali.

18.1 Calculations involving acids and bases

18.1.1 **State** the expression for the ionic product constant of water (K_w).

The ionic product of water

When water is purified by repeated distillation its electrical conductivity falls to a very low, constant value. This is evidence that pure water dissociates to a very small extent to form ions:

$$H_2O(l) \rightleftharpoons H^+(aq) + OH^-(aq) \qquad \text{or} \qquad 2H_2O(l) \rightleftharpoons H_3O^+(aq) + OH^-(aq)$$

If the equilibrium law (Chapter 7) is applied to the first equation:

$$K_w = [H^+(aq)] \times [OH^-(aq)]$$

where K_w represents a constant known as the **ionic product constant of water**. At 298 K (25 °C) the measured concentrations of $H^+(aq)$ and $OH^-(aq)$ in pure water are 1.0×10^{-7} mol dm^{-3}, therefore:

$$K_w = [H^+(aq)] \times [OH^-(aq)] = 1.0 \times 10^{-7} \times 1.0 \times 10^{-7} = 1.0 \times 10^{-14} \text{ mol}^2 \text{ dm}^{-6}$$

This is a key equation in acid–base chemistry. Note that the product of $[H^+(aq)]$ and $[OH^-(aq)]$ is a *constant* at a given temperature. Thus as the hydrogen ion concentration of a solution increases, the hydroxide ion concentration decreases (and *vice versa*).

The solution is described as neutral when the concentration of hydrogen ions equals the concentration of hydroxide ions, so that at 298 K (25 °C) the value of K_w is 1.0×10^{-14} mol^2 dm^{-6}.

Worked example

Calculate the hydroxide ion concentration of an aqueous solution whose hydrogen ion concentration at 25 °C is 5.4×10^{-4} mol dm^{-3}.

$$K_w = [H^+(aq)] \times [OH^-(aq)]$$

$$1.00 \times 10^{-14} \text{ mol}^2 \text{ dm}^{-6} = 5.4 \times 10^{-4} \text{ mol dm}^{-3} \times [OH^-(aq)]$$

$$[OH^-(aq)] = 1.9 \times 10^{-11} \text{ mol dm}^{-3}$$

■ Extension: Degree of dissociation

Consider a 1 dm^3 sample of pure water (Figure 18.1) with a density of 1 kg dm^{-3}. The relative molecular mass of water (M_r) is 18.016, and so this sample contains:

$$\frac{1000\,g}{18.016\,g\,mol^{-1}} = 55.5 \text{ moles of water.}$$

From the value of K_w at 25 °C, we know that
$[H^+(aq)] = [OH^-(aq)] = 1.0 \times 10^{-7}$, so that the proportion
of water molecules that are dissociated at this temperature is:

$$\frac{10^{-7}}{55.5} = 1.8 \times 10^{-9} \text{ (approximately 2 in } 10^9)$$

Figure 18.1 Distilled water (prepared by laboratory distillation apparatus) in a plastic dispenser

18.1.2 Deduce $[H^+(aq)]$ and $[OH^-(aq)]$ for water at different temperatures given K_w values.

The values of K_w at several temperatures are tabulated in Table 18.1.

Table 18.1 Values of the ionic product of water at different temperatures

Temperature/°C	0	25	40	100
K_w/mol^2 dm^{-6} (/10^{-14})	0.11	1.0	2.9	51.3

This data indicates that water becomes increasingly dissociated and hence acidic as the temperature rises (Figure 18.2). The increase in the ionic product of water, K_w, with temperature accounts for the corrosive action of hot pure water on iron pipes: this behaviour is caused by the increase in concentration in hydrogen ions. The pH decreases with an increase in temperature but the solution is described as chemically neutral since the concentrations of hydroxide and hydrogen ions remain *equal*.

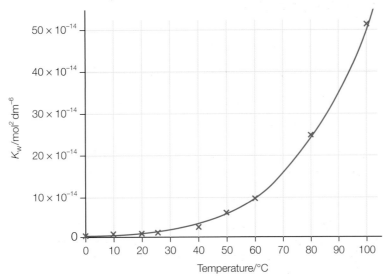

Figure 18.2 The relationship between temperature and the ionic product constant of water

The dissociation of water is expected to be an endothermic process (Chapter 5) because energy is needed to separate oppositely charged ions. It can also be deduced that the dissociation of water is an endothermic process since applying Le Châtelier's principle (Chapter 7) predicts that increasing the temperature will favour the reaction that absorbs the heat, that is, the endothermic reaction. The increasing values of K_w also show that increasing the temperature favours the dissociation of water molecules, thereby producing more hydrogen and hydroxide ions.

The ionic product constant of water can be used to calculate the concentration of hydroxide or hydrogen ions at any specified temperature.

Worked example

At 60 °C, the ionic product constant of water is 9.6×10^{-14} mol^2 dm^{-6}. Calculate the pH of a neutral solution at this temperature. (pH = $-\log_{10}[H^+(aq)]$ – see below.)

$[H^+(aq)][OH^-(aq)] = 9.6 \times 10^{-14}$

If the solution is neutral:

$[H^+(aq)] = [OH^-(aq)] = \sqrt{(9.6 \times 10^{-14})} = 3.1 \times 10^{-7}$

pH = $-\log_{10}[H^+(aq)] = -\log_{10}(3.1 \times 10^{-7}) = 6.5$

Language of Chemistry

The pH of an aqueous solution is a measure of its hydrogen ion concentration; it depends on temperature because the degree of dissociation of an acid does, i.e. K_a changes. Values of the pH of pure water at $0\,°C$ and $100\,°C$ are approximately 7.5 and 6.1. Both the solutions are, however, *neutral*. The definition of this is pH-independent: a neutral solution is one where the concentration of hydrogen ions equals the concentration of hydroxide ions. This is pH 7 only at $25\,°C$. ∎

The pH and pOH scales

18.1.3 Solve problems
involving
$[H^+(aq)]$,
$[OH^-(aq)]$,
pH and pOH.

pH and **pOH** are formally defined as the negative logarithms to the base 10 of the concentration in $mol\,dm^{-3}$ of the hydrogen and hydroxide ion concentrations, respectively, that is:

$$pH = -\log_{10}([H^+(aq)]/mol\,dm^{-3}) \qquad \text{and} \qquad pOH = -\log_{10}([OH^-(aq)]/mol\,dm^{-3})$$

(You can only take logarithms of a pure number and not of a quantity.)

The logarithmic pH and pOH scales reduce the extremely wide variation in concentrations of hydrogen ions, $H^+(aq)$, and hydroxide ions, $OH^-(aq)$, in dilute aqueous solutions of acids and bases (typically 1 to 10^{-14}) to a narrower range of pH (typically 1 to 14) (Figure 18.3). pH values of solutions are measured with a pH meter or narrow-range indicator paper (Figure 18.4).

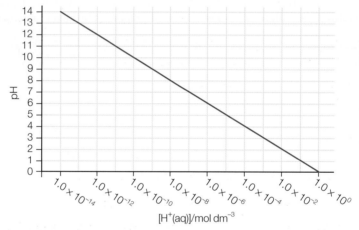

Figure 18.3
The variation of pH
with $[H^+(aq)]$

Figure 18.4
Narrow-range
indicator paper

The *negative* sign of the logarithmic function produces pH and pOH values that are *positive* for most dilute solutions.

It can also be shown that pH + pOH =14:

$$K_w = [H^+(aq)] \times [OH^-(aq)] = 1.00 \times 10^{-14}\,mol^2\,dm^{-6}$$

Taking negative logarithms to the base 10 of both sides:

$$-\log_{10} K_w = -\log_{10}([H^+(aq)] \times [OH^-(aq)]) = -\log_{10}(1.00 \times 10^{-14})$$
$$= -\log_{10}[H^+(aq)] - \log_{10}[OH^-(aq)] = -\log_{10} 10^{-14}$$

$$pK_w = pH + pOH = 14$$

Calculating the pH values of aqueous solutions of strong acids and bases is relatively easy since the concentration of hydrogen ions is *directly related* to the concentration of the acid or base.

Worked example

Calculate the pH of $0.01\,mol\,dm^{-3}$ hydrochloric acid, $HCl(aq)$.

$$HCl(aq) \rightarrow H^+(aq) + Cl^-(aq)$$

$$pH = -\log_{10}[H^+(aq)]$$

$$[H^+(aq)] = 0.01\,mol\,dm^{-3}$$

($[H^+(aq)]$ is equal to the concentration of the acid, because it is a strong acid that is completely ionized.)

$$pH = -\log_{10}(0.01) = 2$$

Language of Chemistry

The term pH was introduced by S. P. L. Sørensen (Chapter 8) in 1909, but was written as PH in his original paper. This is because Sørensen's work dates from the time when printed material was produced using metal type: the typesetting of superscripts and subscripts was difficult and the finished printed text was often difficult to read. Sørensen's convention avoided this problem. 'p-Functions' have also been adopted for other concentrations and concentration-related numbers. For example, pCa = 5.0 means a concentration of calcium ions equal to 10^{-5} mol dm^{-3}, and pK_a = 4.0 means an acid dissociation constant equal to 10^{-4}. ■

Worked examples

Calculate the pH of 0.01 mol dm^{-3} aqueous sodium hydroxide, NaOH(aq).

$NaOH(aq) \rightarrow Na^+(aq) + OH^-(aq)$

$[OH^-(aq)] = 0.01$ mol dm^{-3}

$pOH = -\log_{10}[OH^-(aq)] = -\log_{10}(0.01) = 2$

$pOH + pH = 14$

$2 + pH = 14$

$pH = (14 - 2) = 12$

Calculate the pH of 0.01 mol dm^{-3} aqueous barium hydroxide, Ba(OH)$_2$(aq).

$Ba(OH)_2(aq) \rightarrow Ba^{2+}(aq) + 2OH^-(aq)$

$[OH^-(aq)] = (2 \times 0.01$ mol dm$^{-3}) = 0.02$ mol dm^{-3}

$pOH = -\log_{10}[OH^-(aq)] = -\log_{10}(0.02) = 1.7$

$pOH + pH = 14$

$1.7 + pH = 14$

$pH = (14 - 1.7) = 12.3$

Calculate the pH of 0.01 mol dm^{-3} sulfuric acid, H$_2$SO$_4$(aq).

$H_2SO_4(aq) \rightarrow H^+(aq) + HSO_4^-(aq) \rightleftharpoons SO_4^-(aq) + 2H^+(aq)$

Note that although sulfuric acid is diprotic (dibasic), the dissociation is not complete and the concentration of hydrogen ions is *not* twice that of the acid concentration. In other words, H$^+$(aq) is *not* equal to 0.02 and the pH is *not* therefore 1.7 ($\log_{10} 0.02$).

To simplify the pH calculation, sulfuric acid may be assumed to be monoprotic (monobasic) and the small but significant contribution to hydrogen ion concentration from the second dissociation is ignored.

$pH = -\log_{10}[H^+(aq)]$

$[H^+(aq)] = 0.01$ mol dm^{-3}

$pH = -\log_{10}(0.01) = 2$

(In practice, the second dissociation makes a small but significant contribution to the hydrogen ion concentration and the accurately experimentally determined pH is 1.84.)

To calculate pH or pOH values from hydrogen or hydroxide ion concentrations on a calculator, type in the hydrogen or hydroxide ion concentration, press the log function key, then multiply by −1 to change the sign.

To calculate hydrogen ion concentrations from pH values on a calculator, type in the pH value, insert a negative sign and then press the 10^x or antilog function key. Mathematically this is expressed as:

$[H^+(aq)] = 10^{-pH}$ and $[OH^-(aq)] = 10^{-pOH}$

Worked example

Calculate the hydrogen ion concentration of a solution whose pH is 2.80.

$[H^+(aq)] = 10^{-pH}$

$[H^+(aq)] = 10^{-2.80}$

$[H^+(aq)] = 1.6 \times 10^{-3} \, mol \, dm^{-3}$

Alternatively, you may be asked to calculate the pH given a known mass of a pure anhydrous acid made up to a solution of known volume with distilled water.

Worked examples

A solution of nitric acid, $HNO_3(aq)$, contains $1.26 \, g$ of the pure acid in every $100 \, cm^3$ of aqueous solution. Calculate the pH of the solution.

$$\text{Amount of nitric acid in } 100 \, cm^3 \text{ of solution} = \frac{1.26 \, g}{63 \, g \, mol^{-1}} = 0.020 \, mol$$

Hence, the amount of nitric acid in $1000 \, cm^3$ or $1 \, dm^3$ of solution is $\frac{1000}{100} \times 0.020 \, mol = 0.20 \, mol$.

Nitric acid is a strong acid and completely dissociated or ionized when in dilute aqueous solution:

$$HNO_3(aq) \rightarrow H^+(aq) + NO_3^-(aq)$$

Hence, the concentration of the hydrogen ions is identical to the concentration of the nitric acid. So:

$[H^+(aq)] = 0.20 \, mol \, dm^{-3}$

$pH = -\log_{10}(0.20) = 0.70$

A solution of sodium hydroxide contains $0.40 \, g$ in every $50 \, cm^3$ of aqueous solution. Calculate the pH of the solution.

$$\text{Amount of sodium hydroxide in } 50 \, cm^3 \text{ of solution} = \frac{0.40 \, g}{40 \, g \, mol^{-1}} = 0.010 \, mol$$

Hence, the amount of sodium hydroxide in $1000 \, cm^3$ or $1 \, dm^3$ of solution is:

$$\frac{1000}{50} \times 0.010 \, mol = 0.20 \, mol.$$

Sodium hydroxide is a strong base and completely dissociated or ionized when in dilute aqueous solution:

$$NaOH(aq) \rightarrow Na^+(aq) + OH^-(aq)$$

Hence, the concentration of the hydroxide ions is identical to the concentration of the sodium hydroxide. So:

$[OH^-(aq)] = 0.20 \, mol \, dm^{-3}$

$pOH = -\log_{10}(0.20) = 0.70$

$pH + pOH = 14$

$pH = 14 - 0.70 = 13.30$

Another common type of calculation is calculating the pH of a strong acid after dilution.

Worked example

$10 \, cm^3$ of an aqueous solution of a monoprotic (monobasic) strong acid is added to $990 \, cm^3$ of water. Calculate the change in pH.

$$\text{The dilution factor} = \frac{(10 + 990)}{10} = \times 100$$

Since pH is logarithmic to the base 10 a change in a hydrogen ion concentration of 100 means a change in pH of 2. As pH is the negative logarithm to the base ten, the pH will increase by two.

The pH scale is used to describe *both* alkaline and acidic solutions. Alkaline solutions will contain a low concentration of hydrogen ions derived from the dissociation of water.

The pOH scale can also be used to describe acidic and alkaline solutions, but it is not as widely used.

Worked examples

Calculate the pOH of a solution with a hydroxide ion concentration of $0.1 \, mol \, dm^{-3}$.

$pOH = -log_{10}[OH^-(aq)] = -log_{10}(0.1) = 1$

Calculate the hydroxide ion concentration of a solution with a pOH of 2.

$[OH^-(aq)] = 10^{-pOH} = 10^{-2}$

$[OH^-(aq)] = 0.01 \, mol \, dm^{-3}$

Calculate the pH of a solution with a pOH of 1.

$pH + pOH = 14$
$pH + 1 = 14$
$pH = (14 - 1) = 13$

Calculate the pOH of a solution with a pH of 2.

$pH + pOH = 14$
$2 + pOH = 14$
$pOH = (14 - 2) = 12$

■ Extension: The pH of very dilute solutions

If asked to calculate the pH of an aqueous solution of hydrochloric acid of concentration $1.0 \times 10^{-8} \, mol \, dm^{-3}$, many students will reason as follows: the concentration of hydrogen ions from the acid (assuming complete dissociation since it is a strong acid) is $1.0 \times 10^{-8} \, mol \, dm^{-3}$. $pH = -log_{10}[H^+(aq)]$ and hence the pH is 8. The students have overlooked the fact that, in a solution of acid, there are *two* sources of hydrogen ions: one from the ionization of acid and another from the self-ionization of water. So:

$$[H^+(aq)]_{total} = [H^+(aq)]_{from \, water} + [H^+(aq)]_{from \, acid}$$

At higher concentrations of hydrochloric acid, from $0.1 \, mol \, dm^{-3} \, HCl(aq)$, the hydrogen ion contribution from the water is insignificant and neglected, but at very low concentrations ($< 1 \times 10^{-6} \, mol \, dm^{-3}$), the hydrogen ion contribution from the water becomes significant. So:

$$[H^+(aq)]_{total} = [H^+(aq)]_{from \, water} + [H^+(aq)]_{from \, acid}$$

Let $[H^+(aq)]_{from \, water} = y$ (which is also equal in value to $[OH^-(aq)]_{from \, water}$)

$$[H^+(aq)] \times [OH^-(aq)] = 1.00 \times 10^{-14} \, mol^2 \, dm^{-6}$$

Substituting:

$$[y + (1.0 \times 10^{-8})] \times y = 1.0 \times 10^{-14}$$

and rearranging:

$$y^2 + (1.0 \times 10^{-8})y - (1.00 \times 10^{-14}) = 0$$

This is a quadratic equation and can be solved to give a positive answer for y of $9.5 \times 10^{-8} \, mol \, dm^{-3}$.

$$[H^+(aq)]_{total} = 9.5 \times 10^{-8} \, mol \, dm^{-3} + 1.0 \times 10^{-8} \, mol \, dm^{-3}$$
$$[H^+(aq)]_{total} = 10.5 \times 10^{-8} \, mol \, dm^{-3}$$
$$pH = 6.98$$

so the solution is weakly acidic.

■ **Extension:** Negative pH values

The pH scale has been presented as a scale whose values range from 0 to 14. However, negative values of pH and pH values above 14 are possible. For example, commercially available concentrated hydrochloric acid solution (37% by mass) has a pH of approximately −1.1, while saturated sodium hydroxide solution has a pH of approximately 15.0. If the concentration of hydrogen ions is greater than $1 \, \text{mol} \, \text{dm}^{-3}$ then the solution will have a negative value of pH. For example, a $12 \, \text{mol} \, \text{dm}^{-3}$ hydrochloric acid solution would be expected to have a pH of $-\log_{10} 12 = -1.08$.

However, there are some complications in highly concentrated acid solutions that make pH calculations from acid concentration inaccurate and difficult to verify experimentally.

Because there are so few water molecules per formula unit of acid, the influence of hydrogen ions in the solution is *increased*. The *effective concentration* of hydrogen ions (or the activity) is much higher than the actual concentration. The IB definition of pH as $-\log_{10}[\text{H}^+(\text{aq})]$ is better written as pH $= -\log_{10} a_{\text{H}^+}$ where a_{H^+} represents the activity of hydrogen ions. The activity of hydrogen ions approximates to the numerical value of the concentration at low concentrations.

18.1.4 State the equation for the reaction of any weak acid or weak base with water, and hence **deduce** the expressions for K_a and K_b.

Acid dissociation constant

A weak monobasic acid, HA, reacts with water according to the equation:

$$\text{HA(aq)} \rightleftharpoons \text{H}^+(\text{aq}) + \text{A}^-(\text{aq})$$

The equilibrium constant for this reaction is known as the **acid dissociation constant**, K_a, and has units of $\text{mol} \, \text{dm}^{-3}$.

$$K_a = \frac{[\text{H}^+(\text{aq})] \times [\text{A}^-(\text{aq})]}{[\text{HA(aq)}]}$$

The acid dissociation constant is a measure of the strength of a weak acid. The larger the value of K_a the stronger the acid and the greater the extent of ionization or dissociation.

Since acid dissociation constants, K_a, tend to be small and vary considerably, they are often expressed as pK_a values where:

$$pK_a = -\log_{10} K_a \qquad (\text{cf. } [\text{H}^+(\text{aq})] \text{ and pH})$$

Values of pK_a are also a measure of acid strength, but now the *smaller* the value of pK_a the *stronger* the acid.

A change of 1 in the value of the pK_a means a change in acid strength of a factor of 10 (cf. pH and $[\text{H}^+(\text{aq})]$).

The various factors that determine the strengths of organic acids and hence determine values of K_a and pK_a are discussed in Chapter 27.

(Acid dissociation constants are not usually quoted for strong acids because these effectively undergo *complete* ionization or dissociation in water. Their dissociation constants are very large and tend towards *infinity* in dilute solutions. It is difficult to measure them accurately because the concentration of undissociated acid molecules is so low.)

Values of K_a and pK_a are equilibrium constants and, like other equilibrium constants, are not affected by changes in concentrations, only by changes in temperature. This means that acid strengths vary with temperature and that the order of acid strengths can vary with temperature.

The pH of a solution of a weak acid can only be calculated if the acid dissociation constant, K_a, (or pK_a) is known.

$$K_a = \frac{[\text{H}^+(\text{aq})] \times [\text{A}^-(\text{aq})]}{[\text{HA(aq)}]}$$

but since $[\text{H}^+(\text{aq})] = [\text{A}^-(\text{aq})]$, in a solution where only the acid is present:

$$K_a = \frac{[H^+(aq)]^2}{[HA(aq)]}$$

Rearranging:

$$[H^+(aq)] = \sqrt{[HA(aq)] \times K_a}$$

and then

$$pH = -\log_{10}[H^+(aq)]$$

You can also use this approach to calculate the K_a (and hence pK_a) of a weak acid if you know the pH of the solution and its concentration.

Worked examples

The pH of $0.01\,mol\,dm^{-3}$ benzenecarboxylic acid solution, $C_6H_5COOH(aq)$, is 3.10. Calculate the acid dissociation constant, K_a, at this temperature.

$$pH = -\log_{10}[H^+(aq)]$$

$$[H^+(aq)] = -antilog\,(3.10)\ or\ 10^{-3.10} = 7.94 \times 10^{-4}\,mol\,dm^{-3}$$

$$K_a = \frac{[H^+(aq)] \times [C_6H_5COO^-(aq)]}{[C_6H_5COOH(aq)]}$$

but since $[H^+(aq)] = [C_6H_5COO^-(aq)]$

$$K_a = \frac{[H^+(aq)]^2}{[C_6H_5COOH(aq)]}$$

$$K_a = \frac{(7.94 \times 10^{-4})^2}{0.01} = 6.3 \times 10^{-5}\,mol\,dm^{-3}$$

Calculate the pH value of a $0.1\,mol\,dm^{-3}$ solution of ethanoic acid, $CH_3COOH(aq)$, given that its K_a value is $1.8 \times 10^{-5}\,mol\,dm^{-3}$.

$$K_a = \frac{[H^+(aq)] \times [CH_3COO^-(aq)]}{[CH_3COOH(aq)]}$$

but since $[H^+(aq)] = [CH_3COO^-(aq)]$

$$K_a = \frac{[H^+(aq)]^2}{[CH_3COOH(aq)]}$$

Rearranging:

$$[H^+(aq)] = \sqrt{([CH_3COOH(aq)] \times K_a)}$$
$$= \sqrt{(1.8 \times 10^{-5} \times 0.1)} = 1.34 \times 10^{-3}\,mol\,dm^{-3}$$

and then $pH = -\log_{10}[H^+(aq)] = -\log_{10}(1.34 \times 10^{-3}\,mol\,dm^{-3}) = 2.87$

Assumptions and simplifications

For weak acids we assume that there is no dissociation or ionization and that none of the acid molecules react with water to release ions. This is not far from reality since the dilute acid solutions used in the laboratory are typically about 1% dissociated. However, as indicated previously this becomes less true as the solution is progressively diluted.

We can take the dissociation of the acid into consideration when we perform a calculation with an aqueous solution of a weak acid. However, a quadratic equation results and the slight increase in accuracy rarely justifies the additional mathematical effort required.

Worked example

Calculate the pH of a $1.00\,mol\,dm^{-3}$ aqueous solution of hydrofluoric acid, HF(aq) ($K_a = 7.2 \times 10^{-4}$).

$HF(aq) \rightleftharpoons H^+(aq) + F^-(aq)$

$$K_a = \frac{[H^+(aq)] \times [F^-(aq)]}{[HF(aq)]}$$

Before any dissociation has occurred:

$[HF(aq)] = 1.00\,mol\,dm^{-3}$; $[F^-(aq)] = 0\,mol\,dm^{-3}$ and $[H^+(aq)] = 10^{-7}\,mol\,dm^{-3} \approx 0$

(However, we ignore the very small concentration of hydrogen ions formed from the dissociation of water.) Once equilibrium has been reached and dissociation of the acid has occurred:

$[HF(aq)] = 1.00 - x\,mol\,dm^{-3}$; $[F^-(aq)] = x\,mol\,dm^{-3}$; $[H^+(aq)] = x\,mol\,dm^{-3}$

where x represents the concentration of hydrofluoric acid that dissociates, which at the present is unknown. We now substitute these equilibrium concentrations into the expression for the acid dissociation constant:

$$K_a = 7.2 \times 10^{-4} = \frac{[H^+(aq)] \times [F^-(aq)]}{[HF(aq)]}$$

$$= \frac{x \times x}{1.00 - x}$$

$$= \frac{x^2}{(1.00 - x)}$$

This expression can be rearranged to give a quadratic equation:

$x^2 + (7.2 \times 10^{-4})\,x - (7.2 \times 10^{-4} \times 1.00) = 0$

Comparing this expression with the general form of a quadratic equation:

$ax^2 + bx + c = 0$; $\quad a = 1$; $b = 7.2 \times 10^{-4}$ and $c = -7.2 \times 10^{-4}$

One method of finding the two values of x that satisfy a quadratic equation is to use the quadratic formula:

$$x = \frac{-b \pm \sqrt{(b^2 - 4ac)}}{2a}$$

Substituting the values of a, b and c into the quadratic formula and evaluating the expression will give an accurate value for the concentration of hydrogen ions in this solution of hydrofluoric acid.

This approach is not demanded by the IB Chemistry programme, but you do need to know how to simplify this type of calculation so that a lengthy quadratic formula calculation can be avoided, and when such an approximation is valid.

To simplify the calculation and to avoid the tedious task of solving a quadratic, we assume (as in previous calculations) that the amount of acid dissociated, x, is negligible compared to the concentration of the hydrofluoric acid. In other words:

$1.00 - x \approx 1.00$ (where $\approx$ means approximately equal to).

The equilibrium expression now simplifies to:

$$7.2 \times 10^{-4} = \frac{x^2}{1.00}$$

$x^2 = (7.2 \times 10^{-4}) \times 1.00$

$x = \sqrt{7.2 \times 10^{-4}} = 2.7 \times 10^{-2}$

A quadratic equation is only solved when the two calculated pH values differ by more than 5%, since typically K_a values for weak acids and weak bases are known to an accuracy of about ±5%. The example above with hydrofluoric acid can be subjected to this 5% 'rule'.

Write out the simplified equilibrium expression and make x the subject:

$$K_a = \frac{x^2}{[HF(aq)]}$$

$$x = \sqrt{K_a \times [HF(aq)]}$$

Then compare the sizes of x and $[HF(aq)]$:

$$\frac{x}{[HF(aq)]} \times 100$$

Substituting:

$$\frac{2.7 \times 10^{-2}}{1.00} \times 100 = 2.7\%$$

If the expression above is less than or equal to 5% (as it will be for IB questions), the value of x is such that the approximation below is valid and the pH can be calculated.

$$[HF(aq)] - x \approx [HF(aq)]$$

$$pH = -\log_{10}(2.7 \times 10^{-2}) = 1.6$$

Relationship between K_a for a weak acid and K_b for its conjugate base

The relationship between K_a for a weak acid HA and K_b for its conjugate base, A^- (Chapter 8) is:

$$K_a(HA(aq)) \times K_b(A^-(aq)) = K_w = 1.0 \times 10^{-14}$$

Since $pK_a = -\log_{10} K_a$ and $pK_b = -\log_{10} K_b$, the logarithmic form of the equation above is:

$$pK_a(HA(aq)) + pK_b(A^-(aq)) = pK_w = 14.00$$

The stronger the acid, the larger the value of K_a and the smaller the value of pK_a. Likewise the stronger the base, the larger the value of K_b and the smaller the value of pK_b. The two equations show that as the value of K_a increases (and the value of pK_a decreases), the value of K_b decreases (and the value of pK_b increases). These equations give quantitative support to the statement 'the stronger the acid, the weaker the conjugate base' (Chapter 8).

The justification for the first equation follows from the equations below. Recall from Chapter 7 that if two chemical equilibria are added together then the equilibrium constant for the reaction is the product of the two equilibria.

Weak acid:

$$HA(aq) + H_2O(l) \rightleftharpoons H_3O^+(aq) + A^-(aq) \qquad K_a(HA(aq)) = \frac{[H_3O^+(aq)][A^-(aq)]}{[HA(aq)]}$$

Conjugate base:

$$A^-(aq) + H_2O(l) \rightleftharpoons HA(aq) + OH^-(aq) \qquad K_b(A^-(aq)) = \frac{[HA(aq)][OH^-(aq)]}{[A^-(aq)]}$$

$$\overline{2H_2O(l) \rightleftharpoons H_3O^+(aq) + OH^-(aq)} \qquad K_w = [H_3O^+(aq)][OH^-(aq)]$$

$([H_3O^+(aq)] = [H^+(aq)]$, remembering that each hydrogen ion, H^+, will be attached to a water molecule, H_2O.)

Relationship between K_b for a weak base and K_a for its conjugate acid

Analogous equations can be written to describe the relationship between K_b for a weak base B and K_a for its conjugate acid HB^+:

$$K_b(B(aq)) \times K_a(BH^+(aq)) = K_w = 1.0 \times 10^{-14}$$
$$pK_b(B(aq)) + pK_a(BH^+(aq)) = pK_w = 14.00$$

The equations below provide justification for these results:

Weak base:

$$B(aq) + H_2O(l) \rightleftharpoons BH^+(aq) + OH^-(aq) \qquad K_b(BH^+(aq)) = \frac{[BH^+(aq)][OH^-(aq)]}{[B(aq)]}$$

Conjugate acid:

$$BH^+(aq) + H_2O(l) \rightleftharpoons H_3O^+(aq) + B(aq) \qquad K_a(BH^+(aq)) = \frac{[H_3O^+(aq)][B(aq)]}{[BH^+(aq)]}$$

$$2H_2O(l) \rightleftharpoons H_3O^+(aq) + OH^-(aq) \qquad K_w = [H_3O^+(aq)][OH^-(aq)]$$

■ Extension: Successive acid dissociation constants

A number of important acids, such as carbonic acid, sulfuric acid and phosphoric(V) acid, release more than one proton per molecule and are called **polyprotic** acids. Polyprotic acids always dissociate in a stepwise manner with one proton being released at a time (to varying degrees). For example, phosphoric(V) acid dissociates in the following steps:

$$H_3PO_4(aq) \rightleftharpoons H^+(aq) + H_2PO_4^-(aq)$$

$$K_{a1} = \frac{[H^+(aq)][H_2PO_4^-(aq)]}{[H_3PO_4(aq)]} = 7.5 \times 10^{-3}$$

$$H_2PO_4^-(aq) \rightleftharpoons H^+(aq) + HPO_4^{2-}(aq)$$

$$K_{a2} = \frac{[H^+(aq)][HPO_4^{2-}(aq)]}{[H_2PO_4^-(aq)]} = 6.2 \times 10^{-8}$$

$$HPO_4^{2-}(aq) \rightleftharpoons H^+(aq) + PO_4^{3-}(aq)$$

$$K_{a3} = \frac{[H^+(aq)][PO_4^{3-}(aq)]}{[HPO_4^{2-}(aq)]} = 4.8 \times 10^{-13}$$

These results are typical for a weak polyprotic acid, namely, that the acid dissociation constants become progressively smaller: $K_{a1} > K_{a2} > K_{a3}$.

(The overall dissociation constant for the reaction $H_3PO_4(aq) \rightleftharpoons 3H^+(aq) + PO_4^{3-}(aq)$ is given by the expression $K_{a1} \times K_{a2} \times K_{a3}$.)

These *decreasing* values indicate that the species $H_3PO_4(aq)$, $H_2PO_4^-(aq)$ and $HPO_4^{2-}(aq)$ are increasingly *less acidic*, and present in increasingly lower concentrations, than the preceding species.

It becomes progressively more difficult to remove the positively charged protons as phosphoric acid is dissociated. This is because the protons are being removed from species that become increasingly negatively charged, resulting in stronger electrostatic forces of attraction between the departing proton and anion.

■ Extension: Percent dissociation

It is often useful to calculate the amount of weak acid (or weak base) that has dissociated once equilibrium has been reached in an aqueous solution.

$$\text{percent dissociation} = \frac{\text{concentration of weak acid or base dissociated (mol dm}^{-3})}{\text{initial concentration of weak acid of base (mol dm}^{-3})} \times 100$$

Worked example

A $1.00 \, \text{mol dm}^{-3}$ aqueous solution of hydrofluoric acid, HF(aq), has a concentration of hydrogen ions of $2.7 \times 10^{-2} \, \text{mol dm}^{-3}$. Calculate the percent dissociation.

$$\text{Percent dissociation} = \frac{2.7 \times 10^{-2} \, \text{mol dm}^{-3}}{1.00 \, \text{mol dm}^{-3}} \times 100 = 2.7\%$$

As stated previously the percent dissociation of a weak acid or weak base increases upon dilution. This behaviour can be accounted for in terms of Le Châtelier's principle (Chapter 7). For example consider the dissociation of ethanoic acid in water:

$$CH_3COOH(aq) + H_2O(l) \rightleftharpoons CH_3COO^-(aq) + H_3O^+(aq)$$

A simple (but *incorrect*) argument is to say that the addition of water increases its 'concentration' and that the equilibrium is restored by a shift to the right. However, this argument, although correctly predicting the response of the system, is invalid because the 'concentration' of water in a dilute solution of ethanoic acid remains effectively constant.

The correct argument is as follows. The addition of water decreases the concentrations of ethanoic acid molecules, ethanoate ions and hydrogen ions. Two of these species are on the right side of the equation and only one on the left: a shift to the right is more effective at restoring the original concentrations of molecules and ions than a shift to the left. The decrease in concentration reduces the rate of the backward reaction to a greater extent than the forward reaction.

■ **Extension:** ## Factors controlling the strength of inorganic acids and bases

Electronegativity

When all other factors are kept constant, acids become stronger as the X—H bond becomes more polar. The second period non-metal hydrides, for example, become more acidic as the difference between the electronegativity of the X and H atoms increases. Hydrogen fluoride is the strongest of these four acids, and methane is one of the weakest Brønsted–Lowry acids known (Table 18.2).

Table 18.2 Effect of electronegativity on the acidity of non-metal hydrides from period 2

Increasing acidity	Increasing electronegativity difference
HF	
H$_2$O	↑ (Increasing acidity)
NH$_3$	↑ (Increasing electronegativity difference)
CH$_4$	

When these compounds act as acids in aqueous solution, an H—X bond is broken to form H$^+$(aq) and X$^-$(aq) ions. The more polar this bond, the easier it is to form these ions. Hence, the greater the polarity of the bond, the stronger the acid (at a constant temperature).

The size of the X atom

It might be expected that HF, HCl, HBr and HI would become weaker acids as group 7 is descended because the H—X bond becomes less polar (Chapter 4). Experimentally, the *opposite* trend is found: the acids actually become stronger as the group is descended.

This occurs because the size of the atom X influences the acidity of the H—X bond. Acids become stronger as the H—X bond becomes weaker, and bonds generally become weaker as the atoms become larger.

The K_a data for HF, HCl, HBr and HI reflect the fact that the H—X bond energy becomes smaller as the X atom becomes larger (Table 18.3).

Table 18.3 Effect of the size of the halogen atom on the acidity of hydrides from group 7

Increasing acidity	Decreasing bond energy
HF	
HCl	↓ (Increasing acidity)
HBr	↓ (Decreasing bond energy)
HI	

The charge on the acid or base

The charge on a molecule or ion can influence its ability to act as an acid or a base. This is clearly shown when the pH of $0.1 \, mol \, dm^{-3}$ solutions of H_3PO_4 and the $H_2PO_4^-$, HPO_4^{2-} and PO_4^{3-} ions are compared:

H_3PO_4	pH = 1.5
$H_2PO_4^-$	pH = 4.4
HPO_4^{2-}	pH = 9.3
PO_4^{3-}	pH = 12.0

Compounds become less acidic and more basic as the negative charge increases.

Acidity:	$H_3PO_4 > H_2PO_4^- > HPO_4^{2-}$
Basicity:	$H_2PO_4^- < HPO_4^{2-} < PO_4^{3-}$

The oxidation state of the central atom

There is no difference in the polarity, size or charge when we compare oxoacids of the same element, such as H_2SO_4 and H_2SO_3 or HNO_3 and HNO_2, yet there is a significant difference in the strengths of these acids. Consider the following K_a data, for example.

H_2SO_4	$K_a = 1.0 \times 10^3$	HNO_3	$K_a = 28$
H_2SO_3	$K_a = 1.7 \times 10^{-2}$	HNO_2	$K_a = 5.1 \times 10^{-4}$

The acidity of these oxoacids increases significantly as the oxidation state (Chapter 9) of the central atom becomes larger. H_2SO_4 is a much stronger acid than H_2SO_3, and HNO_3 is a much stronger acid than HNO_2. This trend is easiest to see in the four oxoacids of chlorine.

Oxoacid	K_a	Oxidation number of chlorine
HOCl	2.9×10^{-8}	+1
HOClO	1.1×10^{-2}	+3
$HOClO_2$	5.0×10^2	+5
$HOClO_3$	1.0×10^3	+7

Table 18.4 Effect of the oxidation state of the central atom on the acidity of oxoacids

This factor of 10^{11} difference in the value of K_a for chloric(I) acid (HOCl) and chloric(VII) acid ($HOClO_3$) can be traced to the fact that there is only one value for the electronegativity of an element, but the tendency of an atom to draw electrons toward itself increases as the oxidation number of the atom increases.

As the oxidation number of the chlorine atom increases, the atom becomes more electronegative. This tends to draw electrons away from the oxygen atoms that surround the chlorine, thereby making the oxygen atoms more electronegative as well. As a result, the O—H bond becomes more polar, and the compound becomes more acidic.

History of Chemistry

George Olah (1927–) is a Hungarian-born American chemist who was awarded the 1995 Nobel Prize in Chemistry for his work on carbocations. These are short-lived intermediates in many organic reactions (Chapter 20 and Chapter 27). Olah found that a solution of antimony pentafluoride (a strong Lewis acid) in liquid hydrogen fluoride (a strong Brønsted–Lowry acid) allowed the generation of carbocations from a wide range of hydrocarbons and halogenoalkanes and would preserve them for many months. This allowed their structure to be determined by NMR (Chapter 21). The HF/SbF_5 system is such a strong acid that it can even force a methane molecule to accept a proton:

$$CH_4 + HF + SbF_5 \rightarrow CH_5^+ + SbF_6^-$$

Language of Chemistry
The HF/SbF$_5$ system is a superacid, meaning that its acidity is greater than 100% sulfuric acid. It was also named 'magic acid' by one of Olah's research students who placed a candle in a sample of it. The candle dissolved, showing the ability of the acid to protonate hydrocarbons (which are not basic). ■

18.2 Buffer solutions

18.2.1 Describe the composition of a buffer solution and **explain** its action.

A buffer solution is an aqueous solution whose pH (and hence hydrogen ion concentration) remains unchanged by dilution with water or when relatively small amounts of acid or base are added to it. Buffers *resist* changes in pH.

Types of buffers

There are three types of buffer:

- acidic buffers, which are prepared from a weak acid and a salt of the acid, for example, ethanoic acid and sodium ethanoate
- basic or alkaline buffers, which are prepared from a weak base and a salt of the base, for example, ammonia and ammonium chloride
- neutral buffers, which are prepared from phosphoric acid and its salts.

Many chemical processes require the pH to be controlled. For example, the use of shampoos and other hair treatments, developing photographs, medical injections (Chapter 24) and fermentation. Enzymes in living organisms must be buffered to ensure that they function as catalysts (Chapter 22). Soil also contains a number of important buffers (Chapter 25).

Action of a buffer

Acidic buffers
Since ethanoic acid is only slightly dissociated and sodium ethanoate is completely dissociated, a mixture of the two contains a relatively low concentration of hydrogen ions, but a large proportion of ethanoic acid molecules and ethanoate ions:

$$CH_3COONa(aq) \rightarrow CH_3COO^-(aq) + Na^+(aq)$$
$$CH_3COOH(aq) \rightleftharpoons CH_3COO^-(aq) + H^+(aq)$$

If an acid is added to the buffer, the additional hydrogen ions will be removed by combination with the ethanoate ions to form undissociated acid molecules. The presence of sodium ethanoate ensures there is a large 'reservoir' of ethanoate ions to 'mop up' the additional hydrogen ions from an acid.

If an alkali is added, the hydroxide ions combine with the hydrogen ions to form water molecules:

$$H^+(aq) + OH^-(aq) \rightarrow H_2O(l)$$

The removal of hydrogen ions via neutralization results in the dissociation of ethanoic acid molecules to replenish the hydrogen ions removed. The presence of ethanoic acid ensures that there is a large 'reservoir' of undissociated ethanoic acid molecules that will dissociate following the addition of an alkali.

Basic buffer
Since ammonia is only slightly dissociated and ammonium chloride is completely dissociated, a mixture of the two contains a relatively low concentration of hydroxide ions, but a large proportion of ammonia molecules and ammonium ions:

$$NH_4Cl(aq) \rightarrow NH_4^+(aq) + Cl^-(aq)$$
$$NH_3(aq) + H_2O(l) \rightleftharpoons NH_4^+(aq) + OH^-(aq)$$

If an acid is added, the hydrogen ions will combine with hydroxide ions to form water:

$$H^+(aq) + OH^-(aq) \rightarrow H_2O(l)$$

As a result more ammonia molecules react with water to release hydroxide ions and restore the equilibrium.

If an alkali is added, the hydroxide ions react with the ammonium ions from ammonium chloride to form ammonia and water. The presence of ammonium chloride ensures that there is a large 'reservoir' of ammonium ions to cope with the addition of an alkali.

Two assumptions are made to simplify calculations involving buffers solutions:

- In the buffer solution, the weak acid or weak base is not dissociated. This is because the presence of ions from the dissociation of its salt will prevent dissociation of the acid or base molecules.
- In the buffer solution it is assumed that all the ions present in the solution are produced by the dissolution of the salt: none originate from the acid or base.

Calculations involving buffer solutions

Consider the equilibrium for a weak acid:

$$HA(aq) \rightleftharpoons H^+(aq) + A^-(aq)$$

$$K_a = \frac{[H^+(aq)] \times [A^-(aq)]}{[HA(aq)]}$$

Rearranging:

$$[H^+(aq)] = \frac{K_a \times [HA(aq)]}{[A^-(aq)]}$$

Taking negative logarithms to the base 10 of both sides:

$$pH = pK_a - \log_{10} \frac{[HA(aq)]}{[A^-(aq)]}$$

or

$$pH = pK_a + \log_{10} \frac{[A^-(aq)]}{[HA(aq)]}$$

This equation is often called the **Henderson–Hasselbalch equation**. It indicates that:

- the pH of a buffer solution depends on the K_a of the weak acid
- the pH of a buffer solution depends upon the ratio of the concentrations of the acid and its conjugate base and not on their actual concentrations.

■ **Extension:** **Basic buffers**

The Henderson–Hasselbalch equation described previously can be readily applied to basic buffers since $pK_w = pK_a + pK_b$.

For example, ammonia has a base dissociation constant, K_b, of $1.78 \times 10^{-5}\,mol\,dm^{-3}$. Hence, the pK_b value is 4.75 and the pK_a value is 9.25.

However, the Henderson–Hasselbalch equation can also be applied to a basic buffer:

$$K_b = \frac{[B^+(aq)] \times [OH^-(aq)]}{[BOH(aq)]}$$

Taking negative logarithms of both sides of the equation:

$$-\log_{10} K_b = -\log_{10} [OH^-(aq)] - \log_{10} \frac{[B^+(aq)]}{[BOH(aq)]}$$

$$pK_b = pOH - \log_{10} \frac{[B^+(aq)]}{[BOH(aq)]}$$

$$pOH = pK_b + \log_{10} \frac{[B^+(aq)]}{[BOH(aq)]}$$

Calculating the pH of a buffer system

Worked example

Calculate the pH of a buffer containing 0.20 moles of sodium ethanoate in 500 cm³ of 0.10 mol dm⁻³ ethanoic acid. K_a for ethanoic acid is 1.8×10^{-5}.

(Assume complete dissociation of sodium ethanoate and that the dissociation of ethanoic acid is insignificant, so that the equilibrium concentration of ethanoic acid is the same as the initial concentration.)

$[CH_3COO^-(aq)] = 0.20 \times \dfrac{1000}{500} = 0.40 \text{ mol dm}^{-3}$

$\dfrac{[H^+(aq)] \times [CH_3COO^-(aq)]}{[CH_3COOH(aq)]} = 1.8 \times 10^{-5}$

$1.8 \times 10^{-5} = \dfrac{(x) \times (0.40)}{0.10}$

$x = 4.5 \times 10^{-6} = [H^+(aq)]$

$pH = -\log_{10}[H^+(aq)] = -\log_{10}(4.5 \times 10^{-6}) = 5.3$

Calculating the mass of a salt required to give an acidic buffer solution with a specific pH

Worked example

Calculate the mass of sodium propanoate ($M = 96.07$ g mol⁻¹) that must be dissolved in 1.00 dm³ of 1.00 mol dm⁻³ propanoic acid ($pK_a = 4.87$) to give a buffer solution with a pH of 4.5. (Let x represent the concentration of propanoate ions and y represent the amount of sodium propanoate.)

$[H^+(aq)] = 10^{-pH} = 1 \times 10^{-4.5} = 3.16 \times 10^{-5} \text{ mol dm}^{-3}$

$K_a = 1 \times 10^{-4.87} = 1.35 \times 10^{-5}$

$K_a = \dfrac{[H^+(aq)][CH_3CH_2COO^-(aq)]}{[CH_3CH_2COOH(aq)]} = 1.35 \times 10^{-5}$

$1.35 \times 10^{-5} = \dfrac{(3.16 \times 10^{-5})(x)}{1.00}$

$x = 0.427 \text{ mol dm}^{-3}$

$0.427 \text{ mol dm}^{-3} = \dfrac{y}{1.00 \text{ dm}^3}$

$y = 0.427 \text{ mol}$

$96.07 \text{ g mol}^{-1} \times 0.427 \text{ mol} = 41.0 \text{ g}$

Calculating the pH of a buffer after base is added

Worked example

A buffer contains 0.20 mol of sodium ethanoate (CH_3COONa) in 500 cm³ of 0.10 mol dm⁻³ ethanoic acid. K_a for ethanoic acid is 1.8×10^{-5}. Calculate the pH after 0.025 moles of sodium hydroxide is added.

The addition of hydroxide ions will cause the acid dissociation to shift to the right. So the amount of hydroxide ions added must be *subtracted* from the ethanoic acid and *added* to the amount of ethanoate ions.

$K_a = \dfrac{[H^+(aq)][CH_3COO^-(aq)]}{[CH_3COOH(aq)]} = 1.8 \times 10^{-5}$

$[H^+(aq)] = x$

$[CH_3COO^-(aq)] = \dfrac{(0.20 \text{ mol} + 0.025 \text{ mol})}{(0.500 \text{ dm}^3)} = 0.45 \text{ mol dm}^{-3}$

$[CH_3COOH(aq)] = (0.1 \text{ mol dm}^{-3} - \dfrac{0.025 \text{ mol}}{0.500 \text{ dm}^3}) - x = 0.050 \text{ mol dm}^{-3} - x \approx 0.050 \text{ mol dm}^{-3}$

$1.8 \times 10^{-5} = \dfrac{(x)(0.45)}{0.050} - x \approx \dfrac{(x)(0.45)}{0.050}$

$x = 2.0 \times 10^{-6} = [H^+(aq)]$

$pH = -\log_{10}[H^+(aq)] = -\log_{10}(2.0 \times 10^{-6}) = 5.7$

If acid is added to an acidic buffer composed of ethanoic acid and sodium ethanoate, then the addition of hydrogen ions will cause the acid dissociation to shift to the left. Hence the amount of hydrogen ions must be *added* to the ethanoic acid and *subtracted* from the amount of ethanoate ions.

History of Chemistry

Lawrence Joseph Henderson (1878–1942) was an American scientist, philosopher and sociologist whose major contributions were in the field of biochemistry. He was a professor at Harvard University where he performed physiological research. He discovered that the pH of blood is regulated by buffer systems together with the lungs, red blood cells and kidneys. He wrote an equation describing the use of carbonic acid as a buffer solution.

Karl Albert Hasselbalch (1874–1962) was a physician and chemist. He pioneered the use of pH measurement in medicine with Christian Bohr, father of Niels Bohr (Chapter 2). In 1916 he converted the 1908 equation of Lawrence Henderson to a logarithmic form, which is now known as the Henderson–Hasselbalch equation.

Characteristics of buffer solutions

Dilution

The Henderson–Hasselbalch equation indicates that the pH of a buffer solution will depend only on the ratio of the concentrations of the acid and its conjugate base, so that dilution of the buffer solution should have no effect. This is because when you add distilled water to a buffer solution you dilute both components of the buffer to the same degree.

Buffering capacity

The buffering capacity is the ability of the buffer to resist changes in pH. The buffering capacity increases as the concentration of the buffer salt/acid solution increases. The closer the buffered pH is to the pK_a, the greater the buffering capacity. The buffering capacity is expressed as the concentration of sodium hydroxide required to increase pH by 1.0

Applications of Chemistry

Blood is an important example of a buffered solution. Human blood is slightly basic, with a pH of about 7.4. In a healthy person, the pH is never more or less than 0.2 pH units from the average value. Whenever the pH falls below about 7.4 the condition is known as acidosis; whenever it rises above 7.4 it is known as alkalosis. Acidosis is more common because cells produces several acids.

The body uses three methods to control blood pH. The blood contains several buffers, including $H_2CO_3(aq)/HCO_3^-(aq)$ and $H_2PO_4^-(aq)/HPO_4^{2-}(aq)$ pairs, and hemoglobin-containing acid–base pairs. The kidneys absorb or release $H^+(aq)$ from the blood. The pH of urine is normally about 5.0 to 7.0. Acidosis involves an increase in the loss of body fluids as the kidneys reduce $H^+(aq)$. The concentration of hydrogen ions, $H^+(aq)$, is also altered by the rate at which carbon dioxide is removed from the lungs:

$$H_2CO_3(aq) \rightleftharpoons H_2O(l) + CO_2(aq)$$

Removal of carbon dioxide shifts this equilibrium to the right, thereby decreasing the concentration of hydrogen ions, $H^+(aq)$. Acidosis or alkalosis disrupt the mechanism by which hemoglobin transports oxygen in the blood. Hemoglobin (Hb) is involved in a series of equilibria whose overall result is:

$$HbH^+(aq) + O_2(aq) \rightleftharpoons HbO_2(aq) + H^+(aq)$$

In acidosis, this equilibrium is shifted to the left, and the ability of hemoglobin to form oxyhemoglobin (HbO_2) is decreased. The smaller amount of dissolved oxygen, $O_2(aq)$, available to cells in the body causes fatigue (tiredness) and headaches. Temporary acidosis occurs during rapid and heavy and exercise, when energy demands exceed the oxygen available for complete oxidation of glucose to carbon dioxide (Chapter 22). In this case the glucose is converted to an acidic metabolite, lactic acid, $CH_3CHOHCOOH$.

Acidosis also occurs when glucose is not available to the cells. This situation can arise, for example, during starvation or as a result of diabetes. In diabetics, glucose is unable enter the cells because of inadequate insulin (Chapter 22). When glucose is unavailable, the body obtains energy from stored lipids, which produce acid metabolism products.

18.3 Salt hydrolysis

18.3.1 Deduce whether salts form acidic, alkaline or neutral aqueous solutions.

A **salt** is defined as a compound formed when the hydrogen of an acid is completely or partially replaced by a metal (Table 18.5). If all the hydrogen is replaced a normal salt is formed; if the hydrogen is not completely replaced then an acidic salt is formed. Acid salts can only be formed by diprotic and triprotic acids; monoprotic acids can only form normal salts.

Acid	Salt	Example(s)	Classification
Hydrochloric acid, HCl	Chlorides	Sodium chloride, NaCl	Normal
Nitric acid, HNO_3	Nitrates	Sodium nitrate, $NaNO_3$	Normal
Ethanoic acid	Ethanoates	Sodium ethanoate, CH_3COONa	Normal
Sulfuric acid, H_2SO_4	Sulfates and hydrogensulfates	Sodium sulfate, Na_2SO_4, and sodium hydrogensulfate, $NaHSO_4$	Normal and acidic
Carbonic acid, H_2CO_3	Carbonates and hydrogencarbonates	Sodium carbonate, Na_2CO_3, and sodium hydrogencarbonate, $NaHCO_3$	Normal and acidic
Cyanic acid, HCN	Cyanides	Sodium cyanide, NaCN	Normal

Table 18.5 Salts of the common acids

Acid salts, if they are soluble in water, dissolve to form acidic solutions, for example sodium hydrogensulfate ionizes to release hydrogen, sulfate and sodium ions:

$$NaHSO_4(s) + (aq) \rightarrow Na^+(aq) + H^+(aq) + SO_4^{2-}(aq)$$

The resulting solution is acidic and exhibits typical acidic properties; for example it turns blue litmus red.

Normal salts, if they are soluble in water, often dissolve to form neutral solutions; for example sodium chloride ionizes to release sodium and chloride ions, neither of which react with water:

$$NaCl(s) + (aq) \rightarrow Na^+(aq) + Cl^-(aq)$$

This occurs when the salt has been formed by the neutralization of a strong acid by a strong base. In the case of sodium chloride the corresponding acid is hydrochloric acid, a strong acid, and the base is sodium hydroxide, a strong base or alkali:

$$NaOH(aq) + HCl(aq) \rightarrow NaCl(aq) + H_2O(l)$$

However, some normal salts dissolve in water to form either acidic or alkaline solutions. This is because one of the ions reacts with the water to release an excess of either hydroxide or hydrogen ions. This phenomenon is called **salt hydrolysis** and occurs when the salts are formed from weak acids or weak bases.

An example of the hydrolysis of a salt of a weak acid and a strong base is sodium carbonate solution, $Na_2CO_3(aq)$ (Figure 18.5):

$$CO_3^{2-}(aq) + H_2O(l) \rightleftharpoons H_2CO_3(aq) + OH^-(aq)$$

Figure 18.5 Universal indicator solution added to a concentrated solution of sodium carbonate: the colour shows that it is alkaline

The sodium ions are spectator ions (Chapter 1) and do not participate in the hydrolysis. The resulting solution will contain an excess of hydroxide ions and is alkaline. Its pH will be greater than 7.

Similar reactions will occur between water and the anion of the salt formed from a weak acid and strong base, for example sodium ethanoate, sodium hydrogencarbonate and sodium fluoride.

An example of the hydrolysis of a salt of a strong acid and a weak base is ammonium chloride solution, $NH_4Cl(aq)$ (Figure 18.6):

$$NH_4^+(aq) + H_2O(l) \rightleftharpoons NH_3(aq) + H_3O^+(aq)$$

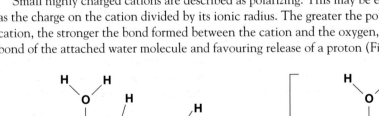

Figure 18.6 Universal indicator solution added to a concentrated solution of aqueous ammonium chloride: the colour shows that it is acidic

The chloride ions are spectator ions and do not participate in the hydrolysis. The resulting solution will contain an excess of hydrogen ions and is acidic. Its pH will be less than 7.

An example of the hydrolysis of a salt of a weak acid and a weak base is ammonium ethanoate, $CH_3COONH_4(aq)$:

$$NH_4^+(aq) + H_2O(l) \rightleftharpoons NH_3(aq) + H_3O^+(aq)$$
and
$$CH_3COO^-(aq) + H_2O(l) \rightleftharpoons CH_3COOH(aq) + OH^-(aq)$$

Both ions react with water and undergo hydrolysis: the final pH of the solution depends on the equilibrium constants for the two reactions. In this example, the two values are approximately the same and the two processes cancel each other out and the solution is neutral.

The two equations above can be summed together:

$$NH_4^+(aq) + CH_3COO^-(aq) + 2H_2O(l) \rightleftharpoons NH_3(aq) + H_3O^+(aq) + CH_3COOH(aq) + OH^-(aq)$$

Additional examples of salt hydrolysis occur with salts whose metal cation is small and highly charged. Examples include copper(II) sulfate, $CuSO_4$, aluminium sulfate, $Al_2(SO_4)_3$ and iron(III) chloride, $FeCl_3$ (Figure 18.7).

For example, in iron(III) chloride solution the hydrated iron(III) ions are $Fe[(H_2O)_6]^{3+}(aq)$. An equilibrium is set up between the hydrated iron(III) ions and water molecules leading to the release of hydrogen ions (oxonium ions).

Figure 18.7 Hydrated iron(III) chloride, $FeCl_3.6H_2O$

$$Fe[(H_2O)_6]^{3+}(aq) + H_2O(l) \rightleftharpoons Fe[(H_2O)_5OH]^{2+}(aq) + H_3O^+(aq)$$

This type of hydrolysis occurs very readily with trivalent cations, e.g. $Fe^{3+}(aq)$, much less with $Cu^{2+}(aq)$ and not at all with unipositive ions, e.g. $Ag^+(aq)$. The acidity also varies with the ionic radius: the smaller the ion, the greater the hydrolysis.

Small highly charged cations are described as polarizing. This may be expressed quantitatively as the charge on the cation divided by its ionic radius. The greater the polarizing power of the cation, the stronger the bond formed between the cation and the oxygen, weakening the O—H bond of the attached water molecule and favouring release of a proton (Figure 18.8).

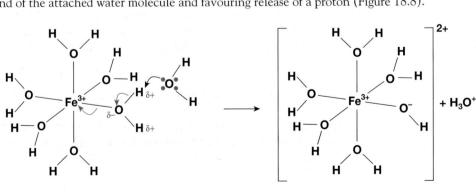

Figure 18.8 The polarization of the hydrated iron(III) ion by a water molecule

Applications of Chemistry

Chlorine is widely used to prevent the growth of bacteria in swimming pools. When chlorine is added to water, it rapidly reacts forming two acids:

$$Cl_2(aq) + H_2O(l) \rightleftharpoons HCl(aq) + HOCl(aq)$$

The hydrochloric acid is a strong acid and hence fully ionized, but the chloric(I) acid is a weak acid, so an equilibrium exists:

$$HOCl(aq) \rightleftharpoons H^+(aq) + ClO^-(aq)$$

The position of this equilibrium is important, as it controls the relative proportions of HOCl, a good bactericide, and ClO⁻, whose effectiveness is much lower (due to its negative charge). HCl is not a bactericide. Chlorine may be introduced directly as a gas in some large swimming pools, though smaller pools normally use a solution of sodium chlorate(I), NaOCl. The total chlorine concentration is about $2\,mg\,dm^{-3}$.

It is important to control the pH of the pool water carefully; if the water is too alkaline then scale can be produced, blocking filters, while if the solution is too acidic mortar and metal pipes are corroded. The pH should be kept in the range 7.4–7.6. It is usually controlled by the addition of hydrochloric acid or sodium hydrogensulfate ($NaHSO_4$) if the pH is too high, or sodium carbonate if it is too low.

18.4 Acid–base titrations

18.4.1 Sketch the general shapes of graphs of pH against volume for titrations involving strong and weak acids and bases, and explain their important features.

Measuring changes in pH during titrations

The changes in pH that occur during an acid–base titration can be measured by the method shown in Figure 18.9.

The pH of the solution in the titration flask is measured with a pH probe and meter (which may be connected to a data logger). The aqueous alkali is added in 0.5 or $1.0\,cm^3$ portions and the pH measured and recorded. A table of pH against volume of alkali is obtained and a graph is drawn.

Figure 18.9 Apparatus to record titration curves: the reaction mixture is stirred magnetically and the change in the pH recorded on the meter

Titration of a strong acid against a strong alkali

If we have $0.1\,mol\,dm^{-3}$ aqueous hydrochloric acid in the flask at the beginning of the titration then the pH will be $-\log_{10}0.1$, i.e. 1.

Let us now calculate the pH after $22.5\,cm^3$ of $0.100\,mol\,dm^{-3}$ aqueous sodium hydroxide is added to $25.0\,cm^3$ of $0.100\,mol\,dm^{-3}$ aqueous hydrochloric acid.

The addition of the alkali has removed $\frac{22.5}{25.0}$ or, if we divide the ratio by 2.5, $\frac{9}{10}$ (90%) of the hydrogen ions. In other words, after the neutralization $\frac{1}{10}$ (10%) of the original number of moles of hydrogen ions is left. This ratio can be termed the reaction factor.

An *approximate* pH for the resulting solution is calculated as follows:

$$[H^+(aq)] = 0.100 \times \tfrac{1}{10} = 0.0100\,mol\,dm^{-3}; \quad pH = -\log_{10}0.0100 = 2.00$$

So when 90% of the hydrogen ions have been neutralized, the pH has only changed by one unit from 1 to 2.

However, this simple calculation ignores the 'dilution effect' which follows from the fact that the addition of aqueous sodium hydroxide not only adds hydroxide ions, but also changes the volume and hence concentration of the resulting solution.

We can modify the previous equation to include the dilution factor and slightly improve the accuracy of the calculation:

'new' $H^+(aq)$ concentration = 'old' $H^+(aq)$ concentration × reaction factor × dilution factor

$$[H^+(aq)] = 0.100 \times \frac{1}{10} \times \frac{25.0}{(25.0 + 22.5)} = 5.26 \times 10^{-3}\,mol\,dm^{-3}$$

pH = 2.28

Now repeat the above calculation, without the dilution factor, with a higher volume of NaOH(aq):

- 24.75 cm³ of 0.100 mol dm⁻³ sodium hydroxide added to 25.0 cm³ of 0.100 mol dm⁻³ hydrochloric acid gives a pH of 3.00.

    $$reaction\ factor = \frac{(25.0 - 24.75)}{25.0} = 0.0100$$

    $[H^+(aq)] = 0.100 \times 0.0100 = 0.001\,00\,mol\,dm^{-3}$
    pH = $-\log_{10} 0.001\,00 = 3.00$

- 24.975 cm³ of 0.100 mol dm⁻³ sodium hydroxide added to 25.0 cm³ of 0.100 mol dm⁻³ hydrochloric acid gives a pH of 4.00.

    $$reaction\ factor = \frac{(25.0 - 24.975)}{25.0} = 0.001\,00$$

    $[H^+(aq)] = 0.100 \times 0.001\,00 = 0.000\,100\,mol\,dm^{-3}$
    pH = $-\log_{10} 0.000\,100 = 4.00$

- 24.9975 cm³ of 0.100 mol dm⁻³ sodium hydroxide added to 25.0 cm³ of 0.100 mol dm⁻³ hydrochloric acid gives a pH of 5.00.

    $$reaction\ factor = \frac{(25.0 - 24.9975)}{25.0} = 0.000\,100$$

    $[H^+(aq)] = 0.100 \times 0.000\,100 = 0.000\,010\,0\,mol\,dm^{-3}$
    pH = $-\log_{10} 0.000\,010\,0 = 5.00$

- 24.9999 cm³ of 0.100 mol dm⁻³ sodium hydroxide added to 25.0 cm³ of 0.100 mol dm⁻³ hydrochloric acid gives a pH of 6.

    $$reaction\ factor = \frac{(25.0 - 24.9999)}{25.0} = 0.000\,010\,0$$

    $[H^+(aq)] = 0.100 \times 0.000\,010\,0 = 0.000\,001\,00\,mol\,dm^{-3}$
    pH = $-\log_{10} 0.000\,001\,00 = 6.00$

These simplified calculations show that the pH rises *very rapidly* near the **end-point** or **equivalence point** where the reacting volumes will be equal. The end-point or equivalence point can also be found using an acid–base indicator.

Finally, when exactly 25 cm³ of sodium hydroxide is added to 25 cm³ of 0.100 mol dm⁻³ hydrochloric acid it gives a pH of exactly 7, since the hydroxide and hydrogen ions are in an exactly reacting molar ratio of 1 : 1. Neither sodium nor chloride ions react with water and the solution of sodium chloride is neutral with a pH of exactly 7.

Once we continue adding aqueous sodium hydroxide beyond the end-point, the hydroxide ions are now present in excess since all the acid has been removed by neutralization. The hydroxide ion concentration begins to rise as quickly after the end-point as the hydrogen ion concentration decreased before the end-point. The titration curve (Figure 18.10) is symmetrical around an imaginary horizontal axis that runs through pH 7 at the equivalence or end-point.

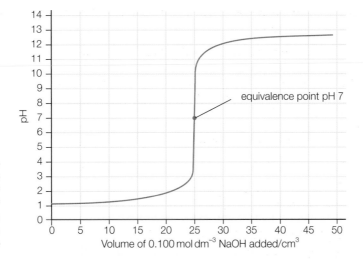

Figure 18.10 The titration curve for the titration of 25.0 cm³ of 0.100 mol dm⁻³ hydrochloric acid with 0.100 mol dm⁻³ sodium hydroxide

Let us calculate the pH after 25.5 cm³ of 0.100 mol dm⁻³ sodium hydroxide has been added to 25.0 cm³ of 0.100 mol dm⁻³ hydrochloric acid.

$$\text{reaction factor} = \frac{(25.5 - 25.0)}{25.0} = 0.02$$

An approximate pH (ignoring the dilution factor) for the resulting solution is calculated as follows:

$[OH^-(aq)] = 0.100 \times 0.0200 = 0.002\,00\,\text{mol dm}^{-3}$
$pOH = -\log_{10}0.002\,00 = 2.69; \qquad pH = 11.3$

Now calculate the pH after 27.5 cm³ of 0.100 mol dm⁻³ sodium hydroxide has been added to 25.0 cm³ of 0.100 mol dm⁻³ hydrochloric acid:

$$\text{reaction factor} = \frac{(27.5 - 25.0)}{25.0} = 0.100$$

An approximate pH (ignoring the dilution factor) for the resulting solution is calculated as follows:

$[OH^-(aq)] = 0.100 \times 0.100 = 0.0100\,\text{mol dm}^{-3}$
$pOH = -\log_{10}0.0100 = 2.00; \qquad pH = 12.0$

Titration of a weak acid against a strong alkali

25.0 cm³ of 0.100 mol dm⁻³ ethanoic acid requires exactly 25.00 cm³ of 0.100 mol dm⁻³ sodium hydroxide to reach the end-point or equivalence point. When these volumes react together 'neutralization' has occurred and only sodium ethanoate and water will be present:

$$CH_3COOH(aq) + NaOH(aq) \rightarrow CH_3COONa(aq) + H_2O(l)$$

However, the resulting solution will *not* be neutral. Ethanoic acid is a weak acid and exists mainly as molecules:

$$CH_3COOH(aq) \rightleftharpoons CH_3COO^-(aq) + H^+(aq)$$

At the beginning of the titration the pH will be about 3 since the acid is a weak acid and is only slightly dissociated into ions. (The exact pH of the ethanoic acid can be calculated if the value of K_a or pK_a is known.)

The addition of sodium hydroxide adds hydroxide ions to this equilibrium, which is 'pulled over' to the right as hydroxide ions remove the hydrogen ions via formation of water:

$$H^+(aq) + OH^-(aq) \rightarrow H_2O(l)$$

As alkali is added the ethanoic acid molecules undergo increasing dissociation to replace the hydrogen ions removed:

$$CH_3COOH(aq) \rightarrow CH_3COO^-(aq) + H^+(aq)$$

The overall reaction as an ionic equation is therefore:

$$CH_3COOH(aq) + OH^-(aq) \rightarrow H_2O(l) + CH_3COO^-(aq)$$

Note: the sodium ions are spectator ions and do not participate in the neutralization reaction, nor do they react with water molecules.

As a consequence of ethanoic acid being a weak acid, the line of the titration curve (Figure 18.11) (compared to hydrochloric acid) starts at a higher value of pH and stays higher because most of the hydrogen ions are kept 'in reserve' in undissociated ethanoic acid molecules. The dissociation of ethanoic acid gradually occurs as the alkali is added, hence the steady increase in pH with total volume of alkali added.

The equivalence point, where there are equal amounts of ethanoic acid and sodium hydroxide, will be at a pH *above* 7 due to salt hydrolysis since sodium ethanoate is the salt of a weak acid and strong base.

Ethanoate ions are a stronger base than water molecules and the following equilibrium is established:

$$CH_3COO^-(aq) + H_2O(l) \rightleftharpoons CH_3COOH(aq) + OH^-(aq)$$

with the forward reaction heavily favoured.

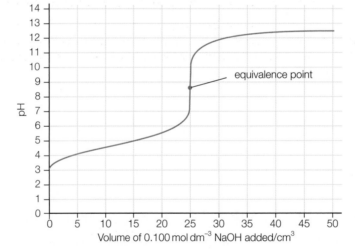

Figure 18.11 The titration curve for the titration of 25.0 cm³ of 0.100 mol dm⁻³ ethanoic acid with 0.100 mol dm⁻³ sodium hydroxide

The relatively flat portions of titration curves are where the pH changes most slowly on addition of acid or alkali. These flat portions are, therefore, where the best buffering action occurs. They are known as the buffer regions.

The production of a titration curve for a weak acid such as ethanoic acid is helpful, as it allows the pK_a and hence K_a to be calculated graphically (Figure 18.12) since the pH of the *half-neutralized* acid (at 12.5 cm³ of alkali) corresponds to the pK_a of the acid.

During the titration of ethanoic acid by sodium hydroxide the hydroxide ions gradually convert ethanoic acid molecules into ethanoate ions, so half-way to the end-point half of the ethanoic acid molecules will have been converted to ethanoate ions.

So specifically for the *half-neutralized* solution:

$$[CH_3COOH(aq)] = [CH_3COO^-(aq)]$$

However, in general:

$$K_a = [H^+(aq)] \times \frac{[CH_3COO^-(aq)]}{[CH_3COOH(aq)]}$$

Hence, $K_a = [H^+(aq)] \times \dfrac{1}{1}$ since the two concentrations are equal

So, $\quad K_a = [H^+(aq)]$

Taking logarithms to the base 10 of both sides:

$\quad pK_a = pH$

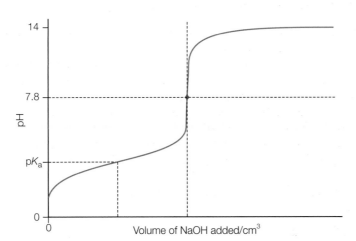

Figure 18.12 The determination of the pK_a of ethanoic acid from its titration curve

Titration of a strong acid against a weak alkali

If $0.100\,mol\,dm^{-3}$ hydrochloric acid, HCl(aq), is titrated against $0.100\,mol\,dm^{-3}$ aqueous ammonia, $NH_3(aq)$, then the pH changes very little until near the equivalence point, when it changes rapidly (Figure 18.13). The pH levels off again, but at a relatively low pH since aqueous ammonia is a weak base.

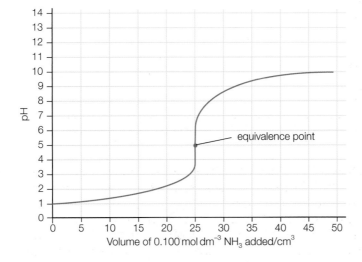

Figure 18.13 The titration curve for the titration of $25.0\,cm^3$ of $0.100\,mol\,dm^{-3}$ hydrochloric acid with $0.100\,mol\,dm^{-3}$ ammonia

Titration of a weak acid against a weak alkali

If $0.100\,mol\,dm^{-3}$ ethanoic acid is titrated against $0.100\,mol\,dm^{-3}$ aqueous ammonia solution then a very different titration curve results (Figure 18.14). There is no sharp or abrupt change in pH and hence no vertical section in the titration curve; the pH changes gradually during the titration process. No indicator is suitable for following this type of neutralization. A pH probe and meter are often used to identify the end-point in this type of titration.

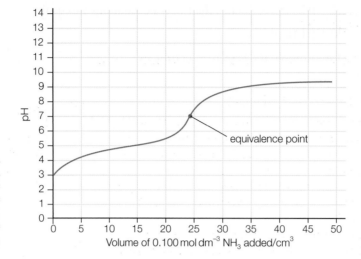

Figure 18.14 The titration curve for the titration of 25.0 cm³ of 0.100 mol dm⁻³ ethanoic acid with 0.100 mol dm⁻³ ammonia

■ Extension: Polybasic acids

Diprotic (dibasic) acids are acids that dissociate in water to release two hydrogen ions per molecule and triprotic (tribasic) acids dissociate in water to release three hydrogen ions per molecule. If the triprotic acid phosphoric(V) acid, H_3PO_4, is titrated against an aqueous solution of strong alkali, a titration curve (Figure 18.15) with three vertical regions is produced. Phosphoric(V) acid undergoes dissociation in three steps:

$$H_3PO_4(aq) \rightarrow H^+(aq) + H_2PO_4^-(aq)$$
$$H_2PO_4^-(aq) \rightarrow H^+(aq) + HPO_4^{2-}(aq)$$
$$HPO_4^{2-}(aq) \rightarrow H^+(aq) + PO_4^{3-}(aq)$$

Each of the step-wise dissociations is characterized by a different vertical section of the titration curve. The acid dissociation constant, K_a, for each of the dissociations can be determined graphically.

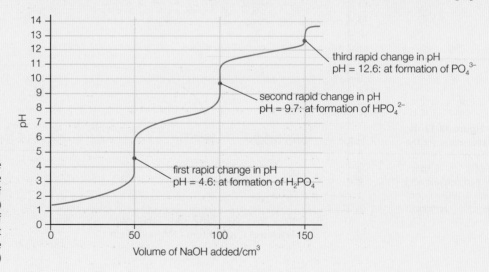

Figure 18.15 The titration curve for titration of phosphoric(V) acid (50.0 cm³ of 0.100 mol dm⁻³) against sodium hydroxide (0.100 mol dm⁻³)

18.5 Indicators

18.5.1 **Describe** qualitatively the action of an acid–base indicator.

Nature of indicators

Acid–base indicators are soluble dyes that change colour according to the hydrogen ion concentration, that is, the pH. They are usually weak acids whose acid and conjugate base are different colours. The wavelength of the light absorbed by the acid changes greatly when a proton is lost to form the conjugate base (Chapter 2).

Common acid–base indicators include litmus, methyl orange, screened methyl orange, bromothymol blue, bromophenol blue, phenol red (Figure 18.16) and phenolphthalein. Figure 18.17 shows the structural formulae of the dissociated or ionized and undissociated or un-ionized forms of the indicator phenolphthalein.

Figure 18.16 Phenol red: acidic (yellow) and alkaline (red) forms

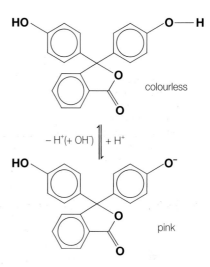

Figure 18.17 Phenolphthalein is colourless in acid, but pink in alkaline conditions

Phenolphthalein is slightly unusual for an indicator as one of its two forms is colourless. (Since they are brightly coloured, only small quantities of indicators need to be used and their addition to a solution will not significantly alter the concentration of hydrogen ions already present.)

Action of indicators

Using HIn(aq) for the acid form of an indicator and In⁻(aq) for its conjugate base, the equilibrium for the dissociation of the indicator can be generalized to:

$$HIn(aq) \rightleftharpoons In^-(aq) + H^+(aq)$$

or $\quad HIn(aq) + H_2O(l) \rightleftharpoons In^-(aq) + H_3O^+(aq)$

Consider the indicator bromophenol blue whose colours are yellow (undissociated form, HIn(aq)) and blue (dissociated form, In⁻(aq)), respectively.

In a neutral solution, very few of the acid molecules dissociate since indicators are weak acids. The solution therefore appears yellow due to the relatively high concentration of HIn(aq). This means nearly all the indicator molecules will exist as the undissociated yellow form.

- **Addition of an acid:** If an excess of an acidic solution is added to a solution of the indicator the increase in hydrogen ion concentration will, according to Le Châtelier's principle, shift the equilibrium above to the left so that the concentration of HIn(aq) is very high. This means almost all the indicator molecules will exist as the undissociated yellow form.
- **Addition of an alkali:** If an excess of an alkaline solution is added to a solution of the indicator the hydroxide ions will combine with the hydrogen ions to form water, thereby removing them from the equilibrium. The removed hydrogen ions will be partly replaced by the dissociation of HIn(aq). A relatively high concentration of In⁻(aq) will be produced and the solution will be blue.

Since indicators are generally weak acids an equilibrium expression for the acid dissociation constant, K_a, can be written for them. In general:

$$K_a = \frac{[H^+(aq)] \times [In^-(aq)]}{[HIn(aq)]} \quad \text{or} \quad K_a = \frac{[H_3O^+(aq)] \times [In^-(aq)]}{[HIn(aq)]}$$

The acid dissociation constant, K_a, is sometimes known as the dissociation constant for the indicator and given the symbol K_{In}.

The equation can be rearranged to make the ratio of the concentrations of the two coloured forms the subject:

$$\frac{[HIn(aq)]}{[In^-(aq)]} = \frac{[H^+(aq)]}{K_a}$$

This equation shows that the colour of an indicator depends not only on the hydrogen ion concentration, that is the pH, but also on the value of the acid dissociation constant, K_a. This means that different indicators change colour over different pH ranges.

$$K_a = \frac{[H^+(aq)] \times [In^-(aq)]}{[HIn(aq)]}$$

The equilibrium expression above can also be transformed into the Henderson–Hasselbalch equation previously derived in Section 18.2, by rearranging to:

$$\frac{1}{[H^+(aq)]} = \frac{1}{K_a} \times \frac{[In^-(aq)]}{[HIn(aq)]}$$

Taking logarithms to the base 10 of both sides:

$$pH = pK_a + \log_{10}\frac{[In^-(aq)]}{[HIn(aq)]}$$

This equation allows the calculation of any of the four variables, given the other three.

pH range of indicators

As shown previously, the colour of a solution to which the indicator has been added depends on the ratio of [HIn(aq)]/[In⁻(aq)] or [yellow]/[blue], which in turn depends on the hydrogen ion concentration or pH.

If the ratio for bromophenol blue is 10, the solution is yellow as the colour of HIn(aq) predominates. This happens, as shown previously, when the pH is low; that is, when the hydrogen ion concentration is high. The human eye cannot detect the small concentration of the blue In⁻(aq) form present. At this point:

$$pH = pK_a + \log_{10}\frac{[In^-(aq)]}{[HIn(aq)]}$$

$$pH = pK_a + \log_{10}\frac{1}{10}$$

$$pH = pK_a - 1$$

If the ratio [HIn(aq)]/[In⁻(aq)], or yellow to blue, is equal to 1 the solution is green, as the blue and yellow forms of the indicator are present in equal concentrations. At this point:

$$pH = pK_a + \log_{10}\frac{[In^-(aq)]}{[HIn(aq)]}$$

$$pH = pK_a + \log_{10}\frac{1}{1}$$

$$pH = pK_a$$

If the ratio [HIn(aq)]/[In⁻(aq)] is less than $\frac{1}{10}$, that is 0.1, then the solution is blue, as the colour due to the In⁻(aq) form predominates. The human eye cannot detect the small concentration of the yellow HIn(aq) form present.

$$pH = pK_a + \log_{10}\frac{[In^-(aq)]}{[HIn(aq)]}$$

$$pH = pK_a + \log_{10}\frac{10}{1}$$

$$pH = pK_a + 1$$

In general for indicators (Table 18.6) the colour change takes place over a range of about 2 pH units, specifically, from pH = pK_a − 1 to pH = pK_a + 1. This generally corresponds to the change described above, that is, going from 10% of one form of the indicator to 10% of the other form (Figure 18.18).

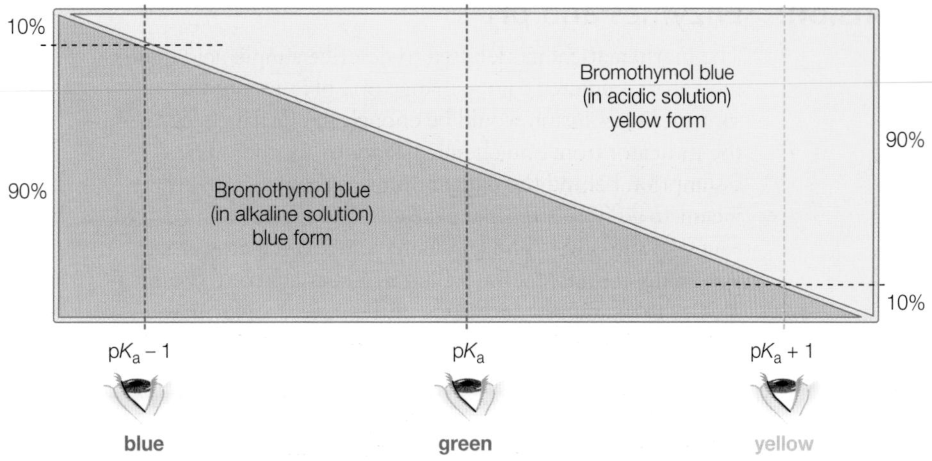

Bromothymol blue
(in acidic solution)
yellow form

Bromothymol blue
(in alkaline solution)
blue form

10%

90%

90%

10%

$pK_a - 1$

pK_a

$pK_a + 1$

blue

green

yellow

Figure 18.18 A diagram illustrating the behaviour of a typical acid–base indicator: bromothymol blue is placed into a transparent plastic box diagonally divided into halves

Indicator	'Acid colour'	'Alkaline colour'	pH range and pK_a
Methyl orange	Red	Yellow	3.2–4.4 and 3.4
Bromothymol blue	Yellow	Blue	6.0–7.6 and 7.1
Bromophenol blue	Yellow	Blue	3.0–4.6 and 4.0
Phenolphthalein	Colourless	Pink	8.2–10.0 and 9.4
Thymol blue	Red	Yellow	1.2–2.8 and 1.6
	Yellow	Blue	8.0–9.6 and 8.9
Methyl red	Red	Yellow	4.8 to 6.0 and 5.0
Litmus	Red	Blue	5.0 to 8.0 and 6.5

Table 18.6 The pH ranges of some common acid–base indicators

A few indicators undergo more than one change. Thymol blue, for example, changes colour in the pH range 1.2 to 2.8 and again in the range 8.0 to 9.6. This is because the undissociated acid is in equilibrium with *two* ionized forms. Universal indicators (Figure 18.19) are composed of mixtures of carefully selected indicators so as to give a series of gradual colour changes over a relatively large range of pH values.

Figure 18.19 Bottle of universal indicator solution

■ Extension: Indicators as weak bases

Methyl orange (Figure 18.20), unlike most indicators, is a weak base which can be represented as BOH(aq). In aqueous solution the following equilibrium is set up:

$$BOH(aq) \rightleftharpoons B^+(aq) + OH^-(aq)$$

yellow   red

base   conjugate acid

Application of Le Châtelier's principle predicts that in alkaline solution the yellow form BOH(aq) will predominate and in acidic solution the red $B^+(aq)$ will predominate (Figure 18.21).

Figure 18.20 Methyl orange indicator in acidic and alkaline solutions

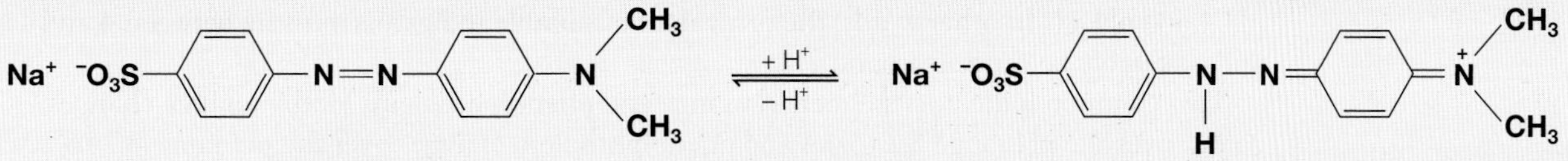

Figure 18.21 The structure of methyl orange in acidic and alkaline conditions

■ Extension: Enzymes and pH

The mathematical model used to describe simple acid–base indicators predicted that a change of 2 pH units in the surrounding solution would be enough to effectively convert the indicator from one chemical state to another. The assumption behind the calculation was that 'to convert' meant to go from a tenfold excess of one species to a tenfold excess of the other. The argument was that a factor of ten was an appropriate factor, which would enable one chemical species to swamp the influence of the other 'colour-wise'.

In fact if there is a situation where any protonated and de-protonated conjugate pair are in equilibrium in a solution, then a 2 pH unit increase in the background pH should be enough to convert the system from 'mostly protonated' (that is, most sites with hydrogen ions attached) to 'mostly de-protonated' (where 'mostly' means $10:1$).

This model can be readily applied to enzymes. Figure 18.22 shows the active site of an enzyme (Chapter 22). The assumption is that the active site includes at least one positively polarized hydrogen ion. Figure 18.22 shows that the substrate is expecting a protonated active site, and if it does not find one, it will not interact and bind to the active site. A protonated enzyme is being viewed as a weak acid in equilibrium with its conjugate base:

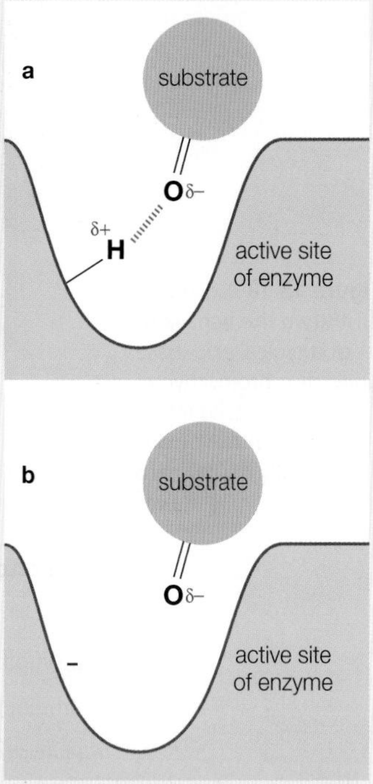

$$\text{active site-H} \rightleftharpoons \text{active site}^- + H^+(aq)$$

This is exactly the same situation as for acid–base indicators, only without the colour change: +2 pH units are enough to convert the enzyme from being $10:1$ protonated (active site-H) to $10:1$ de-protonated (active site$^-$).

The 'indicators' mathematical model is applicable and useful beyond the field of acid–base indicators, extending to any proton exchange equilibrium in solution. It predicts that the proton exchange will be substantially completed in either direction over a range of 2 pH units and therefore that any events dependent on the degree of protonation will change their course over that range. So, for example, enzymes are likely to have quite precise needs concerning the pH of their medium, and within about 2 pH units an enzyme will switch from a functional to a non-functional (denatured) condition.

Figure 18.22 A diagram illustrating the pH dependent action of enzymes. **a** Enzyme at low pH. There is a stabilizing hydrogen bond binding the substrate to the enzyme's active site. **b** Enzyme at high pH. The substrate is no longer able to hydrogen bond to the active site

Choice of indicator for titrations

An indicator used for any acid–base titration should ideally change colour at the pH corresponding to the mid-point of the almost straight portion of the titration curve. (However, there is little loss in accuracy if the indicator changes colour anywhere within the range of the almost vertically straight portion of the curve, since the pH change is relatively *large* for the addition of a relatively *small* amount of acid or base.)

For a strong acid/strong base titration any of the indicators could be used since all of them change colour within the almost vertical straight portion of the titration curve between about pH 4 and pH 11. In other words, the pK_a values of suitable indicators must lie between 4 and 11, and preferably be centred around 7.

Two common indicators are methyl orange and phenolphthalein. Methyl orange changes colour over the pH range 3.2 to 4.4 and phenolphthalein changes over the pH range 8.2 to 10.0. Both indicators are suitable for titrations involving a strong acid and a strong base (Figure 18.23).

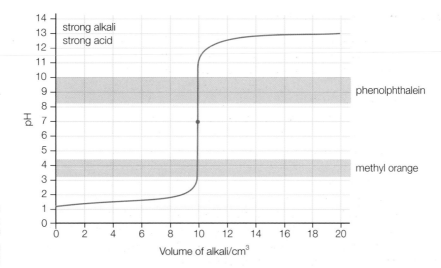

Figure 18.23 Titration curve starting with 100 cm³ of 0.100 mol dm⁻³ strong acid and adding 1.0 mol dm⁻³ strong alkali

However, the choice of indicator is more limited if a weak acid or a weak base is used in the titration, since the pH range of the almost straight portion is much smaller and fewer indicators change colour completely over this range.

For a strong acid/weak base titration, such as that between 0.1 mol dm⁻³ aqueous ammonia and 0.1 mol dm⁻³ hydrochloric acid, the indicator needs to change between pH values 4 and 7. Methyl orange is a suitable indicator, but phenolphthalein is not.

As Figure 18.24 shows, phenolphthalein would *not* be a suitable indicator because it will change colour at the wrong volume (not at the end-point) and over a large volume change of aqueous ammonia solution. It would therefore be impossible to find the end-point accurately using phenolphthalein as the indicator.

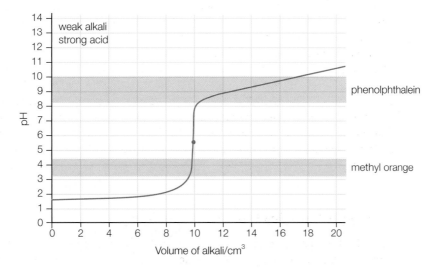

Figure 18.24 Titration curve starting with 100 cm³ of 0.100 mol dm⁻³ strong acid and adding 1.0 mol dm⁻³ weak alkali

For a weak acid/strong base titration, such as that between 0.1 mol dm⁻³ ethanoic acid and 0.1 mol dm⁻³ sodium hydroxide, the indicator needs to change between pH values 6 and 10. Phenolphthalein is a suitable indicator; but methyl orange is not.

Figure 18.25 shows that methyl orange would *not* be a suitable indicator because it will change colour very slowly over a relatively large volume of sodium hydroxide so that it would be very difficult to locate the end-point accurately. In addition, the colour change would occur at the wrong volume.

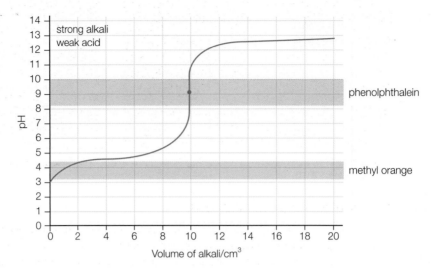

No indicator is suitable for the titration of a weak acid with a weak base, for example, 0.1 mol dm⁻³ ethanoic acid and 0.1 mol dm⁻³ aqueous ammonia, since there is no almost straight portion present in the titration curve (Figure 18.26). In other words, the pH changes *gradually* throughout the titration.

If bromothymol blue, whose pK_a is approximately 7, is used as an indicator, it would change colour over a relatively large volume of ammonia. Hence it would not be possible to find the end-point accurately.

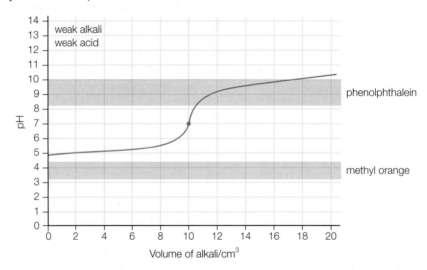

Table 18.7 summarizes the use of phenolphthalein and methyl orange as acid–base indicators. The principles described here can be used to select other suitable indicators for acid–base titrations.

Alkali	Acid	Indicator
Strong	Strong	Methyl orange or phenolphthalein
Strong	Weak	Phenolphthalein (Figure 18.27)
Weak	Strong	Methyl orange
Weak	Weak	None

No indicator can be used for an accurate titration of a weak acid with a weak base since the colour change of any indicator is going to be gradual. Such titrations are therefore often performed using a pH probe and meter or conductivity probe and meter (Figure 18.28).

Figure 18.27 Phenolphthalein in acidic and basic solutions

Figure 18.28 Conductivity probe and meter

■ Extension: Conductometric titrations

The end-point of an acid–base titration can be found by monitoring the conductivity (Figure 18.29a) of the solution as the alkali is progressively neutralized by the addition of acid. For example, during the titration of barium hydroxide and dilute sulfuric acid the electrical conductivity is zero at the end-point (Figure 18.29b).

$$Ba(OH)_2(aq) + H_2SO_4(aq) \rightarrow BaSO_4(s) + 2H_2O(l)$$

or ionically,

$$2OH^-(aq) + 2H^+(aq) \rightarrow 2H_2O(l)$$

At the end-point the electrical conductivity is zero because the ions of barium hydroxide and sulfuric acid are replaced by insoluble barium sulfate and water molecules.

Similar results are obtained when any base (strong or weak) is titrated against any acid (weak or strong). A V-shaped graph is obtained whose trough corresponds to the volume of acid required to achieve neutralization. Weak bases and acids, of course, have *shallower* gradients than strong acids or bases, and finish at *lower* points on conductivity graphs since their concentrations of ions are relatively smaller.

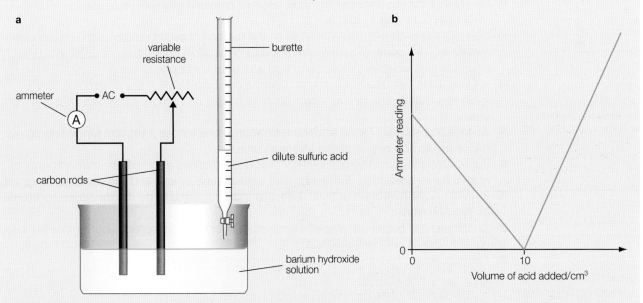

Figure 18.29 Neutralization of barium hydroxide with dilute sulfuric acid: **a** apparatus to monitor the neutralization by measuring conductivity changes and **b** a graph of sample results

SUMMARY OF KNOWLEDGE

- Water molecules are in equilibrium with a very low concentration of hydrogen and hydroxide ions.
- The ionic product constant of water, $K_w = [H^+(aq)] \times [OH^-(aq)] = 1.00 \times 10^{-14}\, mol^2\, dm^{-6}$ (at 298 K).
- The ionic product constant of water is a constant for all aqueous solutions (at constant temperature). When the hydrogen ion concentration increases, then the hydroxide ion concentration decreases (and *vice versa*).
- The strength of an acid is quantified by its acid dissociation constant, K_a, which is the equilibrium constant of its reaction with water:
$$K_a = \frac{[H^+(aq)] \times [A^-(aq)]}{[HA(aq)]}$$
- For a pure solution of a weak acid, $[H^+(aq)] = \sqrt{(K_a \times c)}$ (where c represents concentration of acid).
- The larger the value of the acid dissociation constant, the stronger the acid.
- $pK_a = -\log_{10} K_a$ for a weak acid. The stronger the acid, the smaller the pK_a value.
- The strength of a base is quantified by its base dissociation constant, K_b, which is the equilibrium constant of its reaction with water.
$$K_b = \frac{[BH^+(aq)][OH^-(aq)]}{[B(aq)]}$$
- For a pure solution of a weak base, $[OH^-(aq)] = \sqrt{(K_b \times c)}$ (where c represents concentration of base).
- The larger the value of the base dissociation constant, the stronger the base.
- $pK_b = -\log 10\, K_b$ for a weak base. The stronger the base, the smaller the pK_b value.
- The acidity of an aqueous solution is expressed on the pH scale, which is the negative logarithm (to the base 10) of the hydrogen ion concentration. $pH = -\log_{10} [H^+(aq)]$.
- For strong monoprotic acids, $[H^+(aq)]$ is the same as the concentration of the acid, $[HA(aq)]$. For strong bases $[OH^-(aq)]$ is the same as the concentration of the base, $[BOH(aq)]$.
- The alkalinity of an aqueous solution is expressed on the pOH scale, which is the negative logarithm (to the base 10) of the hydroxide ion concentration. $pOH = -\log_{10}[OH^-(aq)]$.
- Because of the constancy of the ionic product constant in aqueous solutions, $pH + pOH = 14$ and $K_w = K_b \times K_a$.
- During an acid–base titration, the pH of the solution changes rapidly in the region of the end-point. The graph of pH against the volume of alkali added takes a different curve depending on whether the reacting acid and base are strong or weak.
- An acid–base indicator is a weak acid whose conjugate base has a different colour. Indicators change colour over approximately a 2 pH unit range, either side of its own pK_a value.
- An indicator suitable for an acid–base titration is one whose 2 pH unit range falls across the nearly vertical region of the titration curve.
- Buffers are aqueous chemical systems that resist changes in pH when small amounts of acid or based are added. Buffers contain a weak base or acid and the corresponding salt.
- Buffers release hydrogen or hydroxide ions from equilibria to neutralize added hydroxide or hydrogen ions.
- Conjugate-pair buffer systems when mixed in equimolar proportions buffer at a pH value equal to the pK_a of the weak acid.
- The behaviour of acidic buffers is described by the following equation:
$$pH = pK_a + \frac{\log_{10}[A^-(aq)]}{[HA(aq)]}$$
- The behaviour of basic buffers is described by the following equation:
$$pOH = pK_b - \frac{\log_{10}[B(aq)]}{[BH^+(aq)]}$$

■ *Examination questions – a selection*

Paper 1 IB questions and IB style questions

Q1 An aqueous solution has a pH of 10. Which concentrations are correct for the ions below?

	$[H^+(aq)]/mol\,dm^{-3}$	$[OH^-(aq)]/mol\,dm^{-3}$
A	10^4	10^{-10}
B	10^{-4}	10^{-10}
C	10^{-10}	10^{-4}
D	10^{-10}	10^4

Higher Level Paper 1, Nov 05, Q26

Q2 Which compound will dissolve in water to give a solution with a pH greater than 7?
  A rubidium chloride
  B sodium carbonate
  C ammonium nitrate
  D calcium sulfate

Q3 What is the relationship between K_a and pK_a?
  A $pK_a = -\log K_a$
  B $pK_a = 1.0 \times 10^{-7} / K_a$
  C $pK_a = \log K_a$
  D $pK_a = 14 / K_a$

Q4 A buffer solution can be prepared by adding which of the following to $50\,cm^3$ of $0.20\,mol\,dm^{-3}$ $CH_3COOH(aq)$?
  I $50\,cm^3$ of $0.20\,mol\,dm^{-3}$ $CH_3COONa(aq)$
  II $25\,cm^3$ of $0.20\,mol\,dm^{-3}$ $NaOH(aq)$
  III $50\,cm^3$ of $0.20\,mol\,dm^{-3}$ $NaOH(aq)$
  A I only
  B I and II only
  C II and III only
  D I, II and III

Q5 An acid–base indicator, HIn, dissociates according to the following equation:
  $HIn(aq) \rightleftharpoons H^+(aq) + In^-(aq)$
  Colour A   Colour B
  Which statement about this indicator is correct?
  I In a strongly acidic solution colour A would be seen.
  II In a neutral solution the concentrations of $HIn(aq)$ and $In^-(aq)$ must be equal.
  III It is suitable for use in titrations involving weak acids and bases.
  A I only
  B II only
  C III only
  D None of the above

Q6 When the following $2.0\,mol\,dm^{-3}$ aqueous solutions are arranged in order of decreasing pH, which is the correct order?
  I ammonium chloride
  II ammonium ethanoate
  III potassium ethanoate
  A I, II and III
  B II, I and III
  C III, I and II
  D III, II and I

Q7 The hydrogen ion concentration of a $0.050\,mol\,dm^{-3}$ lactic acid (a weak monobasic organic acid) solution is $2.62 \times 10^{-3}\,mol\,dm^{-3}$. The acid dissociation constant, K_a, is 1.37×10^{-4}. What is the pH of this solution?
  A 1.30
  B 2.58
  C 5.94
  D 3.86

Q8 The pH of a $0.10\,mol\,dm^{-3}$ solution of a weak monobasic acid represented by HA is 4.20. What is the value of K_a for HA?
  A 4.0×10^{-8}
  B 2.5×10^{-7}
  C 4.0×10^{-5}
  D 2.5×10^{-9}

Q9 What is the pH in a titration when $20.0\,cm^3$ of $0.011\,mol\,dm^{-3}$ sodium hydroxide, NaOH, has been added to $25.0\,cm^3$ of $0.014\,mol\,dm^{-3}$ hydrochloric acid, HCl?
  A 2.28
  B 1.85
  C 2.54
  D 3.46

Q10 Which one of the following $0.20\,mol\,dm^{-3}$ salt solutions is the most acidic?
  A potassium carbonate, K_2CO_3
  B potassium cyanide, KCN
  C sodium nitrate, $NaNO_3$
  D iron(III) chloride, $FeCl_3$

Q11 Pyridine is a weak base that reacts with water according to the equation:

  $pyridine(aq) + H_2O(l) \rightleftharpoons pyridineH^+(aq) + OH^-(aq)$

  What is the pH of a $0.050\,mol\,dm^{-3}$ solution of pyridine if the base dissociation constant, K_b, is 1.4×10^{-9}?
  A 5.1
  B 4.4
  C 8.9
  D 9.6

Q12 Which compound dissolves in water to form a solution that does **not** conduct electricity?

- **A** HCl
- **B** K_2SO_4
- **C** $CH_3CH_2CH_2OH$
- **D** CH_2ClCH_2COOH

Q13 The following pairs are mixtures of $0.5\,mol\,dm^{-3}$ aqueous solutions. Which pair of chemicals would act as a buffer?

- **A** HCl and NaCl
- **B** HCl and $NaNO_3$
- **C** NH_3 and NH_4NO_3
- **D** NaOH and HCl

Q14 Which of the following statements correctly describes a **weak** acid?

  - **I** It has a low pK_a value.
  - **II** It has a strong conjugate base.
  - **III** It has a relatively high electrical conductivity in dilute aqueous solutions.

- **A** II only
- **B** II and III
- **C** I and III
- **D** I, II and III

Q15 The dissociation of water is an endothermic reaction. Which of the statements is **false** when $1000\,cm^3$ of pure water is heated from $25\,°C$ to $60\,°C$?

- **A** The concentration of hydrogen ions increases.
- **B** The pH of the water decreases.
- **C** The water becomes acidic.
- **D** The concentration of hydroxide ions decreases.

Q16 For which of the following reactions is the equilibrium constant called the basicity constant, K_b?

- **A** $HCOOH(aq) + NH_3(aq) \rightleftharpoons HCOO^-(aq) + NH_4^+(aq)$
- **B** $CH_3CH_2NH_2(aq) + H_2O(l)$
$\rightleftharpoons CH_3CH_2NH_3^+(aq) + OH^-(aq)$
- **C** $H_3O^+(aq) + OH^-(aq) \rightleftharpoons 2H_2O(l)$
- **D** $CH_3CH_2COOH(aq) + OH^-(aq)$
$\rightleftharpoons CH_3CH_2COO^-(aq) + H_2O(l)$

Q17 For a given weak acid, HA, the value of K_a:

- **A** will change with the $[H^+(aq)]$
- **B** will change with the pOH
- **C** will change with temperature
- **D** cannot be less than 10^{-5}

Q18 What is the pH of a solution which is $0.0100\,mol\,dm^{-3}$ in HA (a weak monobasic acid) and also $0.0020\,mol\,dm^{-3}$ in NaA (its sodium salt) ($K_a = 9.0 \times 10^{-6}$)?

- **A** 4.35
- **B** 3.75
- **C** 6.65
- **D** 5.65

Q19 Which one of the following mixtures is suitable for making a buffer with a pH of about 9?

- **A** CH_3CH_2COOH and CH_3CH_2COONa
- **B** HF and NaF
- **C** $Ca(OH)_2$ and NaOH
- **D** NH_4Cl and NH_3

Q20 Which graph shows how the pH changes when a weak base is added to a strong acid?

A
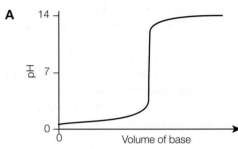

B
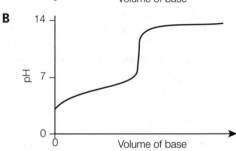

C
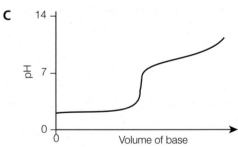

D
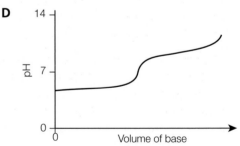

Higher Level Paper 1, Nov 05, Q27

Paper 2 IB questions and IB style questions

Q1 a i Define the term *pH*. [1]
 ii A 25.0 cm³ sample of 0.100 mol dm⁻³
 hydrochloric acid is placed in a conical flask,
 and 0.100 mol dm⁻³ sodium hydroxide is
 added until a total of 50.0 cm³ has been
 added. Sketch a graph of pH against
 volume of NaOH(aq) added, clearly showing
 the volume of NaOH(aq) needed for
 complete reaction and the pH values at the
 start, the equivalence point and finish. [4]
 iii The experiment in **a ii** was repeated but
 with 25.0 cm³ sample of 0.100 mol dm⁻³
 ethanoic acid in the conical flask instead of
 the hydrochloric acid. Use information from
 Table 15 on page 20 of the IB *Chemistry
 data booklet* to calculate the pH at the start
 of the experiment. State the approximate
 pH value at the equivalence point. [5]
 b i Describe how an indicator, HIn, works. [3]
 ii Name a suitable indicator for the reaction
 between ethanoic acid and sodium
 hydroxide. Use the information from Table
 16 on page 22 in the IB *Chemistry data
 booklet* to explain your choice. [2]
 c i Identify **two** substances that can be added
 to water to form a basic buffer solution. [1]
 ii Describe what happens when a small
 amount of acid solution is added to the
 buffer solution prepared in **i**. Use an
 equation to support your explanation. [2]
 d Define the terms *Brønsted–Lowry acid* and
 Lewis acid. For each type of acid, identify
 one example other than water and write an
 equation to illustrate the definition. [5]
 e Predict and explain whether an aqueous
 solution of 0.10 mol dm⁻³ AlCl₃ will be acidic,
 alkaline or neutral. [2]
 Higher Level Paper 2, May 06, Q8

Q2 a The equilibrium reached when propanoic acid
 is added to water can be represented by the
 following equation:
 $$CH_3CH_2COOH(aq) \rightleftharpoons CH_3CH_2COO^-(aq) + H^+(aq)$$
 Identify two Brønsted–Lowry acids and two
 Brønsted–Lowry bases. [2]
 b The pH of a solution is 4.9. Using information
 from Table 16 of the IB *Chemistry data
 booklet*, deduce and explain the colours of the
 indicators bromophenol blue and phenol red in
 this solution. [3]
 c Calculate the pH of a buffer solution
 containing 0.0500 mol dm⁻³ of ethanoic acid
 ($K_a = 1.74 \times 10^{-5}$) and 0.0200 mol dm⁻³ of
 sodium ethanoate. [3]

Q3 A mixture of benzoic acid (C_6H_5COOH) and
 sodium benzoate ($NaC_6H_5CO_2$) can act as a
 buffer solution.
 a Define the term *buffer solution* and describe
 what happens when an acid is added to a
 buffer solution. [5]
 b Calculate the pH of a solution containing 7.2 g
 of sodium benzoate in 1.0 dm³ of
 3.0×10^{-2} mol dm⁻³ benzoic acid,
 ($K_a = 6.3 \times 10^{-5}$ mol dm⁻³) stating any
 assumptions you have made. [5]
 c Benzoic acid is a *weak monoprotic* acid. Explain
 these terms. [2]

Q4 a In the reaction
 $$2H_2O(l) \rightarrow H_3O^+(aq) + OH^-(aq)$$
 use the Brønsted–Lowry theory to discuss the
 acidic and/or basic nature of water. [2]
 b What is the conjugate base of the hydroxide
 ion, OH^-? [1]
 c Define the term *pH* and give the pH of pure
 water (at 25 °C). [2]
 d i Write an expression for the ionic product of
 water, K_w. [1]
 ii The value of K_w increases with temperature.
 Explain with reasoning whether the
 dissociation of water is endothermic or
 exothermic. [3]

Q5 a Define the term *salt hydrolysis*. [2]
 b Predict whether each of the following solutions
 would be acidic, alkaline or neutral. In each
 case explain your reasoning.
 i 1.0 mol dm⁻³ $FeCl_3$(aq)
 ii 1.0 mol dm⁻³ KNO_3(aq)
 iii 1.0 mol dm⁻³ K_2CO_3(aq) [6]
 c Determine the pOH of a solution with an
 ammonia concentration of 0.120 mol dm⁻³.
 (pK_b of ammonia is 4.75.) [4]

19 Oxidation and reduction

STARTING POINTS

- Redox reactions involve electron transfer from a reducing agent to an oxidizing agent.
- Redox reactions, like all other reactions, tend towards a state of chemical equilibrium.
- Redox reactions can be performed in a voltaic cell so that electron transfer takes place along a wire connecting two electrodes. This allows the chemical energy from the redox reaction to produce a potential difference (voltage).
- Any reversible reaction that contains species related by the gain or loss of electrons can act as an electrode in a voltaic cell.
- The potential of a voltaic cell (under standard conditions) can be calculated from standard electrode potentials.
- Standard electrode potentials are measured against the standard hydrogen electrode.
- Standard electrode potentials are a measure of the tendency of a chemical element to form ions or change oxidation number.
- The sign of a cell potential indicates whether a reaction is spontaneous under standard conditions.
- Measurements of the potential differences of voltaic cells allow chemists to calculate free energy changes of redox reactions.
- Electrolysis reactions are made spontaneous by passing an electric current through a reaction mixture containing mobile ions.
- Electrolysis reactions performed in aqueous solution may involve hydrogen and hydroxide ions from the dissociation of water.

19.1 Standard electrode potential

Redox equilibria

A chemical equilibrium (Chapter 7) known as a **redox equilibrium** is established when a piece of metal is placed in an aqueous solution of its ions. The forward reaction involves the atoms of the metal entering the solution as hydrated metal ions, leaving behind a layer of electrons on the surface of the metal. This oxidation process can be represented by:

$$M(s) \rightarrow M^{n+}(aq) + ne^-$$

where n represents 1, for example sodium, 2, for example copper or 3, for example aluminium.

The reverse reaction involves hydrated metal ions in the solution accepting electrons from the surface of the metal and being deposited as metal atoms on the surface of the piece of metal.

This reduction process can be represented by:

$$M^{n+}(aq) + ne^- \rightarrow M(s)$$

Equilibrium (Figure 19.1) is established when the rates of the forward and backward reactions are equal:

$$M(s) \rightleftharpoons M^{n+}(aq) + ne^-$$

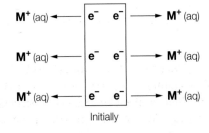
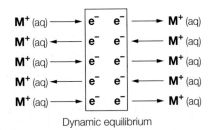

Figure 19.1 The establishment of a redox equilibrium

Initially

Dynamic equilibrium

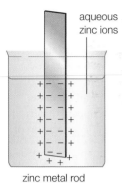

aqueous zinc ions

zinc metal rod

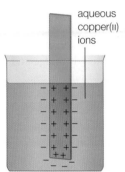

aqueous copper(II) ions

copper metal rod

Figure 19.2 Redox equilibria for zinc in zinc(II) ions and copper in copper(II) ions

The position of equilibrium depends on a number of factors, but in particular, the *reactivity* of the metal. For example, when a piece of zinc, a relatively reactive metal, is placed in an aqueous solution of zinc ions, the equilibrium lies to the right-hand side:

$$Zn(s) \rightleftharpoons Zn^{2+}(aq) + 2e^-$$

with the overall change being the dissolution of zinc, leaving the surface with a *negative* charge due to the presence of a layer of electrons. The mass of the original piece of zinc has decreased slightly and the concentration of zinc ions in solution has increased slightly. (However, there is a clear distinction between reactivity (a kinetic concept) and position of equilibrium (a thermodynamic concept) which is discussed later in this chapter.)

The opposite process occurs when copper, a relatively unreactive metal, is placed in an aqueous solution of copper(II) ions. In this reaction the equilibrium lies to the left-hand side:

$$Cu(s) \rightleftharpoons Cu^{2+}(aq) + 2e^-$$

The overall reaction involves the deposition of copper(II) ions as copper atoms and so as a consequence the surface of the piece of copper gains a *positive* charge. The mass of the original piece of copper has increased slightly and the concentration of copper ions in solution has decreased slightly. These two redox equilibria are illustrated in Figure 19.2. (The number of charges shown in the diagram should not be taken literally, but the numbers of positive and negative charges will be equal.)

In both the redox equilibria described above, the solution and the surface of the metal develop opposite electrical charges. A potential difference (measured as a voltage), known as an **electrode potential**, is said to exist between the surface of the metal and the solution because of this charge separation. The piece of metal dipping in the solution is referred to as an electrode and when immersed in its own ions forms a **half-cell**.

A voltaic cell, such as the Daniell cell (Chapter 9), can be constructed by connecting two half-cells using an external circuit and a salt bridge. The cell potential can be measured by introducing a high-resistance voltmeter into the external circuit.

Types of half-cells

The potential difference between an electrode and a solution of aqueous ions is not limited to metals but also applies to non-metals. There are three types of commonly encountered half-cell:

- Metal immersed in its own ions.
- **Inert electrode** (for example, graphite or platinum which does not take part in the redox equilibrium) immersed in an aqueous solution containing two ions of the same element in *different* oxidation states, e.g. Fe^{2+}/Fe^{3+}.
- Gas bubbling over an inert electrode immersed in an aqueous solution containing the ions of the gas, for example, the standard hydrogen electrode, which has an equilibrium between hydrogen ions and hydrogen molecules.

Any two different half-cells can be combined together to form a voltaic cell and allow the electrons to flow from the reducing agent to the oxidizing agent. The resulting movement of electrons allows useful work to be done and also allows chemists to measure the tendency for a redox reaction to occur.

The electrode potential of an element depends on three factors:

- the nature of the element
- the concentration of its ions in solution
- the temperature of the solution.

Note that the amount of metal present does *not* influence the electrode potential of a metal.

Hence, the concentrations and temperature of the electrolytes have to be stated when comparing the electrode potentials of different elements. Standard thermodynamic conditions are usually stated (standard conditions are 298 K (25 °C), one atmosphere pressure and all concentrations 1 mol dm^{-3}).

■ **Extension:** The salt bridge

A simple salt bridge consists of a filter paper soaked in saturated potassium nitrate solution, $KNO_3(aq)$. The function of the salt bridge is to complete the circuit and to allow for the balancing of ionic charges in the two solutions of the Daniell cell and other simple voltaic cells.

The dissolution of zinc from the zinc electrode in the Daniell cell will result in an increase in the concentration of zinc ions in the zinc sulfate solution. The deposition of copper(II) ions from the copper(II) sulfate solution as copper atoms will cause a decrease in the concentration of copper(II) ions in the copper(II) sulfate solution.

Both of these processes will lead to a *surplus* of positive ions in the zinc sulfate and a *deficiency* of positive ions in the copper(II) sulfate. Unless the concentrations of these positive ions are kept constant then the two redox reactions will gradually slow and stop, and the current will drop to zero.

The imbalances in the concentrations of the two positive ions are restored by flows of ions from the salt bridge. Negatively charged nitrate ions leave the salt bridge and their place is taken up by zinc ions. For every two nitrate ions that enter the zinc sulfate, one zinc ion enters the salt bridge. In the copper(II) sulfate solution positive potassium ions leave the salt bridge and are replaced by sulfate ions. For every *two* potassium ions that enter the copper(II) sulfate, *one* sulfate ion enters the salt bridge (Figure 19.3).

These flows of ions maintain the overall net positive charges of metal ions in both solutions. Potassium nitrate is chosen for the salt bridge because neither potassium ions nor nitrate ions react chemically with the other ions present in the two solutions. They also move at similar speeds to copper(II), zinc and sulfate ions in solution.

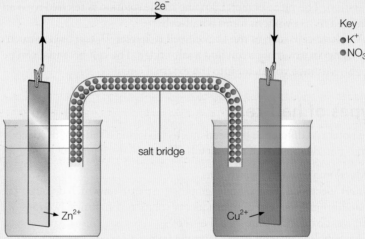

Figure 19.3 Flow of ions in a salt bridge containing potassium and nitrate ions

The standard hydrogen electrode

19.1.1 **Describe** the standard hydrogen electrode.

The voltage of a *single* metal electrode in the half-cell of an electrochemical cell, such as the copper and zinc electrodes of the Daniell cell, cannot be measured. If the metal electrode is connected to a voltmeter using a wire and the wire placed in the solution to complete the circuit, another redox equilibrium and electrode potential will be generated. The voltage will be the difference of the two electrode potentials, not the voltage of the first metal electrode in equilibrium with its ions in aqueous solution.

The solution to this problem is to choose a *standard reference electrode* and measure the potentials of all other electrodes relative to this. In principle any electrode system could be used as a standard reference electrode, but metals, especially the reactive ones, tend to undergo corrosion, which reduces their accuracy.

The internationally agreed reference electrode is the **standard hydrogen electrode** (Figure 19.4), which has hydrogen gas in equilibrium with hydrogen ions. In this system hydrogen is behaving as a metal since positively charged hydrogen ions are present. It is relatively easy to prepare pure hydrogen gas and solutions of hydrogen ions of known concentrations. The standard hydrogen electrode is arbitrarily given an electrode potential of *zero* volts, so all redox systems measured relative to it will be given either a positive or negative value.

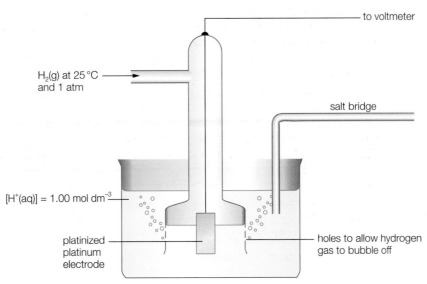

Figure 19.4 The standard hydrogen electrode

The standard hydrogen electrode is maintained by a stream of pure hydrogen gas bubbling over a platinum electrode coated with platinum black (finely divided platinum) immersed in a solution of hydrochloric acid.

The platinum electrode has these functions:

- It acts as an *inert* metal connector to the hydrogen gas/hydrogen ion equilibrium. There is no tendency for the very unreactive platinum itself to ionize and to act as an electrode.
- The platinized surface acts as a heterogeneous *catalyst* (Chapter 16) for the adsorbed hydrogen gas on its surface. This allows standard electrode potentials to be measured quickly.
- An equilibrium is set up between the gas adsorbed on the electrode and the hydrogen ions in the acid solution.

The electrode potential of this electrode is fixed at zero under the following standard state conditions:

- temperature at 298 K (25 °C)
- pressure of hydrogen gas at standard atmospheric pressure (101 325 Pa)
- hydrogen ion concentration at one mole per cubic decimetre (1 mol dm^{-3}).

Standard electrode potentials

19.2.2 **Define** the term *standard electrode potential* ($E^{\ominus}$).

Standard electrode potentials of metals are measured relative to the standard hydrogen electrode. The redox half-equation for the standard hydrogen electrode is:

$$2H^+(aq) + 2e^- \rightleftharpoons H_2(g) \qquad E^{\ominus} = 0.00\,V$$

Thus, if the standard half-cell is connected to a standard hydrogen electrode to form a voltaic or electrochemical cell, the measured voltage, called the electromotive force (e.m.f.), is the standard electrode potential of that half-cell.

The **standard electrode potential** is defined as the potential difference between a standard hydrogen electrode and a metal (the electrode) which is immersed in a solution containing metal ions at 1 mol dm^{-3} concentration at 298 K (25 °C) and 1 atmosphere pressure.

Figure 19.5 shows the arrangement used to measure the standard electrode potential of a zinc half-cell. The e.m.f. is −0.76 V.

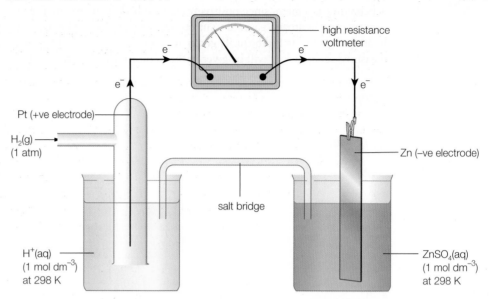

Figure 19.5 Measuring the standard electrode potential of a zinc half-cell

The voltmeter shows that the electrons flow from the zinc electrode to the hydrogen electrode in the external circuit. This means that at the zinc electrode, the following reaction occurs:

$$Zn(s) \rightarrow Zn^{2+}(aq) + 2e^-$$

That is, oxidation occurs and the zinc electrode acts as the negative pole, the anode.

Conversely, the hydrogen electrode acts as the positive pole, the cathode, and the reaction of the half-cell is:

$$2H^+(aq) + 2e^- \rightarrow H_2(g)$$

The overall cell reaction is:

$$Zn(s) + 2H^+(aq) \rightarrow Zn^{2+}(aq) + H_2(g)$$

According to IUPAC (International Union of Pure and Applied Chemistry), the standard electrode potential for the zinc half-cell ($Zn^{2+}(aq)/Zn(s)$) is −0.76 V. The negative sign is used to show that the electrode is the negative pole if it is connected to a standard hydrogen electrode.

Sign convention

By convention, the oxidized species is written first when a particular half-equation and its standard electrode potential are being referred to. In other words, the half-equation is written as a *reduction* process.

$$\text{oxidized species} + ne^- \rightleftharpoons \text{reduced species}$$

Thus, $Ag^+(aq)/Ag(s)$, $E^\ominus = +0.80$ V means that the silver half-cell reaction has a standard electrode potential of +0.80 V.

$$\underset{\text{oxidized species}}{Ag^+(aq) + e^-} \quad \rightarrow \quad \underset{\text{reduced species}}{Ag(s)} \qquad\qquad E^\ominus = +0.80\,V$$

The IB *Chemistry data booklet* contains standard electrode potentials of half-cells recorded in this format in Table 14 on page 18.

In summary:

- If $E^\ominus > 0$, then *reduction* takes place at the electrode (electrons being used up) when it is connected to the standard hydrogen electrode.
- If $E^\ominus < 0$, then *oxidation* takes place at the electrode (electrons being produced) when it is connected to the standard hydrogen electrode.

■ Extension: Cell potential and e.m.f.

The measurement of the electrode potential must be carried out without the flow of electric current so that the concentrations of the solutions in the half-cells do not change. A change in concentration of the solutions will affect the electrode potential of the half-cell. Hence, a high-resistance voltmeter is used for the measurement of the e.m.f. of the electrochemical cell. This means that the current in the external circuit is virtually zero and the electrochemical cell registers its maximum potential difference.

The **cell potential** is the difference in electrode potentials of the two electrodes when the cell is passing current through the circuit. It is measured by a voltmeter, ideally with a high resistance. It is *less* than the maximum voltage obtained from the cell (the e.m.f. of the cell).

The **electromotive force (e.m.f.)** is the potential difference between the two terminals of the voltaic cell when no current is flowing, that is, in an open circuit. The e.m.f. is the *maximum* voltage obtained from the voltaic cell. Strictly speaking, it should only be measured using a potentiometer and not a voltmeter (since voltmeters have a high but finite resistance). However, a good high-resistance voltmeter will give a value very close to the e.m.f.

Electrode reaction	$E^{\ominus}$/volts
$Li^+(aq) + e^- \rightarrow Li(s)$	−3.03
$K^+(aq) + e^- \rightarrow K(s)$	−2.92
$Ca^{2+}(aq) + 2e^- \rightarrow Ca(s)$	−2.87
$Mg^{2+}(aq) + 2e^- \rightarrow Mg(s)$	−2.36
$Al^{3+}(aq) + 3e^- \rightarrow Al(s)$	−1.66
$Mn^{2+}(aq) + 2e^- \rightarrow Mn(s)$	−1.18
$Zn^{2+}(aq) + 2e^- \rightarrow Zn(s)$	−0.76
$Fe^{2+}(aq) + 2e^- \rightarrow Fe(s)$	−0.44
$Ni^{2+}(aq) + 2e^- \rightarrow Ni(s)$	−0.23
$Sn^{2+}(aq) + 2e^- \rightarrow Sn(s)$	−0.14
$Pb^{2+}(aq) + 2e^- \rightarrow Pb(s)$	−0.13
$H^+(aq) + e^- \rightarrow \frac{1}{2}H_2(g)$	0.00
$Cu^{2+}(aq) + 2e^- \rightarrow Cu(s)$	+0.34
$Cu^+(aq) + e^- \rightarrow Cu(s)$	+0.52
$Ag^+(aq) + e^- \rightarrow Ag(s)$	+0.80

Table 19.1 Standard electrode potentials for some common metals (all data taken from the IB *Chemistry data booklet*)

The standard electrode potentials for some common metals are given in Table 19.1. Note that the half-cell reactions are written as *reduction* processes: the metal ions are gaining electrons. The standard electrode potentials are therefore sometimes known as standard reduction potentials.

This arrangement of metals (and hydrogen) in order of decreasing standard electrode potential is known as the **electrochemical series**. It is very similar in arrangement to the reactivity series (Chapter 9).

Metals towards the top of the electrochemical series, that is, those metals with large *negative* electrode potentials, are very reactive and readily give up electrons in solution. In other words, they are powerful *reducing* agents: they have the greatest tendency to form positive ions in aqueous solution.

Towards the bottom of the electrochemical series the metals become progressively weaker reducing agents, *but* conversely their oxidizing power increases. Unreactive metals at the very bottom of the electrochemical series behave as weak oxidizing agents (Figure 19.6).

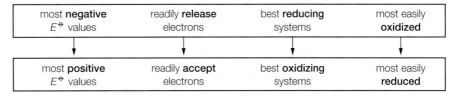

Figure 19.6 Summary of the trends in the electrochemical series

Worked example

Use the standard electrode potentials on page 18 of the IB *Chemistry data booklet* to arrange the following oxidizing agents in increasing order of oxidizing strength (under standard conditions):

Potassium manganate(VII) (in acidic solution)
Iodine
Iron(III) ions
Oxygen (in acidic solution)

Here are the standard electrode potentials:

$$MnO_4^-(aq) + 8H^+(aq) + 5e^- \rightarrow Mn^{2+}(aq) + 4H_2O(l) \qquad E^{\ominus} = +1.51\,V$$

$$Fe^{3+}(aq) + e^- \rightarrow Fe^{2+}(aq) \qquad E^{\ominus} = +0.77\,V$$

$$\tfrac{1}{2}I_2(s) + e^- \rightarrow I^-(aq) \qquad E^{\ominus} = +0.54\,V$$

$$\tfrac{1}{2}O_2(g) + 2H^+(aq) + 2e^- \rightarrow H_2O(l) \qquad E^{\ominus} = +1.23\,V$$

Oxidizing agents undergo reduction (gain of electrons). The more positive the value of the standard electrode potential, the greater the oxidizing power of the chemical species on the left-hand side of the reduction potential.

Hence, the order of increasing oxidizing power (under standard conditions) is:

iodine, iron(III) ions, oxygen (in acidic solution) and potassium manganate(VII) (in acidic solution).

■ **Extension:** ## The reactivity series versus the electrochemical series

Note that the reactivity series (Chapter 9) and electrochemical series, although they are very similar, are based upon different branches of chemistry. The reactivity series is derived from *kinetic* studies and the electrochemical series is based upon *thermodynamic* measurements.

The reactivity series is based upon the reactivities of the pure metals with water, dilute acid or metals ions: the faster the reaction, the more reactive the metal. In contrast, the electrochemical series is based upon the measurement of potentials (voltage – a thermodynamic quantity) of metals against a standard reference electrode.

There is not necessarily a correlation between kinetics and thermodynamics, although there happens to be one for many metals and their reactions. Hence the order of metals in the electrochemical series and activity series is very similar.

The distinction between the reactivity series and the electrochemical series can be illustrated by adding lithium and potassium to water. Lithium reacts more slowly with water than potassium does, despite the former having a more negative electrode potential. The reaction between lithium and water to form lithium ions is more thermodynamically favourable than that between potassium and water, but the reaction is slower.

Predicting cell reactions and voltages

Predictions of cell potentials for spontaneous reactions of any combination of metals and their ions can be easily calculated. The two half-cells are both written as reduction potentials. However, it is not possible for both reactions to accept electrons; one half-cell must be reversed so it releases electrons. The half-cell that is always reversed is the one with *more negative* (or least positive) electrode potential. The cell potential is then the sum of the electrode potentials (including their signs).

Worked example

A voltaic cell is constructed using magnesium and copper electrodes. What is the cell potential for the spontaneous reaction?

The electrode potentials are:

$$Mg^{2+}(aq) + 2e^- \rightarrow Mg(s) \qquad E^\ominus = -2.36\,V$$
$$Cu^{2+}(aq) + 2e^- \rightarrow Cu(s) \qquad E^\ominus = +0.34\,V$$

The magnesium half-cell is the more negative electrode potential, hence the equation and the sign of the electrode potential are reversed:

$$Mg(s) \rightarrow Mg^{2+}(aq) + 2e^- \qquad E^\ominus = +2.36\,V$$
$$Cu^{2+}(aq) + 2e^- \rightarrow Cu(s) \qquad E^\ominus = +0.34\,V$$

The two half-equations are added and the electrons cancelled to generate the ionic equation for the spontaneous reaction:

$$Mg(s) + Cu^{2+}(aq) \rightarrow Mg^{2+}(aq) + Cu(s)$$
$$E^\ominus_{cell} = (2.36\,V) + (0.34\,V) = +2.70\,V$$

Note that this is a larger voltage than that obtained from the Daniell cell (1.1 V). In general, the larger the difference in positions between metals in the electrochemical series, the greater the voltage produced by them in a voltaic cell.

This voltage is only obtained under standard conditions, that is, when $1\,mol\,dm^{-3}$ solutions are used at a temperature of 25 °C and 1 atmosphere pressure. Changing the conditions will alter the voltage and even the direction of the reaction.

Cell diagrams

Electrochemical or voltaic cells can be represented by cell diagrams. For example, the Daniell cell is represented as

$$Zn(s) \mid Zn^{2+}(aq) \parallel Cu^{2+}(aq) \mid Cu(s)$$

The double line in the centre of the cell diagram represents the salt bridge while the single lines represent so-called phase boundaries between the metal electrodes and their ions. It has been agreed that the electrode with the more positive (or less negative) standard electrode potential is placed on the right-hand side of the diagram.

For this cell the electrode potentials are:

$$Zn^{2+}(aq) + 2e^- \rightarrow Zn(s) \qquad E^{\ominus} = -0.76\,V$$
$$Cu^{2+}(aq) + 2e^- \rightarrow Cu(s) \qquad E^{\ominus} = +0.34\,V$$

If this convention is followed then the cell potential or voltage can be calculated using the following equation:

$$E^{\ominus}_{cell} = E^{\ominus} \text{ rhs electrode} - E^{\ominus} \text{ lhs electrode}$$

Hence for the Daniell cell:

$$E^{\ominus}_{cell} = (0.34\,V) - (-0.76\,V) = +1.10\,V$$

A positive value for a cell potential indicates that the cell reaction is thermodynamically spontaneous under standard thermodynamic conditions; in other words, 'it will go'.

The cell reaction that corresponds to this voltage is easily found by replacing the | symbols with arrows and then balancing with electrons:

$$Zn(s) \mid Zn^{2+}(aq) \parallel Cu^{2+}(aq) \mid Cu(s)$$
$$Zn(s) \rightarrow Zn^{2+}(aq) + 2e^- \qquad \text{and} \qquad 2e^- + Cu^{2+}(aq) \rightarrow Cu(s)$$

or

$$Zn(s) + Cu^{2+}(aq) \rightarrow Zn^{2+}(aq) + Cu(s)$$

(Cell potentials are related to values of Gibb's free energy changes, $\Delta G^{\ominus}$ (Chapter 15).)

Non-standard conditions

Electrode potential values can only be used to predict the feasibility of a redox reaction under *standard conditions*. Electrode potentials for oxidizing agents in acidic conditions refer to $1.0\,mol\,dm^{-3}$ concentrations of hydrogen ions, $H^+(aq)$ (pH = 0). Increasing the H^+ concentration increases the oxidizing strength of the oxidizing agent, thus increasing the electrode potential of the half-cell.

Consider the following reaction:

$$MnO_2(s) + 4H^+(aq) + 2Cl^-(aq) \rightarrow Mn^{2+}(aq) + 2H_2O(l) + Cl_2(g)$$
$$E^{\ominus}_{cell} = 1.23 + (-1.36) = -0.13\,V$$

Since the cell potential, $E^{\ominus}_{cell}$, is negative, the reaction is *not* spontaneous under standard conditions. However, when *concentrated* hydrochloric acid is heated with manganese(IV) oxide, the cell potential becomes positive and the reaction can occur and it can oxidize chloride ions. This is the standard laboratory preparation of chlorine. An additional factor is the shifting of the equilibrium to the right by the loss of chlorine gas.

In general, for a redox equilibrium:

$$Ox + ne^- \rightleftharpoons Red$$

increasing the concentration of the oxidized species, [Ox], or decreasing the concentration of the reduced species, [Red], will shift the position of the equilibrium to the right, reducing the number of electrons transferred and hence making the cell potential more positive. Similarly, the cell potential will become more negative if the concentration of the oxidized species, [Ox], is decreased, or the concentration of the reduced species, [Red], increased. These shifts can all be predicted from an application of Le Châtelier's principle (Chapter 7).

A Daniell cell consists of a zinc half-cell connected to a copper half-cell. Under standard conditions electrons spontaneously flow from the zinc electrode to the copper electrode. The cell potential is 1.10 V. However, if the concentration of zinc ions in the zinc half-cell is decreased, then the equilibrium:

$$Zn^{2+}(aq) + 2e^- \rightleftharpoons Zn(s) \qquad E^\ominus = -0.76\,V$$

is shifted to the left and the negative charge on the electrode is increased. This can be predicted from Le Châtelier's principle (Chapter 7): the removal of zinc ions will cause some of the zinc atoms to ionize and replace the zinc ions. This will increase the voltage of the Daniell cell to a value above 1.1 V.

If the concentration of zinc ions in the half-cell of the Daniell cell is increased, then the equilibrium is shifted to the right and the negative charge on the electrode decreased: the addition of zinc ions will cause some of the zinc ions to gain electrons. This will decrease the voltage of the Daniell cell to a value below 1.1 V.

■ Extension: The Nernst equation

The Nernst equation allows chemists to calculate the cell potentials of non-standard half-cells where the concentrations of ions are *not* 1 mol dm^{-3}. The mathematical relationship between the electrode potential and concentration of aqueous ions is known as the Nernst equation. It describes the relationship between cell potential and concentration (at constant temperature). It also describes the relationship between cell potential and temperature (at constant concentration).

For the general case of a metal/metal ion system:

$$M^{n+}(aq) + ne^- \rightleftharpoons M(s)$$

$$E_{cell} = E^\ominus + \frac{2.3RT}{nF} \log_{10} \frac{[M^{n+}(aq)]}{[M(s)]}$$

where R represents the gas constant (8.31 J mol^{-1} K^{-1}), F the Faraday constant (the product of the charge on an electron and the Avogadro constant), T the absolute temperature (in kelvin) and n the number of electrons transferred.

The value of the Faraday constant is 96485 C mol^{-1} (page 537), so at $T = 298\,K$, $2.3RT/F = 0.059$. Since the 'concentration' of a solid is constant (taken as unity, 1) the expression can be simplified to:

$$E_{cell} = E^\ominus + \frac{0.059}{n} \log_{10} [M^{n+}(aq)]\,V$$

which implies a logarithmic relationship between cell potential and concentration.

$$E_{cell} = E^\ominus + \frac{2.3RT}{nF} \log_{10} \frac{[\text{oxidized form}]}{[\text{reduced form}]}$$

is a generalized form of the Nernst equation that can be used to calculate the cell potentials of voltaic cells under non-standard conditions.

Worked example

Use the Nernst equation to calculate the cell potential at 298 K of a Daniell cell where the zinc ion concentration is 0.005 mol dm^{-3} and the copper(II) ion concentration is 1.5 mol dm^{-3}.

$$E_{cell} = 1.10 - \frac{0.059}{2} \log_{10} \frac{0.005}{1.5} = (1.10\,V) - (-0.07\,V) = +1.17\,V$$

■ Extension: Concentration cells

These are voltaic cells that have electrodes of the same element (typically a metal), but different concentrations of the electrolyte in the cathode and anode. The potential difference across the two electrodes is developed because of the difference in the concentrations of electrolytes.

The cell potential can be calculated by applying the Nernst equation:

$$E = \frac{0.059}{n} \log_{10} \frac{C_2}{C_1}$$

where C_2 and C_1 represent the concentrations of electrolyte in the half-cells containing the anode and cathode. For the concentration cell to exhibit a positive value C_2 must be greater than C_1.

History of Chemistry

Walter Hermann Nernst (1864–1941) was a German chemist who was awarded the 1920 Nobel Prize in Chemistry for his work in developing the Third Law of Thermodynamics (Chapter 15). He also made contributions to solid state chemistry and photochemistry, but is best known for his work in electrochemistry and the development of the Nernst equation.

The redox series

The electrochemical series has been extended to give the redox series (Table 19.2) which includes the standard electrode potentials of redox systems in which transition metals are present in different oxidation numbers.

$$Fe^{3+}(aq) + e^- \rightarrow Fe^{2+}(aq)$$

is a half-cell formed by dipping a platinum wire into an aqueous solution containing a mixture of $1\,mol\,dm^{-3}$ iron(II) ions and $1\,mol\,dm^{-3}$ iron(III) ions (Figure 19.7). Electrical contact is made with the mixture of two ions by means of the platinum wire, which acts as an *inert* conductor. As with all half-cells, a redox equilibrium is established:

$$Fe^{2+}(aq) \rightarrow Fe^{3+}(aq) + e^-$$
$$\text{and} \quad Fe^{3+}(aq) + e^- \rightarrow Fe^{2+}(aq)$$

$$\text{or} \quad Fe^{2+}(aq) \rightleftharpoons Fe^{3+}(aq) + e^-$$

Electrons produced by the forward reaction are transferred to the surface of the platinum, making it negatively charged, whereas the backward reaction removes electrons from the surface of the platinum wire. The resultant charge therefore depends on the relative balance between these two opposing processes.

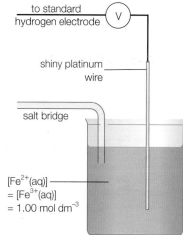

Figure 19.7 The half-cell system used to measure the standard electrode potential for the $Fe^{3+}(aq)/Fe^{2+}(aq)$ system

Electrode reaction	$E^\ominus$/volts
$H^+(aq) + e^- \rightarrow \frac{1}{2}H_2(g)$	0.00
$Cu^{2+}(aq) + 2e^- \rightarrow Cu(s)$	+0.34
$Cu^+(aq) + e^- \rightarrow Cu(s)$	+0.52
$Fe^{3+}(aq) + e^- \rightarrow Fe^{2+}(aq)$	+0.77
$Ag^+(aq) + e^- \rightarrow Ag(s)$	+0.80

Table 19.2 Part of the redox series for metals at 298 K (25 °C)

Worked example

Calculate the cell potential for a voltaic cell constructed from the following half-cells: $Fe^{3+}(aq)/Fe^{2+}(aq)$ and $Ag^+(aq)/Ag(s)$.

$$Fe^{3+}(aq) + e^- \rightarrow Fe^{2+}(aq) \qquad E^\ominus = +0.77\,V$$

$$Ag^+(aq) + e^- \rightarrow Ag(s) \qquad E^\ominus = +0.80\,V$$

The iron(III)/iron(II) half-cell is the least positive, so the half-equation and the sign of the electrode potential are reversed:

$$Fe^{2+}(aq) \rightarrow Fe^{3+}(aq) + e^- \qquad E^\ominus = -0.77\,V$$

$$Ag^+(aq) + e^- \rightarrow Ag(s) \qquad E^\ominus = +0.80\,V$$

The silver half-cell half-equation is multiplied through by two before it is added to the iron(III)/iron(II) half-cell. This is done to make the number of electrons equal, so they cancel to generate the ionic equation:

$$2Ag^+(aq) + Fe^{2+}(aq) \rightarrow Fe^{3+}(aq) + 2Ag(s)$$

The cell potential is then the sum of the electrode potentials (including their signs).

$$E^\ominus{}_{cell} = (-0.77)\,V + (0.80)\,V = +0.03\,V$$

Note that the standard electrode potential is *not* doubled when the stoichiometry is doubled.

The redox series can be extended to include the standard electrode potentials of non-metals and ions (Table 19.3).

Electrode reaction	$E^\ominus$/volts
$Li^+(aq) + e^- \rightarrow Li(s)$	−3.03
$K^+(aq) + e^- \rightarrow K(s)$	−2.92
$Ca^{2+}(aq) + 2e^- \rightarrow Ca(s)$	−2.87
$Mg^{2+}(aq) + 2e^- \rightarrow Mg(s)$	−2.36
$Al^{3+}(aq) + 3e^- \rightarrow Al(s)$	−1.66
$Mn^{2+}(aq) + 2e^- \rightarrow Mn(s)$	−1.18
$H_2O(l) + e^- \rightarrow \frac{1}{2}H_2(g) + OH^-(aq)$	−0.83
$Zn^{2+}(aq) + 2e^- \rightarrow Zn(s)$	−0.76
$Fe^{2+}(aq) + 2e^- \rightarrow Fe(s)$	−0.44
$Ni^{2+}(aq) + 2e^- \rightarrow Ni(s)$	−0.23
$Sn^{2+}(aq) + 2e^- \rightarrow Sn(s)$	−0.14
$Pb^{2+}(aq) + 2e^- \rightarrow Pb(s)$	−0.13
$H^+(aq) + e^- \rightarrow \frac{1}{2}H_2(g)$	0.00

Electrode reaction	$E^\ominus$/volts
$SO_4^{2-}(aq) + 4H^+(aq) + 2e^- \rightarrow H_2SO_3(aq) + H_2O(l)$	+0.17
$Cu^{2+}(aq) + 2e^- \rightarrow Cu(s)$	+0.34
$\frac{1}{2}O_2(g) + H_2O(l) + 2e^- \rightarrow 2OH^-(aq)$	+0.40
$Cu^+(aq) + e^- \rightarrow Cu(s)$	+0.52
$\frac{1}{2}I_2(s) + e^- \rightarrow I^-(aq)$	+0.54
$Fe^{3+}(aq) + e^- \rightarrow Fe^{2+}(aq)$	+0.77
$Ag^+(aq) + e^- \rightarrow Ag(s)$	+0.80
$\frac{1}{2}Br_2(aq) + e^- \rightarrow Br^-(aq)$	+1.09
$\frac{1}{2}O_2(g) + 2H^+(aq) + 2e^- \rightarrow H_2O(l)$	+1.23
$Cr_2O_7^{2-}(aq) + 14H^+(aq) + 6e^- \rightarrow 2Cr^{3+}(aq) + 7H_2O(l)$	+1.33
$\frac{1}{2}Cl_2(aq) + e^- \rightarrow Cl^-(aq)$	+1.36
$MnO_4^-(aq) + 8H^+(aq) + 5e^- \rightarrow Mn^{2+}(aq) + 4H_2O(l)$	+1.51
$\frac{1}{2}F_2(aq) + e^- \rightarrow F^-(aq)$	+2.87

Table 19.3 The redox series

Worked example

Use the standard electrode potential data below to write equations for the two reactions that occur if the half-cells are connected. Write a balanced equation for the overall reaction and hence predict the reaction, if any, when chlorine gas is bubbled into aqueous chromium(III) ions.

$6e^- + Cr_2O_7^{2-}(aq) + 14H^+(aq) \rightarrow 2Cr^{3+}(aq) + 7H_2O(l)$ $\qquad E^\ominus = +1.33\,V$

$\frac{1}{2}Cl_2(aq) + e^- \rightarrow Cl^-(aq)$ $\qquad E^\ominus = +1.36\,V$

The dichromate(VI)/chromium(III) half-cell is the least positive, so the half-equation and the sign of the electrode potential are reversed:

$2Cr^{3+}(aq) + 7H_2O(l) \rightarrow 6e^- + Cr_2O_7^{2-}(aq) + 14H^+(aq)$ $\qquad E^\ominus = -1.33\,V$

$\frac{1}{2}Cl_2(aq) + e^- \rightarrow Cl^-(aq)$ $\qquad E^\ominus = +1.36\,V$

The chlorine/chloride half-cell half-equation is multiplied through by six before it is added to the dichromate(VI)/chromium(III) half-cell.

The cell potential is then the sum of the electrode potentials (including their signs).

$E^\ominus_{cell} = (-1.33)\,V + (1.36)\,V = +0.03\,V$

The $E^\ominus_{cell}$ is positive, so the reaction can take place.

$2Cr^{3+}(aq) + 7H_2O(l) + 3Cl_2(g) \rightarrow Cr_2O_7^{2-}(aq) + 14H^+(aq) + 6Cl^-(aq)$

Applications of Chemistry

The rusting of iron and steel is an important redox reaction. The overall reaction involves the formation of hydrated iron(III) oxide from iron, water and oxygen. However, the first step of the reaction involves the formation of iron(II) hydroxide and can be derived from the following half-equations:

$Fe^{2+}(aq) + 2e^- \rightarrow Fe(s)$ $\qquad E^\ominus = -0.44\,V$

$\frac{1}{2}O_2(g) + H_2O(l) + 2e^- \rightarrow 2OH^-(aq)$ $\qquad E^\ominus = +0.40\,V$

We apply the rule that the more negative half-cell gives up electrons. Hence the iron half-cell is written as an oxidation process (the sign of the electrode potential is reversed) and added to the other half-cell.

$$Fe(s) \rightarrow Fe^{2+}(aq) + 2e^- \qquad E^\ominus = +0.44\,V$$
$$\tfrac{1}{2}O_2(g) + H_2O(l) + 2e^- \rightarrow 2OH^-(aq) \qquad E^\ominus = +0.40\,V$$
$$\tfrac{1}{2}O_2(g) + H_2O(l) + Fe(s) \rightarrow Fe^{2+}(aq) + 2OH^-(aq)$$
$$E^\ominus_{cell} = 0.44\,V + (0.40)\,V = +0.84\,V$$

In this process the iron metal is oxidized to iron(II) ions at the centre of a water drop, where the oxygen concentration is low (due to slow diffusion), and the electrons released reduce the oxygen molecules at the surface of the water, where oxygen concentration is high (Figure 19.8). The iron(II) and hydroxide ions formed diffuse away from the surface of the iron object. Further oxidation by dissolved oxygen in the air results in the formation of rust, hydrated iron(III) oxide.

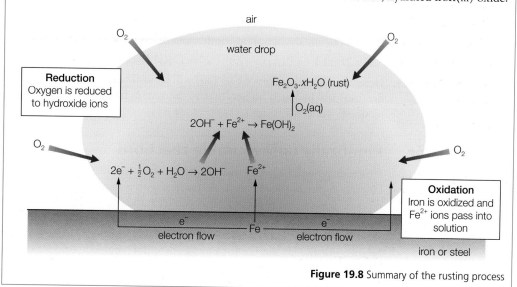

Figure 19.8 Summary of the rusting process

Cell spontaneity

19.1.4 **Predict** whether a reaction will be spontaneous using standard electrode potential values.

An electrochemical cell, such as the Daniell cell, operates by the oxidation reaction producing electrons in the zinc anode, which are then 'pulled round' the external circuit (wires, bulbs, voltmeter, etc.) by the reduction reaction at the copper cathode. As long as the overall reaction is *not* at equilibrium, the oxidation reaction 'pushes' electrons into the external circuit, and the reduction reaction 'pulls' them out. The cell is described as doing work since it produces a force that moves electrons around the external circuit. This work can light a bulb, drive an electric motor, etc.

The amount of work done by an electrochemical cell depends on the cell potential or voltage between its two electrodes: the greater the cell potential, the greater the amount of work the cell can do. A cell in which the overall reaction is at equilibrium can do no work and its cell potential or voltage, as well as its current, are zero.

The maximum amount of electrical work that can be done by an electrochemical cell is equal to the Gibb's energy change (Chapter 15), $\Delta G^\ominus$ (provided the temperature and pressure remain constant). The equation below gives the exact relationship between the Gibb's energy and cell potential:

$$\Delta G^\ominus = -nFE^\ominus_{cell}$$

where n represents the amount of electrons (in moles) transferred between the electrodes for the given equation, F represents the Faraday constant ($96\,485\,C\,mol^{-1}$), the amount of electrical charge carried by one mole of electrons, and $E^\ominus_{cell}$ represents the cell potential or voltage of the cell. For example, in the Daniell cell, the cell reaction is

$$Zn(s) + Cu^{2+}(aq) \rightarrow Zn^{2+}(aq) + Cu(s)$$

and n is two because two moles of electrons are transferred from the zinc atoms to the copper(II) ions in the above equation.

$$Zn(s) + 2e^- \rightarrow Zn^{2+}(aq); \qquad Cu^{2+}(aq) \rightarrow Cu(s) + 2e^-$$

$$\Delta G^{\ominus} = -nFE^{\ominus}$$

$$= -2 \times 96\,485\,C\,mol^{-1} \times 1.1\,V$$

$$= 212\,267\,J\,mol^{-1} = -212\,kJ\,mol^{-1}$$

This relatively large negative value for $\Delta G^{\ominus}$ means that the reaction is thermodynamically *spontaneous* and will take place under standard thermodynamic conditions (namely, 1 atmosphere pressure, 298 K (25 °C) and both solutions with a concentration of 1 mol dm^{-3}).

By contrast, if we apply Hess's law (Chapter 5), the reverse reaction has an equally large, but positive, value for ΔG and is not thermodynamically spontaneous under standard conditions. In other words, 'it will not go'.

$$Zn^{2+}(aq) + Cu(s) \rightarrow Zn(s) + Cu^{2+}(aq) \quad \Delta G^{\ominus} = +212\,kJ\,mol^{-1}$$

For any chemical reaction at equilibrium ΔG is zero, so

$$Zn(s) + Cu^{2+}(aq) \rightleftharpoons Zn^{2+}(aq) + Cu(s) \quad \Delta G = 0\,kJ\,mol^{-1}$$

Here, the concentration of $Cu^{2+}(aq)$ will be well below 1 mol dm^{-3} and the concentration of $Zn^{2+}(aq)$ will be much higher, that is, the system has shifted far enough to the right-hand side to reduce E_{cell} and ΔG to zero.

■ Extension: Kinetically unfavourable reactions

Electrode potentials, when used to predict the feasibility of a reaction, give *no* indication of the kinetics or rate of the reaction. A reaction with a positive $E^{\ominus}_{cell}$ value suggests that the reaction is possible from energy consideration under standard conditions only. However, the reaction may be so slow that it effectively does not occur. This may be due to the reaction having a high activation energy. Such a reaction, which has a positive $E^{\ominus}_{cell}$ but yet occurs very slowly, is said to be energetically favourable but kinetically unfavourable. Consider the following reaction:

$$H_2(g) + Cu^{2+}(aq) \rightarrow Cu(s) + 2H^+(aq) \qquad E^{\ominus}_{cell} = +0.34\,V$$

The positive value suggests that hydrogen gas should displace copper from copper(II) salts in solution under standard conditions. In practice, the rate of reaction is so slow that the reaction is kinetically non-feasible. This is because a relatively large amount of energy is needed to break the strong hydrogen–hydrogen covalent bond before the reaction can start.

19.2 Electrolysis

19.2.1 Predict and **explain** the products of electrolysis of aqueous solutions.

Electrolysis of aqueous solutions

The electrolysis of aqueous solutions of ionic compounds is more complicated than the electrolysis of molten ionic compounds (Chapter 9) since the water itself will undergo electrolysis. This occurs because water is slightly dissociated into hydrogen and hydroxide ions (Chapter 9):

$$H_2O(l) \rightleftharpoons H^+(aq) + OH^-(aq)$$

The hydrogen and hydroxide ions migrate with the ions from the ionic compound and *compete* with them to accept or release electrons at the cathode and anode, respectively. For example, an aqueous solution of sodium chloride contains the following ions:

$H^+(aq)$ and $OH^-(aq)$ from the water

$Na^+(aq)$ and $Cl^-(aq)$ from the sodium chloride

Both positive ions migrate to the negative cathode and both negative ions to the positive anode. At each electrode, depending upon the conditions, one or both of the ions may be discharged as atoms or molecules. Although the concentrations of hydrogen and hydroxide ions from the dissociation of water are very small, they will be rapidly restored via a shifting of the equilibrium if they are removed from the water via reactions with the electrodes.

If the solution of sodium chloride is concentrated, chlorine is produced at the anode and hydrogen is produced at the cathode:

Anode   *Cathode*

$2Cl^- \rightarrow Cl_2 + 2e^-$ $2H^+ + 2e^- \rightarrow H_2$

If the solution of sodium chloride is *dilute* then hydrogen is produced at the cathode and oxygen is produced at the anode:

Anode   *Cathode*

$4OH^- \rightarrow 2H_2O + O_2 + 4e^-$ $2H^+ + 2e^- \rightarrow H_2$

In both of these electrolyses inert graphite (carbon) or platinum electrodes are used and in neither case are sodium ions discharged as sodium metal.

These and other observed results (Table 19.5) suggest the following 'rules' regarding electrolysis (Figure 19.9) of aqueous solutions:

Figure 19.9 Electrolytic cell with graphite electrodes and ignition tubes in which any gases released at the electrodes are collected

Electrolyte	Electrodes	Cathode half-equation	Anode half-equation
Potassium bromide, KBr(aq)	Graphite	$2H^+ + 2e^- \rightarrow H_2$	$2Br^- \rightarrow Br_2 + 2e^-$
Magnesium sulfate, $MgSO_4$(aq)	Graphite	$4H^+ + 4e^- \rightarrow 2H_2$	$4OH^- \rightarrow 2H_2O + O_2 + 4e^-$
Concentrated hydrochloric acid, HCl(aq)	Graphite	$2H^+ + 2e^- \rightarrow H_2$	$2Cl^- \rightarrow Cl_2 + 2e^-$
Dilute sulfuric acid, H_2SO_4(aq)	Graphite	$2H^+ + 2e^- \rightarrow H_2$	$4OH^- \rightarrow 2H_2O + O_2 + 4e^-$
Dilute sodium hydroxide, NaOH(aq)	Graphite	$2H^+ + 2e^- \rightarrow H_2$	$4OH^- \rightarrow 2H_2O + O_2 + 4e^-$
Copper(II) sulfate, $CuSO_4$(aq)	Graphite	$Cu^{2+} + 2e^- \rightarrow Cu$	$4OH^- \rightarrow 2H_2O + O_2 + 4e^-$
Copper(II) sulfate, $CuSO_4$(aq)	Copper	$Cu^{2+} + 2e^- \rightarrow Cu$	$Cu \rightarrow Cu^{2+} + 2e^-$
Copper(II) chloride, $CuCl_2$(aq)	Carbon	$Cu^{2+} + 2e^- \rightarrow Cu$	$2Cl^- \rightarrow Cl_2 + 2e^-$
Potassium iodide, KI(aq)	Carbon	$2H^+ + 2e^- \rightarrow H_2$	$2I^- \rightarrow I_2 + 2e^-$

Table 19.5 Examples of electrolysis of solutions

- Metals, if produced, are discharged at the cathode.
- Hydrogen is produced at the cathode only.
- Non-metals, apart from hydrogen, are produced at the anode.
- Reactive metals, that is, those above hydrogen in the reactivity series (Chapter 9), are not discharged (unless special cathodes are used).
- The products can depend upon the *concentration* of the electrolyte in the solution and the nature of the electrode.
- If halide ions are present in reasonable concentrations they will be discharged more readily than hydroxide ions, *but* if no halide ions are present, hydroxide ions are discharged more readily than other anions.

Applications of electrolysis

Figure 19.10 BAE Hawks (Red Devils Acrobatic Team): the aircraft are composed mainly of aluminium alloy with some magnesium

The use of lightweight alloys containing lithium and magnesium makes the extraction of these metals increasingly important. Aluminium is a particularly useful metal, since in addition to its low density and high tensile strength, it does not suffer corrosion like iron (Figure 19.10).

The conductivity of copper increases by a factor of ten when it is more than 99.9% pure. The impure copper is made the anode of an electrolysis cell and pure copper is the cathode. Impure copper contains small amounts of gold and silver. They drop off the anode as the copper around them dissolves, and fall to the bottom as 'anode sludge'. Gold and silver can be extracted from the filtered sludge.

Important chemicals, such as sodium hydroxide and sodium chlorate(I) (bleach) are made by the electrolysis of brine (saturated salt solution). The electrolysis of brine (Chapter 23) results in the formation of chlorine, hydrogen and sodium hydroxide, useful raw materials for a variety of industrial processes (Figure 19.11).

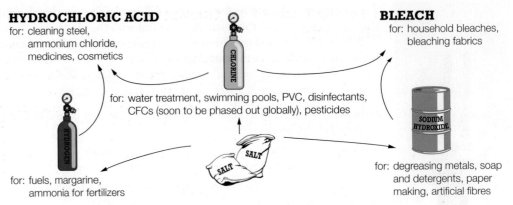

HYDROCHLORIC ACID
for: cleaning steel,
ammonium chloride,
medicines, cosmetics

for: water treatment, swimming pools, PVC, disinfectants,
CFCs (soon to be phased out globally), pesticides

BLEACH
for: household bleaches,
bleaching fabrics

for: fuels, margarine,
ammonia for fertilizers

for: degreasing metals, soap
and detergents, paper
making, artificial fibres

Figure 19.11 Important products from the electrolysis of salt solution

Electroplating (see page 541) has become less important recently, as stainless steel has replaced chromium-plated steel. However, many items are chromium, gold or silver plated. Tin cans are steel cans that have been tin plated.

The electrolysis of water

When very dilute sulfuric acid is electrolysed, one volume of oxygen gas is collected over the anode, and two volumes of hydrogen gas are collected over the cathode (Figure 19.12). At the anode, the hydroxide ions (from the dissociation of water) are discharged in preference to the sulfate ions. They give up electrons and form water and oxygen molecules. At the cathode, hydrogen ions are discharged by accepting electrons to form hydrogen molecules:

Anode: $4OH^-(aq) \rightarrow 2H_2O(l) + 1O_2(g) + 4e^-$

Cathode: $4e^- + 4H^+(aq) \rightarrow 2H_2(g)$

The second half-equation has been adjusted to show that ratio of amounts or volumes of oxygen molecules to hydrogen molecules is $1:2$, so in effect, water is being electrolysed. As the electrolysis proceeds, more water molecules dissociate to replace the ions that have been discharged. Thus, although the *quantity* of sulfuric acid is unchanged, its *concentration* increases as the water is consumed.

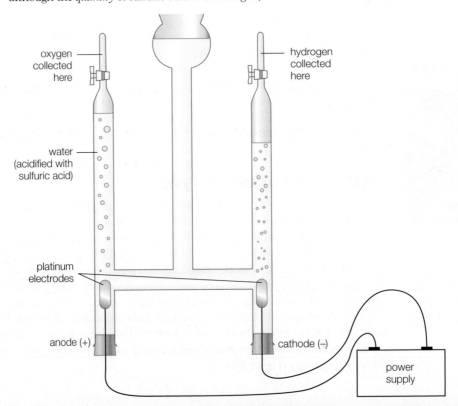

oxygen collected here

hydrogen collected here

water (acidified with sulfuric acid)

platinum electrodes

anode (+)

cathode (−)

power supply

Figure 19.12 Apparatus for the decomposition of water (Hoffman voltameter)

The electrolysis of copper(II) sulfate solution using copper electrodes

No gases are released during this electrolysis. However, if both the anode and cathode are weighed before and after passing the current, it is found that the mass of the anode decreases while that of the cathode increases, the two changes being equal.

At the anode hydroxide ions are present (in low concentration) from the dissociation of water. However, it requires less energy to remove electrons from the copper atoms of the anode than to remove them from the hydroxide ion. Hence, the anode slowly dissolves.

At the cathode, copper(II) ions are discharged in preference to hydrogen ions because hydrogen is below copper in the reactivity series, and so the cathode becomes plated with copper.

Anode: $\quad\quad\quad\quad Cu(s) \rightarrow Cu^{2+}(aq) + 2e^-$

Cathode: $\quad 2e^- + Cu^{2+}(aq) \rightarrow Cu(s)$

The concentration of the copper(II) sulfate solution remains unchanged, but copper is transferred from the anode to the cathode.

Alternative theory to explain the electrolysis of aqueous solutions

In the electrolysis of dilute sulfuric acid using inert electrodes, the formation of oxygen at the anode can be explained in terms of the discharge of hydroxide ions from the dissociation of water molecules. Similarly, the formation of hydrogen gas at the cathode in the electrolysis of sodium chloride solution was accounted for by the discharge of hydrogen ions from the dissociation of water. The degree of ionization in water is extremely small (Chapter 18) and pure water is virtually a non-conductor. The theory assumes that hydrogen and hydroxide ions are discharged from solution much more rapidly than from pure water.

An alternative but equivalent theory (common in North American textbooks) suggests that electrons can be taken or released at the electrodes by water *molecules*. Molecules of water are present in far greater concentration than any of the ions in solution. The observed results for the electrolysis of water can be readily accounted for by this theory:

Anode: $\quad\quad\quad\quad 2H_2O(l) \rightarrow 4H^+(aq) + O_2(g) + 4e^-$

Cathode: $\quad 4H_2O(l) + 4e^- \rightarrow 4OH^-(aq) + 2H_2(g)$

Using standard electrode potentials to explain hydrolysis products

A more rigorous approach to predicting and accounting for electrolysis products uses standard electrode potentials.

During electrolysis, cations are discharged at the cathode:

$$M^{n+}(aq) + ne^- \rightarrow M(s)$$

If hydrogen ions are discharged, then hydrogen gas is produced:

$$2H^+(aq) + 2e^- \rightarrow H_2(g)$$

Since discharge at the cathode involves reduction, ions that accept electrons readily will be reduced first. Therefore, strong oxidizing agents with more positive standard electrode potential values will be preferentially discharged compared to those with less positive values. For example, it is easier to discharge copper(II) ions than zinc ions at the cathode.

$$Zn^{2+}(aq) + 2e^- \rightarrow Zn(s) \quad\quad E^{\ominus} = -0.76\,V$$

$$Cu^{2+}(aq) + 2e^- \rightarrow Cu(s) \quad\quad E^{\ominus} = +0.34\,V$$

Anions are discharged at the anode during electrolysis:

$$2X^{n-}(aq) \rightarrow X_2(g) + 2ne^-$$

Since this is an oxidation reaction, ions that lose electrons readily will be oxidized first. Therefore, an anion with a more negative standard electrode potential will be discharged instead of one with a less negative standard electrode potential.

For example, it is easier to discharge bromide ions than chloride ions:

$$Br_2(aq) + 2e^- \rightarrow 2Br^-(aq) \qquad E^\ominus = +1.09\,V$$
$$Cl_2(aq) + 2e^- \rightarrow 2Cl^-(aq) \qquad E^\ominus = +1.36\,V$$

Another way of understanding this is to look at the *oxidation* potential of both bromide and choride ions. Oxidation potentials are the electrode (reduction) potentials with the *sign reversed*.

$$2Br^-(aq) \rightarrow Br_2(aq) + 2e^- \qquad E^\ominus_{oxidation} = -1.09\,V$$
$$2Cl^-(aq) \rightarrow Cl_2(aq) + 2e^- \qquad E^\ominus_{oxidation} = -1.36\,V$$

As the oxidation potential of bromide is more positive, it also indicates that the oxidation of bromide ions to bromine molecules is energetically more favourable. Therefore, if chloride and bromide ions migrate to an inert platinum electrode, the bromide ions will be preferentially discharged: $2Br^-(aq) \rightarrow Br_2(aq) + 2e^-$.

In summary, when inert electrodes are used during electrolysis:

- Cations with more positive $E^\ominus_{reduction}$ values will be discharged first at the cathode.
- Anions with more negative $E^\ominus_{reduction}$ values will be discharged first at the anode.

or

- Anions with more positive $E^\ominus_{oxidation}$ values will be discharged first at the anode.
- Cations with more negative $E^\ominus_{oxidation}$ values will be discharged first at the cathode.

Worked example

Deduce the relevant half-equation for the electrolysis of copper(II) sulfate solution using inert electrodes.

Step 1 Write the ions present in the electrolyte used:

$Cu^{2+}(aq)$, $SO_4^{2-}(aq)$, $H^+(aq)$ and $OH^-(aq)$ (from the dissociation of water:

$H_2O(l) \rightleftharpoons H^+(aq) + OH^-(aq))$

Step 2 Write the possible reduction reactions that could occur at the cathode. Refer to page 18 of the IB *Chemistry data booklet* and quote the necessary standard electrode potential values, $E^\ominus_{reduction}$ values.

$$Cu^{2+}(aq) + 2e^- \rightarrow Cu(s) \qquad E^\ominus_{reduction} = +0.34\,V$$

For the case of water molecules versus hydrogen ions, the IBO accepts the use of either:

$$2H^+(aq) + 2e^- \rightarrow H_2(g) \qquad E^\ominus_{reduction} = 0.00\,V$$

or

$$H_2O(l) + e^- \rightarrow \tfrac{1}{2}H_2(g) + OH^-(aq)\, E^\ominus_{reduction} = -0.83\,V$$

Step 3 Decide which reaction will take place by comparing the $E^\ominus_{reduction}$ values.

The copper(II) ion discharge half-equation has the more positive electrode potential and hence copper(II) ions will be preferentially discharged.

Step 4 Repeat the procedure for the possible reactions occurring at the anode.

$$S_2O_8^{2-}(aq) + 2e^- \rightarrow 2SO_4^{2-}(aq) \qquad E^\ominus_{reduction} = +2.01\,V$$

For the case of water molecules versus hydroxide ions, the IBO accepts the use of either:

$$\tfrac{1}{2}O_2(g) + H_2O(l) + 2e^- \rightarrow 2OH^-(aq) \qquad E^\ominus_{reduction} = +0.40\,V$$

or

$$\tfrac{1}{2}O_2(g) + 2H^+(aq) + 2e^- \rightarrow H_2O(l) \quad E^\ominus_{reduction} = +1.23\,V$$

Or if you prefer looking at values of $E^{\ominus}_{oxidation}$

$$2SO_4{}^{2-}(aq) \rightarrow S_2O_8{}^{2-}(aq) + 2e^- \quad E^{\ominus}_{oxidation} = -2.01\,V$$

$$4OH^-(aq) \rightarrow O_2(g) + 2H_2O(l) + 4e^- \, E^{\ominus}_{oxidation} = -0.40\,V$$

$$2H_2O(l) \rightarrow O_2(g) + 4H^+(aq) + 4e^- \quad E^{\ominus}_{oxidation} = -1.23\,V$$

Step 5 Decide which reaction will take place by comparing either the $E^{\ominus}_{reduction}$ values or $E^{\ominus}_{oxidation}$ values.

The most negative (or least positive) $E^{\ominus}_{reduction}$ value is the discharge of water molecules or hydroxide ions. They also have the least negative (or most positive) $E^{\ominus}_{oxidation}$ values. Hence, sulfate ions will remain in solution and *not* be discharged.

So the relevant half-equations are:

Cathode: $Cu^{2+}(aq) + 2e^- \rightarrow Cu(s)$

Anode: $2OH^-(aq) \rightarrow \frac{1}{2}O_2(g) + H_2O(l) + 2e^-$

or

$$H_2O(l) \rightarrow \frac{1}{2}O_2(g) + H^+(aq) + 2e^-$$

19.2.2 Determine the relative amounts of the products formed during electrolysis.

Faraday's laws

When a solution of aqueous copper(II) sulfate is electrolysed using copper electrodes (Figure 19.13), the copper anode slowly dissolves away and the copper cathode slowly gains a deposit of copper. Any impurities present in the copper anode collect at the bottom of the electrolytic cell. This method is used on the industrial scale to purify copper.

Cathode: $Cu^{2+} + 2e^- \rightarrow Cu$

Anode: $Cu \rightarrow Cu^{2+} + 2e^-$

Experiments have shown that the amount of copper deposited depends on both the length of time for which the current flows and the size of the current. Results have shown *directly proportional* relationships: if the time is doubled, the mass of copper deposited on the cathode is doubled, and if the size of the current is doubled, then again, the mass of copper deposited is doubled.

The amount of copper deposited therefore depends upon the size of the current and the length of time it is allowed to flow. An electric current is a flow of negatively charged electrons and is measured in units called *amperes* (amps (A), for short). The tiny electrical charge on each electron can be expressed in units called **coulombs** (C).

The total charge carried by an electric current is given by this expression:

charge in coulombs(C) = current(A) × time(s)

Experiments have shown that a mole of electrons carries an approximate charge of 96 500 coulombs. This is known as the **Faraday constant** in honour of the English physicist and chemist, Michael Faraday, and is given the symbol F with a value of $96\,485\,C\,mol^{-1}$. Faraday's investigations into the factors controlling the amounts of products formed during electrolysis are summarized in Faraday's laws of electrolysis.

Faraday's first law

Faraday's first law states that the mass of an element produced during electrolysis is directly proportional to the quantity of electricity (charge) passed during the electrolysis. The quantity of electricity (charge), as measured in coulombs, depends on both the current and time.

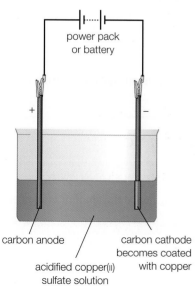

power pack or battery

+   −

carbon anode

carbon cathode becomes coated with copper

acidified copper(II) sulfate solution

Figure 19.13 Apparatus for the electrolysis of copper(II) sulfate solution

Faraday's second law

Faraday's second law states that the masses of different elements produced by the same quantity of electricity form simple whole number ratios when divided by their relative atomic masses.

Here is some experimental data that supports Faraday's second law. During an electrolysis experiment, 2.16 grams of silver are deposited and 0.64 grams of copper (for equal amounts of charge in coulombs). The relative atomic masses of silver and copper are 108 and 64, respectively.

	Silver	*Copper*
Amount	$\dfrac{2.16}{108} = 0.02\,\text{mol}$	$\dfrac{0.64}{64} = 0.01\,\text{mol}$
Divide through by smallest	$\dfrac{0.02}{0.01} = 2$	$\dfrac{0.01}{0.01} = 1$

The results can be accounted for in terms of the relevant half-equations and molar quantities of ions, atoms and electrons.

$$\begin{array}{ccccc}
Ag^+ & + & e^- & \rightarrow & Ag \\
0.02\,\text{mol} & & 0.02\,\text{mol} & & 0.02\,\text{mol}
\end{array}$$

$$\begin{array}{ccccc}
Cu^{2+} & + & 2e^- & \rightarrow & Cu \\
0.01\,\text{mol} & & 0.02\,\text{mol} & & 0.01\,\text{mol}
\end{array}$$

The quantity of electricity consumed in each cell is the same. The amount of copper formed is half that of the amount of silver formed because each mole of copper(II) ions needs two moles of electrons for discharge, whereas each mole of silver ions needs only one mole of electrons for discharge. A modern statement of Faraday's law is therefore that the number of moles of electons required to discharge one mole of an ion at an electrode equals the charge on the ion.

Faraday's laws and the relationships between charge, time, current and amount allow a variety of quantitative calculations involving electrolysis to be solved.

Worked examples

Deduce the charge on an aluminium ion if 5.4 grams of aluminium is deposited by a current of 5.00 A flowing for 3 hours and 13 minutes.

quantity of charge = 5.00 A × 11 580 s = 57 900 C

Hence, 5.4 grams of aluminium is discharged by 57 900 C. One mole of aluminium atoms has a mass of 27 grams.

So 27 grams of aluminium atoms is discharged by $\dfrac{27}{5.4} \times 57\,900 = 289\,500\,\text{C}$.

Experiments have shown that a mole of electrons carries a charge of approximately 96 500 C.

$\dfrac{289\,500\,\text{C}}{96\,500\,\text{C mol}^{-1}} = 3$ moles of electrons, and hence the charge on the aluminium ion is +3.

The equation for discharge is thus: $Al^{3+} + 3e^- \rightarrow Al$.

Calculate the mass of copper that would be plated on the cathode from an aqueous solution of copper(II) sulfate by a current of 2.00 A flowing for 15 minutes.

The relevant half-equation is: $Cu^{2+}(aq) + 2e^- \rightarrow Cu(s)$

quantity of charge = current(A) × time(s) = 2.00 A × (15 × 60 s) = 1800 C

quantity of electrons = $\dfrac{1800\,\text{C}}{96\,500\,\text{C mol}^{-1}} = 0.0187\,\text{mol}$

The molar ratio of the amounts of electrons to copper atoms is 2 : 1, hence the amount of copper atoms deposited is $0.5 \times 0.0187 = 9.33 \times 10^{-3}\,\text{mol}$.

mass of copper = molar mass of copper × amount of copper

mass of copper = $63.5\,\text{g mol}^{-1} \times 9.33 \times 10^{-3}\,\text{mol} = 0.592\,\text{g}$

Calculate the volume of hydrogen gas (in cm³) produced at stp when a current of 4.00 A is passed for 6 minutes and 10 seconds through a solution containing dilute aqueous sulfuric acid.

quantity of charge = $4.00\,A \times [(6 \times 60) + 10]s = 1480\,C$

$$\text{quantity of electrons} = \frac{1480\,C}{96\,500\,C\,mol^{-1}} = 0.015\,mol$$

The relevant half-equation is: $2H^+(aq) + 2e^- \rightarrow H_2(g)$.

2 moles of electrons produces 1 mole of hydrogen gas.

Hence 0.015 moles of electrons will produce 0.0075 moles of hydrogen gas.

The molar gas volume at stp is 22.4 dm³.

Hence, the volume of hydrogen released = $22\,400 \times 0.0075 = 168\,cm^3$.

A current of 3.00 A was passed for 30 minutes through molten lead(II) bromide. Lead of mass 5.60 grams was obtained. Determine the value of the Avogadro constant.

quantity of charge(Q) = current(A) × time(s) = $3.00\,A \times (30 \times 60)\,s = 5400\,C$

To deposit 207 g (1 mole) of lead, we need:

$$\frac{207}{5.6} \times 5400\,C = 199\,607\,C$$

$$Pb^{2+} + 2e^- \rightarrow Pb$$

From the above equation, we see that 2 moles of electrons deposit one mole of lead. One electron carries a charge of e coulombs ($1.6 \times 10^{-19}\,C$), hence two moles of electrons ($2L$) will carry a charge of $2L \times e = 2Le$ coulombs. Therefore:

$$2Le = 199\,607\,C$$

Substituting $1.6 \times 10^{-19}\,C$ for e and solving for L, we get:

$$L = \frac{199\,607\,C}{2 \times 1.6 \times 10^{-19}\,C} = 6.2 \times 10^{23}$$

Language of Chemistry

Michael Faraday, the discoverer of ions, named them after the Greek word *ion*, derived from *ienai*, to go. The terms anode and cathode were derived from the Greek words *anodos* (way up) and *cathodos* (way down). The term electrode was derived from the Greek words *elektron* (meaning amber) and *hodos*, a way. All these terms were developed by Faraday with the Master of Trinity College, Cambridge, William Whewell. ■

History of Chemistry

Michael Faraday (1791–1867) was an English chemist and physicist who made major contributions to electromagnetism and electrochemistry. The SI unit of capacitance, the farad, is named after him, as is the Faraday constant, the charge on a mole of electrons. Faraday was born the son of a blacksmith and was largely self-educated due to a seven-year apprenticeship to a bookbinder and bookseller. In 1812 Faraday attended lectures by Humphry Davy who later appointed him as his assistant. Davy had isolated sodium and potassium by electrolysing their molten hydroxides. Faraday was elected a member of the Royal Society in 1824, appointed director of the laboratory of the Royal Institution in 1825, and in 1833 he was appointed Fullerian Professor of Chemistry.

Figure 19.14 Michael Faraday

In 1847 he discovered that the optical properties of gold colloids differed from those of the corresponding bulk metal. This was probably the first reported observation of the effects of quantum size, and might be considered to be the birth of nanoscience (Chapter 23).

■ Extension: Reversing the flow

If the electrodes of a voltaic cell are connected together, electrons flow from the negative electrode to the positive terminal. If the two electrodes are instead connected to an external voltage supply (power pack) and the applied voltage is increased, electrons still flow as before until the external voltage equals the cell potential, $E^{\ominus}_{cell}$. At this voltage, no current flows, but if the external voltage is increased still further, current flows in the *opposite* direction and electrolysis takes place (Figure 19.15). For example, when an external voltage greater than 1.1 volts is applied to a Daniell cell the reaction runs in *reverse*:

$$Zn^{2+}(aq) + Cu(s) \rightarrow Zn(s) + Cu^{2+}(aq).$$

Figure 19.15 Voltage–current graph for a Daniell cell (copper/zinc voltaic cell)

1.1 volts is the minimum voltage required to bring about electrolysis in the Daniell cell. However, in practice the voltage used for electrolysis is always greater than this minimum. The voltaic cell has resistance and the cell discharge reactions require energy to overcome the activation energy associated with discharge of ions.

Applications of Chemistry

The corrosion resistance of aluminium results from the reaction of the metal with air, which forms a very thin, tough, impermeable oxide layer that protects the metal from further attack by oxygen. If the oxide layer is chemically removed, for example by rubbing the surface of the aluminium foil with mercury(II) chloride solution, the exposed metal reacts very exothermically with air to form aluminium hydroxide.

The thickness of this protective oxide layer can be increased by an electrolytic process known as anodizing (Figure 19.16). The aluminium is used as the anode in an electrolytic cell containing dilute sulfuric acid. Oxygen formed at the metal surface reacts to build up the oxide layer:

$$4OH^-(aq) \rightarrow 2H_2O(l) + O_2(g) + 4e^-$$

Anodized aluminium can be coloured, as the porous oxide layer absorbs a range of dyes.

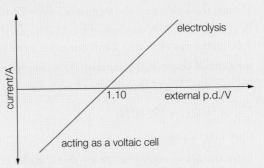

Figure 19.16 A Stirling cycle engine (external combustion engine): the anodized parts are 'gold'

Figure 19.17 Gold-plated spoons

Electroplating

Metals are electroplated to improve their appearance or to prevent corrosion. The most commonly used metals for electroplating are copper, chromium, silver and tin. Familiar examples of electroplated objects include chromium-plated car bumpers and kettles, jewellery, for example gold bracelets, and cutlery (Figure 19.17) including EPNS (electroplated nickel silver) cutlery.

Figure 19.18 shows an electrolytic cell used to perform silver plating. At the cathode the silver ions undergo reduction to form silver atoms: $Ag^+(aq) + e^- \rightarrow Ag(s)$. At the anode the silver atoms undergo oxidation: $Ag(s) \rightarrow Ag^+(aq) + e^-$. As the current flows through the circuit, the anode slowly dissolves and replaces the silver ions in the electrolyte.

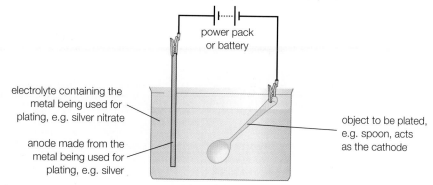

Figure 19.18 Electroplating apparatus: silver plating

The object to be electroplated must be made the cathode. The cathode is the negative electrode and attracts metal ions (cations). The anode must be the metal used for the plating process. The electrolyte solution must contain ions of the metal for plating.

In order to obtain a good coating of metal during electroplating:

- the object to be plated must be clean and free of grease
- the object should be rotated to give an even coating
- the current must not be too large or the 'coating' will form too rapidly and flake off
- the temperature and concentration of the electrolyte must be carefully controlled, otherwise the 'coating' will be deposited too rapidly or too slowly.

Applications of Chemistry

In order to produce a thin, even layer of metal during electroplating, the metal needs to be deposited slowly. This means that the concentration of free ions must be kept low. In chromium plating (Figure 19.19), the electrolyte is a mixture of chromium(III) ions, $Cr^{3+}(aq)$ and chromate(VI) ions, $CrO_4^{2-}(aq)$. As the chromium(III) ions are deposited, they are replenished by the reduction of chromate(VI) ions at the cathode.

Cathode: $\qquad Cr^{3+}(aq) + 3e^- \rightarrow Cr(s)$

$$CrO_4^{2-}(aq) + 8H^+(aq) + 3e^- \rightarrow Cr^{3+}(aq) + 4H_2O(l)$$

In silver plating, a solution containing the complex ion $Ag(CN)_2^-(aq)$ is often used. This is a stable complex and produces a very low concentration of silver ions, $Ag^+(aq)$.

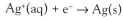

$$Ag(CN)_2^-(aq) \rightleftharpoons Ag^+(aq) + 2CN^-(aq)$$

Cathode: $\qquad Ag^+(aq) + e^- \rightarrow Ag(s)$

Figure 19.19 Chrome-plated bumper from a Renault car

Applications of Chemistry

A common method of rust prevention is to electroplate iron or steel with a metal such as tin or zinc. However, if this metal coating is scratched so that the iron is exposed, and then water enters the scratch, a voltaic couple is set up.

In the case of zinc plating (galvanizing), the zinc, being higher in the electrochemical series (and having a more negative standard electrode potential), acts as the anode and dissolves (Figure 19.20). The zinc ions produced combine with the hydroxide ions from water to form a precipitate of zinc hydroxide, which fills up the scratch and helps to prevent further reaction.

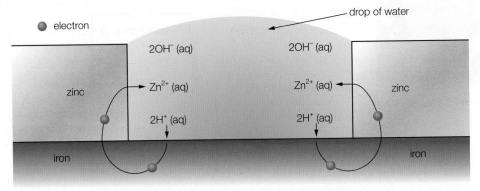

Figure 19.20 How zinc plating prevents further corrosion of iron

Tin is below iron in the electrochemical series (and has a less negative standard electrode potential) and in similar circumstances it will be the iron that acts as the anode and dissolves (Figure 19.21). The presence of tin will *increase* the rate of corrosion (rusting) of the iron if the coating is damaged. Tin is used to protect cans made of iron because zinc would poison the food.

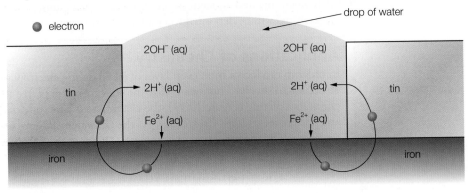

Figure 19.21 If tin plating is damaged it increases the rate at which the iron rusts

Language of Chemistry

Originally, the term galvanization referred to the production of electric shocks. This originated with Luigi Galvani (1737–1798), an Italian physicist who observed that a dead frog's legs twitched when static electricity was discharged through them. The term galvanizing now specifically refers to zinc plating. ∎

SUMMARY OF
KNOWLEDGE

- A metal dipping into a solution of its ions forms a half-cell. This is an example of a simple redox equilibrium.
- Two half-cells connected via a salt bridge and an external circuit form an electrochemical or voltaic cell.
- Under standard conditions, the voltage generated by this cell is known as the standard e.m.f. (electromotive force). Standard conditions are all concentrations at $1 \, mol \, dm^{-3}$, all gases at 1 atm pressure and temperature at 298 K.
- If the cell is connected to the standard hydrogen electrode, the standard e.m.f. is the standard electrode potential, $E^{\ominus}$, of the cell.
- The value of $E^{\ominus}$ measures the oxidizing/reducing power of the half-cell system: the more negative the value of $E^{\ominus}$, the more reducing the half-cell; the more positive the value of $E^{\ominus}$, the more oxidizing the half-cell.
- The electrochemical series is a list of half-cells in order of their standard electrode potentials, with the most negative ions at the top. By convention, half-cells are written with the oxidized form on the left. Hence, they are written as reduction potentials.
- The cell's potential is related to the free energy of the cell reaction, by the following relationship:
  $$\Delta G = -nFE^{\ominus}_{cell}$$
  where n represents the number of electrons being transferred per mole of cell reaction and F is the Faraday constant.
- A redox reaction that is predicted to be spontaneous from $E^{\ominus}_{cell}$ values might be extremely slow because the conditions are non-standard, or because the activation energy barrier is very high.
- Electrolysis is the process of driving a decomposition reaction that is not thermodynamically spontaneous, by passing a continuous electric current through it.
- The ions that are discharged in electrolysis are the ones that require the least amount of energy.
- The more positive/less negative the half-cell potential, the more easily the cation is discharged by reduction at the cathode. The less positive the half-cell potential, the more easily the anion is discharged by oxidation at the anode.
- Concentration (a kinetic effect) also plays a role in determining the products of electrolysis.
- The amount of substance dissolved or deposited during electrolysis is given by:
  $$amount = \frac{I \times t}{nF}$$
  where I represents the current in amps, t the time in seconds, n the charge on the ion and F the Faraday constant (the charge in coulombs carried by one mole of electrons).
- Electrolysis is used to perform electroplating. The object to be electroplated is made the cathode and immersed in an aqueous solution of metal ions.

■ Examination questions – a selection

Paper 1 IB questions and IB style questions

Q1 Which one of the following factors does not influence the voltage in the reaction shown below?

$$Fe(s) + Cu^{2+}(aq) \rightarrow Cu(s) + Fe^{2+}(aq)$$

  A the concentration of copper(II) ions, $Cu^{2+}(aq)$
  B the concentration of iron(II) ions, $Fe^{2+}(aq)$
  C the temperature of the solutions
  D the size of the anode

Q2 A student constructs a voltaic cell using tin and lead electrodes immersed in $1\,mol\,dm^{-3}$ solutions of tin(II) and lead(II) ions. What is the e.m.f. for the spontaneous reaction? The electrode potentials are:

$$Sn^{2+}(aq) + 2e^- \rightarrow Sn(s) \qquad E^{\ominus} = -0.14\,V$$
$$Pb^{2+}(aq) + 2e^- \rightarrow Pb(s) \qquad E^{\ominus} = -0.13\,V$$

  A $-0.27\,V$ **C** $+0.01\,V$
  B $-0.01\,V$ **D** $+0.27\,V$

Q3 For which of the reactions below will $\Delta G^{\ominus}$ be the most negative?
  A $Cu(s) + 2Ag^+(aq) \rightarrow 2Ag(s) + Cu^{2+}(aq)$
                                      $E^{\ominus} = 0.46\,V$
  B $Co(s) + Cu^{2+}(aq) \rightarrow Cu(s) + Co^{2+}(aq)$ $E^{\ominus} = 0.62\,V$
  C $Fe^{2+}(aq) + Cu^{2+}(aq) \rightarrow Fe^{3+}(aq) + Cu^+(aq)$
                                        $E^{\ominus} = -0.61\,V$
  D $H_2(g) + Cr^{2+}(aq) \rightarrow Cr(s) + 2H^+(aq)$ $E^{\ominus} = -0.74\,V$

Q4 According to the standard electrode potentials below, which combination of reagents will produce the greatest voltage (potential difference)?

	$E^{\ominus}$/volts
$Ti^+(aq) + e^- \rightarrow Ti(s)$	-0.336
$Ni^{2+}(aq) + 2e^- \rightarrow Ni(s)$	-0.236
$Mo^{3+}(aq) + 3e^- \rightarrow Mo(s)$	-0.200
$Cu^+(aq) + e^- \rightarrow Cu(s)$	0.518

(Assume that the copper(I) ions are stabilized against disproportionation in aqueous solution.)
  A $Cu^+(aq) + Ti(s)$ **C** $2Cu^+(aq) + Ni(s)$
  B $Mo^{3+}(aq) + 3Ti(s)$ **D** $3Ni(s) + 2Mo^{3+}(aq)$

Q5 The equilibrium constant for the reaction below is 4×10^{16}.

$$Zn(s) + 2Ag^+(aq) \rightleftharpoons Zn^{2+}(aq) + 2Ag(s)$$

What are the signs of the standard free energy change, $\Delta G^{\ominus}$, and the standard electrode potential, $E^{\ominus}$, for the reaction?
  A $\Delta G^{\ominus}$ is $+$; $E^{\ominus}$ is $-$ **C** $\Delta G^{\ominus}$ is $-$; $E^{\ominus}$ is $+$
  B $\Delta G^{\ominus}$ is $+$; $E^{\ominus}$ is $+$ **D** $\Delta G^{\ominus}$ is $-$; $E^{\ominus}$ is $-$

Q6 One mole of electrons was passed through the electrolytic cells in series containing solutions of $Ag^+(aq)$, $Ni^{2+}(aq)$ and $Cr^{3+}(aq)$. What mass of Ag, Ni and Cr will be deposited? [A_r values: Ag = 108, Ni = 59, Cr = 52]
  A $36\,g$, $29.5\,g$ and $52\,g$
  B $108\,g$, $59\,g$ and $52\,g$
  C $108\,g$, $29.5\,g$ and $17.3\,g$
  D $108\,g$, $118\,g$ and $156\,g$
             Higher Level Paper 1, Nov 02, Q34

Q7 The standard potentials for the following half-reactions are:

$$Cl_2(g) + 2e^- \rightarrow 2Cl^-(aq) \qquad E^{\ominus} = 1.36\,V$$
$$Ni^{2+}(aq) + 2e^- \rightarrow Ni(s) \qquad E^{\ominus} = -0.23\,V$$
$$Mg^{2+}(aq) + 2e^- \rightarrow Mg(s) \qquad E^{\ominus} = -2.36\,V$$

What will happen when nickel dust is added to an aqueous solution of magnesium chloride?
  A No reaction will take place.
  B Chlorine gas will be produced.
  C Magnesium metal will form.
  D Nickel chloride will form.

Q8 The standard reduction potentials of tin(II) and copper(I) ions are given below:

$$Sn^{2+}(aq) + 2e^- \rightarrow Sn(s) \qquad -0.141\,V$$
$$Cu^+(aq) + e^- \rightarrow Cu(s) \qquad +0.518\,V$$

What is the e.m.f. for the following reaction?
$$Sn(s) + 2Cu^+(aq) \rightarrow Cu(s) + Sn^{2+}(aq)$$

(Assume that the copper(I) ions are stabilized against disproportionation in aqueous solution.)
  A $0.754\,V$ **C** $0.895\,V$
  B $0.659\,V$ **D** $2.354\,V$

Q9 Given the standard reduction potentials

$$Ni^{2+}(aq) + 2e^- \rightarrow Ni(s) \qquad E^{\ominus} = -0.23\,V$$
$$Cd^{2+}(aq) + 2e^- \rightarrow Cd(s) \qquad E^{\ominus} = -0.40\,V$$
$$2H^+(aq) + 2e^- \rightarrow H_2(g) \qquad E^{\ominus} = 0.00\,V$$

Which pair of substances will react spontaneously?
  A nickel ions with cadmium ions
  B nickel atoms with cadmium ions
  C nickel ions with hydrogen ions
  D cadmium atoms with nickel ions

Q10 From the given standard electrode potentials which statement is correct?

$$Ca^{2+}(aq) + 2e^- \rightleftharpoons Ca(s) \qquad E^{\ominus} = -2.87\,V$$
$$Ni^{2+}(aq) + 2e^- \rightleftharpoons Ni(s) \qquad E^{\ominus} = -0.23\,V$$
$$Fe^{3+}(aq) + e^- \rightleftharpoons Fe^{2+}(aq) \qquad E^{\ominus} = +0.77\,V$$

  A $Ca^{2+}(aq)$ can oxidize Ni(s)
  B $Ni^{2+}(aq)$ can reduce $Ca^{2+}(aq)$

C Fe³⁺(aq) can oxidize Ni(s)
D Fe³⁺(aq) can reduce Ca²⁺(aq)
Higher Level Paper 1, Nov 05, Q32

Q11 Consider an electrochemical cell constructed from standard Ni/Ni²⁺ and Cu/Cu²⁺ half-cells. Which one of the following changes would be expected to increase the measured value of $E^\ominus_{cell}$?

$Cu^{2+}(aq) + 2e^- \rightarrow Cu(s);$ $\quad E^\ominus = +0.34$ volts
$Ni^{2+}(aq) + 2e^- \rightarrow Ni(s);$ $\quad E^\ominus\ -0.23$ volts

A increase the [Ni²⁺(aq)]; increase the [Cu²⁺(aq)]
B decrease the [Ni²⁺(aq)]; increase the [Cu²⁺(aq)]
C decrease the [Ni²⁺(aq)]; decrease the [Cu²⁺(aq)]
D increase the [Ni²⁺(aq)]; decrease the [Cu²⁺(aq)]

Q12 A strip of pure iron is a placed in a 1.00 mol dm⁻³ solution of iron(II) chloride, and a strip of pure copper is placed in a 1.00 mol dm⁻³ solution of copper(II) chloride. The two metal strips are connected by a wire and the two solutions are connected by a salt bridge. Which **one** of the following observations will be made for the system?
A The concentration of iron(II) ions in the iron half-cell is decreased.
B Copper atoms are deposited at the cathode.
C Chlorine gas is produced at the iron electrode.
D Chlorine gas is produced at the copper electrode.

Q13 Which statement is correct for the electrolysis of aqueous copper(II) sulfate using copper electrodes?
A Copper(II) ions move toward the cathode (negative electrode).
B Oxygen molecules are produced at the cathode (negative electrode).
C The anode gains in mass due to deposition of copper atoms.
D The blue colour of the copper(II) sulfate solution becomes lighter.

Q14 Which of the following statements is not correct about the charge on a mole of electrons?
A It always deposits one mole of an element during electrolysis.
B It equals the charge on 6 × 10²³ electrons.
C It is approximately equivalent to 96 500 C of charge.
D It deposits half a mole of copper atoms during electrolysis.

Q15 Which of the following is an **incorrect** statement about the electrolysis of concentrated sodium chloride solution with graphite electrodes?
A The anode (positive electrode) attracts hydroxide and chloride ions.
B Sodium metal is produced at the cathode.

C Hydrogen ions undergo reduction at the cathode.
D More energy is required to discharge chloride ions compared to hydroxide ions.

Q16 A voltaic cell is constructed with pure zinc and copper electrodes in aqueous solutions of Zn(NO₃)₂ and Cu(NO₃)₂, respectively. The voltage of this cell is 0.91 volts, but the value calculated from a table of standard electrode potentials is 1.10 volts. The difference in voltage is probably due to the:
A concentrations of the solutions used
B quantity of solution used
C size of the electrodes used
D diameter of the salt bridge used

Q17 Two standard zinc and silver half-cells are connected via a salt bridge. The two electrodes are connected by a wire and a high-resistance voltmeter. Select the incorrect statement.
A Electrons are flowing along the wire from the zinc electrode to the silver electrode.
B Electrons are flowing through the salt bridge to complete the circuit.
C If the salt bridge is lifted out of the solutions the voltmeter will read zero volts.
D If hydrochloric acid is added to the Ag(s)/Ag⁺(aq) half-cell, the reading on the voltmeter will change.

Q18 According to the following half-reactions and $E^\ominus$ values, which combination of reactions below will result in spontaneous reaction?

	$E^\ominus$/volts
$Mo^{3+}(aq) + 3e^- \rightarrow Mo(aq)$	−0.20
$Pb^{2+}(aq) + 2e^- \rightarrow Pb(s)$	−0.13
$Cu^{2+}(aq) + e^- \rightarrow Cu^+(aq)$	−0.13

(Assume that the copper(I) ions are stabilized against disproportionation in aqueous solution.)
A Cu⁺(aq) + Mo³⁺(aq)
B Pb²⁺(aq) + Mo(s)
C Pb²⁺(aq) + Cu⁺(aq)
D Cu²⁺(aq) + Pb²⁺(aq)

Q19 Which of the following will affect the mass of gold deposited on the cathode (negative electrode) during an electroplating process?
I the concentration of gold ions in the electrolyte
II the size of the current used
III the time of the electroplating process
A I and II
B II and III
C I and III
D I, II and III

Q20 An electrolysis cell connected in series with one which deposits 12 grams of magnesium would, during the same time interval, deposit:
- **A** 23 grams of sodium
- **B** 27 grams of aluminium
- **C** 8 grams of hydrogen gas
- **D** 32 grams of oxygen gas

Paper 2 IB questions and IB style questions

Q1 a Redox equations may be balanced using changes in oxidation number. For the following redox equation calculate the oxidation numbers of copper and nitrogen. Use these values to balance the equation.

$$Cu(s) + HNO_3(aq) \rightarrow Cu(NO_3)_2(aq) + NO(g) + H_2O(l)$$
[5]

b i Draw a diagram for the voltaic cell formed by connecting the following standard half-cells:

$$Ni(s) \mid Ni^{2+}(aq) \parallel Mn^{2+}(aq) \mid Mn(s)$$
[3]

ii Describe the key features of the standard hydrogen electrode. [3]

c i Write an equation for the reaction in each half-cell, identifying the species which is oxidized and the oxidizing agent. [4]

ii State which electrode is the anode and state the direction of electron flow in the external circuit. [2]

iii For the overall cell, calculate its voltage and state the sign of ΔG. [2]

d An aqueous solution of gold nitrate is electrolysed. Predict the product formed at each electrode. [2]

Q2 a When a concentrated aqueous solution of potassium chloride is electrolysed using inert platinum electrodes, two different colourless gases are produced at the electrodes.

i Write half-equations for the oxidation and reduction half-reactions. [2]

ii Explain why potassium is not formed during the electrolysis of aqueous potassium chloride solution. [1]

iii Under what conditions can potassium be formed by electrolysis of potassium chloride? [1]

b Deduce the products formed during the electrolysis of an aqueous solution of potassium fluoride. Write an equation for the reaction at the positive electrode (the anode) and give your reasoning. [4]

Q3 A pale blue aqueous solution of copper(II) sulfate, $CuSO_4$, is electrolysed using copper electrodes.

a Write balanced half-equations for the reactions occurring at the:
i cathode (negative electrode) [1]
ii anode (positive electrode) [1]

b Explain why there is no change in the intensity of the pale blue colour or pH when a current flows. [2]

c Write a balanced equation for the products formed if the reactive copper anode is replaced by an inert platinum or graphite electrode. [2]

d Calculate the mass of copper produced in grams when a current of 0.360 amperes is passed through a $1.0\,mol\,dm^{-3}$ copper(II) sulfate solution for 10 minutes.

Q4 a Define an oxidizing agent in terms of electrons. [1]

b Make a copy of the following representation of a voltaic cell. Label each electrode with a + or − sign, as appropriate, and draw an arrow on the connecting wire to indicate the direction of electron flow (Refer to Table 14 of the IB *Chemistry data booklet*.) [2]

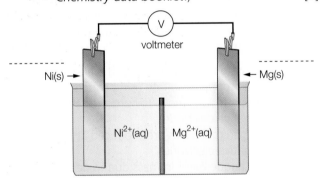

c i Write the balanced equation for the spontaneous reaction in the above cell. [2]

ii Calculate the standard electrode potential. [1]

d Chromium is deposited from an acidic solution containing the dichromate(VI) ion, according to the equation:

$$Cr_2O_7^{2-}(aq) + 14H^+(aq) + 12e^- \rightarrow 2Cr(s) + 7H_2O(l)$$

Calculate the mass of chromium metal (in grams) that can be deposited by a current of 1.00 A flowing for 8.00 hours. [5]

20 Organic chemistry

STARTING POINTS

■ Amines, a homologous series containing the $-NH_2$ group, can be regarded as substituted forms of ammonia. They can be primary, secondary or tertiary amines.

■ Amides are a homologous series derived from carboxylic acids by the substitution of the $-OH$ in the carboxylic acid group by an $-NH_2$ group. The functional group of an amide is a $-CONH_2$ group.

■ The structures and names of esters reflect their origins in the esterification reactions between an alcohol and a carboxylic acid.

■ Nitriles are a homologous series containing the $-CN$ group substituted into the hydrocarbon chain.

■ Nitriles are most readily synthesized by the nucleophilic attack of a cyanide ion, CN^-, on an appropriate halogenoalkane.

■ There is evidence for two very different reaction mechanisms for nucleophilic substitution: S_N1 and S_N2 mechanisms. These mechanisms have their own characteristic kinetics.

■ A nucleophile is a molecular species or negative ion that contains a lone pair of electrons on a highly electronegative atom.

■ Iodoalkanes are the most susceptible compounds to substitution, because of the weakness of the carbon–iodine bond.

■ Tertiary halogenoalkanes are found to undergo an S_N1 mechanism of substitution, while primary halogenoalkanes undergo an S_N2 mechanism.

■ Under certain conditions halogenoalkanes will undergo an elimination reaction with hydroxide ions (OH^-), rather than substitution, to produce an alkene and the appropriate hydrogen halide. In these elimination reactions the hydroxide ion is acting as a base rather than a nucleophile.

■ Condensation reactions are an important type of organic reaction in which two molecules join together with the elimination of a small molecule, usually water, from the structure.

■ Condensation reactions are involved in the synthesis of esters and amides, and of biologically important molecules such as proteins, nucleic acids and polysaccharides.

■ In protein synthesis amino acids are polymerized by the loss of a water molecule at the addition of each new amino acid. The bond formed between adjacent amino acids is the amide or peptide link.

■ Several artificial polymers are now made using condensation reactions. Nylon is an example of a polyamide, while polyesters are made by reacting monomers containing alcohol and carboxylic acid groups.

■ The knowledge of these different functional groups and types of reaction means that further synthetic pathways are available for producing useful organic compounds.

■ Compounds that have the same molecular formula but different structural formulas are stereoisomers.

■ There are two types of stereoisomerism: geometrical isomerism and optical isomerism.

■ Geometrical isomerism arises in molecules where there is restricted rotation built into the structure.

■ Optical isomers (enantiomers) are non-superimposable mirror images of each other and arise because of the presence of a chiral centre in the molecule.

20.1 Introduction

Within the context of our earlier discussion of organic chemistry (Chapter 10) we introduced some of the chemistry that has revolutionized our lives in terms of the materials available for us to use. Addition polymerization has enabled us to develop a whole range of useful practical products that have greater durability than previous natural materials used for the same purpose. There are relatively few natural addition polymers – the exception being rubber. Nature predominantly

Figure 20.1 A schematic representation of the enzyme pepsin showing how the protein chain folds and the regions where there is a helical structure (blue cylinders – alpha helix) and a sheet structure (purple arrows – beta sheet)

uses a different type of chemical reaction to produce its key long-chain molecules. Some of these molecules – DNA, proteins and polysaccharides – are vital to life. Here we introduce this type of polymerization, known as condensation polymerization, and look at some important molecules that are assembled by this type of reaction. Figure 20.1 shows a highly schematic representation of a protein molecule, the enzyme pepsin, which functions in our stomachs to help us digest proteins.

Before discussing this type of polymerization in more detail we must introduce some further homologous series with more complex functional groups. The structures and naming of alkanes, alkenes, alcohols, aldehydes, ketones, carboxylic acids and halogenoalkanes containing up to six carbon atoms have already been covered (Chapter 10). Additionally, the structures of compounds containing the amine group, $-NH_2$, the benzene ring (phenyl group), $-C_6H_5$, group and esters, RCOOR', have also been mentioned. For IB at higher level you should also be able to name compounds containing the amine group and esters in addition to compounds containing two other functional groups: the amides and nitriles.

Further functional groups and their key reactions

20.1.1 Deduce structural formulas for compounds containing up to six carbon atoms with one of the following functional groups: amine, amide, ester and nitrile.

20.1.2 Apply IUPAC rules for naming compounds containing up to six carbon atoms with one of the following functional groups: amine, amide, ester and nitrile.

Amines (R–NH₂)

Amines are compounds derived from ammonia and, as such, have similar properties to ammonia. This relationship to ammonia provides an alternative way of looking at, and classifying, their structures. Rather than regarding the molecules as substituted hydrocarbons, it is possible to view them as compounds based on ammonia in which the hydrogen atoms are progressively replaced by alkyl (R) groups. Amines can then be categorized as one of the three types of organic compound – primary, secondary or tertiary (though these terms have a different meaning in this context than when they were applied to alcohols and halogenoalkanes).

Language of Chemistry

When applied to alcohols and halogenoalkanes, the terms primary, secondary and tertiary refer to the degree of chain branching at the carbon atom to which the functional group (—OH or —X) is attached. This is most easily seen by counting the hydrogen atoms attached to this particular carbon atom in the chain (Figure 20.2).

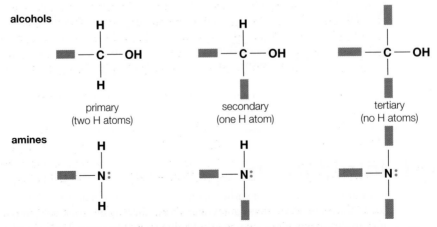

Figure 20.2 A comparison of the use of the terms primary, secondary and tertiary as applied to alcohols and amines

When applying these terms to amines (and indeed to amides) it is the level of branching at the nitrogen that is the key. Again the clue to the type of compound being considered is the number of hydrogen atoms attached, in this case to the nitrogen atom (Figure 20.2). ■

The IUPAC system of naming organic compounds allows for two different ways of naming amines. In the first of these the suffix *-amine* is used to indicate the presence of the $-NH_2$ group. Amines are then named by calling them after the longest hydrocarbon unbranched chain present, with the suffix *-amine*: for example, butan-1-amine, $CH_3CH_2CH_2CH_2NH_2$, and pentan-2-amine, $CH_3CH(NH_2)$ $CH_2CH_2CH_3$. The following system is then applied to naming primary, secondary or tertiary amines.

- If only one hydrogen atom in ammonia has been replaced then a **primary amine** is formed. For example, methanamine or methylamine (Figure 20.3a) – note that it is acceptable to use either the prefix *methan* or *methyl*.

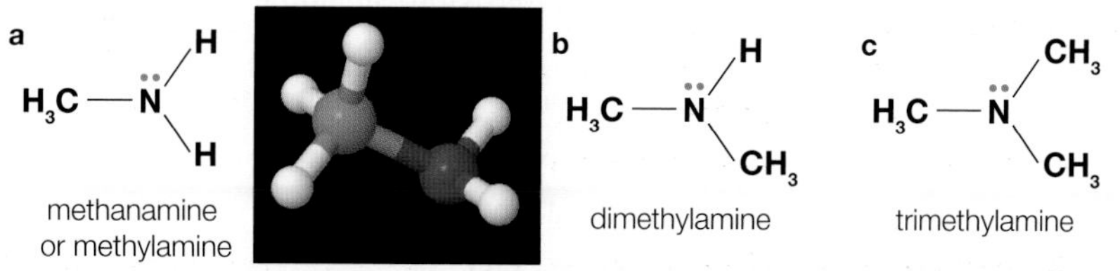

Figure 20.3 The structures of a primary, secondary and tertiary amine

- A **secondary amine** is formed when two of the hydrogen atoms from ammonia are replaced. For example, dimethylamine (Figure 20.3b). The alkyl groups are not necessarily of the same type.
- If alkyl groups replace all three hydrogen atoms then a **tertiary amine** is formed. For example, trimethylamine (Figure 20.3c).

However, the IUPAC system also makes it possible to use the prefix *amino-*, with the location of the $-NH_2$ group being indicated by numbering the carbon atoms in the chain. This method is most often used when the number of carbon atoms in the chain is four or more. Examples (Figure 20.4) of the application of this system are 2-aminopentane, 1,2-diaminoethane (an example of a bidentate ligand, see Chapter 13) and 1,6-diaminohexane.

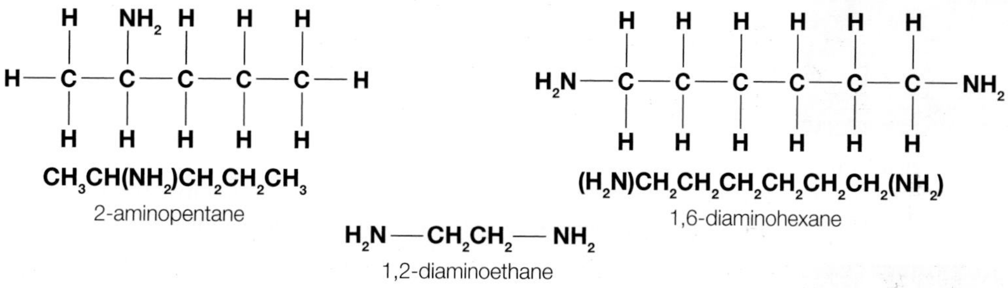

Figure 20.4 The names and structures of some more complex amines

The secondary and tertiary amines shown in Figure 20.3 were simple to name, as the alkyl groups replacing the hydrogen atoms were identical; hence dimethylamine and trimethylamine. In general, when naming a secondary amine the main name of the amine is taken from the longest unbranched carbon chain attached to the nitrogen atom. The other chain is prefixed as an alkyl group, with the location prefix given as an italic *N*. Examples include *N*-methylethylamine and *N*-ethylpropylamine. For tertiary amines there are two prefixes, each with an italic *N*: for example, $CH_3CH_2N(CH_3)_2$ is *N*,*N*-dimethylethylamine (Figure 20.5).

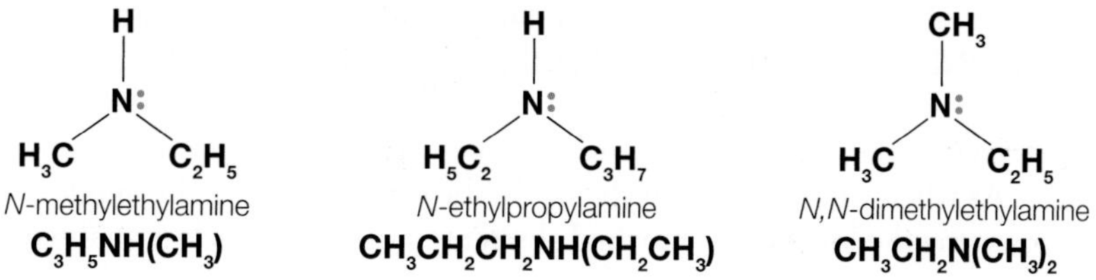

Figure 20.5 Secondary and tertiary amines

Using this systematic way of naming amines, dimethylamine should be named *N*-methylmethylamine and trimethylamine should be named *N*,*N*-dimethylmethylamine.

Aromatic amines are compounds in which an -NH2 group is bonded directly to the benzene ring. The most common is phenylamine, C6H5NH2 (also known as aminobenzene):

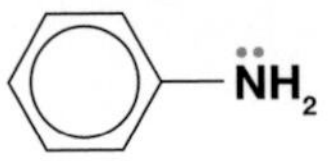

1 Name the following amines:

   a $CH_3CH_2CH_2CH_2NH_2$ **b** $(CH_3CH_2CH_2)_2NH$ **c** $(C_6H_5)_2NH$

2 Give the structural formulas of the following amines.

   a hexane-1,6-diamine   **b** *N*-ethylpropanamine   **c** *N,N*-dimethylpropanamine

1 **a** 1-butylamine, or 1-aminobutane, or 1-butanamine

   b *N*-propylpropanamine, or *N*-propylpropylamine, or dipropylamine

   c diphenylamine or *N*-phenylphenylamine

2 **a** $H_2NCH_2CH_2CH_2CH_2CH_2CH_2NH_2$

   b $CH_3CH_2CH_2NH(CH_2CH_3)$

   c $CH_3CH_2CH_2N(CH_3)_2$

Chemistry and Literature

The different disciplines of intellectual understanding give an insight into one another. Shakespeare's Caliban is a 'wild' uncivilized character in *The Tempest*, the only human inhabitant of the island on which the play is set (Figure 20.6). In Act 2, Scene 2 Trinculo, the court jester, raises questions as to Caliban's nature:

> *What have we here? A man or a fish? Dead or alive?*
> *A fish: he smells like a fish; a very ancient and fish-like smell;*
> *a kind of not-of-the-newest poor-John.*
> *A strange fish!'*

Figure 20.6 Caliban, Prospero's servant in *The Tempest*

It has been suggested that this is an early written description of a human metabolic disorder known as *trimethylaminuria*. This is a rare, genetically inherited condition in which the sufferer gives off a fishy smell due to the fact they cannot oxidize the trimethylamine (Figure 20.7) produced from the digestion of foods such as eggs, liver and various vegetables and grains. Normally the trimethylamine is oxidized in the liver to odourless trimethylamine oxide, which is then excreted from the body in the urine. However, if the ability to oxidize the amine is impaired, large amounts of the unchanged trimethylamine are excreted, producing a highly unpleasant, fishy body odour. It is estimated that one person in 10 000 suffers from what is sometimes known as 'fish odour syndrome'. ∎

$CH_3N(CH_3)_2$

Figure 20.7 The structure of trimethylamine – a tertiary amine

Amides (R–CONH₂)

Amides are derived from carboxylic acids by substitution of the —OH in the carboxylic acid group by an —NH₂ group. As such they retain the >C=O group as part of the amide functional group —CONH₂. It is worth noting, though, that the amide group behaves as a distinctive entity in its own right – not as a ketone and an amine – because the resonance interaction between the oxygen, carbon and nitrogen atoms makes the group function as a unit.

As when we name carboxylic acids, the carbon atom of the amide group is counted when assessing the longest unbranched chain present, and as such is counted as the first in the chain. Thus amides are named after the longest carbon chain, followed by the suffix *-amide*. Figure 20.8 shows the structures of ethanamide (see also Figure 20.9) and 2-methylpropanamide.

Secondary and tertiary amides are named rather like amines, in that the other alkyl group attached to the nitrogen atom is prefixed by an *N*: for example, *N*-methylethanamide (Figure 20.8) is a secondary amide, while *N,N*-dimethylpropanamide, $CH_3CH_2CON(CH_3)_2$, is a tertiary amide.

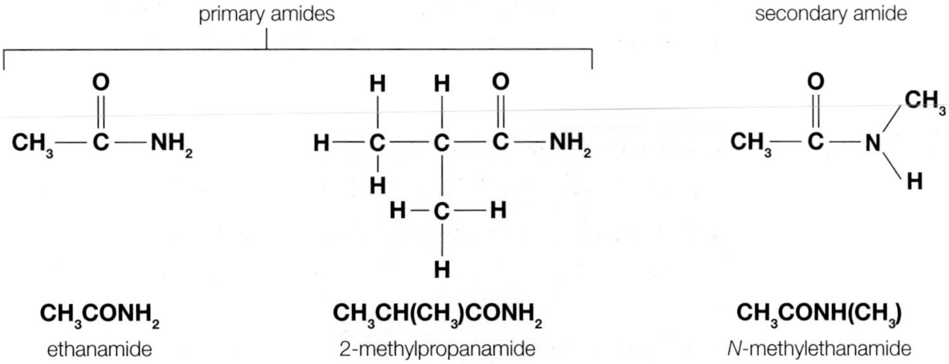

primary amides

secondary amide

CH_3CONH_2
ethanamide

$CH_3CH(CH_3)CONH_2$
2-methylpropanamide

$CH_3CONH(CH_3)$
N-methylethanamide

Figure 20.8 The names and structures of some amides

Figure 20.9 Ethanamide is a white crystalline solid at room temperature. The crystals are water soluble because of the capacity of amide molecules to hydrogen bond to water.

You will note from the structure of *N*-methylethanamide that secondary amides contain the group:

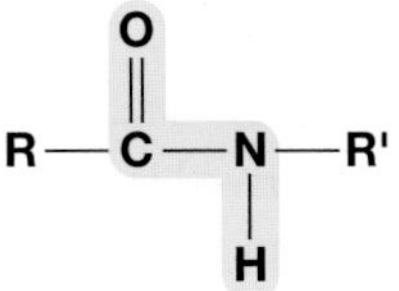

The highlighted group is known as the **amide link** or **peptide bond**. It is highly significant biologically as it is the group linking the amino acid residues together in all proteins. It is also present in some important artificial condensation polymers, the polyamides: most notably, nylon and Kevlar.

Esters (R—COOR′)

Esters are named from the alcohol and carboxylic acid from which they are derived. Unlike the homologous series considered so far, all of which are based on a single carbon chain, esters contain two carbon chains (alkyl or R groups) separated by an oxygen atom. The first part of the name of an ester is taken from the alkyl group of the alcohol from which it was synthesized. There is then a space, followed by the name for the acid anion. This is the part of the structure that contains the carbonyl group >C=O and has a name taken from the carboxylic acid from which it was formed.

Some examples are shown in (Figure 20.10): ethyl methanoate, propyl ethanoate, ethyl propanoate and phenyl ethanoate.

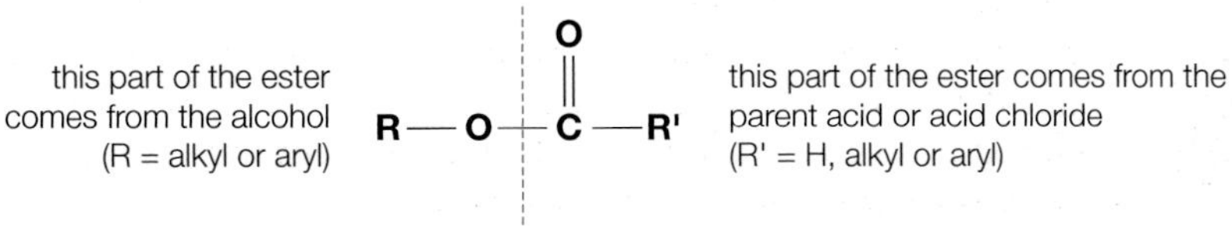

this part of the ester comes from the alcohol (R = alkyl or aryl)

this part of the ester comes from the parent acid or acid chloride (R′ = H, alkyl or aryl)

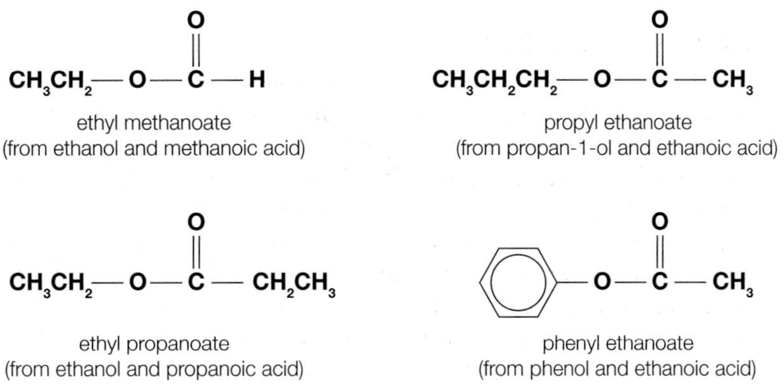

CH_3CH_2—O—C—H
ethyl methanoate
(from ethanol and methanoic acid)

$CH_3CH_2CH_2$—O—C—CH_3
propyl ethanoate
(from propan-1-ol and ethanoic acid)

CH_3CH_2—O—C—CH_2CH_3
ethyl propanoate
(from ethanol and propanoic acid)

—O—C—CH_3
phenyl ethanoate
(from phenol and ethanoic acid)

Figure 20.10 The naming of esters

From Figure 20.10 we can see how the structure of an ester is built up. However, it is more conventional to draw the condensed or full structural formulas of esters the other way round.

Thus, ethyl ethanoate is usually written $CH_3COOC_2H_5$ in chemical equations, for instance. Figure 20.11 shows the structure of some esters (propyl methanoate, methyl ethanoate and ethyl propanoate) written or drawn in this more usual format.

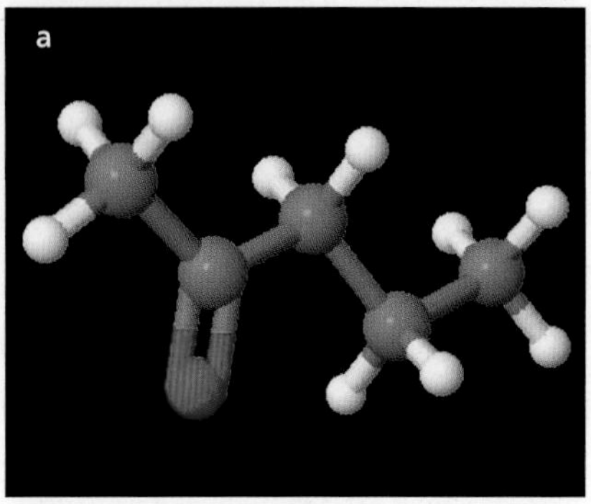

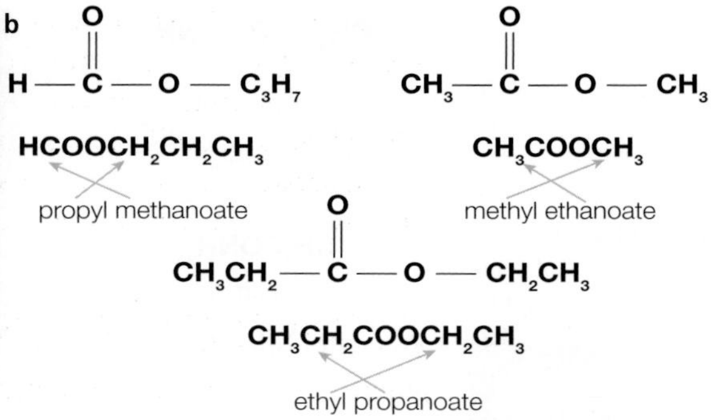

Figure 20.11 a A molecular model of methyl ethanoate **b** The structure of esters – showing the conventional way of writing the structural formulas with the acid grouping first

Nitriles (R–C≡N)

Nitriles are a homologous series derived from carboxylic acids. The early members of the series are pleasant smelling liquids at room temperature. They used to be called cyanides as they contain the —C≡N group which resembles the inorganic cyanide ion, C≡N⁻. So C_2H_5CN used to be known as ethyl cyanide. However, the IUPAC way of naming nitriles is to consider the carboxylic acid from which they are derived, as the —COOH group has been replaced by a —CN group. This means that the carbon atom of the nitrile group counts as the first atom of the hydrocarbon chain of the molecule. The suffix *-nitrile* is added to the name of the hydrocarbon chain. For example, C_2H_5CN now becomes propanenitrile (note there are three carbon atoms in the chain), and ethanenitrile has the formula CH_3CN (Figure 20.12).

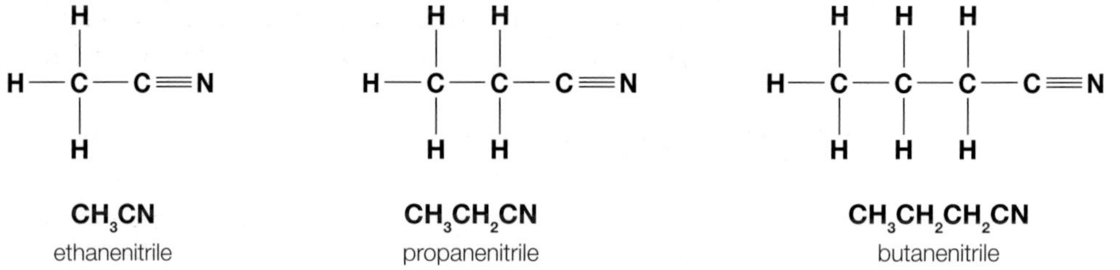

Figure 20.12 The structure and naming of simple nitriles

CH_3CN
ethanenitrile

CH_3CH_2CN
propanenitrile

$CH_3CH_2CH_2CN$
butanenitrile

In certain circumstances the nitrile group must give precedence in the naming of a compound to, say, a carboxylic acid group. In this situation it is referred to in the name by the prefix *cyano-*. So, for example, the compound $CH_2(CN)COOH$ is called cyanoethanoic acid.

Worked example

1 Give the condensed structural formulas of the following organic compounds:

 a ethyl butanoate

 b phenyl propanoate

 c 2-hydroxypropanenitrile

 d 2-cyanopropanoic acid

 e *N,N*-dimethylethanamide

2 Name the following compounds:

 a $CH_3CH_2CH_2COOCH_3$

 b $CH_3CH_2CH_2CH(CH_3)CN$

 c $CH_3CH_2CH(NH_2)CH(NH_2)CH_3$

 d $CH_3CH_2CONH(CH_3)$

1 a $CH_3CH_2CH_2COOCH_2CH_3$

 b $CH_3CH_2COOC_6H_5$

 c $CH_3CH(OH)CN$

 d $CH_3CH(CN)COOH$

 e $CH_3CON(CH_3)_2$

2 a methyl butanoate

 b 2-methylpentanenitrile

 c 2,3-diaminopentane

 d *N*-methylpropanamide

Language of Chemistry

The IUPAC system of naming organic molecules is very useful in connecting the naming of a compound unambiguously with its structure. However, it is helpful to remember that this system is a relatively recent development and that there is a whole set of trivial names for these compounds in the wider world of chemistry.

Thus ethanamide can also be found referred to as acetamide, ethanenitrile as acetonitrile, and phenylamine as aniline, for instance. It is worth being aware of this when you see reagent bottles in a laboratory; but do not be too disturbed, the IB syllabus and examination papers will use the IUPAC names exclusively and consistently. ■

20.2 Nucleophilic substitution reactions

One classic example of a nucleophilic substitution reaction is the hydrolysis of a halogenoalkane by a warm aqueous solution of sodium hydroxide (Chapter 10). In addition to hydroxide ions, other common nucleophiles are water molecules, ammonia molecules, amine molecules and cyanide ions. Each of these nucleophiles reacts to replace the halogen atom, which leaves as the halide ion.

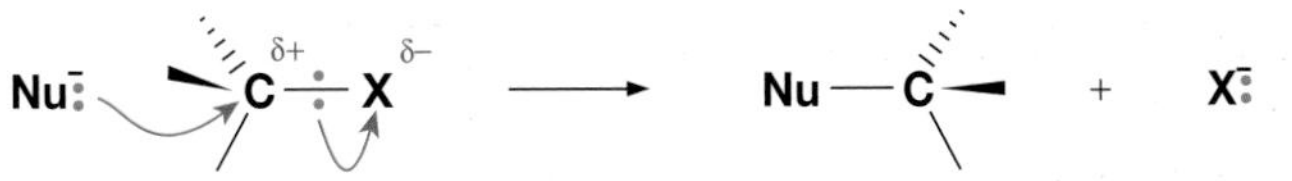

Figure 20.13 A diagrammatic representation of nucleophilic substitution

Here Nu represents a nucleophile (e.g. OH⁻, CN⁻, NH₃ or H₂O), and X represents a halogen atom (Cl, Br or I). We can see from this that a nucleophile is essentially a species with a lone pair of electrons and it can therefore attack an electron-deficient centre in the target molecule (Figure 20.13). As you can see from the list of examples, nucleophiles may also be negatively charged.

We have seen earlier that there are two mechanisms by which nucleophilic substitution can be achieved. The major factor that determines which mechanism takes place in a particular case is thought to depend primarily on the structure of the halogenoalkane.

The S$_N$1 reaction

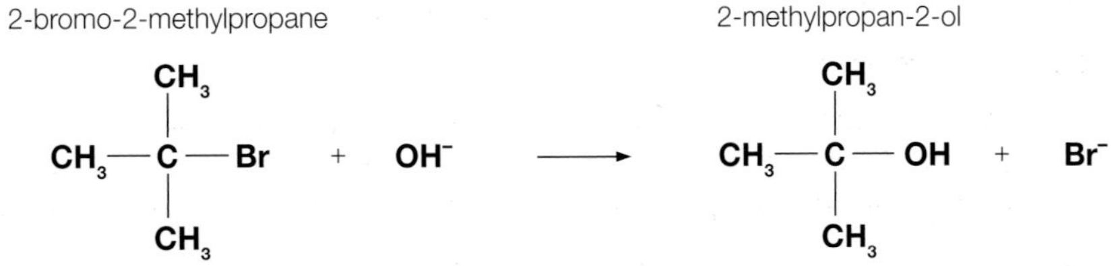

2-bromo-2-methylpropane   2-methylpropan-2-ol

Figure 20.14 An S$_N$1 reaction

rate = k_1[RBr] (where RBr is (CH$_3$)$_3$CBr)

2-Bromo-2-methylpropane is a tertiary halogenoalkane and the hydrolysis of this compound proceeds by an S$_N$1 mechanism (Figure 20.14). The hydrolysis is a first-order reaction, which means that the rate doubles if we double the concentration of the halogenoalkane, for instance. However, if we double the concentration of hydroxide ions, OH⁻, the rate does not change at all. The rate depends only on the concentration of the bromoalkane and is independent of the hydroxide ion concentration.

Kinetically this means that hydroxide ions cannot be involved in the **rate-determining step** – that is the step in the overall reaction that is the slowest and so limits the overall rate of reaction. The mechanism is shown in Figure 20.15.

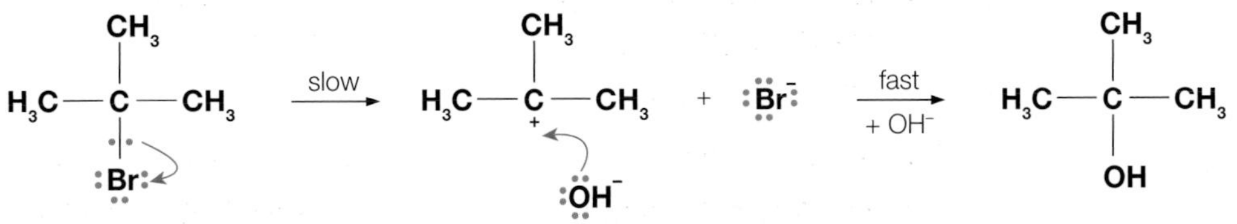

Figure 20.15 The S$_N$1 mechanism for the hydrolysis of 2-bromo-2-methylpropane

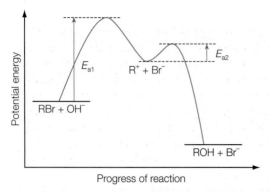

Figure 20.16 The reaction profile for an S_N1 mechanism

The first step of this reaction involves only the heterolysis of the C—Br bond, forming the **carbocation** intermediate and a bromide ion. The original halogenoalkane is tetrahedral in shape around the target carbon atom. This is the slowest (rate-determining) step in the reaction, and the hydroxide ions do not participate in it. If the concentration of hydroxide ions were increased, the rate of the second step would also increase. But this second step is already faster than the first one, so the rate of the overall reaction is unaffected.

Following the detachment of the Br⁻ ion the intermediate carbocation is trigonal planar in shape. This contributes to the speed of the second step – the attachment of the OH⁻ ion – as it can attack from either side of the planar carbocation. The reaction profile of the S_N1 reaction is shown in Figure 20.16.

The first step in this reaction involves bond breaking, which is an endothermic process. This results in the relatively high activation energy of the first step, E_{a1}. The activation energy of the second step, E_{a2}, is lower as oppositely charged ions attract each other strongly.

The S_N2 reaction

Bromomethane is a primary halogenoalkane and as such undergoes an S_N2 mechanism of hydrolysis (Figure 20.17). The hydrolysis of bromomethane is a second-order reaction provided neither reactant is present in large excess.

Figure 20.17 The single-step mechanism envisaged for an S_N2 reaction

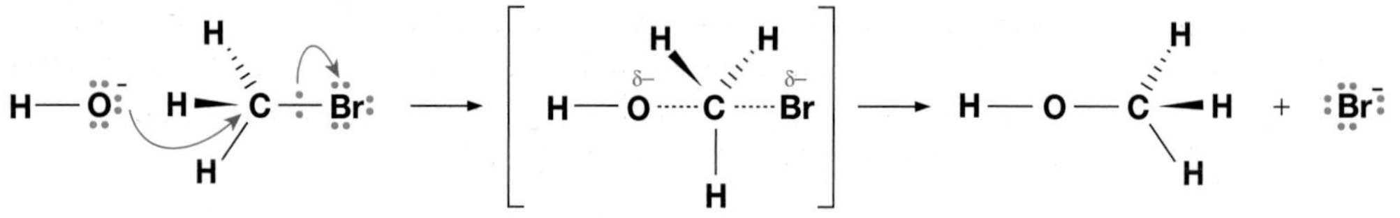

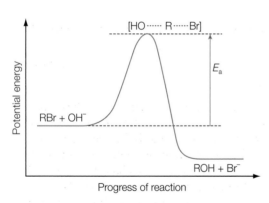

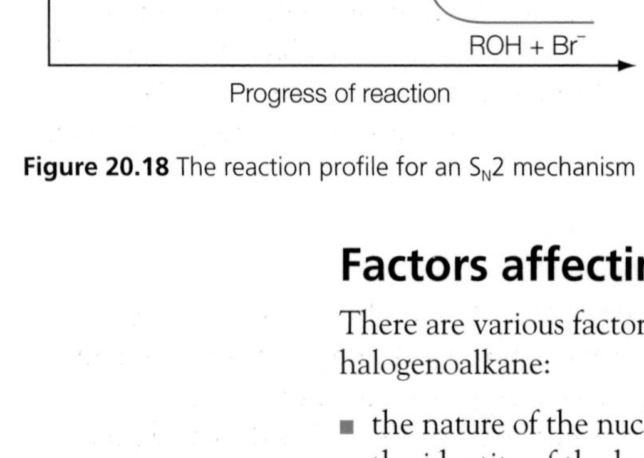

If the concentrations of either bromomethane molecules or hydroxide ions are doubled then the rate of reaction is also doubled. We can therefore deduce that the reaction is first order with respect to both bromomethane molecules and hydroxide ions and that both these species are involved in the rate-determining step. The reaction is thought to be a continuous, one-step process. The chemical structure shown in square brackets is not an intermediate (as the carbocation in the S_N1 reaction is) but a **transition state**. It is a halfway stage in the reaction where covalent bonds on the carbon atom are being simultaneously broken and made. The transition state is believed to have a trigonal bipyramidal shape (see Figure 20.23 and Chapter 14).

The reaction profile of the S_N2 reaction is shown in Figure 20.18. Do note the absence of an intermediate dip in the curve that would have corresponded with the existence of a reaction intermediate.

Figure 20.18 The reaction profile for an S_N2 mechanism

Factors affecting the rate of nucleophilic substitution

There are various factors that affect the rate of a nucleophilic substitution reaction involving a halogenoalkane:

- the nature of the nucleophile
- the identity of the halogen involved
- the structural type of the halogenoalkane
- the solvent involved, which can determine whether a nucleophilic substitution or an elimination reaction takes place.

We will look at the influence of these factors in the next sections.

20.2.1 **Explain** why the hydroxide ion is a better nucleophile than water.

The nature of the nucleophile

In this context a nucleophile is a species that is attracted to the partially positive carbon atom ($\delta+$) that is bonded to the electronegative halogen atom. Effective nucleophiles are neutral or negatively charged structures that have a lone pair of electrons which they can donate to make a new bond to the carbon atom. The more dense the negative charge on the ion or molecule is, the better it will act as a nucleophile. As a result, the negative anions tend to be more reactive than the corresponding neutral species. For example, the rate of substitution with hydroxide ions is faster than with water molecules.

Where species have the same charge, the controlling factor seems to be the electronegativity of the atom carrying the lone pair of electrons. The less electronegative the atom carrying the non-bonded lone pair of electrons, the better the nucleophile. Thus ammonia is a better nucleophile than water, as nitrogen is less electronegative than oxygen.

The reason for this is that the less electronegative the atom is the more easily it can donate its pair of electrons to make the new bond, as they are being held less strongly to that atom. Cyanide ions are the most effective common nucleophile. The effectiveness of nucleophiles follows the order below:

$$CN^- > OH^- > NH_3 > H_2O$$

$\xrightarrow{\hspace{4cm}}$

decreasing nucleophilic strength

20.2.2 **Describe** and explain how the rate of nucleophilic substitution in halogenoalkanes by the hydroxide ion depends on the identity of the halogen.

The identity of the halogen

An essential part of the substitution mechanism is that the halogen group in the target molecule must be able to leave. The carbon–halogen bond must be weak enough to allow this to happen. This consideration seems to be the most important factor in determining which is the best '**leaving group**' of the halogens. The reactivity of halogenoalkanes seems to depend on the ability of the carbon–halogen bond to break heterolytically and for the halogen to then leave as a halide ion. This is the reason the halogen is referred to as the leaving group. Iodine is the best leaving group, and fluorine is the worst (Figure 20.19).

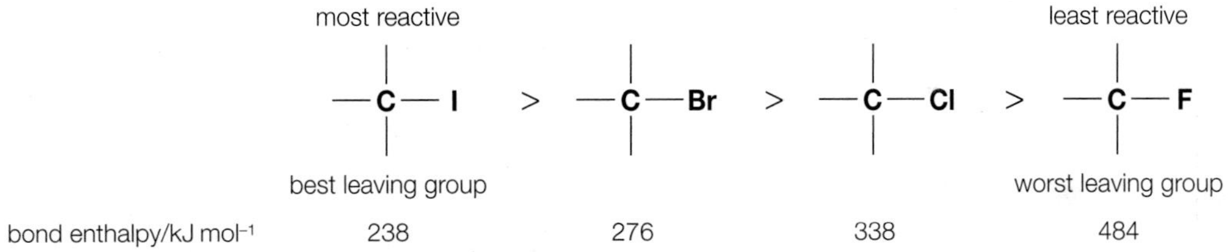

Figure 20.19 The bond enthalpy of the carbon–halogen bond determines how easily the halogen atom leaves the halogenoalkane molecule

most reactive			least reactive
—C—I	—C—Br	—C—Cl	—C—F
best leaving group			worst leaving group

bond enthalpy/kJ mol⁻¹   238   276   338   484

This does raise an interesting point, because fluorine is the most electronegative of the halogens and therefore the carbon–fluorine bond will be the most polar: that is, it will have the most ionic character. Given that nucleophiles are attracted to the electron-deficient ($\delta+$) carbon atom, we might have predicted that fluoroalkanes would in fact be the most reactive. However, the practical evidence that fluoroalkanes are unreactive suggests that the polarity of the carbon–halogen bond in the molecule is not nearly as important a factor as the *strength* of the carbon–halogen bond. This is borne out by the fact that iodoalkanes, where the bonds are the least polarized, are the most reactive, as the carbon–iodine bond is the longest and, therefore, the weakest.

Language of Chemistry

The meaning of the term 'leaving group' is reasonably self-evident in this context: it is the species that detaches from the electron-deficient carbon atom to make way for the nucleophile to bond to that carbon atom. A further factor that influences the ease of 'leaving', other than those that have already been discussed, is the stability of the group once it has detached from the carbon atom. The more stable the molecule or ion the better it functions as a leaving group. The less stable, or more reactive, the leaving group is, the more it is likely to act as a nucleophile. If this happens then the reaction simply goes round in circles.

Neutral molecules are better leaving groups than negative ions. Thus a water molecule is a better leaving group than an OH⁻ ion. Within the halogens, the iodide ion (I⁻) is the most stable ion and therefore the best leaving group. ∎

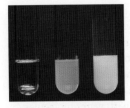

Some of the experimental evidence on which halogenoalkanes are most susceptible to nucleophilic attack has been gained using an aqueous silver nitrate solution as the source of the nucleophile. The solvent, water, is the nucleophile. The usefulness of this approach is that the appearance of the released halide ion is immediately detected by precipitation with the silver ions. Figure 20.20 shows the precipitates produced as a result of adding halobutanes to silver nitrate solution. The precipitates seen here are (left to right) silver chloride, silver bromide and silver iodide. The time taken for the precipitate to appear may be used as a comparative measure of the ease with which the substitution occurs.

The structural type of the halogenoalkane

20.2.3 **Describe** and explain how the rate of nucleophilic substitution in halogenoalkanes by the hydroxide ion depends on whether the halogenoalkane is primary, secondary or tertiary.

Kinetic studies demonstrate that substitution reactions with primary halogenoalkanes proceed by an S_N2 mechanism (Chapter 10). Substitution reactions with tertiary halogenoalkanes, however, proceed by an S_N1 mechanism in which an intermediate carbocation is formed. The reaction is first order with respect to the halogenoalkane, and is independent of the concentration of the nucleophile. It involves two steps. A similar experimental approach to that described earlier using aqueous silver nitrate solution can be used as part of an investigation to determine the relative rate of substitution of primary, secondary or tertiary bromoalkanes.

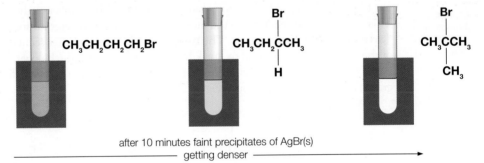

after 10 minutes faint precipitates of AgBr(s)
getting denser ⟶

Figure 20.21 The timing of the appearance of the off-white precipitate of silver bromide indicates the rate of substitution of the halogenoalkane

In these experiments the precipitate of silver bromide appears fastest where the tertiary bromoalkane is being substituted (Figure 20.21). Experimentally, S_N1 reactions are generally found to be faster than S_N2 reactions. This may be because the activation energy required to form the tertiary carbocation intermediate is less than the activation energy required to form the transition state of the S_N2 reaction.

primary   **secondary**   **tertiary** ⟶

carbocations increase in stability

Figure 20.22 The relative stability of primary, secondary and tertiary carbocations

The more alkyl groups there are attached to the positively charged carbon at the centre of the carbocation, the more energetically stable the ion. Therefore the stability of carbocations increases in the order shown in Figure 20.22. As the carbocation becomes more stable, the activation energy for the reaction leading to its formation also decreases (see Figure 20.16). Consequently we would expect that the rate of the S_N1 reaction will increase in the order primary < secondary < tertiary. Table 20.1 gives an indication of the relative rates of reaction for the nucleophilic substitution of different bromoalkanes under identical experimental conditions.

Table 20.1 The relative rates of the S_N1 reactions for different types of bromoalkane

Type of halogenoalkane	Compound	Relative rate of S_N1 reaction
Primary	CH_3CH_2Br	1
Secondary	$(CH_3)_2CHBr$	26
Tertiary	$(CH_3)_3CBr$	60 000 000

■ Extension: Structural influences on nucleophilic substitution

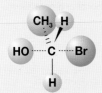

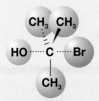

Figure 20.23 Different transition states for S_N2 reactions; the presence of more —CH₃ groups makes the second example more 'crowded' (sterically hindered)

The additional energetic stability of the tertiary carbocation arises because of the electron-donating effect of an alkyl group in such structures. An alkyl group, such as a methyl or ethyl group, affects the electron density of a structure in a way that a hydrogen atom does not. A hydrogen atom has no electrons associated with it other than those in its bond. A methyl group, however, has the three electron pairs in the carbon–hydrogen bonds in the group. In a carbocation, these electrons can be attracted towards the positive charge. Electron density is fed towards the charge, tending to lower the overall value of the charge and stabilizing the ion. Thus a methyl group, and other alkyl groups, are said to be electron-donating. This enhanced stability of the tertiary or secondary carbocation when compared to a primary carbocation is influential not only in the context of nucleophilic substitution mechanisms, but also in determining the major product of electrophilic addition reactions to asymmetric alkenes.

No intermediate carbocations are formed during an S_N2 reaction, so other factors must now come into play in determining the rate of reaction. The transition state in this type of reaction has five groups arranged around the central carbon atom. It is therefore more crowded than either the starting bromoalkane or the alcohol product, each of which have only four groups around the central carbon atom.

Hydrogen atoms are much smaller than alkyl groups. We therefore expect that the more alkyl groups there are around the central carbon atom, the more crowded will be the transition state (Figure 20.23), and the higher will be the activation energy, E_a (Figure 20.19). This will slow down the reaction (Table 20.2).

Type of halogenoalkane	Compound	Relative rate of S_N2 reaction
Primary	CH_3CH_2Br	1000
Secondary	$(CH_3)_2CHBr$	10
Tertiary	$(CH_3)_3CBr$	1

Table 20.2 The relative rates of the S_N2 reactions for different types of bromoalkane

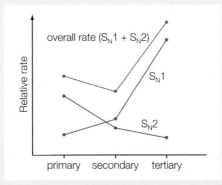

Figure 20.24 The relative rates of reaction for S_N1 and S_N2 reactions for primary, secondary and tertiary halogenoalkanes

These two effects we have discussed here reinforce each other. The S_N1 reaction is faster for tertiary halogenoalkanes than for primary halogenoalkanes, whereas the S_N2 reaction is faster with primary molecules than with tertiary molecules under the same conditions in the same solvent. Overall, therefore, we would expect primary halogenoalkanes to react predominantly by the S_N2 mechanism, tertiary halogenoalkanes to react predominantly by the S_N1 mechanism, and secondary halogenoalkanes to react by a mixture of the two (Figure 20.24).

Nitrogen-containing nucleophiles

> 20.2.4 **Describe,** using equations, the substitution reactions of halogenoalkanes with ammonia and potassium cyanide.

Ammonia as a nucleophile

Ammonia can act as a nucleophile through its lone pair of electrons on the nitrogen atom. Ammonia reacts with bromoethane by an S_N2 mechanism to form ethylamine and hydrogen bromide:

$$CH_3CH_2Br + \overset{..}{N}H_3 \rightarrow CH_3CH_2\overset{..}{N}H_2 + HBr$$
   bromoethane   ethylamine

However, the product ethylamine is also a nucleophile, because the nitrogen atom still has a non-bonding pair of electrons. Thus the reaction can proceed further. The product, ethylamine, can react with more bromoethane to produce diethylamine – a secondary amine.

$$CH_3CH_2Br + CH_3CH_2\overset{..}{N}H_2 \rightarrow (CH_3CH_2)_2\overset{..}{N}H + HBr$$
   bromoethane   ethylamine   diethylamine

Indeed, further substitution is possible to produce the tertiary amine triethylamine, $(CH_3CH_2)_3N$. The fact that these successive substitutions take place means that this is not a 'clean' way to prepare a particular amine and there may be other, preferable, routes if a particular amine is the target of a synthesis.

■ **Extension: Successive substitution reactions**

In fact there are four possible substitution products possible in the reaction between ammonia and a halogenoalkane. Triethylamine itself still has a nitrogen atom with an unbonded pair of electrons at its centre. As such it can again act as a nucleophile and gain a fourth ethyl group to form the equivalent of an ammonium salt, tetraethylammonium bromide, $(CH_3CH_2)_4N^+Br^-$ (Figure 20.25).

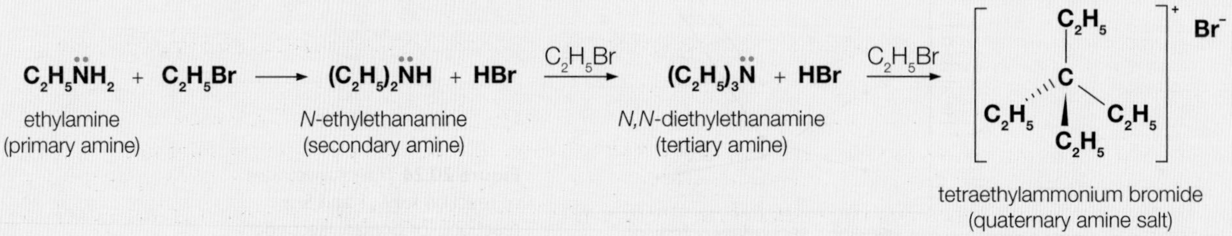

Figure 20.25 The reaction sequence of successive substitutions when ammonia reacts with excess bromoethane. Note the use of alternative IUPAC names for the products

The cyanide ion as a nucleophile

Nucleophilic substitution reactions are important in organic synthesis because the halogen atom on halogenoalkanes can be replaced by other functional groups. The reaction with potassium cyanide is a good illustration of this. The cyanide ion reacts to form a nitrile. For example, bromoethane reacts by an S_N2 mechanism with a solution of potassium cyanide in ethanol to form propanenitrile:

$$CH_3CH_2Br + KCN \rightarrow CH_3CH_2CN + KBr$$
   bromoethane   propanenitrile

A comparison of the names of the reactant and organic product here gives an indication of the potential significance of this particular reaction. With the introduction of the nitrile group into the molecule we have effectively extended the carbon chain.

20.2.5 Explain the reactions of primary halogenoalkanes with ammonia and potassium cyanide in terms of the S_N2 mechanism.

Both the nucleophilic substitution reactions we have just considered proceed by an S_N2 mechanism. Figure 20.26 shows the mechanism involved, with the lone pair on the nitrogen of the ammonia molecule forming a bond to the electron-deficient carbon atom in bromoethane, for instance.

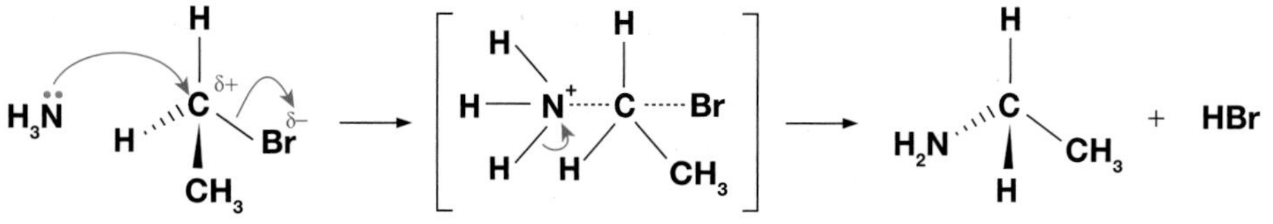

Figure 20.26 The S_N2 mechanism of substitution for the reaction between ammonia and bromoethane

The reaction of cyanide ions with bromoethane is also an S_N2 reaction (Figure 20.27).

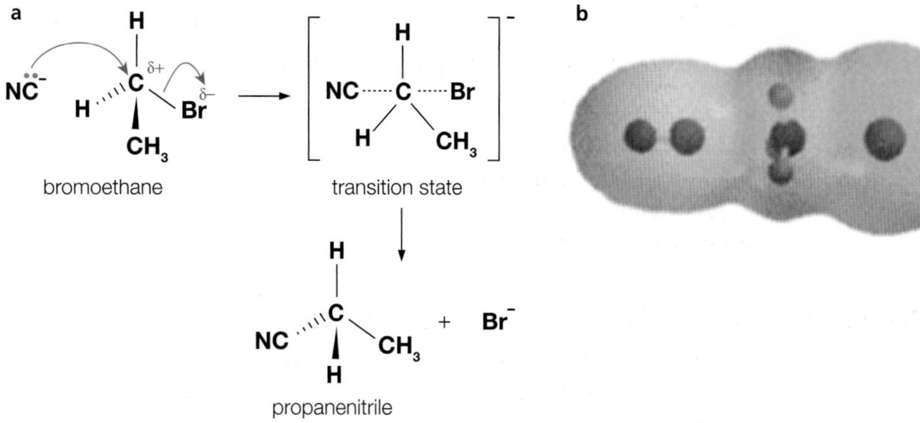

Figure 20.27 a The S_N2 mechanism for the reaction between cyanide ions and bromoethane molecules
b A computerized model of the transition state, showing the central trigonal planar arrangement around the carbon atom

Nitriles as intermediates in organic synthesis

20.2.6 Describe, using equations, the reduction of nitriles using hydrogen and a nickel catalyst.

As we indicated earlier the formation of a nitrile from a halogenoalkane is useful in devising synthetic pathways because it is a way of extending the carbon chain (an alternative method of achieving this uses Grignard reagents, Chapter 27). However, this would be of little use to organic and industrial chemists if the nitrile produced could not be converted into useful organic products. Fortunately, nitriles can be reduced using hydrogen gas and a heated nickel catalyst to form amines.

$$CH_3CH_2CN + 2H_2 \xrightarrow[\text{catalyst}]{\text{Ni}} CH_3CH_2CH_2NH_2$$

They can also be hydrolysed in acid solution to form carboxylic acids, which can then be reacted to form many other compounds (Figure 20.28).

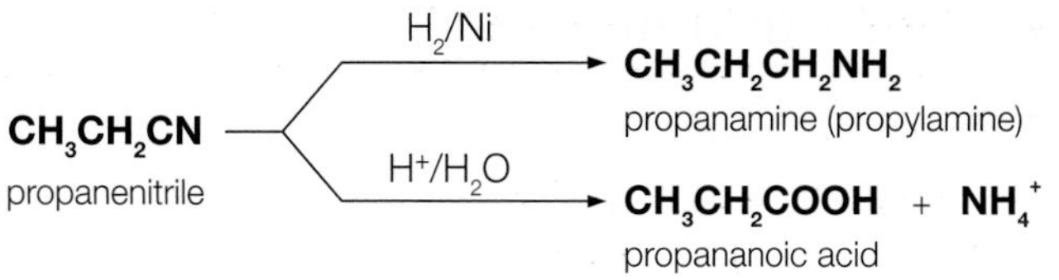

Figure 20.28 Two useful reactions of organic nitriles

H_2/Ni

CH_3CH_2CN propanenitrile

$\longrightarrow CH_3CH_2CH_2NH_2$ propanamine (propylamine)

H^+/H_2O

$\longrightarrow CH_3CH_2COOH + NH_4^+$ propananoic acid

20.3 Elimination reactions

Nucleophilic substitution takes place when a warm dilute aqueous solution of sodium hydroxide reacts with a halogenoalkane to form an alcohol. Hydrolysis of the carbon–halogen bond occurs under these conditions, with the hydroxide ion acting as a nucleophile (an electron pair donor). However, under different conditions, the hydroxide ion can act as a base (a proton acceptor – see Chapter 8) and **elimination** of the hydrogen halide occurs to produce an alkene.

> **20.3.1 Describe,** using equations, the elimination of HBr from bromoalkanes.

The generalized example here is of the elimination of hydrogen bromide (HBr) to produce a carbon–carbon double bond in the alkene:

$$-\overset{\overset{\textstyle H}{|}}{C}-\overset{\overset{\textstyle Br}{|}}{C}- \;+\; OH^- \longrightarrow \; \overset{}{C}=\overset{}{C} \;+\; H_2O \;+\; Br^-$$

The conditions that favour this elimination reaction are where the sodium or potassium hydroxide is dissolved in hot ethanol, a less polar solvent than water. Under these conditions, elimination is favoured by tertiary halogenoalkanes in comparison to secondary and primary molecules, as the intermediate carbocation is stabilized by the presence of the alkyl groups.

The ethanolic solution is added to the halogenoalkane and the mixture is heated under reflux. This provides a good example of how the same reactants can produce completely different products if the conditions are changed; in this case both the solvent and the temperature are changed. Refluxing bromoethane with an ethanolic solution of potassium hydroxide will produce ethene. Propene is produced from either 1-bromopropane or 2-bromopropane.

$$CH_3CH_2Br + KOH \rightarrow CH_2{=}CH_2 + KBr + H_2O$$
bromoethane ethene

$$CH_3CH_2CH_2Br + KOH \rightarrow CH_3CH{=}CH_2 + KBr + H_2O$$
1-bromopropane propene

$$CH_3CHBrCH_3 + KOH \rightarrow CH_3CH{=}CH_2 + KBr + H_2O$$
2-bromopropane propene

The possible elimination products become more complicated with bromoalkanes of increasing chain length. Primary bromoalkanes will always produce an alkene with a terminal carbon–carbon double bond. However, a mixture of products is produced from a secondary bromoalkane such as 2-bromobutane, since the hydrogen atom eliminated can come from either side of the bromine atom concerned.

$$CH_3CH_2CHBrCH_3 + KOH \rightarrow \begin{matrix} CH_3CH{=}CHCH_3 \\ \text{and} \\ CH_3CH_2CH{=}CH_2 \end{matrix} + KBr + H_2O$$

A mixture of but-1-ene and but-2-ene is produced. But-2-ene predominates in the mixture of products.

It is the more energetically stable of the two alkenes as it has two alkyl groups releasing electrons towards the carbon–carbon double bond.

Elimination reaction mechanisms

> **20.3.2 Describe** and explain the mechanism for the elimination of HBr from bromoalkanes.

Language of Chemistry

The mechanisms of these elimination reactions are often referred to by the terms E1 and E2. This is a similar type of 'shorthand' to the terms S_N1 and S_N2 that were applied to nucleophilic substitution reactions. Here the 'E' refers to an elimination reaction, the '1' or '2' to the kinetics of the reaction. Thus an E1 reaction has first-order kinetics, while an E2 reaction has second-order kinetics. ■

Two different mechanisms have been suggested for the elimination of hydrogen bromide from bromoalkanes. Both involve the hydroxide ion acting as a base rather than as a nucleophile. The conditions used, involving a concentrated ethanolic solution of the strong alkali, favour the hydroxide ion removing a proton (H^+ ion) from the bromoalkane. Two possible mechanisms are envisaged.

The first (referred to as the E1 mechanism) is thought to proceed by a carbocation intermediate, as the first step involves the heterolytic fission of the carbon–bromine bond in a similar way to the S_N1 mechanism (Figure 20.29).

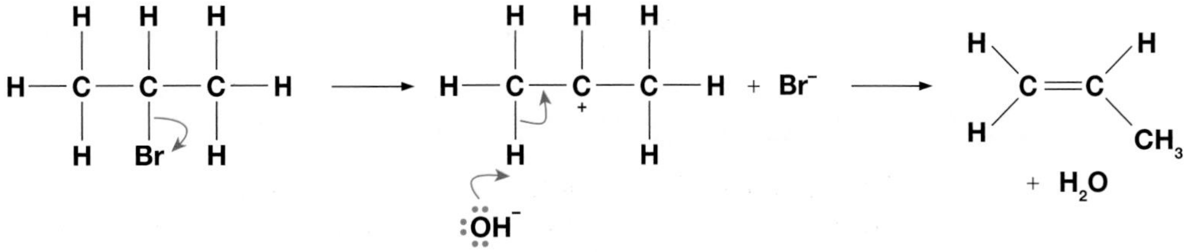

Figure 20.29 The E1 mechanism for elimination of hydrogen bromide from 2-bromopropane

The same factors that favour the production of the carbocation apply here as they did in the substitution reactions. Thus secondary and tertiary bromoalkanes will tend to favour this mechanism.

The alternative mechanism is a concerted process known as the E2 mechanism, because both the hydroxide ion (OH^-) base and the halogenoalkane are involved in the rate-determining step. Figure 20.30 shows the proposed mechanism applied to the elimination of hydrogen bromide from 2-bromopropane. This mechanism suggests that in one single step:

- the hydroxide ion removes an H^+ ion
- the lone pair of electrons left on the carbon atom (left-hand end) shifts to form a double bond between the carbon atoms and
- the bromine atom (Br) leaves as a bromide ion (Br^-).

Thus the products of this bond breaking and making are the alkene, in this case propene, water and the bromide ion.

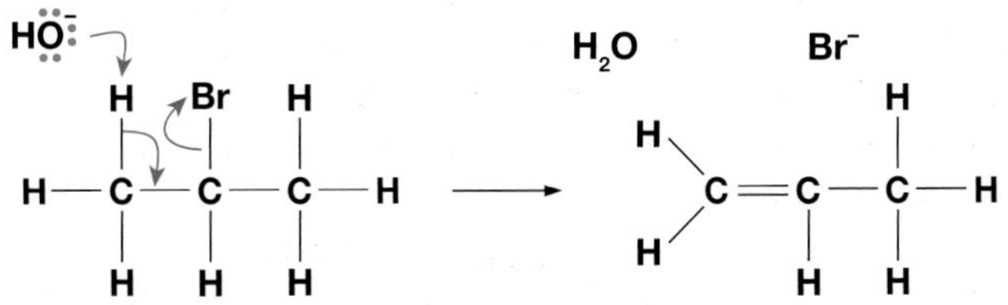

Figure 20.30 The E2 mechanism for the elimination of hydrogen bromide from 2-bromopropane

■ Extension: Competition between nucleophilic substitution and elimination

It is essential to realize that, given the similarity of the reagents involved, the two processes of nucleophilic substitution and elimination are generally in competition with each other. This is particularly true where tertiary halogenoalkanes are involved, as the E1 mechanism favoured in these reactions involves the same first step – the production of the intermediate carbocation.

As a result of this competition between the two processes going on simultaneously it is very difficult to obtain 'clean' products in these reactions. A single product will always have to be separated from a number of by-products.

Worked example

Suggest the structural formulas of the possible alkenes produced when 2-bromo-2-methylbutane is treated with potassium hydroxide dissolved in ethanol. Which of the possible alkenes will be found in greatest yield?

$CH_3CH_2C(CH_3)BrCH_3$
2-bromo-2-methylbutane

$CH_3CH=C(CH_3)_2$ [A]
2-methylbut-2-ene

$CH_3CH_2C(CH_3)=CH_2$ [B]
2-methylbut-1-ene

2-methylbut-2-ene [A] will be found in greater yield than 2-methylbut-1-ene [B] because it is the more substituted alkene.

20.4 Condensation reactions

A condensation reaction is one in which two molecules react together to produce a larger molecule, with the elimination of a small molecule such as water or hydrogen chloride. This type of reaction is one of the most important in terms of the formation of biologically significant molecules.

Esters

> **20.4.1 Describe,** using equations, the reactions of alcohols with carboxylic acids to form esters, and **state** the uses of esters.

One of the best known examples of a condensation reaction is **esterification**. When refluxed in the presence of a small amount of concentrated sulfuric acid, which acts as a catalyst by acting as a source of protons, carboxylic acids react with alcohols to form esters and water. This is an equilibrium reaction, and the sulfuric acid also acts to shift the position of equilibrium towards the products by 'removing' the water (Chapter 7). As an example, ethanol reacts with ethanoic acid to produce ethyl ethanoate and water:

Figure 20.31 The esterification reaction involving ethanoic acid and ethanol, showing how water is eliminated between the two molecules

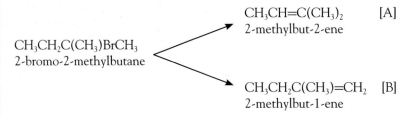

■ Extension: Studying reaction mechanisms

Studies using isotopic labelling with oxygen-18 have shown that it is the carbon–oxygen bond in the carboxylic acid molecule that breaks, not the carbon–oxygen bond in the alcohol molecule (Figure 20.31). Oxygen-18 can be detected by mass spectrometry and it is possible to use ethanol labelled with this isotope to detect in which product the heavier oxygen atom finishes up following the reaction. The finding that the heavier isotope is present in the ester isolated from the reaction mixture indicates which carbon–oxygen bond breaks during the reaction.

$$CH_3COOH + H^{18}OC_2H_5 \rightleftharpoons CH_3CO^{18}OC_2H_5 + H_2O$$

This shows one way in which an organic reaction mechanism can be determined. The method of preparation of a sample of ethyl ethanoate (Figure 20.32) illustrates some of the key practical procedures that are used in organic chemistry. These include reflux, distillation, use of a separating funnel and a drying agent, and final re-distillation to obtain purified product.

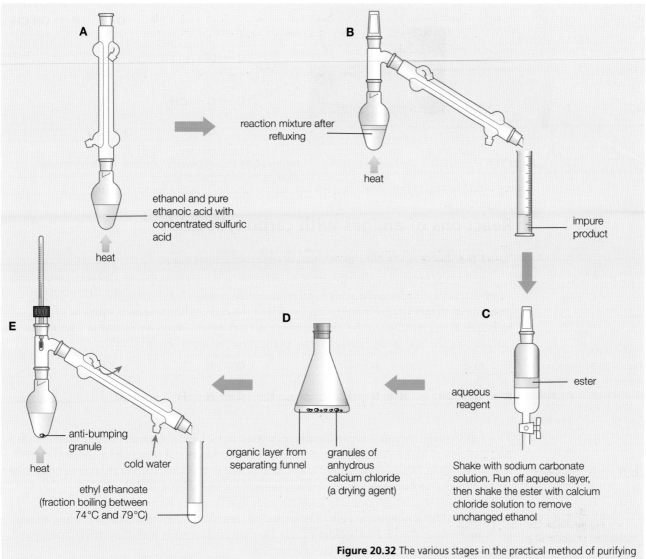

A

reaction mixture after
refluxing

heat

ethanol and pure
ethanoic acid with
concentrated sulfuric
acid

heat

B

heat

impure
product

E

anti-bumping
granule

heat

cold water

ethyl ethanoate
(fraction boiling between
74°C and 79°C)

D

organic layer from
separating funnel

granules of
anhydrous
calcium chloride
(a drying agent)

C

aqueous
reagent

ester

Shake with sodium carbonate
solution. Run off aqueous layer,
then shake the ester with calcium
chloride solution to remove
unchanged ethanol

Figure 20.32 The various stages in the practical method of purifying
the ethyl ethanoate produced by an esterification reaction

Many esters have very distinctive and individual smells. Generally these are quite fruity and
pleasant and so they are used as food flavourings and in perfumes. Table 20.3 lists some esters
responsible for the smells and flavours of various fruit.

Ester	Condensed structural formula	Smell or flavour
ethyl 2-methyl butanoate	$CH_3CH_2CH(CH_3)COOCH_2CH_3$	apple
3-methylbutyl ethanoate	$CH_3COOCH_2CH_2CH(CH_3)CH_3$	pear
1-methylbutyl ethanoate	$CH_3COOCH(CH_3)CH_2CH_2CH_3$	banana
butyl butanoate	$CH_3CH_2CH_2COOCH_2CH_2CH_2CH_3$	pineapple
octyl ethanoate	$CH_3COO(CH_2)_7CH_3$	orange
methylpropyl methanoate	$HCOOCH_2CH(CH_3)CH_3$	raspberry
pentyl butanoate	$CH_3CH_2CH_2COO(CH_2)_4CH_3$	strawberry

Table 20.3 The smells and flavours of esters

Esters can also be used as solvents (for example, in adhesives). Some painkillers (for example,
aspirin) (Figure 20.33) and anaesthetics (benzocaine and procaine, for instance) also contain an
ester functional group (Figure 20.34).

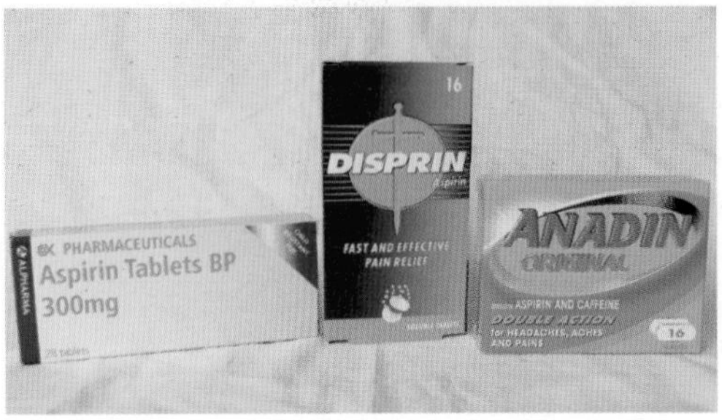

Figure 20.33 Some painkillers contain an ester group

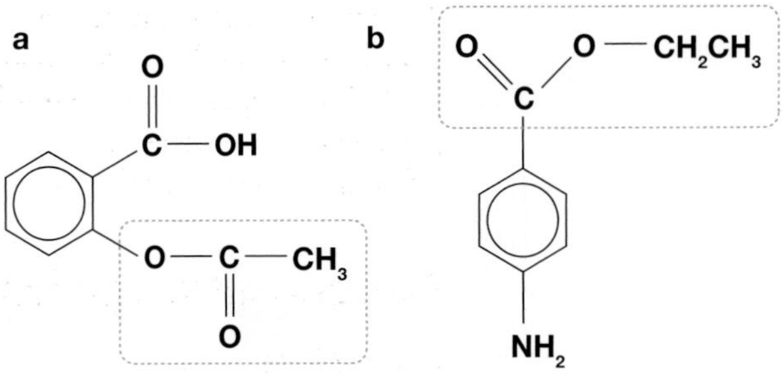

Figure 20.34 The structures of **a** aspirin, and **b** benzocaine, showing the functional groups

Reactions of amines with carboxylic acids

20.4.2 Describe, using equations, the reactions of amines with carboxylic acids.

Amines react with carboxylic acids to form secondary amides that contain the –CO–NH– grouping of atoms. The hydroxyl group (–OH) of the carboxylic acid reacts with one of the hydrogen atoms attached to the nitrogen atom in the amine to form water, and as a result the carbon from the acid group bonds directly to the nitrogen atom (Figure 20.35).

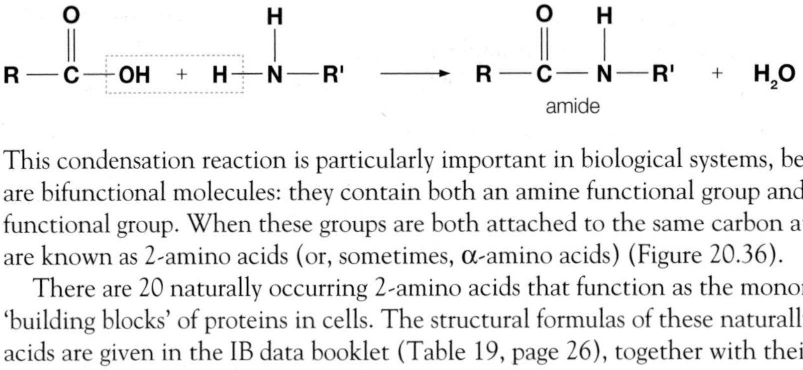

Figure 20.35 The reaction between a primary amine and a carboxylic acid to form a secondary amide and water

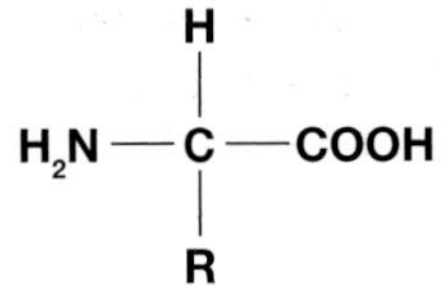

Figure 20.36 The generalized structure of a 2-amino acid

This condensation reaction is particularly important in biological systems, because amino acids are bifunctional molecules: they contain both an amine functional group and a carboxylic acid functional group. When these groups are both attached to the same carbon atom the compounds are known as 2-amino acids (or, sometimes, α-amino acids) (Figure 20.36).

There are 20 naturally occurring 2-amino acids that function as the monomers or basic 'building blocks' of proteins in cells. The structural formulas of these naturally occurring amino acids are given in the IB data booklet (Table 19, page 26), together with their trivial names and abbreviated codes (for example, Ala = alanine, etc).

This type of condensation reaction can be illustrated using 2-aminopropanoic acid (alanine) and 2-aminoethanoic acid (glycine). These two amino acids can react together to form two different organic condensation products, both of which are amides. The fact that the molecules are bifunctional gives rise to the two possible products. As you can see from the equations below, the two products form because of the two possible orientations of the reacting molecules.

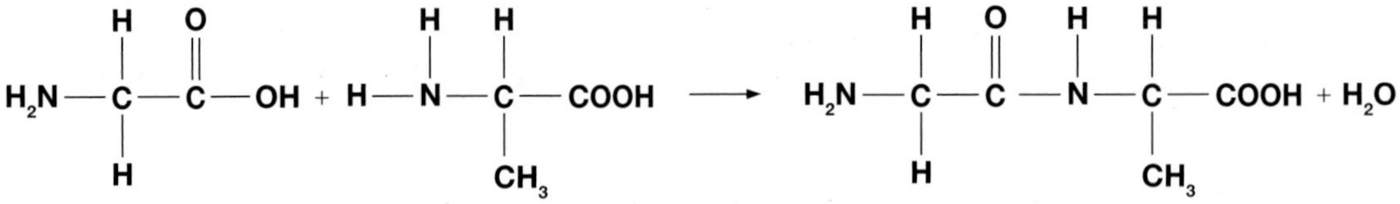

2-aminoethanoic acid glycine — 2-aminopropanoic acid alanine

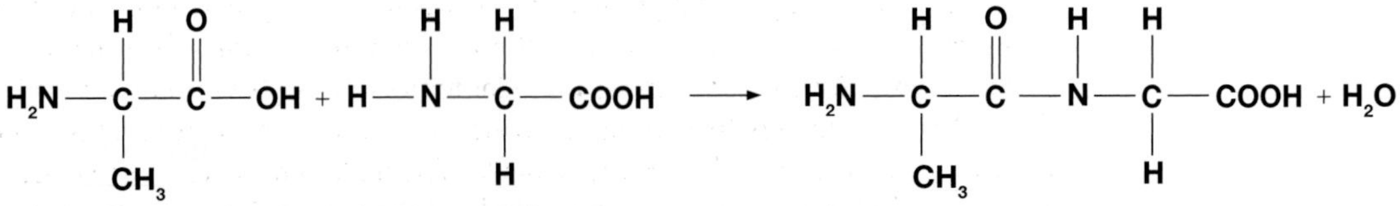

2-aminopropanoic acid alanine — 2-aminoethanoic acid glycine

The amide group (or link) formed in such condensation products from amino acids is also called a **peptide bond**, and the two condensation products are known as **dipeptides**.

Both the dipeptides shown still contain reactive amine and carboxylic acid groups and so it is possible to envisage reactions continuing using these functional groups. Chains built in this way are referred to as polypeptides (or proteins), and such molecules are a type of condensation polymer.

Condensation polymers

Polyesters

> **20.4.3 Deduce** the structures of the polymers formed in the reactions of alcohols with carboxylic acids.

As we saw earlier, the product formed by the condensation reaction between an alcohol and a carboxylic acid is called an ester. If two bifunctional molecules, one containing alcohol groups, the other carboxylic acid groups, react, it should be possible to form a long chain molecule, a **polyester**. Such a molecule would be a **condensation polymer** (or a **step polymer** in modern terminology).

Language of Chemistry

Alcohol nomenclature

Alcohols containing more than one hydroxyl group were briefly mentioned in Chapter 10. The term 'monohydric alcohol' indicates an alcohol possessing one hydroxyl group ($-OH$); methanol, for example. Alcohols also exist with more than one alcohol group. They are termed *polyhydric*; for example, propane-1,2-diol ($C_3H_6(OH)_2$), a dihydric alcohol, and glycerol (propane-1,2,3-triol), $C_3H_5(OH)_3$, a trihydric alcohol (see also Chapters 22 and 26). ∎

For instance, if a diol such as ethane-1,2-diol is reacted with a dicarboxylic acid such as benzene-1,4-dicarboxylic acid then a polyester is produced (Figure 20.37). In its fibre form this polyester is known as Terylene in the UK and as Dacron in the USA. However, it can be produced as a packaging film (Mylar) or in a form suitable for making bottles, when it is referred to as PET (polyethylene terephthalate).

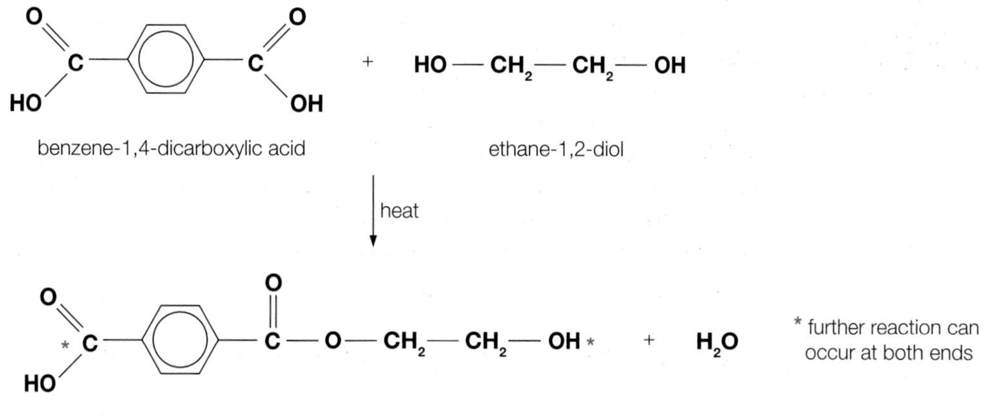

benzene-1,4-dicarboxylic acid   ethane-1,2-diol

heat

* further reaction can occur at both ends

Figure 20.37 The formation of the condensation polymer, Terylene

The different Terylene chains are not particularly strongly attracted together as there is no capacity for hydrogen bonding between the chains. In addition to van der Waals' forces between the chains, weak dipole–dipole interactions can take place between the polar carbonyl $>C=O$ groups as shown in Figure 20.38. The physical properties of a polymer are dependent on three factors. The inter-chain forces mentioned here, the crystallinity of the structure and the orientation of the chains all play their part. Although the intermolecular forces in this polymer are relatively weak, the chain is stiff and rigid, with very little rotation about the bonds. Crystallization of this polyester is slow and so various forms of the polymer can be made. The molten polymer can be extruded to form fibres that are useful for materials for making clothes (Figure 20.39). Since it has the useful property of being able to form permanent creases, it has been used extensively in the production of trousers and skirts. Currently the market is for approximately 60% fibre production but, as we have mentioned, this material can also be used for packaging films. In addition a polyester resin (PET) can be produced which has a low permeability to carbon dioxide and so is used extensively in bottling carbonated drinks.

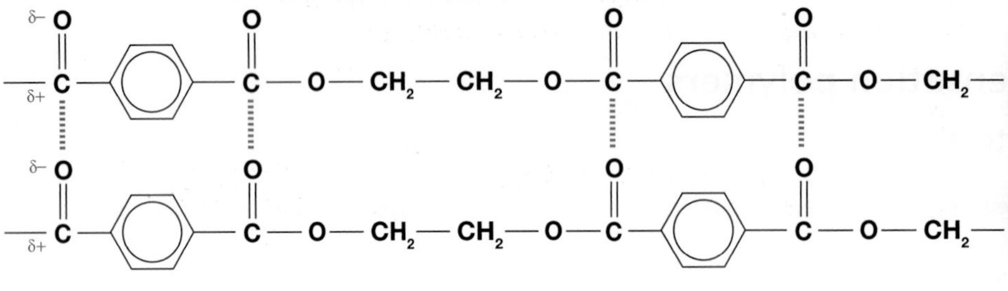

Figure 20.38 The dipole–dipole interactions between adjacent chains in Terylene

Figure 20.39 Terylene (Dacron) finds considerable use in making clothing; shirts, for example. It is usually used as cotton/polyester mixtures

20.4.4 **Deduce** the structures of the polymers formed in the reactions of amines with carboxylic acids.

The production of nylon

History of Chemistry

The American chemist **Wallace Carothers** (1896–1937) is generally accepted as being the inventor of nylon. He had been a member of the team at DuPont in 1930 that created neoprene, the first synthetic rubber. He then went on to create and patent nylon in 1935. The development of nylon was the result of a deliberate intention to build up a polymer chain similar to those protein chains occurring in wool and silk – two natural polyamide polymers. Wallace Carothers suffered from mental illness and took his own life in 1937, just before nylon fibres were first used commercially in 1938.

Language of Chemistry

The name nylon is often incorrectly described as being derived from the names for New York (NY) and London (LON). However, in 1940 John Eckelberry, of DuPont, in answer to the question of the origin of the name, stated that the initial three letters, 'nyl', were arbitrary, and the 'on' was copied from the suffixes of other fibres such as cotton and rayon. ■

Nylon was originally made by the reaction of a diamine with a dicarboxylic acid. The two monomers used initially were 1,6-diaminohexane and hexane-1,6-dioic acid. The polymer chain is made up from the two monomers reacting alternately and results in the chain type:

—A—B—A—B—A—B—A—B—A—B—A—B—A—B—A—

As we saw earlier when looking at the formation of dipeptides, each time a reaction takes place between the two monomers a molecule of water is lost. This is a further example of condensation polymerization (Figure 20.40). The link formed repetitively between the monomers is an amide link and nylon is known as a **polyamide**. This link is that found in proteins, where it is often referred to as a **peptide bond**.

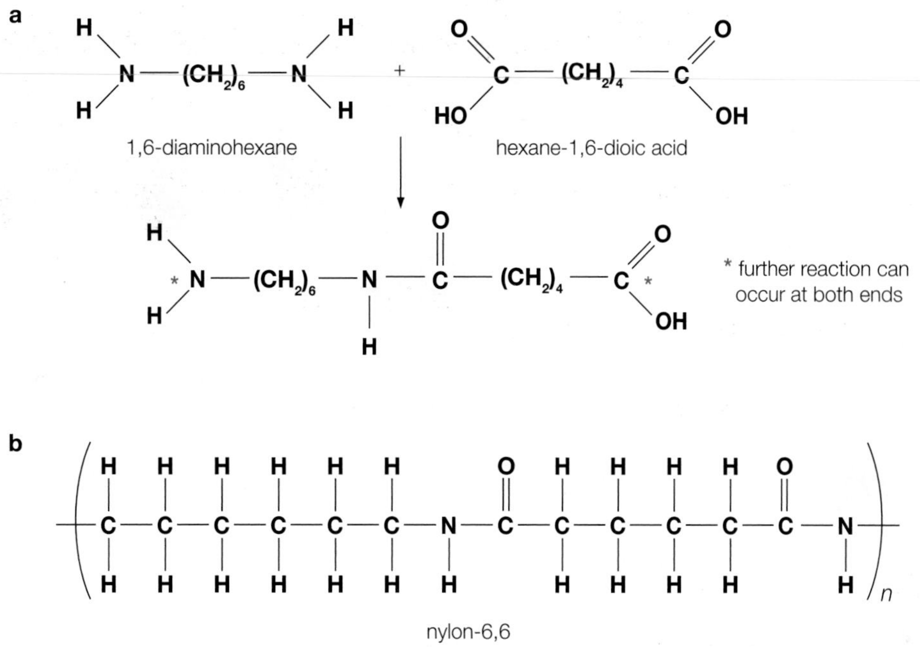

a

1,6-diaminohexane   +   hexane-1,6-dioic acid

* further reaction can
occur at both ends

Figure 20.40
a The polymerization of
1,6-diaminohexane and
hexane-1,6-dioic acid to
form nylon
b Because the two
monomers each contain
a six-carbon chain, this
form of nylon is known as
nylon-6,6

b

nylon-6,6

There are various different forms of nylon, for example, nylon-6,6 (Figure 20.40b) and nylon-6,10. The type of nylon depends on the number of carbon atoms in the monomers used. If the diamine used contains six carbon atoms and the dicarboxylic acid contains a chain of ten carbon atoms, then the resulting nylon is referred to as nylon-6,10. The formation of these different forms of nylon requires high temperatures and the presence of catalysts. The reactions are usually carried out under vacuum conditions to remove the water released.

■ Extension: Nylon-6

After the development of the original forms of nylon, people considered whether it would be possible to make nylon from a single monomer. For this to happen it would be necessary for the monomer to contain an amine group at one end of the molecule and a carboxylic acid group at the other. This would mimic more closely the formation of protein chains, though with only one type of monomer molecule, not the 20 different ones used in cells to make proteins.

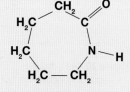

Theoretically, nylon-6 could be made from the monomer $H_2N(CH_2)_5COOH$, 6-aminohexanoic acid, but there are practical difficulties with this. Today nylon-6 is made by heating the cyclic amide caprolactam to 260 °C (Figure 20.41). Usefully, it can then be polymerized in a mould to make small, solid objects directly from the monomer without any moulding process.

caprolactam

Figure 20.41 Caprolactam, the cyclic amide used to make nylon-6

The structure of nylon-6 is $\text{--}[CO(CH_2)_5NH]_n\text{--}$ and is distinctive when compared with the other forms of nylon in that the peptide links are all orientated in the same direction, as in proteins.

When nylon is prepared by the reaction between a diamine and a dicarboxylic acid the reaction is actually quite slow. For a demonstration in the laboratory this can be speeded up by reacting the diamine with an acyl chloride (or acid chloride, a derivative of the acid where the —OH in the carboxylic acid group is replaced by a chlorine atom, Cl). In this case, the condensation reaction is much faster and hydrogen chloride is eliminated between the monomers instead of water (Figure 20.42a). The diamine is dissolved in water and a solution of the acyl chloride in hexane is layered carefully on top of the aqueous solution. The reaction takes place at the *interface* between the two immiscible solutions and the raw nylon can be spooled away as a 'nylon rope' (Figure 20.42b).

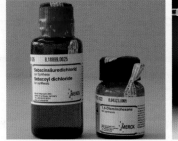

Figure 20.42 a The monomers for the 'nylon rope' experiment (decanedioyl dichloride [or sebacoyl chloride] and 1,6-diaminohexane); **b** the 'nylon rope' can be continuously drawn off from the interface between the solutions of the two monomers. This experiment produces nylon-6,10

Figure 20.43 A nylon climbing rope

When nylon is made in industry it forms a solid, which is melted and then forced through fine jets and extruded. The long filaments cool and the solid nylon fibres produced are stretched to align the polymer molecules and then dried. The resulting yarn can be woven into fabric to make shirts, ties, sheets and parachutes or turned into ropes (Figure 20.43) or racket strings for tennis and badminton rackets.

The molecular chains in nylon fibres interact by hydrogen bonding between the hydrogen atoms of the N—H groups of the amide link of one polymer chain with the $>$C=O groups on adjacent polymer chains (Figure 20.44). Thus nylon fibres contain strong hydrogen bonding between the chains but the chains do show high flexibility and, because nylon crystallizes quickly, the fibres are always semi-crystalline. These factors give nylon its distinctive properties when compared to other polymers.

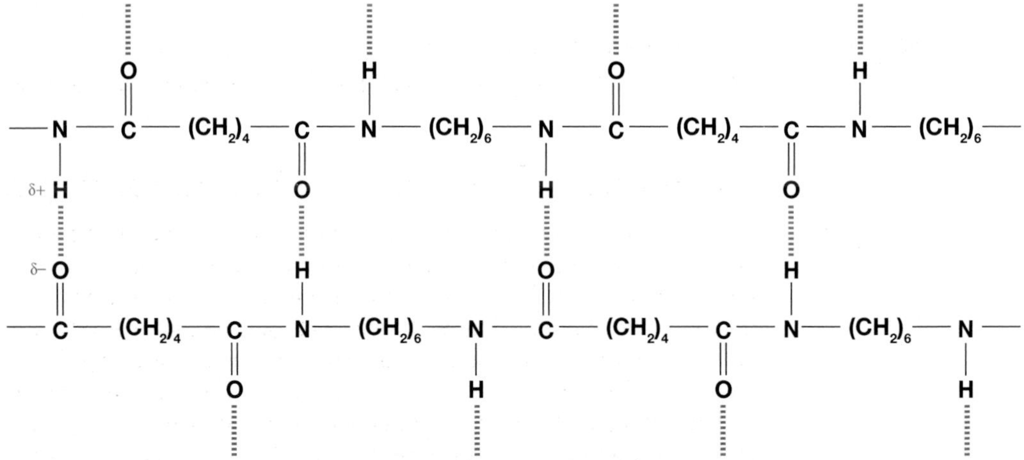

Figure 20.44 Hydrogen bonding between adjacent nylon chains

A recent development in polymer chemistry has been to create a polyamide in which the straight chain hydrocarbon unit within the polymer has been replaced by an aromatic benzene ring. This type of polymer is known as an aramid. The first aramid was made from 3-aminobenzoic acid. However, it was found not to be particularly strong even though it had exceptional fire resistance and could be made into fibres. The starting monomer was modified to create straighter chains in the polymer and a polyaramid was produced with exceptional properties; it is now known as Kevlar (Figure 20.45).

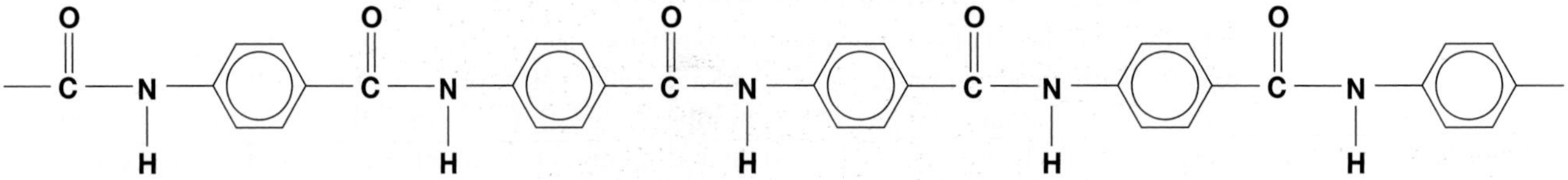

Figure 20.45 A section of the polymer chain in Kevlar

Figure 20.46 This motorcycle helmet is reinforced with Kevlar

In nylon the single covalent bonds within the polymer chain are free to rotate and this tends to make the polymer quite flexible. However, in the case of Kevlar the replacement of the straight hydrocarbon chain parts of the polyamide by rigid benzene rings makes the chains inflexible. This polymer is far more rigid than nylon. Kevlar is exceptionally strong, being five times the strength of steel on a weight for weight basis. In addition it is very fire resistant. These properties have led to a variety of uses in the aircraft and aerospace industry as well as for the manufacture of cables and ropes, and for protective clothing (including bullet-proof vests) (Figure 20.46).

Primary structure of proteins

Proteins are natural polyamides whose assembly from the monomers is under the control of a template known as messenger RNA, copied from the gene for the particular protein chain (Chapter 22). Proteins consist of a long chain of amino acid residues, formed by condensing the amino acids together in the presence of enzymes to form peptide bonds. The folded three-dimensional structure of proteins is complex, but their primary structure is simply the sequence of amino acid residues from the amino(N)-terminus to the carboxy(C)-terminus. One such protein is lysozyme, a relatively simple protein whose structure was one of the first to be determined. Lysozyme is an enzyme, found in many organisms, which breaks down the cell walls of certain types of bacteria. The protein chain in lysozyme consists of 129 amino acids joined together by amide (peptide) links (Figure 20.47).

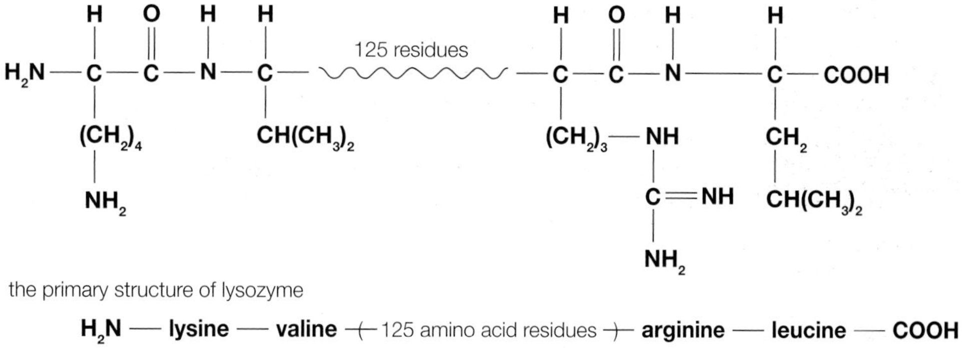

Figure 20.47 The primary structure of the enzyme lysozyme

the primary structure of lysozyme

$$H_2N \text{ — lysine — valine } \text{─}(\text{125 amino acid residues})\text{─} \text{ arginine — leucine — COOH}$$

The economic importance of condensation reactions

20.4.5 Outline the economic importance of condensation reactions.

Polyamides and polyesters are just two examples of condensation polymers. The world production of nylon alone is currently estimated to exceed 5 million tonnes. We are very conscious of polyesters and nylon being used to make clothes and carpets. However, there are many other uses for condensation polymers. Nylon has high strength, resists abrasion and is easy to dye. Nylon fibres are used to make climbing ropes for mountaineering (Figure 20.43), and one of the main uses of nylon is in engineering (gears, etc.). The structure of nylon can be altered to give specific properties for a defined purpose by using fillers, pigments, glass fibre and toughening agents. Nylon film finds many everyday uses, including that of 'boil-in-the-bag' convenience meal packaging. The chemical and physical properties can also be altered by changing the number of carbon atoms in the two condensed carbon chains. Thus nylon-6,10 has the repeating unit $-NH(CH_2)_6NHCO(CH_2)_8CO-$. Kevlar is a polyamide that finds many uses for its strength and rigidity, including reinforcing the chassis of Formula 1 racing cars (Figure 20.48).

Figure 20.48 Mechanics working on a Formula 1 chassis

20.5 Reaction pathways

20.5.1 Deduce reaction pathways given the starting materials and the product.

In Chapter 10 we introduced the idea of linking reactions together in sequence to synthesize an organic compound from starting reagents that would not directly produce the desired compound. Now we have added further reactions (Figure 20.49) and we can use these to expand the possibilities of the synthetic routes available to use.

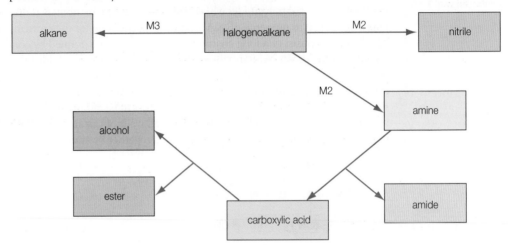

Figure 20.49 The reaction pathways discussed in this chapter: M1 is the free radical substitution mechanism discussed in Chapter 10

M2 & M3: you should know these mechanisms for IB

One useful reaction is the nucleophilic substitution of a halogenoalkane with potassium cyanide, which forms the nitrile. In doing so it effectively extends the carbon chain and ascends the homologous series. Remember to recall not just the interconversions of functional groups but also the relevant equations and conditions involved in the key reactions. When considering the problem posed by a particular conversion remember that working backwards from the required product as well as forward from the reactants may give useful clues to the steps needed.

For example, it is possible to synthesize 1-aminobutane from 1-bromopropane in two stages. First 1-bromopropane can be converted to butanenitrile by reacting the halogenoalkane with potassium cyanide dissolved in ethanol. The nitrile produced can then be reduced to the amine by heating with hydrogen and a nickel catalyst.

1-bromopropane $\xrightarrow[\text{reflux}]{\text{KCN/ethanol}}$ butanenitrile $\xrightarrow[\text{heat}]{\text{H}_2\text{ / nickel catalyst}}$ 1-aminobutane

$CH_3CH_2CH_2Br \xrightarrow{} CH_3CH_2CH_2CN \xrightarrow{} CH_3CH_2CH_2CH_2NH_2$

Worked example

Devise two-step synthetic routes to produce the following products from the given starting compound. Give the necessary conditions and equations for the reactions involved.

a *N*-ethylethanamine from ethene   **c** Pentan-1-ol from pentane
b Propanone from propene   **d** Ethyl ethanoate from ethanol

a

ethene $\xrightarrow{\text{HBr}}$ bromoethane $\xrightarrow[\text{heat}]{\text{CH}_3\text{CH}_2\text{NH}_2}$ *N*-ethylethanamine

$CH_2{=}CH_2 + HBr \rightarrow CH_3CH_2Br$

$CH_3CH_2Br + CH_3CH_2NH_2 \rightarrow CH_3CH_2NHCH_2CH_3 + HBr$

b

$$\text{propene} \xrightarrow{\text{conc. H}_2\text{SO}_4/\text{heat}/\text{H}_2\text{O}} \text{propan-2-ol} \xrightarrow{\text{Na}_2\text{Cr}_2\text{O}_7/\text{H}^+/\text{reflux}} \text{propanone}$$

$$CH_3CH{=}CH_2 + H_2O \rightarrow CH_3CH(OH)CH_3$$

$$CH_3CH(OH)CH_3 + [O] \rightarrow CH_3COCH_3 + H_2O$$

c

$$\text{pentane} \xrightarrow{\text{Cl}_2/\text{uv radiation}} \text{1-chloropentane} \xrightarrow{\text{NaOH(aq) reflux}} \text{pentan-1-ol}$$

$$CH_3(CH_2)_3CH_3 + Cl_2 \rightarrow CH_3(CH_2)_3CH_2Cl + HCl$$

$$CH_3(CH_2)_3CH_2Cl + NaOH \rightarrow CH_3(CH_2)_3CH_2OH + NaCl$$

d

$$\text{ethanol} \xrightarrow{\text{Na}_2\text{Cr}_2\text{O}_7/\text{H}^+/\text{reflux}} \text{ethanoic acid} \xrightarrow{\text{more C}_2\text{H}_5\text{OH}/\text{H}^+/\text{reflux}} \text{ethyl ethanoate}$$

$$C_2H_5OH + 2[O] \rightarrow CH_3COOH + H_2O$$

$$CH_3COOH + C_2H_5OH \rightarrow CH_3COOC_2H_5 + H_2O$$

20.6 Stereoisomerism

> **20.6.1 Describe** stereoisomers as compounds with the same structural formula but with different arrangements of atoms in space.

Where two compounds have the same molecular formula but different structural formulas they are said to be structural isomers. For example, propan-1-ol and propan-2-ol both have the molecular formula C_3H_8O but the carbon atom to which the alcohol functional group (—OH) is attached in the hydrocarbon chain is different in the two molecules. We have also seen that hydrocarbon chains can be branched or straight, giving rise to **structural isomerism**; for example, butane and 2-methylpropane are isomers with the same molecular formula (C_4H_{10}).

Here we introduce a further category of isomerism known as **stereoisomerism** in which the molecules concerned have the same molecular formula *and* structural formula, but their atoms are arranged differently in space. There are two types of stereoisomerism, geometrical isomerism and optical isomerism.

History of Chemistry

Jacobus van't Hoff (1852–1911) was the winner of the first ever Nobel Prize in Chemistry. He was a Dutchman, born in Rotterdam in the Netherlands, and one of the pre-eminent scientists of his day. He was a physical and organic chemist who contributed original ideas in the fields of reaction kinetics and equilibria, osmotic pressure (Chapter 17) and crystallography. He was a professor of chemistry at the University of Amsterdam for almost 18 years before moving to finish his career at the University of Berlin. In 1901 he gained the Nobel Prize for his work on aqueous solutions.

Even before he received his doctorate he had published work on the stereochemistry of organic compounds. He accounted for the phenomenon of optical isomerism with his ideas on the tetrahedral arrangement of the bonding around a carbon atom. He shares the honour of this original idea with the French chemist Joseph Le Bel who independently came up with the same idea.

Joseph Le Bel (1847–1930) was a French chemist best known for his work on stereochemistry. He lived and worked for most of his life in Paris. He put forward his ideas on the spatial arrangement of atoms in molecules in 1874, the same year as van't Hoff. His ideas helped to explain the phenomenon of the optical activity of certain organic chemicals.

If a carbon atom formed four planar (flat) bonds then dichloromethane, for example, would exist in two isomeric forms (Figure 20.50). However, there is only one form of dichloromethane and this type of isomerism is unknown.

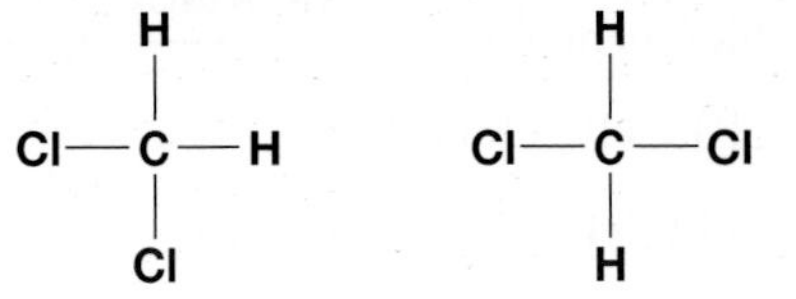

Figure 20.50 Possible isomers of dichloromethane if the bonds from carbon were all in one plane. This is not the case and this type of isomerism does not exist

It is for this reason that van't Hoff and Le Bel independently suggested that the four bonds from a carbon atom were equivalent, and arranged *tetrahedrally*. This was a *paradigm shift* in chemical thinking and this idea was used to explain the existence of optical isomers. Together van't Hoff and Le Bel were responsible for developing a new area of chemistry known as stereochemistry, which deals with the effects produced by the organization of atoms and functional groups in space.

Geometrical isomerism

20.6.2 Describe and explain geometrical isomerism in non-cyclic alkenes.

Geometrical isomerism occurs when the arrangement of the bonds prevents free rotation around the axis through the molecule. This **restricted rotation** occurs in alkenes, and when the two atoms attached to the carbon atoms of the double bond are different we get geometrical isomerism. Consider the structures of but-1-ene and but-2-ene (Figure 20.51).

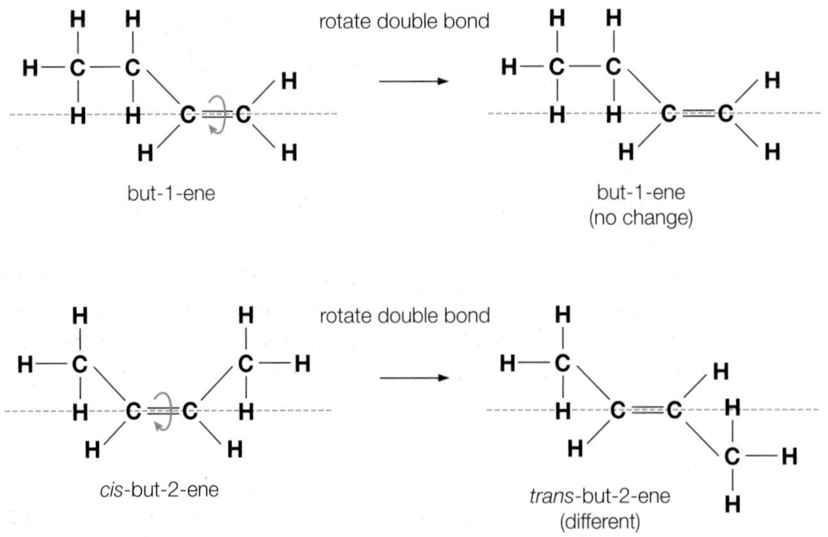

Figure 20.51 The different spatial arrangements possible for but-1-ene and but-2-ene

But-1-ene and but-2-ene are structural isomers because the double bond is in a different place in the hydrocarbon chain. If we were to rotate the double bond in but-1-ene then the outcome would be the same: the two hydrogen atoms on the right-hand carbon atom simply exchange places. But-1-ene, therefore, does not show geometrical isomerism; there is only one structure for but-1-ene.

However, if we were to rotate the double bond in but-2-ene, the resulting structure is different from the original. In the original structure the two methyl groups are on the same, upper, side of the molecule. After rotation the methyl groups are on opposite sides of the axis through the carbon–carbon double bond. Clearly the distance between the two carbon atoms in the methyl groups is different in the two molecules. When the methyl groups are on the same side of the molecule, the compound is known as the *cis* isomer. Conversely, and when they are on opposite sides (across the molecule), the compound is known as the *trans* isomer.

The *cis* and *trans* isomers of but-2-ene exist because there *cannot* be rotation around the carbon–carbon double bond. The carbon atoms involved in the double bond are sp^2 hybridized. One of the bonds between the two carbon atoms is a sigma (σ) bond formed by the end-on overlap of two sp^2 hybrid orbitals. The other bond is a pi (π) bond formed by the sideways overlap of the unused 2p atomic orbitals on each carbon atom (Chapter 14). To achieve this overlap the 2p orbitals must be in the same plane and, once formed, any rotation around the axis through the carbon atoms would mean the breaking of the pi (π) bond (Figure 20.52). As this would require a considerable amount of energy the *cis* and *trans* isomers exist independently and are not easily interconvertible. The high activation energy for the conversion between the two isomeric forms is generally prohibitive except at very high temperatures. The use of molecular models that represent the formation of sigma (σ) and pi (π) bonds illustrates clearly the restriction placed on rotation in alkene-based molecules; 1,2-dichloroethene, for example (Figure 20.53).

both p orbitals must be in the
same plane to overlap and merge

π bond above and below
plane of molecule

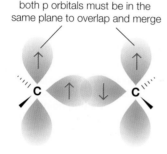

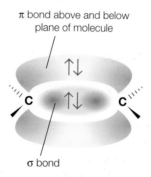

σ bond

Figure 20.52 The nature of the carbon–carbon double bond in an alkene prevents rotation about the carbon–carbon axis. Single bonds are sigma (σ) only, and so rotation can take place without bond breaking (except in cyclic compounds)

Figure 20.53 Molecular models of *cis*- and *trans*-1,2-dichloroethene showing how the presence of the pi (π) bond restricts rotation around the carbon–carbon axis. Single bonds are sigma (σ) only, and so rotation can take place without bond breaking (except in cyclic compounds)

20.6.3 Describe and **explain** geometrical isomerism in C$_3$ and C$_4$ cycloalkanes.

One further situation where rotation about a carbon–carbon axis is restricted occurs in cyclic alkanes. Geometrical isomerism is possible here, even though there are only single bonds between the carbon atoms. The rigid structure of the ring prevents free rotation. Thus, for example, 1,2-dichlorocyclopropane exists as *cis* and *trans* isomers (Figure 20.54).

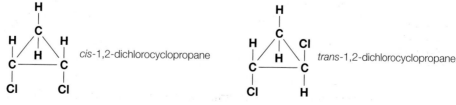

Figure 20.54 Simplified representations of the *cis* and *trans* isomers of 1,2-dichlorocyclopropane

cis-1,2-dichlorocyclopropane

trans-1,2-dichlorocyclopropane

For more complex structures the situation can become quite complicated. Thus there are four isomers of dichlorocyclobutane. Firstly there are two structural isomers of this compound and they are 1,2-dichlorocyclobutane and 1,3-dichlorocyclobutane (Figure 20.55). Each of these structural isomers then has two geometric isomers (Figure 20.55). The *cis* and *trans* isomers of

1,3-dichlorocyclobutane are particularly intriguing as the chlorine atoms involved are not on adjacent carbon atoms. Consideration of these isomers reminds us of how the large number of organic compounds so readily arises.

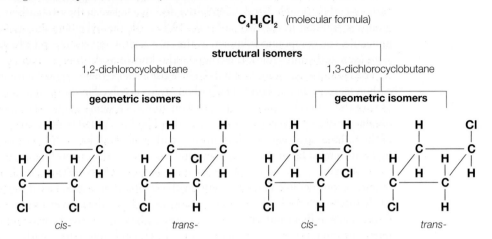

Figure 20.55 The geometric isomers of dichlorocyclobutane

Differences in properties between geometrical isomers

20.6.4 Explain the difference in the physical and chemical properties of geometrical isomers.

A consideration of the melting points and boiling points of the geometric isomers of various compounds illustrates some interesting aspects of the influence of intermolecular forces on these physical properties. Table 20.4 gives the melting and boiling points of two pairs of geometric isomers.

Isomer	Condensed structure	Melting point/K	Boiling point/K
cis-but-2-ene	$CH_3CH{=}CHCH_3$	134	277
trans-but-2-ene	$CH_3CH{=}CHCH_3$	167	274
cis-1,2-dichloroethene	$CHCl{=}CHCl$	193	333
trans-1,2-dichloroethene	$CHCl{=}CHCl$	223	321

Table 20.4 The melting and boiling points of the *cis* and *trans* isomers of but-2-ene and 1,2-dichloroethene

You will note from these figures that in each case:

- the *trans* isomer has the higher melting point, whereas
- the *cis* isomer has the higher boiling point.

As illustration of this, *cis*-1,2-dichloroethene boils at 333 K, which is higher than the boiling point of *trans*-1,2-dichloroethene (321 K). This suggests that there must be stronger intermolecular forces between the molecules of the *cis* isomer than between the *trans* isomer molecules. Both of these isomers have exactly the same atoms bonded together in the same order, which means that the van der Waals' forces between the molecules will be identical. However, where the substituent groups in the isomers contain highly electronegative atoms the geometric isomers can differ significantly in their polarity. *Cis*-1,2-dichloroethene has a dipole moment since the two polar C–Cl bonds are on the same side of the molecule and their effects do not cancel out. There will, therefore, be dipole–dipole attractions between molecules of *cis*-1,2-dichloroethene.

In contrast, *trans*-1,2-dichloroethene is non-polar because the two polar C–Cl bonds are arranged symmetrically on either side of the carbon–carbon double bond. The effects of the two polarized bonds cancel each other in this isomer and the overall molecule is non-polar. The lower boiling point of the *trans* isomer results from this difference in polarity between the two isomers.

In the case of the boiling points of *cis*- and *trans*-but-2-ene a similar argument holds, although the degree of the effect is slighter. The methyl groups in but-2-ene, like other alkyl groups, are electron-releasing. This polarizes the bond to the carbon atoms at either end of the double bond.

The same argument then applies: while *cis*-but-2-ene is a polar molecule, *trans*-but-2-ene is not. Hence the boiling point of the *cis* isomer is higher than that of the *trans* isomer.

The fact that the same pattern is not observed with the melting points indicates that another factor has an effect in this case. When molecules assemble into the solid state they must form a lattice in which the molecules are packed closely together. The shape of the *trans* isomer of a compound means that they pack together better than the *cis* isomer. The poorer packing of the *cis* isomer means that the intermolecular forces are not as effective as they could be, and so the melting point of the *cis* isomer is lower than that of the *trans* isomer.

In some cases there are very marked differences in melting point between two geometric isomers. This is often due to the nature of the functional groups attached to the carbon–carbon double bond and their proximity to each other in the *cis* form. The melting points of *cis*-but-2-ene-1,4-dioic acid and *trans*-but-2-ene-1,4-dioic acid are very different indeed, one being much higher than the other. In the *trans* isomer there is strong intermolecular hydrogen bonding between different molecules, because of the polarity of the carboxylic acid groups. However, in the *cis* isomer these groups are adjacent to each other and much of this hydrogen bonding occurs internally between the two carboxylic acid groups (intramolecular hydrogen bonding). Consequently the intermolecular forces between molecules of the *cis* form are significantly weaker in this case, and so the melting point of this isomer is much lower (Figure 20.56).

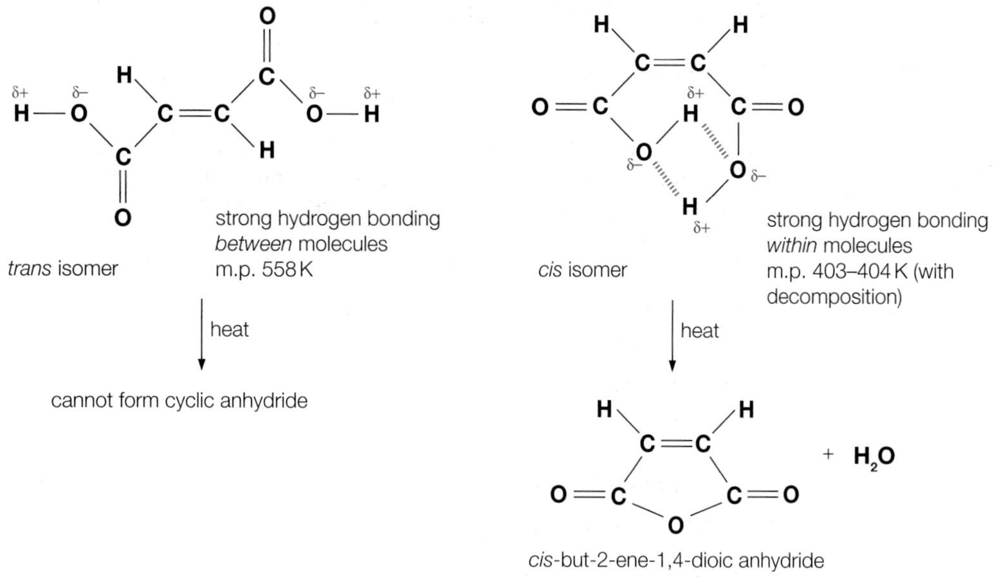

Figure 20.56 The structural difference between the geometric isomers of but-2-ene-1,4-dioic acid, and the consequences for the melting points and chemistry of the isomers

These two geometric isomers of but-2-ene-1,4-dioic acid are also very different chemically. In the *cis* isomer the two carboxylic acid functional groups are close enough together to react. When this isomer is heated a water molecule is eliminated between the two carboxylic acid groups and a cyclic acid anhydride is formed (Figure 20.56). In the *trans* isomer the two groups are distant from each other and no such reaction is possible.

■ Extension: Acidity and geometrical isomerism

Butene-2,3-dioic acid contains two carboxylic acid functional groups and undergoes stepwise ionization according to the following equation:

$$C_2H_2(COOH)_2(aq) \rightleftharpoons C_2H_2(COOH)(COO^-) + H^+(aq) \rightleftharpoons C_2H_2(COO^-)(COO^-) + H^+(aq)$$

The *cis* isomer of butene-2,3-dioic acid is a much stronger acid than the *trans* isomer. The first pK_a of *cis*-butene-2,3-dioic acid is 1.92 whereas the first pK_a of *trans*-butenedioic acid is 3.02 (Chapter 18). This is because the —COO$^-$ group formed in the *cis* isomer can be stabilized by hydrogen bonding from the adjacent —COOH group, which is on the same side of the double bond. This stabilization is not possible in the *trans* isomer because the two groups are too far apart.

■ Extension: The molecular conversion that helps us see

The activation energy needed to convert one geometric isomer to another is generally high enough to prevent interconversion. However, there is one crucially important case where this can be achieved with a photon of light! It relates to the mechanism by which we see.

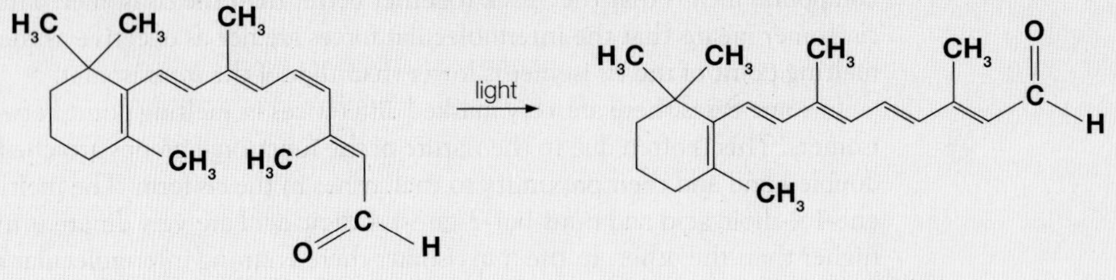

Figure 20.57 The conversion of *cis*-retinal to *trans*-retinal by light

The light-sensitive cells of the retina, the inner lining at the back of the human eye, contain a molecule known as retinal. When light falls on the retina it converts molecules of retinal from the *cis* form to the *trans* form (Figure 20.57).

Light energy promotes this photochemical reaction. The energy is sufficient to cause the homolytic fission of the pi (π) bond to form a free radical. Rotation can then take place around the remaining single sigma (σ) bond and the pi (π) bond is then re-formed, converting one geometric isomer into the other. The energy released by the re-formation of the pi (π) bond causes a nerve impulse to be transmitted, sending a signal to the brain via the optic nerve (Figure 20.58). The *trans*-retinal is then converted back to *cis*-retinal by an enzyme, RPE isomerohydrolase, so that it can function again in response to a light stimulus.

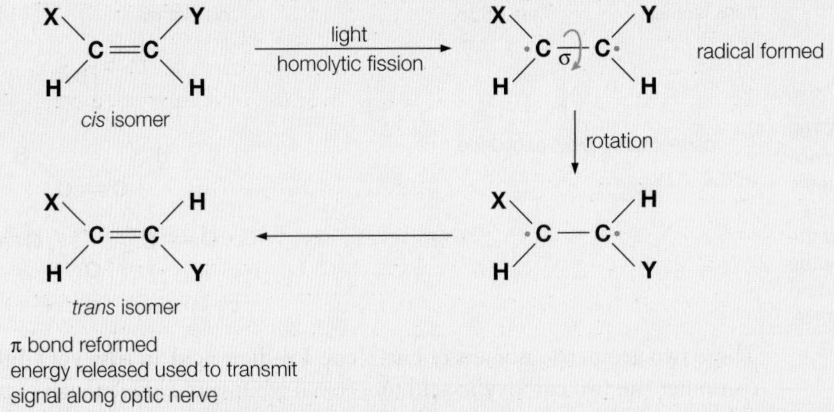

Figure 20.58 Detail of the reaction involved in the conversion of *cis*-retinal to *trans*-retinal by light

■ Extension: The CIP method

Sometimes, in the case of more complicated alkenes, deciding which isomer to label as the *cis* isomer and which the *trans* isomer can be difficult. Three chemists, Cahn, Ingold and Prelog (Chapter 26), devised a method for deciding which group takes priority from the two (or more) functional groups attached to the carbons of the double bond. Their method, sometimes called the CIP method, is based on atomic number priorities, and is as follows. For each group attached to the double bond, apply the following rules, in order:

   Rule 1: If one of the atoms joined to the carbon–carbon double bond has a higher atomic number than the other one, then that atom has the higher priority.

   Rule 2: If both the atoms joined to the carbon–carbon double bond have the same atomic number, the next atom(s) in the groups are compared.

Once the order of priority of the various groups on the double bond has been decided, the letters *E* or *Z* can be assigned, depending on whether the groups of higher priority on each end of the carbon–carbon double bond are on opposite sides (*E*) or the same side (*Z*) of the molecule.

a For the following geometric isomers of but-2-ene, make sure you understand why they have been designated *E* or *Z*.

H₃C, H / H, CH₃

H₃C, CH₃ / H, H

E-but-2-ene
(*trans*-but-2-ene)

Z-but-2-ene
(*cis*-but-2-ene)

b For the following four molecules, make sure you understand why they have been labelled as the *E* or *Z* isomer.

i   ii   iii   iv

i: H₃C, CH₃ / H₃C, CH₃ — *E*

ii: H₃C, CH₂Cl / H₃C, CH₃ — *Z*

iii: H₃C, CH₂Cl / H, HO — *Z*

iv: H₃C, CHO / Cl, HO — *E*

Language of Chemistry

In this chapter we have now come across two systematic terms for naming geometric isomers. The first involves using the terms *cis-* and *trans-* as prefixes. These terms are derived from the Latin: *cis* meaning 'on this side' and *trans* meaning 'across'.

The *E*/*Z* system is most useful for complex alkenes and the letters come from German: *E* stands for *Entgegen*, meaning 'opposite', while *Z* stands for *Zusammen*, meaning 'together'. ■

■ Extension: Conformational isomers

We have seen how important rotation about the axis along a hydrocarbon molecule can be; and indeed the significance of the energy barrier to any possible rotation. Usually there is free rotation around a carbon–carbon single bond, but if the two carbon atoms that are joined by the bond have bulky groups attached to them, not every position of rotation will have the same potential energy. Conformational isomers are the different arrangements that correspond to energy minima as one atom (with its attached groups) is rotated with respect to the other.

Normally, the energy barriers are low enough for molecules to have enough kinetic energy at room temperature to overcome them, and hence one conformer (conformational isomer) easily changes into another. However, if the energy barriers to rotation are high enough, because of very large bulky groups, it is possible to observe the different conformers separately. Different conformations will quite often interact with drug receptors in different ways, or at different rates (Chapter 24). They can also be modelled in computer-based simulations.

The simplest molecule to illustrate conformational isomerism is ethane. If we rotate one methyl group around the carbon–carbon single bond axis, keeping the other fixed, and imagine looking along the axis of the bond, there will be times when the hydrogen atoms on the front carbon atom obscure ('eclipse') the atoms on the far carbon, and times when they do not. The two extremes are, on the one hand, total eclipsing, and on the other hand, the situation where the front hydrogens are exactly in between the back hydrogens, which is the 'staggered' conformation (Figure 20.59).

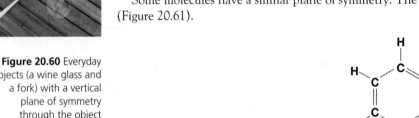

high potential energy   low potential energy

Figure 20.59
The different
conformations of
ethane

displayed view   looking along
the bond

The eclipsed conformer of ethane

displayed view   looking along
the bond

The staggered conformer of ethane

Optical isomerism

20.6.5 **Describe** and explain optical isomerism in simple organic molecules.

Optical isomerism arises from the inherent asymmetry of the molecules that make up each isomer. If you look at the everyday objects pictured in Figure 20.60 you see that these are objects that have a vertical plane of symmetry down through the centre of the object.

Some molecules have a similar plane of symmetry. The benzene molecule is one such example (Figure 20.61).

Figure 20.60 Everyday
objects (a wine glass and
a fork) with a vertical
plane of symmetry
through the object
concerned

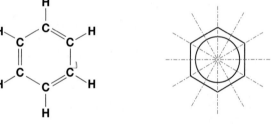

Figure 20.61 Benzene is a molecule with six
planes of symmetry

However, other molecules have no plane or centre of symmetry and they are referred to as **chiral** (pronounced 'kai-ral', from the Greek word meaning 'handed') molecules. The most common cases of chirality occur when a molecule contains a tetrahedral carbon atom surrounded by four different atoms or groups of atoms.

Figure 20.62 shows two three-dimensional models of such a carbon atom. The first model has four different groups around the central carbon atom, and the other model has just two of the groups interchanged. The two models, as you can see, are mirror image forms of each other. The carbon atom at the centre of these models is known as an **asymmetric** carbon atom, and molecules containing an asymmetric carbon atom are said to be chiral molecules. The term 'chiral' is most correctly applied to the whole molecule. However, its use has been extended to refer to the asymmetric carbon atom itself – and so that carbon atom is sometimes referred to as a chiral carbon atom, or a chiral centre.

No matter how we try, it is impossible to superimpose one model upon the other without breaking and remaking bonds. These molecules are said to be non-superimposable. The only way in which you can superimpose them is by reflection in a mirror: they are mirror images of each other. These two mirror images are known as **enantiomers.** Figure 20.63 shows representations on paper of the enantiomers of butan-2-ol and 2-bromobutane. The asymmetric carbon atom in each molecule is indicated here using an asterisk. We saw in Chapter 10 how three-dimensional structures can be represented in two dimensions by using a dotted line for bonds going behind the plane of the paper and wedge-shaped lines for bonds coming out from the plane of the paper.

Figure 20.62 Models of
two tetrahedral carbon-
containing molecules
with four different groups
around the central carbon
atom

Enantiomers are also referred to as **optical isomers** because different enantiomers rotate the plane of polarized light in equal but opposite directions. Optical isomerism is the second form of stereoisomerism. In these molecules the groups in the optical isomers are the same; they differ in their arrangement in space, in this case around the central carbon atom.

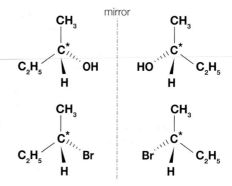

Figure 20.63 The enantiomers of butan-2-ol and 2-bromobutane

Language of Chemistry

Chiral is derived from the Greek word *kheir*, which means 'hand'. The mirror image of the left hand is the right hand, and it cannot be superimposed on the left hand. If molecules (enantiomers) cannot be superimposed on their mirror images, they are showing 'handedness'. They are behaving rather like left and right hands – hence the word 'chiral'. The word 'enantiomer' is derived from the Greek *enantion* meaning 'opposite'.

There is a simple experiment with a plane mirror to show that your hands are chiral. Hold your left hand, outstretched and palm away from you, in front of a mirror. Now hold your right hand, outstretched and palm towards you, next to the mirror. The image of your left hand now looks identical to (is superimposable on) your real right hand (Figure 20.64)! ∎

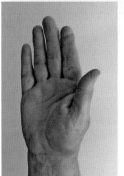

Figure 20.64
Photographs of the 'hand' experiment demonstrating chirality

History of Chemistry

The French chemist **Louis Pasteur** (1822–1895) started his career studying the various isomers of tartaric acid (2,3-dihydroxybutanedioic acid) isolated from the deposits from wine fermentation. Previously it had been found that there were two isomers: one, called racemic acid, had no effect on polarized light. The other, called tartaric acid, rotated polarized light to the right. Chemically, however, the two acids had identical reactions.

Pasteur carefully examined crystals of the sodium-ammonium salt of racemic acid that he had prepared by allowing a saturated solution to evaporate slowly. He noticed there were two types of crystal that were mirror images of each other. Painstakingly, using a hand lens and a pair of tweezers, he separated the crystals into two piles. He then dissolved each sample of crystals in water and examined them in a polarimeter. He found that one pile of crystals rotated the polarized light to the right (just as the same salt of tartaric acid had), but the other rotated the light to the left, to an equal extent (Figure 20.65). He had become the first person to separate the enantiomers from a racemic mixture.

Only later did it become apparent how lucky he had been. Above the transition temperature of 22 °C the racemic salt crystallizes as symmetrical crystals containing equal amounts of both enantiomers. If he had recrystallized the salt in the usual way, by cooling a hot saturated solution, he would never have obtained the mixture of crystals.

Pasteur's remarkably patient isolation of mirror-image crystalline forms of an optically active compound serves to emphasize that chirality is first and foremost a property of the whole molecule. We have since developed that view and applied the language to a specific chiral carbon centre in a molecule.

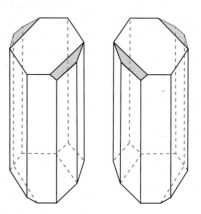

Figure 20.65 Mirror image forms of sodium ammonium tartrate crystals

The defining physical property that distinguishes the optical isomers of a compound from each other is that they rotate the plane of **plane-polarized light** in equal but opposite directions. Normal light is electromagnetic radiation that is oscillating in every plane (Figure 20.66). Light passing through a polarizing filter only oscillates in one particular plane and is referred to as plane-polarized light. The filter shown in Figure 20.66 only allows through light waves oscillating in a vertical plane.

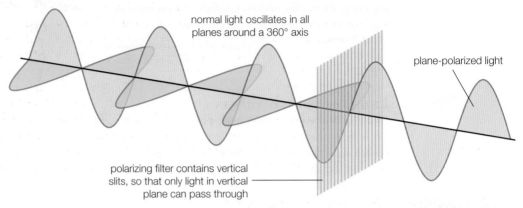

normal light oscillates in all
planes around a 360° axis

plane-polarized light

Figure 20.66 A representation of how a polarizing filter works

polarizing filter contains vertical slits, so that only light in vertical plane can pass through

The polarizing filter acts as if it contains a series of narrow and closely spaced vertical slits, so that only the light oscillating in the same vertical plane can pass through it.

Our eyes are unable to distinguish between normal light and plane-polarized light. However, if another polarizing filter is placed in the path of plane-polarized light, rotated at right-angles relative to the first filter, it will completely block the passage of all light (Figure 20.67). Polarizing filters are used in sunglasses and in camera lenses to cut down glare and reflections.

The ability of enantiomers to rotate the plane of plane-polarized light can be shown in practice by using a polarimeter. This consists of a light source, two polarizing lenses, and a tube to hold the sample of the enantiomer located between the lenses.

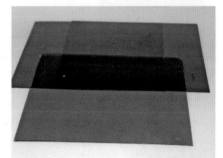

Figure 20.67 If two pieces of Polaroid are laid at right-angles to each other, the light is cut out. This is what is happening at the overlap of the filters here

20.6.6 **Outline** the use of a polarimeter in distinguishing between optical isomers.

A **polarimeter** is an instrument used to distinguish optical isomers. The device consists of six parts, shown in Figure 20.68. Light from a monochromatic light source (which produces light of a single wavelength) passes through a slit to produce a thin beam of light. This beam of light then passes through a Polaroid filter.

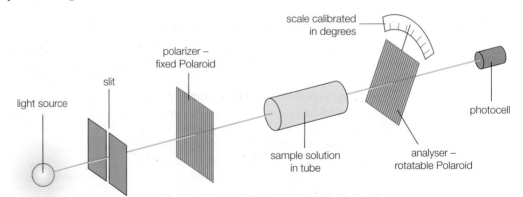

scale calibrated in degrees

polarizer – fixed Polaroid

slit

light source

photocell

sample solution in tube

analyser – rotatable Polaroid

Figure 20.68 The various stages of a polarimeter for analysing optical isomers

The polarized beam now enters and passes through the sample solution being tested. After the sample the beam then passes through another piece of Polaroid, the analyser, and then into the photocell, which produces an electric current proportional to the intensity of light that falls upon it.

The instrument is initially calibrated using pure solvent. The output from the photocell is set to a minimum, at which point the analyser filter is at right angles to the polarizer filter. The angle on the scale is recorded. One of the samples is then placed in the sample tube and the analyser is rotated until the output from the photocell is again a minimum. The new angle on the scale is recorded. The rotation caused by the compound is the difference between the two measured angles.

A sample of the other enantiomer is then analysed in a similar way using a solution of the same concentration. In one case the analyser will have been rotated clockwise. This means that the enantiomer is dextrorotatory (from the Latin *dexter*, meaning 'right'). The other enantiomer will rotate the plane of the light anticlockwise and is said to be laevorotatory (from the Latin *laevus*, meaning 'left'). The two enantiomers rotate the plane of plane-polarized light by the same amount, but in opposite directions. One of the enantiomers is thus known as the D-form and the other as the L-form. If both enantiomers are present in equal amounts the two rotations cancel each other out, and the mixture appears to be optically inactive. Such a mixture is known as a racemic mixture.

■ Extension: Naming enantiomers

There is no general way of predicting which of a pair of enantiomers will rotate the plane of polarized light in a particular direction, and there is no relationship between the configuration of a molecule and the direction or extent of rotation. Various systems for naming the two isomers have been proposed, but it would be useful to have a system that was generally applicable and could be consistently applied. The most widely applicable method of describing the configurations of enantiomers is the R–S convention based on the priority rules of Cahn, Ingold and Prelog (Chapter 26).

One group of compounds that exhibit optical isomerism is the 2-amino acids that are used in cells to synthesize proteins. Of the 20 amino acids utilized for this purpose all but glycine (2-aminoethanoic acid) are chiral molecules. It is interesting to realize that it is the L-form of all these amino acids that is exclusively used by the protein-synthesizing machinery of the cell. Alanine (2-aminopropanoic acid) illustrates the optical isomerism shown by the 2-amino acids (Figure 20.69).

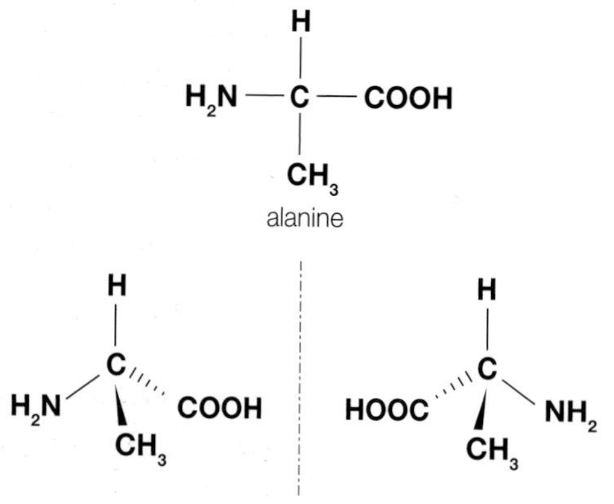

Figure 20.69 Alanine exhibits optical isomerism – the two enantiomers rotate the plane of polarized light in opposite directions

■ Extension: Chiral molecules without chiral centres

Chirality is a property of the entire molecule and there are molecules that are chiral but do not contain a chiral centre. For example, 1,3-dimethylallene is a chiral molecule since it has no plane of symmetry, and can exist as a pair of mirror images (Figure 20.70). The presence of two rigid π bonds prevents free rotation around the three central carbon atoms.

Figure 20.70 The enantiomers of 1,3-dimethylallene

History of Chemistry

Paul Walden (1863–1957) was a Latvian-born chemist who, after studying in Riga and St. Petersburg, moved to Germany and gained his PhD working with Ostwald in Leipzig. He is noted for his work on the 'Walden inversion', which gave insight into the mechanism of nucleophilic substitution reactions, and the invention of the first example of an ionic liquid, ethylammonium nitrate.

TOK Link

The ability to use reasoning based on physical evidence and to link that to our mental models of the microscopic 'world' is an essential part of scientific thinking. Modern scanning tunnel microscopy has recently allowed us to gain a 'view' of atoms, and X-ray crystallography enables us to deduce the distances between atoms, yet we cannot ultimately *prove* the shapes of simple molecules or the mechanisms of organic reactions. The existence of optical isomers does provide good evidence to support our picture of chemistry at the molecular level. The ability to use optical isomerism and its effects on polarized light has provided us with physical evidence that these do represent the true shapes of these molecules and that our ideas regarding reaction mechanism are along the right lines.

Important in this was the application of polarimetry to studies on the S_N2 mechanism for nucleophilic substitution. This was first demonstrated by the Latvian chemist, Paul Walden. Walden reasoned that if he could find an optically active halogenoalkane that underwent nucleophilic substitution by an S_N2 mechanism, then the product should rotate the plane of light in the opposite direction to the starting material because the reaction mechanism suggests that an 'inversion' of the tetrahedral arrangement around the target carbon atom should occur (Figure 20.71).

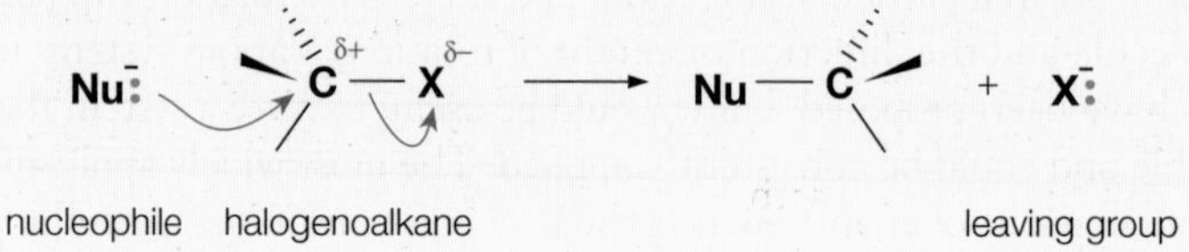

Figure 20.71 The S_N2 mechanism suggests that inversion of the tetrahedral molecule should occur – the umbrella should blow open the wrong way!

In the S_N2 mechanism described in Figure 20.71, the nucleophile is shown approaching the halogenoalkane from the side of the molecule *opposite* the halogen atom – this is known as backside attack (Chapter 6). The nucleophile could also have been shown to be approaching the halogen from the same side. However, experiments strongly support the idea of backside attack. The halogenoalkane used in these experiments needs to have a chiral centre. The secondary halogenoalkane 2-bromooctane is suitable since it has a chiral centre and can exist in two optically active forms (enantiomers) (Figure 20.72).

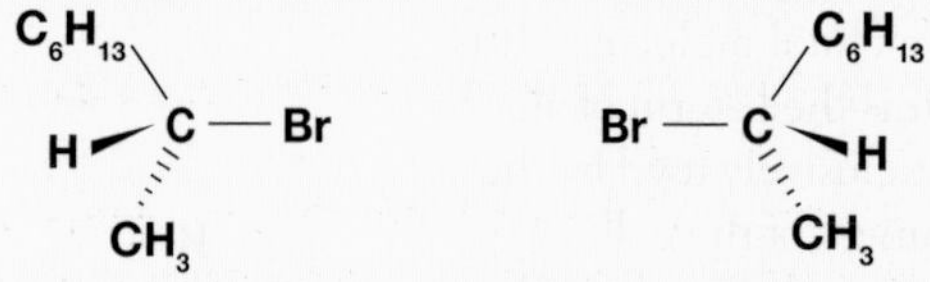

Figure 20.72 The enantiomers of 2-bromooctane

Hydrolysis experiments show that each enantiomer gives the alcohol, octan-2-ol, of *opposite* configuration, showing that substitution involves *inversion* of configuration at the carbon atom that was bonded to the halogen. This observation is consistent with the approach of the hydroxide ion (nucleophile) from the opposite side of the molecule to the halogen atom (Figure 20.73).

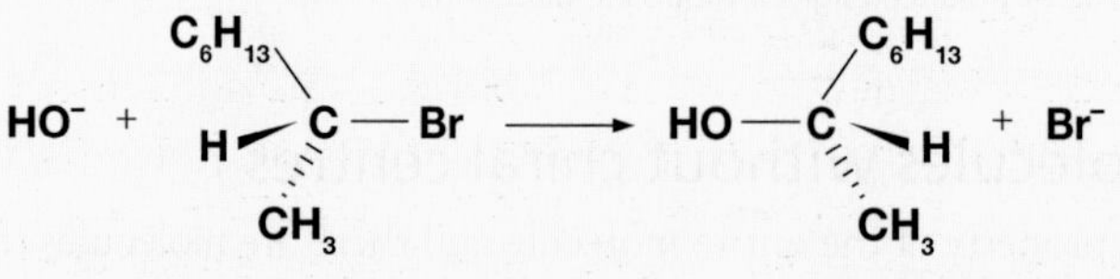

Figure 20.73 The inversion of 2-bromooctane during nucleophilic substitution (sometimes referred to as a Walden inversion)

There is also strong stereochemical evidence in support of the S_N1 mechanism described in Figure 20.74. It has been found that when an optically active sample of a chiral tertiary halogenoalkane undergoes hydrolysis all of the optical activity is lost. This observation is consistent with the suggested mechanism. VSEPR theory (Chapter 4) predicts that the three bonds from the positively charged carbon are in the form of a trigonal planar arrangement (Figure 20.74).

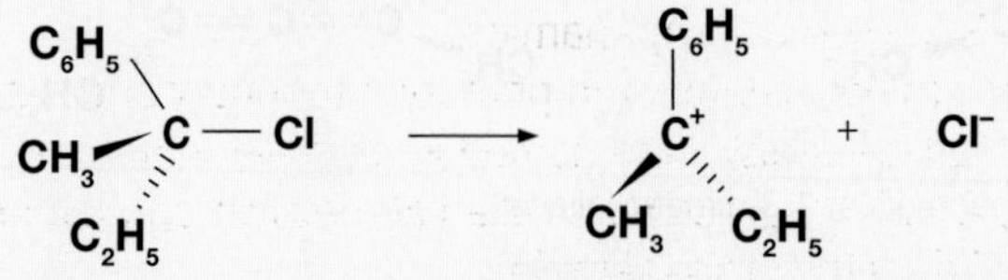

Figure 20.74 The formation of a trigonal planar carbocation intermediate from an optically active halogenoalkane

The planar nature of the intermediate means that when the carbocation from an optically active halogenoalkane is attacked by a nucleophile from *either side* (with equal probability) a mixture of the two enantiomers of the alcohol is obtained (Figure 20.75). This gives a 50:50 racemic mixture of the two optical isomers. The planar shape of carbocations has been confirmed experimentally by X-ray diffraction (Chapter 4).

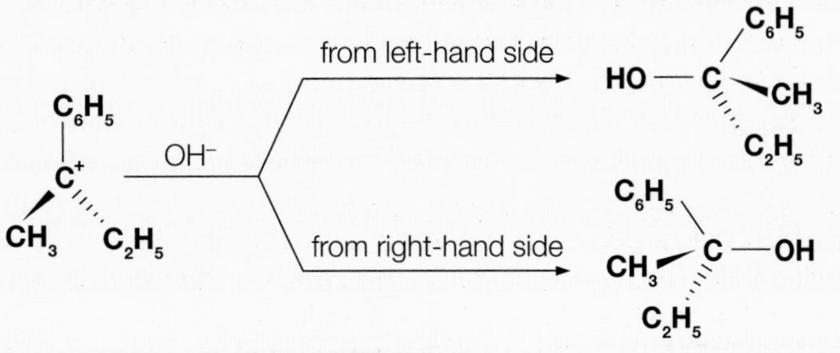

Figure 20.75 The racemization of an optically active halogenoalkane during nucleophilic substitution (S_N1)

Properties of optical isomers

20.6.7 Compare the physical and chemical properties of enantiomers.

The optical isomers of a compound differ in only one defining physical property: they rotate the plane of plane-polarized light in opposite directions. All their other physical properties, such as density and melting point, are identical. Not only are the physical properties of the two enantiomers identical: their general chemical properties are also identical. The exception to this is when they interact with biological sensors. Thus, in the body, the different enantiomers can have completely different physiological effects. For example, one of the enantiomers of the amino acid asparagine (H_2N–$CH(CH_2CONH_2)$–$COOH$) tastes bitter, whereas the other optical isomer tastes sweet. This relates to the fact that biological sensors, and indeed many biochemical reactions, are sensitive to molecular shape.

SUMMARY OF KNOWLEDGE

- Three homologous series involving nitrogen-containing functional groups exist: the amines, the amides and the nitriles.
- The functional groups present in amines are –NH_2 in primary amines, >NH in secondary amines and >N– in tertiary amines.
- The IUPAC naming convention allows two methods for naming amines. In the first of these the prefix *amino-* is used, followed by the name of the hydrocarbon present; hence, 1-aminopropane, $CH_3CH_2CH_2NH_2$ and 2-aminopropane, $CH_3CH(NH_2)CH_3$.
- An alternative way of considering the structure of amines is to view them as substituted forms of ammonia. In a primary amine just one hydrogen atom is replaced by an alkyl group; in a secondary amine two are replaced; while in a tertiary amine all three hydrogen atoms are replaced by alkyl groups.
- The second IUPAC method of naming amines is based on this view of them as substituted ammonia. The suffix *–amine* is added on to the name of the longest hydrocarbon chain in the molecule: hence, ethanamine, $CH_3CH_2NH_2$. This view leads to secondary and tertiary amines having names based on the alkyl groups being substituted at the nitrogen (N) atom. Thus, $CH_3CH_2NHCH_2CH_3$ is known as *N*-ethylethanamine and $CH_3CH_2N(CH_2CH_3)_2$ is N,N-diethylethanamine. IUPAC also allows for the name of the alkyl group to be used instead of that of the hydrocarbon; so *N,N*-diethylethanamine and *N,N*-diethylethylamine are both acceptable.
- Amides contain the –$CONH_2$ group and the suffix *–amide* is added to their names. Hence $CH_3CH_2CONH_2$ is known as propanamide.
- Amides, like the amines, can exist in primary, secondary and tertiary forms.
- Secondary and tertiary amides are named in a similar way to such amines by using the N notation. Thus, $CH_3CH_2CONHCH_3$ is called *N*-methylpropanamide and $CH_3CH_2CON(CH_3)_2$ is known as *N,N*-dimethylpropanamide.

- Members of the homologous series of nitriles contain the nitrile group, $-C\equiv N$.
- In the IUPAC system the suffix *–nitrile* is added to the name for the longest chain of carbon atoms – remembering to include the carbon atom of the nitrile group in the count. Thus, $CH_3CH_2CH_2CN$ is butanenitrile.
- Halogenoalkanes undergo nucleophilic substitution reactions. For example, bromoalkanes react with aqueous sodium hydroxide solution to produce alcohols.
- Two mechanisms (S_N1 and S_N2) have been suggested for this type of reaction .
- The S_N1 mechanism, involving the initial heterolytic fission of the carbon–bromine bond in the halogenoalkane, is favoured by tertiary bromoalkanes.
- The S_N2 mechanism, involving a single step reaction (via a high-energy transition state) in which the carbon–bromine bond is broken simultaneously with the attack of the nucleophile (the hydroxide ion in this case) on the target, electron-deficient, carbon atom, is favoured by primary halogenoalkanes.
- The general equation for an S_N2 reaction is:

- The nucleophilic substitution reactions between primary halogenoalkanes and ammonia and potassium cyanide produce useful products, amines and nitriles respectively.
  $R-Hal + H-NH_2 \rightarrow H-Hal + R-NH_2$ (an amine)
  $R-Hal + CN^- \rightarrow Hal^- + R-CN$ (a nitrile)
- Nitriles are useful intermediates in synthetic reaction pathways as they can be reduced using hydrogen in the presence of a heated nickel catalyst to yield amines.
- If the conditions of the substitution reaction between a bromoalkane and sodium hydroxide are changed then a different reaction, known as elimination, takes place.
- The conditions that favour the elimination reaction with bromoalkanes are a high temperature and the use of an ethanolic solution of sodium hydroxide. In these conditions the hydroxide ion acts as a base rather than a nucleophile.
- The product of the elimination of hydrogen bromide from a bromoalkane is an alkene. For example:
  $CH_3CH_2CH_2Br + OH^- \rightarrow CH_3CH=CH_2 + H_2O + Br^-$
- Two mechanisms (E1 and E2) have been proposed for these elimination reactions.
- The E1 mechanism, involving the initial heterolytic fission of the carbon–bromine bond in the halogenoalkane, is favoured by tertiary bromoalkanes.
- The E2 mechanism, involving the simultaneous removal of an H^+ ion and a Br^- ion from the bromoalkane molecule so that a carbon–carbon double bond is formed, is favoured by primary bromoalkanes.
- Condensation reactions are reactions in which two molecules join together with the elimination of a simple molecule such as water between them.
- The condensation reaction between an alcohol and a carboxylic acid produces an ester.
  $RCOOH + HOR' \rightarrow RCOOR' + H_2O$
- Using suitable bifunctional monomers, such as a diol and a dicarboxylic acid, it is possible to produce a condensation polymer where the monomers are linked by ester groups to form a polyester.
- An example of such a polyester is Terylene/Dacron.
- The condensation reaction between an amine and a carboxylic acid produces a secondary amide.
- If appropriate monomers are used then a polymer linked by amide groups can be produced – such polymers are known as polyamides.
- Proteins are naturally occurring polyamides formed from amino acids.
- Artificial polyamides such as nylon can be produced using appropriate bifunctional monomers.

- The additional functional groups and possible reactions create further useful synthetic pathways for producing useful organic products.
- Stereoisomerism is a form of isomerism where the same atoms and functional groups are present in the molecules, but the atoms are arranged differently in space.
- There are two types of stereoisomerism: geometrical isomerism and optical isomerism.
- Geometrical isomerism gives rise to *cis* and *trans* forms of the compound concerned and occurs most obviously in certain alkenes where the carbon–carbon double bond prevents rotation about the axis of the bond. Consequently the substituent groups are held in specific positions. This type of restriction can also be caused if the molecule has a cyclic structure.
- Optical isomerism arises for chiral molecules. Such molecules usually contain a chiral centre – most commonly a carbon atom with four different substituent groups in a tetrahedral structure.
- Optical isomers (enantiomers) are mirror images of each other and they rotate the plane of plane-polarized light in opposite but equal directions (under the same conditions).
- An instrument known as a polarimeter is used to study and detect optical isomers.
- Some very significant molecules exist as optical isomers – most notably all but one of the 2-amino acids that make up proteins.

■ *Examination questions – a selection*

Paper 1 IB questions and IB style questions

Q1 Which is correct about the order and rate of hydrolysis (relative to 1-bromobutane) for 2-bromo-2-methylpropane $((CH_3)_3CBr)$?

	Order	Rate relative to 1-bromobutane
A	first	faster
B	second	faster
C	first	slower
D	second	slower

Q2 Which one of the following can exist as two optical isomers (enantiomers)?
A 1-chlorobutane
B 2-chloropropane
C 2-chloro-2-methyl propane
D 2-chlorobutane

Q3 Which formula represents an amide?
A $CH_3(CH_2)_2NH_2$
B $CH_3CH_2N(CH_3)_2$
C H_2NCH_2COOH
D CH_3CONH_2

Q4 What is the correct order of reaction types in the following sequence?

$$C_2H_5Br \xrightarrow{\text{I}} C_2H_5OH \xrightarrow{\text{II}} CH_3COOH \xrightarrow{\text{III}} CH_3COOCH_3$$

	I	II	III
A	substitution	oxidation	esterification
B	addition	substitution	substitution
C	substitution	oxidation	substitution
D	oxidation	substitution	addition

Q5 Which of the following products could be formed from the oxidation of ethanol?
I ethanal
II ethanoic acid
III ethene
A I and II only
B I and III only
C I, II and III
D II and III only

Q6 Which group of compounds will polymerize (under appropriate conditions) to form proteins?
A amides
B amino acids
C amines
D alkanes

Q7 Which pair of compounds can be used (in the presence of concentrated sulfuric acid) to prepare the ester CH_3COOCH_3?
A ethanol and methanoic acid
B methanol and ethanoic acid
C ethanol and ethanoic acid
D methanol and methanoic acid

Q8 What is the composition of a racemic mixture?
A equal amounts of *cis* and *trans* isomers
B equal amounts of structural isomers
C equal amounts of a pair of enantiomers (optical isomers)
D equal amounts of structural isomers with different functional groups

Q9 Which one of the following compounds has *cis–trans* isomers?
A $H_2C=C(C_2H_5)_2$
B $BrHC=C(CH_3)_2$
C $CH_3C\equiv CH$
D $CH_3HC=CHCH_3$

Q10 What is the correct name for $CH_3CH_2COOCH_2CH_3$?
A ethyl ethanoate
B diethylcarboxylate
C ethyl propanoate
D propyl ethanoate

Q11 Which one of the following reagents reacts by the donation of a pair of electrons?
A methyl free radical
B electrophile
C methyl carbocation
D nucleophile

Q12 Which one of the following statements will **not** be true for:

$CH_2Br(CH_2)_3CH=CHCOOH$?

A It will react with cyanide ions, CN–(aq).
B It will react with bromine.
C It will be able to exist as *cis* and *trans* isomers.
D It will be able to exist as optical isomers (enantiomers).

Q13 What is the major product when 2-bromobutane is treated with a hot alcoholic solution of sodium hydroxide?
A 2-hydroxybutane
B butan-1-ol
C but-2-ene
D but-1-ene

Q14 What is the correct order of reactivities of the following halogenoalkanes for a S_N2 reaction?
A RCl = RBr = RI
B RCl > RBr > RI
C RBr > RCl > RI
D RI > RBr > RCl

Q15 Which of the following compounds could be the product of a reaction involving a nucleophilic attack by cyanide ions (in aqueous ethanolic solution)?
A C_6H_5CN
B $(CH_3)_2CHCN$
C $H_2C=CHCN$
D $CH_3CH_2CONH_2$

Q16 Which technique can be used to distinguish between optical isomers (enantiomers)?
A chemical tests
B mass spectrometry
C polarimetry using plane-polarized light
D melting point determination

Q17 What is produced when propanenitrile is reacted with hydrogen using a heated nickel catalyst?
A CH_3CONH_2
B $CH_3(CH_2)_2NH_2$
C CH_3NH_2 and CH_4
D $CH_3CH_2NH_2$

Q18 Nylon-6,6 has the repeat unit
$-CO-(CH_2)_4-CO-NH-(CH_2)_6-NH-$

It is made from hexanedioic acid and hexane-1,6-diamine. Which of the following is an incorrect statement?
A Nylon-6,6 is a condensation polymer.
B Amide linkages are formed between the monomers.
C Ammonia is eliminated during the reaction.
D Hydrogen bonding can occur between nylon-6,6 molecules.

Q19 Which one of the following substances is a polyester?
A hexane-16-diamine
B nylon-6,6
C polyethene
D Terylene

Q20 Which of the following will give an amine as the product when reacted together?
  I $(CH_3)_3Cl + NH_3$
  II $C_2H_5CN + H_2$
  III $CH_3COOH + NH_3$
A I, II and III
B I and II only
C I only
D II and III only

Paper 2 IB questions and IB style questions

Q1 a Compounds of formula C_4H_7Br exhibit both geometrical and optical isomerism.
  i Explain why C_4H_7Br shows geometrical isomerism. [1]
  ii Draw the *cis* and *trans* isomers of C_4H_7Br. [2]
  iii Draw the structural formula of C_4H_7Br that shows only one optical isomer. Show the chiral carbon atom with '*'. [2]
b Explain why 1,2-dibromocyclopropane has *cis* and *trans* isomers. Draw the structural formulas of the two isomers. [3]

Q2 One of the alcohols containing four carbon atoms exists as optical isomers.

   a Give the structural formula and name of this alcohol and explain why it exists as optical isomers. Outline how these two isomers could be distinguished. [4]

   b When concentrated phosphoric acid is added to one of these optical isomers and the mixture is warmed, two isomeric organic products are formed. Give the structural formula and name of one of these products and identify the type of reaction occurring. Outline a simple chemical test for the functional group present in this product. [4]

                  Higher Level Paper 2, Nov 05, Q4

Q3 The molecular formula C_4H_9I represents four structural isomers, all of which can undergo nucleophilic substitution reactions with aqueous sodium hydroxide. An equation to represent all these reactions is:

$$C_4H_9I + NaOH \rightarrow C_4H_9OH + NaI$$

   a Explain what is meant by the term *nucleophilic substitution*. [2]

   b The main mechanism for nucleophilic substitution in a tertiary halogenoalkane is S_N1. Give the equations for the substitution reaction of the tertiary isomer of C_4H_9I. Show the structures of the organic reactant and product and use curly arrows to show the movement of electron pairs. [4]

   c The main mechanism for a primary halogenoalkane is S_N2. Give the equations for the substitution reaction of the straight-chain primary isomer of C_4H_9I. Show the structures of the organic reactant and product and use curly arrows to show the movement of electron pairs. [4]

Q4 The molecular formula, $C_3H_4Cl_2$ represents several isomeric compounds. Some isomers are cyclic and some are unsaturated.

   a Draw the structures of two cyclic compounds that are structural isomers and state the names of both isomers. [2]

   b Two of the non-cyclic compounds have geometrical isomers. Draw the structures of these compounds and their geometrical isomers. [2]

            Higher Level Paper 2, Specimen 09, Q5

Q5 **a** The compound 2-bromobutane, $CH_3CHBrCH_2CH_3$, can react with sodium hydroxide to form the compounds **F**, **G** and **H**. Compound **F**, $C_4H_{10}O$, exists as a pair of optical isomers. Compounds **G** and **H**, C_4H_8, are structural isomers, and compound **H** exists as a pair of geometrical isomers.

     i Draw the structures of the two optical isomers of **F**. [2]

     ii Outline the use of a polarimeter in distinguishing between the optical isomers. [2]

     iii Draw diagrams to show the shapes of the two geometrical isomers of **H**. [2]

     iv Draw the mechanism, using curly arrows to represent the movement of electron pairs, to show the formation of **G**. [3]

   b A compound, **J**, has the molecular formula $C_2H_4O_2$ and is obtained from a reaction between methanoic acid and methanol. Write an equation for this reaction and state the name of compound **J**. [3]

Q6 This question refers to the compounds in the following reaction scheme.

$$C_3H_7Br \xrightarrow{NaOH} C_3H_8O \xrightarrow{oxidation} C_3H_6O \xrightarrow{oxidation} C_3H_6O_2$$
$$\textbf{A} \qquad\qquad \textbf{B} \qquad\qquad \textbf{C} \qquad\qquad \textbf{D}$$

   a Give the condensed structural formulas of compounds **C** and **D**. [2]

   b State a suitable reagent for the oxidation of **B** to **C** and **C** to **D**. Explain how the oxidation of **B** to **C** could be achieved without further oxidation to **D**. [3]

   c The conversion of **A** to **B** takes place by an S_N2 mechanism. State what is meant by the term S_N2 and describe, by using 'curly arrows' to show the movement of electron pairs, the mechanism of this conversion. [6]

   d Deduce how the rate of reaction of **A** with NaOH would compare with that of the compound $CH_3CH_2CH_2Cl$ with NaOH. Explain your answer by referring to Table 10 of the IB *Chemistry data booklet*. [2]

   e **B** and **D** react with each other when heated with concentrated sulfuric acid. State the name of this type of reaction and deduce the structure of the product. [2]

   f Write the structure of an ester isomer of **D** and explain why it is less soluble in water than **D**. [3]

Glossary

Entries that are **IB command terms** and entries that are **IB syllabus-required** definitions are coloured red and blue respectively.

A

2-amino acids a group of soluble organic compounds that possess a carboxylic acid group (—COOH) and a primary amine group (—NH$_2$) bonded to a common carbon atom; 2-amino acids are the monomers of proteins

absolute configuration actual three-dimensional structure of a chiral molecule: specified verbally by the Cahn–Ingold–Prelog R, S convention and represented on paper by Fischer projections

absolute temperature a temperature based on the thermodynamic (ideal gas) scale, measured from absolute zero; absolute temperature = (temperature in °C + 273) K; temperature in °C = (absolute temperature – 273)

absolute zero the temperature to which substances cannot be cooled; corresponds to the complete absence of heat; the temperature at which all ionic, molecular and atomic vibrations cease and at which the pressure and volume of an ideal gas are zero

absorbance logarithm of the fractional transmission of light (often at a specified wavelength) through a sample; absorbance, $A = \log_{10}(I_0 / I)$ where, I_0 and I are the intensity of the incident and transmitted radiation, respectively (see Beer–Lambert law)

absorption spectroscopy production and study of the absorption spectra of a material (often gaseous atoms and molecules)

absorption spectrum consists of dark absorption lines superimposed on a bright continuous spectrum; shows the absorption of radiation by a material over a range of wavelengths

accuracy a measure of the agreement between a measurement and the true or correct value

accurate measurement a measurement obtained using accurately calibrated instruments correctly and where no systematic errors arise: it will be close to the true value of the measurement

acid a proton donor (Brønsted–Lowry theory) and/or electron pair acceptor (Lewis theory)

acid anhydride homologous series (RCO(O)COR) containing a functional group characterized by two acyl groups joined by an oxygen atom

acid deposition deposition of acids from the atmosphere in solid or liquid form on the Earth's surface

acid dissociation constant, K_a the equilibrium constant for the reaction in which an acid dissociates and loses its hydrogen in the form of a hydrogen ion (proton) or oxonium (hydronium) ion

acid rain rain water (pH < 5.6) polluted by the presence of a mixture of nitric and sulfuric acids; caused by sulfur dioxide from the burning of coal and oxides of nitrogen from car exhaust emissions

acid salt a salt formed from a polybasic or polyprotic acid in which some of the hydrogen has not been replaced by metal ions

acid–base indicator a weak acid whose dissociated and undissociated forms in aqueous solution are different colours – the proportion of the two coloured forms varies with pH; used to show changes in the pH of aqueous solutions

acid–base reactions the transfer of protons/hydrogen ions from an acid to a base

acidic oxide a covalent oxide, usually an oxide of a non-metal, that reacts and dissolves in water to form a solution of an acid – if insoluble in water, the oxide reacts with alkalis to give salts

acidification fall in pH of soils and water in lakes and rivers, caused by acid rain

activated complex see transition state

activation energy the minimum amount of combined total kinetic energy a colliding pair of ions, atoms or molecules requires for a chemical reaction to occur; the energy barrier that has to be overcome to form the transition state

active site the region of an enzyme, usually a pocket or groove, that binds the substrate molecule(s) and catalyses a reaction

active site (catalyst) site on the surface of a solid catalyst at which catalytic activity occurs

active site (of an enzyme) the region of an enzyme, typically a pocket or groove on the surface, that binds the substrate(s) and performs catalysis

activity series see reactivity series

acylation introduction of an acyl group (RCO—) into an organic molecule

addition polymerization a type of polymerization that occurs when alkene-based monomers undergo repeated addition reactions to form a single molecule

addition reaction a reaction in which two (or more) molecules combine together to form a single molecule

additives (food) substances added to food during manufacture or preparation to improve sensory properties (taste, aroma and colour), nutrient content (vitamins and minerals), shelf-life or safety

adduct the chemical formed by the combination of a Lewis base with a Lewis acid

adenosine triphosphate (ATP) nucleotide formed in photosynthesis and respiration from ADP and phosphate, and functioning as a common intermediate between energy-requiring (endothermic) and energy-yielding (exothermic) reactions

adsorption accumulation, usually temporarily, of gases, liquids or solutes on the surface of a solid or liquid through the formation of weak intermolecular interactions or chemical bonds

aerobic respiration respiration requiring molecular oxygen, involving the oxidation of glucose to carbon dioxide and water

agonist any molecule that improves the activity of a different molecule

AIDS disease caused by a retrovirus, HIV (human immunodeficiency virus), and characterized by failure of the immune system to protect against infections and certain cancers

air pollution unwanted or toxic gases or particles that accumulate in the air as a consequence of natural processes or industrialization

alcohols a homologous series of organic compounds containing the functional group —OH and the general formula $C_nH_{2n+1}OH$

aldehydes a homologous series of compounds with the general formula RCHO, where the —CHO group (the aldehyde group) consists of a carbonyl group attached to a hydrogen atom; R is usually an alkyl or aryl group

aldose any monosaccharide sugar that contains an aldehyde group (—CHO) (in its acyclic (straight chain) form)

algal bloom rapid growth of algae on the surface of freshwaters in response to a supply of nitrogen and/or phosphorus, leading to depletion of light and oxygen below the water surface

aliphatic an organic compound that has carbon atoms arranged in a chain or branched form rather than in an aromatic ring

alkali a strong base which is soluble in water; alkalis are group 1 metal hydroxides and barium hydroxide

alkali metals the group of very reactive metals in group 1 of the periodic table; they react with water to release hydrogen gas and form strongly alkaline solutions

alkaloids basic nitrogen organic compounds (mostly heterocyclic) derived from plants and having a range of pharmacological properties

alkanes saturated hydrocarbons which have the general formula C_nH_{2n+2} (if acyclic)

alkenes unsaturated hydrocarbons containing a carbon–carbon double bond and with the general formula C_nH_{2n} (if acyclic)

alkyl group a group, with the general formula C_nH_{2n+1}, obtained by removing a hydrogen atom from an alkane, and usually represented by R

alkylation introduction of an alkyl group into an organic molecule

alkynes unsaturated hydrocarbons with a carbon–carbon triple bond and with the general formula C_nH_{2n-2} (if acyclic)

allotrope different structures of an element

allotropy the ability of an element to exist in different structural forms or allotropes

alloy mixture with metallic properties made up of two or more metals, or which contains metals and carbon

alpha helix right-handed helical conformation of a protein chain, held together by intra-molecular hydrogen bonding; a common protein secondary structure

alpha particle fast-moving helium nucleus emitted by radioactive nuclei, consisting of two protons and two neutrons tightly bound together; travels only a few centimetres in air and is stopped by thick paper

amalgam alloy that contains mercury

amide a homologous series of organic compounds with the general formula $RCONH_2$

amide link a chemical bond formed between two molecules when the carboxyl group of one molecule reacts with the amine group of the other molecule, thereby releasing a molecule of water

amines organic compounds derived by replacing one or more of the hydrogen atoms in ammonia by alkyl groups

amino acid residue single amino acid within a polypeptide chain

amorphous non-crystalline solid; lacking a definite or regular shape

amount a physical quantity indicating the number of moles of a substance present in a sample

ampere the SI unit of electrical current: a current of one ampere (amp) is a flow of one coulomb of charge per second ($1 A = 1 C s^{-1}$)

amphiprotic a chemical species capable of accepting and donating protons, thus able to behave as both as an acid and a base

amylopectin water-soluble component of starch, consisting of highly branched chains of glucose molecules; the backbone of glycosidic linkages is ($1\alpha \rightarrow 4$), but the branches are ($1\alpha \rightarrow 6$)

amylose water-insoluble component of starch. It consists of between 100 and 1000 linear chains of D-glucose molecules bonded in ($1\alpha \rightarrow 4$) linkages

anabolic steroids group of synthetic hormones that promote the storage of protein and the growth of tissue, sometimes used by athletes to increase muscle size and strength

anaerobic respiration respiration in the absence of molecular oxygen, involving the breakdown of glucose to lactic acid or ethanol

analgesic drug used to alleviate the sensation of pain

analyse interpret data to reach conclusions

anhydrous salt a salt that does not have its water of crystallization

anion a negatively charged ion which migrates to the anode (positive electrode) during electrolysis

annealing process involving heating and cooling, typically used to remove internal stresses or induce softening of e.g. copper or high-carbon steel

annotate add brief notes to a diagram or graph

anode where oxidation (the loss of electrons) occurs during an electrochemical process; in an electrolytic cell the anode is the positive electrode

anomalous data data with unexpected values that does not match the relationship predicted by the hypothesis

anomers two stereoisomers that differ only in the configuration about the carbonyl carbon atom

antacids substances, basic in nature, used to reduce the pH of the gastric juice in the stomach with the aim of relieving indigestion

antagonist molecule that blocks the ability of a given chemical to bind to its receptor, preventing a biological response

anthocyanidins common plant pigments, usually red or blue in colour: the sugar-free counterparts of anthocyanins based on the flavylium ion

anthropogenic originating from the activity of humans

antibiotic substance or a semi-synthetic substance derived from a microorganism, usually a bacterium, and able in dilute solution to inhibit or kill another microorganism, usually a bacterium

antibonding orbital a molecular orbital that is higher in energy than the atomic orbitals from which it is formed

anticodon specific sequence of three nucleotides in a transfer RNA molecule, complementary to a codon for an amino acid in a messenger RNA molecule

antioxidant chemicals that prevents damage caused by oxidation that would normally occur during cell metabolism, or storage of foods, preventing the damaging effects of free radicals

antiparallel (DNA) DNA double helix consists of two linear DNA molecules that are opposite in orientation

antipyretic chemical that reduces fever

antiviral drug that acts against viruses

apply use an idea, equation, principle, theory or law in a new situation

arenes hydrocarbons based on benzene rings

aromatic an organic compound that contains a benzene ring

aromatization process by which a compound forms an aromatic ring

Arrhenius constant a constant that appears in the Arrhenius equation in front of the exponential term: a term that relates to the frequency of collisions and their orientation in space

Arrhenius equation an equation that relates the rate constants (k) for a reaction obtained at different absolute temperatures to the activation energy of the reaction; $\ln k = \ln A = -\dfrac{E_a}{RT}$, where E_a and R refer to the activation energy and gas constant, respectively

Arrhenius plot a plot of the natural logarithm of the rate constant (k) (y-axis) against 1/(absolute temperature) (x-axis) – a straight line will be obtained with a gradient of $\dfrac{E_a}{R}$; the value of the activation energy, E_a, will be in $J mol^{-1}$

Arrhenius temperature dependence for many reactions (with an activation energy approximately equal to $50 kJ mol^{-1}$) a rise in temperature of ten degrees Celsius leads to an approximate doubling of the initial rate

asymmetric carbon atom a carbon atom in a molecule that is attached to four different atoms and/or functional groups

atactic polymer chain in which the stereochemical orientation of the substituents, with respect to each other along the chain, is random

atmosphere total of all the gases surrounding the Earth, extending several hundred kilometres above the surface

atomic absorption spectroscopy instrumental technique for detecting concentrations of atoms to parts per million (or billion) by measuring the amount of light absorbed by atoms or ions vaporized in a flame or an electrical furnace

atomic force microscope instrument able to image surfaces to molecular accuracy by mechanically probing their surface contours

atomic number the number of protons in the nucleus of an atom; in atoms the atomic number is also equal to the number of extra-nuclear electrons

atomic radius half the distance of the closest approach of the nuclei of atoms in the crystal or molecule of a chemical element

Aufbau principle a principle of quantum mechanics which states that the order in which the atomic orbitals of atoms are filled with electrons is the order of increasing energy

autacoid a biological substance secreted by various cells whose physiological activity is restricted to the vicinity of its release; it is often referred to as local hormone

average bond enthalpy enthalpy change per mole when one mole of the same type of covalent bond is broken in the gas phase for many similar molecules

average rate of reaction calculated by dividing the total change in reactant or product concentration by the time for the reaction to end

Avogadro's law at a specified temperature and pressure, equal volumes of (ideal) gases contain equal numbers of moles of particles; there is a directly proportional relationship between the volume of gas, V, and the amount of particles (at constant pressure), n: $V \propto n$

Avogradro constant the number of atoms in exactly 12 grams of carbon-12; it has units of per mol (mol^{-1})

axial overlap overlap of orbitals that occurs directly between the nuclei of atoms – sigma bonds (σ) are formed

B

back titration often two consecutive acid–base titrations, performed when an insoluble and slowly reacting reagent is treated with an excess of an acid or base. The excess acid or base is then titrated with base or acid solution of a primary standard (a back titration may involve redox reactions)

backward reaction the conversion of products into reactants in an equilibrium

bacterium single-celled organism lacking a nucleus

Balmer series a series of lines in the emission spectrum of visible light emitted by excited hydrogen atoms: the lines correspond to the electrons falling down into the second lowest energy level, emitting visible light

barbiturate group of drugs derived from barbituric acid that is used to sedate, to control convulsions, or to induce sleep

base in the Brønsted–Lowry theory a base is a proton acceptor; a Lewis base is an electron pair donor

base (DNA or RNA) one of the nitrogen-containing compounds that occurs attached to the sugar component of DNA or RNA

base dissociation constant the equilibrium constant for the reaction in which base reacts with water to produce the conjugate acid and hydroxide ions; a measure of the extent to which weak bases accept hydrogen ions in solution

base pair two nucleotides in nucleic acid chains that are paired together by intermolecular hydrogen bonding between the bases

basic oxide an ionic oxide, usually an oxide of a metal, that reacts with acids to form a salt and water; some basic oxides react with water to form alkaline or basic solutions

basic oxygen converter vessel in which scrap steel and a small amount of limestone are dissolved in molten iron and pure oxygen is then blown into the molten mixture to remove impurities

Beer–Lambert law absorbance, A, of a given wavelength of light by a solution of a substance is proportional to the concentration of a substance (provided the solution is dilute): $A = \varepsilon cl$, where c is the concentration of the substance, l is the path length for the radiation through the solution and ε is the absorption coefficient; if ε is in $dm^3\,mol^{-1}\,cm^{-1}$ (molar absorption coefficient) and l is in cm then c will be in $mol\,dm^{-3}$

Benedict's test test for a reducing sugar: blue copper(II) ions are converted to a red-brown precipitate of copper(I) oxide

beta particle fast-moving electron emitted by radioactive nucleus which travels a few metres in air or can pass through several millimetres of aluminium; formed when a proton inside a nucleus changes to form a neutron and a fast-moving electron

beta sheet extended, zigzag arrangement of a protein chain; a common secondary structure held together by intramolecular hydrogen bonding

bias a factor of sampling of the variables from an investigation when the conclusions obtained from the investigation do not accurately describe the characteristics of the whole population, i.e. the differences between the

sample and the whole population are not just due to random chance

bimolecular involving the collision between two reactant species in an elementary step in a reaction

biochemical oxygen demand (BOD) amount of oxygen taken up by bacteria that decompose organic waste in water, calculated by keeping a sample of water containing a known amount of oxygen for five days at 20 °C and then measuring the oxygen content; a high BOD value indicates the presence of a large number of bacteria, which suggests a high level of pollution

blast furnace tall furnace allowing continuous production of iron from iron oxide using carbon monoxide as the reducing agent (similar furnaces are used for the extraction of zinc)

blunt ends end of a DNA fragment produced by a restriction enzyme which cuts both strands of DNA at the same point, leaving no single-stranded sections

Bohr's theory a model of the atom that explains emission and absorption of radiation as transitions between states in which the electron orbits the nucleus at a definite distance

boiling the change of a liquid into a gas at constant temperature; occurs when the vapour pressure of the liquid is equal to the external pressure exerted on the liquid – it is characterized by the appearance of bubbles of vapour throughout the liquid which break through the surface of the liquid

boiling point the temperature at which a liquid is converted to a gas at the same temperature; a liquid boils when the vapour pressure of the liquid equals the surrounding pressure

bomb calorimeter device used to measure energy changes (at constant volume) that occur when substances, for example, alcohols or hydrocarbons, are burnt in excess oxygen in a sealed container

bond angle an angle formed by the location of three atoms or two covalent bonds in space, used to describe the shapes of molecules; the angle ABC where atoms A and C are bonded to atom B

bond enthalpy the amount of energy (in kilojoules) required to break one mole of a particular covalent bond in the gaseous state into gaseous atoms (under standard thermodynamic conditions): a measure of the strength of the bond

bonding orbital a molecular orbital that is lower in energy than the atomic orbitals from which it is formed

Born–Haber cycle an enthalpy cycle based upon experimental data commonly used to calculate the lattice enthalpies of ionic solids; it is a series of reactions (and the accompanying enthalpy changes) which, when summed, represents the hypothetical one-step reaction in which elements in their standard states are converted into crystals of an ionic compound (and the accompanying enthalpy change) under standard thermodynamic conditions

bottom-up approach building larger nano-based objects from atoms or molecules

Boyle's law the product of pressure and volume (for a fixed mass of ideal gas at constant temperature) is a constant

branched-chain hydrocarbon a hydrocarbon which has alkyl groups bonded to its longest unbranched chain

broad-spectrum antibiotic antibiotic that is effective against a wide range of strains of bacteria

Brønsted–Lowry theory a theory of acidity that describes an acid as a proton or hydrogen ion donor, and a base as a proton or hydrogen ion acceptor

buffer solution an aqueous solution consisting of a weak base and its conjugate acid, which resists a change in pH when small amounts of either hydroxide ions (from a base) or hydrogen ions (from an acid) are added; buffers typically consist of a weak acid and its corresponding salt (an acidic buffer) or a weak base and its corresponding salt (a basic buffer)

buffering capacity the ability of a buffer to absorb hydrogen ions or hydroxide ions without a significant change in pH

C

C-terminus only amino acid residue at one end of a polypeptide chain that contains a free carboxyl group

calculate find a numerical answer showing the relevant stages in the working (unless instructed not to do so)

calorie notionally the energy required to raise the temperature of 1 g of water by 1 °C; a calorie is approximately equivalent to 4.2 joules

calorific value amount of heat released by a unit mass of a substance, for example a food, (or a unit volume of a gas) being burnt

calorimeter a piece of apparatus (insulated) for measuring the energy released or absorbed during a chemical reaction; in an open vessel at constant pressure, the heat change equates to the enthalpy change (ΔH)

caramelization type of non-enzymatic browning which involves the high-temperature decomposition of sugars, forming brown pigment and releasing volatile products

carbocation an organic ion with a positive charge on an electron-deficient carbon atom

carboxylic acids a homologous series of organic compounds with the general formula RCOOH

carotenoid pigments found in plants, including algae, and bacteria; composed of conjugated molecules containing carbon and hydrogen and sometimes oxygen

carrier gas inert gas used to carry the sample in gas chromatography

catalyst a substance which, when present in relatively small amounts, increases the rate of a chemical reaction but which is not consumed during the overall process – the function of a catalyst is to provide a new reaction pathway with a lower activation energy

catalyst poison a substance that prevents the activity of a catalyst

catalyst selectivity relative activity of a catalyst in reference to a particular compound in a mixture; the relative rate of a single reactant in competing reactions

catalytic converter part of the exhaust system of a modern car running on unleaded petrol, consisting of a platinum/rhodium catalyst in a honeycomb structure which converts carbon monoxide, nitrogen monoxide and unburnt hydrocarbons into carbon dioxide, nitrogen and dinitrogen oxide

catalytic cracking cracking carried out in the presence of a heated catalyst, for example, aluminium oxide (alumina) or silicon dioxide (silica)

catenation the spontaneous linking of atoms of certain chemical elements, such as carbon atoms, to make stable rings or long chains

cathode where reduction (the gain of electrons) occurs during an electrochemical process; in an electrolytic cell the cathode is the negative electrode

cation a positively charged ion attracted to the cathode during electrolysis; metals form cations via loss of one or more electrons

cation-exchange capacity the capacity of a soil to exchange cations with the soil solution, often used as a measure of potential soil fertility: generally expressed in milliequivalents per 100 g of soil

cell diagram a shorthand form of summarizing the electrodes and electrolytes present in a voltaic cell, which traces the path of the electrons; the reduced form of the metal to be oxidized at the anode is written first, followed by its oxidized form, then the oxidized form of the metal to be reduced at the cathode, and finally the reduced form of the metal at the cathode; the salt bridge is indicated by || and the phase boundaries between the electrodes and their ions by |

cell potential the potential difference between the two half-cells (in their standard states) of an electrochemical cell

cellular respiration cellular process by which glucose (and other substances) are broken down, in the presence of enzymes, to release useful energy for other cellular processes

central dogma genetic information flows from DNA to RNA to protein

chain (addition) polymerization chain reaction in which the growth of a polymer chain proceeds exclusively by reaction(s) between monomer(s) and reactive site(s) on the polymer chain, with regeneration of the reactive site(s) at each growth step

chain reaction a reaction intermediate generated in one step reacts in such a way that this intermediate is regenerated

change of state the inter-conversion of a substance between the solid, liquid and gaseous states

charging (battery) process of supplying electrical energy for conversion to stored chemical energy

Charles' law the volume of a fixed mass (at constant pressure) of an ideal gas is directly proportional to absolute temperature

chemical environment number and types of atoms a particular atom within a molecule is bonded to

chemical equilibrium when the rate of the forward reaction is equal to the rate of the backward reaction and consequently there are no changes in the concentrations of reactants and products

chemical feedstock raw materials required for an industrial process

chemical kinetics the branch of physical chemistry concerned with the mechanisms and rates of chemical reactions

chemical library large collection of stored chemicals usually used in screening for drug leads

chemical shift position (in ppm) of a resonance in the NMR spectrum relative to a standard such as TMS (tetramethylsilane); an atomic property that varies depending on the chemical and magnetic properties of an atom and its arrangement within a molecule

chemical weathering breakdown of rock material brought about by the action of chemicals, usually in aqueous solution

chemotherapy use of chemical agents in the treatment or control of disease, particularly cancer, or mental illness

chiral a molecule that has a non-superimposable mirror image

chiral auxiliary a chiral compound that is covalently attached to the substrate as a controlling element in a diastereoselective reaction and is subsequently cleaved from the product

chlor-alkali industry the industrial electrolysis of brine which results in the production of sodium hydroxide and chlorine

chlorofluorocarbons (CFCs) group of compounds in which some or all of the hydrogen atoms of an alkane have been replaced (substituted) by chlorine and fluorine atoms

chlorophyll green pigment found in plants involved in absorbing light for photosynthesis; a magnesium-based porphyrin

chromatogram record obtained from chromatography

chromatography technique for analysing or separating mixtures of gases, liquids or dissolved substances based upon differential solubility in two phases

cis a term used to describe geometrical isomers of 1,2-disubstituted alkenes with functional groups or atoms which are on the same side of the molecule as each other

clock reaction when a mixture of reacting chemical compounds involved in a redox reaction shows a sudden colour change

closed system a system where the amount of matter is fixed but heat can flow in or out in, or the volume can be altered; a prerequisite for the establishment of an equilibrium

co-enzyme organic co-factor required for the action of certain enzymes; often contains a vitamin as a component

codon sequence of three adjacent nucleotides in a nucleic acid that codes for a specific amino acid

coefficients the numbers that appear to the left of chemical formulas in a balanced equation

coke solid material left behind when volatile components in coal have been removed by heating; it contains a high percentage of carbon and is used as a reducing agent and fuel

colligative property a physical property that depends on the number of solute species present, but not on their chemical identity

collision theory a simple model to account for the variation in the rate of reaction with temperature, surface area and concentration; it considers particles to be hard spheres that react with each other when they collide with sufficient kinetic energy

colloid system in which there are two or more phases, with one (the dispersed phase) distributed in the other (the continuous phase)

colorimeter an instrument used to measure the intensity of colour in a solution

column chromatography form of chromatography that uses a column or tube to hold the stationary phase

combinatorial chemistry automated, parallel synthesis of a library of chemical structures, usually drug leads

combined gas law the gas law that combines absolute temperature, pressure and volume, but not the amount of gas
$\frac{PV}{T}$ = constant; $\frac{P_1 \times V_1}{T_1} = \frac{P_2 \times V_2}{T_2}$, where P represents the pressure of the gas, V represents the volume and T represents the absolute temperature

combustion a highly exothermic and rapid chemical reaction in which a substance reacts with oxygen during burning

comment give a judgment based on a given statement or result of a calculation

common ion effect suppression of the solubility of a weak electrolyte by the presence, in the same solution, of a strong electrolyte containing an ion in common with the weak electrolyte

compare give an account of similarities and differences between two (or more) items, referring to both (all) of them throughout

competitive inhibitor generally competes with the substrate for the enzyme's active site: the percentage of competitive inhibition at fixed inhibitor concentration can be decreased by increasing the substrate concentration; at high concentrations of the substrate it is possible to reach V_{max} even in the presence of the inhibitor; however, the value of K_m is decreased

complementary base pairing length of single-stranded DNA whose sequence base pairs with another length of single-stranded DNA

complementary colour one of two coloured lights so related to each other that when blended together they produce white light; so-called because each colour makes up in the other what it lacks to make it white

complex ion a transition metal ion, surrounded by a fixed number of ligands which form dative (or coordinate) covalent bonds with vacant orbitals in the metal ion

compound a substance formed by the chemical combination of two or more chemical elements in fixed proportions

concentrated a solution with a relatively high concentration of solute

concentration the ratio of the amount (in moles) of a substance dissolved in a given volume of solution.; expressed in mol dm^{-3}

condensation polymer a type of polymer made by a process that involves the elimination of small molecules, usually water

condensation reaction an addition reaction immediately followed by an elimination reaction

condensed structural formula a formula in which the single bonds between the atoms are not shown with lines

conducting polymer polymer having high conductivity, approaching that of metals

conduction band energy band in a solid in which electrons are freely mobile and can produce a net electric current

conformational isomers the spatial arrangement of atoms in a molecule that can be adjusted or changed by rotations about sigma bonds

conjugate acid the chemical species formed when a proton or hydrogen ion is accepted by a base

conjugate base the chemical species formed when an acid loses a proton or hydrogen ion

conjugate pair two chemical species related to each other by the loss or gain of a single proton or hydrogen ion

conjugated molecules with double or triple bonds that are separated by one single bond: there is delocalization of electrons in the pi (π) orbitals between the carbon atoms linked by the single bond

construct represent or develop in graphical form

Contact process sulfur dioxide and air are passed over a heated vanadium(v) oxide catalyst to produce sulfur trioxide which is then dissolved in sulfuric acid to form disulfuric acid (oleum) which is diluted to produce concentrated sulfuric acid

continuous spectrum an emission spectrum that exhibits all the wavelengths or frequencies of visible light

control group a group of patients in a clinical trial who are given another drug, a placebo or no treatment at all, to compare with the group receiving the drug under trial

convergence the lines in an emission spectrum become progressively closer to each other (at higher frequency or smaller wavelength) and finally merge

coordinate bond see dative covalent bond

coordination number the number of ligands surrounding a central metal ion

coordination polymerization a form of addition polymerization in which a monomer adds to a growing polymer chain through an organometallic active centre

copolymer a material created by polymerizing a mixture of two (or more) starting compounds

corrosion the process by which a metal undergoes oxidation by air and water

coulomb the SI unit of electrical charge: one coulomb of charge is passed around a circuit when a current of one ampere is allowed to flow for one second

Coulomb's law the force between two charged particles is directly proportional to the product of their charges and inversely proportional to the square of the distance between them

covalent bond the sharing of one or more pairs of electrons between two atoms

cracking process of breaking down long-chain alkanes into smaller alkanes and alkenes using heat, usually in the presence of a catalyst

cross links covalent bonds between adjacent chains in a polymer

cross-tolerance tolerance to a drug that generalizes to drugs that are chemically related or that produce similar effects

crude oil mixture of hydrocarbons formed originally from marine animals, found beneath the ground trapped between layers of sedimentary rock; it is obtained by drilling

crystalline polymer polymer in which sections of adjacent chains are packed in a regular array

curly arrow used to show the notional movement of a pair of electrons in a reaction mechanism; the tail of the arrow shows where the electrons come from and the head where they go to

current the rate of flow of electric charge or electrons through a conductor, measured in coulombs per second or amperes

cyclic molecules having atoms arranged in a ring or closed-chain structure

cyclization formation of a cyclic compound from an open-chain compound

cytochrome class of iron-containing proteins important in cell respiration as catalysts of oxidation–reduction reactions

D

d-block metal metals located between groups 2 and 3 of the periodic table where the d-shell is being filled across the period

d–d splitting a splitting of the d orbitals of the transition metal ion in a complex ion, caused by the ligands

d–d transition an electronic transition between two levels of d orbitals; the energy separation usually corresponds to that of visible light and for this reason, such transitions are responsible for the colours of many transition metal ions

Daniell cell a voltaic cell consisting of a copper cathode immersed in 1 mol dm^{-3} copper(ii)

sulfate solution and a zinc anode immersed in 1 mol dm^{-3} zinc sulfate solution; it has a standard cell potential of 1.10 V

dative covalent bond one of the atoms supplies both electrons of the shared pair

decomposition the formation of new substances from a single substance upon heating

deduce reach a conclusion from the information given

define give the precise meaning of a word, phrase or physical quantity

degeneracy (atomic energy levels) when two or more orbitals have the same potential energy

degeneracy (DNA) ability of codons to code for more than one amino acid; this property of the genetic code makes it more fault-tolerant for mutations

degenerate a group of orbitals with the same energy

delocalized molecules or ions that have p orbitals extending over three or more atoms have delocalized π electrons; metallic bonding involves the delocalization of valence electrons between all the ions within a crystal

denaturation (protein) partial or complete unfolding of a protein chain (or nucleic acid); the process is usually reversible and is brought about by heat and a variety of chemicals, for example, urea and organic solvents

dependent variable the variable that is measured during an investigation

depressant drug used medicinally to relieve anxiety, irritability and tension

derive manipulate a mathematical relationship(s) to give a new equation or relationship

desalination removal of dissolved salts from an aqueous solution

describe give a detailed account

design produce a plan, simulation or model

designer drug drug with properties and effects similar to a known hallucinogen or narcotic but having a slightly altered chemical structure, created in order to evade restrictions against illegal substances

desorption the opposite of adsorption, i.e. a decrease in the amount of adsorbed substance

determine find the only possible answer

dextrorotatory rotates the plane of plane-polarized light clockwise, in the (+) direction

diamagnetic effect weak repulsion by a strong magnetic field

diaphragm cell industrial electrolytic cell in which a porous diaphragm is used to separate the electrodes thereby allowing electrolysis of sodium chloride solution, without allowing the products to react

diatomic a molecule containing two atoms of the same element covalently bonded together

dietary fibre the indigestible carbohydrates (mainly cellulose) found in fruit, vegetables, grain and nuts

dilute a solution containing a relatively low concentration of solute

dimer a molecule formed by the bonding of two identical monomers: the bonds will be relatively strong hydrogen bonds or covalent bonds

diode electronic device that allows current to flow in one direction only; often used to rectify current (convert a.c. to d.c.)

dipeptide a peptide that is composed of two amino acid molecules linked by a peptide bond

dipole a pair of separated opposite electrical charges located on a pair of atoms within a molecule

dipole moment product of the charge at one end of a dipole and the distance between the charges: the larger the dipole moment of a molecule, the greater the polarity of the bond; dipole moments are measured in units called debyes (D)

dipole-dipole forces weak intermolecular forces caused by electrostatic interactions between permanent dipole moments that exist in polar molecules

diprotic an acid which contains two replaceable hydrogen atoms per molecule (also known as dibasic)

directing effect the ability of substituent functional groups to direct further substitution to certain positions on the benzene ring during the electrophilic substitution of benzene derivatives

disaccharide carbohydrate consisting of two covalently joined monosaccharide units

discharge the conversion of ions to atoms or molecules during electrolysis

discuss give an account including, where possible, a range of arguments for and against the relative importance of various factors, or comparisons of alternative hypotheses

displacement reaction a redox reaction in which a more reactive element displaces a less reactive element from a solution of its ions or salt, often in aqueous solution

disproportionation the simultaneous oxidation and reduction of atoms of a single chemical element to produce two products: the element undergoing disproportionation must have a minimum of three stable oxidation states and be in an intermediate oxidation state

distinguish give the differences between two or more different items

disulfide bridge covalent bond (—S—S—) formed (in the presence of an enzyme) between two adjacent polypeptide chains by the reaction between sulfhydryl groups (—SH) of two cysteine residues

diuretic substance that leads to an increase in the discharge of urine

DNA profiling distinctive pattern of fragments of DNA obtained by restriction enzyme digestion and electrophoretic separation of repeated DNA segments from individual people

dopant element introduced into semi-conductor to establish either p-type (acceptors) or n-type (donors) conductivity

doping incorporation of impurities within the crystal lattice of silicon so as to increase its conductivity

double helix coiled structure of double-stranded DNA in which strands linked by hydrogen bonds form a spiral or helical configuration, with the two strands oriented in opposite directions

double-blind study study of the effects of a drug where both the researcher and the participants are not aware of which treatment each participant is receiving

downfield low-field or higher frequency or higher delta ppm value area of the NMR spectrum

draw represent by means of pencil lines

drug chemical intended to affect the structure or any function of the body of humans or other animals

drug candidate chemical compound that has potential to be developed into a therapeutic drug; not all drug candidates become products

drug target gene or protein that plays a role in a disease process and is the intended site of drug activity

ductile the ability of metals to be drawn out under tension and stretched into wires

dynamic equilibrium an equilibrium is described as dynamic because although there is no change in macroscopic properties the forward and backward reactions are occurring (at equal rates)

E

effective nuclear charge the nuclear charge exerted on a particular electron, equal to the actual nuclear charge minus the effect of nuclear shielding

electrochemical series an arrangement of elements and ions (which can undergo redox reactions) arranged in order of their standard reduction potentials, with the most negative (that is, most reducing) at the top of the series

electrode a conductor which dips into the electrolyte of an electrolytic or voltaic cell and allows the current (electrons) to flow to and from the electrodes; electrodes may be inert, functioning only to transfer electrons, or may be active and be involved in the cell reactions

electrode potential the potential formed between a metal and an aqueous solution of its ions or between ions of the same element in different oxidation states

electrolysis a process in which chemical decomposition of an ionic substance, known as the electrolyte, is caused by the passage of an electric current; the conduction in the electrolyte occurs via migration of ions: no free electrons are transferred between the electrodes through the electrolyte

electrolyte the solution or molten liquid in an electrolytic cell or voltaic cell: an ionic compound (a salt, alkali or acid) that will conduct electricity during electrolysis when it is melted or, if soluble, dissolved in water

electromagnetic spectrum entire range of electromagnetic radiation or waves including, in order of decreasing frequency, cosmic-ray photons, gamma rays, X-rays, ultraviolet radiation, visible light, infrared radiation, microwaves and radio waves

electromagnetic wave a wave of oscillating electric and magnetic fields that can move through space; they are transverse in nature and all travel through a vacuum at a speed of $3 \times 10^8\,\mathrm{m\,s^{-1}}$

electromotive force (of a cell) the energy per unit charge that is converted reversibly from chemical energy into electrical energy; the SI unit of electromotive force (e.m.f.) is the volt

electron negatively charged sub-atomic particle present in all atoms and located in shells, or energy levels, outside the nucleus

electron affinity the energy released by the addition of one mole of electrons to one mole of gaseous atoms (under standard thermodynamic conditions) to form one mole of gaseous uninegative ions

electron arrangement the distribution of electrons among the available shells

electron density the probability of finding an electron or pair of electrons in a particular region or volume of an atom or molecule

electron shells the main energy levels of an atom where the electrons are located

electron transport chain a series of electron carriers that transfer high-energy electrons along a redox chain, driving ATP synthesis in the process

electronegativity a measure of the tendency of an atom in a molecule to attract a pair of shared electrons towards itself; electronegativity values increase from left to right across the periodic table and decrease down a group

electrophile a molecule or cation that can act as an electron pair acceptor (lone pair or π pair) or Lewis acid in a reaction with an organic molecule

electrophilic addition addition reaction initiated by the rate-determining attack of an electrophile on the π electrons of the carbon–carbon double bond

electrophilic substitution reaction involving the substitution of an atom or group of atoms in benzene (or derivative) with an electrophile as the attacking species

electrophoresis movement of charged ions in response to an electrical field, often used to separate mixtures of ions, proteins, or nucleic acids

electroplating the coating of a metal with a thin layer of another metal by electrolysis

electrostatic precipitation removal of very fine particles suspended in a gas by electrostatic charging and subsequent precipitation onto a collector in a strong electric field

element a pure substance that cannot be decomposed or broken down into simpler substances by chemical methods

elementary steps many reactions occur in a series of steps each of which involves one or two reacting particles (atoms, ions or molecules)

elimination a reaction in which atoms or small molecules are removed from a single molecule, usually to give a double bond, or between two molecules

Ellingham diagrams diagrams that explain how standard free energies of formation of metal oxides vary with temperature; they allow prediction of the conditions required for metal extraction from oxide ores

elution process of removing an adsorbed material (the adsorbate) from an adsorbent by washing it with a liquid (the eluent): the solution consisting of the adsorbate dissolved in the eluent is the eluate

emission spectroscopy study of emission spectra produced by excited substances (often gaseous atoms or molecules)

empirical formula a formula for a compound which shows the simplest whole number ratio of atoms present

emulsifier substance that enables the mixing of two insoluble liquids

emulsion system consisting of two immiscible liquids, one of which is dispersed in the other in the form of small droplets

enantiomer a chiral compound whose molecular structure is not superimposable on its mirror image

end-point where the indicator changes colour suddenly during a titration: at this stage the acidic and basic forms of the indicator are present in equal concentrations (*see* equivalence point)

endocrine gland gland producing a hormone that passes directly into the bloodstream

endothermic reaction a reaction in which heat energy is absorbed from the surroundings; there is a fall in temperature when the reaction occurs or heat energy has to be continually supplied to make the reaction occur

enthalpy the difference in the enthalpy between the reactants and products can be measured as heat (at constant pressure); it has no absolute values – only enthalpy changes can be measured

enthalpy change of atomization the enthalpy change when one mole of gaseous atoms is formed under standard thermodynamic conditions (with no change in pressure)

enthalpy change of fusion the enthalpy change when one mole of a solid melts to a liquid at its melting point

enthalpy change of hydration the enthalpy change when one mole of aqueous ions is formed from separate ions in the gas phase under standard thermodynamic conditions

enthalpy change of neutralization the enthalpy change when one mole of acid undergoes complete neutralization with a base to form one mole of water under standard thermodynamic conditions:
$H^+(aq) + OH^-(aq) \rightarrow H_2O(l)$

enthalpy change of solution the enthalpy change when one mole of a substance is dissolved in a solvent to infinite dilution (in practice, to form a dilute solution) under standard thermodynamic conditions

enthalpy change of sublimation the sum of the enthalpies of fusion and vaporization at the same temperature

enthalpy change of vaporization the enthalpy change when one mole of a pure liquid is vaporized at its boiling point

enthalpy level diagram a diagram that traces the changes in the enthalpy or potential energy of a chemical system during the course of a reaction

entropy a thermodynamic function that measures how energy is distributed or dispersed among particles (sometimes crudely called disorder or randomness)

entropy change the change in entropy that accompanies a physical or chemical change is given by the sum of the entropies of the products minus the sum of the entropies of the reactants

enzyme a globular protein that acts on a specific substrate molecule and catalyses a specific biochemical reaction

equation of state *see* combined gas law

equilibrium *see* chemical equilibrium

equilibrium constant the value obtained when equilibrium concentrations of the chemical species are substituted in the equilibrium expression – the value indicates the equilibrium position

equilibrium expression the expression obtained by multiplying the product concentrations and dividing by the multiplied reactant concentrations, with each concentration raised to the power of the coefficient in the balanced equation (pure solids and pure liquids are not included in equilibrium expressions involving aqueous solutions)

equilibrium law in any reversible reaction at a state of equilibrium, the rate of the forward reaction equals the rate of the reverse reaction

equilibrium position a particular set of equilibrium concentrations of reactants and products

equivalence point the point in an acid–base titration where the acid and base have been added in stoichiometric amounts, so that neither is present in excess; if a suitable indicator is chosen it will correspond to the end-point

essential fatty acids fatty acids that cannot be synthesized by humans and must be obtained from the diet

esterification the reaction between a carboxylic acid and alcohol to form an ester and water, catalysed by concentrated sulfuric acid

esters organic compounds formed by the condensation reaction between alcohols and acids; esters formed from carboxylic acids have the general formula RCOOR'

estimate find an approximate value for an unknown quantity

eutrophication the process by which lake or pond water becomes rich in mineral and organic nutrients that promote a proliferation of plant life, especially algae, which reduces the dissolved oxygen content and often causes the extinction of other organisms

evaluate assess the implications and limitations

evaporation occurs at the surface of a liquid and involves the liquid changing into a gas at a temperature below the boiling point of the liquid

excess a reactant is in excess when, after the reaction is complete, some of it remains unreacted, i.e. it is present in a molar ratio that exceeds the value implied by the stoichiometric equation

excitation energy minimum energy required to change a system from its ground state to a particular excited state

exothermic reaction a reaction in which heat energy is released to the surroundings from the reactants: the bonds of the products contain less enthalpy than the bonds of the reactants

experimental yield the quantity of a product that is obtained from a chemical reaction (*see* theoretical yield)

explain give a detailed account of causes, reasons or mechanisms

exponential factor the expression $e^{-E_a/RT}$ in the Arrhenius equation

extrapolation to estimate (a value of a variable outside a known range) from values within a known range by assuming that the estimated value follows logically from the known values

F

Faraday constant the quantity of electric charge (in coulombs) transferred by one mole of electrons; it has the (approximate) value of $96\,500\,C\,mol^{-1}$

Faraday's first law of electrolysis the amount of a chemical product formed during electrolysis is directly proportional to the quantity of electricity passed

Faraday's second law of electrolysis the amount of chemical product formed (for a constant quantity of electricity) is proportional to the relative atomic mass of the ion and its charge

fatty acid a group of long monobasic acids, found in animal and vegetable fats and oils, having the general formula $C_nH_{2n+1}COOH$ and containing an even number of carbon atoms

fibrous protein insoluble protein that serves in a protective or structural role

fingerprint region region of the infrared spectrum of a substance between 910 and $1430\,cm^{-1}$ where the pattern of peaks is characteristic of that compound, even though all the peaks might not all have been assigned to specific vibrations of the molecule

first ionization energy the energy required to remove one mole of electrons from one mole of isolated gaseous atoms to form one mole of gaseous unipositive ions under standard thermodynamic conditions

first-order reaction a reaction in which the initial rate of reaction is directly proportional to the concentration of only one reactant

Fischer projection means of depicting the absolute configuration of chiral molecules on a flat page which employs a cross to represent the chiral centre: the horizontal arms of the cross represents bonds coming out of the plane of the page, and the vertical arms of the cross represent bonds going back into the plane of the page

five prime end (5') the end of a DNA or RNA strand with a free 5'-phosphate group

fluidized bed combustion crushed coal and limestone are suspended in the bottom of a boiler by an upward stream of hot air; as the coal burns, sulfur dioxide gas from the coal combines with limestone to form solid calcium sulfate that is recovered with the ash

fly ash finely divided particles of ash present in flue gases resulting from the combustion of a fossil fuel

foam dispersion of a gas in a liquid or solid

formal charge the charge an atom in a Lewis structure would have if the bonding electrons were shared equally

forward reaction the conversion of the reactant into products in an equilibrium reaction

fraction mixture of liquids with similar boiling points collected by fractional distillation

fractional distillation a mixture of liquids is separated into its components by successive vaporizations and condensations in a vertical fractionating column

fractionating column a colum packed with inert beads in which many separate distillations can occur so that a liquid mixture can be separated into its components

free radical a species with one or more unpaired electrons, often produced by photolysis

frequency the number of waves produced every second by the source, measured in units of hertz (Hz) or s⁻¹ (per second) and given the symbol *f*

Friedel–Crafts reaction method for substituting an alkyl or acyl group into a benzene ring and forming a C—C bond, involving the reaction between benzene (or other aromatic compound) with a halogenoalkane or acyl chloride in the presence of a so-called halogen carrier (for example, anhydrous aluminium chloride); it proceeds via electrophilic substitution

fuel cell device which converts chemical energy directly into electrical energy: a gaseous fuel, usually hydrogen or a hydrocarbon, and oxygen are passed over porous electrodes where combustion occurs; this is accompanied by the production of an electric current

full structural formula a formula showing the relative positioning of all the atoms in a molecule and the bonds between them

fullerenes cage-like molecules composed of fused rings of carbon atoms arranged into pentagons and hexagons

functional group an atom or group of atoms (other than hydrogen) that imparts specific physical and chemical properties to a homologous series of organic compounds

G

gamma rays electromagnetic radiation of very short wavelength which can penetrate several metres of concrete; released as a result of energy changes within the nuclei of atoms

gas a state of matter in which there are small attractive forces operating between the particles; the individual particles move at high velocity in straight lines (until they collide with each other or the walls of the container)

gas constant the constant that appears in the ideal gas equation, defined as $R = \frac{PV}{T}$ for one mole of ideal gas; it has the value of $8.314\,J\,mol^{-1}\,K^{-1}$

gas–liquid chromatography (GLC) form of partition or adsorption chromatography in which the mobile phase is a gas and the stationary phase a liquid; solid and liquid samples are vaporized before being introduced to the column

gateway drug habit-forming substance whose use may lead to the abuse of drugs that are more addictive or more dangerous

gene segment of a DNA molecule that contains the genetic code for a single protein molecule

genetically modified (GM) food foods derived from genetically modified organisms which have had specific changes introduced into their DNA by genetic engineering

genome all the genetic material present in a virus, bacterium or nucleus of a cell

geometrical isomerism a form of stereoisomerism that describes the orientation of functional groups at the ends of a bond around which no rotation is possible

giant covalent lattice a regular arrangement, usually three-dimensional, of covalently bonded atoms that extends throughout the substance

giant structure a lattice, usually three-dimensional, of ions or atoms in which the bonding (ionic, covalent or metallic) extends throughout the substance

Gibbs equation $\Delta G = \Delta H - T\Delta S$ where ΔH equals the change in enthalpy, T equals the absolute temperature, and ΔS equals the change in entropy

Gibbs free energy change of formation, ΔG_f the Gibbs free energy of formation of the species from its elements (in their standard states under standard thermodynamic conditions) expressed as Gibbs free energy per mole of the species

Gibbs free energy change, ΔG a thermodynamic function equal to the enthalpy change minus the product of the entropy change and the absolute temperature: a negative sign indicates that the reaction is spontaneous under standard thermodynamic conditions; the energy available to do work (at constant pressure)

global warming increase in the average temperature of the Earth's atmosphere since the Industrial Revolution, believed to be a consequence of rising levels of greenhouse gases, especially carbon dioxide

global warming potential (GWP) the total contribution to global warming resulting from the emission of one unit of a gas relative to one unit of the reference gas, carbon dioxide, which is assigned a value of 1

globular protein soluble protein with a globular or rounded shape

glycolysis an ATP-generating metabolic process that occurs in nearly all living cells, in which glucose is converted in a series of enzyme-controlled steps to pyruvic acid, without the presence of oxygen; it occurs in the cytoplasm of cells

glycosidic bond covalent bond, C—O—C, formed between two reacting sugar molecules; or the bond between the sugar and base in a nucleotide, in the presence of enzymes

grain orderly arrangement of metal atoms in a crystal structure

graphene a single atom thick planar sheet of graphite (in isolation)

gravimetric analysis a method of quantitative analysis for finding the composition and formulas of compounds based on accurate weighing of reactants and products

greenhouse effect heating effect occurring in the atmosphere because of the presence of greenhouse gases that absorb infrared radiation

greenhouse gases gases that contribute to the greenhouse effect and global warming by absorbing infrared energy emitted or reflected from the surface of the Earth

Grignard reagent alkyl- or aryl- magnesium halide; important in the synthesis of carbon–carbon bonds

ground state the lowest possible energy state of an atom or molecule

group a column of the periodic table which contains elements with similar chemical properties: atoms of elements in the same group have the same number of electrons in their outer or valence shell

H

Haber process the industrial manufacture of ammonia from nitrogen and hydrogen, carried out at high pressure (~ 350 atm) and moderate temperature (~ 350 °C) in the presence of an iron catalyst

half-cell an electrode in contact with an aqueous solution of ions: two half-cells can be connected to make a voltaic cell

half-equation the two parts of a redox reaction, one describing the oxidation and the other reduction

half-life (in a chemical reaction) the time taken for the concentration of a reactant to reach a value which is the mean or average of its initial and final values

halide a compound of one of the halogens, group 7

halide ions the halide ions are fluoride (F⁻), chloride (Cl⁻), bromide (Br⁻) and iodide (I⁻); the halides are salts that contain a metal ion combined with the halide ions

hallucinogens drugs that induce alterations in perception, thinking and feeling

halogen carrier electron-deficient metal halide (in its anhydrous form) that acts as a Lewis acid during electrophilic substitution

halogenation any reaction in which a halogen atom (and no other elements) is introduced into a molecule

halogenoalkanes a homologous series of organic compounds in which one (or more) of the hydrogen atoms of an alkane have been substituted or replaced by halogen atoms

halogens a group of reactive non-metals in group 7 of the periodic table, composed of diatomic molecules

hardening (of oils) process of converting unsaturated oils into more solid saturated fats by hydrogenation using a nickel catalyst

heat the form of energy transferred between two objects due to a temperature difference between them; the origin of heat energy is the movement of atoms, ions and molecules, that is, their kinetic energy

heat capacity the amount of heat energy required to raise the temperature of a substance by one kelvin (or one degree Celsius)

heavy metal toxic metals, such as cadmium, mercury and lead, which have relatively high relative atomic masses: they often function as enzyme inhibitors

Heisenberg's uncertainty principle a principle of quantum mechanics which states that it is not possible to determine the position and momentum of a particle at the same instant: the act of observing the particle will in some way affect its behaviour

heme an iron atom complexed in the centre of an organic ring called a porphyrin

Henderson–Hasselbach equation the equation derived from the expression for K_a, that enables the calculation of the pH of a buffer:

$$pH = pK_a + \log_{10}\frac{[A^-(aq)]}{[HA(aq)]}$$

(it is assumed in a calculation that all of the anion, $A^-(aq)$, is derived from the salt and none from the acid)

Henry's law the amount of a gas that dissolves in a liquid is proportional to the partial pressure of the gas over the liquid, provided no chemical reaction takes place between the liquid and the gas

hertz (Hz) a unit of frequency equal to one cycle per second or s^{-1}

Hess's law the total enthalpy change for a reaction is independent of the route taken – it depends only on the initial and final states

heterogeneous catalysis when the catalyst and the reactants are in different phases (or states)

heterogeneous equilibrium where the reactants and products are in more than one phase

heterolytic fission the cleavage of a covalent bond so that one of the atoms or groups separates with both bonding electrons and becomes negatively charged, leaving the other atom or group positively charged

high-density lipoprotein (HDL) smallest and densest lipoproteins, which contain a high proportion of protein; because HDL can remove cholesterol from the arteries, and transport it back to the liver for excretion, they are seen as 'good cholesterol'

high-performance liquid chromatography (HPLC) technique in which the sample is forced through the chromatography column under pressure; the mobile phase is a liquid and the stationary phase is a solid

high-resolution NMR NMR performed in the presence of a strong and stable magnetic field so that spin–spin coupling can be observed

homogeneous catalysis when the catalyst and the reactants are in the same phase (or physical state)

homogeneous equilibrium where the reactants and products are in the same phase

homologous series a group of organic compounds that follow a regular structural pattern and have the same general molecular formula , differing only by the addition of methylene, $-CH_2-$, groups; the members will have almost identical chemical and physical properties, for example, boiling point and viscosity, that increase gradually

homolytic fission the breaking of a covalent bond so that one electron from the bond is left on each fragment, resulting in the formation of two free radicals

homopolymer polymer that is constructed from only one monomer

hormone a molecule that is secreted directly into the bloodstream of an organism through a ductless gland and acts as a chemical messenger, carrying information from one cell or group of cells to another area

humus part of the soil which consists of complex organic matter and is derived from the decayed remains of plants and animals

Hund's rule the electronic configuration in orbitals of the same energy will have the maximum number of unpaired electrons; electrons with parallel spins are lower in energy than a corresponding pair with opposed spins

hybridization the mixing of two or more atomic orbitals to form the equivalent number of hybrid molecular orbitals (all of identical shape and energy) which overlap and form covalent bonds

hydrated ion ions dissolve in water and attract the polar molecules of water, becoming associated with a variable or definite number of water molecules via electrostatic attraction or coordinate bonding

hydrated salt a salt associated with a definite number of molecules of water

hydration a reaction where an unsaturated molecule adds a molecule of water, or where water molecules interact with ions in aqueous solution

hydrocarbon an organic compound containing only hydrogen and carbon atoms

hydrochlorofluorocarbons (HCFCs) compounds containing carbon, hydrogen, chlorine and fluorine; they have a lower ozone-depleting potential than CFCs because the hydrogen makes them less stable and therefore less damaging to the ozone layer.

hydrocracking process by which the hydrocarbon molecules of crude oil or petroleum are broken into simpler molecules, by the addition of hydrogen under high pressure and in the presence of a catalyst

hydrofluorocarbons (HFCs) compounds containing carbon, hydrogen and fluorine which, because they contain no chlorine atoms, do not directly affect stratospheric ozone; however, they are efficient absorbers of infrared radiation

hydrogen bond an unusually strong intermolecular interaction that occurs among molecules possessing permanent dipole moments, for example, O—H, N—H and H—F

hydrogenation the addition of hydrogen across a multiple bond, often performed in the presence of a heated transition metal catalyst

hydrolytic rancidity hydrolytic cleavage of triglycerides in fats and oils to yield free fatty acids – associated with off flavours in dairy products

hydrophilic used to describe molecules or functional groups that are soluble in water

hydrophobic used to describe molecules or functional groups that are poorly soluble or insoluble in water

hypervalent the ability of an atom in a molecule or ion to expand its valence shell beyond the limits of the octet rule

I

ideal gas a hypothetical state that consists of molecules or atoms that occupy negligible space and have no attractive or repulsive forces operating between them, or between themselves and the walls of the container; all collisions between the molecules or atoms are perfectly elastic; the behaviour of an ideal gas is exactly described by the ideal gas equation

ideal gas equation an equation relating the absolute temperature (T), pressure (P), volume (V) and amount (n) of an ideal gas; $PV = nRT$

identify find an answer from a given number of possibilities

incineration the process of burning solid waste under controlled conditions to reduce its weight and volume, and often to produce energy

independent variable the variable that is manipulated or changed during an investigation

individual order of reaction the power to which the concentration of a particular species is raised in the rate expression: if rate = $k[A]^a[B]^b$, then a and b are the individual orders with respect to the reactants; the individual orders indicate how the rate will change when the concentration of that species is changed

induced fit model model of an enzyme–substrate reaction that causes a conformational change in the active site of the enzyme which allows the substrate to fit perfectly

inductive effect effect of a functional group or atom in an organic molecule which attracts sigma electrons towards itself, or repels them, resulting in the formation of a dipole in the molecule

inert electrode an electrode that serves only as a source or sink for electrons without playing a chemical role in the electrode reaction (though it may act as a catalyst)

infrared spectroscopy absorption spectroscopy carried out in the infrared region of the electromagnetic spectrum, generally detecting bond stretching and bending

inhibitor a substance that decreases the rate of a chemical reaction

initial rate the rate of a chemical reaction extrapolated back to the instant the reactants were mixed

initial rates method method of finding the rate-law expression by carrying out a reaction with different initial concentrations and analysing the resultant changes in initial rates

initiation the first elementary step in a free radical reaction; it involves the homolytic cleavage of a bond, typically by ultraviolet radiation or high temperature, to generate free radicals

instantaneous rate of reaction the rate of a reaction at a particular point in time; evaluated from the slope of a plot of concentration versus time

integration trace area under an NMR resonance peak, which is proportional to the number of hydrogens which that resonance peak represents; experimentally, the integrals will appear as a line over the NMR spectrum

inter-nuclear axis the imaginary axis that passes through the two nuclei in a bond

intercalation incorporation of a foreign atom in a crystal lattice (usually in the spaces between atoms)

intermediate a chemical species that is neither an initial reactant nor a final product but is formed and consumed during the overall chemical reaction; intermediates never appear in a rate expression

internal resistance resistance offered by the chemical constituents of a cell or a battery.

interpolation estimating a value between two known values, frequently on a graph

intravenously administering a drug directly into a vein by injection

iodine number mass of iodine in grams which 100 grams of a fat or oil can absorb; it is directly related to the amount of unsaturated fatty acids present in the fat or oil

ion a charged particle formed by the loss or gain of electrons from a single atom (simple or monatomic ion), or a group of atoms (polyatomic ion) (*see* cation and anion)

ion exchange exchange of ions of the same charge between an aqueous solution and a solid in contact with it

ion-exchange chromatography form of chromatography in which ions are retained by oppositely charged groups covalently bonded to a solid support; the analyte ions are retained by displacing ions associated with the bonded functional group

ionic bonding a strong electrostatic force of attraction between all the oppositely charged ions arranged into a lattice

ionic product constant, K_w (of water) the product of the concentrations of hydrogen and hydroxide ions in water under standard thermodynamic conditions

ionic radius the radius of an ion in the crystalline form of a compound, which varies periodically: the ionic radius increases down a group and generally decreases from left to right across a period

ionization energy the enthalpy change when an electron is removed from an atom or ion in the gaseous state, under standard conditions

isoelectric point (of an amino acid) pH at which an amino acid in solution has no overall electrical charge; at this pH the amino acid does not move when placed in an electric field

isoelectronic species two chemical species which have the same number of total electrons or the same number of electrons in the valence shell

isomerization chemical process carried out in a refinery that involves a rearrangement of atoms and bonds within a molecule, without changing the molecular formula

isotactic polymer chain in which the substituents are all sterochemically oriented on the same side of the chain

isotopes two or more atoms of the same element with different numbers of neutrons (and therefore different mass numbers)

K

Kekulé structure a localized description of the structure of benzene in which there is a six-membered ring with alternate double and single bonds

kelvin scale a thermodynamic or an absolute temperature scale starting from absolute zero (−273 °C), the lowest temperature possible, using units of kelvin (K) with the same magnitude as degrees Celsius

ketones a homologous series of compounds with the general formula RCOR', having two alkyl or aryl groups bonded to a carbonyl group

ketose simple sugar that has a ketone as its carbonyl group (in its acyclic (straight chain) form)

kinetic theory explains the physical properties of solids, liquids and gases in terms of the movement of particles (atoms, ions or molecules): the theory also accounts for the changes that occur during a change in state

kwashiorkor form of malnutrition involving an inadequate intake of protein

L

label add labels to a diagram.

Larmor frequency resonance frequency of a spin in a magnetic field; the frequency which will cause a transition between the two spin energy levels of a nucleus

lattice a regular, repeating three-dimensional arrangement of atoms, molecules or ions within a crystal

lattice enthalpy the energy released when one mole of a solid ionic compound is decomposed to form gaseous ions (infinitely far apart) under standard thermodynamic conditions

law of conservation of mass mass is not lost or gained during a chemical reaction – the total mass of the reactants equals the total mass of the products

laxative a medication used to produce bowel movements

Le Châtelier's principle if a constraint is imposed on a system at equilibrium, the system will shift in the direction which tends to partially counteract the change; used to predict which way a reaction will shift if the conditions are changed

least count the smallest division that is marked on an instrument

leaving group an atom or group of atoms which breaks away from a molecule during a substitution or an elimination reaction

lethal dose (50%) dose of a toxicant that will kill 50% of the test organisms within a designated period: the lower the LD_{50}, the more toxic the compound

levorotatory rotates the plane of plane-polarized light anticlockwise, in the (−) direction

Lewis (electron dot diagram) structure a diagram of a molecule showing how the valence electrons are arranged among the atoms in the molecule

Lewis acid a chemical species that can accept an electron pair to form a dative or coordinate covalent bond

Lewis base a chemical species that can donate an electron pair to form a dative or coordinate covalent bond

ligand a molecule or negative ion that donates a pair (or pairs) of electrons to a central metal ion to form a dative or coordinate covalent bond

ligand field splitting removal of a degeneracy of atomic levels in an ion, induced by the bonding or removal of ligands

ligand replacement one or more ligands in a complex ion are replaced, often reversibly, by another

ligand (field) splitting energy ligands complexed to a metal ion will raise the energy of some of its d orbitals and lower the energy of others; the difference in energy is called the ligand field splitting energy

limiting reactant the reactant that is completely consumed, or used up, when a reaction goes to completion; the limiting reagent determines the yield of the reaction

line of best fit the straight or curved line which gives the best approximation to a given set of data

line spectrum an emission spectrum that has only certain wavelengths or frequencies of visible light (it takes the form of a series of bright lines on a back background)

lipids group of organic compounds that contain carbon, hydrogen and oxygen; includes fats, oils, waxes, sterols, and triglycerides, which are insoluble in water but soluble in non-polar organic solvents, for example, ethanol (alcohol)

lipoproteins group of proteins in which at least one of the components is a lipid; lipoproteins, classified according to their densities and chemical qualities, are the means by which lipids are transported in the blood

liquid a state of matter in which particles are loosely attracted by intermolecular forces; a liquid always takes up the shape of the walls of its container and its particles are not arranged into a lattice

liquid crystal substance that flows like a liquid but has some order in its arrangement of molecules

list give a sequence of names or other brief answers with no explanation

literature value a value from the chemical literature of a physical constant or experimental measurement

lobe a rounded region of electron density projecting away from the nucleus

lock and key model model for the mechanism of enzyme activity postulating that the three-dimensional shapes of the substrate and the enzyme are such that they fit together as a key fits into a specific lock

lone pair a lone or non-bonding pair of electrons is a pair of outer or valence shell electrons (that have opposing spins) which are not used to form covalent bonds within the molecule

low-density lipoprotein (LDL) a lipoprotein that carries cholesterol around the body, for use by cells; transports cholesterol to the arteries and increased levels are associated with atherosclerosis, and thus heart attacks and strokes, hence cholesterol inside LDL lipoproteins is called 'bad cholesterol'

low-resolution NMR NMR performed in an inhomogenous magnetic field, where spin–spin coupling cannot be observed

Lyman series emission spectral lines that are caused by the transition of electrons to the ground state of hydrogen; these spectral lines appear at ultraviolet wavelengths

lyotropic liquid crystal liquid crystal prepared by mixing two or more components, one of which is polar in character

M

macronutrients nutrients that the body uses in relatively large amounts – proteins, carbohydrates and fats

macroscopic properties properties of substances in bulk that can be observed or easily measured

magnetic resonance imaging use of a nuclear magnetic resonance spectrometer to produce electronic images of specific atoms and molecular structures in solids, especially human cells, tissues, and organs

Maillard reaction a form of non-enzymatic browning between an amino acid and a reducing sugar, usually requiring the addition of heat – the reactive carbonyl group of the sugar

interacts with the nucleophilic amino group of the amino acid

malleable the ability of metals to be bent and beaten into thin sheets without breaking

marasmus form of malnutrition that involves an insufficient intake of energy (calories)

Markovnikov's rule rule that predicts the major and minor products when a hydrogen halide adds across the double bond in an unsymmetrical alkene: it states that the major product will be the one in which the hydrogen atom attaches itself to the carbon atom with the larger number of hydrogen atoms

mass number the sum of the number of protons and neutrons in the nucleus of the atom or ion

mass spectrometer an instrument (maintained under a hard vacuum) in which gaseous atoms or molecules are fragmented and ionized and then accelerated into a magnetic field where the ions are separated according to their mass-to-charge ratio

Maxwell–Boltzmann distribution curve describes the distribution of velocities or kinetic energies among the atoms or molecules of an ideal gas

measure find a value for a quantity

mechanism a description in terms of bond breaking, bond making and intermediate formation that occurs during the series of elementary steps by which an overall chemical reaction occurs; also describes the movement of electrons

medicine substance intended for use in the diagnosis, cure, mitigation, treatment or prevention of disease

mercury cell electrolytic cell with a flowing mercury cathode used in the industrial production of chlorine from concentrated sodium chloride solution

messenger RNA class of RNA molecules, each of which is complementary (in base pairs) to one strand of DNA; carries the genetic message from the nuclear DNA to the ribosomes

metabolism the chemical reactions that take place in the cells of living organisms

metallic bonding the electrostatic attraction between positively charged nuclei and the sea of delocalized electrons

metals chemical elements which are shiny solids under standard conditions (except mercury) and are good conductors of heat and electricity when solid; they form positive ions (cations)

Michaelis constant the substrate concentration at which an enzyme-catalysed reaction occurs at one half of its maximum rate, V_{max}; it is an approximate measure of the substrate affinity for the enzyme; in general, a lower value of the Michaelis constant K_m means tighter substrate binding

micronutrients nutrients that the body uses in relatively small amounts – vitamins and minerals

mineral ore impure mineral from which a metal can be profitably mined or extracted

mobile phase liquid or gas which percolates through or along the stationary phase during chromatography

moiety part of a molecule

molar absorption coefficient absorbance of light per unit path length (usually the centimetre) and per unit concentration (moles per cubic decimetre): the proportionality coefficient in the Beer–Lambert law

molar gas volume one mole of an ideal gas occupies 22.4 cubic decimetres (dm^3) at 0 °C (273 K) and one atmosphere pressure (stp)

molar mass the mass in grams of one mole of molecules or the formula units of an ionic compound; numerically equal to the relative molecular or atomic mass of a substance, but has units of grams per mole ($g\,mol^{-1}$)

molar solution a solution that contains one mole of solute per cubic decimetre of solution

mole the measure of the amount of a substance; one mole of a substance contains 6.02×10^{23} (Avogadro's constant) of atoms, ions or molecules

molecular formula a chemical formula which shows the actual number of atoms of each element present in a molecule of a covalent compound

molecular ion unipositive ion formed by an unfragmented molecule losing one electron following electron bombardment

molecular orbital theory a theory of chemical bonding based upon the postulated existence of molecular orbitals

molecular orbitals formed in molecules when atomic orbitals combine and merge as atoms bond together – σ and π bonds are molecular orbitals

molecularity the number of chemical species or particles participating in an elementary step of the mechanism: each step has its own molecularity but the overall reaction has no molecularity (unless it is an elementary process)

molecule a group of atoms held together by covalent bonds; a molecule is the smallest unit of a compound that can exist by itself and retain all of its chemical properties

monodentate a ligand that forms a single dative covalent bond to a central metal ion

monomer a small molecule, a large number of which can be polymerized via the formation of covalent bonds to form a polymer

monoprotic an acid which contains one replaceable hydrogen atom per molecule, sometimes also known as monobasic

monosaccharide sugar that cannot be hydrolysed to simpler sugars

multi-stage flash distillation form of distillation in which water is heated then discharged into a chamber maintained slightly below the saturation vapour pressure of the incoming water, so that a fraction of the water content flashes into steam; the steam condenses and the unflashed brine enters another chamber at a lower pressure, where a portion flashes to steam; each evaporation and condensation chamber is called a stage

multiplet pattern of multiple resonances (NMR peaks) observed when the initially single frequency of a given nucleus is split by interactions with neighbouring spins through spin–spin coupling

N

N-terminus amino acid residue at one end of a polypeptide chain that contains a free amino group

n-type semi-conductor formed when the impurities added to silicon donate electrons which enter the unoccupied energy level

nanotechnology engineering of functional systems at the 1–100 nm scale

nanotube a one-dimensional fullerene (a convex cage of atoms with only hexagonal and/ or pentagonal faces) with a cylindrical shape

naphtha general term used to describe a light hydrocarbon fraction from crude oil distillation with a boiling point between 40 °C and 150 °C; an important feedstock used to manufacture other substances

narcotic addictive drug that reduces pain, alters mood and behaviour, and usually induces sleep

narrow-spectrum antibiotic antibiotic that is effective against only a small number of bacterial strains

native state occurrence of an element in an uncombined or free state in nature

natural product a chemical compound produced by a living organism – usually one that has a pharmacological or biological activity

nematic phase phase composed of rod-shaped molecular aggregates that are arranged with parallel but not lateral order

net ionic equation the simplified equation for a reaction involving ionic substances – only those ions that actually participate in the reaction are included in the ionic equation; spectator ions are not included

neutralization a chemical reaction between an acid and a base to produce a salt and water only

neutron neutral sub-atomic particle found in the nucleus of all atoms (except that of the most abundant isotope of hydrogen) – it has approximately the same mass as the proton

nitrating mixture 1:2 molar mixture of concentrated nitric and sulfuric acids used to nitrate some aromatic organic compounds: produces the electrophilic nitronium cation, NO_2^+

nitration type of reaction in which a nitro group ($-NO_2$) is introduced into an aromatic compound, usually via the use of a nitrating mixture

nitrification process in which ammonia in plant and animal wastes and dead remains is oxidized, first to nitrites and then to nitrates; the reactions are catalysed by the bacteria *Nitrosomonas* and *Nitrobacter*

nitrile a compound of the form RCN, where R is an alkyl group or aryl group

noble gases a group of very unreactive gases found in group 0 of the periodic table; they exist as single atoms and all have filled outer or valence shells

non-competitive inhibitor a substance that will bind to a site on the enzyme, other than the active site, which reduces the ability of the enzyme to form the enzyme–substrate complex; V_{max} is decreased by the inhibitor and cannot be restored by increasing the substrate concentration; non-competitive inhibition is thus dependent on the concentration of the inhibitor and the affinity of the enzyme for the inhibitor

non-metals chemical elements that are typically poor conductors of heat and electricity; they form covalent bonds and/or form negative ions (anions)

normal boiling point the temperature at which the vapour pressure of a liquid is exactly one atmosphere

normal phase form of chromatography whereby retention on a sorbent bed increases with the polarity of the sorbent

normal salt formed when all of the replaceable hydrogens of an acid have been replaced by metal ions

nuclear charge the total charge of all the protons in the nucleus

nuclear fission splitting of relatively large atoms or nuclei to release two or more smaller atoms or nuclei, accompanied by production of large amounts of heat and ionizing radiation

nuclear magnetic resonance (NMR) absorption of radio waves at a precise frequency by nuclei with an odd nucleon number in an external magnetic field

nuclear spin a property of certain nuclei (those with odd numbers of protons and/or neutrons in their nucleus) which gives them a magnetic moment; nuclei that do not exhibit this characteristic will not produce an NMR signal

nucleophile a species (molecule or anion) which contains a lone pair of electrons that can be donated to an electron-deficient centre in an organic molecule to form a coordinate (dative) bond

nucleophile a species (molecule or anion) which contains a lone pair of electrons that can be donated to an electron-deficient centre in an organic molecule to form a coordinate (dative) bond

nucleophilic addition type of reaction in which the rate-determining step is the attachment of a nucleophile to a positive (electron-deficient) part of the molecule (often a carbon–oxygen bond in a carbonyl compound)

nucleophilic substitution the substitution of an atom or group of atoms with a nucleophile as the attacking species; can occur via an S_N1 or S_N2 mechanism

nucleotide monomer of the nucleic acids (DNA and RNA) composed of a five-carbon sugar (a pentose), a nitrogen-containing base, and phosphoric(v) acid

nuclide the general term for a unique nucleus

nutrient a substance needed by living organisms for metabolism

nutrient depletion where the nutrients in the soil are being consumed faster than they are being produced or added, so that the amount of nutrients is decreasing

nylon a hard synthetic polymer with a long chain of carbon atoms in which amide groups ($-NH-CO-$) are combined at regular intervals; there is extensive hydrogen bonding between the chains

O

octane number a value used to indicate the resistance of a motor fuel to knock, based on a scale on which isooctane is 100 (minimal knock) and heptane is 0 (bad knock)

octet a set of eight electrons in the valence shell of an atom or ion

octet rule atoms (with an atomic number greater than five) fill their valence or electron shell with eight electrons (an octet) when they form compounds

opiate a medication (or illegal drug) derived from the opium poppy

opium dried latex containing morphine, obtained from the unripe seed pods of the opium poppy (*Papaver somniferum*)

optical isomerism occurs when a molecule has no plane of symmetry and can exist in left- and right-handed forms that are non-superimposable mirror images of each other: the molecule must possess a chiral centre; optical isomers rotate plane-polarized light

orbital a region in space in which an electron may be found in an atom or molecule; each atomic orbital can hold up to a maximum of two electrons with opposite spins

orbits the path of an electron as it travels round the nucleus of an atom (*see* Bohr's theory)

order of reaction *see* overall order or individual order

organic chemistry the study of carbon-containing compounds with the exception of the allotropes of the element itself, metal carbonates and its oxides and halides

organometallic compound organic compounds that contain a metal, particularly compounds in which the metal atom has a direct bond with a carbon atom

osmosis a process where solvent molecules move through a semi-permeable membrane from a dilute solution into a more concentrated solution (which becomes more dilute)

outline give a brief account or summary

overall order of reaction the sum of the individual orders with respect to each of the reactants in the rate expression

oxidation an increase in oxidation number or the loss of electrons

oxidation number a number (usually an integer), positive or negative, given to indicate whether an element in a compound has been reduced or oxidized during a redox reaction

oxidative rancidity reaction of unsaturated fat with excited molecular oxygen to produce undesirable flavours

oxidizing agent a substance that brings about oxidation; it accepts electrons from the reactant or one of the reactants, being itself reduced

oxoanion an anion containing oxygen

ozone depletion the production of a 'hole' in the ozone layer by the action of chlorine atoms released from chlorofluorocarbons (CFCs), which destroy ozone by reactions on the surface of ice crystals

ozone layer layer of ozone in the stratosphere (between 15 and 30 km altitude) which prevents harmful ultraviolet radiation from reaching the Earth's surface

ozone layer part of the Earth's atmosphere that has a high concentration of ozone, O_3, which absorbs UV-B radiation from the Sun, which is harmful to life

P

p-type semi-conductor formed when the impurities added to silicon withdraw electrons from the occupied energy level leaving positive 'holes' which allow conduction to occur

PANs peroxyacetyl nitrates – compounds formed in photochemical smog by the addition of nitrogen dioxide to the peroxyacyl radical

paper chromatography chromatography carried out using a special grade of filter paper as the stationary phase

parallax error an error in reading an instrument when the eye of the observer and the pointer are not in a line perpendicular to the plane of the scale

parallel synthesis single-batch method that uses a mixture of reagents at each step of a synthesis to generate a large number of different products

parenteral drug (or nutrients) taken into the body or administered in a manner other than through the digestive tract, as by intravenous or intramuscular injection

partial pressure the pressure a gas in a mixture of (ideal) gases would exert on the container if it were the only gas in the container; equal to the mole fraction of that gas multiplied by the total pressure

particulates any type of solid particle or droplet in the air in the form of haze, smoke or dust, which can remain suspended in the air or atmosphere for extended periods

partition distribution of a solute between two immiscible solvents

Pauli exclusion principle no two electrons in an atom can have the same set of four quantum numbers: the Pauli principle states that a maximum of two electrons can occupy an atomic orbital – and these electrons must have opposite spins

Pauling scale a common comparative measure of electronegativity which runs from 0 (least electronegative or most electropositive) to 4 (most electronegative or least electropositive)

penicillinase enzyme produced by some bacterial species that inactivates the antimicrobial activity of certain penicillins

penicillins a group of antibiotics, originally derived from the mould *Pencillium notatum*,

which work by disrupting synthesis of the bacterial cell wall

peptic ulcer a hole in the lining of the stomach, duodenum, or esophagus

peptide bond an amide bond resulting from the condensation reaction between the amine group of one amino acid and the carboxylic acid group of another

peptidoglycan carbohydrate polymer cross-linked by proteins, found in the cell walls of bacteria

percentage error an error expressed as a percentage of the value measured or the true value

percentage uncertainty

$$\frac{\text{random uncertainty}}{\text{actual measurement}} \times 100$$

percentage yield the actual or experimental yield as a percentage of the theoretical or calculated yield

$$\text{percentage yield} = \frac{\text{experimental yield}}{\text{theoretical yield}} \times 100$$

period a horizontal row in the periodic table which contains elements with same number of shells, and with an increasing number of electrons in the outer or valence shell

periodic table a table of the chemical elements arranged in order of increasing atomic (proton) number to show the similar chemical properties of elements

periodicity the regular repetition of chemical and physical properties as you move across (from left to right) and down the periodic table

pH the negative logarithm (to the base ten) of the hydrogen ion concentration (in $mol\,dm^{-3}$)

pH probe an electrode that can be used to accurately measure (via voltage) the pH of an aqueous solution

pH scale typically ranges from 0 to 14 and is used to describe the acidity or alkalinity of an aqueous solution

pharmacology science of studying both the mechanisms and the actions of drugs, usually in animal models of disease, to evaluate their potential therapeutic value

pharmacophore molecular framework that carries the essential features responsible for a drug's activity

phase a physically or chemically distinct part of a chemical equilibrium; a phase is homogenous throughout and is separated from other phases by a phase boundary

phase change a change in the nature of a phase or in the number of phases as a result of some variation in externally imposed conditions, such as temperature or pressure

phase diagram a graph of pressure versus temperature showing the regions where each phase of a system is thermodynamically the most stable

phase equilibrium an equilibrium in which the identities of the phases are fixed, with transfer from one phase to another feasible

phenol group of organic compounds in which at least one hydroxyl group is bonded to directly to one of the carbon atoms of a benzene ring

phosphodiester linkage the covalent bond that holds together the polynucleotide chains of

RNA and DNA by joining a carbon atom in the pentose sugar of one nucleotide to a carbon atom in the pentose sugar of the adjacent nucleotide.

phospholipids lipids consisting of glycerol to which two fatty acids and one ionized phosphate group are chemically bonded

photochemical reaction a reaction initiated by light

photochemical smog form of local atmospheric pollution found in large cities in which oxides of nitrogen and unburnt hydrocarbons react in the presence of light to produce a range of harmful products including ozone and PAN

photoconduction change in the electrical conductivity of a substance as a result of absorbing electromagnetic radiation

photoelectric emission ejection of electrons from the surface of a conductor by incident electromagnetic radiation, especially by visible light

photon a 'packet' or quantum of light, or other electromagnetic radiation

photovoltaic device system that converts solar radiation directly into electric energy

physical change a change that does not involve changing any substances into new chemical substances

physical weathering breakdown of rock and minerals into small sized particles through mechanical stress

pi bond a bond formed by the sideways overlap of two p-orbitals; in a π bond the electron density lies to either side of a plane through the nuclei of the two atoms

pi delocalization a situation in a molecule where the π electrons in a molecule are not localized between a pair of atoms but can move over several adjacent atoms

pig iron impure, high-carbon iron produced by reduction of iron ore in a blast furnace

pigment substance that has a colour resulting from selective colour absorption

placebo an inactive substance or treatment that looks the same as, and is given the same way as, an active drug or treatment being tested; the effects of the active drug or treatment are compared to the effects of the placebo

placebo effect improvement in the condition of a sick person that occurs in response to treatment but that cannot be considered due to the specific treatment used

Planck's constant the constant of proportionality relating the energy of a photon to the frequency of that photon; approximately 6.63×10^{-34} joule-seconds (Js)

plane-polarized light light which vibrates in one plane, formed by passing unpolarized light through a polarizer; plane-polarized light is used for the detection of optical activity

plasmids small circular DNA molecules occurring in bacteria, which can be exchanged between different cells under natural conditions

plasticizer substance added to a plastic to make it flexible

pOH the negative logarithm (to the base ten) of the hydroxide ion concentration (in $mol\,dm^{-3}$)

polar covalent bond formed when electrons are shared unequally between two atoms due to a difference in electronegativity: one atom has a partial positive charge, and the other atom has an equal but opposite partial negative charge; the larger the difference in electronegativities the greater the polarity and the larger the partial or fractional charges

polar molecule an unsymmetrical molecule whose individual dipoles do not sum to zero or cancel

polar stratospheric clouds clouds in the stratosphere that form under extremely cold conditions, when nitric acid, water vapour and other trace chemicals freeze to form ice crystals; they provide a solid surface on which chlorine-containing reservoir compounds can gather, and on this surface the chemical reactions involved in the depletion of ozone are greatly increased

polar vortex a phenomenon that occurs during the polar winter in which stratospheric air moves in a circular motion, with a volume of relatively still air at its centre; the temperature in the vortex is approximately $-80\,°C$, which assists in the formation of polar stratospheric clouds

polarimeter a device used to study optically active substances

polarization the separation of charge in a polar bond

pollutant a waste material that pollutes air, water or soil

pollution modification of the environment caused by human activities

polyamide a polymer in which the monomer molecules are linked by amide bonds

polydentate a ligand that forms multiple dative covalent bonds to a central metal ion

polyester a synthetic polymer formed by reacting alcohols with acids, so that the monomers are linked by the group, —O—CO—

polymer a compound containing very large molecules composed of repeating units called monomers

polymerase chain reaction (PCR) repetitive enzyme-based procedure, using a thermostable DNA polymerase, which results in the production of many copies of a DNA sequence

polymorphic natural variations in a gene, DNA sequence or chromosome that have no adverse effects on the individual and occur with fairly high frequency in the general population

polynucleotide a chain made up of many nucleotides which forms nucleic acids: DNA strands are long polynucleotide chains

polypeptide long linear chain of between 10 and 100 amino acids linked via peptide bonds

polysaccharide carbohydrate whose molecules contain chains of monosaccharide molecules

polysome a group of several ribosomes simultaneously translating the same messenger RNA: shortened form of the term polyribosome

potential difference the work done per coulomb in moving charge from one point to another; if the potential difference between two points is 1 volt, then the passage of 1 coulomb of charge between these points involves 1 joule of energy ($1\,V = 1\,J\,C^{-1}$)

precipitate an insoluble substance formed by a chemical reaction in solution

precipitation a reaction in which an insoluble substance is prepared from two solutions of two soluble substances

precision total amount of random error present in a measurement: indicates how close together a series of measurements are

predict give an expected result

pressure law the gas law stating that the pressure of a fixed mass (at constant volume) of an ideal gas is directly proportional to absolute temperature

primary alcohol an alcohol in which the hydroxyl group is bonded to a primary carbon

primary battery group of voltaic cells in which the chemical reaction producing the voltage is not reversible; cannot be recharged

primary carbon atom a carbon atom that is singly bonded to only one other carbon atom

primary pollutant pollutant that enters the atmosphere directly from various sources

primary smog smog containing soot particles and aqueous sulfur dioxide from the combustion of coal in power stations

primary standard a reagent that is extremely pure, stable, has no water of crystallization, and has a high molar mass; can be used to prepare a solution of accurately known concentration (*see* standard solution)

primary structure order or sequence of the amino acids in a polypeptide chain

principal quantum number (n) a quantum number that can be regarded as a 'label' for the shells of an atom

probe radio-labelled fragment of nucleic acid containing a nucleotide sequence complementary to a section of a gene

prodrug inactive form of a drug; the drug is converted to its active form by processes in the body of the person who has taken it

promoter a substance added to a solid catalyst to improve its performance

propagation an elementary reaction involving one radical causing the formation of another radical

prostaglandins group of organic compounds derived from essential fatty acids and causing a range of actions, including inflammation

protein a long sequence of amino acid residues combined together via peptide bonds, which takes up a particular shape or conformation when folded

proticity the number of replaceable hydrogen atoms a molecule of acid contains

proton positively charged sub-atomic particle found in the nuclei of all atoms – it has approximately the same mass as the neutron

pseudo first-order rate constant if the rate expression is rate = $k[A][B]$, then the function $k[A]$ is the pseudo first-order rate constant with respect to B; similarly, $k[B]$ is the pseudo first-order rate constant with respect to A: the term is used when the concentration of one reactant greatly exceeds that of another

purine nitrogen-containing base found in nucleotides and nucleic acids; contains one six-membered ring fused with a five-membered ring

pyrimidine nitrogen-containing base found in nucleotides and nucleic acids; contains one six-membered ring

pyrolysis chemical decomposition occurring as a result of high temperature in inert (non-oxidizing) conditions

Q

qualitative analysis analysis used to determine the nature but not the concentration of the constituents of a material or substance

quantitative analysis analysis used to determine the amount of each component in a material or substance

quantum number a label for a state of a quantized system, such as a sub-atomic particle, atom, ion or molecule

quaternary structure overall three-dimensional structure of a protein composed of two or more polypeptide chains

quenching process of rapidly cooling steel from a temperature above the critical temperature

R

racemic mixture an equimolar mixture of two enantiomers of the same compound; as their rotation of plane-polarized light is equal but opposite, the mixture is not optically active

radical a species with one or more unpaired electrons, often produced by photolysis

radical cation radical with a positive charge

radioactive a substance that contains unstable radioactive atoms that release ionizing radiation as they decay

radioactivity spontaneous disintegration of the nuclei of certain elements or atoms with the emission of alpha or beta particles, sometimes accompanied by gamma radiation

radiocarbon dating the determination of the approximate age of an object by measuring the amount of carbon-14 it contains

radioisotope a naturally or artificially produced radioactive isotope of an element

radiotherapy treatment of disease with radiation, especially by selective irradiation with X-rays or other ionizing radiation and by ingestion of radioisotopes

random uncertainty an error which is present every time a measurement is recorded – the effects can be reduced by repeating the measurement and averaging

rate constant, k the constant of proportionality in a rate expression; the rate constant is unaffected by changes in the concentrations of the reactants, and is affected only by changes in temperature

rate expression the experimentally determined relationship between the rate of a reaction and the concentrations of the chemical species that occur in the reaction

rate of reaction how fast reactants are being converted to products during a chemical reaction – either the rate of formation of a product or the rate of consumption of a reactant may be used: they are related to each other by dividing the corresponding coefficient in the stoichiometric equation; the rate of a reaction has units of $mol\,dm^{-3}\,s^{-1}$

rate-determining step the slowest elementary step in a reaction mechanism that controls the rate of the overall reaction

reaction kinetics *see* chemical kinetics

reaction pathway a sequence of reactions involving the conversion of organic compounds

reaction quotient a value obtained by applying the equilibrium law to any concentrations other than to equilibrium concentrations

reactivity series an order of metal reactivity based on the relative rates of reactions of metals with oxygen, water, dilute aqueous acid and solutions of metal ions or salts (*see* electrochemical series)

real gas a real gas does not obey the gas laws and exhibits non-ideal behaviour; its molecules have a finite size and there are intermolecular or interatomic forces of attraction operating between the molecules or atoms

redox equation an equation constructed by combining two half-equations so the numbers of electrons on both sides of the equation cancel

redox equilibrium an equilibrium involving a redox (electron transfer) reaction

redox reaction a reaction involving reduction and oxidation and which results in one or more electrons being transferred and is accompanied by changes in oxidation numbers

redox titration a titration used to determine the concentration of a solution of an oxidizing agent or of a reducing agent

reducing agent a chemical that brings about reduction; it donates electrons to a reactant, being itself oxidized

reducing sugar a type of sugar with an aldehyde or ketone group, which allows the sugar to act as a reducing agent, for example in the Maillard reaction and Benedict's reaction

reduction a decrease in oxidation number or gain of electrons

reflux process of boiling a liquid in a flask connected to a condenser so that the condensed liquid runs back into the flask

reforming cyclization of straight-chain hydrocarbons from crude oil by heating under pressure with a catalyst

relative atomic mass the weighted average mass (according to relative abundances) of all the naturally occurring isotopes of an element compared with an atom of the $^{12}_{6}C$ carbon isotope, which has a mass of exactly 12

relative formula mass the sum of the relative atomic masses of the atomic species as given in the formula of an ionic compound; it is a pure number without units

relative isotopic mass the mass of a particular isotope of an element compared to the mass of one twelfth of a carbon-12 atom; measured using a mass spectrometer

relative molecular mass the relative molecular mass of a compound is the sum of the relative atomic masses of the atomic species in one molecule; it is a pure number without units

repeatability the closeness of agreement between independent results obtained with the same method, under the same conditions (same operator, same apparatus, same laboratory and after short intervals of time) (*see* precision)

replication (DNA) synthesis of double-stranded DNA (daughter DNA) identical in sequence to the original (parent) DNA

reproducibility the ability to duplicate measurements over long periods of time by different laboratories

residence time the amount of time a substance can remain in a reservoir: the reservoir can be aquatic, atmospheric or terrestrial

resonance occurs when more than one valid Lewis structure can be written for a particular molecule or ion: the actual electronic structure is not represented by any one of the Lewis structures but by a weighted average of them

resonance energy the difference in energy between the resonance hybrid and the individual resonance structures

resonance hybrid a molecule or ion described by a single localized structure, often using dotted lines to indicate partial bonds

resonance structure the Lewis structures that can be drawn for a resonance-stabilized molecule: their sum (according to their weightings) or blend describes the resonance hybrid

restricted rotation the phenomenon where bonded atoms cannot rotate relative to one another because either the bond between them prevents it or the two atoms are part of a ring that restricts rotation; restricted rotation gives rise to geometrical isomerism in organic compounds

restriction enzyme enzyme that causes cleavage of both strands of double-stranded DNA at or near specific base sequences

retention factor, R_f distance travelled by a compound on a chromatography column or chromatography paper divided by the distance travelled by the solvent or eluent

retention time time between injection of a substance onto a chromatography column and the appearance of the peak

retrovirus RNA virus containing reverse transcriptase

reverse osmosis method for removing dissolved salts from water, for example, in the desalination of sea water, by the use of a semi-permeable membrane and high pressure

reverse phase form of chromatography in which non-polar to moderately polar analytes are extracted from a polar solution using a non-polar sorbent

reverse reaction *see* backward reaction

reverse transcriptase enzyme found in retroviruses capable of synthesizing DNA complementary to RNA

reversible reaction a physical or chemical reaction that can go either backwards or forwards depending on the conditions; when a reversible reaction has equal forward and reverse rates the reaction is at equilibrium

Reye's syndrome potentially fatal condition that causes damage to the brain and liver – associated with aspirin consumption by children during a viral illness

ribosome complex of proteins and RNA, the site of protein synthesis in cells; receives messenger RNA from the nucleus

ring-opening polymerization polymerization in which a cyclic monomer is converted into a polymer which does not contain rings

RNA polymerase an enzyme that links the growing chain of ribonucleotides during transcription

S

salinization the increase in salt concentration in soil

salt an ionic compound formed by the reaction of an acid with a base, in which the hydrogen of the acid has been replaced by a metal ion

salt bridge an electrical connection made between two half-cells that contains an electrolyte with ions that do not cause precipitation of the ions in the two half-cells; it allows ions to flow to maintain electroneutrality while preventing the two solutions from mixing, and thus prevents a build up of charge which would stop the flow of current

salt hydrolysis occurs when one or both ions derived from a soluble salt undergo a chemical reaction with water molecules, leading to production of an acidic or an alkaline solution

saturated used to describe an organic molecule, for example an alkane, that contains no carbon–carbon multiple bonds and contains only carbon–carbon single bonds

saturated vapour pressure the pressure exerted by a pure substance (at a given temperature) in a system containing only the vapour and liquid of the substance

scanning tunnelling microscope an instrument able to image conducting surfaces to atomic accuracy

Schrödinger wave equation a mathematical equation explaining the movement of an electron around an atom in the form of a standing wave

scission breaking of a molecular bond causing the loss of a side-group or shortening of the overall chain

second electron affinity the addition of one mole of electrons to one mole of gaseous singly negative ions (under standard thermodynamic conditions)

Second Law of Thermodynamics spontaneous processes are accompanied by an increase in the total entropy of the universe

second messenger a method of cellular signalling when a signalling molecule is rapidly produced/secreted, which can then go on to activate proteins within the cell to produce a cellular response

second-order reaction a reaction in which the rate of reaction is proportional to the product of the concentrations of two of the reactants or to the square of the concentration of one of the reactants

secondary alcohol an alcohol in which hydroxyl group is bonded to a secondary carbon atom

secondary battery group of voltaic cells in which the chemical reaction producing the voltage is reversible; it can be recharged

secondary carbon atom a carbon atom that is singly bonded to two other carbon atoms

secondary pollutant formed when primary pollutants react with each other or with other compounds present in the atmosphere

secondary smog *see* photochemical smog

secondary structure (of a protein) the three-dimensional conformation of sections of polypeptide chains; common protein secondary structures include the α-helix, the β-sheet and the random coil

sedative drug that depresses the central nervous system, producing a calming effect

semi-conductor crystalline material with a conductivity intermediate between that of a conductor and an insulator

semi-crystalline polymer thermoplastic with both amorphous and crystalline regions

semi-synthetic synthesized using natural chemicals as starting materials

shelf-life the period of time which a product can be stored, under specified conditions, and remain in optimum condition and suitable for consumption

shielding screening of an applied magnetic field experienced by a nucleus due to the electron cloud around an atom or molecule

shielding effect shielding electrons are the electrons in the energy levels between the nucleus and the outer or valence electrons, described as 'shielding' electrons because they 'shield' the valence electrons from the nuclear charge and reduce the attractive force on them by the protons in the nucleus

show give the steps in a calculation or derivation

side-effect secondary and usually adverse effect of a drug

sigma bond formed by the head-on overlap between atomic orbitals along an imaginary line joining the two nuclei (the inter-nuclear axis); the electron density is concentrated around the two nuclei

significant figures the figures in a number that are meaningful

sink a natural removal process for pollutants in the troposphere such as dissolution and removal via rain

skeletal formula a formula representing organic molecules where the carbon and hydrogen atoms are omitted and only the bonds between them are shown

sketch represent by means of a graph showing a line and labelled but unscaled axes but with important features (for example, intercept) clearly indicated

slag mixture of molten non-metallic oxides produced during the extraction of iron in the blast furnace

smelting melting or fusing an ore, often with a reducing agent, in order to extract a metal

smog form of air pollution consisting of a combination of smoke and fog

S_N1 (halogenoalkanes) a nucleophilic substitution in which a carbocation intermediate is formed in the rate-determining step which then reacts with the nucleophile

S_N2 (halogenoalkanes) a nucleophilic substitution where a concerted reaction occurs in which the nucleophile begins to bond with the carbon bearing the halogen as the halogen begins to leave the molecule

soil degradation decline in soil quality caused by its improper use

soil organic matter (SOM) carbon-containing material in the soil that is derived from living organisms

soil solution aqueous liquid found within a soil; normally contains ions released from mineral particles, organic matter or plant roots and leaves

solid a state of matter whose particles are in fixed positions and are not able to move from one location to another; the particles are held in a lattice by chemical bonds or intermolecular forces

solid phase chemistry carrying out a synthesis with one of the reactant molecules attached to an insoluble material known as a solid support

solubility product for a sparingly soluble salt, MX $[M^+X^-]$ in water, the solubility product, K_{sp}, is defined as $[M^+(aq)] \times [X^-(aq)]$. The solubility product is the equilibrium constant for the dissolution process: $MX(s) \rightleftharpoons M^+(aq) + X^-(aq)$. In general for a salt M_aX_b, $K_{sp} = [M^{b+}]^a [X^{a-}]^b$ with units of $(mol\,dm^{-3})^{a+b}$

solute the solid, liquid or gas that has been dissolved to form a solution

solution formed when a solid, liquid or gas is dissolved into a solvent

solve obtain an answer using algebraic and/or numerical methods

solvent a liquid that dissolves solids, liquids or gases to form a solution

source areas or sites where a pollutant is released into the atmosphere

Southern blotting DNA that has been separated by electrophoresis, transferred from the gel to nitrocellulose in single-strand form for hybridization

specific heat capacity the amount of heat energy required to raise the temperature of a given mass of substance (which can be 1 g or 1 kg), by one kelvin (or degree Celsius), equivalent to the heat capacity divided by mass

spectator ions ions present in solution that do not participate directly in a reaction

spectrochemical series arrangement of ligands in order of increasing ability to produce d-orbital splitting

spectroscope optical instrument that produces a spectrum for obervation by the eye

spin pair two electrons with opposite spins, usually occupying the same orbital

spin resonance exchange of energy between two systems at a specific frequency

spin–spin splitting/coupling interaction between the spin magnetic moments of different nuclei: in high-resolution NMR spectroscopy it gives rise to multiplet patterns

spontaneous reaction a reaction which will occur when the reactants are mixed together under standard conditions, accompanied by a decrease in free energy: the Gibbs free energy change, $\Delta G^{\ominus}$, is negative; spontaneous reactions can do work (note that a spontaneous reaction may be fast or slow)

standard a measurement, specification or object against which other measurements are compared

standard electrode potential the potential difference generated by a half-cell under standard conditions when an electrode is connected by a salt bridge and an external circuit to a standard hydrogen electrode

standard enthalpy change of combustion the enthalpy change when one mole of a compound is completely combusted (burned) in excess oxygen under standard thermodynamic conditions (with no change in pressure)

standard enthalpy change of formation the enthalpy change when one mole of a compound is formed under standard thermodynamic conditions (with no change in pressure) from its elements in their standard states

standard enthalpy change of reaction the enthalpy change for the amounts of reactants in a specified stoichiometric equation under standard thermodynamic conditions (pressure 101.3 kPa and 298 K)

standard hydrogen electrode a reference half-cell used to measure standard electrode potentials; it consists of hydrogen gas (at a pressure of one atmosphere) bubbled over a platinum electrode in a one molar aqueous solution of hydrochloric acid at 298 K and is assigned a voltage of zero

standard reference electrode an electrode measuring system which provides a constant stable voltage regardless of the composition of the external solution: this voltage is used as the base from which standard electrode potentials are measured

standard solution a solution whose concentration is accurately known and does not change with time

standard state a reference state for a specific element according to standard thermodynamic conditions

state give a specific name, value or other brief answer without explanation or calculation

state symbol indicates whether a substance shown in an equation is a solid (s), liquid (l), gas (g) or in aqueous solution (dissolved in water) (aq)

states of matter solid, liquid and gas are the three states of matter in which all substances can exist, depending on the conditions of temperature and pressure

static equilibrium a system in which the total amounts do not change, and the starting materials are no longer turning into products, and the products do not turn back into starting materials

stationary phase one of the two phases forming a chromatographic system, which may be a solid, a gel or a liquid; if a liquid, it may be distributed on a solid – the liquid may also be chemically bonded to the solid (bonded phase) or immobilized on it (immobilized phase)

steady state term used in reaction kinetics to describe the near constant (steady) concentration of a reactive intermediate species in a reacting system when it is formed at a rate which is approximately equal to its rate of removal

steam cracking cracking in the presence of steam at very high temperatures

step polymer a type of polymer formed when all molecules in the system can react with each other, and the growth of polymer chains takes place by condensation reactions between molecules of all degrees of polymerization

stereoisomerism isomerism arising from differences in the shapes of molecules; includes geometrical and optical isomerism

stereoregular polymer regular polymer, the molecules of which can be described in terms of only one species of stereo-repeating unit in a single sequential arrangement

steric factor an attempt to account for deviations of reaction rates from collision theory which incorporates the idea that molecules colliding will react only in certain orientations

steric hindrance the prevention or slowing down of a reaction by atoms or functional groups blocking the access of an attacking molecule or ion

steroid group of lipids with a characteristic fused carbon-ring structure, which includes cholesterol and the sex hormones

sticky ends two DNA molecules with short overhanging single-stranded sequences that are complementary to one another, facilitating the sealing of the ends

stimulant drug that temporarily arouses or accelerates physiological activity and prevents sleep

Stock notation a naming system for inorganic compounds based on the indication of the oxidation number (as a Roman numeral, in brackets) of each of the major elements in the compound

stoichiometric quantities refers to a reaction where amounts of reactants are reacted together so that all are consumed at exactly the same time

straight-chain hydrocarbon a hydrocarbon with an open chain of atoms with no side-chains

stratosphere layer of the atmosphere between 20 and 50 kilometres above the surface of the Earth, above the troposphere

strong acid an acid completely dissociated or ionized when dissolved in water

strong base a base completely dissociated or ionized when dissolved in water

structural isomerism a type of isomerism in which the connectivity of atoms in the isomeric compounds differs

sub-atomic particles the particles – electrons, protons and neutrons – from which all atoms are made

sub-shell a sub-division of a shell; atoms can contain up to four different types of sub-shell: s, p, d and f

substitution reaction a reaction in which one atom or group of atoms is replaced by another atom or functional group

substrate compound acted upon by an enzyme

successive ionization energies the energies involved for the successive removal of electrons from an atom

suggest propose a hypothesis or other possible answer

suppository a small plug of medication designed for insertion into the rectum or vagina, where it melts

suspension mixture in which small solid or liquid particles are suspended in gas or liquid

sympathomimetic drugs drugs that mimic the effects of the sympathetic nervous system

syndiotactic polymer chain in which the stereochemical orientation of the substituents, with respect to each other along the chain, alternates

synergistic effect condition in which the result of the combined action of two or more drugs is greater than the sum of their separate, individual effects

synthesis (direct) a reaction in which a compound is formed from its elements

system the mixture of chemicals being studied – everything outside the system is the 'surroundings'; the system and the surroundings (the rest of the universe) are separated by a boundary or interface

systematic error an error which biases your measurements in some predictable way – the effects cannot be reduced by repeating the measurement and averaging

T

tacticity any type of regular or symmetrical molecular arrangement in a polymer structure

temperature inversion occurs when cold dense air is near the ground, with a layer of warmer and therefore lighter air above it; often occurs in valleys, trapping pollutants formed during the day in towns and cities located there

tempering process of reheating quench-hardened steel which relieves internal stresses and imparts toughness and ductility (*see* annealing)

teratogenic able to disturb the growth and development of an embryo or fetus

termination an elementary step in a chain reaction involving the combination of two radicals to form a molecule

tertiary alcohol an alcohol in which a hydroxyl group is bonded to a tertiary carbon

tertiary carbon atom a carbon atom that is directly attached to three other carbon atoms

tertiary structure overall three-dimensional folded shape of a protein composed of a single polypeptide chain

tetramethylsilane (TMS) reference standard for proton (and carbon) nuclear magnetic resonance

theoretical yield the maximum amount or mass of a particular product that can be formed when the limiting reactant is completely consumed and there are no losses or side reactions

therapeutic window range of a drug's concentration in the blood between its therapeutic level and its toxic level

thermal cracking cracking carried out a high temperature in the absence of a catalyst

thermal decomposition the chemical decomposition of a substance into simpler substances at high temperatures

thermal pollution industrial discharge of heated water into a river, lake or other body of water, causing a rise in temperature that endangers aquatic life by decreasing the solubility of dissolved oxygen

thermochemical equation a balanced stoichiometric chemical equation that includes the enthalpy change

thermoplastics plastics that soften when heated and can then be re-moulded

thermosetting plastics plastics that do not soften on heating but only char and decompose – they cannot be re-moulded

thermotropic liquid crystal liquid crystal prepared by heating the substance

thin-layer chromatography form of chromatography in which compounds are separated by a suitable solvent or solvent mixture on a thin layer of adsorbent material coated onto a flat support

three prime end (3') the end of a nucleic acid that lacks a nucleotide bound at the 3' position of the terminal residue

threshold frequency minimum frequency of electromagnetic radiation, typically ultraviolet radiation, capable of generating photoelectric emission in a metal

titration a chemical technique in which one solution is used to analyse another solution and find its concentration or amount

titration curve a plot showing the pH of a solution being analysed as a function of the amount of acid or base added

tolerance capacity of the body to endure or become less responsive to a drug with repeated use or exposure

top-down approach making nano-scale structures by machining and etching techniques

trans a prefix typically used to describe the geometrical isomer of a 1,2-disubstituted alkene with atoms or functional groups on opposite sides of a double bond or ring

trans fat common name for unsaturated fat with *trans*-isomer fatty acid(s). *Trans* fats may be monounsaturated or polyunsaturated but never saturated

transcription enzyme-controlled process where the sequence of bases in one strand of the DNA of a gene is used to specify a complementary sequence of bases in a messenger RNA chain

transfer RNA small RNA molecules that find specific amino acids and attach them to a growing polypeptide chain in the sequence specified by the codons in the messenger RNA

transition element a metal in the d-block of the periodic table which has at least one of its ions with a partly filled d-sub shell

transition metal *see* transition element

transition state the partially bonded chemical species located at the top of the activation energy barrier as a reaction proceeds from reactants to products

transition-state theory explains the reaction rates of elementary steps: the theory assumes an equilibrium between the reactants and the transition state

translation the process by which messenger RNA directs the amino acid sequence of a growing polypeptide during protein synthesis

triglyceride a molecule of triglyceride is composed of a molecule of glycerol in which each of the three carbons is linked through an ester bond to a fatty acid

triose monosaccharide with three carbon atoms

triple point the unique temperature and pressure at which all three phases of a substance (solid, liquid and gas) are in equilibrium

troposphere layer of the atmosphere closest to the ground and extending upwards (15–30 kilometres) to the stratosphere

twisted nematic type of liquid-crystal display where the liquid-crystal fluid rotates the plane of polarization 90°

U

ultraviolet/visible spectroscopy the absorption of ultraviolet/visible light by a molecule causing the promotion of an electron from a ground electronic state to an excited electronic state

unimolecular involving one particle dissociating into two or more particles in an elementary step in a reaction mechanism

universal indicator a mixture of several acid–base indicators which change colour as the pH increases

unsaturated used to describe a molecule, such as an alkene, containing one or more carbon–carbon double bonds

V

valence band band of allowed energy states that exists in solids and is completely filled with electrons at zero degrees kelvin; the valence band is separated from the conduction band by an energy gap in insulators and semi-conductors and overlaps the conduction band in metals

valence shell electron pair repulsion (VSEPR) a model which states that the shape of bonds around the central atom of a molecule is determined by minimizing the repulsion between the centres of negative charge (lone pairs and bonding pairs)

van der Waals' forces intermolecular forces that operate between noble gas atoms and between non-polar molecules in the solid and liquid states – believed to occur when the electrons within an atom or molecule induce a temporary dipole in an adjacent atom or molecule (the term van der Waals' forces can be used to collectively describe all interactions between non-polar molecules)

vapour a vapour may be described as a gas in contact with its liquid at a temperature below its boiling point; it may be formed by the sublimation of a solid or the evaporation of a liquid

vapour pressure the pressure of the vapour over a liquid where the liquid and vapour are in equilibrium

virus simple organism that consists essentially of a core of RNA or DNA surrounded by a protein coat; only able to replicate inside a host cell

vitamin various unrelated fat-soluble or water-soluble organic substances essential in minute amounts for normal growth and activity of the body and obtained naturally from plant and animal foods

volatile organic compounds (VOCs) organic chemicals that have a very high vapour pressure and low solubility in water under normal atmospheric conditions

volatility a qualitative measure of the how readily a liquid or solid is vaporized upon heating or evaporation

volt the unit of potential difference defined as one joule of work per coulomb of charge transferred (1 volt = 1 joule per coloumb)

voltaic cell a voltaic cell contains two half-cells, each of which is composed of an electrode in contact with an electrolyte

W

water of crystallization water molecules that are chemically incorporated into the crystal lattice of many inorganic salts when they crystallize from aqueous solution

wave equation the velocity (v) of a wave is given by the product of the wavelength (λ) and the frequency (f), that is, $v = f \times \lambda$

wave–particle duality light sometimes acts like a wave, and sometimes acts like a particle, depending on the experiment you are performing

wavelength the distance between corresponding points, for example, between two peaks or two troughs, on a wave; denoted by the Greek letter lambda (λ)

wavenumber reciprocal of the wavelength of radiation (in cm), and thus the number of complete waves within 1 cm: commonly used in infrared spectroscopy

weak acid an acid that is only partially dissociated or ionized when dissolved in water

weak base a base that is only partially dissociated or ionized when dissolved in water

weathering chemical and physical breakdown of rock into small fragments

work the force multiplied by the distance through which the force is applied

wrought iron formed when carbon is removed from pig iron; it is relatively pure iron and is easily shaped

Z

zeolite natural or synthetic hydrated aluminosilicate with an open three-dimensional structure, in which water molecules are held in cavities in the lattice

zero error when the scale on the measuring instrument does not start from exactly zero

zero-order reaction a chemical reaction in which the rate of reaction is independent of the concentration of a reactant

Ziegler–Natta process method for the manufacture of high-density polyethene and isotactic polypropene using a catalyst of titanium(IV) chloride and an alkyl aluminium compound

zwitterions ion whose positive and negative charges are separated from each other within the chemical species; zwitterions are responsible for the relatively high melting points of amino acids and their solubility in water

Acknowledgements

The publishers have made every effort to contact copyright holders. If any have been inadvertently overlooked, they will be pleased to make the necessary arrangements at the earliest opportunity.

Photo credits

Thanks are due to the following for permission to reproduce copyright photographs.

All photos by kind permission of Cesar Reyes except:
p.58 Science Photo Library/NASA; **p.62** *t* Science Photo Library/Lynette Cook, *bl* Science Photo Library/Andrew Lambert, *bc* David Talbot, *br* Robert Balcer; **p.89** Science Photo Library/Andrew Lambert; **p.91** David Talbot; **p.120** *t* Chun Mei Kin; *b* copyright Harry Kroto and used with permission of the Sussex Fullerene Research Centre (photographer Nicholas Sinclair); **p.133** Robert Balcer; **p.139** David Talbot; **p.146** NASA; **p.156** Dr Colin Baker; **p.175** *t* Science Photo Library/J. C. Revy, *c* Dr Colin Baker; **p.180** Alamy/Nagelstock; **p.181** *t* Corbis, *b* Richard Harwood; **p.182** *t,b* Richard Harwood; **p.188** Richard Harwood; **p.193** Richard Harwood; **p.195** *t,b* Richard Harwood; **p.201** Corbis; **p.211** Science Photo Library/Andrew Lambert; **p.216** *r* David Talbot; **p.223** David Talbot; **p.227** *r* David Talbot; **p.244** Dr Colin Baker; **p.256** Science Photo Library/Klaus Boller; **p.257** Chris Talbot, *b* Rasmol Library/Richard Harwood; **p.258** *t,c,b* Rasmol Library/Richard Harwood; **p.262** *t* Science Photo Library/Martin Bond, *b* David Talbot; **p.264** Richard Harwood; **p.265** Richard Harwood; **p.270** Rasmol Library/Richard Harwood; **p.276** Chris Talbot; **p.277** Science Photo Library/Martyn f Chillmaid: **p.278** *tl* Science Photo Library/Cordelia Molly, *tr* Science Photo Library/David Taylor, *b* Science Photo Library/Paul Rapson; **p.279** Science Photo Library/CSIRO; **p.280** Science Photo Library/Paul Rapson; **p.281** Chris Talbot; **p.282** David Talbot; **p.284** *t,b* Chris Talbot; **p.286** Science Photo Library/Andrew Lambert; **p.287** David Talbot; **p.290** *l* Science Photo Library/Roger Job, *r* Science Photo Library/Vanessa Vick; **p.291** *l,r* Science Photo Library/Andrew Lambert; **p.293** *l,r* Chris Talbot; **p.294** *l,r* Science Photo Library/Andrew Lambert, *b* Chris Talbot; **p.333** Angus Labs; **p.349** *t* Science Photo Library/Charles D. Winters; **p.350** Science Photo Library/Andrew Lambert; **p.365** *t* Science Photo Library/Andrew Lambert; **p.412** *l* David Talbot; **p.417** David Talbot; **p.423** *t* photography by Dr Thomas Schneider, Center for Cancer Research Nanobiology Program(CCRNP), National Institutes of Health in Frederick, Maryland; **p.455** Frederick Tudgay, The *Dunedin* off the English Coast. 1875, oil on canvas, Acc No: 02/1. Hocken Collections, Uare o Taoka Hakena, University of Otago, Dunedin, New Zealand.; **p.456** *t* Chris Talbot, *b* Rex Features; **p.460** *tl* Chris Talbot, *tr* David Talbot, *b* Richard Harwood; **p.463** *t,b* Chris Talbot; **p.464** Chris Talbot; **p.466** Science Photo Library/Andrew Lambert; **p.475** *t* Science Photo Library/Sovereign ISM, *b* Chris Talbot; **p.477** David Talbot; **p.503** Richard Harwood; **p.533** *b* David Talbot; **p.540** David Talbot; **p.541** *t,b* David Talbot; **p.548** Richard Harwood; **p.549** Richard Harwood; **p.550** *t* Arenapal/Nigel Norrington, *b* Richard Harwood; **p.551** Chris Talbot; **p.552** Richard Harwood; **p.556** Science Photo Library/Andrew Lambert; **p.564** David Talbot; **p.566** Richard Harwood; **p.568** *tl* Chris Talbot, *tr* Science Photo Library/Charles Winters, *b* David Talbot; **p.569** *t* David Talbot, *b* Science Photo Library/Gustoimages; **p.573** Richard Harwood; **p.578** *t* David Talbot, *b* Chris Talbot; **p.579** David Talbot; **p.580** David Talbot; **p.589** Dr Jon Hare; **p.590** Chris Talbot; **p.591** *t,b* Chris Talbot; **p.594** Chris Talbot; **p.595** Chris Talbot; **p.598** Dr Jon Hare; **p.599** Chris Talbot; **p.606** Dr Jon Hare; **p.612** Istock; **p.613** Dr Jon Hare; **p.618** Istock; **p.619** Dr Jon Hare; **p.625** Chris Talbot; **p.626** Chris Talbot; **p.630** *t* Chris Talbot, *b* David Talbot;

p.634 *t* David Talbot, *b* Dr. Jon Hare; **p.635** *t* Robert Balcer, *b* Chris Talbot; **p.641** Chris Talbot; **p.657** David Talbot; **p.667** *tr* Science Photo Library/James King-Holmes, *bl* Chris Talbot; **p.681** David Talbot; **p.690** David Talbot; **p.692** Chris Talbot; **p.721** Prof. Alan Fersht, University of Cambridge; **p.724** Chris Talbot; **p.726** Science Photo Library/Sovereign; **p.740** David Talbot; **p.741** *t* Dr Calvin Davidson, *b* David Talbot; **p.742** *all* David Talbot; **p.746** David Talbot; **p.747** *t* Istock, *b* David Talbot; **p.748** Richard Harwood; **p.751** *t,b* Chris Talbot; **p.752** Corbis; **p.758** David Talbot; **p.759** David Talbot; **p.768** Rex Features; **p.772** David Talbot; **p.773** David Talbot; **p.775** *t* Science Photo Library/James Bell, *b* Chris Talbot; **p.776** David Talbot; **p.778** Chris Talbot; **p.779** *t,b* Chris Talbot; **p.780** *t* IBM Almaden Research Centre, *b* Chris Talbot; **p.781** Chris Talbot; **p.783** IBM Almaden Research Centre; **p.784** Dr Chris Ewels; **p.786** *t,b* Chris Talbot, *c* Corbis; **p.789** Chris Talbot; **p.789** Chris Talbot; **p.796** Chris Talbot; **p.813** Chris Talbot; **p.818** National Cancer Institute; **p.819** Istock; **p.823** Istock; **p.824** Chris Talbot; **p.825** Chris Talbot; **p.827** *t* Science Photo Library/A. B. Dowsett, *b* Chris Talbot; **p.829** David Talbot; **p.832** Istock; **p.837** David Talbot; **p.842** Chris Talbot; **p.846** Dr Ryota Kuroki; **p.853** David Talbot; **p.855** Science Photo Library/Pat & Tom Leeson; **p.858** Istock; **p.859** Istock; **p.860** Istock; **p.867** Richard Harwood; **p.873** Istock; **p.879** Philippe Sainte-Laudy; **p.882** NASA; **p.892** David Talbot; **p.895** Chris Talbot; **p.897** David Talbot; **p.898** Istock; **p.903** Science Photo Library/David Hay Jones; **p.904** David Talbot; **p.905** Christ Talbot; **p.931** *t* Chris Talbot, *b* David Talbot; **p.934** David Talbot; **p.937** Chris Talbot; **p.942** Chris Talbot; **p.943** David Talbot; **p.945** *all* David Talbot; **p.948** Press Association; **p.949** David Talbot; **p.957** David Talbot; **p.959** David Talbot; **p.969** Chris Talbot; **p.982** Richard Harwood; **p.987** Richard Harwood; **p.988** *t* C. Coates, *b* David Talbot; **p.1008** Peter O'Bryne; **p.1009** Peter O'Bryne; **p.1011** Peter O'Bryne.

Artwork credits

The publishers would like to thank the following for permission to reproduce copyright material:

p.10 Fig. 1.17 Kim Gyeoul; **p.21** Fig. 1.24 Saito Susumu; **p.29** Fig. 1.33 Kim Gyeoul; **p.50** Fig. 2.2 Jon Harwood; **p.51** Fig. 2.4 Kirstie Gannaway; **p.59** Fig. 2.25, **p.77** Fig. 3.16 Jon Harwood; **p.152** Fig. 5.25 Jon Harwood with calligraphy by Ding Gang Lee; **p.201** Fig. 7.19 Jon Harwood; **p.210** Fig. 8.1 Kim Gyeoul; **p.212** Fig. 8.5, **p.223** Fig. 8.26, **p.256** Fig. 10.2, **p.332** Fig. 12.14 Jon Harwood; **p.360** Fig. 13.33 Kim Gyeoul; **p.379** Figs 14.17, 14.20, **p.391** Fig. 14.59 Jon Harwood; **p.411** Fig. 15.31 Kim Gyeoul; **p.444** Fig. 16.17, **p.539** Fig. 19.1 , **p.618** Fig. 21.71 Jon Harwood; **p.887** Fig. 22.15 Theresa Harwood; **p.903** Fig. 25.32 Kirstie Gannaway; **p.986** Fig. 27.33 Jon Harwood; **p.956** Fig. 26.58 Kirstie Gannaway.

Text credits

p.86 Mario Petrucci's poem *Last Wish* is published in *Heavy Water: a poem for Chernobyl* (Enitharmon Press, 2004).

Examination questions credits

Examination questions have been reproduced with kind permission from the International Baccalaureate Organization.

Index

Please note: page numbers in blue refer to chapters on the accompanying CD-ROM.